Nonvolatile Semiconductor Memory Technology

IEEE Press
445 Hoes Lane, P.O. Box 1331
Piscataway, NJ 08855-1331

Nonvolatile Semiconductor Memory Technology

A Comprehensive Guide to Understanding and Using NVSM Devices

Edited by

William D. Brown
University of Arkansas

Joe E. Brewer
Northrop Grumman Corporation

IEEE Press Series on Microelectronic Systems
Stu Tewksbury, *Series Editor*

IEEE Solid-State Circuits Council, *Sponsor*

IEEE Components, Packaging, and Manufacturing Technology Society, *Sponsor*

The Institute of Electrical and Electronics Engineers, Inc., New York

This book and other books may be purchased at a discount
from the publisher when ordered in bulk quantities. Contact:

IEEE Press Marketing
Attn: Special Sales
445 Hoes Lane, P. O. Box 1331
Piscataway, NJ 08855-1331
Fax: (732) 981-9334

For more information on the IEEE Press,
visit the IEEE home page: http://www.ieee.org/

Printed in the United States of America
10 9 8 7 6 5 4 3 2

ISBN 0-7803-1173-6
IEEE Order Number: PC5644

Library of Congress Cataloging-in-Publication Data

Nonvolatile semiconductor memory technology : a comprehensive guide to understanding and using NVSM devices / edited by W. D. Brown, Joe E. Brewer.
p. cm. – (IEEE Press series on microelectronic systems)
Includes bibliographical references and index.
ISBN 0-7803-1173-6 (cloth)
1. Semiconductor storage devices. I. Brown, W. D. (William D.). (date) . II. Brewer, Joe (Joe E.) III. Series.
TK7895.M4N634 1997
621.39′732–dc21 97-19691

CIP

Contents

List of Contributors

Joe E. Brewer
Electronic Sensors and Systems Division, Northrop Grumman, Maryland

William D. Brown
Department of Electrical Engineering, University of Arkansas, Arkansas

Manzur Gill
Intel Corporation, California

G. Grueseneken
Interuniversity Microelectronics Center, Belgium

Yukun Hsia
YHL Consultants, California

Stefan Lai
Intel Corporation, California

Frank R. Libsch
IBM Research Division, New York

H. C. Lin
Department of Electrical Engineering, University of Maryland, Maryland

H. E. Mass
Interuniversity Microelectronics Center, Belgium

George Messenger
Messenger and Associates, Nevada

William Owens
Xicor Corporation, California

Ravi Ramaswaml
Hewlett-Packard Company, Oregon

Dave Sweetman
Silicon Storage Technology, California

Vance C. Tyree
Information Sciences Institute, University of Southern California, California

J. Van Houdt
Interuniversity Microelectronics Center, Belgium

H. A. Richard Wegener
Xicor Corporation, California

Marvin H. White
Sherman Fairchild Laboratory, Lehigh University, Pennsylvania

J. S. Witters
MIETEC, Belgium

List of Acronyms

ACEE - Array Contactless Electrically Erasable
AMLCD - Active Matrix Liquid Crystal Display
AOQ - Average Outgoing Quality
AOQL - Average Outgoing Quality Lot
AQL - Acceptable Quality Level
ASIC - Application Specific Integrated Circuit
ASTM - American Society for Testing and Measurements
BBT - Band-to-Band Tunneling
BEM - Breakdown Energy of Metal
BiCMOS - Bipolar CMOS
BIOS - Basic Input Output System
BIST - Built-In Self-Test
BL - BitLine
BOE - Buffered Oxide Etch
BORAM - Block-Oriented Random Access Memory
C-V - Capacitance-Voltage
CDF - Cumulative Distribution Function
CHE - Channel Hot Electron

CMOS - Complementary-Metal-Oxide-Semiconductor
CPU - Central Processing Unit
CVD - Chemical Vapor Deposition
CZ - Czochralski
DEIS - Dual Electron Injection Structure
DI - DeIonized
DIFLOX - DIfused-layer FLoating gate thin Oxide
DINOR - DIvided bitline NOR
DOD - Department Of Defense
DOE - Department Of Energy
DRAM - Dynamic Random Access Memory
DSP - Digital Signal Processing or Drain-Source Protected
E/W - Erase/Write
EAROM - Electrically-Alterable-Read-Only Memory
ECC - Error Correction Circuit
ECR - Electron Cyclotron Resonance
EDAC - Error Detection And Correction
EEPROM - Electrically-Erasable Programmable Read-Only Memory
EM - ElectroMigration
EOR - Exclusive OR
EOS - Electrical OverStress
EPLD - Electrically Programmable Logic Device
EPROM - Electrically Programmable Read-Only Memory
ESD - ElectroStatic Discharge
ETOX - Electron Tunneling OXide
FAMOS - Floating gate-Avalanche-Metal-Oxide-Semiconductor
FET - Field-Effect Transistor
FETMOS - Floating gate-Electron Tunneling-Metal-Oxide-Semiconductor
FF-EEPROM - Full-Feature EEPROM
FG - Floating Gate
FIT - Failure unIT = 1 failure/10E9 device-hours
FITS - Failures In Time
FLOTOX - FLOating gate Thin Oxide
FN - Fowler–Nordheim
FPGA - Field Programmable Gate Array
FPLD - Field Programmable Logic Device
FR - Failure Rate
FRAM - Ferroelectric Random Access Memory
FZ - Float Zone
G-V - Conductance-Voltage
HiCR - High Capacitive-coupling Ratio
HIMOS - High Injection Metal-Oxide-Semiconductor
IC - Integrated Circuit
IDE - Integrated Drive Electronics
IEDM - International Electron Devices Meeting
IEMP - Internal ElectroMagnetic Pulse

JEDEC - Joint Electron Device Engineering Council
JTAG - Joint Testability Action Group
LAT - Large Angle Tilt
LDD - Lightly Doped Drain
LEO - Low Earth Orbit
LET - Linear Energy Transfer
LINAC - LINear ACcelerator
LOCA - Loss Of Coolant Accident
LOCOS - LOCal Oxidation of Silicon
LPCVD - Low Pressure Chemical Vapor Deposition
LTPD - Lot Tolerant Percent Defective
MAOS - Metal-Aluminum oxide-silicon Oxide-Semiconductor
MIMIS - Metal-Insulator-Metal-Insulator-Semiconductor
MIS - Metal-Insulator-Semiconductor
MNOS - Metal-Nitride–Oxide-Semiconductor
MOS - Metal-Oxide-Semiconductor
MTOS - Metal-Tantalum oxide-silicon Oxide-Semiconductor
MTTF - Mean Time To Failure
NASA - National Aeronautics and Space Administration
NCHE - N-channel Channel Hot Electron
NFN - N-channel Fowler–Nordheim
NIST - National Institute of Standards and Technology
NMOS - N-channel Metal-Oxide-Semiconductor
NOVRAM - NOnVolatile Random Access Memory
NV - NonVolatile
NVM - NonVolatile Memory
NVSM - NonVolatile Semiconductor Memory
OEM - Original Equipment Manufacturer
ONO - Oxide–Nitride–Oxide
OTP - One Time Programmable
PCM - Process Control Monitor
PCMCIA/AT - Personal Computer Memory Card International Association/Advanced Technology
PDA - Personal Digital Assistant
PIN - Personal Identification Number
PLD - Programmable Logic Device
PMOS - P-channel Metal–Oxide–Semiconductor
POA - Post Oxidation Anneal
PPB - Parts Per Billion
PPM - Parts Per Billion
PROM - Programmable Read-Only Memory
QCI - Quality Conformance Inspection
RAM - Random Access Memory
RIE - Reactive Ion Etch
ROM - Read Only Memory
RTONO - Rapid Thermal OxyNitradation of Oxide

SAMOS - Stacked gate-Avalanche-Metal-Oxide-Semiconductor
SATO - Self-Aligned Thick Oxide
SCSG - Source-Coupled Split-Gate
SCSI - Small Computer Systems Interface
SEC - Standard Evaluation Circuit
SEM - Scanning Electron Microscopy
SEP - Single Event Phenomena
SEU - Single Event Upset
SIMOS - Stacked gate-Injection-Metal-Oxide-Semiconductor
SMD - Standardized Microcircuit Drawings
SNOS - Silicon-Nitride–Oxide-Semiconductor
SOI - Silicon On Insulator
SONOS - Silicon-Oxide-Nitride-Oxide-Semiconductor
SOS - Silicon On Sapphire
SRAM - Static Random Access Memory
SSI - Source-Side Injection
SSIMOS - Shielded-Substrate-Injection-Metal-Oxide-Semiconductor
SSTR - Stacked-Self-aligned-Tunnel Region
SWEAT - Standard Wafer-level Electromigration Acceleration Test
TDDB - Time Dependent Dielectric Breakdown
TEFET - Trench Embedded Field Enhanced Tunneling
TEM - Transmission Electron Microscope
TFT - Thin Film Transistor
TIPS - Tilt angle Implanted Punchthrough Stopper
TLD - ThermoLuminescent Dosimeter
TPFG - Textured Poly Floating Gate
TRIGA - Training-Research-Isotopes-General-Atomic
TTL - Transistor-Transistor Logic
UTO - Ultra-Thin Oxide
UV - UltraViolet
VGA - Virtual Ground Array
WKB - Wentzel–Kramers–Brillouin
WL - WordLine
WLR - Wafer Level Reliability

Foreword

Nonvolatile Semiconductor Memory (NVSM) has for some time been an important part of modern information processing systems, and its application in systems is expected to grow as device capabilities increase. Engineers who are faced with any of the tasks of selecting, specifying, testing, or using NVSM must cope with the unique features of this useful class of devices and, therefore, must have a working knowledge of the technology. In particular, the retention and endurance characteristics of NVSM pose some interesting challenges.

Since the mid-1960s, development has proceeded on a variety of NVSM technologies, and a large body of technical literature has evolved. Individual NVSM device manufacturers supplement this literature by providing application notes and device data sheets.

Unfortunately for the typical device user, the NVSM technical literature is by and large fragmented and directed toward the needs and interests of device scientists and engineers. Consequently, considerable effort is required to pull together a reasonably complete set of references, read and comprehend the materials, and hopefully obtain a sufficiently complete perspective of the subject adequate to support system-level decision making.

Vendor application notes and device data sheets may be more useful. These publications can help the engineer understand how a given vendor approaches the design, fabrication, and test of given parts. Well-prepared vendor data can give the working engineer the basic information required to procure and apply specific parts. However, the caveat "well prepared" is an important matter. In the past, surveys taken of vendor data sheets revealed a number of serious shortcomings.

It is the purpose of this book to provide a reasonably comprehensive, readable, integrated overview of the NVSM field which gives an engineer sufficient background knowledge to make informed decisions regarding NVSM devices. The goal is to give the engineer a one-volume source book that provides a technically accurate statement of fundamentals for the major NVSM technology options available today. The word "readable" is used to describe the level of the text because it is not written for a device scientist (although the device scientist may find it useful). This does not mean that the information has been simplified to the point of being nontechnical. Rather, it is intended to provide technical information at a level sufficient for an applications engineer. For example, it is necessary that the materials used for charge storage, their properties, and the charge conduction mechanisms used in specific device types be described. For the nonspecialist such information can be daunting, but the level of treatment in this book should be comfortably within the grasp of the average electronics engineer.

The material is organized to allow the reader to pursue particular interests without having to read the entire book. However, it is recommended that all readers read Chapter 1 before pursuing other specific subject matter. Chapter 1, "Basics of Nonvolatile Semiconductor Memory Devices," provides a condensed treatment of all NVSM technologies with the intent of helping the reader to arrive at some perspective of the field. Subsequent chapters deal with specific nonvolatile semiconductor memory technologies. The closing chapters deal with important considerations for general device usage: reliability, radiation characteristics, and procurement considerations.

We the editors have been involved in the development and application of various nonvolatile semiconductor memory technologies since the inception of the NVSM concept in the 1960s. Over and over again, we have lamented the lack of a suitable comprehensive one-volume treatment of the subject. During the mid-1980s, we shared our concern with many of our colleagues and found a warm reception for our concern. In fact, we were able to establish a volunteer NVSM Study Group. At the peak of that effort more than 60 highly qualified individuals participated in the group activities. The study led to the submission of a book proposal to the IEEE PRESS.

Implementation of the book required that we approach the effort in a more focused way than the loose organization of the study group would allow. We then contacted some senior people, recognized leaders who have worked in the NVSM technology area for many years, and asked them to assume responsibility for the preparation of specific chapters. We were delighted with their response to our request and the text material they created. Their willingness to participate, their talents, and the support of industry in general have made our task easy.

We, along with the contributors, are convinced that this book is a significant contribution to the field of microelectronics which can help the proliferation of NVSM technology. We sincerely hope that you find it useful in your own NVSM activity.

Joe E. Brewer
Northrop Grumman
Baltimore, Maryland

William D. Brown
The University of Arkansas
Fayetteville, Arkansas

Chapter 1

G. Groeseneken,
H. E. Maes,
J. Van Houdt,
and J. S. Witters

Basics of Nonvolatile Semiconductor Memory Devices

1.0. INTRODUCTION

Since the very first days of the mid-1960s, when the potential of metal–oxide semiconductor (MOS) technology to realize semiconductor memories with superior density and performance than would ever be achievable with the then commonly used magnetic core memories became known, chip makers have thought of solutions to overcome the main drawback of the MOS memory concept, that is, its intrinsic volatility. The first sound solutions to this problem, with applicability beyond the mere read-only memory (ROM) function, were the floating gate concept [1.1] and the metal–nitride–oxide–semiconductor (MNOS) memory device [1.2], both of which were proposed in 1967. A 1 Kbit UV-erasable programmable read-only memory (PROM) (EPROM) part, based on the floating gate concept, became readily available in 1971, shortly after 1 Kbit random access memories (RAM) came on the market.

The ultimate solution—a genuine nonvolatile RAM that retains data without external power, can be read from or programmed like a static or dynamic RAM, and still achieve high-speed, high-density, and low-power consumption at an acceptable cost—remains unfeasible to this day. Yet tremendous progress has been made over the years in realizing the "alternative best" idea of a reliable, high-density, user-friendly reprogrammable ROM memory. During the last decade, these reprogrammable memories have constituted an almost steady 10% of the total semiconductor memory market. This can be seen from Fig. 1.1, which shows the increase in the

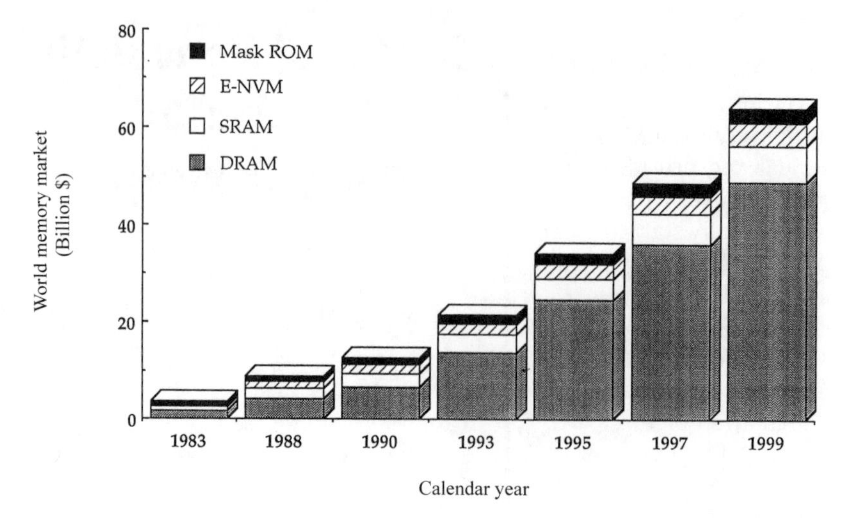

Figure 1.1 The increase of the world memory market during the last decade, the forecast for 1997 and 1999, and the share in this market of the different major classes of memories.

world memory market during the last decade, the forecast for 1997 and 1999, and the share of this market for the different major classes of memories. Until 1992, this 10% share came almost entirely from the least sophisticated and least functional version of this class of memories—that is, EPROMs, which do not allow in-system reprogrammability and are used mainly for standard program storage.

Reprogrammable nonvolatile memories can be subdivided into the following classes:

1. UV-erasable EPROM and one-time programmable (OTP) devices.
2. EEPROM memories, which can be further subdivided into full-feature electrically erasable programmable read-only memory (FF-EEPROMs) and Flash EEPROMs.
3. Nonvolatile RAM (NOVRAM), which combines the nonvolatility of EEPROM with the ease of use and fast programming characteristics of static RAM.
4. Ferroelectric RAM (FRAM).

Adding in-system reprogrammability to PROM memories (leading to FF-EEPROMs and to Flash EEPROMs), however, yields increased system flexibility and opens a broad new range of applications such as intelligent controllers; self-adaptive, reconfiguring, and remotely adjustable systems; programmable/adaptable logic; artificial intelligence; and numerous others [1.3]. The term *Flash* refers to the fact that the contents of the whole memory array, or of a memory block (sector), is erased in one step.

In 1983, 16 Kbit EEPROMs based on both the MNOS [1.4] and the floating gate concept [1.5] were introduced, many analysts projected that EEPROMs would grow into a high-volume market and gradually even replace EPROM as the standard program storage medium in microprocessor-controlled systems. Figure 1.2 shows the actual and projected growth of the EPROM, FF-EEPROM, and Flash EEPROM markets over a 16-year period. In 1984, it was forecast that the EEPROM market would really start to take off around 1985, with projected global sales on the order of $2.5 billion by 1988. It is clear from Fig. 1.2 that this predicted significant increase in the EEPROM market was delayed by more than six years. Moreover, the increase has not been as strong as it was then anticipated. The growth of the EPROM market has, however, slowed down and recently reversed. In 1994, and certainly 1995, the EEPROM market surpassed that of EPROM. Figure 1.2 also shows the emerging domination of Flash EEPROMs for the programmable ROMs for the next generations. They are at present the fastest growing MOS memory segment, and it is expected that they will eventually constitute the third largest segment behind DRAM and SRAM.

The EEPROM market did not grow as previously predicted because of their high cost per bit as compared to EPROM, the lack of large-scale applications for full-featured EEPROMs, and the poorly understood reliability of these components. The reliability issues of EEPROMs and Flash memories have, however, recently been thoroughly investigated and are now much better known and documented. In addition, recent lower pricing and increased performance of Flash memories have stirred new interest in these parts. New large-scale applications are emerging

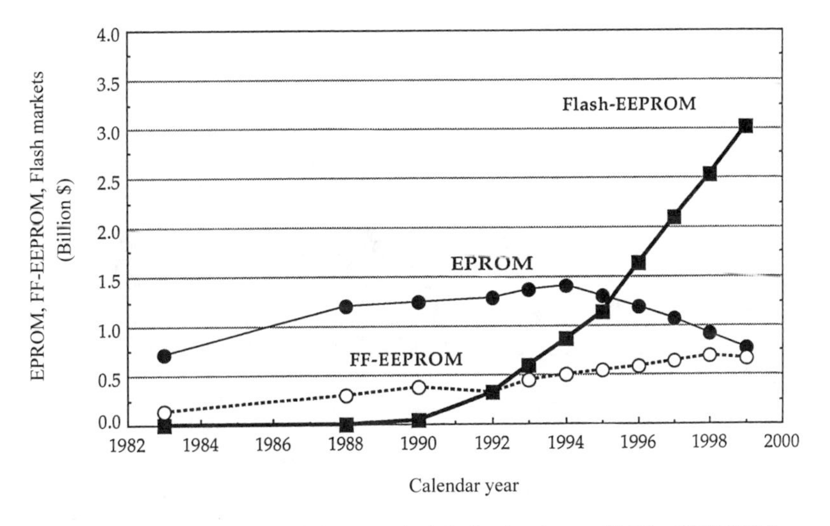

Figure 1.2 Comparison of the actual global sales (up to 1995) of EPROM, EEPROM, and Flash EEPROM, and forecasted market evolution (after 1995).

(i.e., memory cards, small, compact, and portable memories). These Flash EEPROMs were themselves developed in the late 1980s and introduced around 1990 when manufacturers were searching for nonvolatile devices that were still electrically erasable, but that could become nearly as cost effective as EPROMs. They combined the best concepts of EPROM and traditional EEPROM into a single-transistor Flash EEPROM.

This chapter presents the basic concepts and physics of operation of all the nonvolatile semiconductor memory types and classes listed previously. It is intended as a solid introduction to all the following chapters in this book. We will first present the basic principles and history of nonvolatile memory (NVM) devices in Section 1.1. The different programming mechanisms used in the various devices are discussed in Section 1.2, and the basic NVM memory products are presented in Section 1.3. A review of the major NVM devices in use today is given in Section 1.4 and is concluded by a rather general comparison of the different types of memory concepts. The basic equations and models specific to these NVM devices are presented in Section 1.5, which is followed by a detailed discussion in Section 1.6 of the NVM device characteristics and reliability issues for the different types of devices. Finally, Section 1.8 discusses some specific radiation aspects of NVM devices.

1.1. BASIC PRINCIPLES AND HISTORY OF NVM DEVICES

1.1.1 Basic Operating Principle

The basic operating principle of nonvolatile semiconductor memory devices is the storage of charges in the gate insulator of a MOSFET, as illustrated in Fig. 1.3. If one can store charges in the insulator of a MOSFET, the threshold voltage of the transistor can be modified to switch between two distinct values, conventionally

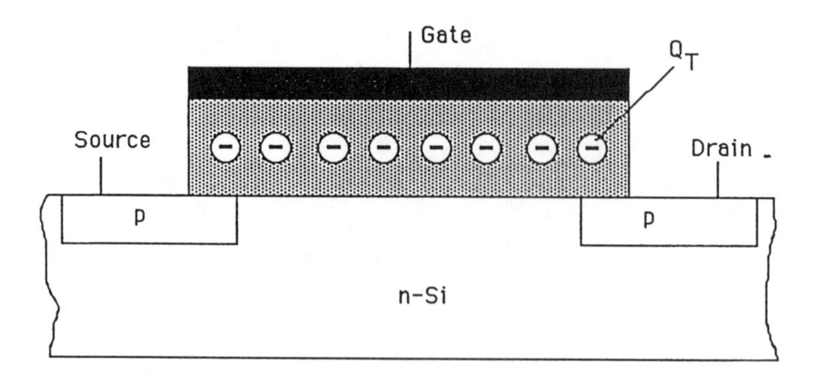

Figure 1.3 Basic operating principle of nonvolatile semiconductor memory: the storage of charges in the gate insulator of a MOSFET.

defined as the "0" or erased state and the "1" or written (programmed) state, as illustrated in Fig. 1.4.

From the basic theory of the MOS transistor, the threshold voltage is given by

$$V_{TH} = 2\phi_F + \phi_{ms} - \frac{Q_I}{C_I} - \frac{Q_D}{C_I} - \frac{Q_T}{\epsilon_I} d_I \tag{1.1}$$

where ϕ_{MS} = the work function difference between the gate and the bulk material
ϕ_F = the Fermipotential of the semiconductor at the surface
Q_I = the fixed charge at the silicon/insulator interface
Q_D = the charge in the silicon depletion layer
Q_T = the charge stored in the gate insulator at a distance d_I from the gate
C_I = the capacitance of the insulator layer
ϵ_I = the dielectric constant of the insulator

Thus, the threshold voltage shift, caused by the storage of the charge Q_T is given by

$$\Delta V_{TH} = -\frac{Q_T}{\epsilon_I} d_I \tag{1.2}$$

The information content of the device is detected by applying a gate voltage V_{read} with a value between the two possible threshold voltages. In one state, the transistor is conducting current, while, in the other, the transistor is cut off. When the power supply is interrupted, the charge should, of course, remain stored in the gate insulator in order to provide a nonvolatile device.

The storage of charges in the gate insulator of a MOSFET can be realized in two ways, which has led to the subdivision of nonvolatile semiconductor memory devices into two main classes.

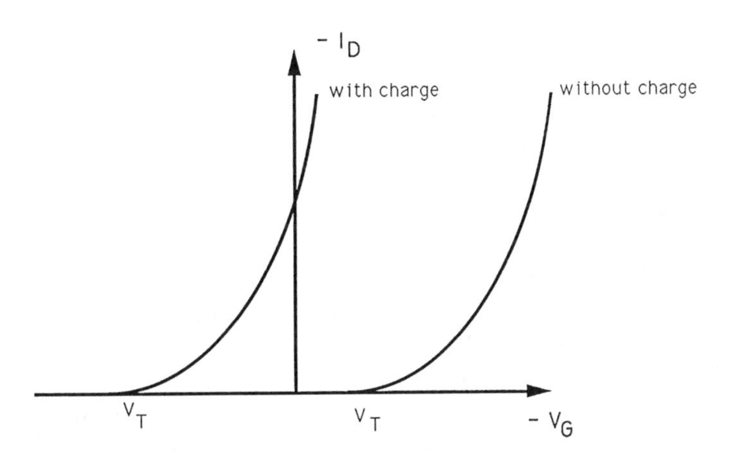

Figure 1.4 Influence of charge in the gate dielectric on the threshold of a p-channel transistor.

The first class of devices is based on the storage of charge on a conducting or semiconducting layer that is completely surrounded by a dielectric, usually thermal oxide, as shown on Fig. 1.5*a*. Since this layer acts as a completely electrically isolated gate, this type of device is commonly referred to as a floating gate device [1.6, 1.7].

In the second class of devices, the charge is stored in discrete trapping centers of an appropriate dielectric layer. These devices are, therefore, usually referred to as charge-trapping devices. The most successful device in this category is the MNOS device (metal–nitride–oxide–semiconductor) structure [1.2, 1.8], in which the insulator consists of a silicon nitride layer on top of a very thin silicon oxide layer, as shown in Fig. 1.5*b*. Other possibilities, such as Al_2O_3 (MAOS) and Ta_2O_5 (MTOS) [1.9, 1.10], have never been successfully exploited.

Further details on the cell types, features, and new developments, as well as a comparison of these classes of nonvolatile memory cells, are given in Section 1.4.

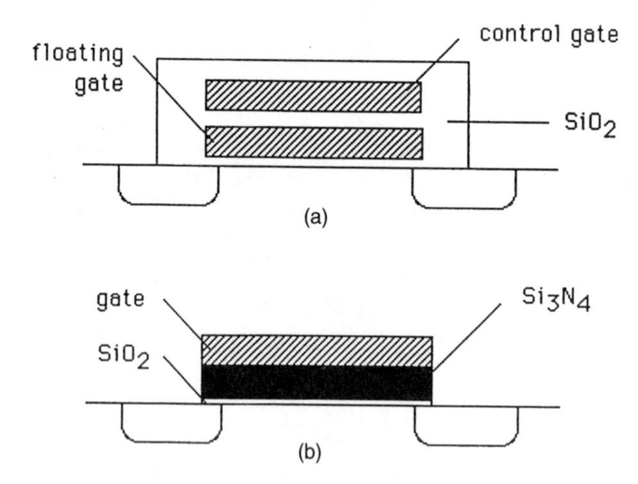

Figure 1.5 Two classes of nonvolatile semiconductor memory devices: (*a*) floating gate devices; (*b*) charge-trapping devices (MNOS device).

1.1.2 Short Historical Review

The idea of using a floating gate device to obtain a nonvolatile memory device was suggested for the first time in 1967 by D. Kahng and S. M. Sze [1.1]. This was also the first time that the possibility of nonvolatile MOS memory devices was recognized.

The memory transistor that they proposed started from a basic MOS structure, where the gate structure is replaced by a layered structure of a thin oxide I_1, a floating but conducting metal layer M_1, a thick oxide I_2, and an external metal gate M_2, as shown in Fig. 1.6. This device is referred to as the MIMIS (metal–insulator–metal–insulator–semiconductor) cell. The first dielectric I_1 has to be extremely thin in order to obtain a sufficiently high electric field to allow tunneling of electrons toward the floating gate. These electrons are then "captured" in the conduction band of the floating gate M_1, if the dielectric I_2 is thick enough to prevent discharging. When the gate voltage is removed, the field in I_1 is too small to allow

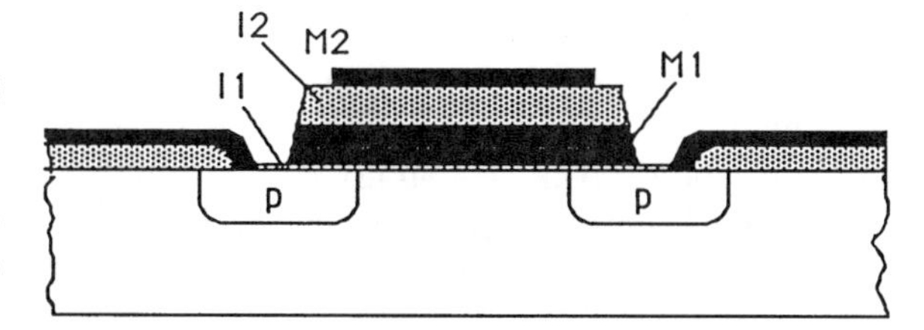

Figure 1.6 Introduction of the floating gate principle: the MIMIS structure, introduced by Kahng and Sze [1.1]. Writing and erasing the device is performed by direct tunneling of electrons through the thin oxide I1.

backtunneling. The injection mechanism to bring electrons to the floating gate is direct tunneling. To discharge the floating gate, a negative voltage pulse is applied at M_2, removing the electrons from the floating gate by the same direct tunneling mechanism.

The direct tunneling programming mechanism imposes the use of very thin oxide layers ($< 5\,\text{nm}$), which are difficult to achieve without defects. Any pinhole in I_1 will cause all the charge stored on M_1 to leak off. Because of this technological constraint, the MIMIS cell could not be reliably built at that time. Therefore, the importance of this device is merely historical, not only because it introduced the basic concept of nonvolatile memory devices in general, but also because it contained several essential concepts that have led to the development of both classes of nonvolatile memory devices: the direct tunneling concept has been used in charge-trapping devices, while the floating gate concept has led to a whole range of floating gate memory types.

In order to solve the technological constraint of the MIMIS cell, two types of improvements are possible: (1) replace the conducting layer on top of I_1 by a dielectric layer without losing the capture possibilities, which is actually the approach utilized in charge-trapping devices, or (2) increase the thickness of the tunneling dielectric I_1, which implies the need for other injection mechanisms.

The first solution was used in the MNOS cell, introduced in 1967 by Wegener et al. [1.2], almost simultaneously with the MIMIS cell. In the MNOS cell, the M_1 and I_2 layers are replaced by a nitride layer, as shown in Fig. 1.5*b*, which contains a lot of trapping centers in which holes and electrons can be captured. These traps fulfill the storage function of M_1 with the important difference that an eventual pinhole in the thin tunneling oxide (I_1) will not lead to a complete discharge of the cell since the individual traps are isolated from each other by the nitride. The device is programmed by applying a high voltage to the gate such that electrons tunnel from the silicon conduction band to the nitride conduction band and are then trapped in the nitride traps. This results in a positive threshold voltage shift. Erasing is achieved by applying a high negative voltage to the gate, so that holes tunnel from the silicon valence band into the nitride traps, resulting in a negative threshold voltage shift. The MNOS device has the intrinsic advantage that both programming and erasing operations can be performed electrically. The concept has been used widely in several kinds of applications, specifically in a class of memory products called EEPROM, which are further discussed in Section 1.3. At present, however, this class of memory cells is used only for military and applications that must be resistant to radiation, and only marginally in commercial high-density nonvolatile memory circuits.

The second solution has been used in a wide range of nonvolatile memory devices. The first operating floating gate device, shown in Fig. 1.7, was introduced in 1971 by Frohman–Bentchkowsky and is known as the *F*loating gate *A*valanche injection *MOS* (FAMOS) device [1.6–1.7, 1.11–1.12]. In the original p-channel FAMOS cell, a polysilicon floating gate is completely surrounded by a thick ($\approx 100\,\mathrm{nm}$) oxide. Here, the problem of possible shorting paths is obviated, but, at the same time, direct tunneling is excluded as the programming mechanism. In the FAMOS cell, the charging mechanism is based on injection of highly energetic electrons from an avalanche plasma in the drain region underneath the gate. This avalanche plasma is created by applying a high negative voltage ($> 30\,\mathrm{V}$) at the drain. The injected electrons are drifted toward the floating gate by the positive field in the oxide induced by capacitive-coupling between the floating gate and the drain. The FAMOS device has found wide applications and was the first cell to reach volume manufacturing levels comparable to other semiconductor memory types. FAMOS devices have evolved into a class of memory products called EPROM, and are further discussed in Section 1.3. The original FAMOS device, however, had several drawbacks, with the inefficiency of the programming process as the most salient one. In addition, no mechanism for electrical erasure existed since no field emission is possible due to the lack of an external gate. Therefore, erasure was possible only by UV or X-ray irradiation.

The drawbacks of the FAMOS device were alleviated in several adapted concepts. In the *S*tacked gate *A*valanche injection *MOS* (SAMOS) [1.13, 1.14], an external gate is added, as shown in Fig. 1.8, in order to improve the writing efficiency, and thus, the programming speed by an increased drift velocity of the electrons in the oxide, a field-induced energy barrier lowering at the Si–SiO_2 interface, and a decreased drain breakdown voltage. Electrical erasure also became possible by field emission through the top dielectric due to polyoxide conduction. Consequently, EEPROM products became feasible.

These first floating gate memory devices were all p-channel devices. In n-channel devices, avalanching the drain yields hole injection, which is much less efficient. Several alternative injection mechanisms have been proposed, most of which, however, were not sufficiently adequate for large-volume applications. Out of the various proposed injection mechanisms, only a few have proven feasible in floating gate applications for large production volumes. These programming mechanisms are discussed in the next section, and the cells that have emerged are the subject of Section 1.4.

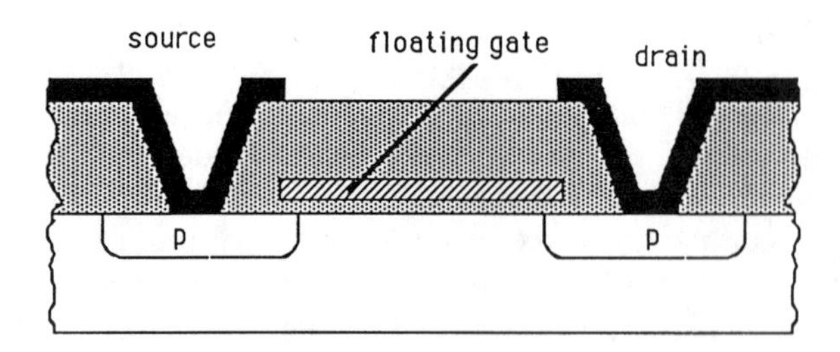

Figure 1.7 First operating floating gate device: the FAMOS (**F**loating gate **A**valanche injection **MOS**) device, introduced by Frohman–Bentchkowsky [1.6]. Writing the device is performed by injection of high energetic electrons created in the drain avalanche plasma. Erasure is possible by UV or X-ray radiation.

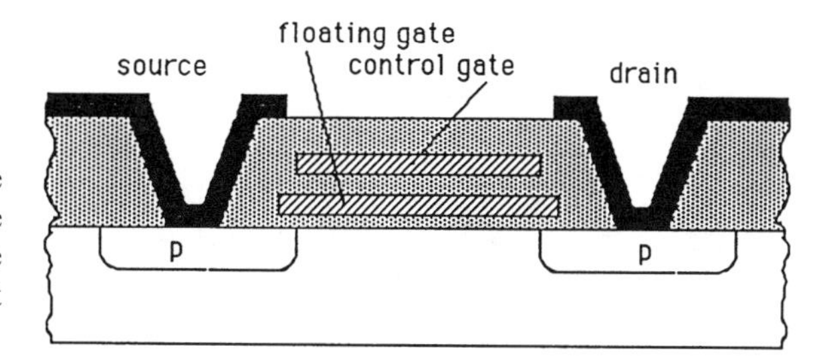

Figure 1.8 The SAMOS (Stacked gate Avalanche injection MOS) device [1.13]. The device is written like the FAMOS device. Several different erasure mechanisms are possible.

1.2. BASIC PROGRAMMING MECHANISMS

Electrical conduction through thin dielectric layers has been studied extensively in the past. It is generally understood that the electrical current behavior through dielectrics can be divided into two main classes: *bulk-limited conduction* and *electrode-limited conduction*. In the bulk-limited class, the current is determined mainly by the characteristics of the dielectric itself, and is independent of the electrodes from which the current originates. In the class of electrode-limited current, on the other hand, the conduction is determined by the characteristics of the electrodes, that is, the interface from which the current originates.

Many dielectrics, such as silicon nitride (Si_3N_4) or tantalum oxide (Ta_2O_5), belong to the bulk-limited conduction class. The current through silicon nitride is determined by Schottky emission from trapping centers in the nitride bulk and is commonly referred to as Poole–Frenkel conduction [1.15]. Thin nitride layers (< 30 nm), however, also show a strong electrode-limited contribution.

In silicon oxide, on the other hand, the current is determined mainly by the electrode characteristics, more specifically by the characteristics of the injection interface. This is due to the fact that SiO_2 has a large energy gap (about 9 eV compared to 5 eV for Si_3N_4) and a high energy barrier at its interface with aluminum or silicon. For example, the barrier of SiO_2 is about 3.2 eV for electrons in the conduction band of silicon and 4.8 eV for holes in the valence band, compared to 2 eV for holes and electrons in Si_3N_4, as shown in Fig. 1.9, which gives a comparison of the band structure for both materials. This means that conduction through SiO_2 will be determined primarily by electron injection, while, in Si_3N_4, both holes and electrons can contribute to the injection currents [1.16].

In both classes of nonvolatile memory devices, charge-trapping and floating gate devices, the charge needed to program the device has to be injected into an oxide layer, either to store it in the isolated traps in the nitride for the case of MNOS devices or to collect it at the floating gate in floating gate devices.

During the last two decades, various mechanisms for charge injection into the oxide have been considered. In order to change the charge content of floating gate devices, four mechanisms have been shown to be viable: Fowler–Nordheim tunneling (F–N) through thin oxides (< 12 nm) [1.5, 1.17], enhanced Fowler–Nordheim tunneling through polyoxides [1.18, 1.19], channel hot-electron injection (CHE)

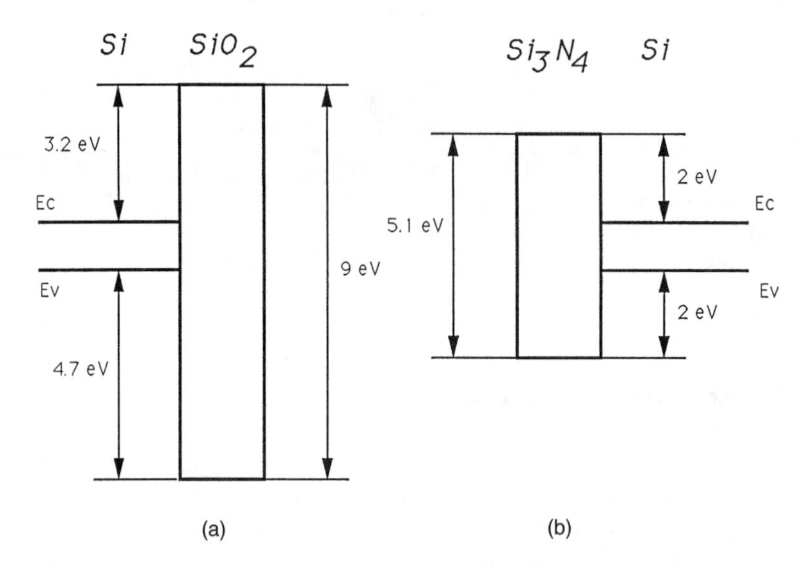

Figure 1.9 Energy band structures of (*a*) the Si–SiO_2 system and (*b*) the Si–Si_3N_4 system.

[1.20, 1.21], and source-side injection (SSI) [1.22, 1.23]. The first two are based on a quantum mechanical tunneling mechanism through an oxide layer, whereas the last two are based on injection of carriers that are heated in a large electric field in the silicon, followed by injection over the energy barrier of SiO_2. In order to change the charge content in charge-trapping devices, direct band-to-band tunneling and modified Fowler–Nordheim tunneling mechanisms are used. In the following sections, these six mechanisms are discussed briefly.

1.2.1 Fowler–Nordheim Tunneling

One of the most important injection mechanisms used in floating gate devices is the so-called Fowler–Nordheim tunneling, which, in fact, is a field-assisted electron tunneling mechanism [1.24]. When a large voltage is applied across a polysilicon–SiO_2–silicon structure, its band structure will be influenced as indicated in Fig. 1.10. Due to the high electrical field, electrons in the silicon conduction band see a triangular energy barrier with a width dependent on the applied field. The height of the barrier is determined by the electrode material and the band structure of SiO_2. At sufficiently high fields, the width of the barrier becomes small enough that electrons can tunnel through the barrier from the silicon conduction band into the oxide conduction band. This mechanism had already been identified by Fowler and Nordheim for the case of electrons tunneling through a vacuum barrier, and was later described by Lenzlinger and Snow for oxide tunneling. The Fowler–Nordheim current density is given by [1.24]:

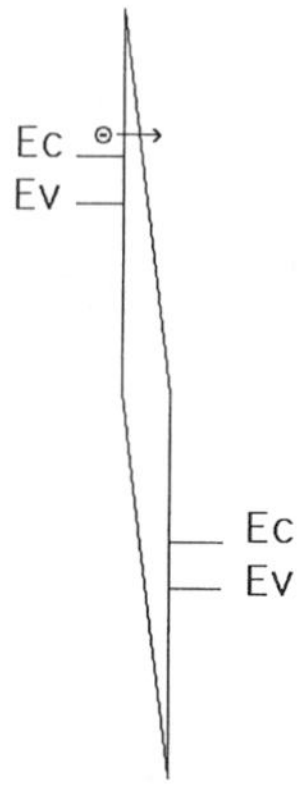

Figure 1.10 Energy band representation of Fowler–Nordheim tunneling through thin oxides: the injection field equals the average thin oxide field. Electrons in the silicon conduction band tunnel through the triangular energy barrier.

$$J = \alpha\, E_{inj}^2 \exp\left[\frac{-E_c}{E_{inj}}\right] \tag{1.3}$$

with

$$\alpha = \frac{q^3}{8\pi h \phi_b} \frac{m}{m^*} \tag{1.4}$$

and

$$E_c = 4\sqrt{2m^*} \frac{\phi_b^{3/2}}{3\hbar q} \tag{1.5}$$

where h = Planck's constant
ϕ_b = the energy barrier at the injecting interface (3.2 eV for Si–SiO_2)
E_{inj} = the electric field at the injecting interface
q = the charge of a single electron (1.6×10^{-19}C)
m = the mass of a free electron (9.1×10^{-31} kg)
m^* = the effective mass of an electron in the band gap of SiO_2 (0.42m [1.22])
$\hbar$ = $h/2\pi$

Equation (1.3) is the simplest form for the Fowler–Nordheim tunnel current density and is quite adequate for use with nonvolatile memory devices. A complete expression for the tunnel current density takes into account two second-order effects: image force barrier lowering and the influence of temperature.

The image force lowers the effective barrier height due to the electrostatic influence of an electron approaching the interface. Two correction factors $t(\Delta\phi_b)$ and $v(\Delta\phi_b)$, have to be introduced into Eq. (1.3), both of which are tabulated elliptic integrals and slowly varying functions. The reduction in energy barrier height ($\Delta\phi_b$) is given by [1.24]:

$$\Delta\phi_b = \frac{1}{\phi_b}\sqrt{\frac{q^3\,E_{inj}}{4\pi\epsilon_{ox}}} \tag{1.6}$$

Although tunneling is essentially independent of temperature, the number of electrons in the conduction band, available for tunneling, is dependent on the temperature. This dependence can be taken into account by a correction factor f(T), given by [1.24]:

$$f(T) = \frac{\pi ckT}{\sin(\pi ckt)} \tag{1.7}$$

with

$$c = \frac{2\sqrt{2m^*}\;t(\Delta\phi_b)}{hqE_{inj}} \tag{1.8}$$

Taking these two corrections into account, we see that the expression for the Fowler–Nordheim tunnel current density becomes:

$$J = \alpha\;E_{inj}^2\;\frac{1}{t^2(\Delta\phi_b)}\;f(T)\;\exp\left[\frac{-E_c}{E_{inj}}\;v(\Delta\phi_b)\right] \tag{1.9}$$

The influence of the correction factors is small, however, and, for most practical calculations, the basic Eq. (1.3) is sufficiently accurate.

The Fowler–Nordheim tunnel current density is, thus, almost exponentially dependent on the applied field. This dependence is shown in Fig. 1.11*a* for the monocrystalline silicon–SiO_2 interface. The Fowler–Nordheim current is usually plotted as $\log(J/E^2)$ versus $1/E$, which should yield a straight line with a slope proportional to the oxide barrier, as shown in Fig. 1.11*b*. In this case, the numerical expression is

$$J\;[A/m^2] = 1.15\;10^{-6}\;E_{inj}^2\;\exp\left[\frac{-2.54\;10^{10}}{E_{inj}}\right] \tag{1.10}$$

which, at an injection field of 10 MV/cm, leads to a current density of approximately 10^7 A/m^2 or 10^7 pA/μm^2. This high value of injection field is of the order of that needed across the oxide during the programming of a nonvolatile memory device. The breakdown field of these oxides should, of course, be significantly larger than this value. In order to reach these high-field values and limit the voltages needed during programming, very thin tunnel oxides are used; an injection field of 10 MV/cm is attained by applying a voltage of 10 V across an oxide of 10 nm thickness. In order to reduce the programming voltage, the tunnel oxide should become even thinner. A thickness of 6 nm, however, is the lower limit for good retention behavior. But these thin oxides are difficult to grow with low defect densities, as is required for floating gate devices. Moreover, below these values, other injection mechanisms, such as direct tunneling, can become important. Yield considerations now limit the usable oxide thicknesses to 8 to 10 nm [1.23].

It should be noted that the tunnel current density is totally controlled by the field at the injecting interface, and not by the characteristics of the bulk oxide. Once

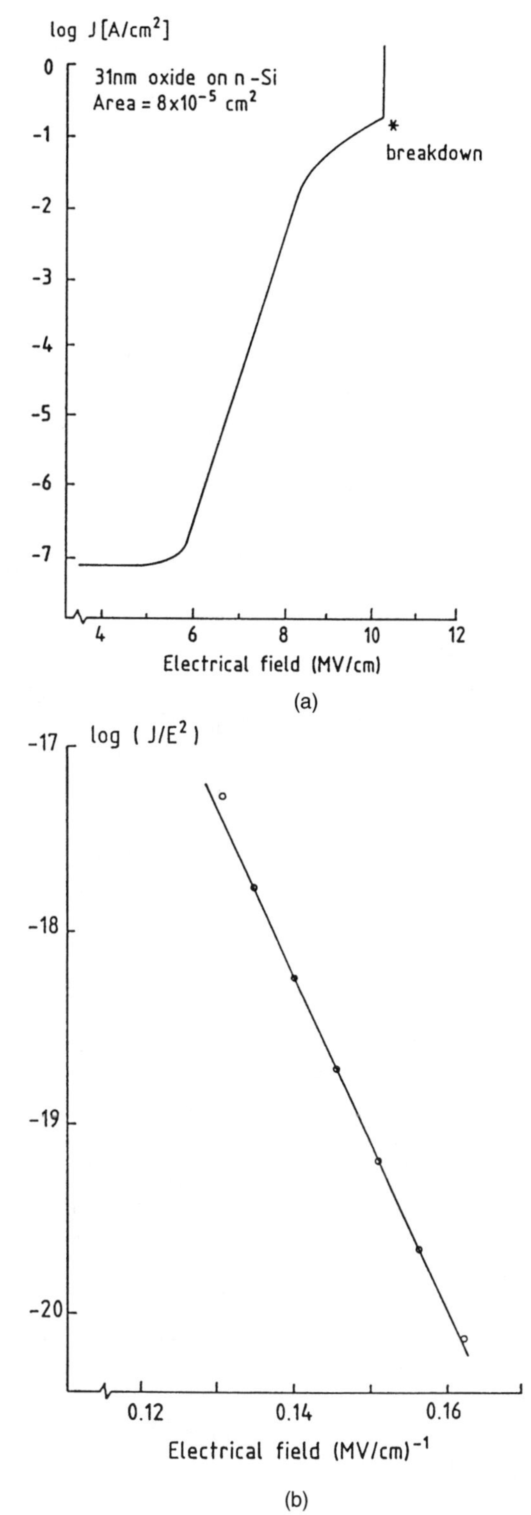

Figure 1.11 (*a*) Fowler–Nordheim tunneling current as a function of applied field across the oxide. The current is exponentially dependent on the field. Breakdown occurs around 10 MV/cm. (*b*) Fowler–Nordheim plot: J/E^2 as a function of $1/E$, extracted from the data (*a*). A straight line is obtained.

the electrons have tunneled through the barrier, they are traveling in the conduction band of the oxide with a rather high saturated drift velocity of about 10^7 cm/s [1.26].

For the calculation of the injection field at a silicon–SiO_2 interface, however, the flatband voltage has to be taken into account as seen by

$$E_{inj} = \frac{V_{app} - V_{fb}}{t_{ox}} \tag{1.11}$$

where V_{app} = the voltage applied across the oxide
V_{fb} = the flatband voltage
t_{ox} = the thickness of the oxide

When voltages are applied so that the silicon is driven into depletion, a voltage drop in the induced depletion layer must be accounted for in the calculation of the oxide field.

The tunnel current for a given applied voltage can be calculated as the product of the tunnel current density and the injecting area only if the tunnel current density has the same value over the whole injecting surface—that is, if the injection occurs uniformly over the area of the tunnel oxide. This assumes a perfectly plane injecting interface which, in many practical devices, will not be the case. Special cases of nonuniform injection are discussed in Sections 1.2.2 and 1.2.3.

1.2.2 Polyoxide Conduction

Fowler–Nordheim tunneling requires injection fields on the order of 10 MV/cm to narrow the Si–SiO_2 energy barrier so that electrons can tunnel from the silicon into the SiO_2 conduction band, as discussed in the previous section.

In oxides thermally grown on monocrystalline silicon, the injection field is equal to the average field in the SiO_2; therefore, thin oxides have to be used to achieve large injection fields at moderate voltages. Oxides thermally grown on polysilicon, called polyoxides, however, show an interface covered with asperities due to the rough texture of the polysilicon surface [1.27, 1.28]. This has led to the name "textured polyoxide." These asperities give rise to a local field enhancement at the interface and an enhanced tunneling of electrons [1.29, 1.30]. In polyoxides, the field at the injecting interface is, therefore, much larger than the average oxide field. Consequently, the band diagram of a polysilicon–polyoxide interface is as shown schematically in Fig. 1.12. Average oxide fields of the order of 2 MV/cm are sufficient to yield injection fields of the order of 10 MV/cm. This has the big advantage that large injection fields at the interface can be obtained at moderate voltages using relatively thick oxides, which can be grown much more reliably than the thin oxides necessary for Fowler–Nordheim injection from monocrystalline silicon.

A quantitative analysis of the tunnel current–voltage relations for polyoxides is rather complex. Although the tunnel mechanism itself is described by the same formula (Eq. 1.3), discussed in a previous section, the difficulty lies in the accurate determination of the injection fields to be used. It is no longer possible to use a single value for this injection field because of the nonuniformity of the field enhancement

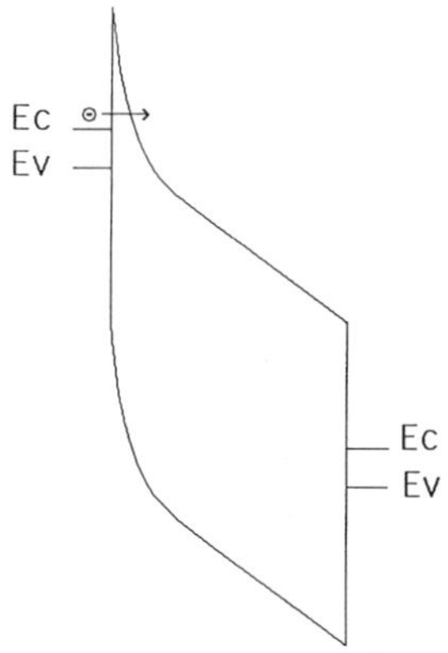

Figure 1.12 Energy band representation of Fowler–Nordheim tunneling through oxides thermally grown on polysilicon: the injection field is much higher than the average oxide field. The high injection field is due to local field enhancement at polysilicon–oxide interface asperities.

over the injecting interface. Indeed, the field enhancement factor is not uniform over the surface of one asperity bump [1.31–1.33]: the factor is maximum at the top of the asperity and decreases strongly down the slope on the bump surface. In addition, variations of the bump shape may be another cause for the nonuniformity.

In the past, attempts have been made to model the current through the polyoxide by use of some mean field enhancement factor [1.34], but as was proven in [1.32], this method always leads to incorrect results. A complete model for the current conduction must be based on [1.32, 1.33]:

- the Fowler–Nordheim expression for tunnel current density
- a model for the distribution of the field enhancement factors over the total injecting area
- a model for the charge-trapping behavior of the oxide under current injection

The last-named model has to be taken into account because charge trapping is of much more importance in polyoxides than in oxides grown on monocrystalline material. This is again due to the strong nonuniform field enhancement. Initially, the injection current originates almost completely from the regions of maximum field enhancement. Extremely large current densities occur at these injection points, leading to strong local trapping of electrons near these sites. This trapping reduces the injection locally. Consequently, the current is taken over by regions with a slightly lower field enhancement. This process proceeds gradually so that the injection current, which initially is extremely localized, becomes more and more uniform and decreases continuously. Unlike conventional Fowler–Nordheim injection in which the trapping occurs only after some critical current level has been reached, charge trapping and current injection occur simultaneously over the whole current range during polyoxide injection. The nonuniform field enhancement at the polysilicon–polyoxide interface makes it impossible to use a closed analytical expression for polyoxide conduction. A complete model for polyoxide conduction, based on the principles indicated above, can be found in [1.32, 1.33].

An example is given in Fig. 1.13 where the injection current is shown during a ramped voltage experiment for two consecutive runs on the same polyoxide capacitor. The dashed lines represent the experimental currents, while the solid lines are simulations based on the above referenced model [1.32]. As can be seen, during the first run the current is increasing less rapidly than expected from the conventional Fowler–Nordheim mechanism. This is due to the gradual decrease of the mean enhancement factor of the polyoxide surface, which is caused by the local electron trapping and accompanying shielding of the sites of maximum field enhancement. The second ramp shows a large shift with respect to the first one. Unlike the case of uniform Fowler–Nordheim injection, however, *this shift cannot be interpreted in terms of trapped charge only*, but is due mainly to a decrease in the mean enhancement factor after the first run.

Another consequence of the fast decrease in mean enhancement factor due to local trapping in polyoxides is the fast current decrease observed during measurement of the time behavior of the current after application of a voltage step across the polyoxide. This decrease can be described by a time power law ($I = C\, t^{-n}$), with a decay factor n. Whereas this decay factor is expected to be 1 for uniform trapping, the decay factor is found to be smaller than 1 for polyoxides [1.29]. Again, this is because the decay is due not only to charge trapping but also to the decrease in the mean enhancement factor of the polyoxide interface. An example is shown in Fig. 1.14, where, again, experimental results are compared with results from the abovementioned model [1.32].

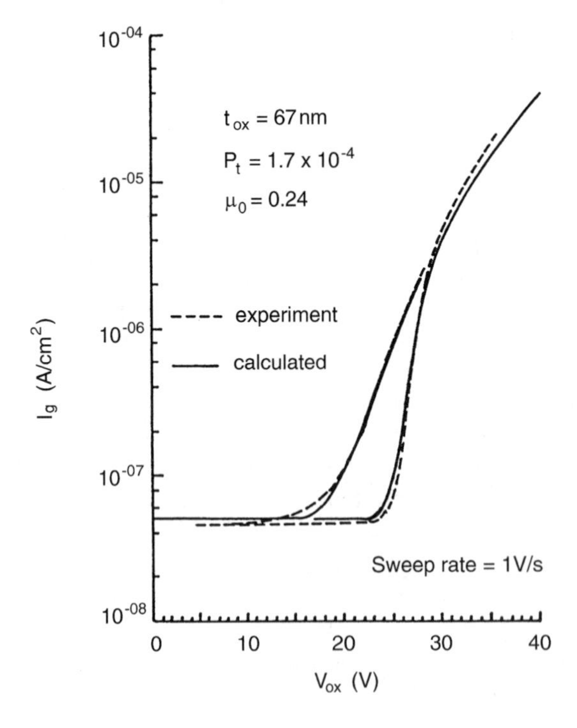

Figure 1.13 Injection current in polyoxides as a function of applied voltage during a ramped voltage experiment for two consecutive runs on the same polyoxide capacitor. The dashed lines represent experimental currents, while the solid lines are simulations based on the model by Groeseneken et al. [1.32].

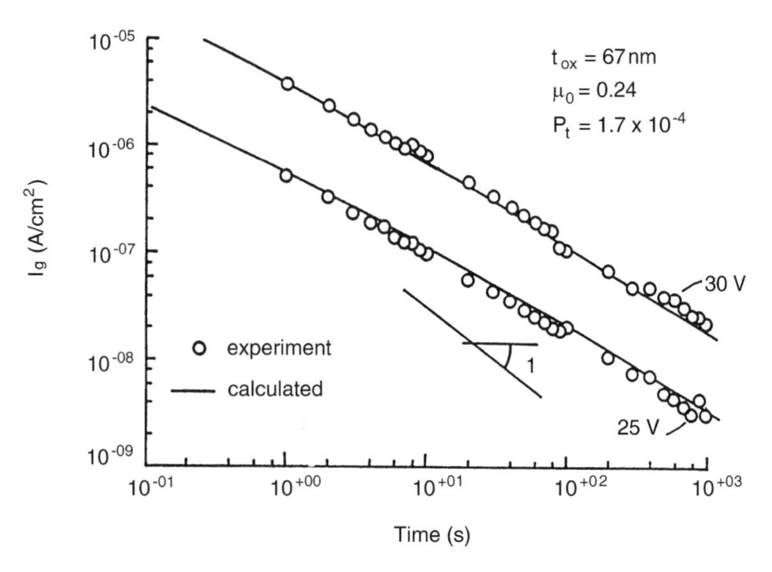

Figure 1.14 Time dependence of the polyoxide current for two values of applied voltage. The circles are experimental values, while the solid lines are results, based on the model of [1.32].

Polyoxide conduction has the advantage that considerable current levels can be attained at moderate average oxide fields, and thus, moderate applied voltages. The need for thin polyoxides is, therefore, not so stringent. From a reliability point of view, this is an advantage since the oxides are not stressed at large fields during programming so that dielectric breakdown failures are avoided [1.25]. On the other hand, the growth of textured polyoxides has to be carefully controlled in order to obtain the desired interface features (shape and size of the asperities) that determine the injection current and reliability characteristics. For this reason, reproducibility may be a problem for this kind of injection mechanism. Another disadvantage is that the injection is asymmetric with respect to polarity. For injection from a top polysilicon layer, the currents are much smaller. Finally, the strong change in injection currents due to a decrease in mean enhancement factor during current injection can pose severe constraints on the number of programming cycles that can be allowed if this mechanism is used for programming a memory cell [1.32]. In nonvolatile memories, polyoxides of 25 nm to 60 nm are used with programming voltages from 12 V up to 20 V.

1.2.3 Hot-Electron Injection

At large drain biases, the minority carriers that flow in the channel of a MOS transistor are heated by the large electric fields seen at the drain side of the channel and their energy distribution is shifted higher. This phenomenon gives rise to impact-ionization at the drain, by which both minority and majority carriers are generated. The highly energetic majority carriers are normally collected at the substrate contact and form the so-called substrate current. The minority carriers, on the other hand, are collected at the drain. A second consequence of carrier heating occurs when some of the minority carriers gain enough energy to allow them to surmount the SiO_2 energy barrier. If the oxide field favors injection, these carriers are injected over the barrier into the gate insulator and give rise to the so-called hot-carrier injection gate current [1.35, 1.36]. This mechanism is schematically represented for the case of an n-channel transistor in the energy band diagram shown in Fig. 1.15.

For nonvolatile memory applications, n-channel transistors are generally used, and therefore, the discussion here will be limited to n-channel devices. In case of an n-channel transistor, the gate current of the transistor consists of those channel hot electrons that actually reach the gate of the transistor. In a floating gate transistor, these electrons change the charge content of the floating gate. An important difference between hot-carrier injection and the two previously discussed injection mechanisms is that, with hot-electron injection, it is only possible to bring electrons onto the floating gate. They cannot be removed from the floating gate by the same mechanism. Although the use of hot-hole injection as a compensating programming mechanism has been tried [1.37], it has never found application due to the very small current levels that can be attained in this way.

In the past, several models have been used to describe the gate current due to channel hot-electron injection. In contrast to the Fowler–Nordheim tunneling case, no closed form analytical expression exists for the channel hot-electron injection current due to the complex two-dimensional nature of the phenomenon and many unknown physical parameters. Therefore, the models are merely qualitative. They can be divided into three main categories: the lucky electron models, the effective electron temperature models, and the physical models.

The lucky electron model [1.38, 1.39] assumes that an electron is injected into the gate insulator if it can gain enough energy in the large lateral electric field with-

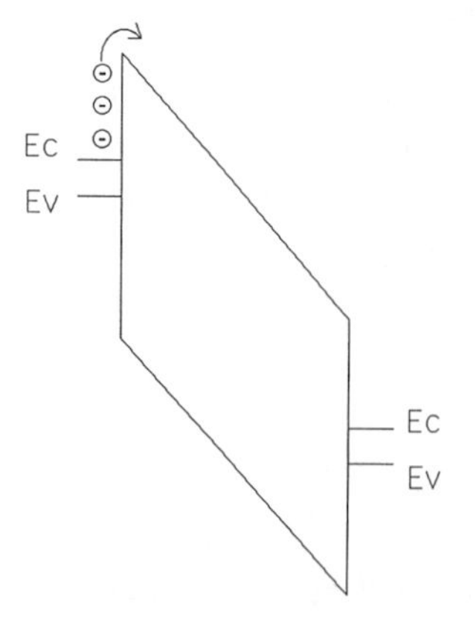

Figure 1.15 Energy band representation of hot-electron injection in the oxide; the oxide field is low, but the electrons are heated by the high lateral fields at the drain in the channel. Some of them acquire enough energy to overcome the interface energy barrier.

out undergoing a collision, by which energy could be lost. By phonon scattering, the electrons are then redirected toward the Si–SiO_2 interface. If these electrons can reach the interface and still have enough energy to surmount the Si–SiO_2 energy barrier (and eventually also a repulsive field), they will be injected into the gate insulator.

The effective electron temperature model [1.40] assumes that the electrons forming the channel current are heated and become an electron gas with a Maxwellian distribution, with an effective temperature, T_e, that is dependent on the electric field. The gate current can then be calculated as the thermionic emission of heated electrons over the interface barrier energy.

The physical models [1.41] attempt to calculate the gate currents based on a more physical treatment and an accurate solution of the two-dimensional electric field distribution at the drain side of the channel. Then, the gate current is calculated based on an injection efficiency that is dependent on the interface barrier energy and the lateral electric fields.

For all the above-mentioned models, we always have to keep in mind that a difference exists between the number of injected electrons and the number of electrons actually reaching the gate. Indeed, due to a repulsive oxide field, all or part of the injected electrons can be repelled into the silicon [1.42].

Qualitatively, it can be stated that the gate current is determined on the one hand by the number of hot electrons and their energy distribution (which is largely dependent on the electric fields occurring in the channel of the transistor) and on the other hand by the oxide field (which determines the fraction of hot electrons that can actually reach the gate).

The magnitude of the gate current is dependent on both the applied gate and drain voltages. A characteristic gate current, as a function of the applied gate voltage and with the drain voltage as a parameter, is shown in Fig. 1.16 for an n-channel transistor. It is important to notice that the hot-electron gate current shows a maximum at approximately $V_g = V_d$, and thus, is not a monotonically increasing function of the applied gate voltage, as is the case for both Fowler–Nordheim and polyoxide conduction.

This typical shape is explained by both determining factors—the gate and drain voltages [1.36, 1.42]. For gate voltages greater than the drain voltage, the oxide field is always favorable for charge collection at the gate, which means that the gate current is limited by the number of hot electrons that are injected. The lateral electric field, and thus, the number of hot electrons that can be injected into the oxide, increases with decreasing gate voltage. Therefore, for $V_g > V_d$, the gate current increases with decreasing gate voltage. For gate voltages smaller than the drain voltage, however, the oxide field becomes repulsive for the injected electrons. Therefore, part of the injected electrons are repelled into the channel. Although the number of hot electrons that are available to be injected still increases with decreasing gate voltage, the gate current now drops rapidly with decreasing gate voltage. Due to this typical gate voltage dependence of the hot-electron gate current, the gate voltage during programming of a nonvolatile memory cell, using hot-electron injection, has to be chosen carefully in relation to the applied drain voltage.

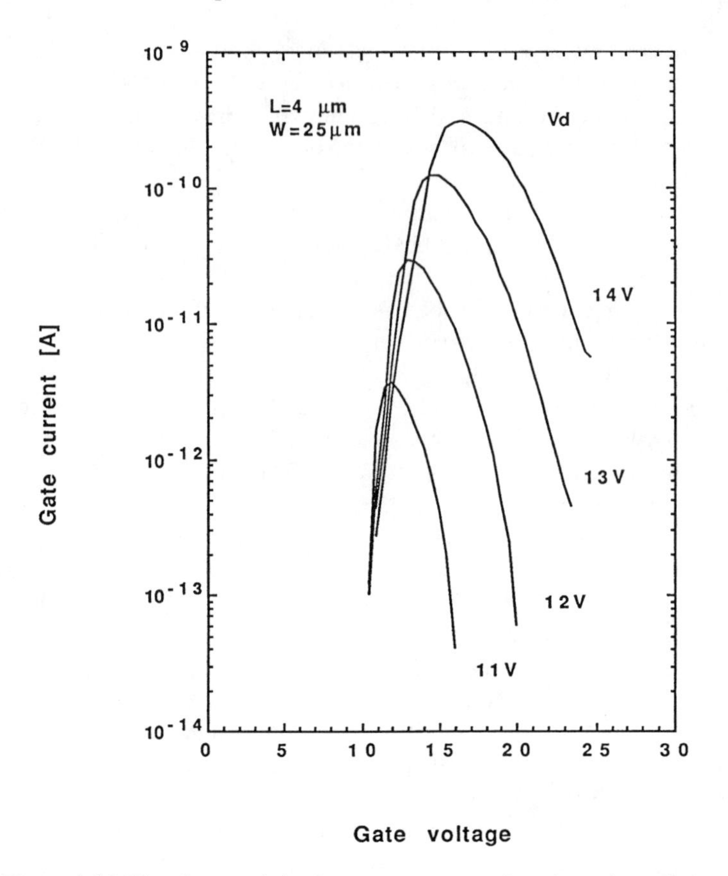

Figure 1.16 Hot-electron injection currents as a function of applied gate voltage with the drain voltage as a parameter. The maximum current occurs when $V_g = V_d$ and is exponentially dependent on the drain voltage.

In order to evaluate the dependence of the injection current on processing and geometrical parameters, a simplified expression for the lateral electric field in the channel can be used [1.43]:

$$E \approx \frac{V_d - V_{dsat}}{L} \tag{1.12}$$

with

$$L \approx 0.22\, t_{ox}^{1/3}\, x_j^{1/2} \tag{1.13}$$

and with V_{dsat} expressed as [1.44]:

$$V_{dsat} = \frac{(V_g - V_t)\, L_{eff}\, E_{sat}}{V_g - V_t + L_{eff}\, E_{sat}} \tag{1.14}$$

where E_{sat} is the electric field at which the electron mobility saturates.

From these formulas, it can be concluded that the gate current increases with thinner gate oxides, shallower junctions, smaller effective channel lengths, and higher substrate doping levels (through the influence on the threshold voltage at the drain through the body effect).

1.2.4 Source-Side Injection

The main disadvantage of the conventional channel hot-electron injection mechanism for programming a nonvolatile element stems from its low injection efficiency, and consequently, its high power consumption. This is due to the incompatibility of having a high lateral field and a high vertical field, favorable for electron injection, at fixed bias conditions, as explained in the previous section. Indeed, the lateral field in a conventional MOS device is a decreasing function of the gate voltage, while the vertical field increases with the gate voltage. Therefore, in order to generate a large number of hot electrons, a low gate voltage is required, combined with a high drain voltage. However, for electron injection and collection on the floating gate of the memory device, a high gate voltage and a low drain voltage are required (see Fig. 1.17). In practice, both gate and drain voltages are kept high as a compromise. The main drawback is clearly the high drain current (on the order of mA's) and the correspondingly high power consumption.

Therefore, a novel injection scheme, now commonly referred to as source-side injection (SSI), has been proposed to overcome this problem [1.45]. In most cases, the MOS channel between the source and drain regions is split into two "subchannels" controlled by two different gates. The gate on the source side of the channel is biased at the condition for maximum hot-electron generation, that

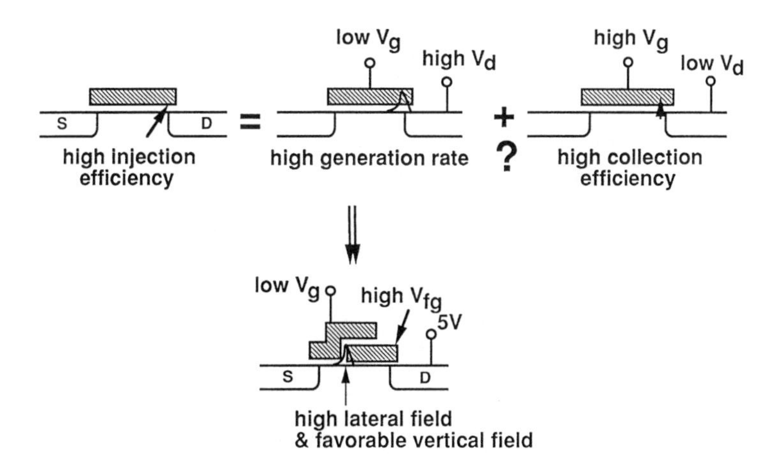

Figure 1.17 Schematic representation of the problem of low-injection efficiency for channel hot-electron injection and the principle of source-side injection.

is, very close to the threshold voltage of this channel [1.42]. The gate at the drain side, which is the floating gate of the cell, is capacitively coupled to a potential that is comparable to or higher than the drain voltage in order to establish a vertical field component that is favorable for hot-electron injection in the direction of the floating gate. The latter condition can be accomplished either by implementing an additional gate with a high coupling ratio toward the floating gate [1.23, 1.45], or by using a high drain-coupling ratio [1.22]. As a result, the drain potential is entirely or partially extended toward the region between the gates that control the MOS channel. This effect is referred to as the virtual drain effect since the inversion layer under the floating gate merely acts as a drain extension, while the effective transistor channel is formed by the subchannel at the source side of the device [1.46]. Consequently, a high lateral field peak is obtained in the gap between both subchannels (Fig. 1.17). The hot electrons are thus generated inside the MOS channel and not at the drain junction of the cell. Because of the high floating gate potential, the vertical field at the injection point is favorable for electrons, and most of the generated hot electrons that overcome the potential barrier between the channel and the oxide layer are effectively collected on the floating gate.

The main advantage of this injection mechanism is the much higher injection efficiency (on the order of 10^{-3} and higher) which allows for fast 5 V-only and even 3.3 V-only operation combined with a low power consumption [1.23, 1.45]. This is illustrated in Fig. 1.18 where the gate currents of the conventional channel hot-electron injection and the source-side injection mechanisms are shown for comparable devices [same channel width to length ratio (W/L), same drain voltage, same technology]. It is clear that the SSI mechanism provides a gate current that is more than three orders of magnitude higher than conventional hot-electron injection [1.46]. At the same time, the drain current in the SSI case is also reduced by a factor

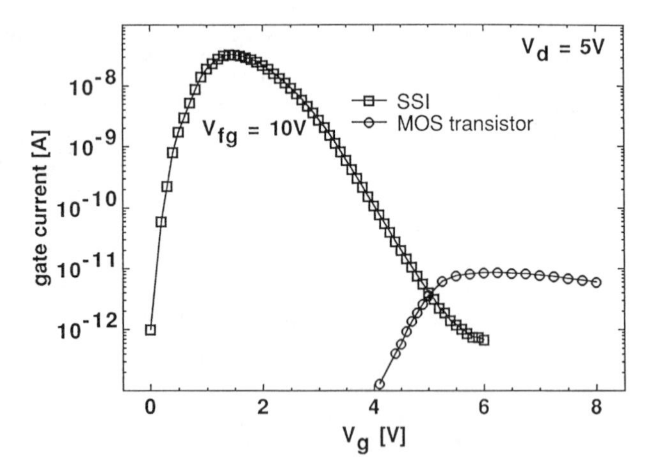

Figure 1.18 Comparison of injected gate current for source-side injection (SSI) and channel hot-electron injection, both measured at the same drain voltage of 5V.

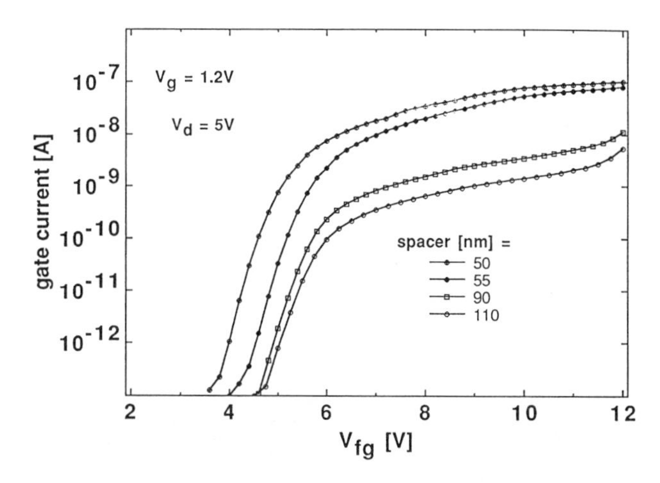

Figure 1.19 Source-side injection current versus floating gate voltage for various values of the interpoly spacer thickness. This characteristic also gives the evolution of the programming current during a programming operation.

of 40 with respect to the conventional case [1.45]. Figure 1.19 shows typical gate current characteristics for the SSI mechanism, but this time as a function of the floating gate voltage. The gate current tends to saturate with increasing floating gate voltage because the virtual drain potential (i.e., the channel potential at the injection point) approaches the externally applied drain voltage. Since the floating gate voltage changes during programming, the maximum observed in the conventional gate current characteristic (Fig. 1.16) is no longer relevant in the SSI case. The gate current decreases monotonically, and only slightly, while programming an SSI cell. Figure 1.19 also shows that the gate current is a strong function of the interpoly width between the gates that control the subchannels. Furthermore, the SSI mechanism is no function of the drain profile and is instead only a smooth linear function of the channel length of the device. This is in strong contrast to conventional hot-electron injection where the injection is strongly dependent on both drain profile and channel length.

1.2.5 Direct Band-to-Band Tunneling and Modified Fowler–Nordheim Tunneling

Section 1.2.1 treated Fowler–Nordheim tunneling, which is field-assisted electron tunneling from the silicon band into the silicon dioxide band through the triangular energy barrier. In MNOS devices with ultra-thin oxides ($< 3\,\text{nm}$), the injection current can either be direct silicon band to nitride band tunneling only through the oxide barrier, as illustrated in Fig. 1.20*a*, or modified Fowler–Nordheim tunneling through the oxide barrier and a nitride barrier, as shown in

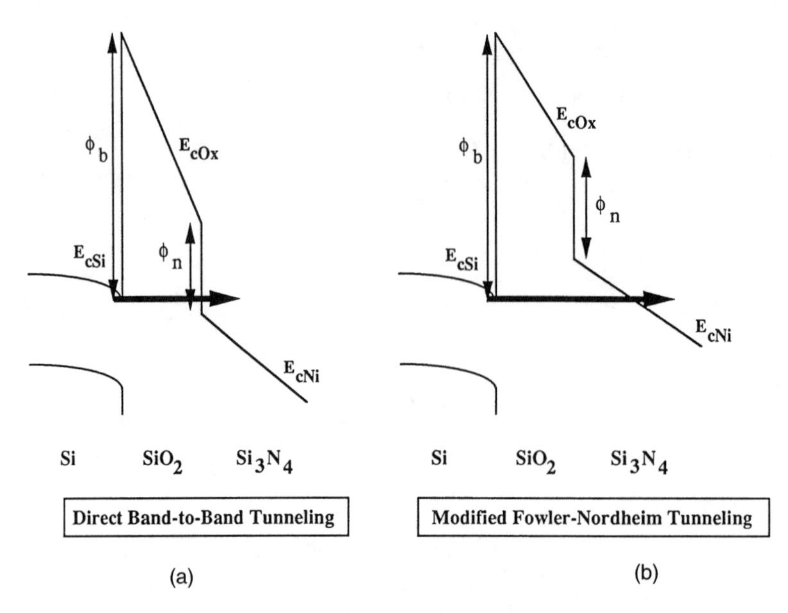

Figure 1.20 Energy band representation of (*a*) direct band-to-band tunneling and (*b*) modified Fowler–Nordheim tunneling between the silicon and the nitride conduction band.

Fig. 1.20*b*. Whether one or the other of these conditions applies depends strongly on the values of the oxide field and oxide thickness. These currents can be expressed as in [1.47]:

$$J = C_{FN}\, E_{ox}^2\, P_{ox}\, P_n \tag{1.15}$$

where C_{FN} is a constant with a similar meaning as α in Eq. (1.3), and P_{ox} and P_n represent the tunneling probabilities through the oxide and the nitride barriers, respectively, and are given by

$$P_{ox} = \exp\left\{-\frac{4}{3h}\sqrt{2qm_{ox}^*}\,\frac{\left[\phi_b^{3/2} - (\phi_b - E_{ox}t_{ox})^{3/2}\right]}{E_{ox}}\right\} \tag{1.16}$$

and

$$P_n = \exp\left\{\frac{4}{3h}\sqrt{2qm_n^*}\,\frac{(\phi_b - \phi_n - E_{ox}t_{ox})^{3/2}}{E_n}\right\} \tag{1.17}$$

where E_{ox} is the field in the oxide and E_n is the field in the nitride, m_{ox}^* and m_n^* are the effective masses in oxide and nitride, respectively, and ϕ_n is the oxide–nitride barrier. In Eqs. (1.16) and (1.17), a negative term within a radical must be replaced by zero.

If $(\phi_b - \phi_n - E_{ox}t_{ox}) < 0$, direct tunneling occurs only through the oxide potential barrier and $P_n = 1$ (see Fig. 1.20*a*). If $(\phi_b - E_{ox} \cdot t_{ox}) < 0$, Eq. (1.16) reduces to the expression for Fowler–Nordheim tunneling [Eq. (1.9)]. Since $E_n \approx (\epsilon_{ox}/\epsilon_n)E_{ox}$ with ϵ_{ox} and ϵ_n, the dielectric constants of oxide and nitride, respectively, Eq. (1.15) provides the current–oxide field relation for tunneling through these double potential barriers. The oxide field, E_{ox}, for a double insulator system (oxide thickness, t_{ox}, and nitride thickness, t_n) and for a given V_{app} and V_{fb} can be obtained from an expression similar to Eq. (1.11) for the case of a single insulating layer from

$$E_{ox} = \frac{V_{app} - V_{fb}}{t_{ox} + \frac{\epsilon_{ox}}{\epsilon_n} t_n} \tag{1.18}$$

1.3. BASIC NVSM MEMORY PRODUCTS

The nonvolatile memory cell concept has been used in several kinds of applications, and many different products have emerged in recent years. The core of the applications are high-volume stand-alone nonvolatile memories, but besides these, it has also been applied to other purposes such as electrically programmable logic devices (EPLD), application specific integrated circuits (ASIC) (embedded memories), and redundancy. Before discussing the different types of nonvolatile cells, this section first treats some of the important features of the main nonvolatile memory products. The aim is not to be complete, however, for the application fields and the product range for nonvolatile memory cells are so large that they cannot be covered within the limited focus of this chapter. A more general overview of the trends in nonvolatile memories can be found in [1.48, 1.49].

This section discusses the main features of EPROM and OTP, EEPROM, Flash EEPROM, and NOVRAM memory products, as well as a new type of concept, FRAM. These classes of nonvolatile memory products emerged under the influence of three main factors: (1) the limitations posed by the available cells (EPROM/OTP), (2) the requirements of the users (EEPROM, NOVRAM), and (3) market and price considerations (Flash EEPROM).

1.3.1 EPROM/OTP

The electrically programmable read-only memory (EPROM) was, in fact, the first nonvolatile memory that could be electrically programmed by the user and that could be erased afterward. All EPROM products rely on the floating gate cell concept. Present EPROM devices all use channel hot-electron injection. Since this mechanism can only supply electrons to the floating gate, EPROM memories are not electrically erasable. UV light is used to erase the memory. For programming and reading, the memory is byte addressable, and each byte can be addressed separately. Obviously, the erase operation always affects the whole memory. It is the user who can perform both operations. For both operations, however, some additional tools are needed—an EPROM programmer and UV light for erasing.

The channel length of EPROM cells has been steadily decreasing, reaching values down to 0.8 μm [1.50]. Since the end of the 1980s, however, EPROMs have been gradually replaced by Flash EEPROM products. Channel hot-electron programming requires high currents and high voltages. Consequently, EPROM memory products require an external supply voltage of typically 12 V for programming. The programming time ranges from 1 ms down to 100 μs per byte of information. Erasing typically takes 20 minutes of UV light exposure. During erasure, the component is not powered on. Because the EPROM functionality does not need addressing down to byte level during an erase operation, the memory cell can be kept fairly simple. One floating gate transistor suffices to build an EPROM memory cell, as illustrated in Fig. 1.21, and is, therefore, called a single-transistor memory cell. This allows for small cell sizes in the range of 8 μm^2 for 0.8 μm technologies, and bit densities comparable to those of DRAMs. At present, the highest bit densities available are 4 and 8 Mbit. The evolution of EPROM cell size and bit density is shown in Figs. 1.22 and 1.23 [1.49].

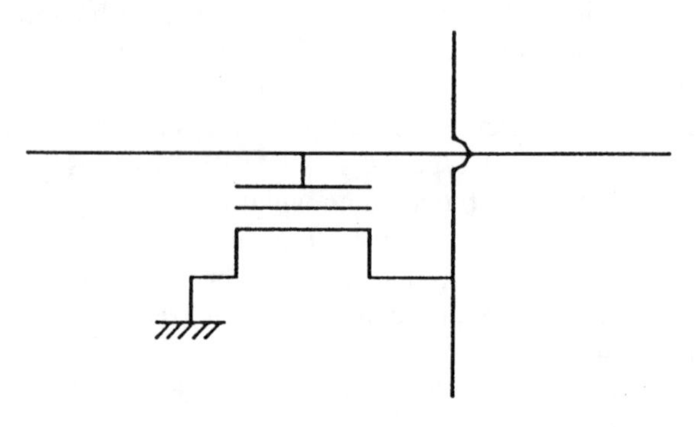

Figure 1.21 Single-transistor EPROM floating gate memory cell.

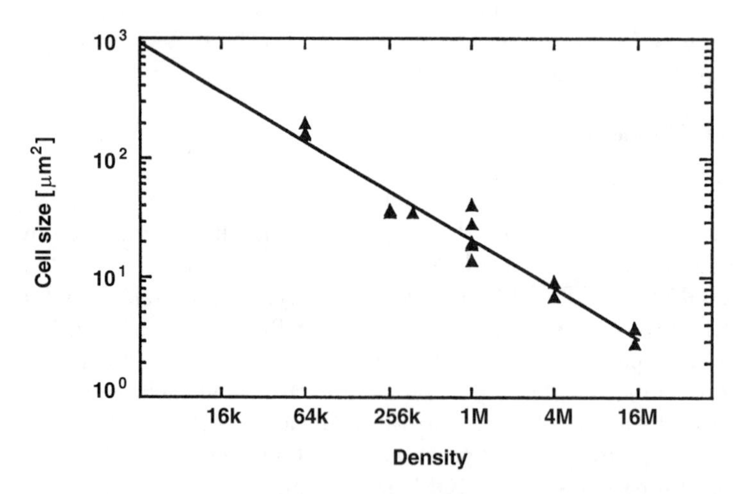

Figure 1.22 Evolution of the cell size of EPROM memory cells as a function of memory bit density.

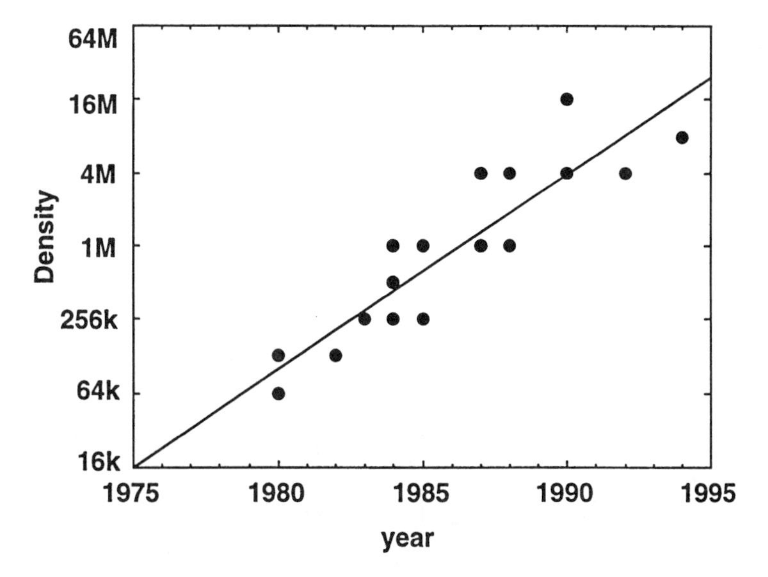

Figure 1.23 Evolution of bit density as a function of time showing a new generation of bit densities every three years.

Since UV light is used for erasure, a quartz window has to be provided in the EPROM package, which makes this package quite expensive. The package also has to be taken out of the circuit board in order to erase or reprogram the memory. In order to avoid these problems, a new product, which actually uses the same chips as the EPROM, called the one-time programmable (OTP) memory, was developed. This device can be written only once and is used like a PROM. Since no erasure of the memory is intended, the quartz window is not necessary and the device can be housed in a cheaper plastic package.

1.3.2 EEPROM

Although EPROM memories are reprogrammable, the reprogramming of the device is not user-friendly. The circuit has to be taken off the circuit board. The erase operation takes about 20 minutes, and then the whole memory circuit has to be reprogrammed byte by byte. This rather tedious erase procedure must be performed even if the content of a single byte has to be changed. These drawbacks have been obviated in the electrically erasable programmable read-only memory (EEPROM). In this type of nonvolatile memory circuit, all operations are controlled by electrical signals. The circuit can be reprogrammed while residing on the circuit board. Each operation, including erasing, can be performed in a byte-addressable way.

This higher level of functionality results in a larger memory cell. Since the EEPROM is byte addressable for reading and for all programming operations, the memory cell has to consist of a memory transistor and a select transistor [1.51] as shown in Fig. 1.24, thus leading to the so-called two-transistor memory cell. As a result, this memory cell is larger than the EPROM cell. Consequently, the densities of EEPROM products have always lagged behind EPROM densities by one to two generations, as illustrated in Fig. 1.25, with 1 Mbit to 2 Mbit memories as the present state-of-the-art EEPROM densities. Typical cell sizes of 1 Mbit parts range between 30 and 50 μm^2.

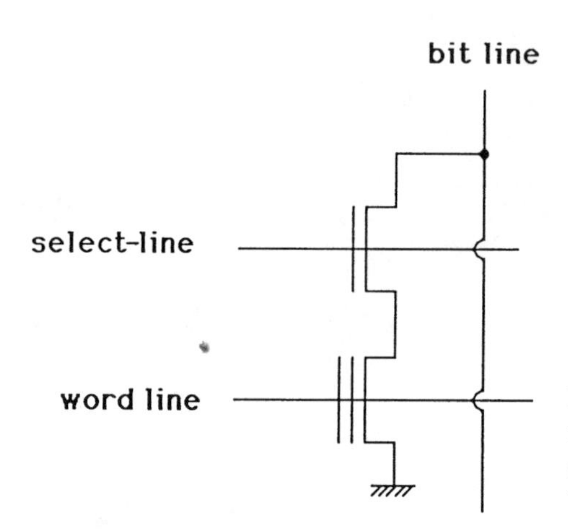

Figure 1.24 Two-transistor EEPROM memory cell. In order to allow byte selective write and erase, a select transistor is added to the memory cell.

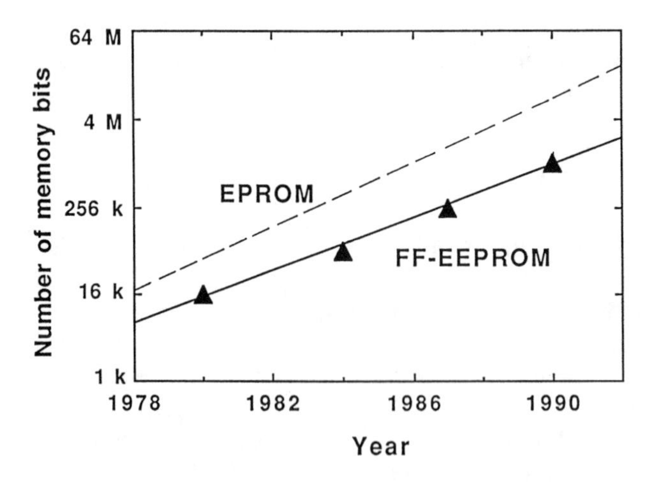

Figure 1.25 Evolution of the bit density as a function of time for EEPROM, and comparison with EPROM. EEPROM is lagging behind about one to two generations.

Charge-trapping as well as floating gate cells are used for EEPROM products. As mentioned previously, the MNOS cell already is inherently electrically-erasable. The floating gate cells usually rely on Fowler–Nordheim tunneling or polyoxide conduction in order to achieve electrical programmability for both operations.

The continuing search for a nonvolatile memory part that is as easy to use as a RAM has led to the incorporation of more and more features on the memory chip, which had to be provided externally in earlier generations of EEPROMs. Considering the EEPROM evolution to this point in time, we can mention three generations. The first generation required an external, high-voltage power supply and wave-shaped write signals with critical specifications for rise, overshoot, and pulse times. In the second generation, the wave-shaping and high-voltage external power supply were eliminated, leading to the first 5 V-only EEPROM products [1.52–1.54]. The third generation is even more complex, with features such as data and address latching, internal timing for the programming operation, page-mode programming capabilities, on-chip pulse shaping, complete transistor-transistor logic (TTL) compatibility, power on/off protection circuitry, on-chip error checking and correcting circuits, data polling possibilities, and 5 V-only operation [1.55, 1.56]. This evolution has made the present EEPROM products completely compatible with other types of memories like SRAM and DRAM.

The EEPROM circuits have become 5 V-only by generating programming voltages on the chip by means of voltage multiplier circuits [1.52–1.54]. This is possible only because programming mechanisms rely on tunneling (direct, Fowler–Nordheim, or polyoxide tunneling), which does not require large programming currents. The erase operation that has to be performed before a byte can be written into a new state is made invisible to the user. Every write request is automatically preceded by the proper erase operation, which is totally controlled by circuits incorporated on the memory chip.

The properly shaped control signals for programming in all recent circuits are generated on chip. The memory just needs a TTL-compatible pulse to initiate a write operation. The timing and application of the different voltage levels are controlled by on-chip circuits [1.55]. Moreover, by using an intelligent data polling feature, the external circuitry can find out if the data have been successfully written into the memory, or if the internal write operation is still busy [1.55], which can be used to reduce the effective programming time.

In order to shorten the quite lengthy programming operation, so-called page-mode programming has been added [1.55, 1.56]. The user writes a whole page (typically 16 to 64 bytes) to the EEPROM as if it were a RAM. On the EEPROM chip, the information and the appropriate addresses are stored, and afterward, the whole page is written in parallel into the nonvolatile memory cells. This effectively reduces the programming time per byte by a factor of 16 to 64.

Devices incorporating all the above-mentioned features are called full-featured EEPROMs. Another class of EEPROM devices is high-speed EEPROMs. These circuits, though not as user-friendly, have a read access time in the range of 30 to 50 ns, comparable to SRAM devices and to bipolar products [1.57].

1.3.3 Flash EEPROM

During the 1980s, a novel nonvolatile memory product was introduced, referred to as the Flash EEPROM [1.58]. The general idea was to combine the fast programming capability and high density of EPROMs with the electrical erasability of EEPROMs. The first products were merely the result of adapting EPROMs in such a way that the cell could be erased electrically. Consequently, these devices used channel hot-electron injection for programming and Fowler–Nordheim tunneling through a thin gate oxide or through a polyoxide for erasure.

All Flash EEPROM products are based on the floating gate concept. The memory can be erased electrically but not selectively. The content of the whole memory chip is always cleared in one step. The advantages over the EPROM are the faster (electrical) erasure and the in-circuit reprogrammability, which leads to a cheaper package. Its cost is lower than that of EEPROM devices, and the part was introduced partially to cope with the low volumes of the market that could be reached with the full-featured EEPROM, until recently.

In the 1990s, Flash memory has become the largest market in nonvolatile technology due to a highly competitive tradeoff between functionality and cost/bit. Since the cell size of Flash devices has the potential to track that of DRAM cells, competitively priced Flash concepts are expected to find a huge market and even to become one of the main technology drivers of the semiconductor industry. Apart from the replacement of EPROMs and EEPROMs, novel application fields have also arisen, such as solid-state disks for portable and handheld computers, and smart cards. Also, novel device structures have been proposed based on Fowler–Nordheim tunneling for both programming and erasure in order to allow operation from a single supply voltage. Moreover, source-side injection Flash devices (see Section 1.2.4) have gained considerable interest because of their unique combination of very fast programming capabilities with low power consumption.

Currently, Flash products up to 32 Mbit are commercially available. The cell size attained in a 32 Mbit product is on the order of 1.5 μm^2 [1.59–1.61]. Finally, there is a strong demand for embedded Flash memory on ASICs, digital signal processing (DSP) chips, microcontrollers, and so on. In the case of microcontrollers, process compatibility, development cost, and single-supply voltage operation are more stringent requirements than high density and high performance.

In 1995, Flash's cost/MB had already become smaller than DRAM's, and additional improvements are still to be expected because of its high scalability. Furthermore, due to the demand for ever higher Flash memory densities (also in embedded applications such as smart cards), the multilevel charge storage (MLCS) option has recently gained considerable interest. The MLCS principle is based on the relatively high stability of the charge level that can be stored inside the floating gate memory cell in a virtually continuous (or analog) manner. In this way, more than two levels, and hence, more than 1 bit, can be stored inside a single memory transistor. This further increases memory capacity without the need for considerable changes in die size or aggressive technology scaling, hence drastically decreasing the cost/MB even further.

1.3.4 NOVRAM

The most complex nonvolatile memory device is the *NO*n *V*olatile *RAM* (NOVRAM) in which an EEPROM memory acts as a shadow memory for a static (or dynamic) RAM [1.62–1.64]. Each memory bit, therefore, consists of a RAM memory cell and an EEPROM element, as shown in Fig. 1.26. Some vendors provide nonvolatile RAM memories by using a battery backup included in the chip package. Battery backup NOVRAM parts up to 512 K are presently available. Other manufacturers provide inherent nonvolatility; that is, all data input/output (I/O) occurs through the RAM (thus allowing fast read and write operations). Data can be written into the EEPROM by copying the entire RAM content in parallel within 10 ms, the normal programming time of an EEPROM memory. Both charge-trapping and floating gate type cells have been used. The nonvolatile element can be based on polyoxide [1.64] or on thin oxide Fowler–Nordheim tunneling. This type of nonvolatile memory with a shadow SRAM is available in densities up to 16 Kbit. This density has not been increased, indicating that there is seemingly no need for larger devices, mainly because of its very high cost. Nevertheless, a 64 Kbit

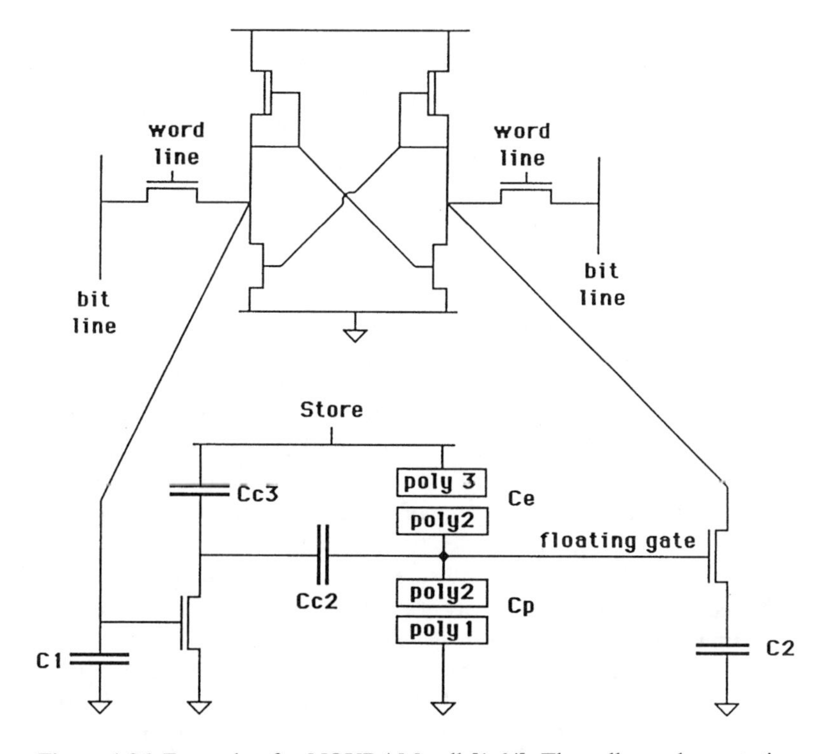

Figure 1.26 Example of a NOVRAM cell [1.64]. The cell couples a static RAM cell with a nonvolatile TPFG memory cell (see also Section 1.4.1.3).

NOVRAM, making use of two SNOS transistors in combination with a static four-transistor RAM cell, has been reported [1.61]. A nonvolatile DRAM memory has also been reported which uses the combination of a conventional high-density DRAM cell and an EEPROM, leading to a shadow DRAM [1.66–1.68]. This results in a much smaller cell size than is obtained in commercially available products.

1.4. BASIC NVSM DEVICES PRESENTLY IN USE

Following our discussion of different programming mechanisms that have shown feasibility for use in nonvolatile memory cells, and of the different nonvolatile memory products, this section is concerned with the different basic nonvolatile memory devices. As already mentioned, nonvolatile memory devices can be subdivided into two main classes: floating gate devices and charge-trapping devices. Floating gate devices are used for EPROM as well as for EEPROM, Flash EEPROM, and NOVRAM, whereas charge-trapping devices are more suited for EEPROM or NOVRAM applications because of their inherent electrical erasability. The following subsections discuss the different basic floating gate device concepts, charge-trapping devices, and ferroelectric RAM device concepts, and presents a brief comparison between the various classes of devices. This discussion is concerned mainly with the basic configuration and operation principles of the different cells and with some of the most important variations and latest improvements. For a more detailed examination of the various types of cells and of more recent developments, the reader is referred to the corresponding chapters in this book.

1.4.1 Floating Gate Devices

Basically, all floating gate memory cells have the same generic cell structure. They consist of a stacked gate MOS transistor, as was shown in Fig. 1.5*a*. The first gate is the *floating gate*, since it is completely embedded inside the dielectric. The second gate, which is usually referred to as the *control gate*, acts as the external gate of the memory transistor. Between the floating gate and the substrate, and between the floating gate and the control gate, a dielectric layer is provided for isolating the floating gate from the external nodes. These dielectric layers can be oxides, nitrides, oxynitrides, or stacked layers of oxide and nitride (ONO). Furthermore, special features are provided in order to implement the selected programming mechanism inside the cell. In fact, the differences between the various classes of floating gate devices are based on the programming mechanisms that are used for writing or erasing the cell.

When channel hot-electron injection is used as the programming mechanism, the cell is referred to as a SIMOS (*S*tacked gate *I*njection *MOS*), and it is used mainly for EPROM purposes. When the programming mechanism is Fowler–Nordheim tunneling, the cell is often called FLOTOX (*FLO*ating gate *T*hin *OX*ide), which is used primarily in EEPROM, Flash EEPROM, and NOVRAM applications. When polyoxide conduction is used for writing and erasing the memory, the cell is called TPFG (*T*extured *P*oly *F*loating *G*ate), which is used in both EEPROM and NOVRAM applications. Finally, there are cells in which a combination of two programming mechanisms is used to write or erase the device. In the following paragraphs, these different classes of nonvolatile memory cells are briefly discussed.

1.4.1.1 SIMOS (EPROM, FLASH EEPROM). The SIMOS (*S*tacked gate *I*njection *MOS*) cell is the n-channel version of the previously discussed SAMOS cell [1.20]. It consists of a double-polysilicon stacked gate device, as shown in Fig. 1.27, and is the basic cell configuration for almost all EPROM memories.

The SAMOS device, discussed in Section 1.1, is a p-channel device using drain avalanche electron injection as the programming mechanism, and consequently, a programmed device would behave as a normally-on device. Proper operation in a memory array, therefore, implies inclusion of an additional select transistor in each memory cell.

This is no longer the case for an n-channel transistor. Its threshold voltage increases due to electron injection, and it offers higher electron mobilities in comparison to p-channel devices. Avalanching the drain of an n-MOS transistor, however, only yields hole injection, which is even less efficient than drain avalanche electron injection in a p-MOS device, due to a higher oxide barrier for hole injection and a larger hole-trapping probability in silicon oxide.

Therefore, in the SIMOS device, channel hot-electron injection is used as the programming mechanism. However, as discussed in Section 1.2, this mechanism is inefficient, and therefore, programming the device is very power consuming, which has prevented the realization of 5 V-only devices. Therefore, all EPROM products require an external supply voltage, which is typically 12 V for programming, and the eventual use of on-chip high-voltage multipliers is excluded. Since channel hot-electron injection is only capable of putting electrons onto the floating gate, UV light is used for erasure.

Typical conditions during operation of the SIMOS cell are shown in Table 1.1. During the read operation, the control gate of the device, which is connected to the wordline of the memory array, is brought to V_{cc} (5 V), while the drain, connected to the bitline of the array, is held at 1 to 2 V, and the source is grounded. If the cell is programmed (high-threshold voltage), no current is detected, whereas an erased cell

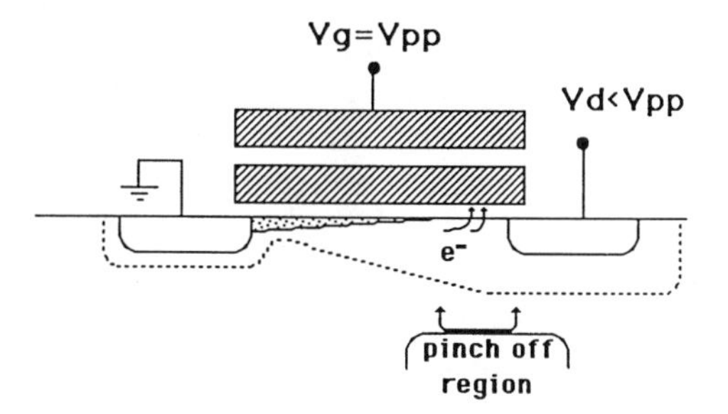

Figure 1.27 The SIMOS cell (Stacked-gate **I**njection **MOS**) is the n-channel version of the SAMOS cell [1.20]. Programming occurs through channel hot-electron injection. Erasing is done by UV irradiation or polyoxide conduction. The cell is used mainly for EPROM applications.

TABLE 1.1 SIMOS OPERATION CONDITIONS

	V_{cg}	V_d	V_s
Read	V_{cc} (5 V)	2 V	GND
Write	V_{pp} (12 V)	8–9 V	GND

will conduct a high current. During programming, the control gate or wordline is brought to the programming voltage, V_{pp}, which is typically 12 V, while the drain is held at 8 to 9 V and the source is again grounded. It is important to notice that the reading and programming configurations are the same. The difference lies only in the voltage levels. The programming gate voltage can be generated on-chip since it is not consuming any current. The programming drain voltage, however, has to be supplied externally. It is also important to mention that selection of the bit during programming is automatically performed by raising the control-line and the bitline. Cells that are connected to the same wordline but to a different bitline, as well as cells connected to the same bitline but to a different wordline, will not be programmed. Consequently, no additional select transistor is needed, and a minimum-size one-transistor cell can be used.

The SIMOS cell has been and, in fact, still is the workhorse of almost all EPROM memories available on the market. Since it became clear that channel hot-electron (CHE) programmable transistors are to be used in high-density EPROMs, technology has been adapted in order to yield fast-programmable, high-density floating gate EPROM structures. The efficiency of the channel hot-electron programming process is largely dependent on substrate (and drain junction) doping level, effective channel length, and floating gate–drain junction overlap. High density can be achieved by using the self-aligned double-polysilicon stacked gate structure [1.69], in which the drain-source junctions are self-aligned with respect to the floating gate, and in which the double-polysilicon stacked gate structure is etched in one step, usually by means of an anisotropic dry etching process [1.69].

Although the conventional self-aligned stacked gate double-polysilicon transistor is still used at the 4 Mbit density level, several new architectures are reported to allow a further decrease in the effective cell size. A first approach is the contactless self-aligned EPROM cell [1.70, 1.71] which consists of a cross-point array configuration defined by continuous buried n^+ diffusions (forming the bitlines) and WSi_2/Poly control gate wordlines. Metal is used to contact the bitline every sixteenth wordline in order to reduce bitline resistance [1.72]. In fact, this is only one example of the use of a virtual ground array. Figure 1.28 shows this virtual ground array architecture in which the common ground line in the array, as well as the drain contact in each memory cell, is eliminated. This technique has been introduced to obtain small cell sizes [1.73, 1.74]. The array architecture relies on the use of asymmetrical floating gate transistors [1.73, 1.74], or on proper source and drain decoding [1.70].

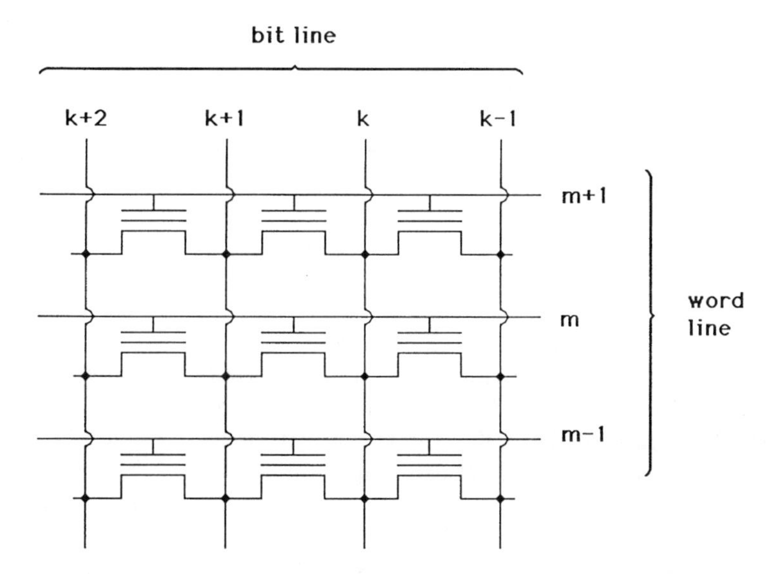

Figure 1.28 The virtual ground array architecture [1.70, 1.73, 1.74]. In this architecture, the common ground line in the array, as well as the drain contact in each memory cell, is eliminated.

Many new device structures, all relying on the original SIMOS concept, have been presented. These variations generally serve one common purpose—namely, to increase the injection current without increasing the programming voltage. The injection of hot electrons is a very inefficient process inasmuch as a proper biasing condition for hot-electron generation does not go together with a favorable condition for injection into the oxide. In practice, very high drain and gate voltages are needed, and the injection efficiency is very low. Alternatives should be based on creating a high hot-electron generating electric field in the channel, and simultaneously, a favorable oxide injection field at the site of the hot-electron generation. Several solutions that meet these requirements have been proposed.

One possibility is the use of a split gate EPROM cell [1.73, 1.75]. In this cell, shown in Fig. 1.29, the series transistor ensures a high immunity to drain turn-on, which otherwise can constitute a serious problem, and eventually, can put a limit on the minimum effective channel length of the EPROM transistor [1.76]. The series transistor also eliminates source-drain punch-through problems. This allows the use of a very small length for the floating gate, thus realizing a high programming speed and a high read current [1.73].

Alternative concepts rely on the source-side injection mechanism described in Section 1.2.4. In the dual gate structure [1.22], shown in Fig. 1.30*a*, a strong potential drop is induced in the center of the channel where neither of the gates is controlling the channel potential. The injection occurs at this site. The injection efficiency can be increased from 10^{-7}, for conventional hot-electron injection, to 10^{-3} for this cell.

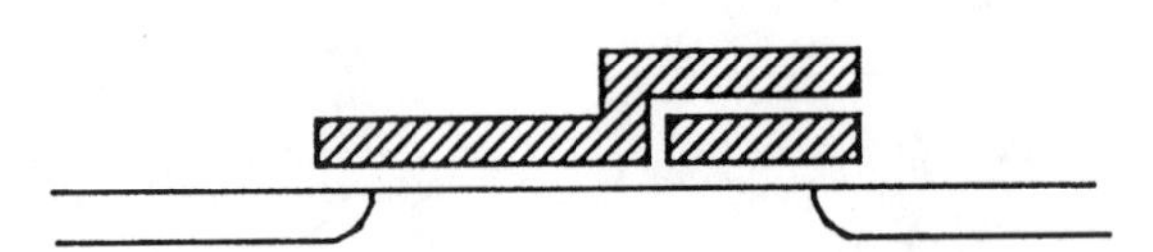

Figure 1.29 The split gate EPROM cell [1.73, 1.75]. The series transistor, incorporated in the memory transistor, allows the use of a floating gate with minimal effective length.

Another cell, based on source-side injection [1.23], is shown in Fig. 1.30*b*. It uses a side-wall gate and a conventional stacked gate structure. Under the spacer oxide between the side-wall gate and the stacked gate, a weak gate control region is formed. This creates a high channel field, located near the source, where the oxide field is highly favorable for injection. The injection efficiency of this cell is on the order of 10^{-5} to 10^{-6}. Other alternatives are the side-wall floating gate structure [1.77], shown in Fig. 1.30*c*, the trench gate–oxide structure [1.78], shown in Fig. 1.30*d*, and the focused ion-beam implantation cell [1.79], shown in Fig. 1.30*e*.

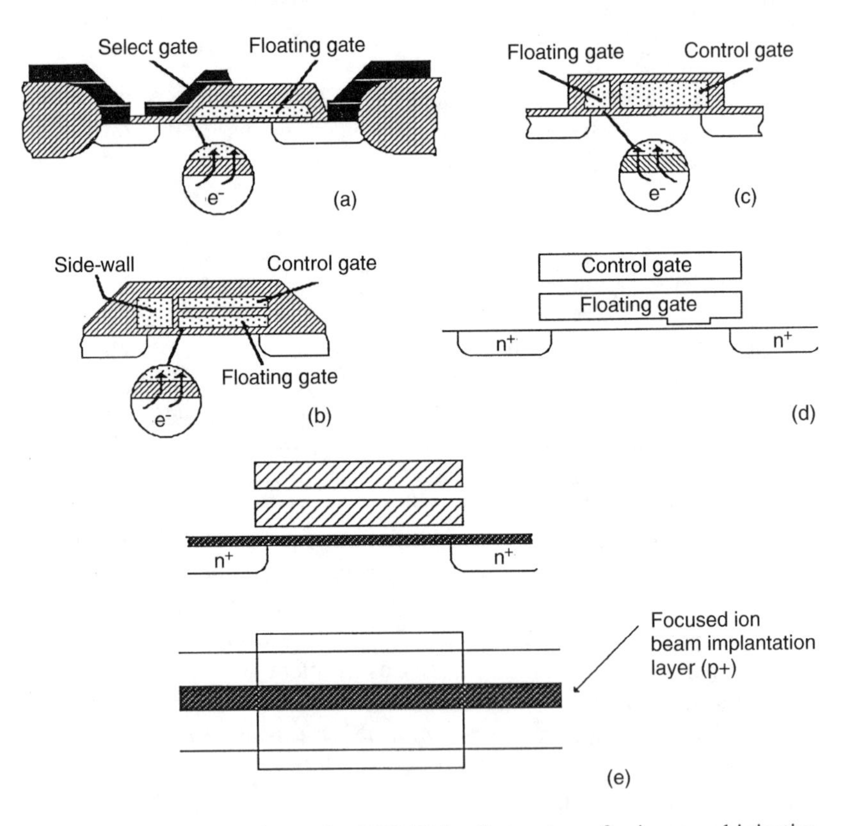

Figure 1.30 Five alternative EPROM cell structures for increased injection efficiency (*a*) PACMOS cell [1.22], (*b*) side-wall floating gate cell [1.23], (*c*) source-side injection cell [1.77], (*d*) trench–gate–oxide structure [1.78], and (*e*) focused ion beam implanted cell [1.79].

Although the SIMOS cell has been used mainly in EPROM devices, the hot-electron injection mechanism is also used in most Flash EEPROM devices. The cells that are used for these applications are primarily combinations of the SIMOS and the FLOTOX cell (to be discussed in the next section), in either a split gate or a stacked-cell configuration. This is discussed further in Section 1.4.1.4.

1.4.1.2 FLOTOX (EEPROM, Flash EEPROM, NOVRAM). The first nonvolatile memory device relying on Fowler–Nordheim tunneling for both writing and erasing was proposed by Harari et al. in 1978 [1.62], and was used in a non-volatile RAM cell. This device incorporates a small, thin oxide region over the drain and has, in fact, exactly the same structure as the FLOTOX (*FLO*ating gate *T*hin *OX*ide) device, shown in Fig. 1.31. The FLOTOX approach [1.5] relies on Fowler–Nordheim tunneling through a thin oxide (8–10 nm) for both programming and erasure, and was introduced in 1980 as an EEPROM memory transistor [1.5].

In order to increase the threshold voltage of the cell, a high voltage is applied at the control gate of the device (typical 14 V), while source, drain, and substrate are grounded. This high voltage is capacitively coupled to the floating gate, by which the necessary high field appears across the thin oxide, leading to tunneling of electrons from the drain to the floating gate. When the threshold voltage has to be decreased, a large voltage is applied at the drain, while the control gate and substrate are grounded and the source is left open. By grounding the control gate and substrate, the floating gate is capacitively coupled to near ground, and again, a high field appears across the thin oxide, this time inducing electrons to tunnel from the floating gate to the drain.

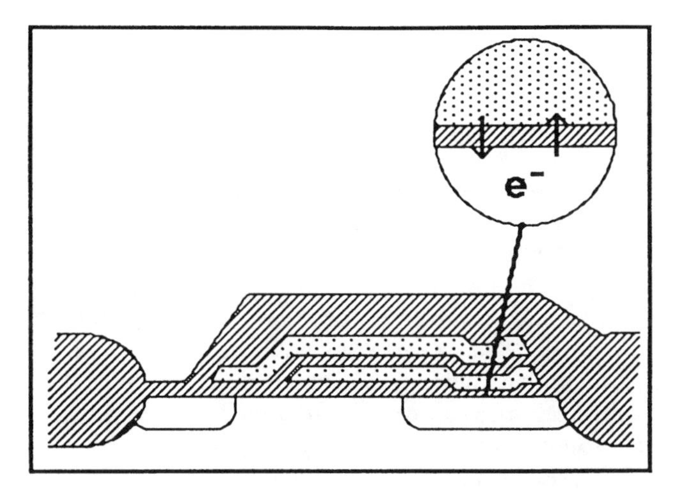

Figure 1.31 Cross section of the FLOTOX device (**FLO**ating gate Thin **OX**ide) [1.5]. The device is written and erased by Fowler–Nordheim tunneling of electrons through the thin oxide and is used for EEPROM and Flash EEPROM applications.

Special attention should be paid when incorporating the FLOTOX cell into a memory array in order not to erase or write a cell on the same wordline or bitline when trying to program another cell. Indeed, unlike the SIMOS cell for which programming involves the application of two voltages (one at the wordline and one at the bitline), which is sufficient to make an appropriate selection of the cell to be programmed, programming of a FLOTOX-type cell involves the use of only one voltage, which is not sufficient to select one single cell. Cells that are connected to the same wordline, but at a different bitline, will also be programmed when the control-line is raised, while cells connected to the same bitline will lose charge from the floating gate when a high voltage is applied to the bitline.

Therefore, since the EEPROM should be byte addressable for read as well as for programming operations, a select transistor becomes necessary, as shown in Fig. 1.32. Each byte has its own erase/write control transistor. First, the control gates of the memory transistors and the select-line of the select transistor are raised to a high voltage while grounding the column lines, as shown in Fig. 1.32*a*. In this way, the floating gates are all charged with electrons. Then, the bits to be programmed are selectively purged of electrons by raising the drain to the programming voltage and grounding the control gates, while the select gates are raised to a high voltage, as shown in Fig. 1.32*b*.

During the read operation, the presence or absence of charge on the floating gate is detected by applying a positive voltage to the control gates and select gates, while biasing the column lines to about 2 V. Like the case of the SIMOS cell, the cells that have electrons on their floating gate do not conduct current, while the cells that are depleted of electrons on the floating gate are in a high conductive state.

For memory cells that use thin oxides, the injection field equals the average oxide field, as discussed in Section 1.2. Consequently, these cells need strong coupling between the floating gate and externally controlled terminals of the device in order to couple the high external voltages onto the floating gate, thereby inducing a high electric field across the thin oxide. The high gate capacitance of the floating gate to substrate transistor can be used for this purpose during the programming operation that lowers the threshold voltage (electrons tunneling off the floating gate). If both programming operations use F–N tunneling, large coupling-capacitance areas between the control gate and the floating gate are necessary.

The FLOTOX cell is used in many commercial products. One of the main reasons why is the low development-entry cost. The only additional process step required in standard double-poly processes is the growth of the thin oxide. It is even possible to realize this type of nonvolatile memory in single-poly processes. Hence, this cell is highly suitable for ASIC and logic applications [1.80, 1.81]. The drawback of the larger cell area is not so important for these applications. Scaling is often difficult because of the complex cell layouts used. To allow the programming voltages to decrease, the tunnel oxides should become even thinner. Thicknesses of 6 nm, however, are the limit for good retention behavior. But these oxides are hard to grow with low defect densities. Yield considerations now limit the oxide thickness to 8 to 10 nm [1.25]. Thinner tunnel oxides imply higher capacitances and thus, larger coupling areas. Really small floating gate tunnel oxide (FLOTOX) cells are, therefore,

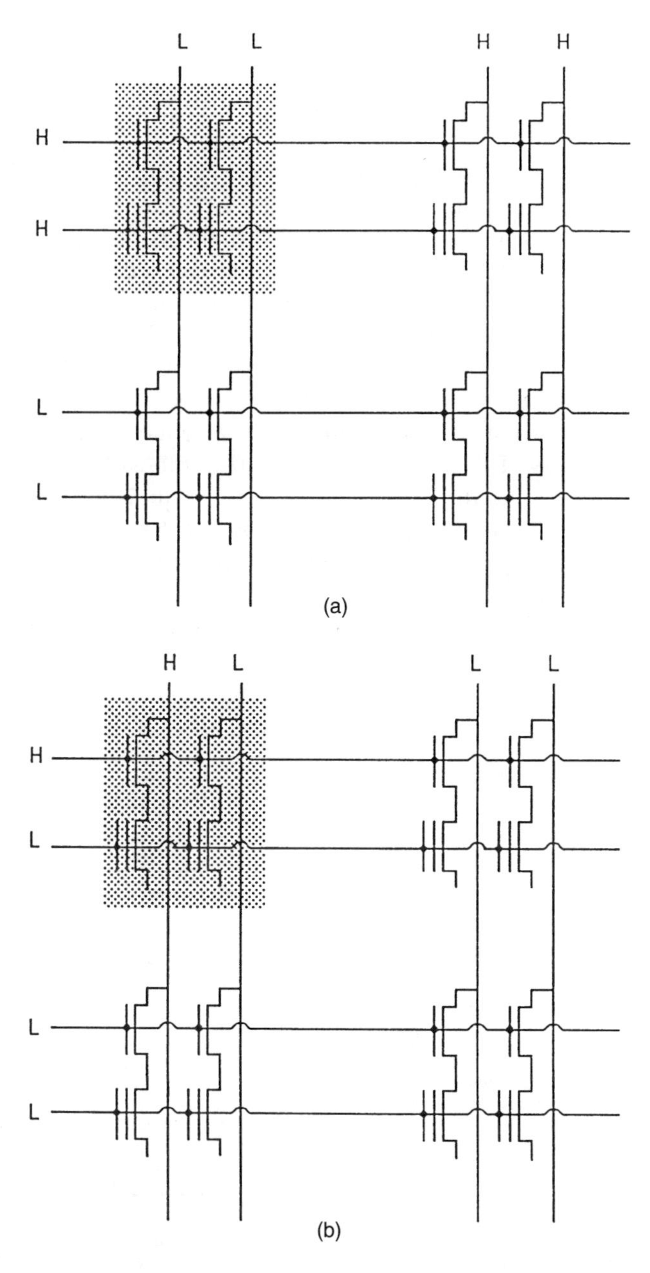

Figure 1.32 Configurations of the voltages in the EEPROM matrix during erase (*a*) and write (*b*) of a memory byte. The shaded parts in the figure are the locations that are being programmed.

hard to achieve. For large memories, however, further scaling and the use of thin oxides become mandatory [1.25, 1.82]. The use of new tunnel materials can obviate some problems. Oxynitrides or nitrided oxides offer better endurance [1.83, 1.84], while oxides grown on highly doped injection regions show higher tunnel current conductance [1.85, 1.86].

The advantage of using low power Fowler–Nordheim tunneling for both programming operations is the possibility of using on-chip high-voltage multipliers to generate the programming voltages. In this way, one of the main drawbacks of the SIMOS cell, namely, the need for an external power supply, can be avoided.

For the FLOTOX concept, several variations of the mechanisms themselves, or to the cell design, have been proposed in order to create alternative memory cells with improved performance. The thin oxide can cover the entire channel area of the floating gate transistor, as shown in Fig. 1.33. This type of nonvolatile memory transistor is called a *F*loating gate *E*lectron *T*unneling *MOS*, or FETMOS [1.17]. The device structure is very simple, and to realize it, only a few processing steps are required in addition to a standard *C*omplementary *M*etal *O*xide *S*emiconductor (CMOS) process. This makes small cell sizes impossible. The writing operation is done uniformly over the entire channel area by applying a high voltage at the gate (Fig. 1.33*a*), whereas the erasing operation can be done either uniformly, using a negative gate voltage, while grounding the other electrodes (Fig. 1.33*b*), or locally, at the drain, by grounding the gate and using a high voltage at the drain (Fig. 1.33*c*). The drawback of this structure is the large capacitance of the thin gate oxide, which

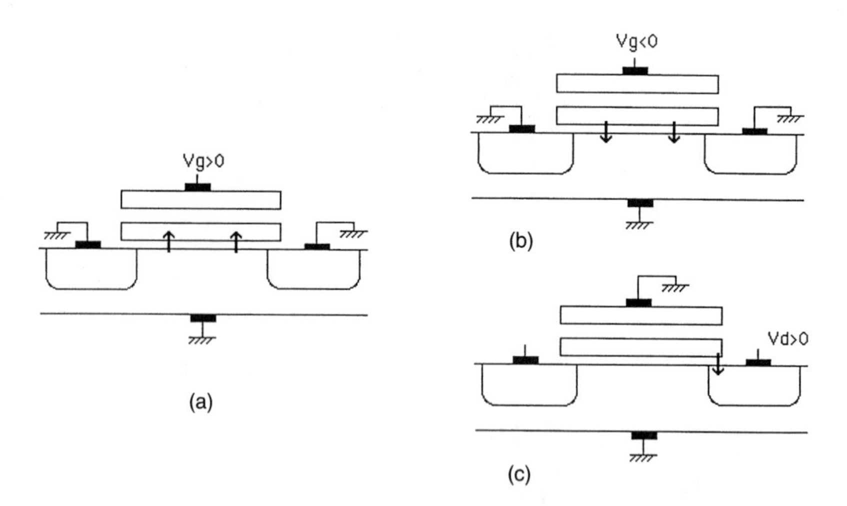

Figure 1.33 Cross section of the FETMOS device (**F**loating gate **E**lectron **T**unneling **MOS**) [1.17]. The thin oxide is covering the entire channel area. (*a*) When the control gate voltage is high, electrons tunnel from the channel to the floating gate. The opposite programming operation can be performed applying either (*b*) a negative voltage at the control gate (uniform) or (*c*) a positive voltage at the drain (nonuniform).

counteracts the coupling between the control gate and floating gate, and which necessitates the use of large coupling capacitors.

Other variations of the FLOTOX device generally have one common purpose, namely, to increase the injection current without increasing the programming voltage. Whereas some of them are merely new ideas with a questionable chance for future application, others have already been implemented in real memory circuits.

For the tunneling mechanism, several possibilities can be considered to increase the injection efficiency without changing the programming voltage: (1) reducing the oxide thickness, (2) reducing the oxide injection barrier, or (3) increasing the coupling factor between the control and the floating gate without requiring cells with too large an area.

Reducing the tunnel oxide thickness, however, puts severe constraints on device reliability due to direct tunneling leakage problems and the enhanced problem of layer integrity. Reducing the effective tunnel oxide barrier was attempted in the past, for example, by using Si-rich SiO_2 [1.87] or nitrides [1.88]. In fact, the use of textured polyoxide (to be discussed in the next section) also falls within this category. Another possibility is to grow the thin tunnel oxide on a highly doped, n-type silicon substrate. It was reported that, in this way, the effective energy barrier for tunneling could be reduced from about 3.2 eV to 1.8 eV, allowing programming voltages of only 12 V with tunnel oxides of 14 nm [1.85, 1.86].

Increasing the coupling factor without sacrificing too much chip area can also be realized in several ways. One possibility is illustrated in Fig. 1.34 (*S*hielded *S*ubstrate *I*njection *MOS*). The floating gate is shielded from the substrate by the control gate [1.89]. This type of cell only takes about a quarter of the area of the larger cells (FLOTOX). Another version of this idea, the *S*tacked *S*elf-aligned *T*unnel *R*egion (SSTR) cell [1.90], is shown in Fig. 1.35. Another approach to increasing the coupling factor is to replace the coupling capacitor of the conventional stacked gate structure by one that is formed by a tunnel oxide MOS capacitor, as shown in Fig. 1.36 [1.80]. This cell can be realized in a single-poly process.

The NAND (Not AND) structure cell has also been proposed for reducing EEPROM cell size [1.91]. The main disadvantage of this approach is the very long read access time [1.92].

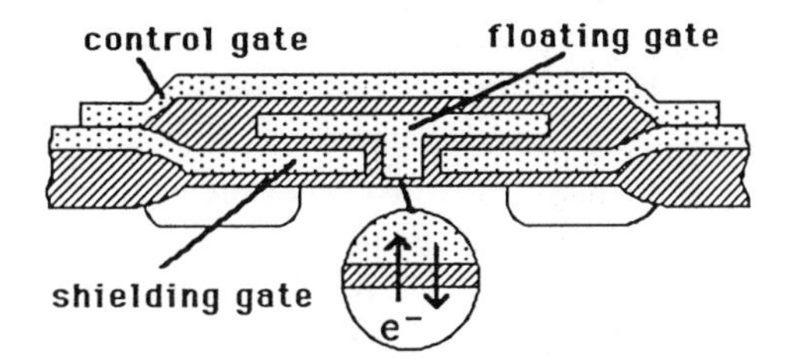

Figure 1.34 Cross section of the Shielded Substrate Injection MOS (SSIMOS) memory cell [1.89]. The floating gate is totally surrounded by the control gate.

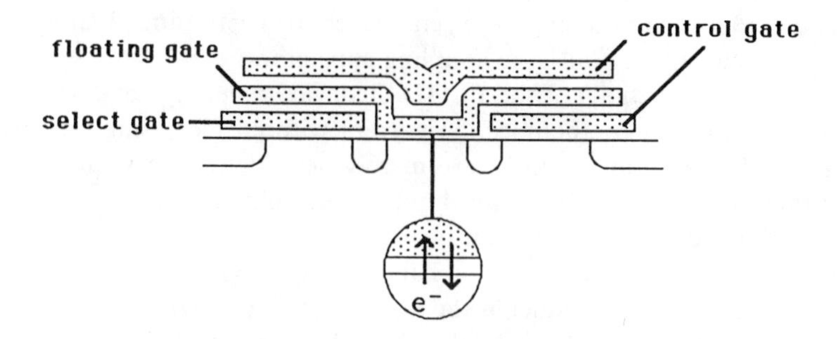

Figure 1.35 Cross section of the **S**tacked **S**elf-aligned **T**unnel **R**egion (SSTR) cell [1.90]. The control gate shields the floating gate from the substrate, and the select transistor is incorporated in the cell.

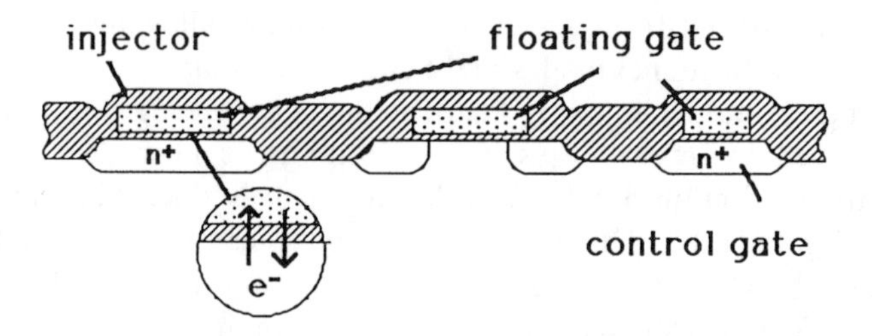

Figure 1.36 Cross section of the single-poly floating gate transistor [1.80]. The control gate is formed by a n+ doped region, which is coupled to the floating gate through a thin oxide region.

1.4.1.3 TPFG (EEPROM, NOVRAM). As discussed in Section 1.2, programming operations can also be based on tunneling through oxides grown thermally on polysilicon. These oxides feature current conduction at lower average oxide fields due to field enhancement at the asperities on the polysilicon surface. This allows the use of much thicker oxides for the same externally applied voltages during programming. Since injection will be enhanced in just one direction, a memory transistor, relying on this enhanced tunnel current for both programming operations, must incorporate two distinct injection regions.

A triple-poly nonvolatile memory transistor, called the *T*extured *P*oly *F*loating *G*ate (TPFG) transistor, was first reported in 1979 [1.18] when it was used in a nonvolatile RAM, and in 1980, as the key element of an EEPROM [1.19]. The device structure is shown in Fig. 1.37. Some other implementations of this enhanced

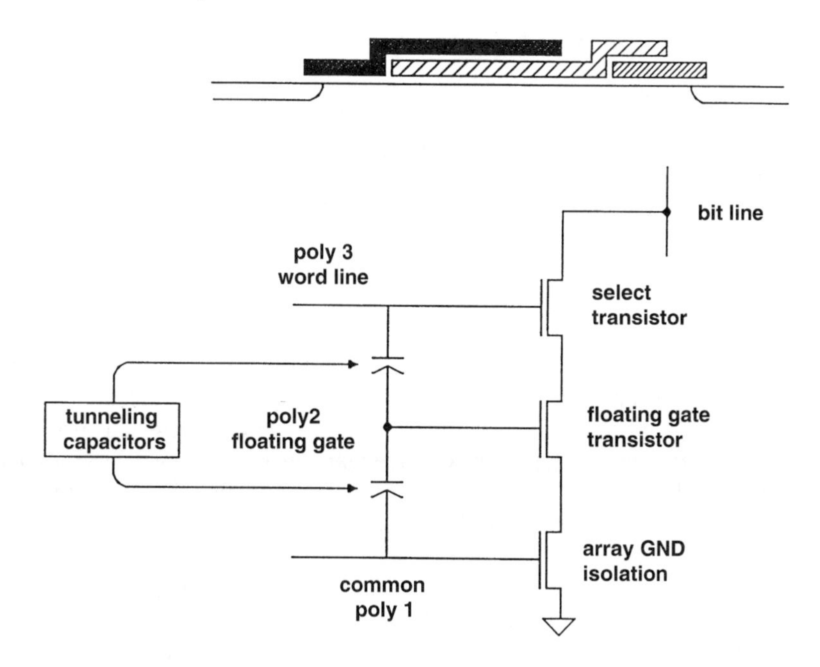

Figure 1.37 Cross section of the **T**extured **P**oly **F**loating **G**ate (TPFG) device [1.18, 1.19]. A select transistor is incorporated; writing and erasing the device is performed by means of Fowler–Nordheim tunneling through the polyoxide in two different areas. Therefore, a triple-polysilicon device is needed.

tunnel–current principle have been proposed. In the DEIS (*D*ual *E*lectron *I*njection *S*tructure) [1.87], tunnel–current enhancement was achieved by using Si-rich SiO_2 layers. Another approach is based on an improved technology for depositing polysilicon layers, resulting in symmetrical enhanced tunnel–current characteristics for both injection directions in polyoxide [1.93].

TPFG devices are used in large-density memory circuits. Again, as in the case of FLOTOX cells, no external power supply is needed, and on-chip high-voltage multiplier circuitry can be used. The main difficulty in this approach is the growth of polyoxides with desired interface features (shape and size of the asperities), which determine the injection current characteristics. Wearout features are quite dependent on the quality of the polysilicon–SiO_2 interface. The structure is rather complex. Three polysilicon layers [1.19], or two layers and an additional buried contact [1.94], are needed. The accurate alignment of these layers requires precise lithography. This has delayed use of this cell in ASIC or logic applications. The need for thin polyoxides is not that stringent. The injection current through these oxides is determined primarily by the shape and size of asperities, and not by the oxide thickness [1.95, 1.96]. The smaller number of injection points of a scaled TPFG memory cell can, however, aggravate intrinsic wearout as a result of trapping electrons in the polyoxide, a process known as trap-up.

1.4.1.4 Combinations. As was previously discussed in Section 1.2, four different mechanisms are presently used to change the amount of charge on a floating gate: Fowler–Nordheim (F–N) tunneling through thin oxides ($< 12\,\text{nm}$), enhanced Fowler–Nordheim tunneling through polyoxides, channel hot-electron injection (CHE), and source-side injection (SSI). Of these, only channel hot-electron injection and source-side injection can be used to bring electrons to the floating gate. The SSI mechanism is, in fact, a special case of channel hot-electron injection and will, therefore, be treated together with CHE in the following discussion. Therefore, a total of six combinations of main classes of floating gate cells can be defined.

Figure 1.38 shows these six possible combinations with an indication of how these devices are programmed. Each column corresponds to a mechanism that allows electrons to be brought to the floating gate. These are, from left to right, F–N tunneling, polyoxide conduction, and channel hot-electron injection (CHE or SSI). Each row corresponds to a mechanism that allows electrons to be removed from the floating gate. The upper row is for F–N tunneling, and the lower one is for polyoxide conduction. Table 1.2 summarizes the main advantages and drawbacks of the obtained cells.

Of these six combinations, two have already been discussed, namely, the FLOTOX cell (a) and the TPFG cell (e). The combinations (c) and (f) are, in fact, extensions of the SIMOS structure, where special features have been provided to allow an electrical erasure of the cell in order to obtain a Flash EEPROM cell. In the cell shown in (c), electrons are injected onto the floating gate by channel hot-electron injection, as in the SIMOS case. But in order to be able to remove the electrons from the floating gate electrically, either the gate oxide underneath the floating gate is kept thin, as in the case of the FETMOS cell, or a separate thin oxide is provided above the drain, as in the case of the FLOTOX [1.97]. This concept is the basic cell of most of the present Flash EEPROM technologies and is also referred to as electron tunneling oxide (ETOX).

For the cell shown in (f), again channel hot-electron injection is used for programming, but the electrons can now be removed from the floating gate by polyoxide conduction through the interpoly dielectric [1.98]. Because, for cells (c) and (f), electron injection toward the floating gate is performed by channel hot-electron injection, no large coupling areas between the control and floating gate are necessary. This leads to an electrically erasable cell but with a much smaller cell size. Therefore, these cells have been utilized primarily in Flash EEPROM applications [1.98]. The main disadvantage of both cells is still the high programming power, with the necessity of an external power supply, which again excludes the use of on-chip voltage multipliers.

Scaling structures that use channel hot-electron injection can lead to new opportunities for this kind of memory. At small channel lengths, a drain voltage of 5 V can be sufficient to generate hot electrons. Only the gate will then need a higher voltage for programming. Nonvolatile memories with CHE programming could thus operate from a 5 V supply voltage. But the high programming current remains, and therefore, these types of cells are not used in conventional EEPROMs (which are reprogrammed frequently). They are, however, the main cell for Flash EEPROM products.

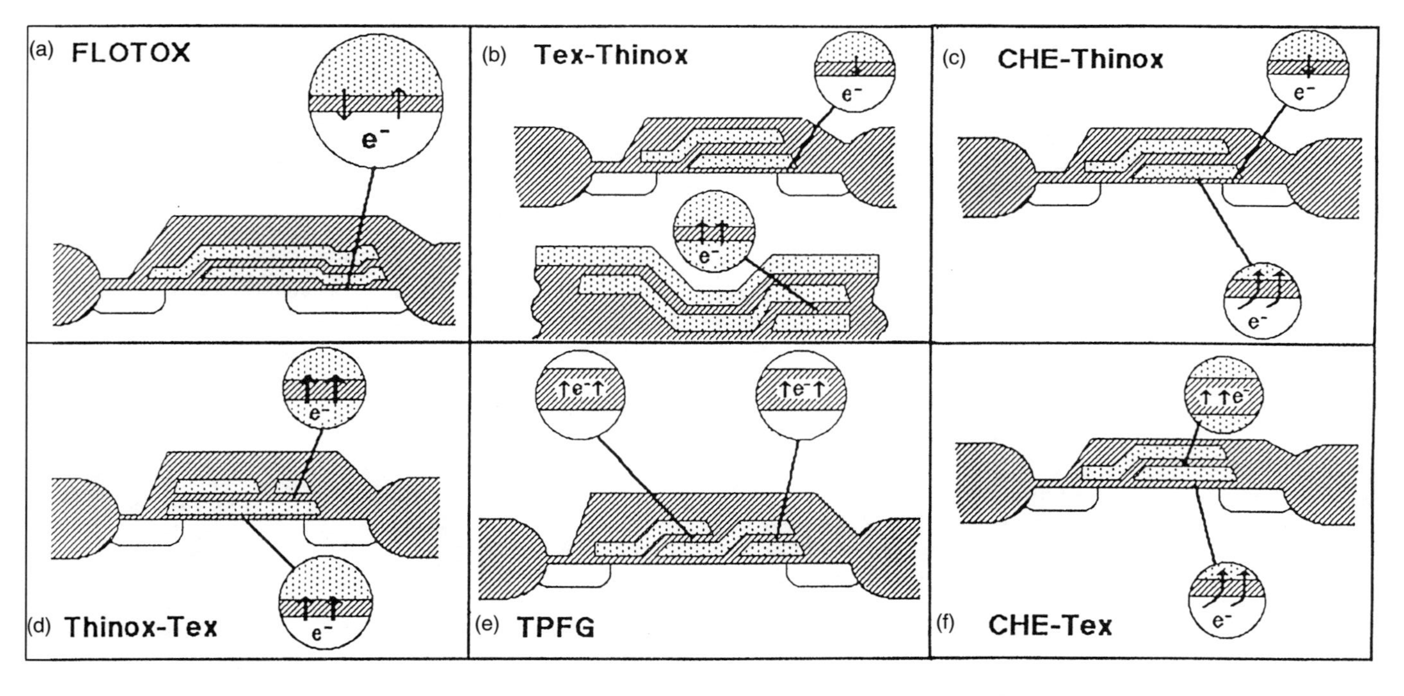

Figure 1.38 Six possible combinations of floating gate memory cells. The three columns represent the three mechanisms to bring electrons to the floating gate (thin oxide Fowler–Nordheim tunneling, polyoxide conduction, hot-electron injection), while the two rows represent the two mechanisms to remove electrons from the floating gate (thin oxide Fowler-Nordheim tunneling and polyoxide conduction). Hot-electron injection cannot be used to remove electrons from the floating gate.

TABLE 1.2 COMPARISON OF VARIOUS POSSIBLE EEPROM CONCEPTS

(a) FLOTOX EEPROM/ASIC/Logic		(b) Tex-Thinox EEPROM/F-EEPROM		(c) CHE-Thinox EEPROM/F-EEPROM/ASIC	
+	–	+	–	+	–
• compatibility • low development entry cost • possible with 1 poly layer	• large cell • difficult scaling • thinox defect density	• small cell • easily scalable • low program power	• complex cell • 2 program mechanisms –thinox (defects) –polyoxide (wearout)	• small cell • relatively simple cell • easily scalable	• high program power • 2 program mechanisms –thinox (defects) –CHE (degradation) • ?? endurance ??
(d) Thinox-Tex EEPROM/F-EEPROM		**(e) TPFG EEPROM/ASIC**		**(f) CHE-Tex EEPROM/EPROM/ASIC**	
+	–	+	–	+	–
• low program power	• large cell • scaling difficulties • 2 program mechanisms –thinox (defects) –polyoxide (wearout)	• thick oxide • small cell • low program power • direct write • easily scalable	• complex cell • higher program voltage –trap-up –window variation • critical tunnel oxide	• small cell • thick oxide • easily scalable	• high program power • 2 program mechanisms –polyoxide (wearout) –CHE (degradation)

Of the six possible floating gate memory cells, two have never been used in commercial products, namely, the Thinox–Tex, shown as (d), and the Tex–Thinox, shown as (b). The Thinox–Tex approach seems to have only drawbacks (cf. Table 1.2). For example, large cell areas are needed for the thin oxide Fowler–Nordheim tunneling used to remove electrons from the floating gate. This also makes the cell difficult to scale. The two different programming mechanisms enhance the technological problems. The Tex–Thinox cell has a rather complex structure. Three polysilicon layers are needed, and the two different programming mechanisms have to be optimized separately. This concept has never been used in commercial products.

1.4.2 Charge-Trapping Devices

1.4.2.1 MNOS and SNOS devices (EEPROM, NOVRAM). MNOS (metal–nitride–oxide–silicon) devices were invented in 1967 [1.2] and were the first electrically alterable semiconductor (EAROM) devices. The nonvolatile function of these devices is based on the storage of charges in discrete traps in the nitride layer. These charges (electrons or holes) are injected from the channel region into the nitride by quantum mechanical tunneling through an ultra-thin oxide (UTO, typically 1.5 to 3 nm). These trapped charges cause a significant shift in the threshold voltage of the transistor [see Eq. (1.2) with Q_T the trapped charge in the nitride layer]. Although over time some of these charges will be lost after programming is completed and will, therefore, result in a gradual decrease of the threshold voltage, the programmed state of the device can typically be maintained for at least 10 years.

By 1975, metal gate, p-channel EAROM with densities up to 8 Kbit were available. They employed a 1 transistor per bit configuration based on the so-called trigate transistor cell concept [1.99]. In this transistor, shown in Fig. 1.39, only the center part of the channel contained the programmable UTO–nitride sandwich structure. At both drain and source, a thicker oxide–nitride sandwich was used, which induced a fixed threshold voltage in the erased state and prevented the device from going into the depletion mode. These memory devices suffered from low-speed, limited-density, inherent read disturbance (sensing the device-required application of a small read voltage to the gate), and the need for 2 to 3 voltage supplies to operate the memory.

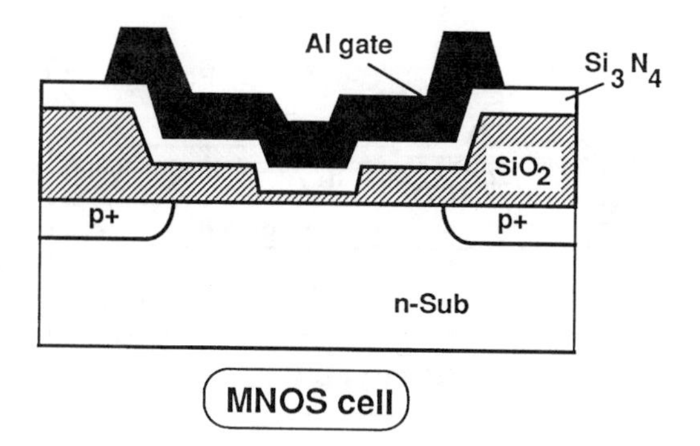

Figure 1.39 Cross section of the p-channel tri-gate MNOS device [1.99]. The thin tunneling oxide (1.5–3 nm) is present only at the center of the channel. At source and drain, a thicker oxide–nitride sandwich acts as a select transistor.

An important breakthrough was achieved for MNOS in 1980 with the development of the Si-gate n-channel SNOS (silicon–nitride–oxide semiconductor) process [1.100], which resulted in the first 16 Kbit SNOS–EEPROM [1.4]. The reliability of the SNOS technology was based mainly on the use of *L*ow-*P*ressure *C*hemical *V*apor *D*eposition (LPCVD) silicon nitride and a pre-metallization high-temperature hydrogen anneal to improve the quality of the nitride–UTO–silicon interface [1.101–1.104]. A cross-sectional diagram of the SNOS cell, which is still used in today's commercially available 256 Kbit SNOS–EEPROM memories and the announced 1 Mbit parts, is shown in Fig. 1.40. A two transistor per bit configuration is used where the MOS transistor acts as the select device whose implementation has allowed the complete elimination of the problem of read disturbance [1.4]. The SNOS transistor consists of a silicon nitride layer (20–40 nm) on top of the UTO on silicon.

Because the SNOS transistor is, in fact, a two-polarity device, necessitating the application of memory bulk voltages, isolation of the memory bulk from the peripheral circuitry bulk is required. The most common approach is the use of separate p-wells for the peripheral MOS circuits and the memory array. Providing separate wells within the memory array then allows for full byte function. In LPCVD nitrides, net positive and negative charge can be stored in almost equal amounts.

The programming of the cell is as follows: during the write operation, a high (positive) voltage is applied to the gate with the well grounded. Electrons tunnel from the silicon conduction band into the silicon nitride conduction band through the modified Fowler–Nordheim tunneling process discussed in Section 1.2.5 and are trapped in the nitride traps, resulting in a positive threshold voltage shift. Erasing is achieved by grounding the gate and applying a high (positive) voltage to the well. This induces direct tunneling of holes from the silicon valence band into the nitride valence band, or the nitride traps [1.105, 1.106], resulting in a negative threshold voltage. During the off-state, the gate is grounded and the select transistor is required for proper operation within the array. Reading of the cell is accomplished by addressing the cell through the select transistor and by sensing the state of the SNOS transistor. Although the gate is grounded, the charge content within the nitride will be modified in time primarily by backtunneling charges through the UTO.

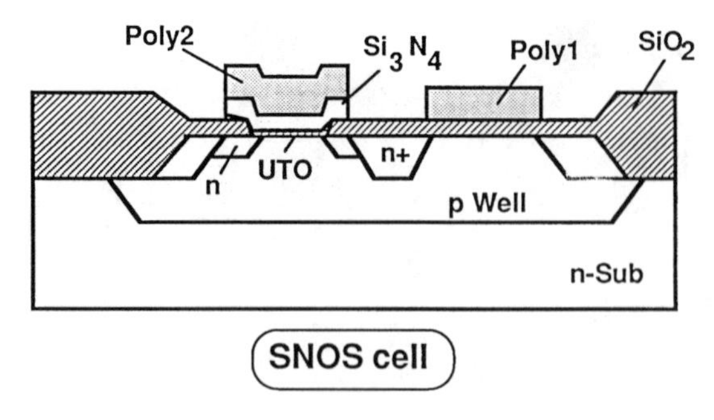

Figure 1.40 Cross section of the two-transistor n-channel SNOS memory cell [1.4, 1.100] consisting of a MOS select transistor and a SNOS memory transistor, both located in a p-well that allows full byte programmability.

Hagiwara et al. [1.106] showed that the integrity of nitride layers can be guaranteed to thicknesses below 20 nm. Scaling of the devices, which must be carried out in parallel with the scaling of the peripheral MOS transistors, is straightforward up to memory densities of 1 Mbit. Yatsuda et al [1.105–1.107] proposed a scaling scheme for SNOS, which is based on a reduction in the dielectric layer thicknesses almost in proportion to the program voltage, except for the UTO. In order to conserve a constant programming time, the UTO thickness has to be reduced slightly. Minami et al. [1.108] showed that the written-state retentivity of these conventional SNOS devices is improved when the nitride thickness is further reduced. For a 1 Mbit SNOS–EEPROM, a nitride thickness of about 20 nm and a UTO thickness of 1.6 nm are used. The programming voltage is about 10 volts.

Hole injection from the gate will, however, limit the memory window [1.105, 1.109–1.111], a problem that becomes more severe for thinner nitride layers. An efficient way to cope with this problem is described in the next section (1.4.2.2). Another solution is the use of a silicon oxynitride layer instead of a pure silicon nitride layer [1.112, 1.113]. Although these layers require slightly larger programming voltages because of their higher energy barriers, it has been shown that retention and endurance properties of SNOS parts using oxynitrides with a moderate oxygen content are markedly better than their nitride counterparts. In particular, the improved endurance characteristics point to a significant reduction in gate injection. The optimum composition for the oxynitride layer has been found to be $[O]/([O]+[N]) \approx 0.17$ [1.113].

SNOS memories have two features that are worth mentioning and that have made this technology the first choice in military and space applications requiring nonvolatility. The first one is their inherent radiation hardness, which is discussed in Section 1.7. The second feature is the ability of SNOS devices to be adjusted to the envisaged application: very slow programming (1–100 ms) for long nonvolatile retention (years, EEPROM function) or fast programming (1–10 μs) for limited nonvolatile retention (hours, days, NOVRAM function) [1.114].

1.4.2.2 SONOS Devices. A reduction of the injection from the gate can be ensured by providing a thin oxide (2–3 nm) on top of the nitride, yielding the so-called SONOS (*S*ilicon–*O*xide–*N*itride–*O*xide *S*emiconductor) device [1.115], as illustrated in Fig. 1.41*a* (SONOS1). This oxide can be formed by steam oxidation of the nitride at the expense of the nitride thickness or by deposition. The blocking efficiency of this top layer has been proven for both types of oxide [1.105, 1.115, 1.116]. However, when thinning the nitride below 20 nm, another problem arises. It is known that the trapping length in nitrides is larger for holes than for electrons, that is, 15 to 20 nm for holes [1.117, 1.118] and 5 to 10 nm for electrons [1.118, 1.119]. If the nitride is reduced in thickness, holes will be trapped close to the gate electrode and will be mostly lost through the gate electrode, even in the presence of this thin oxide. This results in a significant reduction of the threshold voltage in the erased state.

Therefore, further scaling of the SONOS device for higher density memories and lower programming voltages requires a new concept. This concept was first

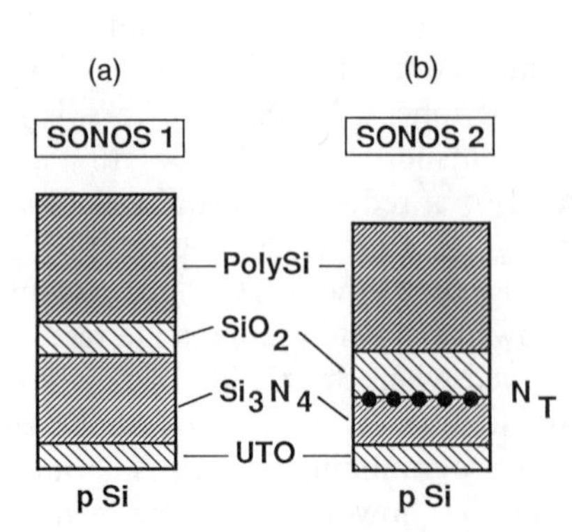

Figure 1.41 Schematical representation of the gate dielectric for two types of SONOS devices. In order to prevent hole injection from the gate, an oxide layer is added on top of the nitride layer. (*a*) The SONOS1 configuration [1.115] consists of a thin oxide (1–2 nm), a thicker nitride, and again a thin oxide (2–3 nm). (*b*) The SONOS2 configuration [1.120–1.122] consists of a thin oxide (1–2 nm), a thin nitride (< 10 nm), and a thicker oxide layer (> 3 nm).

proposed by Suzuki et al. [1.120] and was refined by others [1.121, 1.122]. This new SONOS concept is shown schematically in Fig. 1.41*b* (SONOS2). It consists of a UTO of the same thickness as before, a thin nitride layer (< 10 nm), and a thicker top oxide layer (> 3 nm). The aim of the top oxide is not only to inhibit gate injection, but also to block the charges injected from the silicon at the top oxide–nitride interface, resulting in a higher trapping efficiency and thus, a reduction in the problem related to nitride layer reduction. In this way, the total thickness of the insulator structure can be reduced, and consequently, the programming voltage can be reduced.

This device shows additional advantages. First, large oxygen-related electron trap densities are obtained at the nitride–top oxide interface due to the oxidation of the nitride [1.123]. This results in a larger memory window in spite of the decreased nitride thickness. For a constant top oxide layer thickness, this would eventually make the threshold of the written state independent of the nitride thickness [1.122]. Next, if pinholes are present in the thinner nitride layer, they can be filled with oxide afterward, during oxidation of the nitride. Finally, the retention and degradation behavior are improved, as is discussed in Section 1.6. Low-voltage operation, down to 5 V, has been demonstrated [1.124] for a nitride of 3 nm thickness and a blocking oxide thickness of 5.5 nm. Although optimization of the process and structure is still required, the application of this SONOS cell concept has allowed realization of memories with densities in the Mbit range.

1.4.3 Ferroelectric Devices

A new type of nonvolatile memory is emerging whose operation is based on the ferroelectric effect [1.125]. Certain crystalline materials show the tendency to polarize spontaneously under the influence of an external field and to remain polarized after the external field is removed. The polarization can simply be reversed by applying a field of opposite polarity. As such, a bistable nonvolatile capacitor is obtained in which stored information is based on polarization state rather than on stored charge.

The data stored in a capacitor can be read by sensing the interaction of a "read field" with the polarization state of the element. If a read voltage is applied to the ferroelectric capacitor of polarity opposite to the previous write voltage, the polarization state will switch, giving rise to a large displacement charge that can be sensed by proper circuitry.

The ferroelectric material used in nonvolatile memory applications is a lead–zirconate–titanate compound ($Pb[Zr,Ti]O_3$, PZT), which is a perovskite-type ceramic. Different configurations can be envisioned for a nonvolatile RAM which uses the polarizable medium as the storage element. These configurations can be divided into "backup"-type memories or "primary storage"-type memories.

For the first type, the configuration is, in fact, similar to that described for NOVRAMs in Section 1.3.4. It consists of a DRAM or SRAM configuration for which each memory element has a shadow ferroelectric capacitor backup cell. Only upon power failure or after an intentional store signal is the information present in the RAM transferred to the backup nonvolatile element. The cell itself does not affect the RAM performance during normal operation. When power comes up, or after a recall command, information stored in the ferroelectric capacitor is destructively read out and stored in the corresponding RAM cell. The advantages of this type of NOVRAM over the conventional concept discussed previously are first, the high programming speed of the ferroelectric elements, which allows a very fast transfer of the data content from the volatile to the nonvolatile part (typically well below 200 ns), and second, the high density that can be achieved since the nonvolatile capacitors are built above the conventional memory circuitry.

In order to achieve high-density, nonvolatile RAMs, however, the ferroelectric capacitor should be used as the primary storage element in an advanced DRAM-type configuration in which a single transistor and the ferroelectric capacitor make up the cell. This configuration is referred to as the *F*erroelectric *RAM*, or FRAM, and is the first true nonvolatile read/write memory. The FRAM configuration no longer needs a store or power-failure detection, but each write and access cycle is directed toward the capacitor. However, since each read operation is destructive and implies a rewrite, the FRAM concept can become successful only if very high endurance (more than 10^{15}) is assured and if program times are small enough, that is, below 50 ns. Figure 1.42 shows a schematic of a 2 capacitor/bit FRAM configuration. The memory bit consists of a wordline (WL) controlling two pass transistors, a

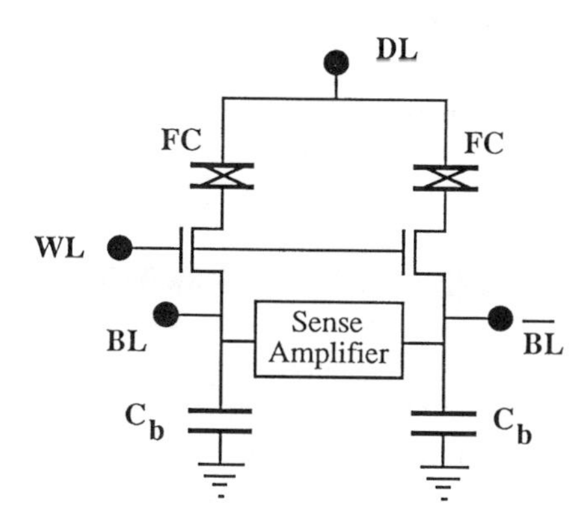

Figure 1.42 Schematic of a 2 capacitor/bit FRAM (**F**erroelectric **RAM**) configuration.

bitline (BL), a $\overline{\text{bitline}}$ ($\overline{\text{BL}}$) to collect charge from the capacitors, and a driveline (DL) to drive the capacitors. A sense amplifier connects both bitlines.

The combination of fast write and read with nonvolatility, high-density, simple cell structure, better endurance, and radiation hardness makes this approach highly promising.

1.4.4 Comparison of the Floating Gate, Charge-Trapping, and Ferroelectric Devices

As the previous sections have made clear, the different nonvolatile technologies and approaches have their merits and drawbacks. The progress made in the physical understanding of the different programming mechanisms, the mastering of the technology, and the capability to adapt the technology and device to a specific application enable any technology or approach to be used or engineered for almost any application, provided sufficient effort is put into the development. However, other criteria, such as development-entry cost, compatibility with standard technology, experience with the technology, and environmental requirements, greatly influence and determine the selection of a particular technology or approach for a specific nonvolatile memory application. In Table 1.3, technologies or approaches (SNOS, TPFG, FLOTOX, Flash EEPROM, FRAM) are compared against a number of criteria.

TABLE 1.3 COMPARISON OF NONVOLATILE APPROACHES

Criteria	SNOS	TPFG	FLOTOX	Flash	FRAM
Scaled cell size	+	+	–	++	++
Voltage scaling	++	+	–	+	++
Complexity of technology	H	H	L	L	H
Compatibility	o	+	++	++	o
Complexity of cell	L	H	M	M	L
Retention	+	++	++	++	++
Endurance	+	+	+	o	++
Radiation hardness	++	–	–	–	++
Development entry cost	H	H	L	L	H

++: very good; +: good; o: medium; –: poor
H: high; M: medium; L: low

1.5. BASIC NVSM DEVICE EQUATIONS AND MODELS

Before discussing the most important nonvolatile memory device characteristics, special attention should be paid to some aspects that are typical for the understanding and modeling of floating gate memory devices. Indeed, apart from the dual dielectric layer that is used in charge-trapping devices, there are no fundamental differences between these devices and conventional MOSFETs from the point of view of typical MOSFET characteristics (I_d-V_g, I_d-V_d, etc.). For the modeling of their memory behavior, some basic considerations are made in Section 1.5.5. For floating gate devices, however, the presence of a floating, but conductive, gate inside the gate dielectric has some important consequences from a device modeling point of view. These consequences are the subject of the present section. First, the capacitor model is discussed, defining the important coupling factors for a floating gate cell. Then, the influence on I-V characteristics of a floating gate cell is examined, followed by the experimental procedures for determining the model parameters and the basic equations for modeling the memory behavior of floating gate (FG) devices.

1.5.1 The Capacitor Model

Obviously, in a floating gate memory cell, the floating gate itself cannot be accessed. Its voltage is controlled through capacitive-coupling with the external nodes of the device. Often, the floating gate transistor is modeled by a capacitor equivalent circuit [1.126–1.135] called the capacitor model.

In this model, shown in Fig. 1.43, all capacitors present in a typical double-poly floating gate transistor are represented. C_k is the total capacitance between the control and floating gates, while C_s and C_d are the floating gate source and drain capacitance, respectively. In the SIMOS device, C_d is determined by the floating gate–drain overlap, while, in the FLOTOX device, it is dominated by the thin oxide injection region capacitance. C_g and C_f are the floating gate channel and field region capacitance, respectively. C_t is then defined as the total capacitance of the floating gate: $C_t = C_k + C_d + C_s + C_g + C_f$.

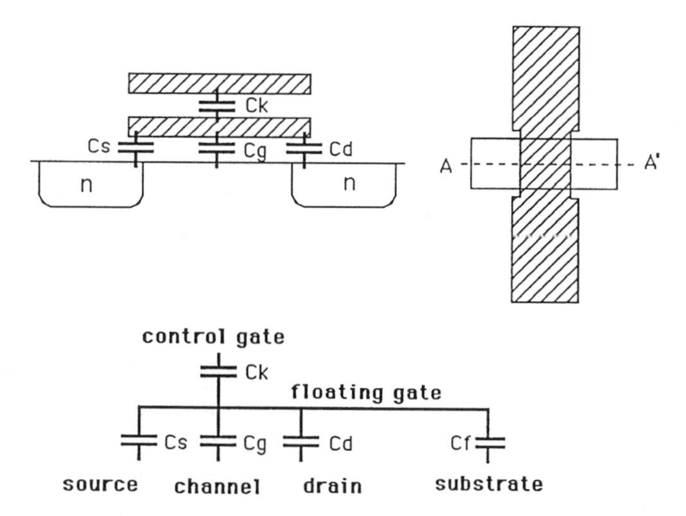

Figure 1.43 The capacitor model showing the various capacitances of the floating gate to the external nodes.

Two important coupling ratios can now be defined: k, the control gate coupling factor, and d, the drain-coupling factor:

$$k = \frac{C_k}{C_t} \qquad d = \frac{C_d}{C_t} \tag{1.19}$$

These capacitances determine the fraction of the control gate and the drain voltage, respectively, that is capacitively-coupled to the floating gate. Therefore, they are important parameters in the design, modeling, and study of floating gate memory cells.

A capacitance that is often omitted in determining the coupling factors is the coupling capacitor C_f between the floating gate and the substrate through the field oxide. In a floating gate transistor, this capacitor can be of importance because of the large coupling area between the floating gate and the control gate located over the field oxide as, for example, in the case of a FLOTOX cell.

In order to estimate the error that is made by neglecting this field capacitance in the capacitor model, assume that t_k and t_f represent the effective oxide thickness of the coupling oxide and the field oxide, respectively, and k' and d' are the coupling ratios if the field oxide capacitor C_f is neglected. The correct coupling factors, k and d, can then be expressed as

$$\begin{cases} k = \dfrac{k'}{1 + k'\dfrac{t_k}{t_f}} \\ d = \dfrac{d'}{1 + k'\dfrac{t_k}{t_f}} \end{cases} \tag{1.20}$$

It can be concluded that, by neglecting the field oxide capacitance, there is an error in both d and k of the same relative importance. This error is larger if the coupling factor k' is large and if the ratio t_k/t_f is large. The error is the same for FLOTOX and SIMOS devices: the gate oxide thickness has no influence. Table 1.4 shows some examples that illustrate the error that can be made by neglecting the field oxide capacitance for the case when t_k/t_f is 1/12.

In most cases, the capacitor values are estimated by using a simple parallel-plate model. However, fringing capacitances should also be taken into account [1.128, 1.129]. These capacitances are due to coupling between the different terminals at the edges of the polysilicon layers and to the fringing fields existing in the substrate of the device. In thin oxide devices, normally the parallel-plate approximation is fairly good, and these fringing effects cause a deviation of only a few percentage points.

The capacitor model can be used to calculate the potential of the floating gate if the voltages at the external nodes and the charge on the floating gate are known. Once the floating gate potential is known, it can then be fitted into the conventional MOS models or equations as a replacement for the conventional gate voltage in order to describe the conventional MOS characteristics. Conversely, the capacitor model is also used to calculate the model parameters, such as the control gate and drain-coupling factors from the measured MOS characteristics.

TABLE 1.4 INFLUENCE OF THE FIELD CAPACITOR ON THE COUPLING FACTORS

$\frac{t_k}{t_f} = \frac{1}{12}$			
k′ [%]	d′ [%]	k [%]	d [%]
50	10	48	9.60
60	10	57	9.52
70	10	66	9.45
80	10	75	9.38

In most cases, this procedure is applied under the assumption that the model's capacitors are formed by identical ideal conductors. In reality, however, the capacitor model, as depicted in Fig. 1.43, can only account for the electrostatics in the floating gate transistor. Therefore, the capacitor model, by itself, can lead to wrong conclusions. Indeed, in a MOS transistor, distinction should be made between the externally applied voltages and the internal electrostatic potential. In this section, the correct equations will be used, that is, the electrostatic potentials instead of the externally applied voltages.

Figure 1.44 shows an energy band diagram for a floating gate transistor at the onset of inversion—that is, when the band bending at the silicon surface equals twice the Fermipotential of the substrate. This situation clearly defines the threshold voltage as measured at the control gate.

The threshold voltage of the floating gate–oxide–substrate transistor is given by V_{t0}:

$$V_{t0} = \phi_{ms} + 2\phi_F - \frac{Q_{ox}}{C_{ox}} - \frac{Q_d}{C_{ox}} \tag{1.21}$$

where ϕ_F = the Fermipotential of the substrate (which is positive for the p-type substrate)

ϕ_{ms} = the work function difference between the gate material and the bulk material

Q_{ox} = the equivalent fixed oxide charge, located at the oxide substrate interface

Q_d = the charge in the depletion layer

First, it has to be noted that the influence of fixed oxide charges is taken into account by means of an equivalent oxide charge Q_{ox} located at the oxide–substrate interface. In order to account for the influence of this charge as sensed at the control gate, the exact distribution of the charges within the oxide must be known. Thus, charges located at the floating gate–oxide interface have no influence at all on the threshold voltage of the floating gate–oxide–substrate transistor, but will certainly be detected at the control gate. Since the exact distribution of the charges within the

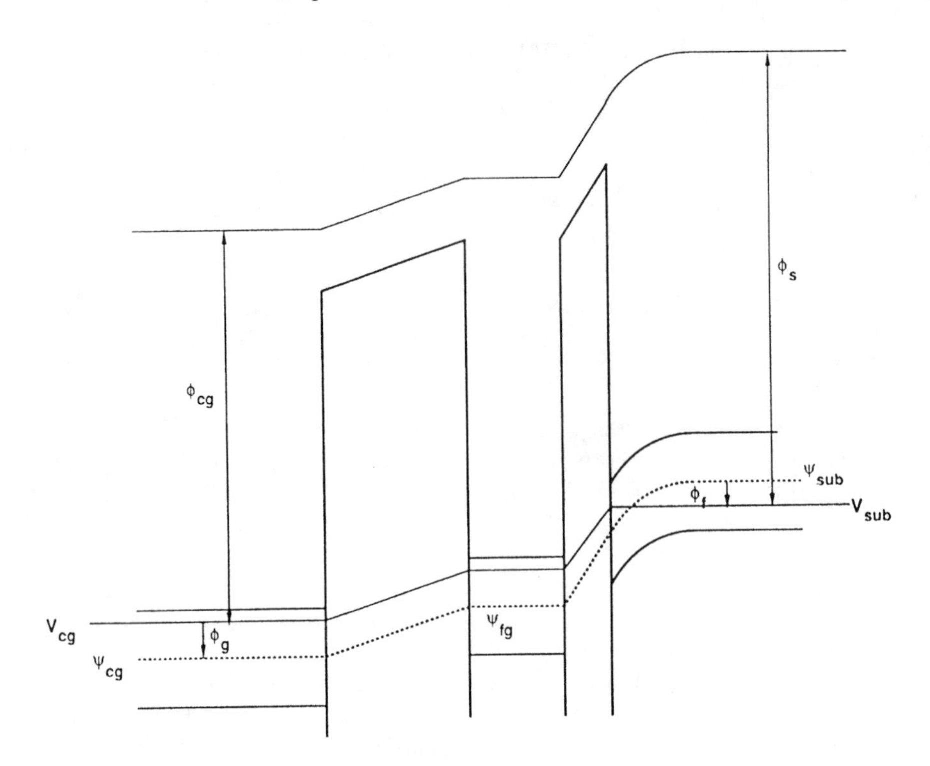

Figure 1.44 Energy band diagram for a floating gate transistor at the onset of inversion (with no charge on the floating gate).

oxide is not known, the assumption is made that the oxide-trapped charge is located at the oxide–substrate interface.

To define the threshold voltage at the control gate, the capacitor model [1.126] can be used to make the charge balance of the floating gate, using the electrostatic potential Ψ in the various regions of the cell (referred to as E_i; see Fig. 1.44):

$$Q_{fg} = (\Psi_{fg} - \Psi_{cg})\, C_k + (\Psi_{fg} - \Psi_{sub})\, C_g + (\Psi_{fg} - \Psi_d)\, C_d + (\Psi_{fg} - \Psi_f)\, C_f \quad (1.22)$$

or

$$\Psi_{fg} = \frac{C_k}{C_t}\, \Psi_{cg} + \frac{C_g}{C_t}\, \Psi_{sub} + \frac{Q_{fg}}{C_t} + \frac{C_d}{C_t}\, \Psi_d + \frac{C_f}{C_t}\, \Psi_f \quad (1.23)$$

with Q_{fg} being the charge on the floating gate. By taking $V_{sub} = 0$ as the reference, and with Eq. (1.19), this yields the floating gate voltage, V_{fg}:

$$V_{fg} + \phi_g = kV_{cg} + dV_d + \frac{Q_{fg}}{C_t} + k\,\phi_g + d\,\phi_d + \frac{C_g}{C_t}\,\phi_f - \frac{C_f}{C_t}\,\phi_f \quad (1.24)$$

It is assumed that the substrate underneath the field capacitor is in accumulation, and thus, $\Psi_f = \phi_f$. Taking into account that the drain is n^+-doped like the gate, and thus, $\phi_g \approx \phi_d$, the threshold voltage at the control gate can be expressed as the control gate voltage for which $V_{fg} = V_{tO}$:

$$V_{tcg} = \frac{V_{tO}}{k} - \frac{Q_{fg}}{C_k} - \frac{d}{k} V_d - \frac{C_g}{C_k} (\phi_{ms} + 2\,\Phi_f) + \frac{C_f}{C_k}\,\phi_f \tag{1.25}$$

In most reports, only the first three terms of this formula are used, based on the capacitor model. The last two terms are usually omitted [1.127–1.132]. The fourth term accounts for the difference in work function between the p-doped substrate and the n^+-doped polysilicon (ϕ_{ms}) and band bending in the substrate ($2\phi_f$). The last term accounts for the influence of the field capacitor. Equation (1.25) is derived for the case of an n^+-doped polysilicon gate. If other gate materials are used (e.g., silicides), the fourth term of Eq. (1.25) will change due to a different work function.

The error introduced when neglecting the last terms in Eq. (1.25) can be significant. If we try to measure the coupling factor, k, by just dividing the measured threshold voltages at the floating gate and the control gate, as was proposed in [1.129] and based on the capacitor model, the resulting coupling factor for n-MOS floating gate devices with n^+-doped polysilicon gates is always too small. The error becomes relatively more important for floating gate transistors that have small coupling factors (as used in EPROM devices). This is shown in Table 1.5 where the ratio V_{tO}/V_{tcg} [Eq. (1.25)] is compared to the real coupling factor, k.

TABLE 1.5 COMPARISON OF k AND $\frac{V_{to}}{V_{tcg}}$

$$V_{t0} = \phi_{ms} + 2.\phi_f - \frac{Q_{ox}}{C_{ox}} + \frac{Q_d}{C_{ox}}$$
$$= -0.9 + 0.6 + 1.1 = 0.8\,\mathrm{V}$$

$d = 0.05 \quad k = 0.9.k'$

k	V_{tcg}	$\frac{V_{t0}}{V_{tcg}}$
0.5	1.87	0.43
0.6	1.51	0.53
0.7	1.25	0.64
0.8	1.06	0,75

1.5.2 I-V Characteristics of Floating Gate Devices

The floating gate forms an equipotential plane between the control gate and the substrate and is parallel to both of them. This kind of equipotential plane does not exist in an ordinary MOS transistor, and thus, there are some important consequences when examining the I-V characteristics of a floating gate memory cell [1.131]. Indeed, as discussed in the previous section, the floating gate voltage is determined by capacitive-coupling between the floating gate and the externally applied voltages, more specifically, the control gate and drain voltages. From the definitions of the coupling factors, the influence of external voltages on the I-V characteristics can be calculated using the capacitor model.

When calculating the I-V characteristics of the floating gate transistor, we can start from the I-V characteristics of the floating gate–oxide substrate transistor and substitute the floating gate voltage using Eq. (1.24). In the linear region, the current is calculated as

$$\begin{aligned} I_{ds} &= \frac{\mu C_{ox} W}{L}\left[V_{fg} - V_{tO} - \frac{V_d}{2}\right]V_d \\ &= \frac{\mu C_{ox} W}{L}\left[V_{cg} - V_{tcg} - \frac{V_d}{2k}\right]V_d \end{aligned} \tag{1.26}$$

In the saturation region, the current is

$$\begin{aligned} I_{ds} &= \frac{\mu C_{ox} W}{2L}(V_{fg} - V_{tO})^2 \\ &= k^2 \frac{\mu C_{ox} W}{2L}[V_{cg} - V_{tcg}]^2 \end{aligned} \tag{1.27}$$

In both formulas, which actually are valid for long channel transistors, of course, the correct value of V_{tcg}, taking into account the influence of the applied drain voltage [Eq. (1.25)], must be used.

As an example, in Fig. 1.45, the output characteristics of a conventional MOS transistor are compared to those of an equivalent floating gate transistor. The characteristics differ in three ways: the threshold voltage of the floating gate transistor is higher, while the transconductance and the output resistance are lower. The first two effects are due to capacitive-coupling between the control gate and the floating gate. The threshold voltage of the floating gate transistor is given by Eq. (1.25) and is, therefore, roughly a factor of $1/k$ higher than that of the corresponding MOS device if there is no charge on the floating gate. The transconductance decreases by the factor k, as given by Eqs. (1.26) and (1.27).

The third distortion is the increase of the output current with drain voltage for the floating gate device. This is due to the capacitive-coupling between the drain and the floating gate, and is described by the drain voltage dependent V_{tcg}, as given by Eq. (1.25), in Eqs. (1.26) and (1.27). This effect can even lead to the turn-on of the transistor at high-drain voltages, even if the transistor is off at low-drain voltages.

1.5.3 Experimental Determination of the Coupling Factors k and d

The values of the different capacitors of the model can be calculated from the layout of the cell and the different oxide thicknesses, which can be determined from capacitor measurements. In these calculations, the effect of the field oxide capacitor and the influence of the stray capacitances must be taken into account. Therefore, this calculation will not be very accurate, and deviations from the experimentally obtained values can be as large as 10 to 15% [1.134].

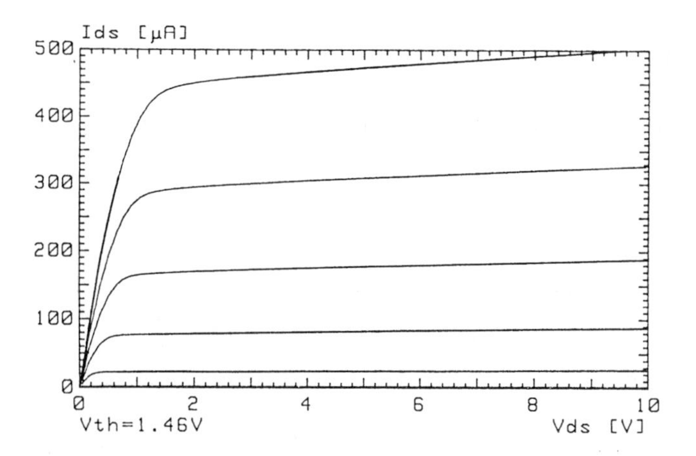

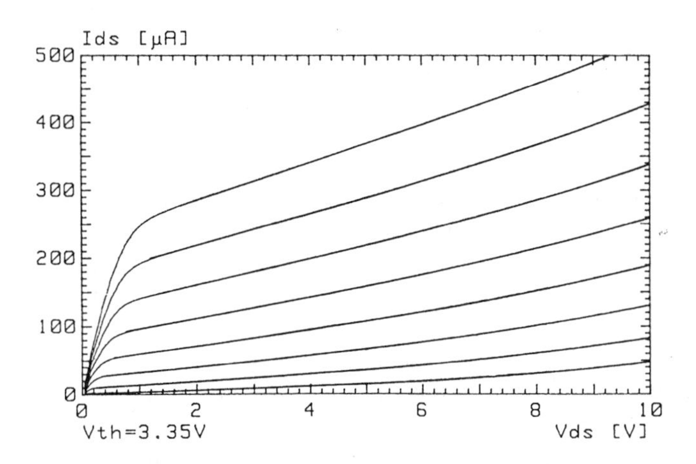

Figure 1.45 Comparison of the $I_{ds} - V_{ds}$ output characteristics of a conventional MOS transistor (upper) and its equivalent floating gate transistor with the same MOS geometry (lower). Upper figure: V_{gs}=2 V, 2.5 V, 3 V, 3.5 V, and 4 V. Lower figure: V_{gs} = 3.5 V, 4 V, 4.5 V, 5 V, 5.5 V, 6 V, 6.5 V, and 7 V.

The value of the capacitors can be measured directly on the floating gate structures and on contacted floating gate structures [1.128]. The problem in this case is that the capacitance measurements on the small floating gate transistors are very difficult and limited in accuracy by the accuracy of the measurement equipment and setup.

A third method of determining the coupling factors is to compare the threshold voltage value of the floating gate device, V_{tcg} (measured at the control gate), to that of the equivalent contacted floating gate, V_{tO} (measured at the contacted floating gate). Apart from the uncertainty about charges stored on the floating gate, the conventionally used formulas [1.129] are also not valid because they only account for electrostatics in the structure, as was discussed previously. The use of the correct formulas introduces other unknown parameters, and the determination of the coupling factors again is not accurate.

The control gate coupling factor can also be determined from a comparison of the subthreshold I-V characteristics of floating gate devices and contacted floating gate devices [1.129]. But, as shown in reference [1.126], this will always result in a coupling factor k that is larger than the real one.

Another method is based on the comparison of the I-V characteristics of floating gate devices and contacted floating gate devices [1.128]. Equations (1.25–1.27) indeed show that the coupling factors k and d can be calculated from the ratio of the currents and from the dependence of the threshold voltage V_{tcg} on the drain voltage. This measurement also directly incorporates the influence of the field oxide capacitor. The most accurate methods for determining coupling factors therefore rely on the comparison of I-V characteristics of floating gate transistors and their contacted floating gate counterparts.

As an example, Fig. 1.46 illustrates the determination of k in both the linear (upper figure) and saturation (lower figure) region, where I_{ds}-V_{gs} curves are compared for a floating gate transistor and its equivalent MOS counterpart. In order to obtain an accurate result for k, it is mandatory that the structures indeed be identical. Not only must the layout of the transistors themselves be the same, but also the contacts made at the different terminals of the transistor and, in particular, the substrate contact, must be identical for the two structures that are being compared. Any difference in layout can cause different series resistances for the terminals of the transistors and can give rise to inaccuracies. If equivalent current ranges are considered and the same fitting method is used for the results from both the contacted floating gate transistor and the memory structure, the determined coupling factor is weakly dependent on these elements. For this example, k is found to be 0.487 in the linear region and 0.532 in the saturation region. As indicated in [1.128], the measured coupling factor k is dependent on the applied drain voltage by the dependence of the gate capacitance C_g on this drain voltage. The value of C_g is indeed smaller when the transistor operates in saturation than when it operates in the linear regime. The coupling factor k will, therefore, always be larger in the saturation regime than in the linear region.

The drain-coupling factor can be calculated from the dependence of the threshold voltage V_{tcg} on the drain voltage using Eq. (1.25). In principle, this dependence can be measured in both the linear and saturation regimes. But the range of drain voltages allowed to operate the transistor in the linear regime is so small that the drain-coupling factor cannot be determined accurately. Therefore, it is preferable that the I-V characteristics for different drain voltages be measured in the saturation regime and that the V_{tcg} values extracted from these results be used to calculate the drain-coupling factor d.

Figure 1.47 shows an example of the determination of the drain-coupling factor. In the upper figure, the $\sqrt{I_{ds}} - V_{gs}$ characteristics are shown, measured for several drain voltages in saturation (between 4 V and 10 V). It becomes clear from these curves, as expected from Eq. (1.27), that the transconductance remains constant, independent of the drain voltage, but that the threshold voltage decreased with drain

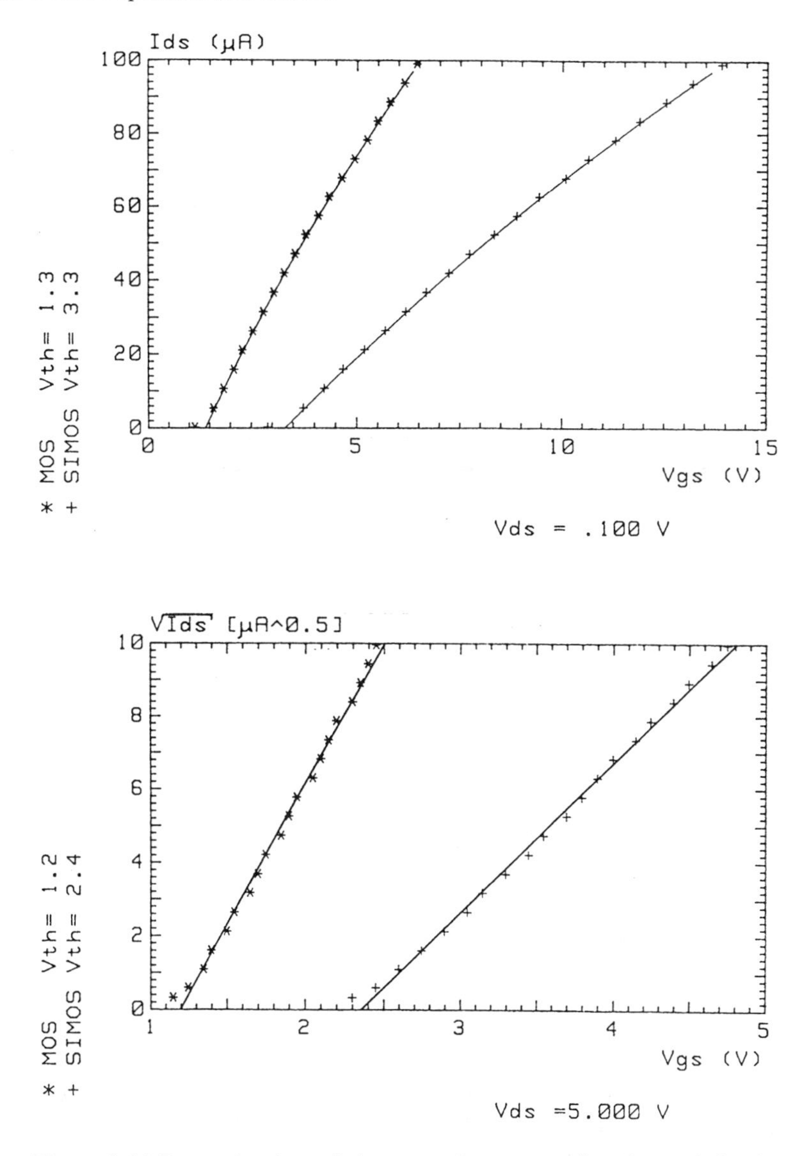

Figure 1.46 Determination of the control gate coupling factor k in the linear (upper) and in the saturation region (lower) from comparison of the $I_{ds} - V_{gs}$ characteristics of the floating gate device and its equivalent MOS counterpart.

voltage, as expected from Eq. (1.25). From Eq. (1.25), it can be concluded that the threshold voltage decreases linearly with drain voltage, with a slope of d/k. The extrapolated threshold voltage is plotted as a function of the drain voltage (lower figure), yielding a straight line. The slope of this line yields d/k, which is found to be 0.119 in this case.

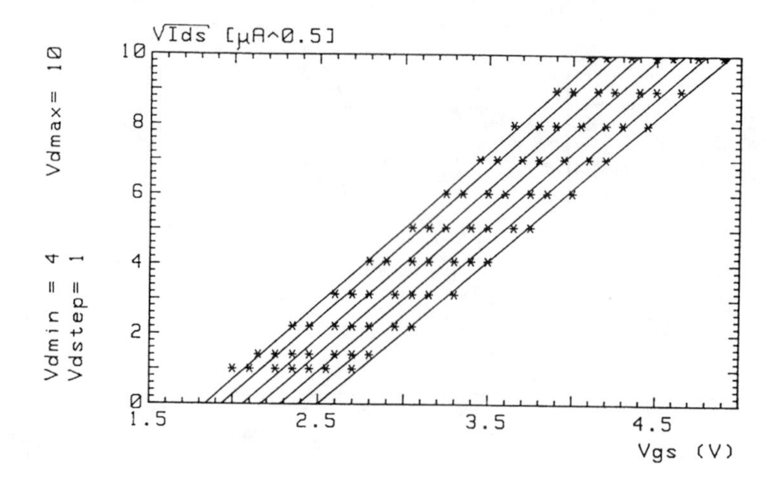

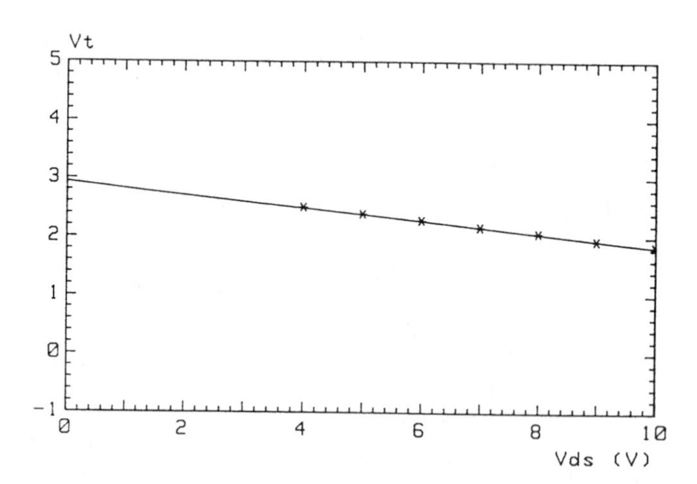

Figure 1.47 Determination of the drain-coupling factor d from the relationship between V_{tcg} and V_{ds}. The upper figure shows the influence of the drain voltage on the $\sqrt{I_{ds}} - V_{gs}$ characteristics. The lower figure shows the extrapolated threshold voltage as a function of V_{ds} yielding a straight line with slope d/k.

1.5.4 Modeling of the Memory Characteristics of FG Cells

Besides the use of the capacitor model and the I-V characteristics of floating gate devices for design and analysis purposes, they can also be applied to model the memory behavior of the cell. This model allows the calculation of the programming or transient characteristics, as well as the retention characteristics, which are discussed in Section 1.6. If a degradation model such as, for example, an electron trapping and generation or interface trap generation model is used, the endurance characteristics can also be calculated and predicted [1.135]. A comparison of measured and calculated characteristics reveals the validity of the assumptions made concerning the physical mechanisms governing the programming or degradation behavior. In this section, the basic equations that are needed for modeling the memory behavior of FG cells are discussed. The models themselves are strongly dependent on the type of cell (programming mechanism, geometry, and design) and are not treated here.

When charge is injected to or emitted from the floating gate, the floating gate potential will change, as given by Eq. (1.23). Due to the change of the floating gate potential, the electric fields in the surrounding oxides change as well, by which the injected currents increase or decrease. This process continues until a steady state is reached. Thus, it is important to remember that the injection currents during a programming operation are not constant, but change rather rapidly as a function of time. The modeling of the memory behavior of floating gate cells is, therefore, based on the following elements.

1. The basic memory model starts from the expression for the charge on the floating gate:

$$\frac{dQ_{fg}(t)}{dt} = \int_{A_{fg}} J_{fg}(t)dA \tag{1.28}$$

   In this expression, the integral over the complete floating gate area can usually be replaced by a summation over the different oxides or dielectrics, "i", that surround the floating gate:

$$\frac{dQ_{fg}(t)}{dt} = \sum_{A_i} J_i(t) \tag{1.29}$$

   where A_i is the injection area to the floating gate. The currents, J_i, through the different dielectrics are a function of time because the electric fields, E_i, in the dielectrics are changing with time. By integrating Eq. (1.29) with respect to time, an expression for the charge that is accumulated on the floating gate is obtained:

$$Q_{fg}(t) = \int_0^t \sum_{A_i} J_i[E_i(t)]dt \tag{1.30}$$

2. This expression cannot be solved as such because the electric fields, E_i, are dependent on the floating gate potential, and thus, also on the floating gate charge, $Q_{fg}(t)$. Therefore, a model is needed to allow the calculation of the fields occurring inside the device during programming. For floating gate transistors, this model is the capacitor model for the device described in Section 1.5.1.

The relation between the floating gate potential and the charge on the floating gate is given by Eq. (1.24), which is often simplified by neglecting the correction factors due to the work function differences and the Fermipotentials:

$$V_{fg}(t) = k\ V_{cg}(t) + d\ V_d(t) + \frac{Q_{fg}(t)}{C_t} \tag{1.31}$$

where $V_{cg}(t)$ and $V_d(t)$ are the control gate and drain voltages, respectively, as a function of time during charge injection (e.g., programming). A more general expression can be obtained by also incorporating the other external potentials that are able to couple voltages to the floating gate such as, for example, the source voltage, V_s, and the channel potential, V_{ch}, together with their respective coupling factors. In most cases, however, these nodes are grounded, so they will not be considered further here.

From V_{fg} and the external potentials V_{cg}, V_d, V_s, and V_{ch}, the electrical fields, E_i, inside the respective dielectrics surrounding the floating gate can be calculated.

3. The next step is to implement a model for charge injection into the gate insulator under programming conditions based on the above calculated electrical fields. These models have been described in Section 1.2. As was discussed there, for some injection mechanisms like Fowler–Nordheim tunneling, a closed form expression for the current as a function of the electrical field exists. If the externally applied voltages are also time independent, Eq. (1.30) can be solved analytically. For other injection mechanisms, however, such as hot-electron injection or polyoxide conduction, as well as for time-dependent external voltages, no analytical expressions can be obtained. For these cases, Eq. (1.30) must be solved numerically.
4. If Eq. (1.30) has been solved, the external threshold voltage of the transistor cell can easily be found using expression (1.25), again given by the capacitor model. This expression links the threshold voltage of the nonvolatile memory transistor to the amount of charge stored on the floating gate. If the correction factors due to work function differences and Fermipotentials are neglected, this yields:

$$V_{tcg}(t) = \frac{V_{tO}}{k} - \frac{Q_{fg}(t)}{C_k} - \frac{d}{k}\ V_d \tag{1.32}$$

where, in this case, V_d is the drain voltage applied during sensing (i.e., reading) of the contents of the memory transistor.

This expression has to be used in combination with a clear definition of the threshold voltage as employed in the measurement of the transient programming characteristics. A commonly used definition is the voltage applied at the externally accessible gate of the transistor in order to allow a predefined current to flow from drain to source in the memory transistor. This imposes the problem that the I–V characteristics of the memory transistor have to be known. In most cases, the expression for the threshold voltage of the nonvolatile memory transistor is based on a physical criterion (like channel inversion) that must be related to the threshold voltage definition based on drain-source currents. In many cases, this distinction is not clearly made, but the error introduced by neglecting this difference is normally small since it is threshold voltage shifts that are considered in the first place, and the applied drain voltage during measurement of the threshold voltage and predefined current levels are kept small. If only the shift has to be predicted, Eq. (1.32) simplifies to

$$\Delta V_{tcg}(t) = -\frac{\Delta Q_{fg}(t)}{C_k} \tag{1.33}$$

5. In case of an eventual degradation of the oxide—such as electron trapping or interface trap generation, which will influence the injected currents J_i—an additional model that describes this degradation has to be implemented. This is needed not only to simulate the endurance characteristics but also in other cases since degradation of the injection currents can occur, even within the time needed for one programming operation. This is the case for programming operations based on polyoxide conduction, an injection mechanism that induces rapid degradation due to charge trapping in the polyoxide, as has already been discussed in Section 1.2.2. Numerous degradation models have been proposed in the past, but it is impossible to discuss these models within the focus of this introductory chapter. Instead, it is important to note that most of the degradation models used are based on experiments that were carried out on capacitors, based on either constant voltage or constant current injection experiments. In floating gate cells, however, neither constant voltage nor constant current conditions exist; therefore, care should be taken to extrapolate the results from capacitors to real situations in memory cells. These extrapolations can lead to erroneous models and conclusions concerning the degradation mechanisms that occur in transistor cells.

In principle, with the help of the above-described models, a complete transient programming characteristic (i.e., the shift of the threshold voltage of the nonvolatile memory transistor as a function of programming time during programming) can be calculated. This can be repeated for different programming conditions (e.g., applied voltages or voltage pulse shapes) and for different transistor geometries. In addition, parameters that cannot be measured directly can be calculated. It is important to know the fields occurring across the gate insulator during programming in order to avoid oxide breakdown, as well as the fields across the insulator between the floating gate and control gate in floating gate devices so that the leakage current through this insulator during programming can be calculated. The same equations can be used to predict or calculate the retention or endurance characteristics, provided that the correct physical mechanisms responsible for these characteristics (charge loss or gain mechanisms for retention and degradation mechanisms for endurance) are taken into account by an appropriate model.

1.5.5 Modeling of the Memory Characteristics of SNOS and SONOS Devices

Modeling of the memory behavior (transient and programming characteristics) of SNOS and SONOS devices is based on numerical integration of the continuity equation which relates the current gradient to the rate of change of the local stored charge content in the nitride layer. The model, therefore, requires knowledge of the injection current mechanisms (at both interfaces and in both energy bands) and an appropriate trap model. Finally, the Poisson equation is used to determine the electric field distributions in all layers. Again, it is not the purpose of this chapter to treat these techniques and results in detail. The advantage of such a rigorous treatment is that charge distributions in the nitride layer can be obtained. The shape of the distribution seriously affects the retention behavior of these devices [1.108, 1.136]. This retention behavior can be computed numerically, as was done by Williams et al. [1.137], Heyns and Maes [1.138, 1.139], and Libsch et al. [1.140], once the trapped charge distribution is known. The effects of reducing the nitride thickness of SNOS or SONOS devices on the retention characteristics [1.108, 1.120] can then be studied and simulated [1.137], and will be useful in the further scaling of these devices. Numerous models have been proposed for the memory traps in silicon nitride. A review of these models is presented in [1.140]. Charge distributions obtained from computations based on two-carrier conduction and two or three trap-level models (electron trap, hole trap, and recombination center) were obtained by Remmerie et al. [1.116], whereas calculations based on two-carrier conduction and a single deep-level amphoteric trap model were presented by Libsch et al. [1.124, 1.140].

1.6. BASIC NVSM MEMORY CHARACTERISTICS

Besides conventional device characteristics, nonvolatile memory cells also have some important functional memory characteristics, which are used to evaluate the memory performance of the cell. These characteristics can be divided into three main classes.

1. The transient characteristics describe the time dependence of the threshold voltage during programming.
2. The endurance characteristics, which are meaningful only for EEPROM cells, give the memory threshold window, which is the difference between the threshold voltages in the written and the erased states, as a function of the number of programming cycles; they are characteristic for the intrinsic number of write/erase cycles that can be endured before both programmed states are no longer distinguishable.
3. Finally, the retention characteristics give the threshold voltage in either programmed state as a function of the time after programming, and indicate the intrinsic ability of the memory cell to retain its content over long periods of time.

These basic memory characteristics are discussed in more detail in the following sections.

1.6.1 Transient Characteristics

The functioning of nonvolatile memory devices is based on the possibility of bringing charges onto the floating gate or into the gate insulator and removing them again in order to change the threshold voltage of the nonvolatile transistor. In the ideal nonvolatile cell, programming can be performed with externally applied voltages that are as low as possible. However, this programming operation should be as fast as possible.

The transient programming characteristic of a nonvolatile memory transistor is the shifting of the threshold voltage of the transistor as a function of time during programming. The exact knowledge of these characteristics allows the determination of the programming voltages and times needed to obtain a useful threshold voltage window.

Some remarks have been made in the preceding sections concerning the meanings of write, erase, program, and clear. Indeed, some confusing terminology is often used with respect to the different programming operations. In fact, two possible choices can be made:

- In a FLOTOX-type EEPROM memory matrix organization, 8 adjacent bits make up a byte. Due to the connection between these 8 bits (i.e., the sources of the memory transistors are connected), it is impossible for one programming operation to be performed selectively for the bits within one byte. Therefore, a byte in this EEPROM matrix architecture is programmed in two steps: first, the threshold voltages of the 8 bits are all brought to a high level simultaneously, and then some of the bits are selectively written to a low threshold voltage within one byte. Therefore, calling the operation that is performed simultaneously on all bits within one byte, "erase," and the selective operation, "write," can be defended. Then, "erasing" in this case means bringing the threshold voltage to a high (positive) level, while writing means bringing it to a low level. This convention is used in most EEPROM products.
- On the other hand, the output voltage on the data line of a memory device being equal to 0 V corresponds to the low-threshold voltage of the memory cell. Indeed, the memory chip will be designed so that the access time for output high and output low are the same. Because it takes more time for the output buffer to drive the data-line high, one will choose this to correspond to the memory cell threshold voltage that will be detected the quickest by the sense amplifier. For a memory cell with a high-threshold voltage, the bitline voltage (and thus, the input voltage of the sense amplifier) will not change, or will change only slightly, during read-out, and the sense amplifier does not need to switch. Therefore, a high-memory cell threshold voltage corresponds to an output voltage on the data-line $V_{out} = 5$ V, and the operation to achieve

a high-memory threshold voltage is called "write." This choice also corresponds to the situation for EPROM memories where "programming" by means of hot carriers induces an increase in the threshold voltage. "Erasure" of EPROM memories is done with UV light and brings the threshold voltage low.

Since the transient programming characteristic consists of a shift of the threshold voltages during a programming operation, the transient programming measurement can be performed in two ways.

In the first approach, the programming operation is stopped before the programming is completed, and the threshold voltage is measured. Then, the memory transistor is erased until the initial threshold voltage is reached again. The programming operation is repeated under the same conditions as before, but with a longer time before the programming is halted. By repeating this procedure, a complete transient characteristic can be recorded. The drawback of this method of measurement is that it implicitly assumes that neither the programming operation nor the erase operation introduces any degradation. This degradation would then be superimposed on the real transient programming characteristic. A simple check on whether this assumption is valid consists of recording the transient programming characteristic twice: both transient programming characteristics should coincide. An advantage of this measurement method is that, in contrast with the second method, it can easily be implemented for different voltage pulse shapes.

A second measurement method for the transient programming characteristic divides the total considered programming time into short time periods. The programming voltages are applied during each small time period, and in between, the threshold voltage is measured. The simplest example is programming with constant voltages. During the measurement, voltage pulses of short duration are applied and the threshold voltage is measured after every pulse. In the transient characteristic, it is the cumulative effective programming time that is put on the x-axis. For this method, the assumption is made that the frequent interruption of the programming operation (in most cases, the application and disconnection of the programming voltages) has no influence on the results. This assumption can easily be checked by repeating the transient measurement where, this time, the programming conditions have to be applied during the whole programming time without interruption. Again, the two results should coincide. A drawback of this measurement method is that it cannot easily be implemented in the case where shaped programming voltage pulses are used.

A special sort of programming characteristic is the so-called soft-write. This terminology is used primarily for EPROM devices and is related to the conditions that occur at the different memory cells in a memory circuit during the programming of the whole memory. The soft-write characteristic describes the shift in threshold voltage of one cell under conditions that are present when another cell is being programmed. For EPROM (or Flash EEPROM) memories, it is mandatory that all cells of the circuit be written without disturbing the information content of any previously written cell. For all recent EEPROM circuits, the danger for soft-write does not exist because of the use of an isolating select transistor per memory cell.

Another special case of programming is the "read-disturb." During read-out of the information, voltages have to be applied to the nonvolatile memory transistor. These voltages can induce a threshold voltage shift in the cell that is addressed, as well as in some other cells. Read-disturb is mainly a concern for EPROM (and Flash EEPROM) memories for the same reason as noted above.

As an example, Fig. 1.48 shows the programming (both write and erase) characteristics of a FLOTOX-type nonvolatile memory transistor. For more details on the transient characteristics of the various cells, the reader is referred to the other chapters of this book.

1.6.2 Endurance Characteristics

With respect to overall reliability, two different features of the nonvolatile memory have to be considered. Nonvolatile memories can be reprogrammed frequently, but, in contrast to RAM memories, each write operation introduces some sort of permanent damage. This implies that the total number of write operations is limited; for example, most commercially available EEPROM products are guaranteed to withstand, at most, 10^4 programming cycles. The damaging of the memory cell during cycling is normally referred to as "degradation" and the number of cycles the memory can withstand is normally called its "endurance." Another failure mode of the nonvolatile memory is retention failure. This failure mode is discussed in the next section.

1.6.2.1 Floating Gate Devices. The program/erase endurance of floating gate devices is determined by four phenomena: tunnel oxide breakdown, gate oxide breakdown, trap-up, and degradation of the sense transistor characteristics. Whereas the first two are self-explanatory, trap-up is defined as the trapping of electrons in the oxide during programming operations. These trapped charges change the injection fields and thus, the amount of charge transferred to and from the floating gate during programming. This eventually leads to a situation where the difference in threshold voltage in the two possible memory states is so small that the sense circuit can no longer discern the two states. Degradation of the transistor characteristics occurs when CHE injection is used for programming. CHE injection is used primarily in EPROMs where endurance is restricted to a low number of cycles. Fowler–Nordheim injection also causes degradation of the transistor characteristics, but most devices make use of a separate tunnel area, leaving the sense transistor unaffected by programming operations.

As described by Mielke et al. [1.25], the main cause of endurance failure in TPFG devices is believed to be the trapping of electrons in the tunnel oxide (called trap-up), while thin oxide devices fail mainly because of thin oxide breakdown induced by very high oxide fields during programming. Trap-up also occurs in

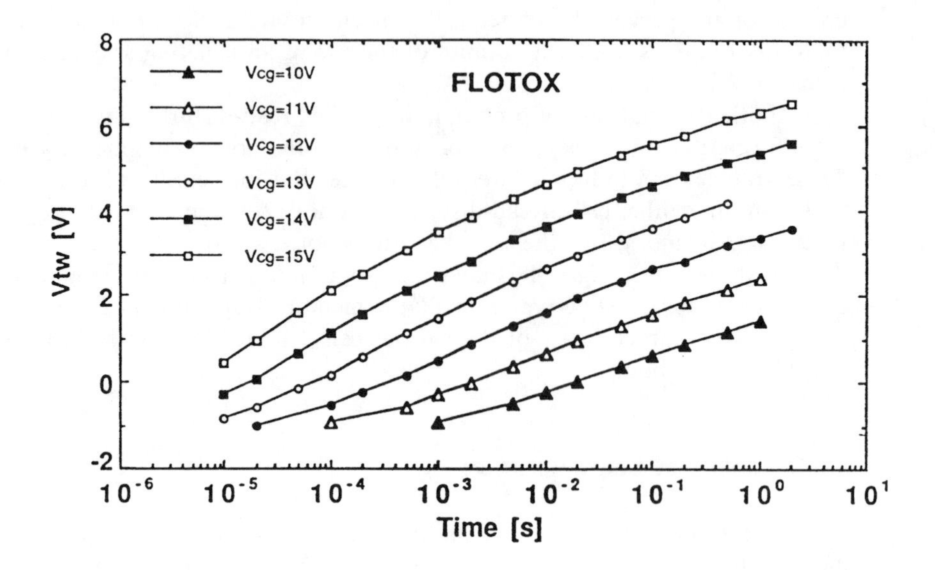

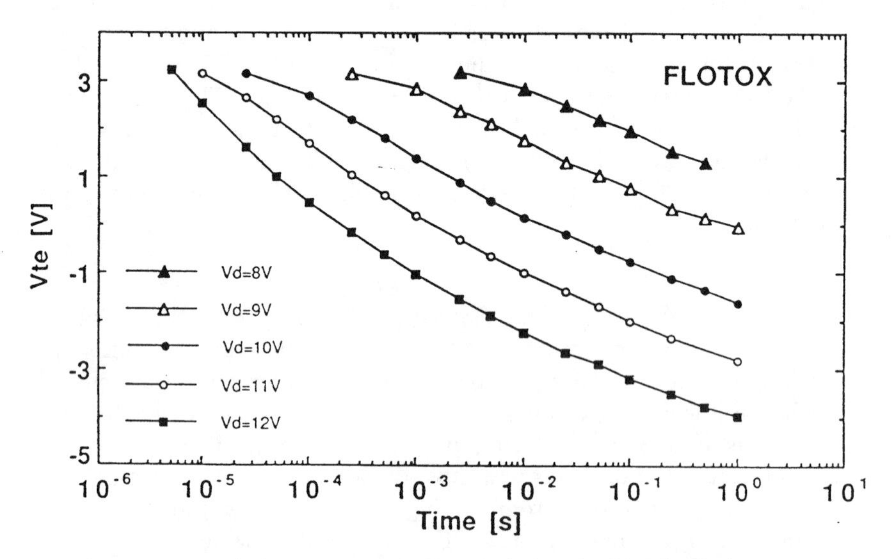

Figure 1.48 Typical programming characteristics of a FLOTOX-type memory cell showing the external threshold voltage as a function of the programming time. Upper figure = erasing, lower figure = writing.

thin oxide devices, but it is far less important than for TPFG transistors. Thin oxide devices are limited primarily by oxide breakdown.

The oxide breakdown phenomenon associated with thin oxide devices manifests itself slightly differently in device operation than during breakdown tests. As suggested in reference [1.25], the amount of energy available inside the memory transistor is too small to cause immediate and total breakdown. Therefore, this breakdown is generally regarded as being caused by a weak spot in the oxide that is "activated" by high electric fields and eventually becomes so leaky that the charges stored on the floating gate can no longer be retained. This is detected as a retention failure.

As explained above, thin oxide nonvolatile devices fail mainly because of the presence of defects (weak spots) in the thin oxide, which become shorts under the influence of high fields. As the total area of thin oxide per chip increases with increasing memory density, the probability that one memory cell will fail (and thus, that the circuit will fail) increases with increasing memory density. It was even predicted that thin oxide FLOTOX-type EEPROM circuits with densities larger than 16 KB would be less reliable than their TPFG counterparts, so that they would never become important [1.25]. As has been argued more than once [1.141], however, the density of weak spots in the thin oxide layer is largely dependent on processing conditions and is constantly decreasing by the use of advanced processing techniques. Recently, 1 Mbit EEPROM products, making use of thin oxide, have been achieved [1.142]. These circuits generally incorporate some sort of redundancy or error correcting circuitry, which can obviate the defect-related yield and reliability problems of thin oxide devices.

The fact that defects in the thin oxide limit the endurance features of thin oxide nonvolatile devices explains why little attention has been paid to the degradation behavior of these devices caused by charge trapping. This behavior, however, is important because it determines the endurance limit of the optimum memory cell (i.e., without defects) or its equivalent, constructed by means of redundancy. The study of degradation behavior also reveals some phenomena that would otherwise not be detected and that are responsible for some mismatches between experimentally obtained and calculated characteristics (e.g., some features of transient programming characteristics).

With respect to endurance, two different characteristics are important. The first one describes the degradation of the memory during cycling and gives the value of the threshold voltage for one memory transistor as a function of applied programming cycles, all with the same programming conditions. An example of such a characteristic for a FLOTOX cell is given in Fig. 1.49. A threshold voltage window opening in the first tens of cycles is observed, followed by a severe window closing after 10^5–10^6 cycles.

The results of threshold voltage degradation measurements are frequently presented in the literature. It is assumed that the results obtained on MOS capacitors can be used to predict the behavior of MOS transistors during high-field stressing. For stressing with alternating field polarities, as is the case for EEPROM devices, the degradation mechanisms have not been well understood until recently [1.25, 1.86, 1.126, 1.143, 1.144]. The explanation of observed memory degradation behavior is restricted to the general statement that threshold voltage window opening is caused by positive charge trapping, whereas window closing is caused by electron trapping in the oxide [1.25, 1.86, 1.143, 1.144].

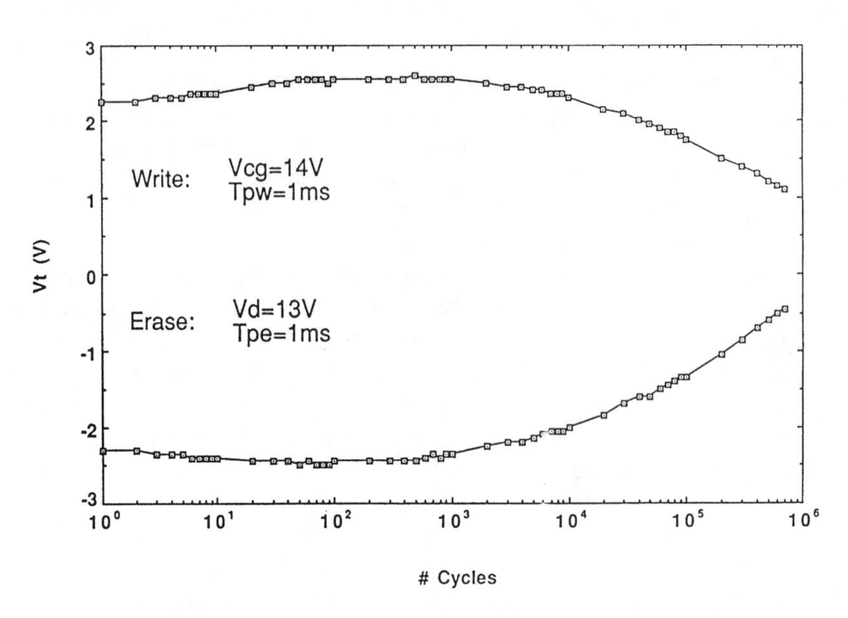

Figure 1.49 Typical endurance characteristics of a FLOTOX-type memory cell showing the threshold voltage in the written and the erased state as a function of the number of applied write/erase cycles. The threshold voltage shows a threshold voltage window opening during the first tens of cycles, followed by a window closure after 10^5–10^6 cycles.

For an understanding of the threshold voltage degradation characteristics, it is important that the threshold voltages be given from the first cycle on. In many cases, application of a number of cycles before the start of the degradation measurement in order to get a stable threshold voltage hides this valuable information.

A second concern about endurance is related to statistical information regarding the distribution of the cell characteristics. Specifically, the cumulative failure probability as a function of applied programming cycles for a large number of nonvolatile memory devices is an important indicator for the overall endurance reliability of memory circuits containing many memory cells. An example is shown in Fig. 1.50 [1.25]. For endurance failures originating from oxide breakdown, a broad random-life failure rate distribution with an almost flat failure rate is normally observed. For endurance failures caused by trap-up, which is a more intrinsic feature, a sharp wearout beyond some number of cycles is recorded.

In order to fully characterize the endurance behavior of a memory circuit, knowledge of the distribution of the endurance failures is not sufficient. The worst case endurance of any given cell in the array defines the endurance of the chip. Within a given chip, the endurances of individual bits have a small range of values, and it is pure chance whether a highly cycled bit has an endurance on the high end or low end of that distribution. Knowledge of the statistical distribution of the worst

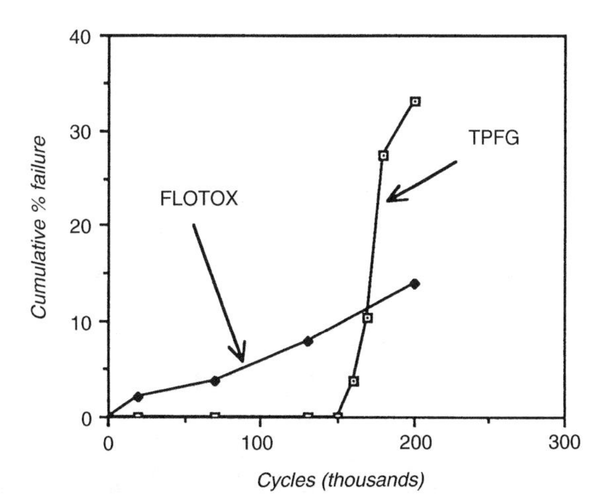

Figure 1.50 Cumulative failure distribution as a function of the number of cycles for two types of floating gate technologies (FLOTOX and TPFG) [1.25].

case bit from each of thousands of devices makes the endurance of a given chip statistically predictable.

1.6.2.2 Charge-Trapping Devices. For charge-trapping devices, program/erase endurance is determined by charge transport through the UTO layer. In conventional SNOS devices, it has been shown that the degradation is caused primarily by hole transport through the UTO layer [1.138, 1.145] by which hole traps are created in the oxide and interface traps are generated [1.112, 1.146], resulting in shifts of the threshold voltage and reduced retention. It was shown recently that hole transport toward the silicon is the most damaging [1.138]. This explains why a reduction in hole injection from the gate in SNOS devices with an oxynitride storage layer [1.113], or in SONOS devices, has given rise to improved endurance [1.105]. For the scaled-down SONOS structure, hole injection is almost eliminated, and consequently, even better endurance can be expected. This has been confirmed on ultra-thin SONOS structures (2 nm SiO_2, 8.5 nm Si_3N_4, 5 nm SiO_2), which showed no noticeable degradation after 10^7 10 V erase/write (E/W) pulse cycles [1.147].

1.6.3 Retention Characteristics

Another failure mode of the nonvolatile memory is retention failure. In this case, the memory loses its information and is, therefore, no longer nonvolatile. Most commercially available products are guaranteed to retain their information for at least 10 years after programming, either when operating or with power turned off. The retention feature of the memory device is influenced by degradation. It is possible that, after a number of programming cycles, the memory cell can still be reprogrammed but can no longer meet its retention limit.

1.6.3.1 Floating Gate Devices. For floating gate devices, there is no intrinsic retention problem since retention is limited only by defect densities. Defects can be activated by the stress that the oxide layer undergoes in a high-temperature bake or from a large number of programming cycles. Thus, many endurance failures are actually retention failures.

Since the retention time to be guaranteed by the manufacturer is generally quite high, all retention tests use a combination of accelerating conditions. For example, higher voltages during read-out is one possible acceleration. The most commonly used retention test is storage of the programmed devices at a high temperature (up to 250°C). In any case, a complete reliability evaluation of a nonvolatile device must be a combination of endurance and retention tests.

1.6.3.2 Charge-Trapping Devices. For charge-trapping devices, there is an intrinsic retention problem. The threshold voltage of programmed SNOS devices decreases with time. This decrease is due either to *loss of charge* from the nitride layer by backtunneling to the silicon bands [1.136, 1.148] or by injection from the silicon into the nitride layer of carriers of the opposite type [1.139] or to *redistribution of charge* in the nitride layer [1.137]. The loss by backtunneling can be reduced through an appropriate hydrogen anneal step [1.103, 1.106], by which the interface trap density is reduced substantially [1.146]. For conventional SNOS and for SONOS devices, the threshold voltage decay is logarithmic in time, and, from extrapolation of the data taken over several decades of time, retention times of well over 10 years can be expected, even at elevated temperatures. Furthermore, a slowdown of the decay rate is observed for longer times [1.122, 1.139]. As an example, Fig. 1.51 shows the threshold voltage decay of a p-channel MNOS transistor and compares its behavior for annealed and nonannealed devices after storage at room temperature and at high temperature (125°C) [1.146]. The hydrogen anneal conditions used in this experiment were 800°C and 15 minutes [1.104]. As the figure shows, the decay is logarithmic in time for all cases and the hydrogen anneal results in a reduction of the decay rate by 25% [1.103]. Recently, Minami et al. [1.108] found that scaled-down SNOS devices show improved retention behavior: the written state decay rates do, in fact, decrease with nitride thickness, whereas the erased-state retentivity is almost independent of nitride thickness.

For the scaled-down SONOS device, it is not clear at present whether further scaling will improve the retention behavior as has been predicted [1.122]. If, in this scaled device, the major contribution to the threshold voltage shift is a result of charge stored at the top oxide–nitride interface, backtunneling can indeed be expected to decrease significantly. However, in this case, injection from the silicon and compensation of the stored charge could become a major issue of concern [1.139] in view of the high ($> 2\,\mathrm{MV/cm}$) fields that will be reached in the tunneling oxide for stored charges on the order of 10^{12}–$10^{13}\,\mathrm{cm}^{-2}$.

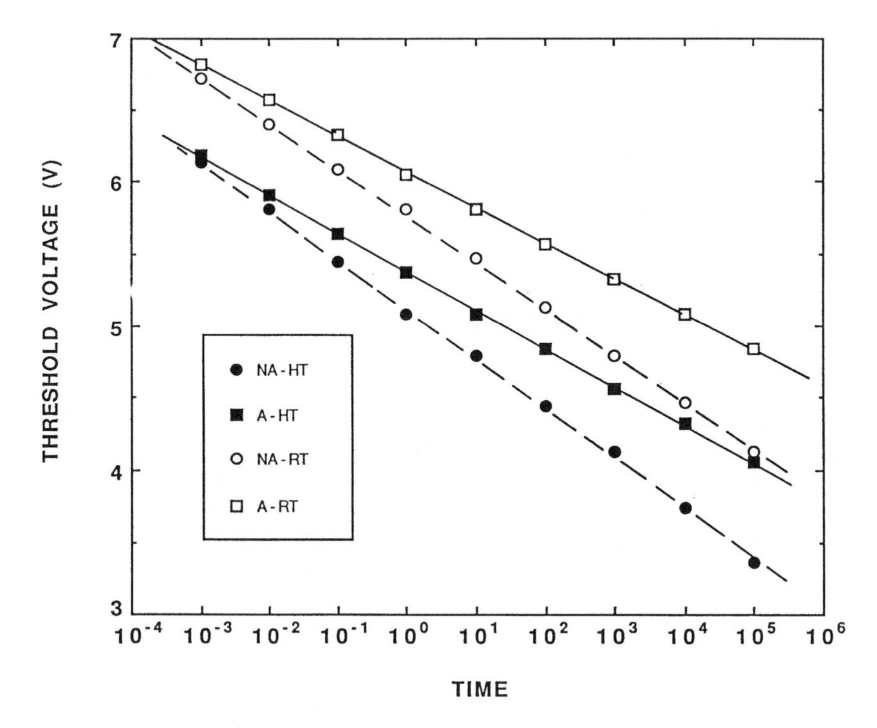

Figure 1.51 Threshold voltage decay in a p-channel MNOS transistor after a write pulse of 1 ms, $V_W = -20$ V, $t_{ox} = 2.1$ nm, $t_n = 52$ nm. Writing was always performed at room temperature. This figure shows the effect of a high-temperature hydrogen anneal (800°C, 15 min; A = annealed, NA = not annealed) and of the temperature during read (RT = room temperature, HT = high temperature = 125°C).

1.7. RADIATION ASPECTS OF NONVOLATILE MEMORIES

Finally, some issues related to the sensitivity of the different nonvolatile memory technologies to ionizing radiation are briefly discussed in this section.

1.7.1 SNOS Technology

SNOS memory devices are significantly harder than MOS structures because, unlike silicon dioxide, the mobilities of electrons and holes are not much different in nitrides. When exposed to ionizing radiation, both generated carriers can be rapidly swept out of the insulator, resulting in a negligible amount of trapped charge [1.149].

Acceptable shifts for a total radiation dose of up to a Megarad (Si) at 77 K have been obtained for SNOS structures [1.150]. A reduction of the programming voltages and the absence of thick oxide parts in the newer SNOS devices have improved their radiation hardness. For SNOS/CMOS technologies, special techniques are, however, required to harden the peripheral circuitry [1.151, 1.152]. For scaled SONOS devices with a thick top oxide, radiation hardness might become a matter of concern in view of the increasing dominance of SiO_2 parts in the device.

1.7.2 Floating Gate Technology

Floating gate memories suffer from a lower failure rate due to α-particles than volatile memories since the charge is stored on a floating gate that is less susceptible to α-particle-induced electron-hole pairs [1.153]. As a result, the memory cell itself is not susceptible to α-particle upset. However, the peripheral circuitry is.

An important issue in radiation hardness of memory devices is their total dose characteristic. Since floating gate technology is essentially a MOS technology, the total dose radiation hardness of these technologies is expected to be rather weak if no special precautions are taken.

Recently, radiation characteristics of a floating gate memory technology were reported [1.154, 1.155]. Figure 1.52 shows an example of the memory threshold voltage window as a function of radiation dose [1.154]. The radiation was carried out with a Cobalt-60 source using a dose rate of 142 rad(Si)/sec. It can be seen that the high-V_t state decays with dose, whereas the low-V_t state remains nearly constant.

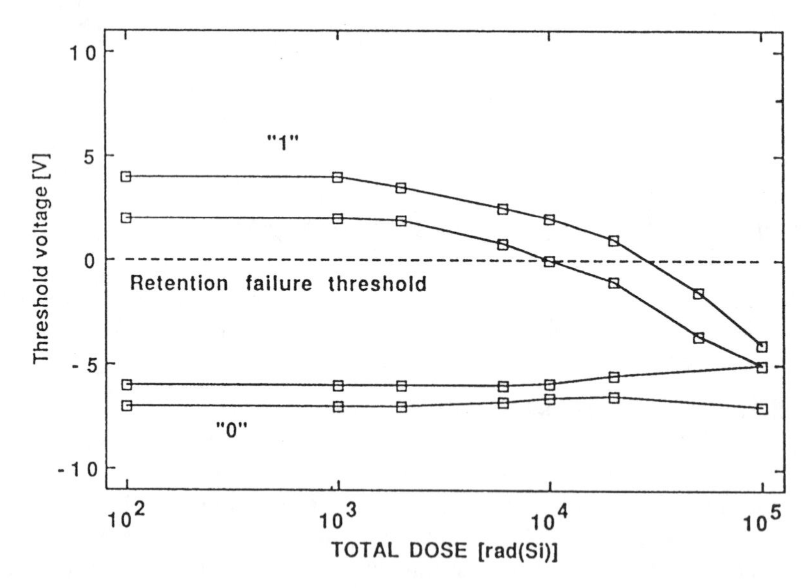

Figure 1.52 Memory threshold voltage window as a function of total radiation dose under Cobalt-60 radiation [1.154] for the floating gate technology.

For a fixed reference sense amplifier, total dose radiation is limited to values of about 10 to 30 Krad(Si), depending on the initial (pre-radiation) high-V_t state of the memory cell. This indicates a tradeoff between the cell write time and the radiation retention failure. Moreover, by using a differential sense amplifier, maximum total dose values can be increased up to values of 100 Krad, at the cost, however, of memory density. It was shown [1.154] that the decay of the high-V_t state of the memory cell is caused by three main mechanisms, illustrated in Fig. 1.53: (1) holes (or electrons) generated in the oxide layers are injected onto the floating gate and decrease the stored charge (the oxide layers involved can be the tunnel oxide, the interpoly oxide, and, often the most important one, the field oxide [1.156]; (2) holes generated in the oxide can also be trapped in the oxide; and (3) electrons stored on the floating gate can be emitted over the energy barriers toward the substrate or the control gate.

Among the different cell types, floating gate memory cells manufactured in a single-polysilicon technology have recently been reported to show a much higher radiation tolerance than double-polysilicon cells [1.157]. Model calculations have indicated that the oxide thickness of the coupling capacitor of the cells appears to be the dominant parameter affecting their radiation response. The thinner oxide of this capacitor in single-polysilicon cells is the primary reason for their better hardness. Hardness levels of more than 110 Krad(SiO_2) have been achieved [1.157].

The radiation hardness of floating gate memory cells can be improved in different ways. The use of thinner oxides can increase the radiation hardness because the volume for carrier generation decreases [1.154, 1.157]. Another way to improve the radiation hardness is to perform an additional threshold voltage implant in order to increase the high-V_t state of the memory cell. The cost, however, is a reduction in speed of the cell. As already mentioned above, differential sensing can increase the maximum total dose to values of 100 to 200 Krad(Si). Finally, refreshing techniques, as in DRAMs, could be used to extend the hardness levels of the memory cells.

Another important point is that, in practical cases, it will not necessarily be the memory cell that causes the failure due to total dose radiation, but rather the peripheral circuits [1.155]. It was reported recently that, for a 256 K EEPROM, the total dose failure levels are limited to values of 10 to 30 Krad(Si) due to loss of drive

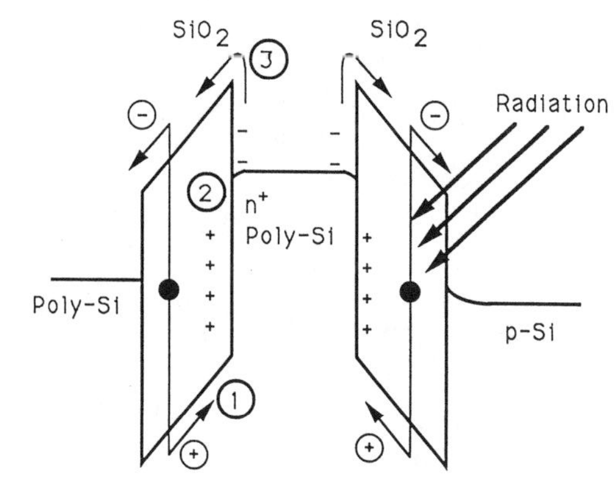

Figure 1.53 Schematic representation of the three main mechanisms that can cause the threshold voltage decay for the high V_t state of a floating gate memory cell: (1) hole injection onto the floating gate, (2) hole trapping in the oxides, and (3) electron emission from the floating gate.

capability of the peripheral circuitry, and not charge loss from the memory transistors. Moreover, the operating mode during radiation (reading or writing) and the exposure time of radiation influence the total dose of memory circuits [1.155]. Finally, it was found that dose rates less than 10^8 rad(Si)/s can be allowed without generating dose-rate upset or latchup of the memory.

In conclusion, and based on the few reports that have been published, it can be stated that floating gate technologies can be used in low total dose applications [maximum 30 Krad(Si)] and with dose rates below 10^8 rad(Si)/s.

1.7.3 Ferroelectric Technology

Ferroelectric capacitors are much more radiation hard than MOS or SNOS capacitors. The data storage mechanism, polarization state, is the result of a net ionic displacement in the unit cells of the material. Thus, high energy particles, gamma radiation or neutrons, have to move ions in the lattice in order to distort the existing polarization states, which would require very high doses. Typically, these ferroelectric capacitors can function satisfactorily with little degradation to doses up to 10 Mrads, with or without applied bias. However, whereas the memory cell based on ferroelectric materials exhibits a high hardness, the ultimate hardness of the FRAM depends on that of the MOS parts in the memory cell and the control circuity.

References

[1.1] D. Kahng and S. M. Sze, "A floating gate and its application to memory devices," *Bell Syst. Tech. J.*, vol. 46, p. 1288, 1967.

[1.2] H. A. R. Wegener, A. J. Lincoln, H. C. Pao, M. R. O'Connell, and R. E. Oleksiak, "The variable threshold transistor, a new electrically alterable, non-destructive read-only storage device," *IEEE IEDM Tech. Dig.*, Washington, D.C., 1967.

[1.3] H. E. Maes, "Recent developments in non-volatile semiconductor memories," *Digest of Technical Papers ESSCIRC83*, p. 1, 1983.

[1.4] T. Hagiwara, Y. Yatsuda, R. Kondo, S. Minami, T. Aoto, and Y. Itoh, "A 16kbit electrically erasable PROM using n-channel Si-gate MNOS technology," *IEEE J. Solid State Circuits*, vol. SC-15, p. 346, 1980.

[1.5] W. Johnson, G. Perlegos, A. Renninger, G. Kuhn, and T. Ranganath, "A 16Kb electrically erasable nonvolatile memory," *IEEE ISSCC Dig. Tech. Pap.*, p. 152, 1980.

[1.6] D. Frohman-Bentchkowsky, "A fully decoded 2048-bit electrically programmable MOS-ROM," *IEEE ISSCC Dig. Tech. Pap.*, p. 80, 1971.

[1.7] D. Frohman-Bentchkowsky, "Memory behaviour in a floating gate avalanche injection MOS (FAMOS) structure," *Appl. Phys. Lett.*, vol. 18, p. 332, 1971.

[1.8] D. Frohman-Bentchkowsky, "The metal–nitride–oxide–silicon (MNOS)-transistor—Characteristics and applications," *Proc. IEEE*, vol. 58, p. 1207, 1970.

[1.9] S. Sato and T. Yamaguchi, "Study of charge behaviour in metal–alumina–silicon dioxide–silicon (MAOS) field effect transistor," *Solid State Electr.*, vol. 17, p. 367, 1974.

[1.10] R. L. Angle and H. E. Talley, "Electrical and charge storage characteristics of the tantalum oxide–silicon dioxide device," *IEEE Trans. Elect. Dev.*, vol. ED-25, p. 1277, 1978.

[1.11] D. Frohman-Bentchkowsky, "A fully decoded 2048 bit electrically programmable FAMOS read-only memory," *IEEE J. Sol. St. Circ.*, vol. SC-6, p. 301, 1971.

[1.12] D. Frohman-Bentchkowsky, "FAMOS—A new semiconductor charge storage device," *Sol. St. Electr.*, vol. 17, p. 517, 1974.

[1.13] H. Iizuka, T. Sato, F. Masuoka, K. Ohuchi, H. Hara, H. Tango, M. Ishikawa, and Y. Takeishi, "Stacked gate avalanche injection type MOS (SAMOS) memory," Proc. 4th Conf. Sol. St. Dev., Tokyo, 1972; *J. Japan. Soc. Appl. Phys.*, vol. 42, p. 158, 1973.

[1.14] H. Iizuka, F. Masuoka, T. Sato, and M. Ishikawa, "Electrically alterable avalanche injection type MOS read-only memory with stacked gate structure," *IEEE Trans. Elect. Dev.*, vol. ED-23, p. 379, 1976.

[1.15] S. M. Sze, "Current transport and maximum dielectric strength of silicon nitride films," *J. Appl. Phys.*, vol. 38, p. 2951, 1967.

[1.16] H. E. Maes and G. Heyns, "Two-carrier conduction in amorphous chemically vapour deposited (CVD) silicon nitride layers," *Proc. of the Int. Conference on Insulating Films on Semiconductors*, North Holland Publ. Co., p. 215, 1983.

[1.17] J. Yeargain and K. Kuo, "A high density floating gate EEPROM cell," *IEEE IEDM Tech. Dig.*, p. 24, 1981.

[1.18] R. Klein, W. Owen, R. Simko, and W. Tchon, "5-V-only, nonvolatile RAM owes it all to polysilicon," *Electronics*, October 11, p. 111, 1979.

[1.19] G. Landers, "5-V-only EEPROM mimics static RAM timing," *Electronics*, June 30, p. 127, 1980.

[1.20] B. Rössler and R. Müller, "Electrically erasable and reprogrammable read-only memory using the n-channel SIMOS one-transistor cell," *IEEE Trans. Elect. Dev.*, vol. ED-24, p. 806, 1977.

[1.21] D. Guterman, I. Rimawi, T. Chiu, R. Halvorson, and D. McElroy, "An electrically alterable nonvolatile memory cell using a floating gate structure," *IEEE Trans. Elect. Dev.*, vol. ED-26, p. 576, 1979.

[1.22] M. Kamiya, Y. Kojima, Y. Kato, K. Tanaka, and Y. Hayashi, "EPROM cell with high gate injection efficiency," *IEEE IEDM Tech. Dig.*, p. 741, 1981.

[1.23] A. Wu, T. Chan, P. Ko, and C. Hu, "A novel high-speed, 5-V programming EPROM structure with source-side injection," *IEEE IEDM Tech. Dig.*, p. 584, 1986.

[1.24] M. Lenzlinger and E. H. Snow, "Fowler–Nordheim tunneling in thermally grown SiO_2," *J. Appl. Phys.*, vol. 40, p. 278, 1969.

[1.25] N. Mielke, A. Fazio, and H.-C. Liou, "Reliability comparison of FLOTOX and textured polysilicon EEPROM's," *Proc. Int. Rel. Phys. Symp. (IRPS)*, p. 85, 1987.

[1.26] R. C. Hughes, "High field electronic properties of SiO_2," *Sol. St. Electr.*, vol. 21, p. 251, 1978.

[1.27] D. J. DiMaria and D. R. Kerr, "Interface effects and high conductivity in oxides grown from polycrystalline silicon," *Appl. Phys. Lett.*, vol. 27, p. 505, 1975.

[1.28] R. M. Anderson and D. R. Kerr, "Evidence for surface asperity mechanism of conductivity in oxide grown on polycrystalline silicon," *J. Appl. Phys.* vol. 48, p. 4834, 1977.

[1.29] H. R. Huff, R. D. Halvorson, T. L. Chiu, and D. Guterman, "Experimental observations on conduction through polysilicon oxide," *J. Electrochem. Soc.*, vol. 127, p. 2482, 1980.

[1.30] P. A. Heimann, S. P. Murarka, and T. T. Sheng, "Electrical conduction and breakdown in oxides of polycrystalline silicon and their correlation with interface texture," *J. Appl. Phys.*, vol. 53, p. 6240, 1982.

[1.31] R. K. Ellis, "Fowler–Nordheim emission from non-planar surfaces," *IEEE Elect. Dev. Lett.*, vol. EDL-3, p. 330, 1982.

[1.32] G. Groeseneken and H. E. Maes, "A quantitative model for the conduction in oxides thermally grown from polycrystalline silicon," *IEEE Trans. Elect. Dev.*, vol. ED-33, p. 1028, 1986.

[1.33] J. Bisschop, E. J. Korma, E. F. F. Botta, and J. F. Verwey, "A model for the electrical conduction in polysilicon oxide," *IEEE Trans. Elect. Dev.*, vol. ED-33, p. 1809, 1986.

[1.34] R. D. Jolly, H. R. Grinolds, and R. Groth, "A model for conduction in floating gate EEPROM's," *IEEE Trans. Elect. Dev.*, vol. ED-31, p. 767, 1984.

[1.35] P. E. Cottrell, R. R. Troutman, and T. H. Ning, "Hot electron emission in n-channel IGFET's," *IEEE J. Sol. St. Circ.*, vol. SC-14, p. 442, 1979.

[1.36] B. Eitan and D. Frohman-Bentchkowsky, "Hot electron injection into the oxide in n-channel MOS-devices," *IEEE Trans. Elect. Dev.*, vol. ED-28, p. 328, 1981.

[1.37] Y. Tarui, Y. Hayashi, and K. Nagai, "Electrically reprogrammable non-volatile semiconductor memory," *IEEE J. Sol. St. Circ.*, vol. SC-7, p. 369, 1972.

[1.38] C. Hu, "Lucky electron model of hot electron emission," *IEEE IEDM Tech. Dig.*, p. 22, 1979.

[1.39] P. Ko, R. Müller, and C. Hu, "A unified model for hot electron currents in MOSFET's," *IEEE IEDM Tech. Dig.*, p. 600, 1981.

[1.40] E. Takeda, H. Kume, T. Toyabe, and S. Asai, "Submicrometer MOSFET structure for minimizing hot carrier generation," *IEEE Trans. Elect. Dev.*, vol. ED-29, p. 611, 1982.

[1.41] S. Tanaka and M. Ishikawa, "One-dimensional writing model of n-channel floating gate ionization-injection MOS (FIMOS)," *IEEE Trans. Elect. Dev.*, vol. ED-28, p. 1190, 1981.

[1.42] K. R. Hoffmann, C. Werner, W. Weber, and G. Dorda, "Hot-electron and hole emission effects in short n-channel MOSFET's," *IEEE Trans. Elect. Dev.*, vol. ED-32, p. 691, 1985.

[1.43] T. Y. Chan, P. K. Ko, and C. Hu, "Dependence of channel electric field on device scaling," *IEEE Elect. Dev. Lett.*, vol. EDL-6, p. 551, 1985.

[1.44] C. Hu, "Hot electron effects in MOSFET's," *IEEE IEDM Tech. Dig.*, p. 176, 1983.

[1.45] J. Van Houdt, L. Haspeslagh, D. Wellekens, L. Deferm, G. Groeseneken, and H. E. Maes, "HIMOS—a high efficiency Flash EEPROM cell for embedded memory applications," *IEEE Trans. Elect. Dev.*, vol. ED-40, p. 2255, 1993.

[1.46] J. Van Houdt, P. Heremans, L. Deferm, G. Groeseneken, and H. E. Maes, "Analysis of the enhanced hot-electron injection in split-gate transistors useful for EEPROM applications," *IEEE Trans. Elect. Dev.*, vol. ED-39, p. 1150, 1992.

[1.47] K. I. Lundström and C. M. Svensson, "Properties of MNOS structures," *IEEE Trans. Elect. Dev.*, ED-19, p. 826, 1972.

[1.48] H. E. Maes, J. Witters, and G. Groeseneken, "Trends in non-volatile memory devices and technologies," *Proc. 17th European Solid State Device Research Conference*, p. 743, 1987.

[1.49] H. E. Maes, G. Groeseneken, H. Lebon, and J. Witters, "Trends in semiconductor memories," *Microelectr. J.*, vol. 20, p. 9, 1989.

[1.50] S. Atsumi, S. Tanaka, S. Saito, N. Ohtsuka, N. Matsukawa, S. Mori, Y. Kaneko, K. Yoshikawa, J. Matsunaga, and T. Iizuka, "A 120ns 4Mb CMOS EPROM," *IEEE ISSCC Dig. Tech. Pap.*, p. 74, 1987.

[1.51] G. Yaron, S. Prasad, M. Ebel, and B. Leong, "A 16K EEPROM employing new array architecture and designed-in reliability features," *IEEE J. Sol. St. Circ.*, vol. SC-17, no. 5, p. 833, 1982.

[1.52] G. Landers, "5-V only EEPROM mimics static RAM timing," *Electronics*, June 30, p. 127, 1980.

[1.53] B. Gerber and J. Fellrath, "Low voltage single supply CMOS electrically erasable read-only memory," *IEEE Trans. Elect. Dev.*, vol. ED-27, p. 1211, 1980.

[1.54] D. H. Oto, V. K. Dham, K. H. Gudger, M. Reitsma, G. S. Gongwer, Y. W. Hu, J. Olund, H. Jones, and S. Nieh, "High-voltage regulation and process considerations for high-density 5V-only EEPROM's," *IEEE J. Sol. St. Circ.*, vol. SC-18, p. 532, 1983.

[1.55] D. Cioaca, T. Lin, A. Chan, L. Chen, and A. Milhnea, "A million-cycles CMOS 256K EEPROM," *IEEE J. Sol. St. Circ.*, vol. SC-22, p. 684, 1987.

[1.56] P. I. Suciu, M. Briner, C. S. Bill, and D. Rinerson, "A 64K EEPROM with extended temperature range and page mode operation," *IEEE ISSCC Dig. Tech. Pap.*, p. 170, 1985.

[1.57] R. Vancu, L. Chen, R. L. Wan, T. Nguyen, C.-Y. Yang, W.-P. Lai, K.-F. Tang, A. Mihnea, A. Renninger, and G. Smarandoiu, "A 35ns 256K CMOS EEPROM with error correcting circuitry," *IEEE ISSCC Dig. Tech. Pap.*, p. 64, 1990.

[1.58] F. Masuoka, M. Asano, H. Iwahashi, and T. Komuro, "A new Flash EEPROM cell using triple polysilicon technology," *IEEE IEDM Tech. Dig.*, p. 464, 1984.

[1.59] K. Imamiya, Y. Iwata, Y. Sugiura, H. Nakamura, H. Oodaira, M. Momodomi, Y. Ito, T. Watanabe, H. Araki, K. Narita, K. Masuda, and J. Miyamoto, "A 35 ns-cycle-time 3.3V-only 32 Mb NAND Flash EEPROM," *IEEE ISSCC Dig. Tech.* Pap., p. 130, 1995.

[1.60] A. Nozoe, T. Yamazaki, H. Sato, H. Kotani, S. Kubono, K. Manita, T. Tanaka, T. Kawahara, M. Kato, K. Kimura, H. Kume, R. Hori, T. Nishimoto, S. Shukuri, A. Ahba, Y. Kouro, O. Sakamoto, A. Fukumoto, and M. Nakajima, "A 3.3V high density AND Flash memory with 1 ms/512B Erase and Program Time," *IEEE ISSCC Dig. Tech. Pap.*, p. 124, 1995.

[1.61] K.-D. Suh, B.-H Suh, Y.-H. Lim, J.-K. Kim, Y.-J. Choi, Y.-N. Koh, S.-S. Lee, S.-C. Kwon, B.-S. Choi, J.-S. Yum, J.-H. Choi, J.-R. Kim, and H.-K. Lim, "A 3.3V 32 Mb NAND Flash memory with incremental step pulse programming scheme," *IEEE ISSCC Dig. Tech. Pap.*, p. 128, 1995.

[1.62] E. Harari, L. Schmitz, B. Troutman, and S. Wang, "A 256bit non-volatile static RAM," *IEEE ISSCC Dig. Tech. Pap.*, p. 108, 1978.

[1.63] R. Klein, W. Owen, R. Simko, and W. Tchon, "5-V-only, nonvolatile RAM owes it all to polysilicon," *Electronics*, October 11, p. 111, 1979.

[1.64] J. Drori, S. Jewell-Larsen, R. Klein, and W. Owen, "A single 5V supply non-volatile static RAM," *IEEE ISSCC Dig. Tech. Pap.*, p. 148, 1981.

[1.65] A. Weiner, C. Herdt, J. Tiede, and B. Geston, "A 64K Non-volatile SRAM," Non-volatile Semiconductor Memory Workshop, Vail, Colo., 1989.

[1.66] Y. Terada, K. Kobayashi, T. Nakayama, H. Arima, and T. Yoshihara, "A new architecture for the NVRAM—An EEPROM backed-up dynamic RAM," *IEEE J. Sol. St. Circ.*, vol. SC-23, no. 1, p. 86, 1988.

[1.67] Y. Yamauchi, K. Tanaka, and K. Sakiyama, "A novel NVRAM cell technology for high density applications," *IEEE IEDM Tech. Dig.*, p. 416, 1988.

[1.68] Y. Yamauchi, H. Ishihara, K. Tanaka, K. Sakiyama, and R. Miyake, "A versatile stacked storage capacitor on FLOTOX cell for megabit NVRAM applications," *IEEE IEDM Tech. Dig.*, p. 595, 1989.

[1.69] G. Gerosa, C. Hart, S. Harris, R. Kung, J. Weihmeir, and J. Yeargain, "A high performance CMOS technology for 256K/1Mb EPROMs," *IEEE IEDM Tech. Dig.*, p. 631, 1985.

[1.70] J. Esquivel, A. Mitchell, J. Paterson, B. Riemenschneider, H. Tiegelaar, T. Coffman, D. Dolby, M. Gill, R. Lahiry, S. Lin, D. McElroy, J. Schreck, and P. Shah, "High density contactless self-aligned EPROM cell array technology," *IEDM Tech. Dig.*, p. 592, 1986.

[1.71] T. Coffman, D. Boyd, D. Dolby, M. Gill, S. Kady, R. Lahiry, S. Lin, D. McElroy, A. Mitchell, J. Paterson, J. Schreck, P. Shah, and F. Takeda, "A

1M CMOS EPROM with a 13.5μ^2 cell," *IEEE ISSCC Dig. Tech. Pap.*, p. 72, 1987.

[1.72] A. Esquivel, B. Riemenschneider, J. Paterson, H. Tiegelaar, J. Mitchell, R. Lahiry, M. Gill, J. Schreck, D. Dolby, T. Coffman, S. Lin, and P. Shah, "A 8.6 μm^2 cell technology for a 35.5mm^2 Megabit EPROM," *IEEE IEDM Tech. Dig.*, p. 859, 1987.

[1.73] B. Eitan, Y. Ma, C. Hu, R. Kazerounian, K. Sinai, and S. Ali, "A self-aligned split-gate EPROM," presented at the 9th NVSM Workshop, Monterey, Calif. 1988.

[1.74] K. Yoshikawa, S. Mori, K. Narita, N. Arai, Y. Oshima, Y. Kaneko, and H. Araki, "An asymmetrical lightly-doped source (ALDS) cell for virtual ground high density EPROMs," *IEEE IEDM Tech. Dig.*, p. 432, 1988.

[1.75] S. Ali, B. Sani, A. Shubat, K. Sinai, R. Kazerounian, C.-J. Hu, Y. Ma, and B. Eitan, "A 50ns 256K CMOS split-gate EPROM," *IEEE J. Sol. St. Circ.*, vol. SC-23, no. 1, p. 79, 1988.

[1.76] K. Prall, W. Kinney, and J. Macro, "Characterization and suppression of drain coupling in submicrometer EPROM cells," *IEEE Trans. Elect. Dev.*, vol. ED-34, no. 12, p. 2463, 1987.

[1.77] Y. Mizutani and K. Makita, "A new EPROM cell with a side-wall floating gate for high density and high performance device," *IEEE IEDM Tech. Dig.*, p. 63, 1985.

[1.78] S. Chu and A. Steckl, "The effect of trench-gate-oxide structure on EPROM device operation," *IEEE Elect. Dev. Lett.*, vol. EDL-9, no. 6, p. 284, 1988.

[1.79] S. Shukiri, Y. Wada, T. Hagiwara, K. Komori, and M. Tamura, "A novel EPROM device fabricated using focused Boron ion beam implantation," *IEEE Trans. Elect. Dev.*, vol. ED-34, p. 1264, 1987.

[1.80] R. Cuppens, C. Hartgring, J. Verwey, H. Peek, F. Vollebregt, E. Devens, and I. Sens, "A EEPROM for microprocessors and custom logic," *IEEE J. Sol. St. Circ.*, vol. SC-20, no. 2, p. 603, 1985.

[1.81] R. Yoshikawa, and N. Matsukawa, "EPROM and EEPROM cell structures for EPLD's compatible with single poly gate process," presented at the 8th NVSM Workshop, Vail, Colo., 1986.

[1.82] N. Mielke, L. Purvis, and H. Wegener, "Reliability comparison of FLOTOX and textured-poly EEPROMs," presented at the 8th NVSM Workshop, Vail, Colo., 1986.

[1.83] S. Lai, J. Lee, and V. Dham, "Electrical properties of nitrided-oxide systems for use in gate dielectrics and EEPROM," *IEEE IEDM Tech. Dig.*, p. 190, 1983.

[1.84] C. Jenq, T. Chiu, B. Joshi, and J. Hu, "Properties of thin oxynitride films used as floating gate tunneling dielectrics," *IEEE IEDM Tech. Dig.*, p. 309, 1982.

[1.85] N. Matsukawa, S. Morita, and H. Nozawa, "High performance EEPROM using low barrier height tunnel oxide," Ext. Abstr. Int. Conf. on Sol. St. Dev. and Mat., p. 261, 1984.

[1.86] H. Nozawa, N. Matsukawa, and S. Morita, "An EEPROM cell using a low barrier height tunnel oxide," *IEEE Trans. Elect. Dev.*, vol. ED-33, no. 2, p. 275, 1986.

[1.87] D. DiMaria, K. De Meyer, C. Serrano, and D. Dong, "Electrically alterable read-only-memory using Si-rich SiO2 injectors and a polycrystalline silicon storage layer," *J. Appl. Phys.*, vol. 52, p. 4825, 1982.

[1.88] T. Ito, S. Hijiya, T. Nozaki, H. Arakawa, H. Ishikawa, and M. Shinoda, "Low voltage alterable EAROM cells with nitride barrier avalanche injection MIS (NAMIS)," *IEEE Trans. Elect. Dev.*, vol. ED-26, p. 906, 1979.

[1.89] R. Stewart, A. Ipri, L. Faraone, J. Cartwright, and K. Schlesier, "A shielded substrate injector MOS (SSIMOS) EEPROM cell," *IEEE IEDM Tech. Dig.*, p. 472, 1984.

[1.90] H. Arima, N. Ajika, H. Morita, T. Shibano, and T. Matsukawa, "A novel process technology and cell structure for megabit EEPROM," *IEEE IEDM Tech. Dig.*, p. 420, 1988.

[1.91] M. Momodomi, R. Kirasawa, R. Nakayama, S. Aritome, T. Endoh, Y. Itoh, Y. Iwata, H. Oodaira, T. Tanaka, M. Chiba, R. Shirota, and F. Masuoka, "New device technologies for 5V-only 4Mb EEPROM with NAND structure cell," *IEEE IEDM Tech. Dig.*, p. 412, 1988.

[1.92] Y. Itoh, M. Momodomi, R. Shirota, Y. Iwata, R. Nakayama, R. Kirasawa, T. Tanaka, K. Toita, S. Inoue, and F. Masuoka, "An experimental 4Mb CMOS EEPROM with a NAND structure cell," *IEEE ISSCC Dig. Tech. Pap.*, p. 134, 1989.

[1.93] R. Stewart, A. Ipri, L. Faraone, and J. Cartwright, "A low voltage high density EEPROM memory cell using symmetrical poly-to-poly conduction," presented at the 6th NVSM Workshop, Vail, Colo., August 1983.

[1.94] K. Sarma, A. Owens, D. Pan, B. Rosier, and L. Yeh, "Double poly EEPROM cell utilizing interpoly tunneling," presented at the 8th NVSM Workshop, Vail, Colo., 1986.

[1.95] H. Maes and G. Groeseneken, "Conduction in thermal oxides grown on polysilicon and its influence on floating gate EEPROM degradation," *IEEE IEDM Tech. Dig.*, p. 476, 1984.

[1.96] H. Wegener, "Build up of trapped charge in textured poly tunnel structures," presented at the 8th NVSM Workshop, Vail, Colo., 1986.

[1.97] G. Verma, and N. Mielke, "Reliability of ETOX based flash Memories," Proc. IRPS, p. 158, 1988.

[1.98] R. Kazerounian, S. Ali, Y. Ma, and B. Eitan, "A 5 volt high density poly-poly erase flash EEPROM," *IEEE IEDM Tech. Dig.*, p. 436, 1988.

[1.99] J. R. Cricchi, F. C. Blaha, and M. D. Fitzpatrick, "The drain-source protected MNOS memory device and memory endurance," *IEEE IEDM Tech. Dig.*, p. 126, 1973.

[1.100] Y. Yatsuda, T. Hagiwara, R. Kondo, S. Minami, and Y. Itoh, "N-channel Si-gate MNOS device for high speed EAROM," *Proc. 10th Conf. Solid State Devices*, p. 11, 1979.

[1.101] G. Schols, H. E. Maes, G. Declerck, and R. Van Overstraeten, "High temperature hydrogen anneal of MNOS structures," *Revue de Phys. Appl. 13*, p. 825, 1978.

[1.102] Y. Yatsuda, S. Minami, R. Kondo, T. Hagiwara, and Y. Itoh, "Effects of high temperature annealing on n-channel Si-gate MNOS devices," *Jap. J. Appl. Phys.*, vol. 19, S19-1, p. 219, 1980.

[1.103] H. E. Maes and G. Heyns, "Influence of a high temperature hydrogen anneal on the memory characteristics of p-channel MNOS transistors," *J. Appl. Phys.*, vol. 51, p. 2706, 1980.

[1.104] G. Schols and H. E. Maes, "High temperature hydrogen anneal of MNOS structures," Proc. ECS, vol. 83-8, Eds. V. J. Kapoor and H. Stein, p. 94, 1983.

[1.105] Y. Yatsuda, T. Hagiwara, S. Minami, R. Kondo, K. Uchida, and K. Uchiumi, "Scaling down MNOS nonvolatile memory devices," *Jap. J. Appl. Phys.*, vol. 21, S21-1, p. 85, 1982.

[1.106] T. Hagiwara, Y. Yatsuda, S. Minami, S. Naketani, K. Uchida, and T. Yasui, "A 5V only 64k MNOS EEPROM," 6th NVSM, Vail, Colo., 1983.

[1.107] Y. Yatsuda, S. Minami, T. Hagiwara, T. Toyabe, S. Asai, and K. Uchida, "An advanced MNOS memory device for highly integrated byte erasable 5V only EEPROMs," *IEEE IEDM Tech. Dig.*, p. 733, 1982.

[1.108] S. Minami, Y. Kamigaki, K. Uchida, K. Furusawa, and T. Hagiwara, "Improvement of written-state retentivity by scaling down MNOS memory devices," *Jap. J. Appl. Phys.*, vol. 27, p. L2168, 1988.

[1.109] D. K. Schroder and M. H. White, "Characterization of current transport in MNOS structures with complementary tunneling emitter bipolar transistors," *IEEE Trans. Elect. Dev.*, vol. ED-26, p. 899, 1979.

[1.110] H. E. Maes and G. Heyns, "Carrier transport in LPCVD silicon nitride," 4th NVSM, Vail, Colo., 1980.

[1.111] H. E Maes and G. L. Heyns, "Two-carrier conduction in amorphous chemically vapour deposited (CVD) silicon nitride layers," Insulating Films on Semiconductor 83, Eds. J. Verwey and D. Wolters, p. 215, 1983.

[1.112] H. E. Maes and E. Vandekerckhove, "Non-volatile memory characteristics of polysilicon-oxynitride-oxide-silicon devices and circuits," *Proc. ECS*, vol. 87–10, Eds. V. J. Kapoor and K. T. Hankins, p. 28, 1987.

[1.113] H. E. Maes, "The use of oxynitride layers in non-volatile S-OxN-OS (silicon–oxynitride–oxide–silicon) memory devices," Chapter 6 in *LPCVD Silicon Nitride and Oxynitride Films*, pp. 127–146, Springer-Verlag, Berlin, Ed. F. H. Habraken, 1991.

[1.114] W. D. Brown, R. V. Jones, and R. D. Nasby, "The MONOS memory transistor: application in a radiation-hard nonvolatile RAM," *Sol. St. Electr.*, vol. 29, p. 877, 1985.

[1.115] P. C. Chen, "Threshold-alterable Si-gate MOS devices," *IEEE Trans. Elect. Dev.*, ED-24, p. 584, 1977.

[1.116] J. Remmerie, H. E. Maes, J. Witters, and W. Beullens, "Two carrier transport in MNOS devices," *Proc. ECS*, vol. 87-10, Eds. V. J. Kapoor and K. T. Hankins, p. 93, 1987.

[1.117] H. E. Maes and R. Van Overstraeten, "Simple technique for determination of the centroid of nitride charge in MNOS structures," *Appl. Phys. Lett.*, vol. 27, p. 282, 1975.

[1.118] F. L. Hampton and J. R. Cricchi, "Space charge distribution limitations on scale down of MNOS memory devices," *IEEE IEDM Tech. Dig.*, p. 374, 1979.

[1.119] B. H. Yun, "Electron and hole transport in CVD nitride films," *Appl. Phys. Lett.*, vol. 27, p. 256, 1975.

[1.120] E. Suzuki, H. Hiraishi, K. Ishi, and Y. Hayashi, "A low voltage alterable EEPROM with metal–oxide–nitride–oxide–semiconductor (MONOS) structure," *IEEE Trans. Elect. Dev.*, vol. ED-30, p. 122, 1983.

[1.121] C. C. Chao and M. H. White, "Characterization of charge injection and trapping in scaled SONOS/MONOS memory devices," *Sol. St. Electr.*, vol. 30, p. 307, 1987.

[1.122] T. A. Dellin and P. J. McWhorter, "Scaling of MONOS nonvolatile memory transistors," *Proc. ECS*, vol. 87-10, Eds. V. J. Kapoor and K. T. Hankins, p. 3, 1987.

[1.123] V. J. Kapoor and S. B. Bibyk, "Energy distribution of electron trapping defects in thick-oxide MNOS structures," *Physics of MOS Insulators*, Eds. G. Lucovsky, S. Pantelides, and G. Galeener, p. 117, 1980.

[1.124] F. R. Libsch, A. Roy, and M. H. White, "Amphoteric trap modeling of multidielectric scaled SONOS nonvolatile memory structures," 8th NVSM, Vail, Colo., 1986.

[1.125] J. Evans and R. Womack, "An experimental 512-bit nonvolatile memory with ferroelectric storage cell," *IEEE J. Sol. St. Circ.*, vol. SC-23, p. 1171, 1998.

[1.126] J. Witters, "Characteristics and reliability of thin oxide floating gate memory transistors and their supporting programming circuits," Ph.D. Thesis, K. U. Leuven, 1989.

[1.127] A. Bhattacharyya, "Modelling of write/erase and charge retention characteristics of floating gate EEPROM devices," *Sol. St. Electr.*, p. 899, 1984.

[1.128] G. Groeseneken, "Programming behaviour and degradation phenomena in electrically erasable programmable floating gate memory devices," Ph.D. Thesis, K. U. Leuven, 1986.

[1.129] K. Prall, W. Kinney, and J. Macro, "Characterization and suppression of drain coupling in submicron EPROM cells," *IEEE Trans. Elect. Dev.*, vol. ED-34, p. 2463, 1987.

[1.130] P. I. Suciu, B. P. Cox, D. D. Rinerson, and S. F. Cagnina, "Cell model for EEPROM floating-gate memories," *IEEE IEDM Tech. Dig.*, p. 737, 1982.

[1.131] S. T. Wang, "On the I-V characteristics of floating gate MOS transistors," *IEEE Trans. Elect. Dev.*, vol. ED-26, p. 1292, 1979.

[1.132] S. T. Wang, "Charge retention of floating gate transistors under applied bias conditions," *IEEE Trans. Elect. Dev.*, vol. ED-27, p. 297, 1980.

[1.133] A. Kolodny, S. Nieh, B. Eitan, and J. Shappir, "Analysis and modeling of floating gate EEPROM cells," *IEEE Trans. Elect. Dev.*, vol. ED-33, p. 835, 1986.

[1.134] R. Bez, D. Cantarelli, P. Cappelletti, and F. Maggione, "SPICE model for transient analysis of EEPROM cells," in *Proc. of ESSDERC88*, p. 677, 1988.

[1.135] J. T. Mantey, "Degradation of thin silicon dioxide films and EEPROM cells," Ph.D. Thesis, Ecole Polytechnique Federale de Lausanne, 1990.

[1.136] L. Lundkvist, I. Lundström, and C. Svensson, "Discharge of MNOS structures," *Sol. St. Electr.*, vol. 16, p. 811, 1973.

[1.137] R. A. Williams and M. E. Beguwala, "The effect of electrical conduction of nitride on the discharge of MNOS memory transistors," *IEEE Trans. Elect. Dev.*, vol. ED-25, p. 1019, 1978.

[1.138] G. Heyns, "Bijdrage von Electronen en Gaten tot de Conductie, Retentie en de Degradatie von MNOS Niet–Vluchtige Geheugens," Ph.D. Thesis, K. U. Leuven, Belgium, 1986, unpublished.

[1.139] G. Heyns and H. E. Maes, "A new model for the discharge behaviour of MNOS non-volatile memory devices," *Appl. Surf. Sci.*, vol. 30, p. 153, 1987.

[1.140] F. R. Libsch, A. Roy, and M. H. White, "A computer simulation program for erase/write characterization of ultra-thin nitride scaled SONOS/MONOS memory transistors," Ext. Abstr. Fall Meeting ECS 1986, vol. 86–2, Abstract 560, 1986.

[1.141] L. Blauner, "Study in reliability improvement of tunnel oxide for FLOTOX cell," presented at the 8th NVSM Workshop, Vail, Colo., August 1986.

[1.142] Y. Terada, K. Kobayashi, T. Nakayama, M. Hayashikoshi, Y. Miyawaki, N. Ajika, H. Arima, T. Matsukawa, and T. Yoshihara, "120ns 128kx8b / 64kx 16b CMOS EEPROM's," *IEEE ISSCC Dig. Tech. Pap.*, p. 136, 1989.

[1.143] B. Euzent, N. Boruta, J. Lee, and C. Jenq, "Reliability aspects of a floating gate EEPROM," *Procs. IRPS*, p. 11, 1981.

[1.144] K. Hieda, M. Wada, T. Shibata, I. Inoue, M. Momodomi, and H. Iizuka, "Optimum design of dual control gate cell for high density EEPROM's," *IEEE IEDM Tech. Dig.*, p. 593, 1983.

[1.145] E. Suzuki and Y. Hayashi, "Degradation properties of MNOS structures," *J. Appl. Phys.*, vol. 52, p. 6377, 1981.

[1.146] H. E. Maes and S. Usmani, "Charge pumping measurements on stepped-gate MNOS memory transistors," *J. Appl. Phys.*, vol. 53, p. 7106, 1981.

[1.147] A. Roy, F. R. Libsch, and M. H. White, "Investigations on ultrathin silicon nitride and silicon dioxide films in nonvolatile semiconductor memory transistors," *Proc. ECS*, vol. 87-10, Eds. V. J. Kapoor and K. T. Hankins, p. 38, 1987.

[1.148] H. E. Maes and R. J. Van Overstraeten, "Memory loss in MNOS capacitors," *J. Appl. Phys.*, vol. 47, p. 667, 1976.

[1.149] J. R. Cricchi, M. D. Fitzpatrick, F. C. Blaha, and B. T. Ahlport, "1 MRad hard MNOS structures," *IEEE Trans. Nucl. Sci.*, vol. NS-24, p. 2185, 1977.

[1.150] M. C. Peckerar and N. Bluzer, "Hydrogen annealed nitride/oxide dielectric structures for radiation hardness," *IEEE Trans. Nucl. Sci.*, vol. NS-27, p. 1753, 1980.

[1.151] P. Vail, "Radiation hardened MNOS : a review," *Proc. ECS*, vol. 83-8, p. 207, 1983.

[1.152] M. G. Knoll, T. A. Dellin, and R. V. Jones, "A radiation-hardened 16 kbit MNOS EAROM," *IEEE Trans. Nucl. Sci.*, vol. NS-30, p. 4224, 1983.

[1.153] J. M. Caywood and B. L. Prickett, "Radiation induced soft errors in floating gate memories," Xicor Inc. Report.

[1.154] E. S. Snyder et al., "Radiation response of floating gate EEPROM memory cells," *IEEE Trans. Nucl. Sci.*, vol. NS-36, p. 2131, 1989.

[1.155] T. F. Wrobel, "Radiation characterization of a 28C256 EEPROM," *IEEE Trans. Nucl. Sci.*, vol. NS-36, p. 2247, 1989.

[1.156] D. Wellekens, G. Groeseneken, J. Van Houdt, and H. E. Maes, "On the total dose radiation hardness of floating gate EEPROM cells," Proc. NVMTR, IEEE Cat. #93TH0547-0, p. 54, 1993.

[1.157] D. Wellekens, G. Groeseneken, J. Van Houdt, and H. E. Maes, "Single poly floating gate cell as the best choice for radiation-hard EEPROM technology," *IEEE Trans. Nucl. Sci.*, vol. NS-40, p. 1619, 1993.

Chapter 2

Ravi Ramaswami
and H. C. Lin

Floating Gate Planar Devices

2.0. INTRODUCTION

The rapid growth of portable computer systems and measurement instrumentation in recent years has resulted in a large market for fast nonvolatile semiconductor memories, and since the system requirements include on-board programmability, bit/byte programmability, and short programming times, UV-erasable PROMs (EPROMs) are falling out of favor. EEPROMs are increasingly finding use as hard disk replacements in portable computer systems. To meet these requirements, different technologies have been developed, which were reviewed in Chapter 1.

In floating gate technology, electrons are stored in an electrically isolated polysilicon layer (called the floating gate) surrounded by a high-quality dielectric. Two mechanisms are used to transport electrons to and from the floating gate in EEPROMs—high-field tunneling and channel hot-electron (CHE) injection. Tunneling requires an electric field of the order of 10 MV/cm to lower the oxide conduction band edge sufficiently so that it lines up with the silicon conduction band within a distance of 30 to 50 Å from the Si–SiO_2 interface. Otherwise, the electron-tunneling probability rapidly falls to zero. To keep programming voltages low, the tunnel dielectric thickness needs to be small, consistent with reliability and manufacturability requirements. The high fields present during the tunneling operation result in severe degradation of the tunnel dielectric, which is one of the failure mechanisms of the memory cell.

An alternative programming mechanism is channel hot-electron injection. Channel hot electrons are created in the high-field region near the drain junction, and, if the oxide field favors injection, electrons can be transported from the substrate to the floating gate. Since the high energy electrons jump over the oxide conduction band at the Si–SiO_2 interface, all that is required is a small field in the tunnel oxide so that the oxide conduction band slopes toward the floating gate. Hence, the high-field stress faced by the oxide during tunnel injection is avoided in this method. The only problem is that the CHE injection mechanism can only bring electrons to the floating gate, which must now be removed by tunneling.

Memory cell technology falls into two basic categories: planar and nonplanar. The difference between the two is that, in the planar the electron injection field is equal to the average tunnel oxide field, while in the nonplanar the average tunnel oxide field is much smaller than the injection field. (The electric field enhancement due to surface topography on the cathode can bend the oxide conduction band sufficiently for electron tunneling at low voltages.) The terms *write* and *erase* are used to denote operations that charge and discharge the floating gate. A write operation brings electrons to the floating gate, while the erase operation removes electrons from the floating gate; programming means either write or erase. The programming voltage, V_{pp}, is typically in the range of 10 to 15 V. The repeated transport of charges across the thin tunnel dielectric results in memory cell degradation, which in turn produces an upper limit for cell programmability, referred to as cell endurance. The endurance of the cell (number of write/erase cycles before failure) is around 10^5 cycles.

A comparison between three planar cell technologies—Flash EEPROM, UV-EPROM, and conventional EEPROM—is shown in Fig. 2.1 [2.1]. Flash EEPROMs use a single transistor per bit, whereas conventional full-feature EEPROMs require two transistors per bit, resulting in a higher cost per bit. The basic functional difference between conventional and Flash EEPROMs is that a user can write or erase any bit of a conventional EEPROM, whereas only block erase is allowed in Flash EEPROM. Presently, conventional EEPROMs typically find use in specialized applications, ASICs, or PLDs (programmable logic devices) rather than as stand-alone memories. Cell size then depends on the specific technology used; examples are illustrated later in this chapter. Conventional floating gate EEPROM technology is the focus of this chapter, which includes discussions of physical electron-transport processes to charge/discharge the floating gate, fabrication technology, memory array implementation, and degradation and failure mechanisms of these cells.

2.1. CELL STRUCTURES AND OPERATION

The FLOTOX cell, shown in cross section in Fig. 2.2*a*, may be considered a forerunner to many of the present-day planar technology structures; it uses electron tunneling for programming operations. In 1980, Intel broke ground with an efficient 16 Kbit EEPROM (a cell area of 0.85 mil^2 or 548.3 μm^2 and a channel length of 3.5 μm), which was fabricated using n-channel double-polysilicon gate technology.

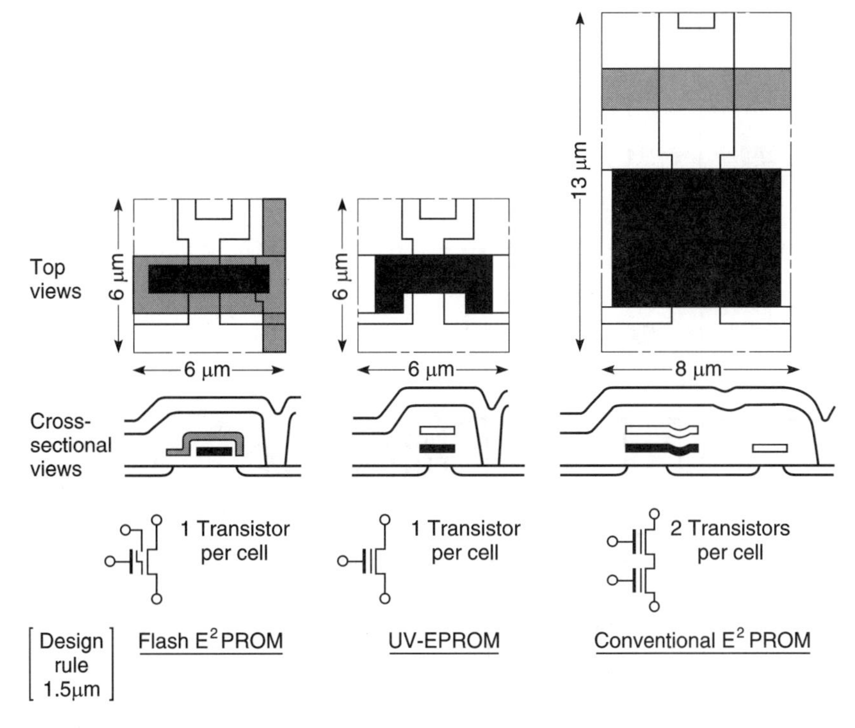

Figure 2.1 Comparison between UV-EPROM and Flash, Full-functional EEPROM [2.1].

Programming is accomplished by electron tunneling between the floating gate and drain across the thin tunnel oxide (< 200 Å). To write, $V_{pp} = 21$ V is applied to the control gate with the drain/source/substrate grounded so that electrons tunnel from the drain to the floating gate. To erase, the drain is connected to V_{pp} with the control gate/substrate grounded and the source floating, causing electrons to tunnel from the floating gate to the drain. Both operations typically require 10 msec for completion to achieve a threshold voltage shift of ≈ 4 V, and they are self-limiting since the field in the tunnel oxide region decreases as the floating gate charges or discharges. The cell endurance is about 10^5 cycles.

Using 1 μm technology reduces the cell area to 94.5 μm^2. The cell layout, shown in Fig. 2.2*b*, tries to minimize the cell area while maximizing the fraction of the applied V_{pp} voltage dropped across the tunnel dielectric. The floating gate (poly1) to control gate (poly2) capacitance is increased by using an oxide/nitride dual dielectric of 40 nm equivalent oxide thickness and extending the floating gate over the field oxide area. The control gate covers the floating gate completely. The tunnel oxide area is integrated with the drain of the sense transistor using the following process flow. The tunnel oxide area is defined by two mask steps. A tunnel oxide mask is used to define a rectangular opening in the first gate oxide, arsenic is implanted through the tunnel oxide opening, and the tunnel oxide is then grown. Poly1 is

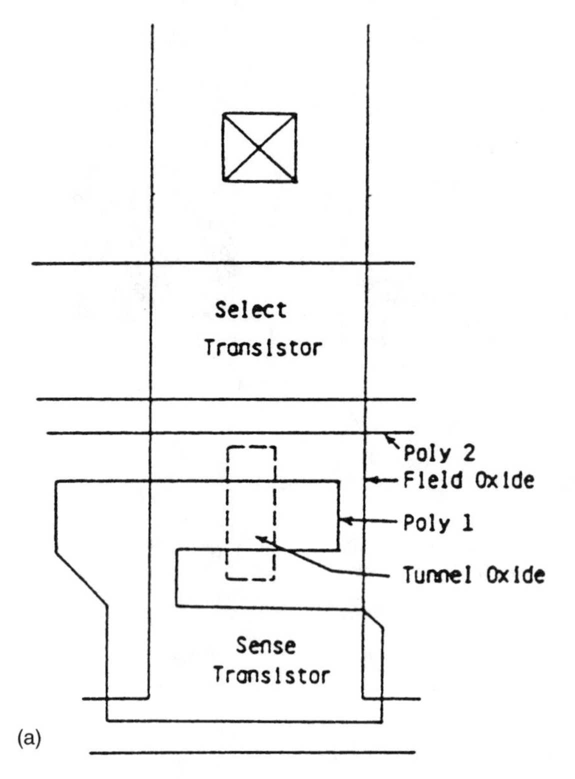

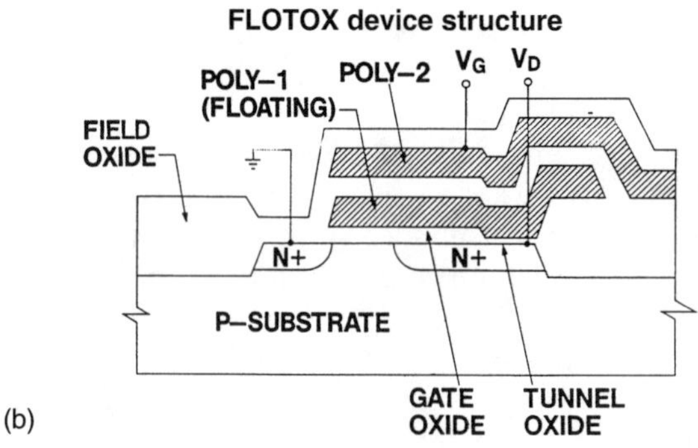

Figure 2.2 (*a*) Floating gate tunnel oxide cell cross section. (*b*) Schematic layout: the top transistor is the select transistor, and the bottom transistor is the sense transistor. The tunnel oxide is connected at the top to the floating gate through the finger structure and at the bottom to the drain of the sense transistor through the tunnel implant extension. Poly2 covers the floating gate completely [2.2].

defined as a narrow finger perpendicular to the length of tunnel oxide leaving implanted areas not covered by poly1. The tunnel oxide area of about 1 μm^2 is defined by the product of the width of the tunnel oxide opening and the poly1 finger width. In order to have the rest of the drain region connected electrically, the poly1 definition is followed by a low-dose arsenic implant self-aligned to poly1. The implant dose is lower than that used for the source-drain region. The regions form a depletion device with a very negative threshold.

Instead of growing the thin tunnel oxide over the drain region, the active channel area can be used [2.3]. The tunneling efficiency is determined by the fraction of the voltage applied at the cell terminals which appears across the tunnel oxide. This can be increased without increasing the cell area by increasing the capacitive-coupling between the control gate and the floating gate using the *S*hielded *S*ubstrate *I*njection *MOS* (SSIMOS) structure shown in Fig. 2.3 [2.4]. Not only is the coupling increased, but also the control gate between the floating gate and the substrate acts as a serial transistor without any increase in cell area. Fabricated using p-well CMOS technology, the cell area is 135 μm^2 using 3 μm layout rules, except for a 2 μm poly-to-poly spacing in the tunnel window. The process sequence is shown in Fig. 2.4. The memory contains an array of multiple wells, each well being 8 bits wide and as long as necessary. To write, 16 V is applied to the control-line with the source drain and well at ground. The erase procedure is to ground the wordline and connect the well and bitlines to a 12 V erase voltage.

A stacked floating gate with Self-Aligned Tunnel Region (SSTR) cell has been implemented in a 1 Mbit EEPROM, fabricated using a twin-well, triple-poly, double-metal CMOS process with 1 μm lithography as shown in Fig. 2.5*a* [2.5, 2.6]. The cell achieves a high coupling ratio of 0.83 in a small cell size of 30.4 μm^2, and has a merged select transistor. The process steps are shown in Fig. 2.5*b*. The first poly layer is used as the select gate and lower control gate, and these two gates form the poly window. The second poly layer, which makes the floating gate, is on the underlying first poly layer except in the poly window region (tunnel oxide region). The third

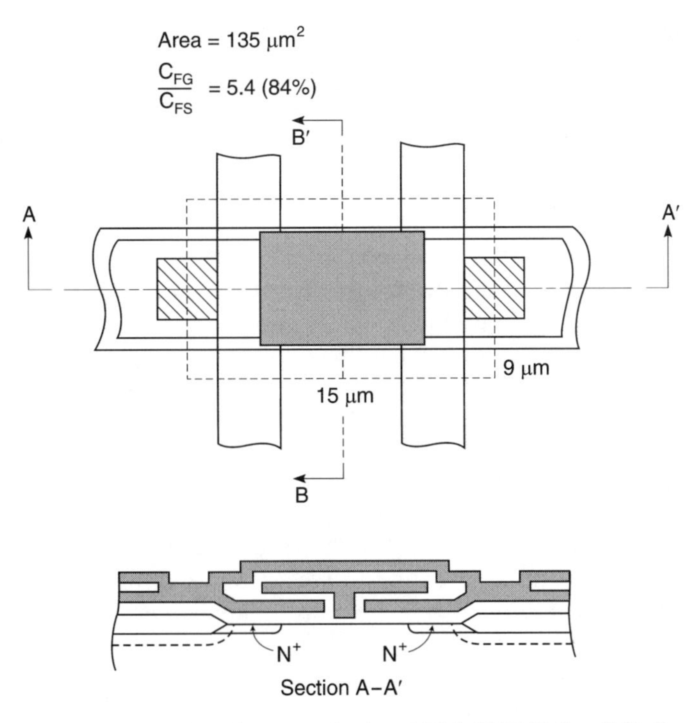

Figure 2.3 Shielded Substrate Injection MOS (SSIMOS) cell [2.4].

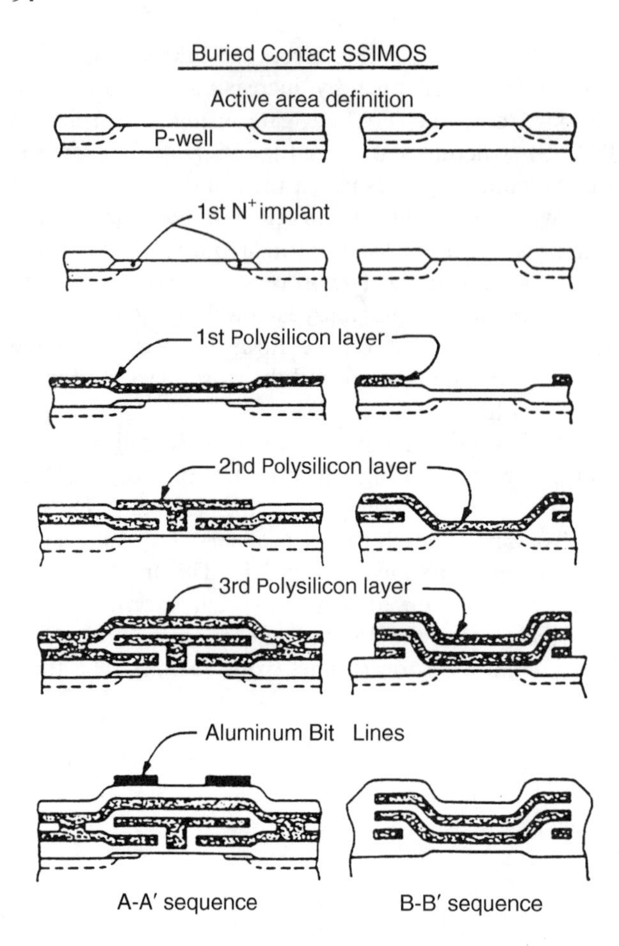

Figure 2.4 SSIMOS cell process flow showing cross sections parallel and perpendicular to the cell [2.4].

poly layer forms the upper control gate and is electrically connected to the lower control gate. The first metal layer is connected to the n^+ drains of the select gate transistor which forms the bitline. The poly wordline (control gate) is stitched by the second metal layer to reduce the wordline resistance. The floating gate transistor resides between the select gate and lower control gate. The source region of the select gate transistor is merged with the tunnel n^+ region of the floating gate device. The width of the tunnel n^+ region is not limited by the minimum lithography feature size and is defined to less than a quarter micron using a newly developed self-aligned technique. The erase and write operation is similar to the conventional FLOTOX cell. To erase, 16 V is applied to the select gate and control gate, with the bitline and source at ground. For writing, the select gate and bitline are raised to 16 V, with the control gate at ground and the source-line floating.

An improved version of the SSTR cell, shown in Fig. 2.6*a*, has been implemented in 4 Mbit memory resulting in a cell area of 9.72 μm^2 with 0.8 μm CMOS technology and a high coupling ratio [2.7]. The first poly layer is used as the select gate. The tunnel n^+ region is merged with the source region of the select transistor and the drain region

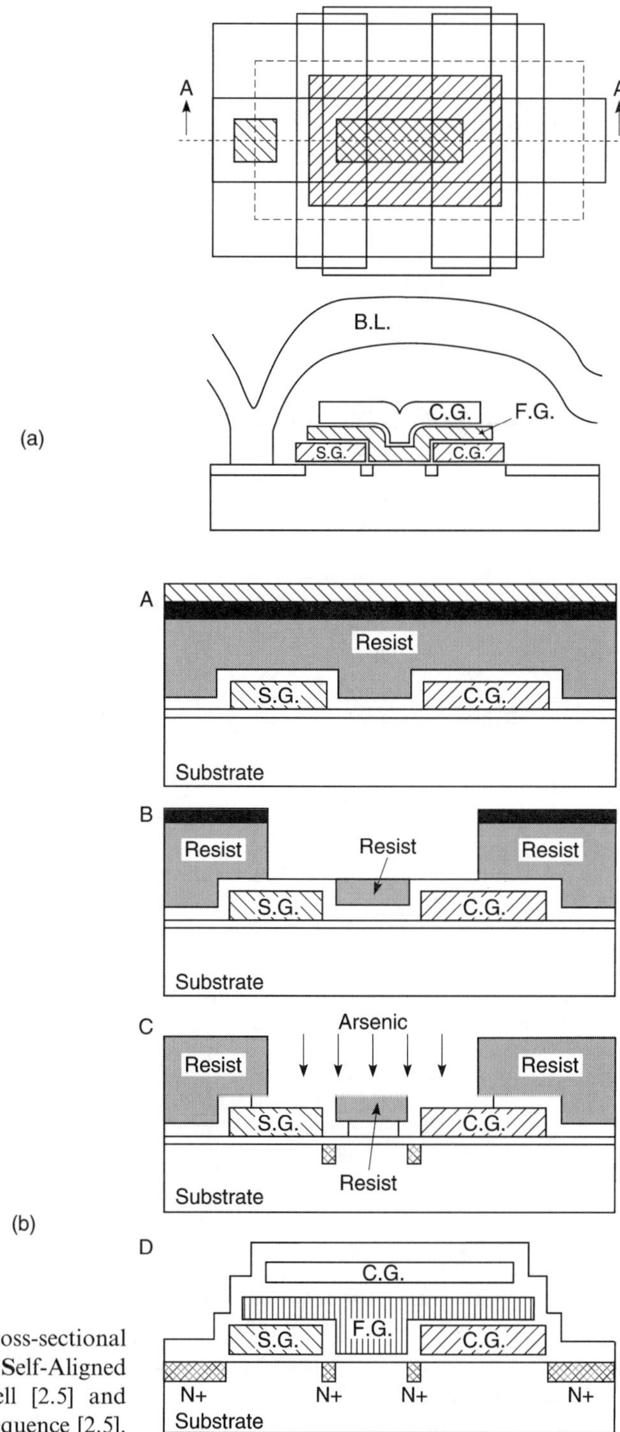

Figure 2.5 (*a*) Top and cross-sectional view of the **S**tacked **S**elf-Aligned **T**unnel **R**egion (SSTR) cell [2.5] and (*b*) SSTR cell fabrication sequence [2.5].

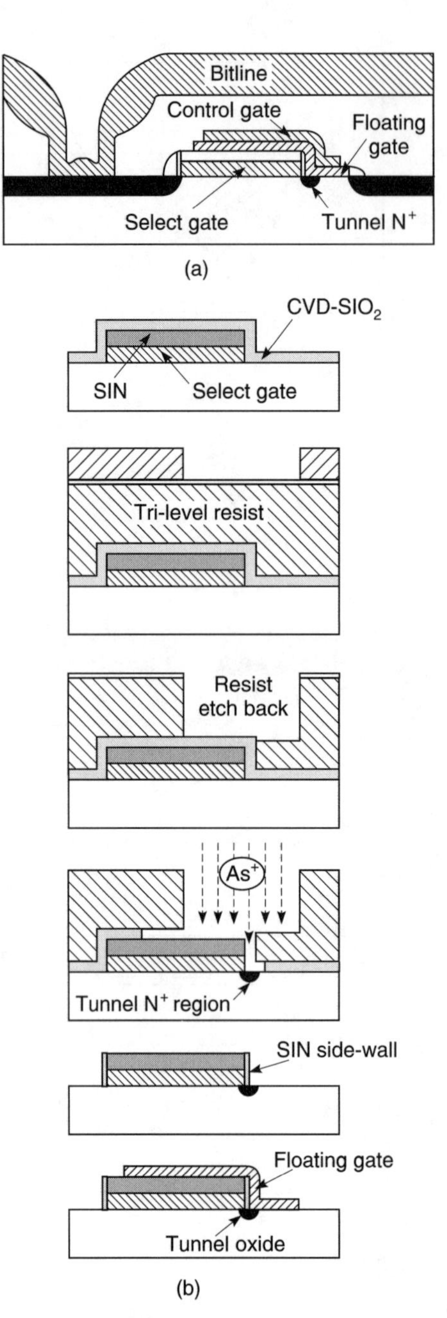

Figure 2.6 (*a*) Modified SSTR cell cross section and (*b*) process flow [2.7].

of the floating gate transistor at the same time. The tunnel n^+ region is formed by self-aligned implants, so it is not affected by lithographic limitations. The second and third poly layers form the floating gate and control gate, respectively, and are stacked on the select gate. The capacitance between the floating gate and substrate is reduced by the shielding effect of the select gate. The select gate, tunnel region, and floating gate are arranged in series between the bitline contact and the source-line to form a merged one-transistor structure, with no electrically useless regions. The first metal layer is used as the bitline. The key process steps are shown in Fig. 2.6*b*.

Instead of a polysilicon layer acting as a control gate, the cell shown in Fig. 2.7 uses a diffusion layer [2.8]. The 256 Kbit *D*iffused *L*ayer *F*loating *G*ate cell with *T*hin

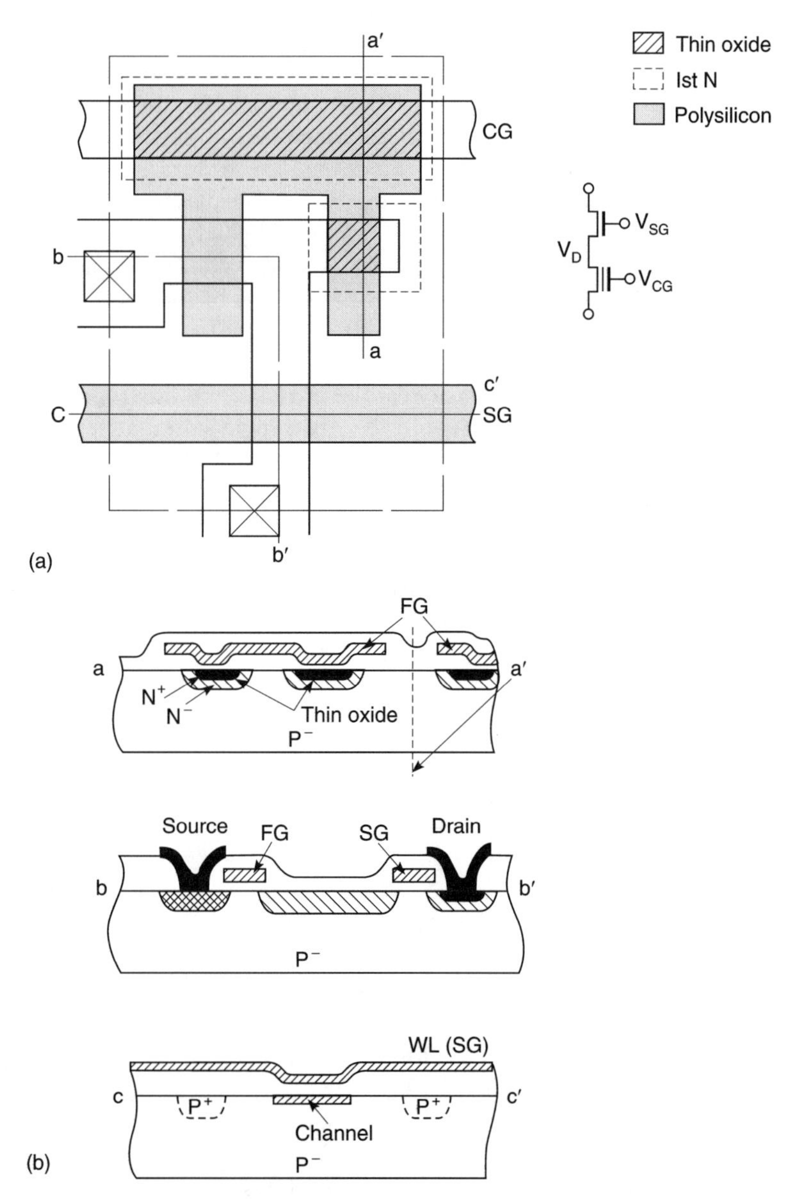

Figure 2.7 The DIFLOX cell [2.8]. (*a*) Cell layout; the cell consists of a select gate and a floating gate transistor and (*b*) cross-sectional views along lines a-a′, b-b′, and c-c′.

*O*xide (DIFLOX) EEPROM is fabricated with a single-poly, single-metal CMOS process yielding a cell area of 86.25 μm^2 with a 1.4 μm gate length. The tunnel oxide thickness is 85 Å. The floating gate is controlled by two thin oxide capacitors coupled with two n^+ diffused layers—the drain and control gate. To write, $V_{pp} = 15$ V is applied to the control gate and the drain is grounded, while, for erase, the connections are reversed. The floating gate is controlled by the capacitive-coupling of the n^+ diffused layers and the drain through the thin oxide. The n^+ implant dose must be carefully chosen since a low concentration suppresses the effective applied voltage across the thin oxide due to the voltage drop in the diffusion layer, while a high concentration lowers the oxide interface barrier height. To increase junction breakdown voltage, the n^+ region under the thin oxide is surrounded with a lower doped n^- region. Selective polyoxidation (SEPOX) isolation technology was used to minimize active area separation [2.9]. Threshold voltage shifts of +3.5 V and −1.5 V have been obtained using a $V_{pp} = 15$ V supply and 2 msec pulses. To obtain a wider threshold window or to eliminate degradation by the select gate, the wordline voltage should be above 20 V. A heavy channel stop implantation was used between the select gates.

The effective bit density can be increased by using a multilevel approach, with each physical bit having more than two logic states. A four-state EEPROM using floating gate technology has been reported [2.10]. This cell, shown in Fig. 2.8, con-

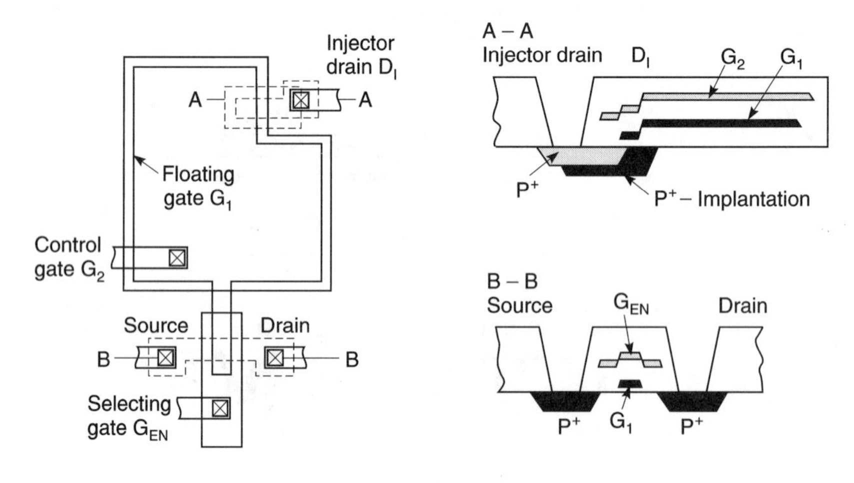

Table 1
Terminal Configurations for the Operation Modes

Mode	Logic-state	V_{G2}	V_{DI}	V_{EN}	Source	A_0	A_1	CLK
Read	x	0	0	V_{SS}	0	x	x	off
Erase	0	V_{ER}	0	x	x	x	x	off
Write	1	0	V_p	V_{SS}	0	1	0	on
Write	2	0	V_p	V_{SS}	0	0	1	on
Write	3	0	V_p	V_{SS}	0	1	1	on

Figure 2.8 Four-state memory cell [2.10].

sists of a control gate G_2 for erasing, an injector drain D_1 for writing, and a gate G_{EN} for selecting the read transistor. Programming is performed by applying a high negative voltage to the injector drain D_1 or the control gate G_2. This causes tunneling current to flow through the thin tunnel oxide and leads to increase or decrease of the charge stored on the floating gate. In the write mode, the cell transistor is used to monitor the actual state of the memory and control the charge on the floating gate, since this charge affects the drain current. Each of the four levels is characterized by its lower and upper limits, as well as by a nominal current that is reached after programming the memory cell. A write/erase voltage of −20 V was used. Depending on the memory state of the cell, the write time was about 30 msec and the erase time was 100 msec.

2.1.1 Structures for Custom IC Applications

Being more cost effective per memory bit, Flash EEPROMs have almost completely displaced conventional full-feature EEPROMs in the high-density NVM market. The latter EEPROM technology is finding increasing use in ASICs to provide improved product functionality, such as EEPROM trimmed data converters, *f*ield *p*rogrammable *l*ogic *d*evices (FPLDs), smart modems, and analog memories.

Integrating EEPROMs with a standard CMOS/BiCMOS process becomes challenging because of incompatible process requirements. EEPROMs require high-voltage transistors and field isolation to sustain even higher voltages, whereas the rest of the process is geared for small geometries and two voltages. The trick is to minimize extra process steps in order to realize the EEPROM. Basic requirements for incorporating conventional EEPROM cells are: high junction breakdown voltage, high-field threshold voltage, high-voltage transistors, and robust thin tunnel dielectrics.

A two-poly 1 μm CMOS-EEPROM process describes ways to realize the above requirements without degrading the performance of the high-speed low-voltage devices [2.11]. The cell is programmed using $V_{pp} = 16$ V for 10 msec with a select gate voltage $V_{sg} = 20$ V. High field V_{th} is obtained by a separate field implant (6E13 cm^{-2} @ 120 keV, BF_2) in the EEPROM areas. This implant is retracted 1 μm from the *l*ocal *o*xidation of *s*ilicon (LOCOS) nitride mask edge to increase the junction breakdown voltage. A 6500 Å thick field oxide and a minimum field width of 4.2 μm was found to be necessary for a field V_{th} of 22.5 V. The high-voltage transistors have a 400 Å gate oxide over the channel region. BV_{dss} is increased by using a graded oxide structure (see Fig. 2.9). Instead of using a *l*ightly *d*oped *d*rain (LDD) process (which would affect the low-voltage transistors), an n^- arsenic (As) implant is used to form extended source-drain regions of the high-voltage transistor. A 2700 Å thick decoupling oxide is grown over the n^- regions, which lifts the edges of the poly-Si gates. This process flow allows the low-voltage and high-voltage transistors to be optimized separately.

The tunnel oxide area is defined by the nitride at the same time as the high-voltage drain region. When the decoupling oxide is grown, the tunnel area is protected by the nitride. Subsequently, when the tunnel window is to be opened, it is

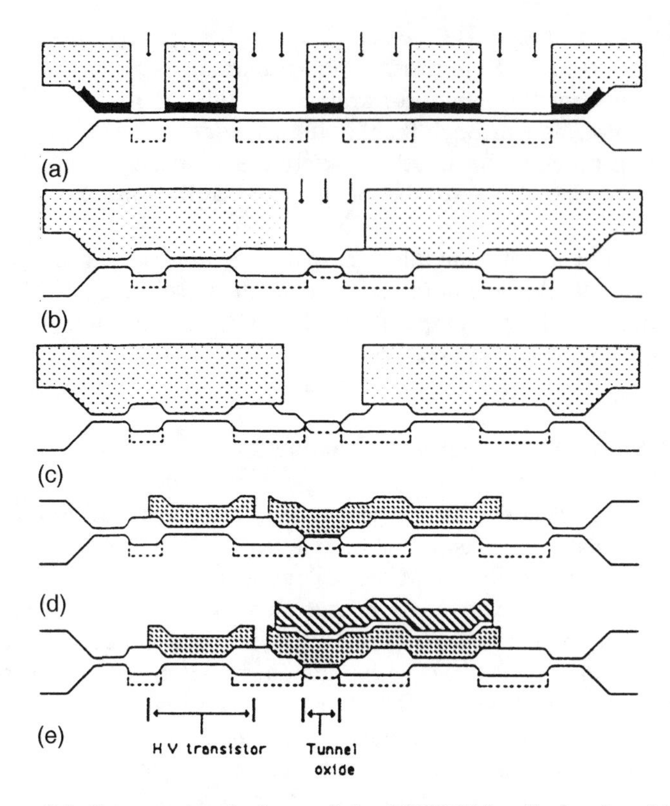

Figure 2.9 Cross-sectional views of the EEPROM cell showing (*a*) the nitride mask and n^- implant used for the graded oxide transistor module, (*b*) the tunnel oxide mask and tunnel implant, (*c*) after wet etch to open the tunnel oxide area, (*d*) after tunnel oxide growth and first poly definition, and (*e*) after interpoly oxide growth and second poly definition [2.11].

defined by the edge of the thick decoupling oxide and not by the tunnel window lithography, which can be larger than the tunnel area. To open the tunnel window, the 400 Å oxide is wet etched [6:1 BOE (*b*uffered *o*xide *e*tch)] to avoid plasma damage to the substrate.

Integration of an EEPROM in a 0.6 μm single-polycide, double-metal, twin-well CMOS process has been achieved using two extra process steps [2.12]: a buried n-type diffusion area for the bottom electrode of the EEPROM cell and tunnel oxide definition/growth. An EEPROM program voltage $V_{pp} = 12.5\,V$ is needed. The integrated process flow steps are given in Table 2.1, and a CMOS-EEPROM cross section is shown in Fig. 2.10. An aggressive field isolation pitch (1.5 μm) was obtained by optimizing the LOCOS/nitride/pad oxide thickness to reduce bird's beak encroachment and gate oxide thinning at the LOCOS edge, and field channel stop implantation through the LOCOS using the same masking step as the NMOS

TABLE 2.1 CMOS-EEPROM PROCESS FLOW FOR 0.6 μm SINGLE-POLY TECHNOLOGY [2.12].

Logic (core)	EEPROM module
Twin-well	
Active area	
LOCOS	
	Buried N+ diffusion
Field and Channel I–I	
Gate oxide	
	Tunnel oxide
Gate poly definition	
CMOS LDD I–I	
NMOS S/D I–I: Shallow $75As^+$ and Deep $31P^+$	
PMOS S/D I–I	
Contact and metallization	

channel threshold voltage implant. A junction breakdown voltage of more than 16 V was obtained. Vertical scaling reduced the topography and line-to-line capacitance. A modified LDD MOSFET used a deep P implant together with a conventional shallow As ion implant during the source-drain formation to yield a more gradual junction profile that partly compensated the p-well concentration at the bottom of the n^+/p junction. This implant increased the junction breakdown voltage by 2 V and, with the field channel stop after LOCOS, reduced the peripheral component of junction capacitance by 5X. In addition, the field implant penetrated deep under the NMOS channel region, creating a retrograde p-well profile that improved punch-through characteristics.

A low-density 5 V only EEPROM memory for custom and ASIC applications, fabricated using a modified 2.5 μm, n-well, CMOS process with a cell area of

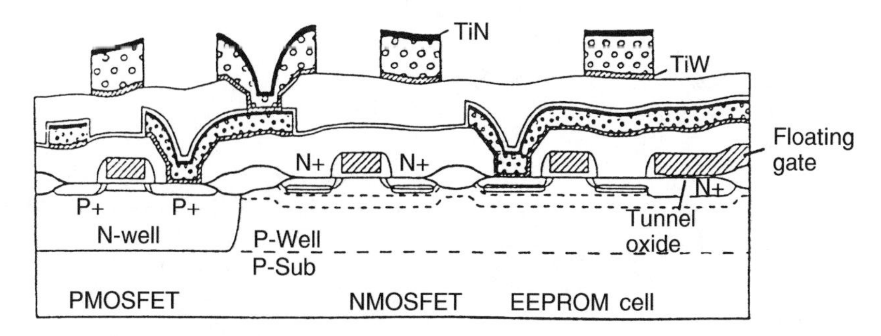

Figure 2.10 Schematic cross-sectional view of the single-poly EEPROM-CMOS process [2.12].

440 μm² and an 80 Å tunnel oxide, is shown in Fig. 2.11 [2.13]. The control gate is an n^+ diffusion that is shared by two adjacent memory cells and is accessed via a pass transistor. The coupling between the floating gate and the control gate is made much larger than between the floating gate and drain (D). As a consequence, only a small fraction of the program voltage applied between the control gate and drain will be over the coupling capacitor, and tunneling through this capacitor can be neglected. Instead of a capacitor between the two poly layers, the cell uses a tunnel oxide (80 Å) MOS capacitor to capacitively couple the floating gate and control gate. The programming voltage is 13 V for this cell. Incorporating this cell in a standard MOS process requires one extra mask for defining the thin oxide regions and additional processing steps.

An EEPROM cell built with standard 0.8 μm CMOS processing using 150 Å gate oxides for custom logic applications is shown in Fig. 2.12 [2.14]. The cell consists of NMOS and P-channel MOS (PMOS) transistors with a common gate used as the floating gate. The inversion layer under the PMOS gate and the p^+ diffusion are

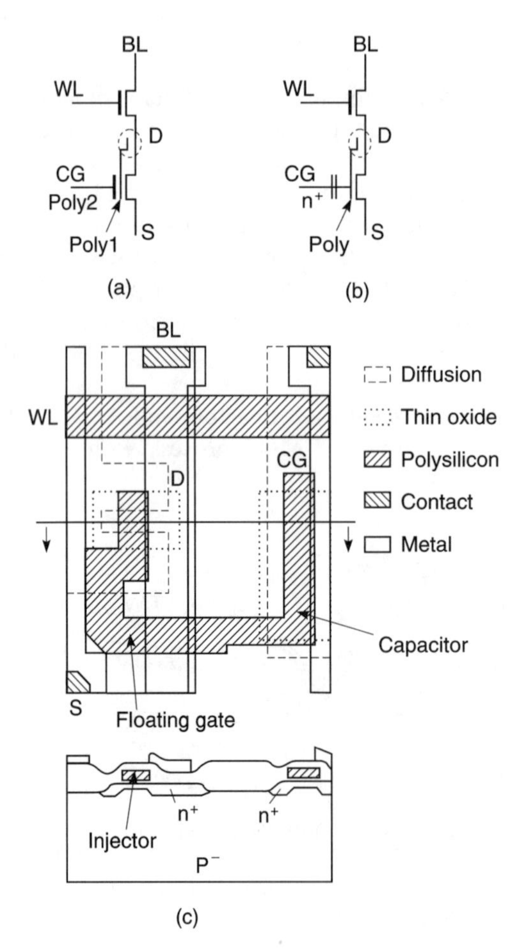

Figure 2.11 (*a*) Schematic representation of double-poly EEPROM cell, (*b*) single-poly EEPROM cell with a thin oxide capacitor as coupling capacitor (circled area in figure is the injector), and (*c*) layout and cross section of single-poly CMOS memory cell usable in ASIC and custom applications [2.13].

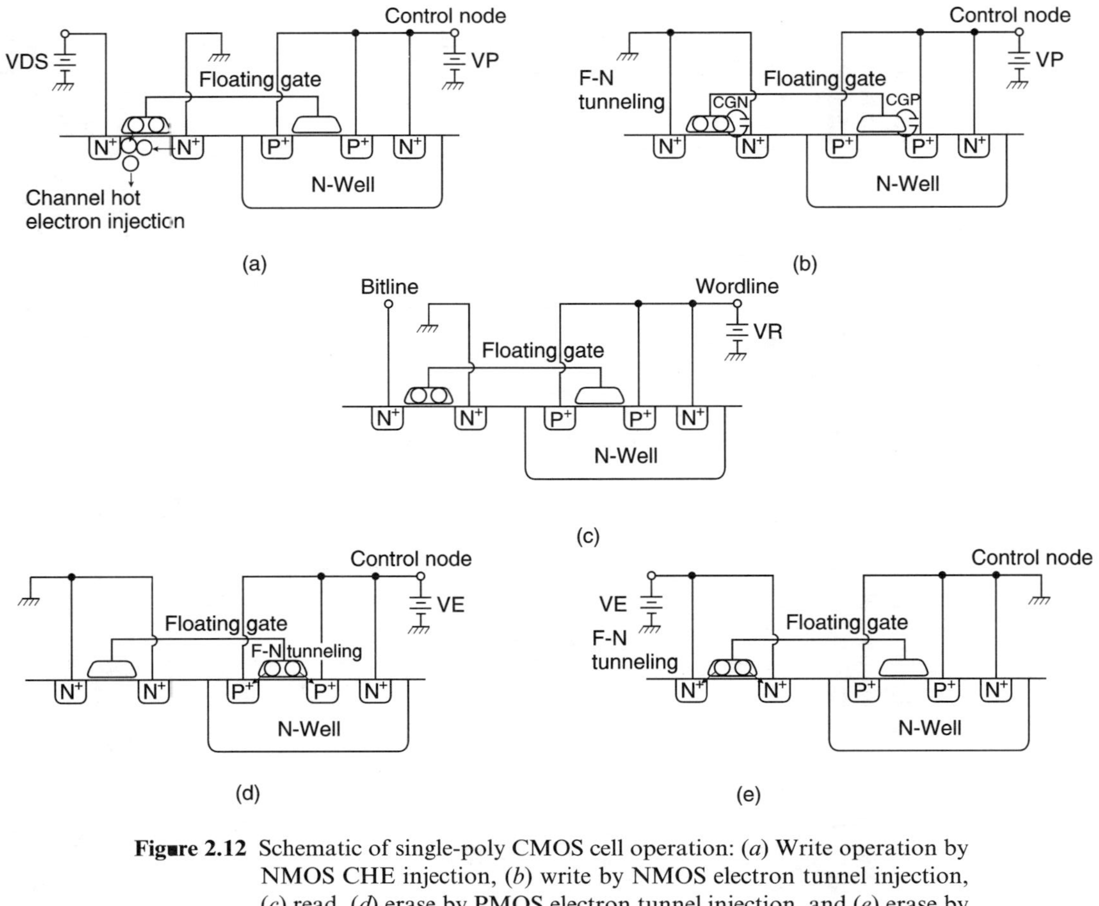

Figure 2.12 Schematic of single-poly CMOS cell operation: (*a*) Write operation by NMOS CHE injection, (*b*) write by NMOS electron tunnel injection, (*c*) read, (*d*) erase by PMOS electron tunnel injection, and (*e*) erase by NMOS electron tunnel injection [2.14].

used as the control gate. The cell can be written by N-channel hot-electron injection (NCHE write) or N-channel Fowler–Nordheim tunneling between the floating gate and the n^+ diffusion (NFN write) in the NMOS transistor. Erase is accomplished by tunneling between the floating gate and the p^+ diffusion of the PMOS transistor, P-channel Fowler–Nordheim, injection (PFN erase), or the floating gate and n^+ diffusion of the NMOS transistor (NFN erase). Selection of a specific mode depends on the PMOS gate (C_{gp}) to NMOS gate capacitance (C_{gn}) ratio. When $C_{gp}/C_{gn} < 1.0$, NCHE write and PFN erase are used. When $1.0 < C_{gp}/C_{gn} < 3.0$, NCHE write and NFN erase are used. When $C_{gp}/C_{gn} > 3.0$, NFN write and NFN erase are used. For the mode combination of NCHE write and PFN erase, the capacitance ratio $C_{gp}/C_{gn} = 2/3$. The write operation is done at $V_p = 15\,V$, $V_{ds} = 9\,V$ with 10 msec pulse duration. Erase is accomplished at $V_e = 18\,V$ with a duration of 100 msec. For the mode combination of NFN write and NFN erase, the gate capacitance ratio $C_{gp}/C_{gn} = 5.0$. The write is performed at $V_p = 19\,V$ and the erase is accomplished at $V_e = 14\,V$, both with a pulse duration of 100 msec.

Twin polysilicon thin film transistor cells have been implemented as shown in Fig. 2.13 using standard self-aligned, n-channel, polysilicon MOS thin film transistors for potential use in active matrix liquid crystal display (AMLCD) image storage and 3-D integrated circuit applications [2.15]. The gate electrodes of the two thin film transistors (TFTs) are connected to form the floating gate. The cell is written and erased by electron tunneling through the gate oxide of the smaller transistor T1 with area W_1L_1. The source and drain of the larger transistor (T2), with area W_2L_2, are connected to form the control gate. The 200 Å gate oxide was deposited by *L*ow-*P*ressure *CVD* (LPCVD) and densified at 600°C for six hours. To write, 15 V is applied to the control gate for 10 msec with the source-drain of T1 at ground. To erase, −15 V is applied to the control gate for 10 msec, resulting in electron tunneling out of the floating gate. A threshold voltage shift of 8 V was obtained for a capacitance ratio W_2L_2/W_1L_1 of 4.

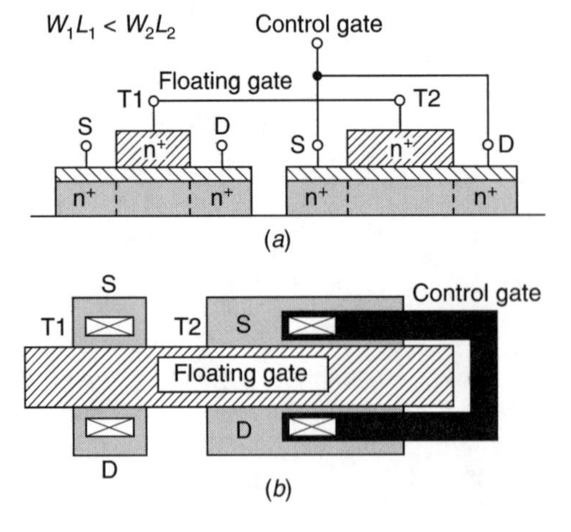

Figure 2.13 Twin polysilicon thin-film transistor memory cell [2.15]. (*a*) Cross section and (*b*) top view. The gate electrodes of T1 and T2 are connected to form the floating gate, while the source and drain of T2 are connected to form the control gate. The cell is programmed and erased by Fowler–Nordheim tunneling through the gate oxide of T1. The channel area of T1 is smaller than the channel area of T2.

2.2. MEMORY ARRAY CIRCUITRY

Incorporating EEPROM cells into large-scale memory arrays presents some unique problems because of the high voltages needed during programming operations. Although the overall memory organization of EEPROMs is similar to that of other volatile and nonvolatile memories—consisting of address/data buffers, row/column decoders, sense-amps, and other control circuits—some key issues have to be resolved before successful matrix implementation is possible—for instance, cell-disturb problems, high-voltage load circuits, over-erase protection circuits, programming timer circuits, and on-chip charge pumps with waveform shaping for 5 V-only operation. In the following discussion, V_{pp} refers to the programming voltage. Its exact value depends on the implementation technology.

When the individual cells (discussed in Section 2.1) are placed in a large memory matrix, the high electric field present during the programming operation can cause *disturb problems* in which programming of a selected cell affects the state of other unselected cells in the matrix. The simplest organization is shown in Fig. 2.14, where a memory transistor has been placed at each cross-point. To write the selected cell, V_{pp} is applied to the selected wordline and the column is grounded. Unselected wordlines are grounded, and columns are connected to V_{pp}, which will cause erasing of the unselected cell. Apart from the disturb problems, a large I_{pp} current will flow from the charge pump in unselected columns.

The disturb problem can be avoided if two transistors are used per memory bit—select and memory transistors, as shown in Fig. 2.15. To write the selected transistor, the selected wordline and column are connected to V_{pp} and ground, respectively, and the unselected wordlines are grounded and the columns pulled to V_{pp}. The select transistor disconnects the drain of the memory transistors of the

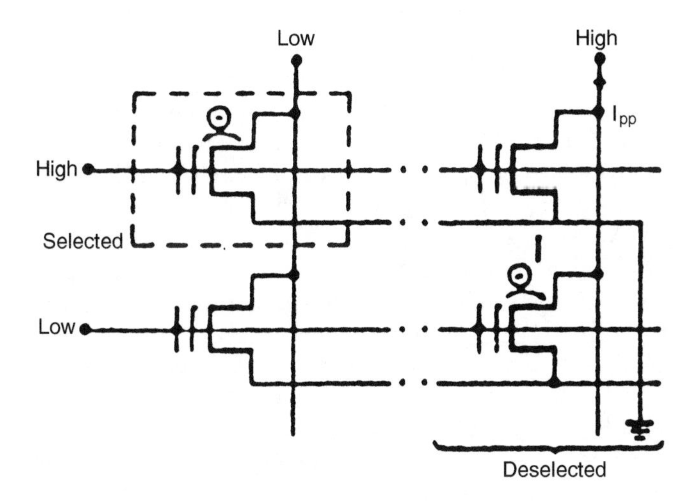

Figure 2.14 Disturb problems in single-transistor EEPROM arrays [2.16].

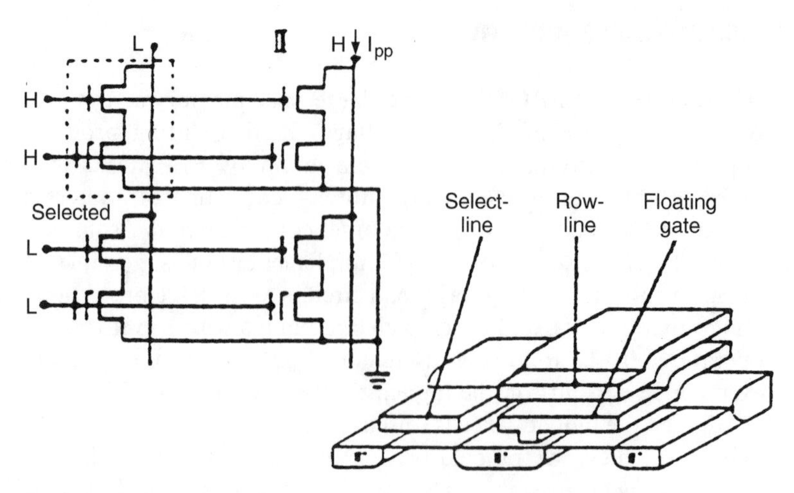

Figure 2.15 (*a*) Schematic of two-transistor cell, and (*b*) implementation of the two-transistor bit into an array configuration [2.16].

unselected rows from the high voltages applied to the unselected columns. The potential problem of high I_{pp} current drawn from the charge pump still remains. In the selected row, all the select transistors are on, and, if memory cells in some column are in the erased state, then large I_{pp} currents can flow.

This situation can be avoided if the sources are disconnected from ground as shown in Fig. 2.16 [2.16]. To write, column pullup V_{cp} is high, all I/O lines are grounded, and Y_j of the selected byte is high (unselected Y_j are grounded). Since the pullup is weak in comparison to the column select, this results in a near-ground potential on all selected columns and near V_{pp} potential on unselected columns. The select-lines X_i and row-lines P_i of the unselected rows are grounded. The select-line of the selected row X_i is held at V_{pp}, transferring the column potential to the drain of the memory transistors that are along the selected row. This results in a near V_{pp} potential on all unselected memory cell drains and near ground potential on selected cell drains. Applying V_{pp} to the selected row results in electron injection only to the floating gate of devices on selected rows and columns. Disconnecting source-lines of adjacent bytes is essential to prevent I_{pp} currents through all unselected bytes that are on the selected row. Erasing is similarly accomplished. Y_j of the selected byte is pulled to V_{pp} with select-lines of the unselected bytes grounded. Thus, voltages on internal I/O lines are transferred only to columns of the selected byte. Applying V_{pp} to a given I/O line results in V_{pp} voltage on the column of the corresponding bit in the selected byte ($V_{cp} = 0$). Grounding the unselected X_i select-lines and unselected P_i row-lines eliminates the program disturbs. Applying V_{pp} to select-lines X1 of the selected byte transfers the high and low I/O voltages to the drains of corresponding memory cells. Grounding the row-lines of the selected bit results in the erasing of all bits whose memory cell drains are at high voltage. Devices whose memory cell drains are grounded remain undisturbed. Note that writing the byte prior to erasing eliminates I_{pp} currents between selected and unselected columns in the selected byte.

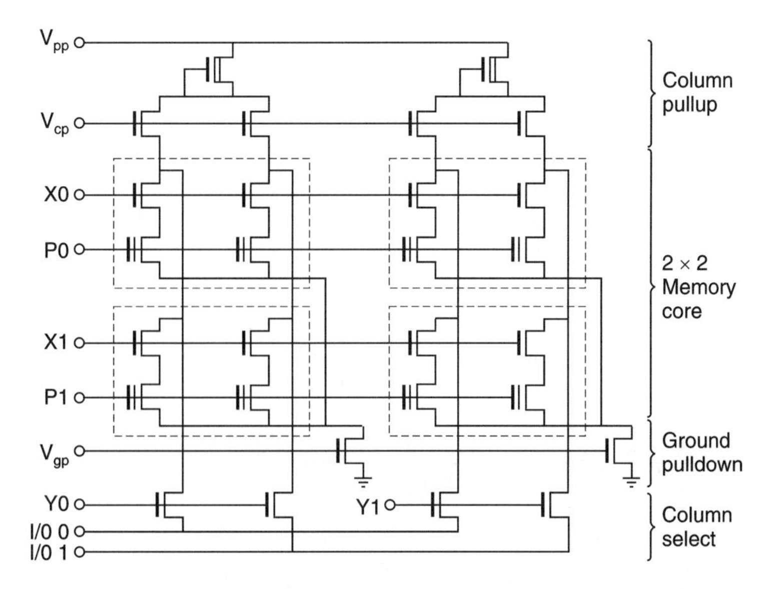

Figure 2.16 Four-byte EEPROM array organized in two rows, each constructed of two bytes (only two bits of the byte are shown) [2.16].

To minimize disturb problems encountered in the simple array given in Fig. 2.14, the half-voltage programming technique can be used, as shown in Fig. 2.17*a*. Here, half voltages, $V_{pp}/2$, are applied to unselected columns and rows, which results in only half the voltage necessary for tunneling being applied across the tunnel oxides of the unselected cells. Since the tunneling current depends strongly on the oxide field, the disturb problem is significantly reduced using this method. The three disturbs that can occur during programming are (a) DC erase, (b) DC program, (c) and program-disturb [2.17]. DC erase occurs on already written cells (i.e., with electrons in the floating gate) sharing the same wordline as the cell being written. During the write operation, the common wordline is high, and, if the electric field across the interpoly oxide is large enough, electron tunneling can occur, resulting in threshold voltage reduction. DC program occurs on cells in the erased state. Raising the wordline voltage of these cells creates a high field across the tunnel gate dielectric, which may cause electron tunneling to the floating gate. A written cell, sharing a column with another cell being written, will experience high electric fields between the floating gate and drain, which may cause electron tunneling from the floating gate to the drain. This effect, called program disturb, leads to reduced cell threshold voltage. This stress occurs at lower voltages (6.5 V) compared to those during DC program and DC erase.

A recent memory matrix implementation using this technique is shown in Fig. 2.17*b* [2.18, 2.19]. Applied voltages for write and erase are given in the figure. All cells are erased simultaneously by applying a negative high-voltage pulse (−11 V) to all wordlines and 5 V bitlines. By applying a negative voltage to the wordline during erase, instead of a high positive voltage to the source-drain junction, graded junc-

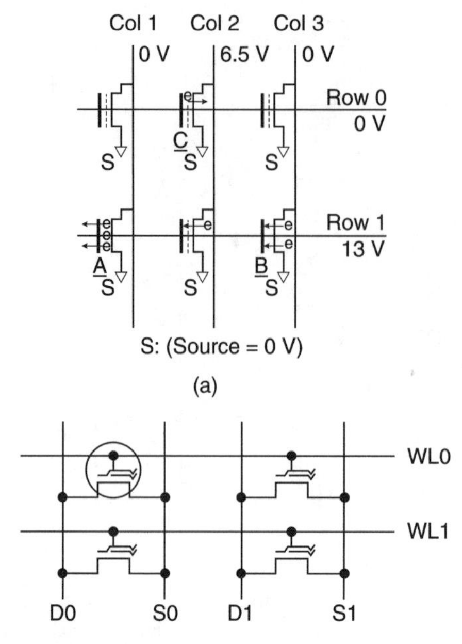

	WLO	WL1	SO	DO	S1	D1
Erase	–11 V	–11 V	5 V	Float	5 V	Float
Program	17 V	7 V	0 V	Float	7 V	Float
Read	3.3 V	0 V	0 V	1.5 V	0 V	1.5 V

(b)

Figure 2.17 (*a*) Schematic description of the disturb mechanism in single-transistor arrays using the half-voltage scheme: Cell A experiences DC erase disturb, Cell B experiences DC program-disturb, and Cell C experiences program-disturb [2.17], and (*b*) matrix implementation of the single-transistor memory cell [2.18].

tions are not needed. (A graded junction is required for a high breakdown voltage, but its drawback is that the cell area cannot be scaled easily.)

2.2.1 Charge-Pump Circuits

The recent development of 5 V-only memories requires that the programming voltage V_{pp} be generated using on-chip charge pumps. The basic voltage multiplier circuit is shown in Fig. 2.18 [2.20]. The nodes of the diode chain are coupled to the inputs via capacitors in parallel so that the capacitors have to withstand the full voltages developed across them. Efficient multiplication can be achieved with relatively high values of stray capacitance, and the current drive capability is independent of the number of multiplier stages. The multiplier operates by pumping charge packets along the diode chains as the coupling capacitors are successively charged and discharged each half-clock cycle. The voltages are not reset after each pumping cycle, so that the average node potentials increase progressively from the input to the output of the diode chain. The circuit implementation is shown in Fig. 2.19, where MOS transistors are used in the diode configuration. The clocks are generated by two inverters driven from an oscillator circuit. Typically, the output is limited by MOS diodes to prevent output voltages from exceeding process limits, such as junc-

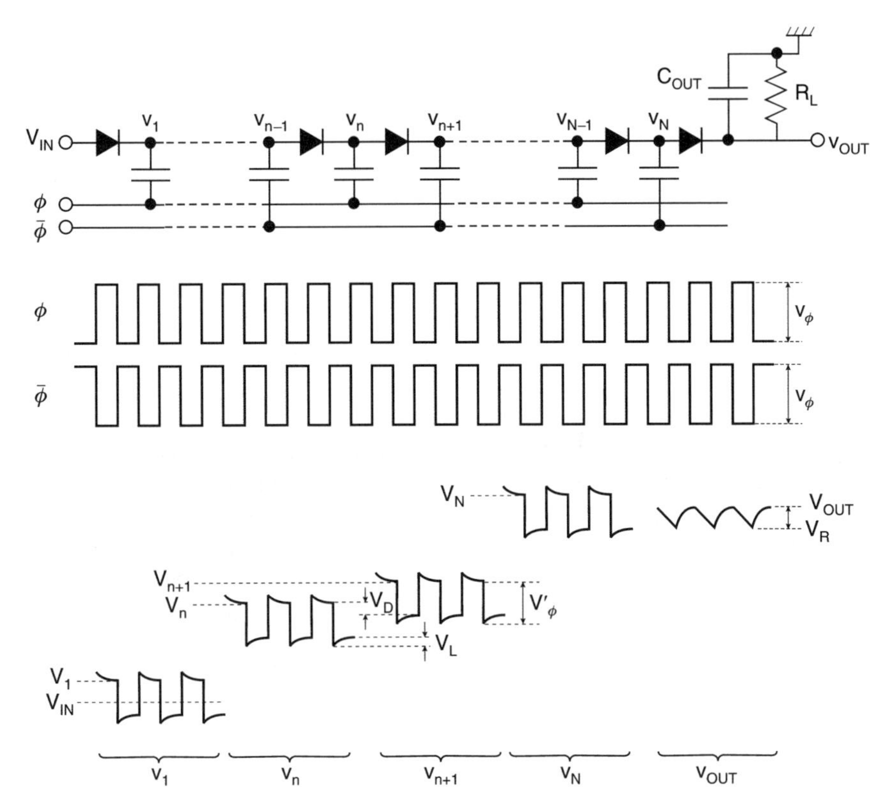

Figure 2.18 Voltage waveforms in an N-stage multiplier showing the voltage relationship between successive nodes of the diode chain [2.20].

tion breakdown voltages, if the supply voltage V_{dd} increases above the specified range.

To reduce the peak fields in the tunnel oxide of cells using electron tunneling-based write/erase operations, the rise time of the programming pulse needs to be well controlled. For an exponential ramp, RC = 600 μsec, and for a linear ramp, the rise time = 40 μs [2.16]. This can be achieved by utilizing a charge pump whose current output is low enough to attain the desired rise time using the capacitance on the high-voltage node. However, this does not compensate for the temperature dependence of junction leakage current. As the temperature increases, leakage currents increase, and the charge pump may fail to achieve the desired output voltage. A closed-loop implementation, shown in Fig. 2.20, uses a feedback network to regulate the charge-pump output and control the programming pulse [2.21]. The feedback loop consists of a clock driver (whose output swing is controllable), a diode-capacitor chain, a voltage divider, and a differential amplifier. The negative feedback forces the output voltage to be equal to the voltage applied to node A multiplied by the reciprocal of the divider ratio. The voltage at A is generated by an RC network that is allowed to charge to V_{ref}. By making the time constant of the RC network

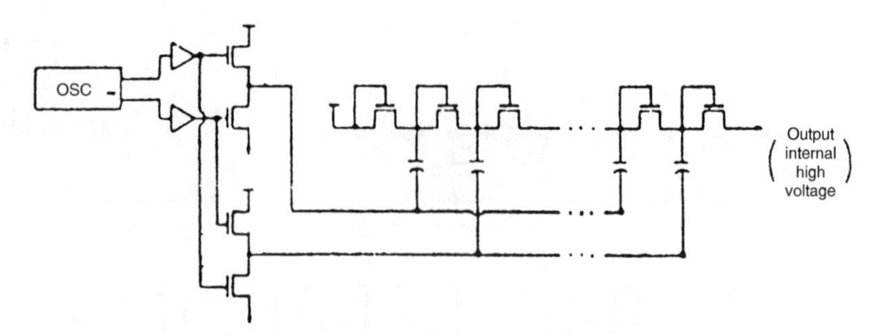

Figure 2.19 Typical circuit implementation in NMOS technology [2.21].

equal to 600 μsec and setting V_{ref} to the optimal programming voltage multiplied by the divider ratio, the desired programming pulse is produced. The feedback loop allows the basic charge pump to be designed to work in the worst case corner condition (high temperature, low V_{cc}) without resulting in overstress in the best case corner (low temperature and high V_{cc}). A nonclocking high-voltage regulation and output shaping circuit is given in Fig. 2.21 [2.22].

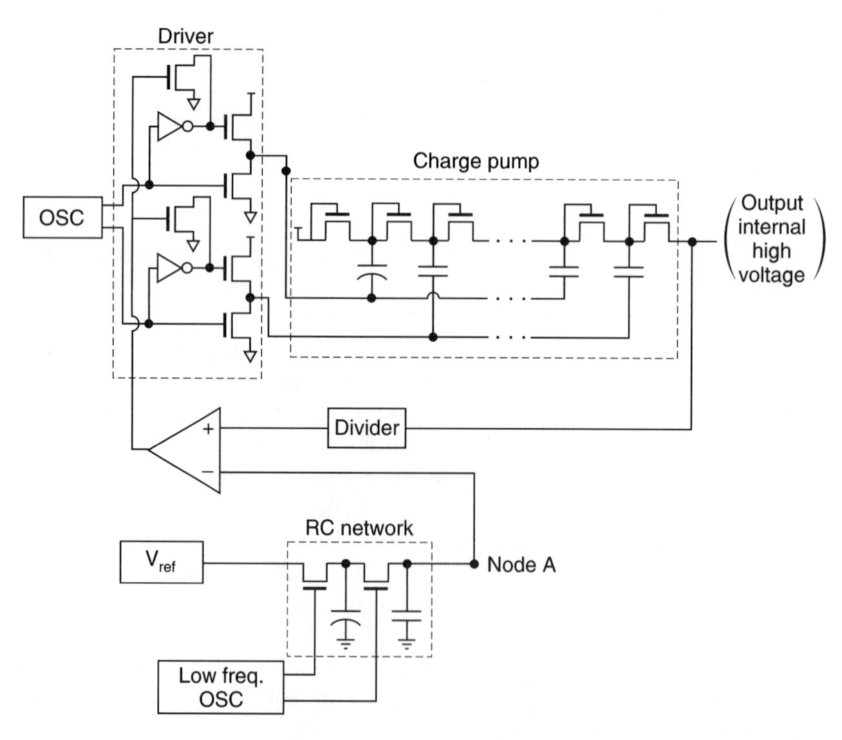

Figure 2.20 Simplified schematic of a high-voltage regulation circuit [2.21].

A quick boosting charge-pump circuit has been proposed for use in memories requiring fast programming times [2.23]. Rather than sequentially charging each bootstrap capacitor, as in the above circuits, the basic concept here is to charge several capacitors in parallel and then connect them in series, which is done by changing the number of stages and capacitance used for charge pumping. The number of stages is controlled to be small, while the capacitance for charge pumping is large, which increases the charge-transfer efficiency (ratio of the average output to input current).

Large negative voltages can be generated (from a 5 V supply) using negative charge pumps. Operating principles are similar to positive charge pumps, except the direction of the diode chain (see Fig. 2.18) is reversed. Typical implementation, in NMOS technology, used for negative substrate bias generation is shown in Fig. 2.22 [2.24]. Node V_x is kept below ground using bootstrap techniques. Deep p-well and n-well implants would be needed to isolate the n^+ junctions connected to large negative voltages. When the oscillator output is high, the charge-pump capacitor CD2 charges to $V_{dd} - V_{t,QD2}$. The enhancement pulldown transistor connected to CD1 is held off by the output of the internal stage of the inverting push–pull buffer connected to CD1. When the oscillator output becomes low, the charge-pump capacitor CD2 couples node B below ground, to $-(V_{dd} - V_t)$. So, the source node of device QD1 is not at ground, but is coupled to $-(V_{dd} - V_t)$. The output of the push–pull buffer now goes high. The voltage across capacitor CD1 charges to $2(V_{dd} - V_t)$. When the oscillator output returns to the high level, the charging circuit for the charge-pump capacitor CD2 is isolated from the rest of the generator by enhancement mode diode QD1, and the voltage at node A is coupled by capacitor CD1 and the falling output voltage of the push–pull buffer to approximately $-2(V_{dd} - V_t)$.

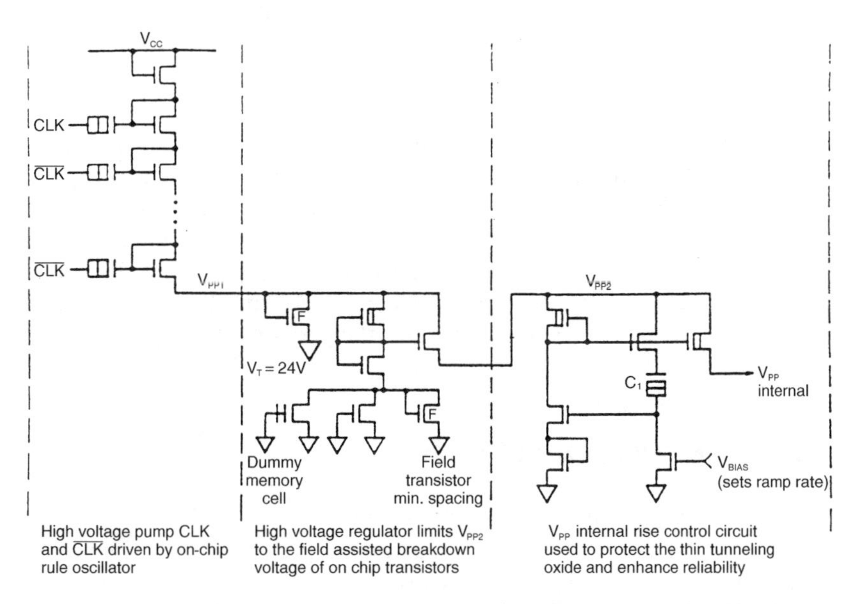

Figure 2.21 High-voltage (V_{pp}) generator for 5V-only operation [2.22].

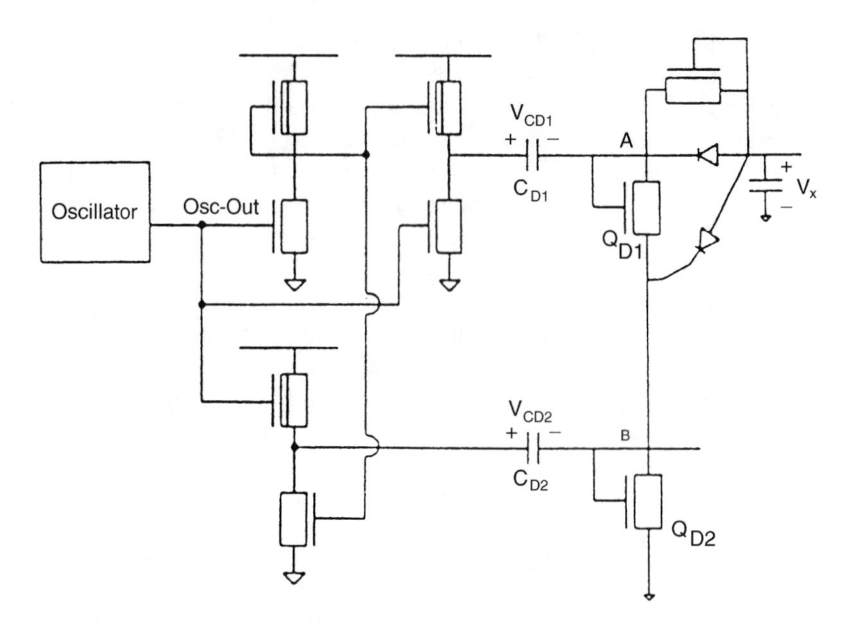

Figure 2.22 NMOS negative charge pump circuit [2.32].

In CMOS implementation, negative charge pumps are typically implemented with PMOS transistors, as shown in Fig. 2.23 [2.18, 2.25, 2.32]. The charge-pump circuit has six cascaded voltage multiplier stages, and each stage consists of one PMOS transistor, one resistor, and two capacitors. The capacitor is formed with a gate capacitor of a PMOS depletion transistor. P7 is a depletion-type PMOS and is used to provide the ground level at V_{ERA}, when the negative pump is not used. With node A initially at ground, both capacitors connected to node A will be charged to $\approx V_{dd}$, that is, 5 V. When PA1 goes low, because node A is isolated, it tracks PA1 and will initially go to about -5 V. This turns P1 on, causing node B to be discharged. The final node A and node B voltages will depend on the capacitor values and the PMOS threshold voltage because the charge on the two capacitors connected to PA1 and PA2 is shared between the capacitors connected to node A and node B. This charge sharing reduces the gate voltage on P1, making it more resistive. Next, PA2 becomes low, P1 is turned on hard since the gate node of P1 becomes lower than node A due to the RC network between node A and PA2, and the node B voltage is lowered further. Each stage discharges the previous stage, and the output node V_{ERA} becomes -13 V.

2.2.2 High-Voltage Load Circuits

Since the amount of current that can be generated by an on-chip charge pump is limited, it is essential that the programming currents be minimized. The memory transistor and decoding circuitry are the main parts of the chip that must be connected to the charge pump. During write, electrons tunnel to the floating gate, and

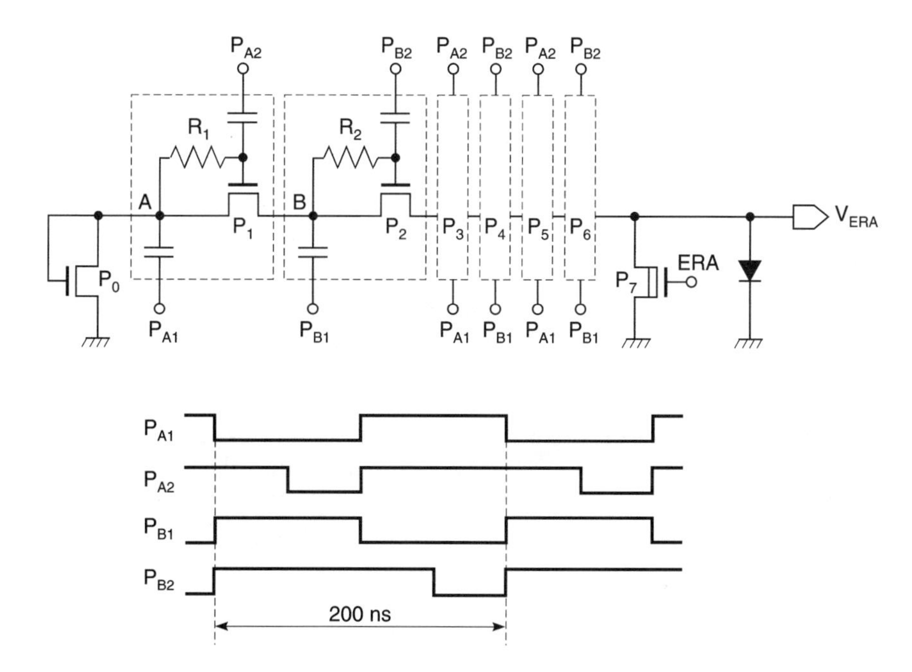

Figure 2.23 Negative high-voltage charge pump using PMOS transistors [2.32].

the loading on the charge pump by the cell is due to the interpoly capacitance. During erase, the drain and select gates are at V_{pp}, and the loading on the charge pump (with source floating) includes the tunnel oxide capacitance, and the capacitance and junction leakage of the diffusions in the memory cell. The select-lines of the array must be able to switch between V_{pp} and ground during programming, which is usually done by connecting a depletion load to the charge pump. Unfortunately, the charge-pump output is quite limited in that it cannot maintain the necessary high voltage while supplying the current required by all the depletion devices on the unselected wordlines. A conventional load circuit and a modified load circuit used to eliminate this problem are shown in Fig. 2.24 [2.21].

Prior to the output going high, the output node is charged to V_{cc}. T3 serves as the transfer device to charge the clocked capacitor T2. T1 directs the current from T2 onto the output node. With each clock cycle, the output node voltage rises until it is at a threshold above V_{pp}. The pumped output voltage is used to eliminate the effects of threshold drops in the decoding circuits and memory cell. Since the output node limits itself to a threshold voltage above V_{pp}, this reduces the chance of overstress in the peripheral circuitry. When the output is held low, T_3 shuts off and no current is drawn from V_{pp}.

This circuit is costly in die area because of the use of a bootstrap capacitor and routing of the clock to all high-voltage nodes. In addition, the rise time of the load

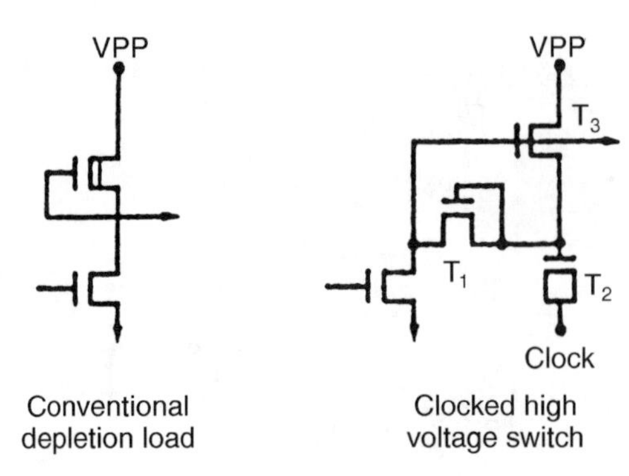

Figure 2.24 Clocked high-voltage switch (T1–T3) minimizes the dc current load on V_{pp} as compared to a standard depletion device [2.21].

depends on the clock frequency and the bootstrap-to-load capacitor ratio. A high-voltage inverter circuit that is used to provide a 0 to the V_{pp} swing without sinking any dc current from the charge pump and that avoids the use of switching transistors, clock signals, and so on, is shown in Fig. 2.25 [2.22]. When $V_{in} = V_{cc}$, T4 is on and pulls the output (node A) to ground. The width/length (W/L) ratio of T1 is much larger than that of T2, so node B is held at nearly V_{cc}. The threshold voltage of T1, T2, and T3 is -3 V. Hence, with $V_{cc} > 4$ V, $V_{pp} = 22$ V, T3 is off since $V_{gs3} \approx -V_{cc}$ (-4 V min). With T3 off, there is no dc current flowing from V_{pp}. When $V_{in} = 0$, T4 is off, and node B and load capacitor C_L (on node A) are charged to V_{pp} through T2 and T3. Again, there is no dc current flowing from V_{pp} because T1 is also off, since $V_{gs1} = -V_{cc}$. The implementation of the high-voltage inverter into a row NOR decoder driver circuit is shown in Fig. 2.26.

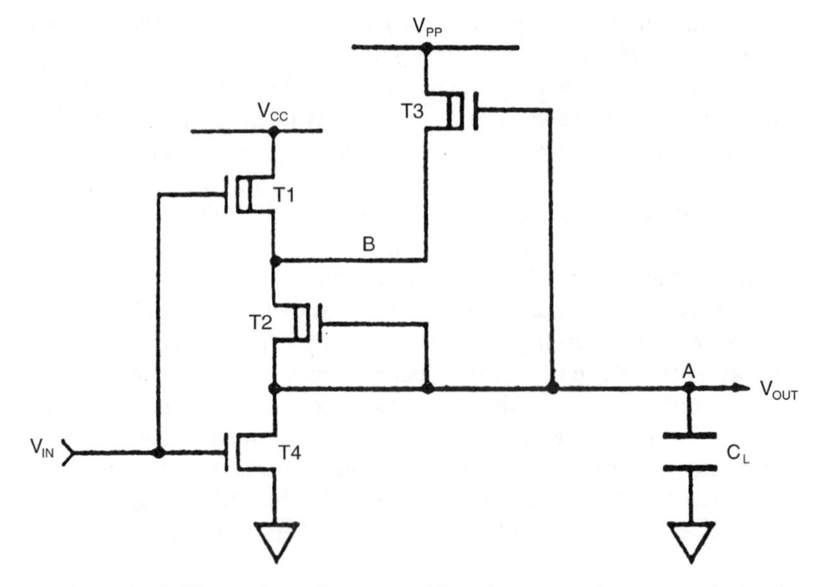

Figure 2.25 High-voltage inverter which draws no dc current from the high-voltage supply V_{pp} [2.22].

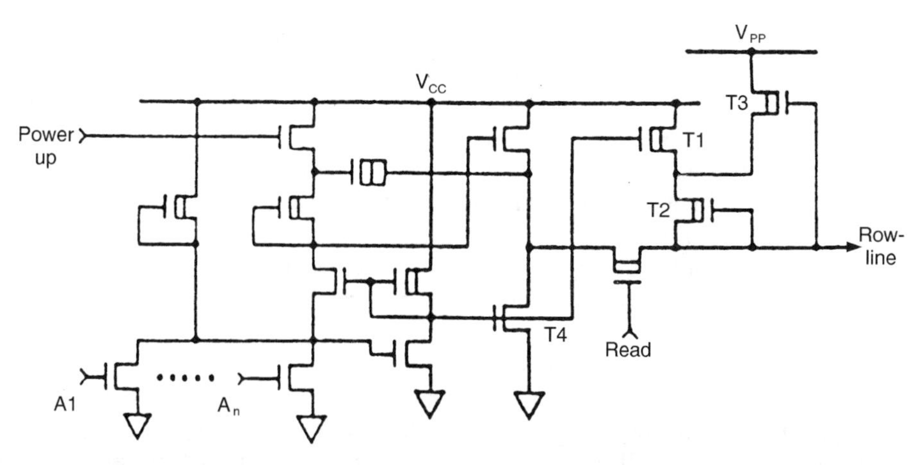

Figure 2.26 Implementation of a high-voltage inverter into a row NOR decoder [2.22].

The limited-current sourcing capability of the charge pump means that small parasitic leakage paths could clamp the output voltage to a level lower than that required for programming operations. This is more of a problem with memory cells in 5 V supply-based custom and ASIC circuits, usually single-poly CMOS technology, where the process is optimized for 5 V logic and leakage paths occur at voltages below the 13 to 15 V needed for programming. Modified design rules for high-voltage circuit layout have been used to avoid critical parasitic leakage paths, punch-through, snapback of short-channel MOSFETs, and turn-on of parasitic field transistors [2.13]. The punch-through voltage can be increased by increasing the minimum spacing between diffusions and increasing the effective channel length. Snapback can be eliminated by biasing the NMOS transistors so that no current flows in the device at high voltages—that is, no gates are switched when high voltages are applied. Parasitic field turn-on can be suppressed by using a poly plane connected to ground or V_{dd} or using the source-drain diffusion of PMOS transistors as a channel stop.

2.2.3 Peripheral Control Circuits

Peripheral circuit blocks include control signal generation and clock generation circuits, decoding circuits, sense amplifiers, and special circuits for improving the EEPROM endurance. Better endurance (in the million cycle range) can be obtained by clever circuit implementation, such as program timer circuits, error correction circuits (ECCs), and cell redundancy.

Timer circuits allow programming operations to be timed internally and are used to correct for cell-to-cell process variations in the memory matrix or tunnel oxide degradation due to charge trapping from repetitive write/erase cycling.

Basically, programming pulses of predetermined duration are repeatedly applied to the cell until the stored data match the input data with adequate threshold margin.

Several methods have been used to increase the memory endurance beyond the 10^5 cycle limit using error correction and memory transistor redundancy. On-chip ECCs have been used to detect and correct single-bit errors or data retention failures and thereby increase endurance to about 10^7 cycles [2.26]. During programming, four parity bits computed using a modified Hamming code are stored along with each byte of data. During subsequent read operations, these parity bits are used to detect and correct single-bit errors. The drawbacks are increased chip area and propagation delays.

An alternative method uses bit-level redundancy in which each memory bit consists of a pair of identical two-transistor cells programmed simultaneously with the same data through two adjacent bitlines, as shown in Fig. 2.27 [2.27]. These cells are read out through two separate sense amps, which, in turn, are combined in a NOR gate to get the final output. By using a diffusion for V_{SS}, two metal lines, BL_1 and BL_2, can be maintained per bit. An array V_{SS} diffusion is shared by four neighboring bytes, two in the X-direction and two in the Y-direction, and is strapped in the middle by an array V_{SS} metal line separating two adjacent byte columns.

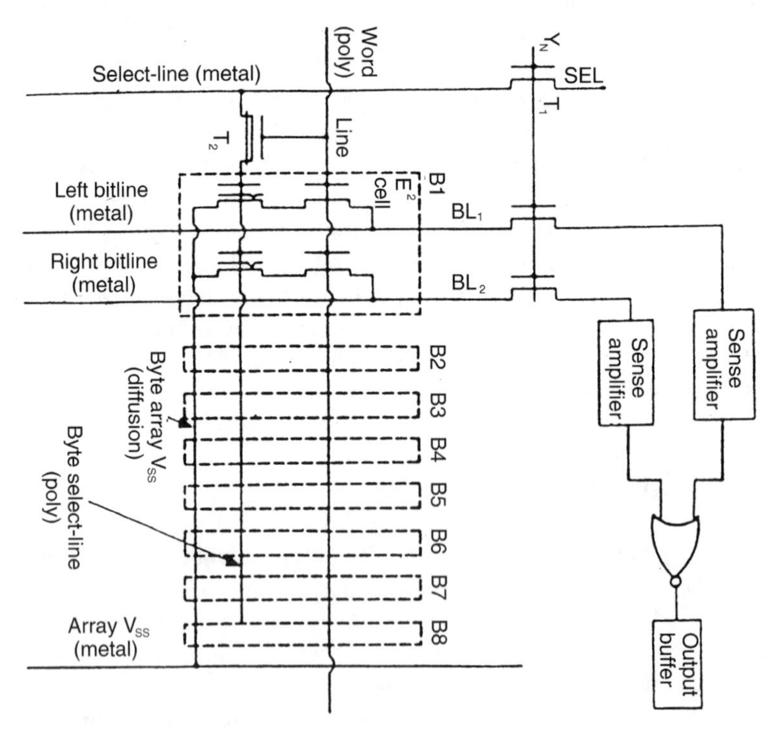

Figure 2.27 Byte organization using the Q-cell concept. The byte has 17.5 metal lines and a common array V_{SS} diffusion [2.27].

Although the erase operation in most cells is self-limiting, whereby the tunnel current decreases as the floating gate gets charged/discharged, it usually continues beyond the neutral point to leave a net positive charge on the floating gate. Erasing without prior writing may even cause the memory cell to become a depletion mode device. At the expense of a larger area, the over-erase problem can be avoided by using a select transistor so that, as long as a cell is readable, a wide variation in the threshold voltage can be tolerated.

In single-transistor memory arrays, this over-erase will cause problems during a read to the column and during subsequent programming operations due to excessive charge-pump loading. One technique used to avoid such problems is to write to the cell first and then erase using an iterative technique, which is a series of partial erase operations followed by measurements and adjustment steps to control the threshold voltage.

Instead of the iterative approach, an "adaptive erase control circuit" has also been used to control the erase operation [2.28, 2.29]. A recent paper suggests the use of feedback techniques to control the programmed threshold voltage of the memory cell [2.30, 2.31]. The basic concept is shown in Fig. 2.28, where V_{pp} is the erase voltage, V_D and V_{FG} are the drain and floating gate voltages, respectively, V_{CGE} is the control gate voltage during erase, and C is the decoupling or programming capacitor assumed to be initially uncharged. The floating gate is charged negatively, resulting in a high-threshold voltage, V_{th}. When V_D becomes high enough (because V_{pp} is raised and no current flows through the transistor), tunnel current I_{tun} flows reducing the floating gate charge and V_{th} decreases. Finally, when the MOSFET is turned on, its channel current becomes high enough to charge the capacitor and $I_{tun} \rightarrow 0$.

The V_{th} measured after erase for cells connected to different programming capacitors is shown in Fig. 2.29 where $V_{pp} = 14.3$ V, $V_{CGE} = V_S = 0$ V was applied. Large programming capacitor values result in a faster V_{th} settling time. Because of capacitive-coupling between the floating gate and drain, when the drain goes high, the floating gate itself tends to rise. Hence, the cell V_{th} at a high-drain voltage is lower than that measured at a low-drain voltage; with $V_{CGE} = V_S = 0$ V during erase, $V_{th} = 4.5$ V at a low-drain voltage is obtained. So, for a target V_{th} at a low-

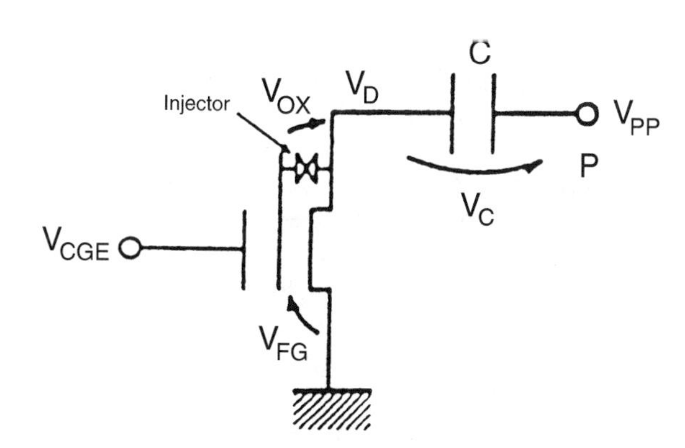

Figure 2.28 Schematic circuit diagram for controlled erase [2.30].

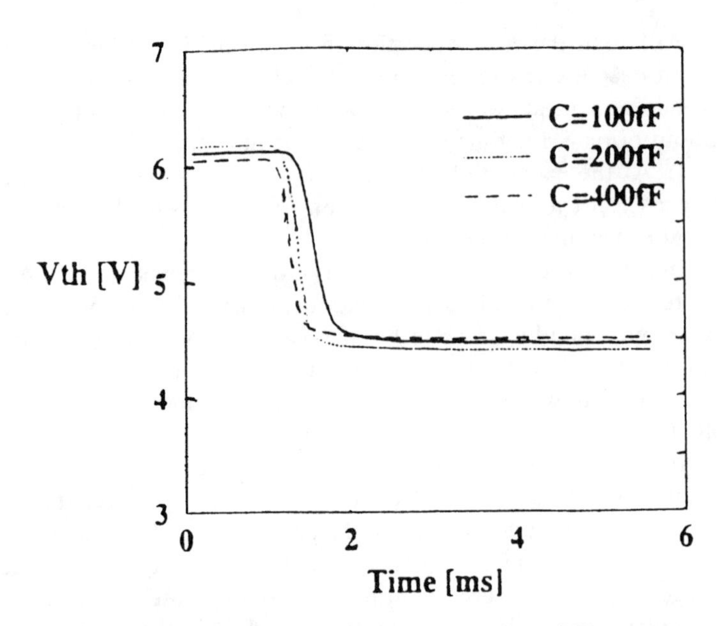

Figure 2.29 Erase characteristics of cells connected to different programming capacitors. The programming pulse applied to the erase circuit has $V_{ppmax} = 14.3\,V$, 5.5 ms duration, and 1.7 ms leading edge duration. The data of this figure have been obtained with $V_{CGE} = V_s = 0\,V$. The final V_{TH} has different values due to the spread in the capacitive-couplings in different cells. The value of the programming capacitor changes the programming dynamics but has no effect on the final V_{TH} [2.31].

drain voltage, the effect of high-drain biasing must be compensated. In particular, a positive source voltage, $V_S \approx 3\,V$, with $V_{CGE} = 0$ was used to achieve $V_{th} = 0\,V$.

2.3. PROCESS TECHNOLOGY

As the previous discussion on EEPROM cell designs shows, as a minimum, a double-polysilicon NMOS process is needed to fabricate high-density memories. The first poly layer is used as the floating gate, and the second layer is used as the control gate. (Since the control gate is connected to wordlines, a polycide layer is generally used to reduce interconnect resistance.) Single-polysilicon cells using diffused control gates require large cell areas and higher programming voltages, so they have only found use in ASIC and other custom applications.

Nonvolatility and long-term storage requires electrical isolation of the floating gate; that is, the process technology should provide high-quality tunnel and interpoly dielectrics. Since on-chip programmability requires the generation and routing of high voltages, the process must provide an effective isolation technology with minimum real estate use, low junction leakage, and high breakdown voltage junctions/ transistors.

Because of the complexity of arraying the cells in a memory matrix, with the associated high-speed peripheral circuits and low power dissipation requirements, memories are fabricated using a double-poly, double-metal CMOS process (typically n-well CMOS) with three oxide thicknesses—a thin tunnel oxide (~ 100 Å) used in memory cells, a thin gate oxide (~ 200 Å) used in the high-speed peripheral circuits, and a thick gate oxide (~ 500 Å) used in the high-voltage transistors. Some recently introduced Flash memories have gone to a triple-well, double-poly, double-metal CMOS process with a 110 Å tunnel oxide, and 200 Å and 300 Å gate oxides for normal and high-voltage transistors. The deep n-well, shown in Fig. 2.30, is used to isolate the p-well when using high negative voltages during erase [2.32].

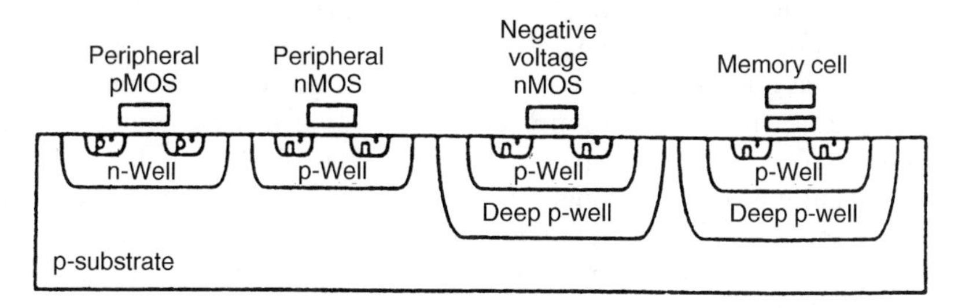

Figure 2.30 Schematic cross-sectional view of the triple-well CMOS structure [2.32].

2.3.1 Ultrathin Tunnel Dielectric Technology

Growth of defect-free ultra-thin (≤ 100 Å) tunnel dielectrics reproducibly, typically SiO_2 and oxynitrides, is necessary for the development of floating gate memories which can operate at low programming voltages and provide reliable long-term storage. The key physical properties of interest include high-breakdown filelds, a low electron trap density for high-cell endurance, and low leakage. Oxide quality is typically measured as the total injected charge per unit area (C/cm^2) required to cause breakdown, referred to as Q_{bd}, and the change in interface trap density, D_{it}, with injected charge. These topics are again discussed in Section 2.4 on oxide degradation and breakdown mechanisms.

A concise review of thin oxide growth technology has been given by Wolf [2.33]. Relatively slow growth rates must be used to reproducibly grow thin oxide films. A number of approaches have been reported, including dry oxidation at atmospheric pressures and relatively low temperatures (800–900°C), dry oxidation with HCl, oxidation at reduced total pressures, oxidation at reduced partial pressures of O_2 (such as the use of 5 or 10% O_2/N_2 ambient), and use of composite oxide films (i.e., using gate oxide films consisting of thermally grown SiO_2 and an overlay of CVD SiO_2).

Factors adversely affecting oxide thickness uniformity include pre-oxidation cleaning effects, residual water content of the O_2 gas in a dry-oxidation process, silicon surface orientation, the presence of native oxide films prior to gate oxide growth, processing conditions, and oxidation during ramp up/ramp down. For dry O_2 oxidation, the trace amount of water should be tightly controlled since a relatively small water concentration in O_2 is sufficient to significantly increase the oxidation rate. It should be kept in mind that the oxidation rate is lower at lower temperatures and pressures, and that oxide density increases as the oxidation temperature is reduced.

Thin oxides are typically grown at atmospheric pressure from 850 to 900°C, in dry O_2 with Cl containing gases, such as HCl, to passivate ionic sodium and improve the breakdown voltage. Instead of using pure O_2, thin oxides have been successfully grown in ozone in order to obtain improved breakdown characteristics and a low surface state density [2.34]. Dry O_2-grown oxides have a high tolerance to hot-carrier degradation but poor breakdown characteristics because of a large hole trap density due to oxygen vacancies. Wet oxides exhibit good breakdown characteristics, but poor hot-carrier degradation because of hydrogen-related bonds. Oxidation in an ozone ambient reduces trap concentrations, and all silicon dangling bonds are tied to oxygen atoms. During oxidation, oxygen radicals, provided by the ozone molecules, tie up the silicon dangling bonds. Figure 2.31*a* shows the D_{it}, after a 1 C/cm^2 stress, for the various oxidation methods: a 7 nm thick oxide grown in ultra-dry O_2, ozone, and HCl/O_2 ambients. The D_{it} of ozone-grown oxides are seen to be lower than those of the other groups. HCl-grown oxides contain Si–H and Si–Cl bonds of binding energies 318 kJ/mole and 397 kJ/mole, respectively, whereas the Si–O binding energy is 622 kJ/mole. Figure 2.31*b* shows the improvement in Q_{bd} for ozone-grown oxides.

A recent study has found that oxidation which proceeds in competition with reduction by active hydrogen (like the hydrogen radical H^+) results in better oxide quality [2.35]. When the oxide layer is formed, mechanical stresses at the interface result in imperfections in the Si–O network, giving rise to weak spots. Crystal defects, surface contaminants, or impurities can also cause such oxide defects. Unlike conventional oxidation, in which only oxidation occurs, the addition of hydrogen radicals reduces the number of weak Si–O bonds. The H^+/H_2O oxides were formed in an ultra-clean $H^+/H_2O/Ar$ ambient at 900°C. However, the H and H_2O remaining in the H^+/H_2O oxide can lead to significant hot-carrier trapping. A post-oxidation anneal step (discussed next) in Ar gas at 900°C was necessary to reduce the electron-hole traps. The breakdown field of H^+/H_2O oxide was about 1 MV/cm higher than that of dry oxides. The improvement in Q_{bd} of the H^+/H_2O oxides, measured by constant-current stressing, is shown in Fig. 2.32.

Post-oxidation anneal (POA) is typically carried out in an inert N_2 or Ar ambient at temperatures of about 1000°C, the purpose being to relieve oxide stress. Some oxygen in the N_2 ambient is thought to bridge nonbridging, positively charged oxygen atoms and to increase the probability of oxidizing free silicon trivalent atoms at the Si–SiO_2 interface. Furthermore, the nitrogen atoms are thought to terminate dangling bonds at the silicon surface by forming Si–N bonds.

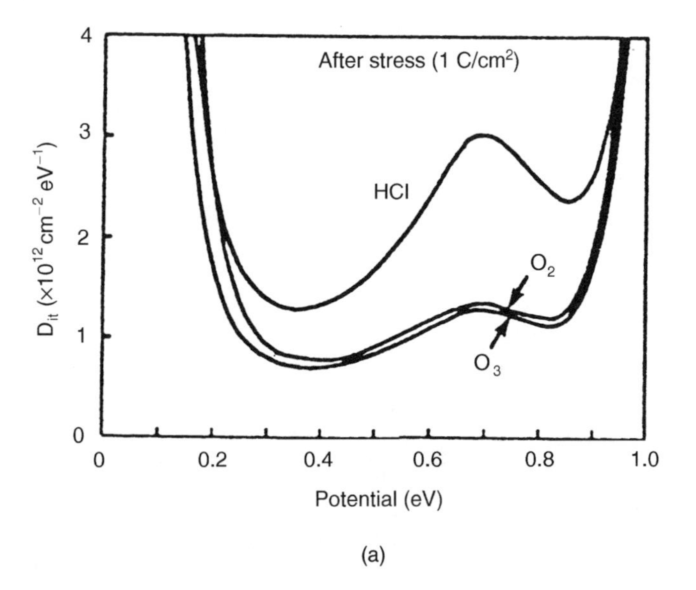

(a)

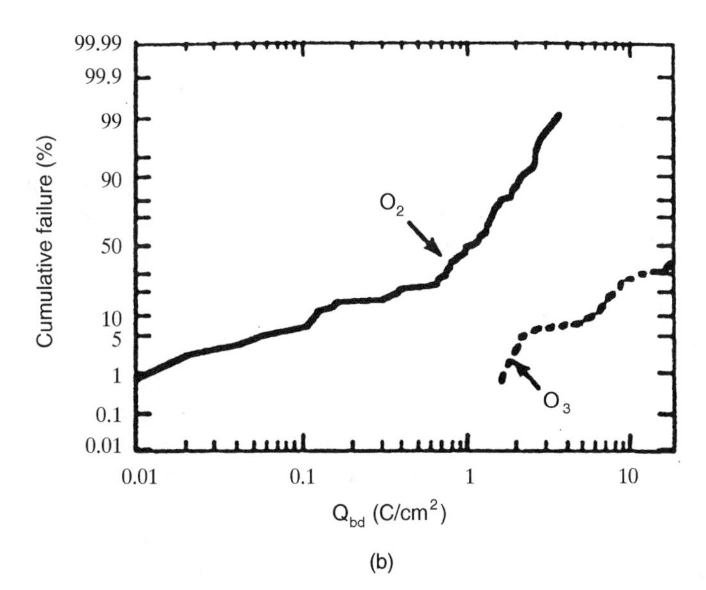

(b)

Figure 2.31 Characteristics of thin oxides grown in ozone: (*a*) surface state density, D_{it}, of 7 nm thick oxides after 1 C/cm^2 stress, (*b*) charge-to-breakdown characteristics of 4 nm oxides under constant 0.1 A/cm^2 stress [2.34].

Thin thermal oxides with an electron barrier height, $\Phi_{bn} < 3.2$ eV, can be obtained by growing oxides on heavily doped silicon. Effective Φ_{bn} values of ~ 1.7 eV have been obtained at the tunnel oxide interface by using phosphorous or arsenic implantation, followed by oxidation in oxygen diluted with Ar at 900°C

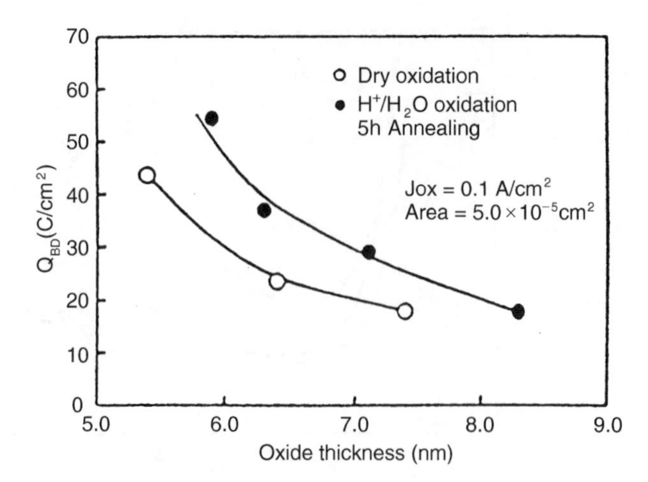

Figure 2.32 Hydrogen radical balanced steam oxidation: dependence of Q_{bd} at 50% cumulative failure on oxide thickness, under constant $0.1\,A/cm^2$ stress, gate area is $5E\text{-}5\,cm^2$. The tunneling injection technique using an n^+ polysilicon gate n-MOSFET structure was performed [2.35].

[2.36]. The variation of Φ_{bn} with implant specie type, dose, and anneal conditions is shown in Fig. 2.33*a*. The lowest barrier height is achieved using an As implant followed by a wet oxidation. However, the oxide breakdown field for wet oxidation is lower than that for dry oxidation and decreases with increasing dose, as shown in Fig. 2.33*b*. In EEPROM cells, phosphorous and arsenic implants are commonly used to create a double-diffused region, followed by tunnel oxide growth.

Instead of using thermal oxides as the tunnel dielectric, most modern EEPROMs use ultra-thin nitrided oxides to improve cell endurance and reliability [2.29, 2.37]. This is because, in oxynitrides, a higher fraction of the Si dangling bonds are terminated by Si–N bonds as opposed to weaker Si–H bonds.

Thermal nitridation of oxide in NH_3 results in nitrogen atom incorporation into the SiO_2 lattice and Si_3N_4 growth at the silicon–oxide interface, but the presence of hydrogen in NH_3 introduces a fixed-charge density and electron traps into the dielectric. Oxynitridation using N_2O has been studied as an alternative process to avoid incorporating H as a potential charge-trapping site. Both conventional furnace oxidation [2.38] and rapid thermal processing (RTP) have yielded high-quality dielectrics [2.39, 2.40]. RTP oxynitridation is typically carried out in a pure N_2O (99.998%) ambient at 1200°C for 5 to 20 seconds in a RTP chamber. Under constant-current injection and hot-carrier stressing, oxynitrides show an improved Q_{bd}, smaller D_{it} shift, and negligible subthreshold degradation compared to control thermal oxides.

Recent papers describe using nitrous oxide (NO) for a lower thermal budget compared to the N_2O process. Oxynitride films have been grown in NO in a conventional furnace using a lower thermal budget, with the films having carrier immu-

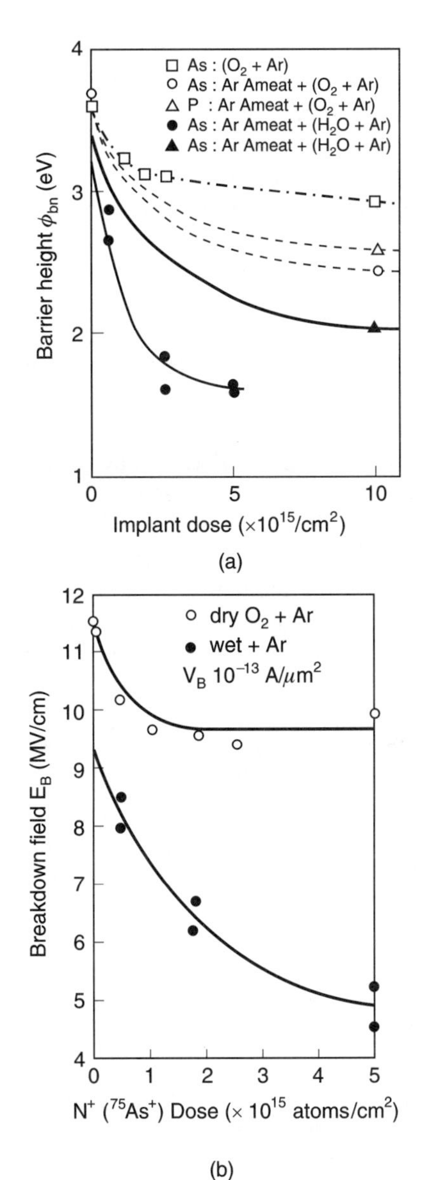

Figure 2.33 Thermal oxides grown on heavily doped silicon: (*a*) barrier height for several kinds of oxidation processing as a function of implanted impurity dose, and (*b*) breakdown field of MOS capacitors for two kinds of oxidation atmosphere, dry oxygen and steam diluted with Ar, as a function of ion-implantation dose [2.36].

nity similar to N_2O-grown films [2.41, 2.42]. Q_{bd} and hot-carrier immunity of NO oxynitride is found to be similar to N_2O oxynitrides.

Silicon surface preparation becomes extremely critical for high-quality thermal oxides in the 50 Å thickness range. The surface needs to be free of native oxide, microscopic reactive impurities, and particles. Native oxides contain more pinholes and defects than thermal oxides and will degrade the final quality of a thermally grown oxide. Surface micro-roughness at the Si–SiO_2 interface correlates with lower breakdown fields, lower Q_{bd}, and increased leakage currents. In addition, studies have shown that the presence of a thin oxide layer prior to carrying out an oxidation is critical for obtaining good oxide quality. This fact appears to be due to roughening of the silicon surface. This surface micro-roughness has been attributed to the etching of the silicon surface by trace O_2 present during temperature ramps in inert ambients. A comparison of oxide quality for various pre-cleans and ambient conditions during ramp-up supports these observations [2.43].

A recent study shows that surface micro-roughness can be reduced by optimizing the NH_4OH content in the cleaning solution NH_4OH-H_2O_2-H_2O; the suggested ratio was (0.05:1:5) instead of the conventional (1:1:5) [2.44]. After a native oxide etch by buffered HF (BHF:$NH_4F + HF + H_2$) with surfactants, the wafers were cleaned in $NH_4OH + H_2O_2 + H_2O$. Another key step was the introduction of a room-temperature DI water rinse prior to a hot DI water rinse. The reduced NH_4OH content did not, however, result in a reduction in metallic contaminant removal or particle removal efficiency.

Surface micro-roughness has also been found to correlate with silicon vacancy concentration [2.45]. A comparison of czochralski (CZ), float zone (FZ), and epitaxial wafers shows that epi-wafers have better breakdown characteristics, lower surface micro-pitting during temperature ramps, and lower Dit. It has been found that CZ wafer quality can be significantly improved by hydrogen denuding; this is achieved by subjecting the wafers to a short hydrogen denudation pre-process step at temperatures between 1050°C and 1200°C for 15 to 30 minutes. Junction leakage was also observed to decrease significantly for H_2 annealed wafers relative to bulk non-epi CZ wafers, resulting in bulk CZ wafers with surface properties similar to epi-wafers without the added cost [2.46].

For better surface passivation, a proposed cleaning method uses a two-dip step (aqueous HF and methanol/HF) after a standard RCA clean and without a final DI rinse [2.47]. The optimum methanol/HF dip time is found to be about 30 seconds. This method is believed to provide more hydrogen passivation on the Si surface due to reduced H_2O species in the methanol/HF solution. The absence of a final DI water rinse prevents native oxide growth and surface micro-roughness. A hydrogen-passivated surface with a low-metal and particle density has been achieved with a modified RCA clean followed by a dip in 0.5% HF + 0.1% IPA + 0.1% chloroacetic acid for 5 minutes [2.48].

In some EEPROM cell designs, silicon surface roughness has been used to an advantage. A textured silicon surface has been obtained by various processing techniques to yield low programming voltages by field enhancement at surface asperities. However, surface roughness typically results in lower quality thermal oxides subsequently grown on the surface, and thicker tunnel oxides have to be grown to minimize leakage currents.

Texturing of single-crystal silicon has been carried out in a plasma etcher using a gas mixture of $CCl_4/O_2/H_2$ [2.49]. Several factors affect surface roughness—O_2 content, the presence of sacrificial oxide, and silicon doping. The I-V characteristics for various process conditions are shown in Fig. 2.34. Etching through a 170 Å sacrificial oxide increased the surface roughness because of preferential etching of silicon over oxide; that is, any nonuniformity, when etching the oxide, is magnified when the silicon is etched.

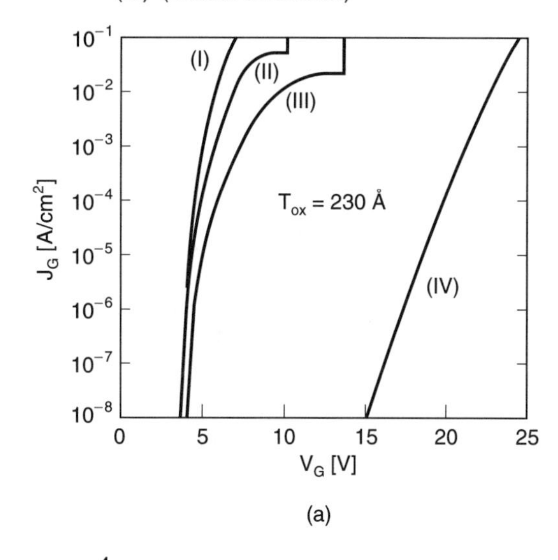

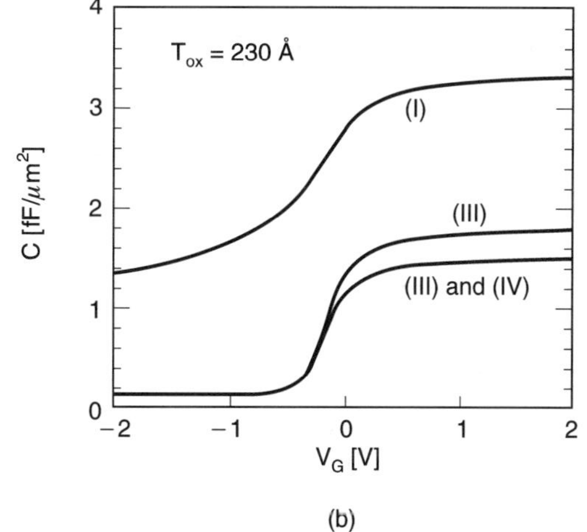

Figure 2.34 (*a*) I-V and (*b*) high-frequency C-V characteristics showing the effects of etching the surface through a sacrificial oxide (I) with and (II) without an As implant. Also shown are (III) results from etching the bare silicon and (IV) the characteristics of a normal oxide [2.49].

A two-stage plasma etch process has been used to control the silicon surface roughness [2.50]. The wafers are etched in a reactive ion etch (RIE) system for 30 seconds in Cl_2 at a pressure of 20 Pa to produce an extremely rough surface, followed by an isotropic plasma etch in SF_6 at 30 Pa to reduce the surface roughness. After the Cl_2 etch, columnar structures roughly 2000 Å high and 400 Å wide are formed. Various reasons have been given for this phenomenon, namely, residual moisture in the vacuum system, possible polymer nucleation on the silicon surface retarding the etching reaction locally, or a combination of residual oxide on the silicon acting as a micromask and the high selectivity of the silicon/oxide etch.

Another technique is the oxidation of thin polysilicon films on single-crystal silicon [2.51]. Thin amorphous silicon film ~ 100 Å thick was LPCVD deposited on silicon wafers and loaded into a furnace. During the temperature ramp, the amorphous films crystallized to polysilicon. The thin tunnel oxide was grown by completely oxidizing the thin poly film in a dry O_2 ambient. Due to the rapid diffusion of O_2 through the grain boundaries of the polysilicon film into the silicon substrate and an enhanced oxidation rate at grain boundaries, a textured Si–SiO_2 interface is formed.

2.3.1.1 Floating Gate Doping and Etching. Doping, annealing, and etching of the polysilicon floating gate strongly affect the barrier height and leakage properties of a memory cell. Degenerate doping of the floating gate polysilicon appears to increase the electron injection barrier height at the floating gate–tunnel oxide interface by up to 250 meV, measured by dark current–voltage characteristics [2.52]. This has been attributed to band-gap narrowing effects. Other studies show that incorporation of dopant atoms from the polysilicon into the surrounding oxide results in a barrier height reduction due to a reduction in the oxide band gap at the interface.

High-temperature anneals after poly deposition result in degradation of tunnel oxide quality. This has been attributed to an increase in roughness at the poly–tunnel oxide interface from the growth of poly-grains. Stress caused by grain growth and oxidation is relieved by deformation of the tunnel oxide due to viscous flow at the anneal temperature, which can result in extremely thin tunnel oxide areas being susceptible to early failure [2.53].

Tunnel dielectrics and gate oxides can be damaged by plasma etching processes [2.54]. Nonuniformities in the plasma across the wafer surface can cause local charge buildup, which can damage the tunnel dielectric. Methods of reducing oxide damage from dry etching processes include modification to the reactor design and operation, and clever layout techniques to minimize charging (antenna effect). A recent study describes the use of an electron cyclotron resonance (ECR) plasma etching technique for reducing gate charging during poly patterning [2.55].

2.3.2 Interpoly Dielectrics

In planar floating gate cells, interpoly dielectrics are used to insulate the floating gate from other electrodes and hence, should be defect-free to prevent charge leakage from the floating gate. Since the floating gate is a polysilicon layer, it is commonly oxidized in conventional processing technologies. Polyoxides can be thermally grown at 800°C in a mixture of pyrogenic steam, dry oxygen, and anhydrous HCl [2.56]. Oxidization of polysilicon modifies the surface topology due to enhanced oxidation at grain boundaries, forming interface protuberances and inclusions [2.57]. The localized field enhancement at these surface nonuniformities lowers the effective interface barrier height, resulting in higher leakage currents, which is a drawback as far as insulating properties are concerned. The extent of field enhancement is dependent on the polysilicon doping process, oxidation temperature, ambient, and polysilicon deposition temperature [2.56, 2.58–2.63]. Incorporation of fluorine into polyoxide appears to improve leakage currents and breakdown strengths [2.64, 2.65], possibly due to a reduction of high stress in the polyoxide by fluorine incorporation, or a reduction in the interface traps by the formation of stronger Si–F bonds in place of Si–H/Si–OH bonds.

Scaling of the interpoly dielectric thickness in high-density memories has led to the use of multiple dielectric stacks, such as oxide–nitride–oxide (ONO), instead of polyoxide for better leakage and breakdown properties [2.66]. Typically, a 50 to 100 Å bottom oxide in the ONO stack is grown by partial pressure, 800 to 850°C dry thermal oxidation or N_2 diluted oxidation of the phosphorous-doped floating gate. Next, a 200 Å CVD silicon nitride is deposited. Lastly, a 30 Å top oxide is formed by a 920°C steam oxidation. One advantage of the ONO stack is that electrons leaking from the floating gate get trapped at the nitride interface, building up an electric field that opposes further charge loss [2.67]. A high-performance, superthin, ONO stacked dielectric has been achieved by oxidizing thin nitride films (LPCVD nitride deposited at 750°C) in low-pressure dry O_2 (0.5 Torr) at 850°C for 30 minutes [2.68].

2.3.3 High-Voltage Technology

Floating gate EEPROMs require high voltages (~ 13 V) for programming operations, which, in present technology, is generated by on-chip charge pumps. With the constant requirement for high-density memories, and correspondingly scaled geometries, on-chip generation and routing of high voltages necessary for programming memory cells pose two basic problems: (1) small parasitic leakage currents can clamp the charge-pump output voltage to low values, making the memory inoperable, and (2) unless carefully laid out, the higher voltages can cause problems with operation of the standard logic present in the memory chip.

In Section 2.2, the problems associated with parasitic leakage paths were discussed, namely, punch-through, snapback, and turn-on of field transistor. The high-voltage design rules given there can easily increase the chip area by more than 20%, which is unacceptable for high-density memories.

Conventional LOCOS isolation with ~ 8500 Å thick field oxide and modified LOCOS processes have been used in 4 Mbit memories [2.5, 2.7]. Effects like bird's beak punch-through and field oxide thinning limit the field oxide thickness (to about 6000 Å) and minimum geometry (to about 0.8 μm) achievable with LOCOS-based technology. Trench and oxide refill technologies, currently used in DRAMs, will also likely be applied to high-density EEPROMs [2.69, 2.70]. Here, deep trenches are etched in the silicon substrate and refilled with an oxide film. Channel stop boron ions are also implanted into the trench side-wall to increase the field turn-on voltage.

Typical processes use at least two gate oxide thicknesses, one for high-speed logic devices (about 150 Å) and another for the high-voltage (HV) devices (about 500 Å to avoid gate oxide breakdown). In order to obtain a high junction breakdown voltage, source-drain junctions of HV transistors are typically formed by a double diffusion. For n-channel field effect transistors (FETs), the n^+ diffusion is surrounded by a deep n$^-$ diffusion. The n$^-$ drift region increases the on-resistance of the transistor.

Increased drift region sheet rho increases the "on-resistance" and breakdown voltage (V_{BD}). A decrease in drift region length decreases V_{BD}. For short channel lengths, V_{BD} is dominated by punch-through. A reduced channel length also results in reduced parasitic bipolar base width, which results in higher gain and increased susceptibility to snapback.

Surface and subsurface punch-through can be suppressed by use of increased channel length, increased minimum spacing of diffused lines, and punch-through stop implants. The tilt angle implanted punch-through stopper (TIPS) process sequence, shown in Fig. 2.35, has been proposed for deep submicron MOSFETS to realize punch-through protection with a minimum area penalty and a lower bottom junction capacitance [2.71]. Snapback can be eliminated by ensuring that negligibly small currents flow through the transistors at high voltages. Fortunately, on-chip charge pumps generate high voltages by increasing the output impedance; hence, they cannot supply large currents needed for initiating snapback.

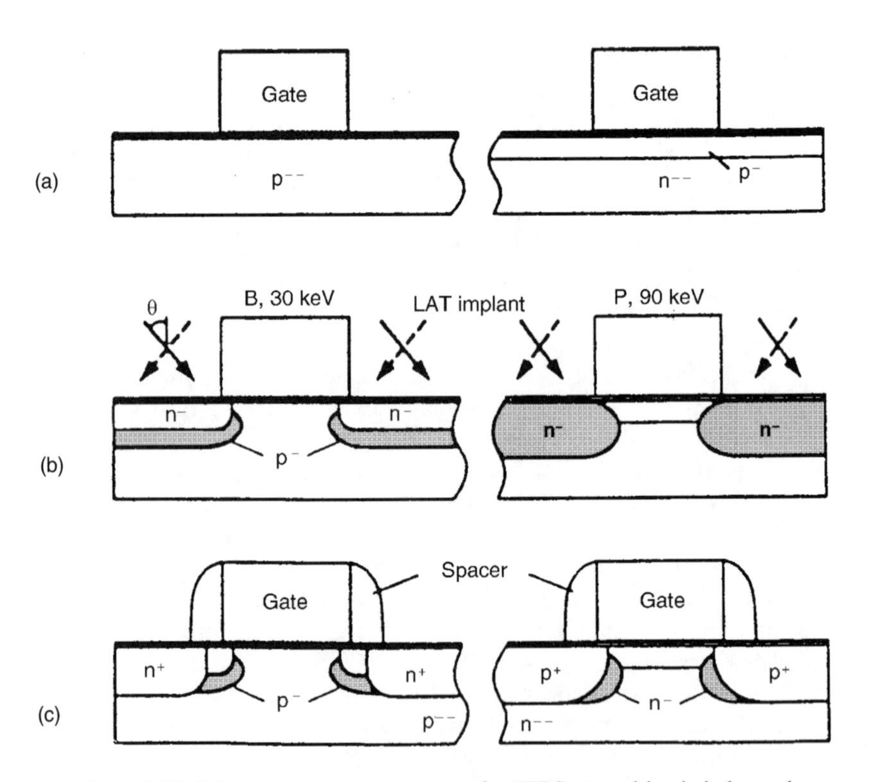

Figure 2.35 Schematic process sequence for TIPS n$^-$ and buried channel p-FETs: (*a*) gate definition, (*b*) LAT implant to form TIPS structures (after n-LDD implant for n-FET), and (*c*) source-drain implant after side-wall spacer formation [2.71].

2.3.4 EEPROM Scaling Considerations

MOS scaling is an approach to minimization whereby the physical dimensions of the device are reduced by a "scaling factor κ." To keep the electric fields in the device constant, the operating voltages as well as the physical dimensions are scaled equally. Thus, current and capacitance are scaled by $1/\kappa$, packing density increases by κ^2, power dissipation decreases by $1/\kappa^2$, and so on. These results fail to hold as the device gets smaller, and secondary effects, like short and narrow channel effects, ballistic transport, and band-gap narrowing, affect the device characteristics. In addition, if the terminal voltages cannot be scaled, then high electric fields will be present, which means that phenomena such as hot-electron effects and velocity saturation have to be taken into account to predict device behavior and reliability.

Scaling effects on EEPROM cell reliability and performance have been investigated using the scaling factor κ for lateral dimensions length, L, and width, W, and the junction depth, x_j [2.72, 2.73]. Scaling ratios for interpoly and tunnel dielectric thickness were derived as $\kappa^{-0.5}$ and $\kappa^{-0.25}$, respectively.

For write/erase by electron tunneling, voltages should be reduced by $\kappa^{-0.25}$ to keep the tunneling field around 11 MV/cm. In cells using hot-electron injection, the maximum channel electric field, E_{mp}, should be maintained as in previous generations of devices because hot-electron generation varies exponentially with $-E_{mp}^{-1}$; $E_{mp} \propto (V_{dp} - V_{dpsat})/(T_{gox} * x_j)^{0.33}$, where T_{gox} is the gate oxide thickness and V_{dp} is the program drain voltage, so V_{dp} scales as $1/k^{0.42}$. Read-disturb has the same channel electric field dependence; hence, the drain read voltage, V_{dr}, should also be scaled as $1/k^{0.42}$.

Oxide–nitride–oxide (ONO) interpoly dielectric thickness scaling guidelines have been established on the basis of charge retention and threshold voltage stability [2.74]. The bottom oxide was thermally grown on p-doped polysilicon using the diluted oxygen method, the SiN layer was deposited by LPCVD, and the top oxide was formed by pyrogenic oxidation. Charge retention was evaluated by a bake test at 300°C for 20 hours. The threshold voltage instability due to charge injected into the ONO dielectric was evaluated by performing a UV erase after the first bake and then re-baking at the same condition. A bottom oxide thickness of about 4 nm is necessary to prevent electron injection into the ONO film, which causes V_{th} instability. The effect of bottom oxide scaling is strongly dependent on the top oxide thickness. Scaling of the bottom oxide thickness down to 5 nm does not lead to a marked degradation of the charge retention characteristics when the top oxide thickness is 3 to 4 nm, which is thick enough to block hole injection from the control gate. A top oxide thickness of about 6 or 7 nm can minimize degradation due to defects. The nitride scaling is mainly dominated by oxidation resistance. The thickness should be around 4–5 nm. Therefore, effective oxide thickness scaling to around 13 nm (bottom/SiN/top = 4.5/4-5/6 nm) should be possible. If a V_{th} instability of 0.2–0.3 V is allowable for actual device operation, scaling down to about 10 nm is possible because a bottom oxide free structure can be adopted.

The above derivation indicates that dielectric fields and maximum channel fields during program/read operations can be kept comparable to previous generation devices. The following considerations indicate that scaling has a physical limit. Because of direct tunneling and an increase in stress-induced leakage currents, tunnel oxides cannot be scaled much below 60 Å, while yield considerations limit oxide thickness to 80 Å. The electron barrier height at the Si–SiO_2 interface of about 3.2 eV prevents significant reduction in the programming voltages. Without a reduction in tunnel oxide thickness, the erase voltage V_{pp} cannot be reduced. Since operating voltages cannot be reduced, in order to prevent high-voltage breakdown/punch-through and parasitic FET turn-on, the minimum channel length and isolation spacing between devices cannot be reduced [2.75].

2.4. DEGRADATION MECHANISMS

Charge injection through dielectrics, whether by tunneling or hot-carrier injection, causes degradation due to broken bonds, generation of interface and bulk traps, and charge trapping, eventually leading to breakdown. (Breakdown is defined as a sudden drop in the dc-resistance of the dielectric.) Dielectric degradation can be measured as the change in the threshold voltage, transconductance, interface state density, or frequency response of MOS capacitor or transistor structures [capacitance–voltage (C–V) or conductance–voltage (G–V)]. Both degradation and breakdown depend on the injected current density, electric field stress, duration of the stress, and total amount of charge injected.

In floating gate memories, electron trapping in the tunnel oxide builds up a permanent negative charge, thereby reducing the electric field and the injected tunnel current for the same applied terminal voltages. For a constant program voltage or program time, this reduces the programmed threshold voltage of the cell, resulting in the threshold window closure problem, as shown in Fig. 2.36. Significant window closure typically occurs after about 10^5 cycles in FLOTOX cells. However, the life can be extended into the 10^6 range by using error correction codes (ECCs) and increasing the sensitivity of the sense amps to detect smaller voltage shifts.

Retention failure refers to the inability of a memory cell to retain the stored charge over a wide range of temperature variations and operating voltages. Floating gate retention failures are due primarily to oxide defects which may greatly reduce the interface barrier height, allowing electrons to leak off [2.76, 2.77]. Possible data-loss mechanisms include electron transport to and from the floating gate through oxide defects, compensation of stored charge by ionic contamination in tunnel/interpoly dielectric, ionic contamination through passivation defects, and intrinsic mechanisms that cause even nondefective cells to lose charge over time. Apart from these process-related mechanisms, voltages applied during programming and read operations may disturb the stored charges in the cell, resulting in retention failure [2.17, 2.78].

Low-field leakage current-induced by high-field stress has been observed in thin oxides (< 100 Å) [2.80–2.83]. Typical characteristics for 65 Å to 130 Å oxides are shown in Fig. 2.37. Positive charge generation and charge-assisted electron tunneling was postulated by Maserjian [2.80]. Olivio et al. [2.82] conclude that leakage current

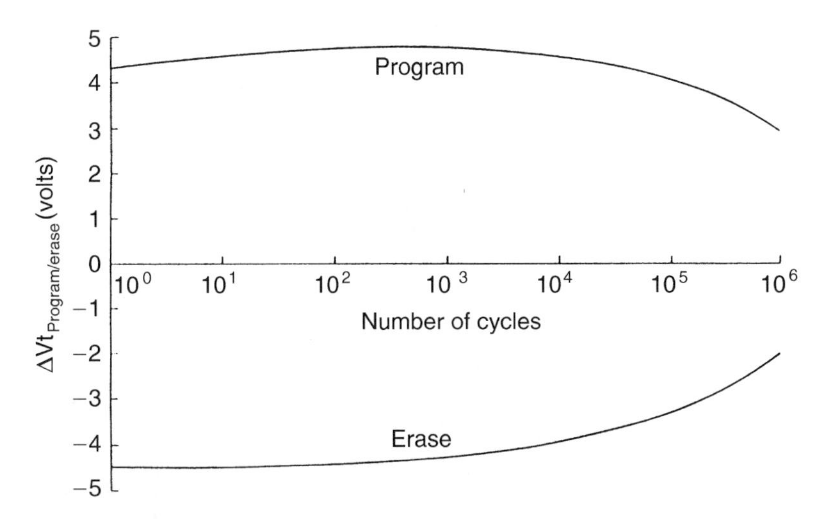

Figure 2.36 Change in write and erase threshold voltage window as a function of the number of write/erase cycles from experiments. The window closing after 10^4 cycles was due to electron trapping in the tunnel oxide [2.109].

originates not from positive charge generation, but from thermally assisted tunneling through a locally reduced injection barrier resulting from defect-related weak spots where the insulator has experienced significant deterioration from electrical stress. Recent studies using oxides with thicknesses ranging from 65 to 130 Å show trap-assisted tunneling to be the possible low-field leakage mechanism [2.83]. In oxides thicker than 100 Å, stress-induced current is seen to decay as traps are filled without significant tunneling out of traps; in thinner oxides, steady-state current flows when there is equilibrium between trap filling and emptying processes.

In actual EEPROMs, the thin dielectric is stressed by pulsed alternating polarity voltages during write and erase operations. Several studies have concluded that the bipolarity stress results in significantly higher Q_{bd} and time-to-breakdown (T_{bd}), compared to unipolarity/DC stressing [2.84–2.86]. The dynamic stress reduces localized hole trapping in weak oxide areas (see the section on oxide breakdown), due to significant hole detrapping, resulting in higher Q_{bd} and T_{bd}.

Oxide degradation from channel hot-electron injection causes a reduction in programming speed and transconductance of the memory cell with write/erase cycles, which results in lower endurance. Charge-pumping and transconductance measurements were made as a function of stressing time to determine the nature of the degradation mechanism [2.87]. The reduction in electron injection into the floating gate is found to be caused mainly by interface states located in the drain overlap region, which are created during the initial step of the write operation, not by charges trapped in the oxide. Reduction in the transconductance is caused mainly by interface states located around the drain edge. These states are created during the final step of the write operation.

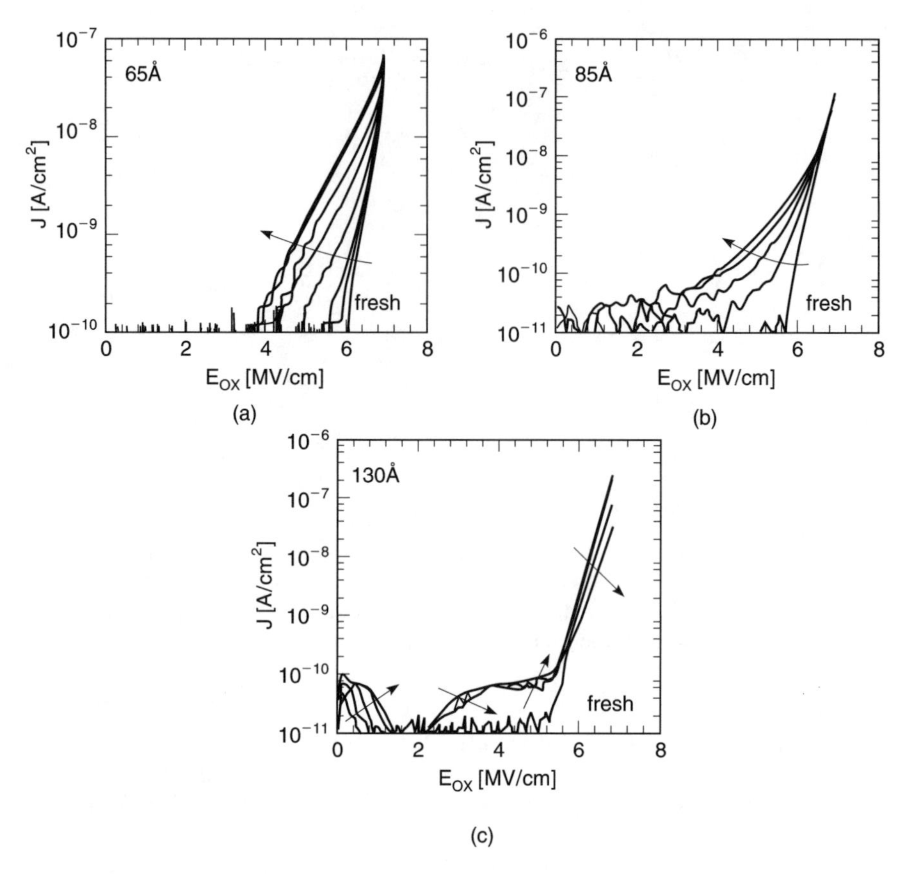

Figure 2.37 Conduction characteristics following Fowler–Nordheim stressing of (*a*) 65 Å, (*b*) 85 Å, and (*c*) 130 Å oxides [2.83]. The capacitors were stressed with a constant-voltage equivalent to 9.5 MV/cm until near breakdown (total injected charge > 20 C/cm^2), and I-V is measured between 0 and 7 MV/cm. Arrows highlight the evolution of the current-voltage characteristic during stress. The initial current observed below 1 MV/cm is attributed to filling of slow interface states generated during stress [2.83].

2.4.1 Electron Capture and Emission

The electron capture process in thermal oxides can be described by first-order kinetics [2.88, 2.89]. Assuming no re-emission of trapped electrons, the rate equation governing the electron trapping process is

$$\frac{dn_{tp}}{dt} = \frac{J(t)}{q_e} \sigma_p \left(N_0 - n_{tp}\right) \tag{2.1}$$

where n_{tp} is the trapped electron density, N_0 is the density of pre-stress deep trap centers, σ_p is the cross section of these centers, and J is the injected current density at time t. New trap centers can also be generated due to stressing, and assuming a constant trap generation rate g (probability of trap creation per electron flow), trapping at these sites is usually modeled by a first-order rate equation

$$\frac{dn_{tg}}{dt} = \frac{J(t)}{q_e}\sigma_g\left[g\int_0^t \frac{J(t)}{q_e}dt - n_{tg}\right] \quad (2.2)$$

Interface traps generated by electron injection into the oxide can be described by the following equation [2.90], which shows that the generation rate saturates for high injection levels. Since the cross section of interface traps is much larger than that of bulk traps, trapping at these sites dominates during the initial stages of injection,

$$\Delta D_{it} = D_{it}^{sat}\frac{\sigma_{it}N_{inj}}{1+\sigma_{it}N_{inj}} \quad (2.3)$$

where ΔD_{it} is the change in interface trap density, D_{it}^{sat} is the saturation value of interface trap density obtained from charge-pumping experiments, σ_{it} is the generation cross section, and N_{inj} is the number of electrons injected.

Detrapping or removal of electrons from traps can occur by several mechanisms—impact-ionization, Frenkel–Poole emission, or tunneling. Detrapping by impact-ionization, that is, collision of free electrons from the conduction band with filled traps, is given by the following rate equation [2.91],

$$\frac{\partial n_t}{\partial t} = -\frac{J(t)\beta n_t}{q_e} \quad (2.4)$$

where n_t is the trapped electron density and β is the field-dependent probability of a passing electron performing an ionizing collision. Frenkel–Poole emission or field-assisted thermal excitation of electrons from traps into the conduction band of the insulator is highly probable only for shallow traps. The electron emission rate $e_n(T)$ at temperature T from a trap at energy E has the form [2.92]

$$e_n(E,T) = \nu(T)e^{(E-E_c)/kT} \quad (2.5)$$

where E_c is the conduction band edge and the pre-exponential factor ν gives the frequency of escape attempts.

Because of the wave nature of electrons, trapped electrons can cross the oxide barrier by tunneling. Factors influencing probability for electron tunneling are — trap depth, applied electric field, and barrier width (distance between trap and anode). Triangular barrier (Fowler–Nordheim tunneling) or trapezoidal barrier (direct tunneling) models may be used to calculate tunneling current I-V characteristics. At high-applied electric fields, electron tunneling occurs from trap to oxide conduction band through a triangular energy barrier, with current density J given by

$$J = AE_{ox}\exp(-B/E_{ox})$$
$$A = \frac{q^3 m}{16\pi^2 \hbar m_{ox}\phi_b} = 1.54 \times 10^{-6}\left(\frac{m}{m_{ox}}\right)\frac{1}{\phi_b} \tag{2.6}$$
$$B = \frac{4(2m_{ox})^{1/2}}{3q\hbar}\phi_b^{3/2}$$

where, q, m, m_{ox}, ϕ_b are electron charge, free space mass, oxide effective mass, and barrier height respectively. At low fields, electron tunneling occurs directly between the trap to available states in the anode through a trapezoidal energy barrier. Direct tunneling has been analyzed with Airy functions and numerical integration. A closed form expression using WKB and other approximations has been derived to predict leakage currents in ultra thin oxides [2.110]. The current density given in Eq (2.6), increases by the following factor,

$$\frac{1}{\left\{1 - \left(\frac{\phi_b - V_{ox}}{\phi_b}\right)^{1/2}\right\}^2}\exp\left(\frac{B\left(\frac{\phi_b - V_{ox}}{\phi_b}\right)^{3/2}}{E_{ox}}\right) \tag{2.7}$$

where, V_{ox} is the applied voltage across the oxide.

It should be noted, however, that these equations are obtained by assuming that electrons in the emitting electrode can be described as free electron gas, and taking into account the normal component of electron momentum only to derive the tunneling probability. As trapped electrons are localized particles with quantized energy states, equations for factors A and B may have to be modified, to accurately calculate the detrapped electron current.

2.4.2 Hole Generation, Capture, and Emission

Because of high fields necessary for injection, holes may be generated by an impact-ionization mechanism [2.93] and get trapped in the oxide. The generated hole concentration Θ_{it}^+ at time t is given by

$$\Theta_{it}^+ = \int_0^t \int_{T_t}^{T_{ox}} \frac{J(t)}{q_e}\alpha_0 e^{-H/E_{ox}} dx dt \tag{2.8}$$

where T_t and T_{ox} are the tunneling distance and oxide thickness, respectively, measured from the cathode/oxide interface, E_{ox} is the electric field in the region from T_t to T_{ox} which may be spatially varying, α_0 and H are the ionization constants [2.94], and J is the injected current density.

Assuming all the generated holes are trapped, these trapped holes may be lost due to recombination [2.95] or tunneling [2.96]. Recombination of trapped holes with injected electrons is again handled as a first-order process using a recombination cross section σ_i.

The rate equation for the density of incoming electrons that recombine (n_i) is given as

$$\frac{dn_i}{dt} = \frac{\sigma_i J(t)}{q_e}(\Theta_{it}^{+} - n_i) \tag{2.9}$$

Since the unrecombined hole density $\Theta_i^{+} = \Theta_{it}^{+} - n_i$, the relation for the unrecombined hole density at time t is given by

$$\frac{d\Theta_i^{+}}{dt} = \frac{d\Theta_{it}^{+}}{dt} - \frac{\sigma_i J}{q_e}\Theta_i^{+} \tag{2.10}$$

The rate equation for tunneling discharge of holes, trapped at a distance x from the interface, is given by

$$\frac{dp}{dt} = -\frac{p(x,t)}{\tau(x)} = -\frac{p(x,t)}{\tau_0 \exp\left[\frac{4}{3}\frac{\sqrt{2m_h^*}}{hqE_{ox}}\left\{E_{th}^{1.5} - (E_{th} - qE_{ox}x)^{3/2}\right\}\right]} \tag{2.11}$$

where p is the hole concentration at x, $\tau(x)$ is the inverse transition rate for hole tunneling from an oxide trap to the valence band, E_{ox} is the oxide field, m_h^* is the hole effective mass in the oxide, E_{th} is the trapped hole energy measured from the oxide valence band edge, and τ_0 is the characteristic tunneling time which is weakly dependent on E_{th} and E_{ox}.

2.4.3 Oxide Breakdown

Although the actual mechanisms leading to the intrinsic breakdown of dielectrics are not fully understood, the breakdown process may be divided into two stages: a buildup stage during which localized high-field/current regions are formed, and a second stage during which the runaway electrical or thermal processes quickly break down the oxide. In MOS structures, constant-current injection studies show that oxide breakdown occurs at a lower injected charge for electron injection from the gate, compared to substrate injection [2.79]. Several models have been postulated to explain the pre-breakdown buildup stage. Most of the models assume that charge trapping at oxide defect sites leads to high local fields and currents, and to a positive feedback condition, resulting in thermal runaway. However, both electron and hole trapping have been proposed as the cause of breakdown; these will be reviewed in this section. A model which proposes that breakdown originates at dielectric defects, but does not make the a priori requirement of charge trapping, is also presented. The hole trapping breakdown theory appears to provide the best physical explanation for the events leading to oxide breakdown.

2.4.3.1 Electron Trapping Breakdown Model. Based on extensive data collected from constant-current stress experiments on thin oxide MOS capacitors, Harari [2.97] suggested that electron trapping in deep traps near the cathode leads to internal oxide fields sufficient to break Si–O bonds, resulting in oxide breakdown.

The band diagram at onset of breakdown is shown in Fig. 2.38. The oxide field is enhanced in the region between the trapped electron centroid and the anode, eventually reaching 3×10^7 V/cm, initiating breakdown. Preexisting oxide defects and strained or broken Si–O bonds can serve as trapping centers. A hole trapping model was rejected since time-to-breakdown (under constant current stress) of oxides stressed at 77 K was higher than at 300 K, which would not be expected, for hole mobility is much lower at 77 K and time-to-breakdown should be correspondingly smaller.

A related model assumes that, because of multiple interactions in the dielectric, any single electron cannot gain enough kinetic energy from the field to cause significant damage, but the collective damage causes breakdown. That is, for a given dielectric, breakdown occurs after a fixed amount of charge per unit area, Q_{bd}, is injected [2.98, 2.99]. Q_{bd} has been shown to be a function of applied fields [2.100].

The scenario for breakdown is as follows. In the initial stages, damage due to charge injection near the anode may consist of broken bonds or displaced atoms that produce a kind of cavity of atomic dimensions. Because cavities have less polarization and higher electric fields, the electrons moving through cavities gain more energy than those within the dielectric since the mean free path is longer. This energy

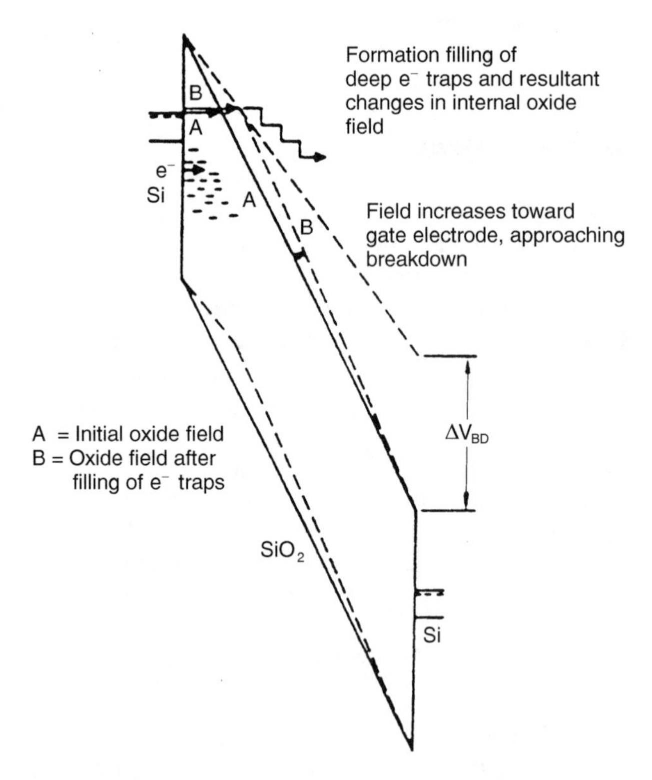

Figure 2.38 Buildup of internal fields due to generation and filling of electron traps close to the injecting electrode [2.97].

is partially dissipated at the wall of the cavity, thereby extending it and providing positive feedback. After sufficient injection, the atoms at the wall may be ionized. The positive ions attract electrons, which preferably move through the cavity. The pathways between the cavities will have a high local current density and, therefore, will erode rapidly. After sufficient injection, the cavities will be interconnected. Such cavities with filamentary interconnection have been observed in stressed dielectrics. When injection continues, the growing feature will interconnect the two electrodes, resulting in breakdown.

2.4.3.2 Hole Trapping Breakdown Model. This model assumes that hole trapping in the oxide leads to current runaway, resulting in breakdown [2.101]. At large fields, electron injection from the cathode results in some electrons gaining enough energy to create electron hole pairs in the oxide by impact-ionization through various mechanisms such as trap-to-band or band-to-band impact-ionization as shown in Fig. 2.39 [2.102, 2.103].

The postulated scenario is as follows. The high-field injected electrons have a large kinetic energy near the anode, which is evident from the band diagram. The generated holes drift to the cathode interface and get trapped at localized defects sites, referred to as "weak spots." The localized trapped holes lead to increased localized fields and conduction, which further accelerates the local hole trapping and results in breakdown when the localized current density reaches a critical value. The energy band diagram at the weak spots is shown in Fig. 2.40. The weak-to-robust area ratio is on the order of 10^{-6}. The trapped charges are shown as sheet charges located at the appropriate centroids. The conclusion from this model is that hole trapping at the interface needs to be minimized to improve oxide breakdown characteristics.

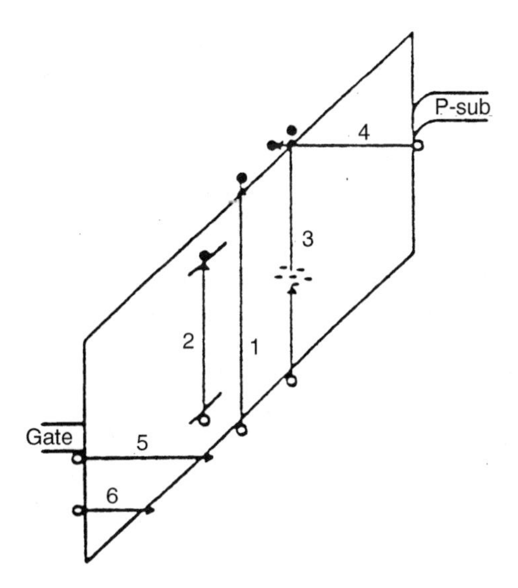

Figure 2.39 Several mechanisms for hole generation in oxides [2.102].

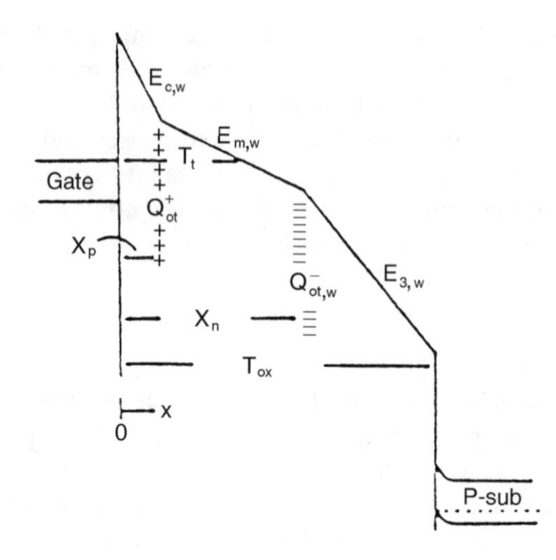

Figure 2.40 Energy band diagram for a weak area oxide susceptible to hole trapping near the cathode. For a fixed oxide voltage, V_{ox}, both Q^+_{ot} and Q^-_{ot} will affect the cathode field [2.102].

Key experimental details and observed data supporting this model are described next. To separate the effect of trapped electrons and trapped positive charges on breakdown, devices were stressed to approximately 90% of the Q_{bd} value and then annealed at 450°C to detrap electrons. A comparison of the gate injection I-V characteristics before and after stress/anneal shows that the post–stress/anneal gate injection I-V characteristic recovers to the pre-stress value. (The I-V curve is shifted to the left due to residual positive charge which was not annealed.) For substrate injection, however, the post–stress/anneal I-V curve showed both an increased current and a lowered slope. This is attributed to hole trapping near the cathode which increases the electric field, resulting in enhanced current injection. Since the trapped hole density, extracted from the C-V shift, was too small to explain the observed slope lowering, hole trapping was postulated to occur in a very small fraction of the oxide area (approximately 10^{-6} of the total area).

After annealing, devices were subjected to constant-current stress of the same and reverse polarity. The data are given in Table 2.2. When the polarity is

TABLE 2.2 ADDITIONAL CHARGE PER UNIT AREA NECESSARY FOR BREAKDOWN Q_{bd} AS A FUNCTION OF INITIAL CHARGE PER UNIT AREA INJECTED INTO THE OXIDE, Q_i, AND THE ANNEALING TEMPERATURE. (NOTE: + REFERS TO ELECTRON INJECTION FROM THE SILICON SUBSTRATE, AND – REFERS TO ELECTRON INJECTION FROM THE POLY-Si GATE. J = 33 mA/cm^2) [2.101].

		Final Q_{bd} C/cm^2		
	Initial Q_i C/cm^2	No anneal	350C anneal	450C anneal
1	0	−19.1	−16.3	−18.8
2	−17.0	−1.4	−0.8	−1.2
3	−17.0	+15.4	+13.9	+16.2

unchanged, the cathode, during final stress, has already experienced field enhancement as a result of the initial stress, and the device breaks down quickly. When the polarity of the current injection is reversed after annealing, the cathode, during the final stress, has not undergone any significant field enhancement. Anneal steps detrapped the electrons but did not increase Q_{bd}. The conclusion is that field enhancement at the cathode interface, due to hole trapping at localized weak spot areas, causes oxide breakdown.

2.4.3.3 Noncharge-Trapping Breakdown Model. Based on breakdown experiments on Al–SiO–Al capacitors, this model does not make the a priori requirement of charge trapping [2.104]. Breakdown is found to originate at randomly distributed inhomogenities in the dielectric (about 0.5 μm in diameter) referred to as "dark spots" and is accompanied by growth of crystalline silicon. Breakdown is postulated to occur by the following reaction, in the presence of a local breakdown field:

$$\text{Si–O–Si} + (\text{local breakdown field}) \rightarrow \text{Si–Si}^+ + \text{O} + e^- \tag{2.12}$$

where Si–Si^+ means that the two atoms are in crystalline form and one is singly ionized. When high fields are applied, the O-atom sees a potential energy as shown in Fig. 2.41, with the length δ becoming smaller with increasing fields. Thus, the O-atom may tunnel out of the barrier, whereby the above reaction can occur and breakdown begins. The above reaction may occur at lower fields if the Si–O–Si bonds are distorted in angle or longitudinally strained due to the influence of neighboring atoms—for instance, from the presence of extra silicon atoms attempting to form a local lattice, which would serve as a nucleating spot for initiation of breakdown. The emitted electrons are injected into the SiO conduction band, thereby increasing the conductivity of the dielectric. The sequence of events leading to breakdown, from a nucleating dark spot, is given in Fig. 2.42. Upon application of breakdown voltage, V_{max}, the above reaction is assumed to start at the dark spot, and the injected electrons decrease the resistance of the shaded region A, which results in a lower field in region A but a higher field in region B. When the field in region B reaches the breakdown field, the above reaction occurs throughout region B. The resistance is again lowered, with corresponding electric field redistribution. This results in growth of silicon as the active region moves from cathode to anode. Once the conductive path in the dielectric is sufficiently established, the entire energy is dissipated in a thin dielectric layer, completing the breakdown.

2.4.4 Parameter Extraction

Parameters used to characterize oxide bulk degradation, like N_t, σ_t, g, and σ_g, can be extracted from constant-current stress, constant-voltage stress, or ramped voltage stress I-V characteristics. Interface state density is extracted from charge-pumping or quasistatic C-V measurements. Trapped charge centroids are obtained by bidirectional I-V curves. Flatband shifts are measured from high-frequency C-V curves.

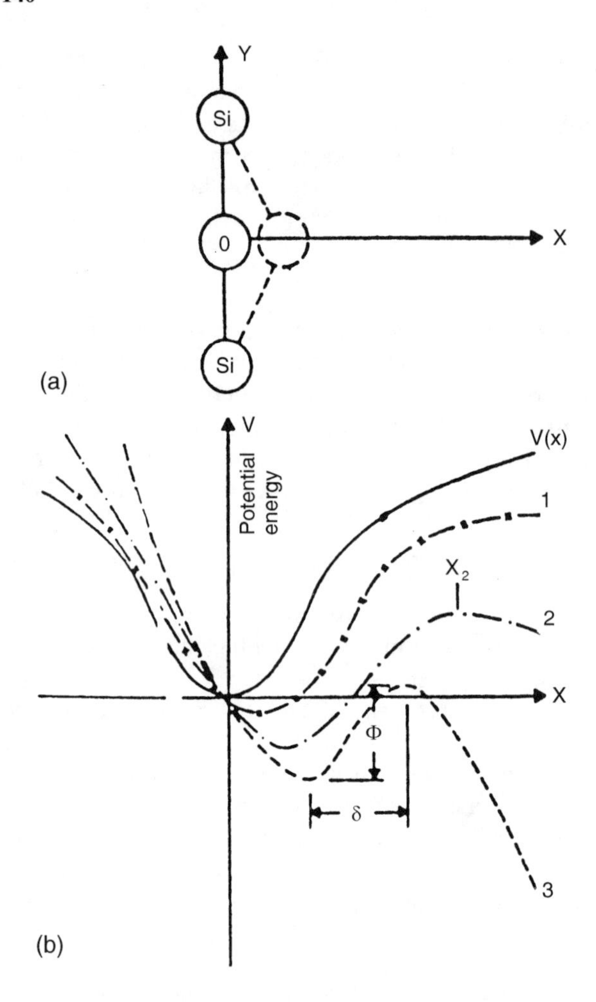

Figure 2.41 (*a*) Linear Si–O–Si bonding arrangement (solid lines) in SiO. Polarization of the Si–O–Si system due to an applied field in the negative X-direction is shown with broken lines. (*b*) Potential energy diagram of the oxygen atom due to the two silicon atoms and an applied field. The curve V(x) represents the potential energy in the plane $y = 0$ when the applied field is zero. Curves labeled 1, 2, 3 represent the total potential energy with progressively increasing electric fields. Beyond X_2 in curve 2, the oxygen atom is no longer bound to the silicon atoms. Curve 3 is believed to represent the situation described by the SiO breakdown equation [2.104].

A good summary of oxide parameter extraction techniques and extensive simulation results have been reported [2.95, 2.101, 2.105]. Typical constant-current stress and ramped voltage stress I-V characteristics are shown in Fig. 2.43.

In general, flatband voltage shift can be written as

$$\Delta V_{FB}^{+} = \frac{q}{\epsilon_{ox}}\left[n_t(T_{ox} - \bar{X}_n) - \Theta_i^{+}(T_{ox} - \bar{X}_p)\right] \tag{2.13}$$

$$\Delta V_{FB}^{-} = \frac{-q\Theta_{it}^{+}}{\epsilon_{ox}}(T_{ox} - \bar{X}_p) \tag{2.14}$$

where $\bar{X}_n$, $\bar{X}_p$ are the negative and positive trapped charge centroids, and ΔV_{FB}^{+}, ΔV_{FB}^{-} are the positive and negative flatband shifts. Since the tunnel current is mainly from electron emission (because of the larger barrier height for holes), electron trapping dominates. So, ΔV_{FB}^{+} reduces to

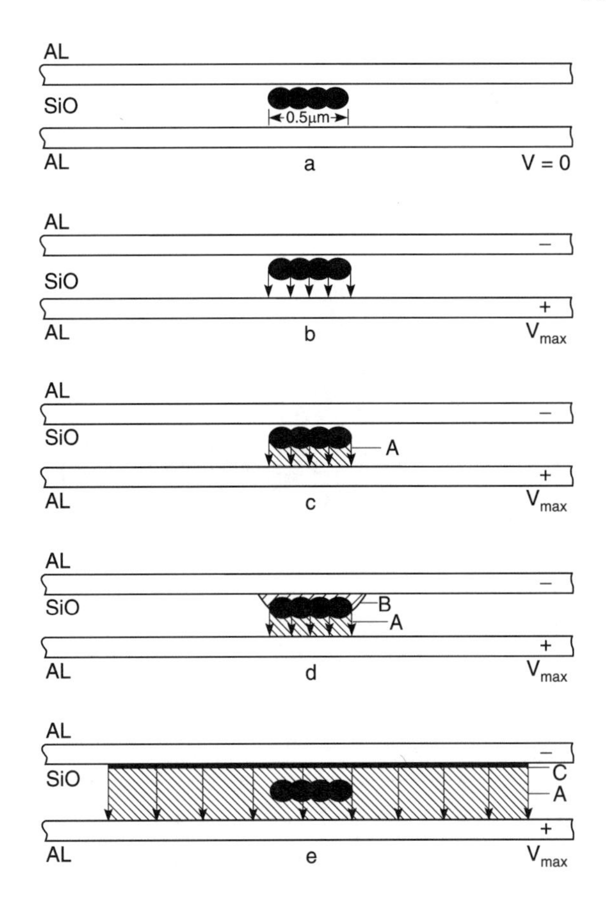

Figure 2.42 Development of breakdown. (*a*) Capacitor cross section showing a "dark spot" when the applied field is zero, (*b*) voltage, V_{max}, has been applied, and the breakdown reaction has started at the dark spot (arrows indicate electrons are injected into the dielectric about the dark spot), (*c*) injected electrons increase the conductivity in the shaded region A, (*d*) voltage across the capacitor is very close to V_{max}, so the increase in conductivity in region A causes the field in this region to decrease and in region B to increase (the field in region B is assumed to at least equal the breakdown electric field, so the breakdown reaction starts in B), and (*e*) the active region rapidly falls back to the thin dielectric layer C immediately adjacent to the cathode (this region must spread laterally to the diameter of the observed breakdown pattern) [2.104].

$$\Delta V_{FB}^{+} = \frac{q}{\epsilon_{ox}} n_t (T_{ox} - \bar{X}_n) \tag{2.15}$$

X_n is determined from bidirectional I-V measurements [2.106]. For the first-order approximation, charges located at the interfaces have no effect on I-V characteristics (but affect C-V shifts). Trapped bulk oxide charge density and the centroid are calculated from a bidirectional I-V measurement using

$$\bar{X}_n = T_{ox} \frac{\Delta V_g^{+}}{\Delta V_g^{+} + \Delta V_g^{-}} \tag{2.16}$$

$$n_t = \Delta V_g^{+} \frac{\epsilon_{ox}}{\bar{X}_n} \tag{2.17}$$

where ΔV_g^{+} and ΔV_g^{-} are voltage shifts for positive and negative gate bias. The net oxide trapped charge density, which is primarily electrons, that is, n_t, and the centroid $\bar{X}_n$ from the gate can be extracted. To minimize undesirable charge trapping, the measurement current should be much smaller than the stressing current.

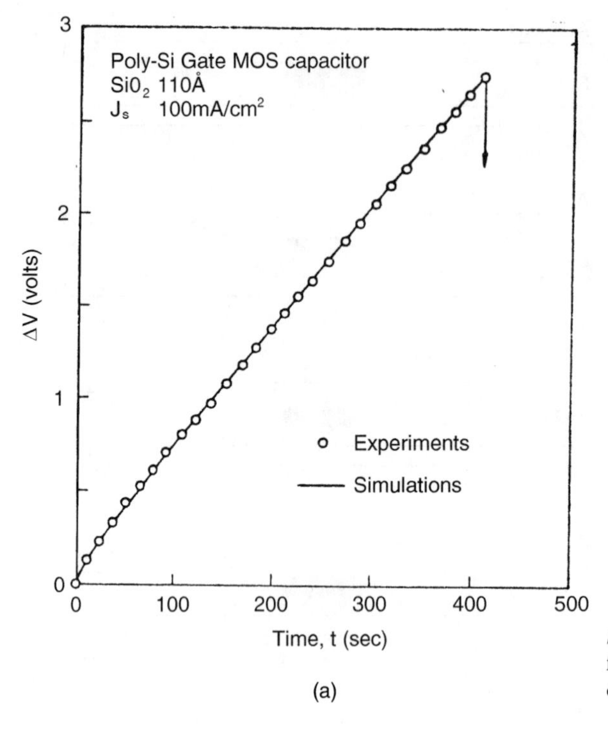

Source: Reprinted (abstracted) with permission from [2.95]. Copyright 1986 American Institute of Physics.

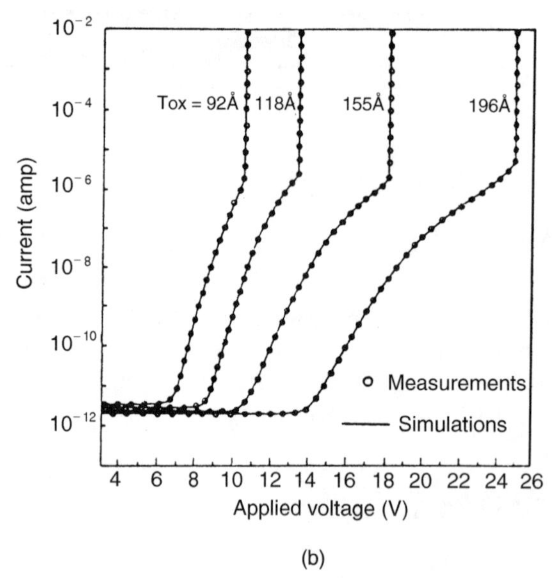

Figure 2.43 (*a*) V-t curve for constant current stressed poly-Si gate capacitor [2.95], and (b) I-V curves for 92 Å, 110 Å, 155 Å, and 196 Å poly-Si gate capacitors under ramped voltage stress [2.105].

The trap generation rate, g, can be determined when the electron fluence, $F = Jt/q$, is large. Under constant-current stress, the voltage drop across the oxide increases due to electron trapping. Neglecting hole trapping in the oxide, the voltage drop variation, $\Delta V(t)$, is given by

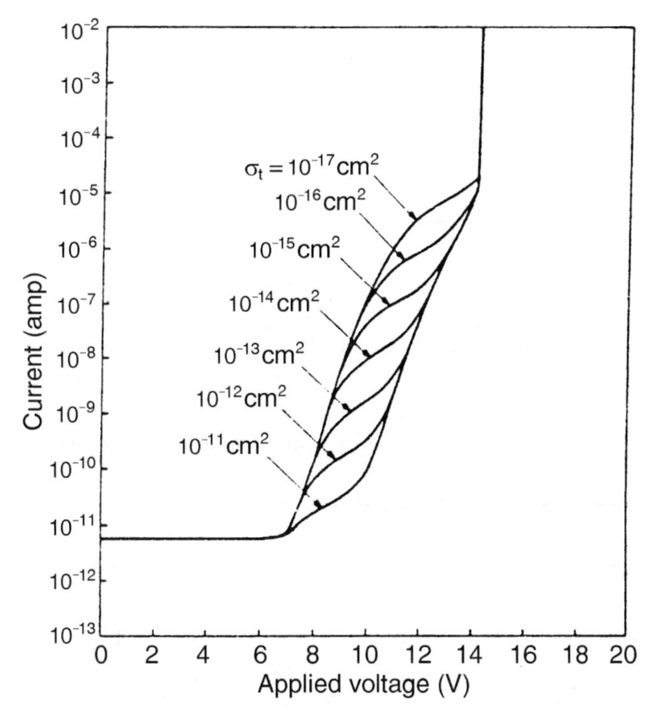

Figure 2.44 Ramped voltage I-V curves simulated for 100 Å poly-Si gate capacitor with electron trap capture cross section, σ_t, as a parameter [2.105].

$$\Delta V(t) = \frac{q}{\epsilon_{ox}}(T_{ox} - \bar{X}_n)\left\{\frac{Jg}{q}t - \frac{g}{\sigma_g}\left[1 - \exp\left(-\frac{\sigma_g J}{q}t\right) + N_t[1 - \exp\left(-\frac{\sigma_t J}{q}t\right)\right]\right\} \tag{2.18}$$

For large F,

$$\Delta V(F) \approx \frac{q}{\epsilon_{ox}}(T_{ox} - \bar{X}_n)\left[Fg - \frac{g}{\sigma_g} + N_t\right] \tag{2.19}$$

$$\frac{\partial \Delta V(F)}{\partial F} \approx \frac{q}{\epsilon_{ox}}(T_{ox} - \bar{X}_n)g \tag{2.20}$$

For large F and small J,

$$\Delta V(F) \approx \frac{q}{\epsilon_{ox}}(T_{ox} - \bar{X}_n)\left(N_t - \frac{g}{\sigma_g}\right) \tag{2.21}$$

For small F, Eq. (2.18) becomes

$$\Delta V(F) \approx \frac{q}{\epsilon_{ox}}(T_{ox} - \bar{X}_n)F\sigma_t N_t \tag{2.22}$$

Ramped voltage stress I-V curves can be used to extract σ_t [2.105]. The effect of σ_t variation on I-V characteristics is shown in Fig. 2.44. This can then be used to obtain σ_g, N_t from the above equations.

2.5. TYPICAL CURRENT–VOLTAGE (I–V) CHARACTERISTICS

The memory cells described in Section 2.1 all use electron tunneling across a thin dielectric to charge/discharge the floating gate. The write operation brings electrons to the floating gate (which increases the threshold voltage of the cell), while the erase operation removes electrons from the floating gate (which decreases the threshold voltage of the cell). Key variables of interest are the cell threshold voltage, V_{th}, after programming, and variations of V_{th} after repeated write/erase cycles.

2.5.1 Programming I-V Characteristics

The programming I-V characteristics can be modeled, in general, by considering the capacitive equivalent circuit given in Fig. 2.45 where C_{pp}, C_{tx}, C_{gate}, C_{gd}, C_{gs}, and C_{fld} are the interpoly (control gate and floating gate), tunnel oxide, gate oxide, floating gate to drain overlap, floating gate to source, and field oxide capacitances, respectively. The corresponding voltages are V_g, V_d', Ψ_s, V_d, V_s, and V_{sub}. The reason for defining V_d (applied drain voltage) and V_d' (effective drain voltage in the tunnel oxide region) separately will be discussed later. For now they can be assumed to be the same. Next, define the total capacitance, C_T, as the sum of the above capacitances, Q_{fg} as the charge on the floating gate, and the initial threshold voltage ($Q_{fg} = 0$) as V_{ti}. To write, a high-voltage pulse ($\sim$ 10 msec wide) is applied to V_g with all other voltages at ground; to erase, a high-voltage pulse is applied to V_d with all other voltages at ground, with V_s floating.

For the purpose of phenomenological discussion, the FLOTOX cell given in Fig. 2.2 will be used [2.107–2.109]. The tunneling current density, J_{tx}, is assumed to follow the Fowler–Nordheim relation and depends on the cathode electric field, E_k, so that

$$J_{tx} = \alpha E_k^2 \exp\left(\frac{-\beta}{E_k}\right) \tag{2.23}$$

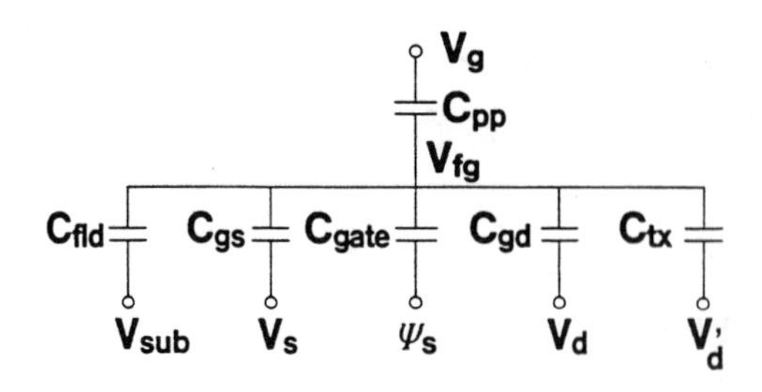

Figure 2.45 Capacitor network representing the basic operation of the cell during high-voltage write and erase operation [2.107, 2.109].

where α and β are experimental constants. The first-order model can be derived assuming no charge trapping in the tunnel dielectric, and neglecting all capacitances except C_{pp}, C_{gate}, and C_{tx}. Then, the cathode field $E_k = V_{tx}/T_{tx}$ where T_{tx} is the tunnel dielectric thickness and V_{tx} is the voltage dropped across this dielectric. V_{tx} is given as

$$V_{tx}(\text{write}) = V_g \frac{C_{pp}}{C_T} + \frac{Q_{fg}}{C_T} \tag{2.24}$$

$$V_{tx}(\text{erase}) = V_d \left\{ 1 - \frac{C_{tx}}{C_T} \right\} - \frac{Q_{fg}}{C_T} \tag{2.25}$$

For a programming pulse of width t, stored charge in the floating gate shifts the write and erase threshold voltage V_{tw} and V_{te} according to

$$V_{tw}(t) = V_{ti} + V_g - \frac{1}{K_w} \frac{B}{\ln(ABt + E_1)} \tag{2.26}$$

$$V_{te}(t) = V_{ti} - V_d' \frac{K_e}{K_w} + \frac{1}{K_w} \frac{B}{\ln(ABt + E_2)} \tag{2.27}$$

where

$$A = \frac{\alpha A_{tx}}{T_{tx} C_T} \tag{2.28}$$

$$B = \beta T_{tx} \tag{2.29}$$

$$E_1 = \exp\left\{ \frac{B}{K_w(V_g + V_{ti} - V_t(0))} \right\} \tag{2.30}$$

$$E_2 = \exp\left\{ \frac{B}{V_d' K_e + K_w V_t(0) - K_w V_{ti}} \right\} \tag{2.31}$$

where $K_w = (C_{pp}/C_T)$, $K_e = (1 - C_{tx}/C_T)$ are referred to as write and erase coupling ratios and have to be maximized for efficient operation; $V_t(0)$ is the cell threshold voltage at time $t = 0$; and A_{tx} is the tunnel area.

The above equations can be illustrated using typical cell parameters: $T_{tx} = 83$ Å, $K_w = 0.73$, $K_e = 0.8$, $V_g = 15$ V (write), $V_d = 16$ V (erase), program pulse duration $= 10$ msec. Electron tunneling is an extremely sensitive function of cathode field and effectively stops when $E_k \approx 10^7$ V/cm. Assuming the write/erase pulse width is long enough to reach $E_k \approx 10^7$ V/cm simplifies the calculations,

$$\Delta V_{th}(\text{write}) = V_g \left(1 - \frac{10^7 T_{tx}}{K_w V_g} \right) = +3.14\,\text{V} \tag{2.32}$$

$$\Delta V_{th}(\text{erase}) = 3.14 - V_d \left(\frac{K_e}{K_w} - \frac{10^7 T_{tx}}{K_w V_d} \right) = -3.3\,\text{V} \tag{2.33}$$

Figure 2.46 shows the calculated threshold voltage shift, ΔV_{th}, with programming pulse width $t = 10$ msec, for various tunnel oxide thicknesses. The effect of charge trapping was shown in Fig. 2.36. ΔV_{th} decreases with the number of write/erase cycles because electron trapping screens the applied V_{pp}, causing a smaller resultant cathode field. The coupling ratios can be obtained by comparing the floating gate cell to an equivalent MOS transistor, and can be extracted from the threshold voltage, body coefficients, or transconductance. The strong capacitive-coupling between the drain and floating gate through the tunnel oxide is apparent from the slope in the saturation region of the I_d–V_d characteristics shown in Fig. 2.47.

Several second-order effects need to be included to correctly model the erase operation: (a) depletion in the channel, (b) deep depletion under the tunnel oxide, and (c) hole flow into substrate, which are shown in Fig. 2.48.

(a) During erase, high voltage is applied to the drain, with the control gate grounded and the source floating. The floating gate potential becomes positive due to coupling of the drain voltage and reduction in the negative stored charge. As a consequence, the channel region gets depleted, reducing C_{gate} and K_e.

(b) At an electric field of 10^7 V/cm, required for significant tunnel current, the n^+ region beneath the tunnel oxide is inverted or depleted depending on its doping and the availability of holes in this region. Hence, a voltage drop appears across the depletion layer in the n^+ region. The effective voltage V'_d will be lower than V_d. Although the deep depletion should collapse as holes are generated in the depletion layer by avalanche multiplication, band-to-band tunneling, or pair generation by impact-ionization in the silicon by the tunnel electrons, the continuous removal of holes from the inversion layer by the surface channel (see Fig. 2.48) forces a deep depletion condition.

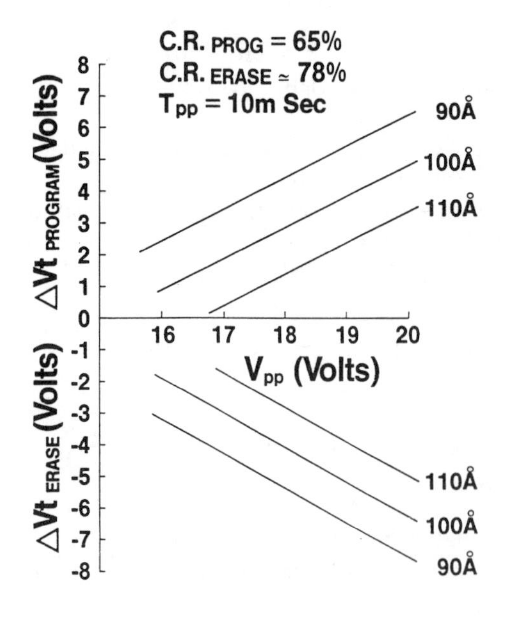

Figure 2.46 Calculated write and erase threshold voltage shifts as a function of V_{pp} for different tunnel oxide thicknesses [2.109].

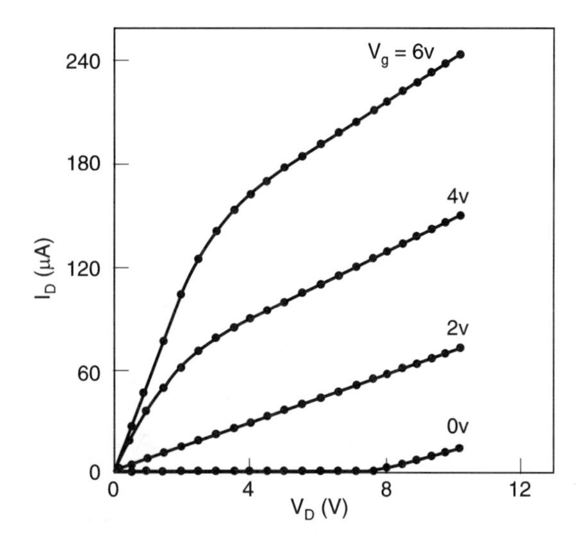

Figure 2.47 I_d-V_d characteristics of FLOTOX cell demonstrating strong coupling from the drain to the floating gate [2.107].

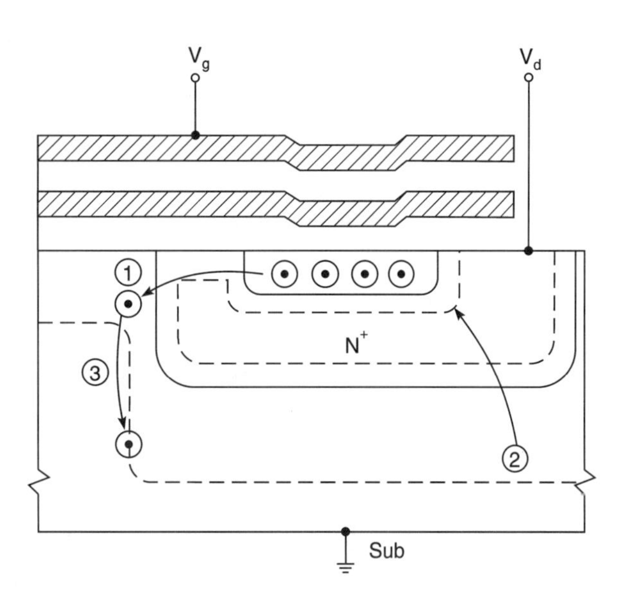

Figure 2.48 Schematic illustration of three mechanisms affecting the erase operation: (1) deep depletion in the channel, (2) deep depletion under the tunnel oxide, and (3) current path for holes from under the tunnel oxide to the substrate [2.107].

2.5.2 Retention Characteristics

The floating gate charge should ideally be stored for more than 10 years under normal chip operating conditions. Leakage through the tunnel dielectric or interpoly dielectric is the basic charge-loss mechanism. During cell read operation, the drain

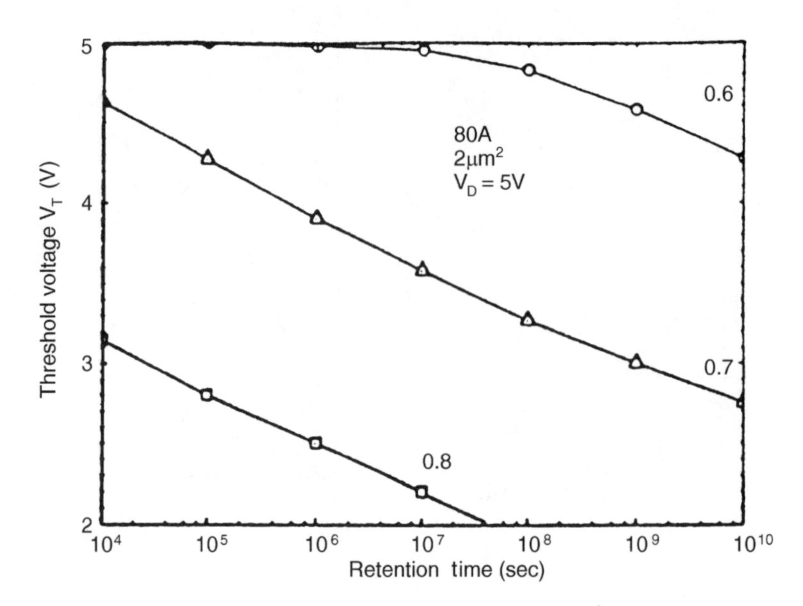

Figure 2.49 Threshold voltage shift versus time for an 80 Å tunnel oxide for different values of the coupling factor with the drain voltage kept at +5 V [2.108].

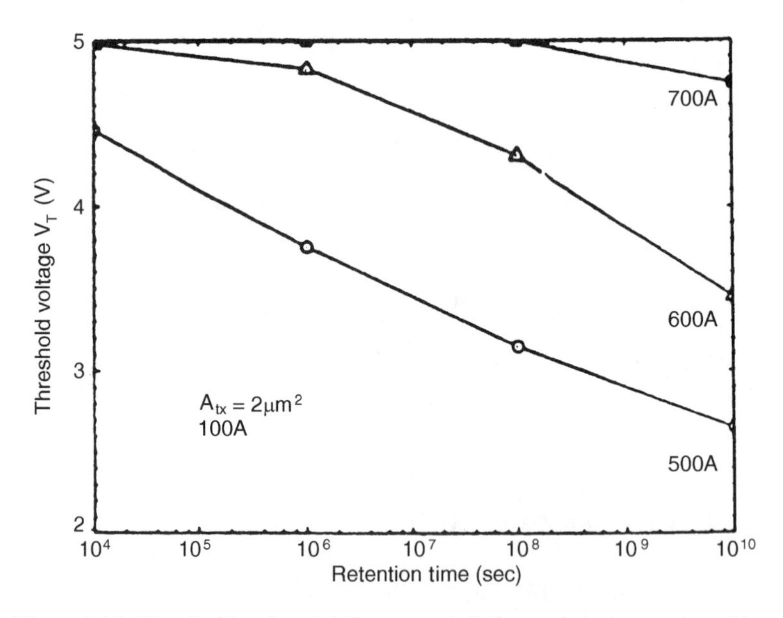

Figure 2.50 Threshold voltage shift versus time for various interpoly oxide thicknesses. The tunnel oxide thickness is 100 Å [2.108].

and control gate will be either high or low depending on row and column selection. From the point of view of a floating gate charge loss, the two worst case conditions are control gate high and drain low (leakage through the interpoly dielectric) or control gate low and drain high (leakage through the tunnel dielectric). Although tunneling is independent of temperature, high temperatures modify the electron energy distribution and the average electron energy increases, resulting in higher leakage currents. The equations given in the previous section can be used to calculate the approximate threshold voltage shifts under the two worst case biasing conditions. More detailed calculations are needed to take temperature effects into account. Figure 2.49 shows the variation in threshold voltage with $A_{tx} = 2\,\mu m^2$ and $V_d = 5\,V$, for an 80 Å tunnel oxide, for different coupling ratios [2.108]. This shows that the interpoly oxide thickness needs to be a minimum to give good retention (from the tunnel oxide leakage point of view) and a reasonable programming voltage. On the other hand, very thin interpoly oxides also cause retention problems due to interpoly leakage. Figure 2.50 shows the effect of interpoly leakage on retention; here $V_g = 5\,V$, $V_d = 0\,V$, $A_{tx} = 2\,\mu m^2$, and $T_{tx} = 100$ Å. Interpoly oxides less than 600 Å have serious retention problems. Compared to polyoxides for interpoly dielectric, oxide–nitride–oxide (ONO) stacked dielectrics (discussed in Section 2.3.2) provide significantly better retention behavior [2.66], with a total thickness down to about 150 Å.

References

[2.1] F. Masuoka, "Technology trends of Flash EEPROMs," *Symp. VLSI Technol. Dig. Tech. Pap.*, p. 6, 1992.

[2.2] W. Johnson et al., "A 16 Kb electrically erasable nonvolatile memory," ISSCC Dig. Tech. Pap., p. 152, 1980; S. K. Lai et al., "Design of an EEPROM memory cell less than 100 square microns using 1 micron technology," *IEEE IEDM Tech. Dig.*, p. 468, 1984.

[2.3] J. R. Yeargain and C. Kuo, "A high density floating gate EEPROM cell," *IEEE IEDM Tech. Dig.*, p. 24, 1981.

[2.4] R. Stewart et al., "A shielded substrate injector MOS (SSIMOS) EEPROM cell," *IEEE IEDM Tech. Dig.*, p. 472, 1984.

[2.5] H. Arima et al., "A novel process technology and cell structure for megabit EEPROM," *IEEE IEDM Tech. Dig.*, p. 420, 1988.

[2.6] Y. Terada et al., "120 ns 128 K × 8/64K × 16 bit CMOS EEPROM," *IEEE J. Sol. St. Cir.*, vol. 24, p. 1244, 1989.

[2.7] N. Ajika et al., "A novel cell structure for 4 Mbit full feature EEPROM and beyond," *IEEE IEDM Tech. Dig.*, p. 295, 1991.

[2.8] J. I. Miyamoto et al., "An experimental 5 V only 256 Kbit CMOS EEPROM with a high performance single polysilicon cell," *IEEE J. Sol. St. Cir.*, vol. 21, p. 852, 1986.

[2.9] N. Matsukawa et al., "Selective polysilicon oxidation technology for VLSI isolation," *IEEE Trans. Elect. Dev.*, vol. ED-29, p. 561, 1982.

[2.10] C. Bleiker and H. Melchior, "A four state EEPROM using floating gate memory cells," *IEEE J. Sol. St. Cir.*, vol. 22, p. 460, 1987.

[2.11] S. Nariani et al., "An ASIC compatible EEPROM technology," IEEE Custom Int. Cir. Conf., p. 9.5.1, 1992.

[2.12] A. O. Adan et al., "A scaled 0.6 μ*m* high speed PLD technology using single-poly EEPROMs," IEEE Custom Int. Cir. Conf., p. 55, 1995.

[2.13] R. Cuppens et al., "An EEPROM for microprocessors and custom logic," *IEEE J. Sol. St. Cir.*, vol. 20, p. 603, 1985.

[2.14] K. Ohsaki, N. Asamoto, and S. Takagaki, "A single poly EEPROM cell structure for use in standard CMOS processes," *IEEE J. Sol. St. Cir.*, vol. 29, p. 311, 1994.

[2.15] M. Ciao et al., "A simple EEPROM cell using twin polysilicon thin film transistors," *IEEE Elect. Dev. Lett.*, vol. 15, p. 304, 1994.

[2.16] G. Yaron et al., "A 16K EEPROM employing new array architecture and designed-in reliability features," *IEEE J. Sol. St. Cir.*, vol. 17, p. 833, 1982.

[2.17] G. Verma and N. Mielke, "Reliability performance of ETOX based flash memories," *IEEE Int. Rel. Phys. Symp.*, p. 158, 1988.

[2.18] M. Gill et al., "A novel sublithographic tunnel diode based 5 V-only flash memory," *IEEE IEDM Tech. Dig.*, p. 119, 1990.

[2.19] S. D'Arrigo et al., "A 5 V only 256 Kbit CMOS flash EEPROM," *ISSCC Dig. Tech. Pap.*, p. 132, 1989.

[2.20] J. Dickson, "On-chip high voltage generation in MNOS integrated circuits using an improved voltage multiplier technique," *IEEE J. Sol. St. Cir.*, vol. 11, p. 374, 1976.

[2.21] D. H. Oto et al., "High voltage regulation and process considerations for high density 5 V-only EEPROMs," *IEEE J. Sol. St. Cir.*, vol. 18, p. 532, 1983.

[2.22] E. M. Lucero, "A 16 Kbit smart 5 V-only EEPROM with redundancy," *IEEE J. Sol. St. Cir.*, vol. 18, p. 539, 1983.

[2.23] T. Tanzawa et al., "A quick boosting charge pump circuit for high density and low voltage flash memories," *Symp. VLSI Technol. Dig. Tech. Pap.*, p. 65, 1994.

[2.24] T. Dillinger, *VLSI Engineering*, Englewood Cliffs, N.J., Prentice-Hall, p. 498, 1988.

[2.25] A. Umezawa et al., "A 5V-only operation 0.6 μm Flash EEPROM with row decoder scheme in triple well structure," *IEEE J. Sol. St. Cir.*, vol. 27, p. 1540 (1992).

[2.26] S. Mehrotra et al., "A 64 Kb CMOS EEPROM with on-chip EEC," *ISSCC Dig. Tech. Pap.*, p. 142, 1984.

[2.27] D. Cioaca et al., "A million cycle CMOS 256 K EEPROM," *IEEE J. Sol. St. Cir.*, vol. 22, p. 684, 1987.

[2.28] J. Kupec et al., "Triple level polysilicon E^2PROM with single transistor per bit," *IEEE IEDM Tech. Dig.*, p. 602, 1980.

[2.29] S. Mukherjee et al., "A single transistor EEPROM cell and its implementation in a 512K CMOS EEPROM," *IEEE IEDM Tech. Dig.*, p. 616, 1985.

[2.30] M. Lanzoni et al., "A novel approach to controlled programming of tunnel based floating gate MOSFETs," *IEEE J. Sol. St. Cir.*, vol. 29, p. 147, 1994.

[2.31] M. Lanzoni and B. Ricco, "Experimental characterization of circuits for controlled programming of floating gate MOSFETs," *IEEE J. Sol. St. Cir.*, vol. 30, p. 706, 1995.

[2.32] T. Jinbo et al., "A 5 V only 16 Mb flash memory with sector erase mode," *IEEE J. Sol. St. Cir.*, vol. 27, p. 1547, 1992.

[2.33] S. Wolf, *Silicon Processing for the VLSI Era—Submicron MOSFET*, vol. 3, Chap. 7, p. 495, Lattice Press, 1995.

[2.34] T. Nakanishi et al., "Improvement in MOS reliability by oxidation in ozone," *Symp. on VLSI Technol. Dig. Tech. Pap.*, p. 45, 1994.

[2.35] K. Ohmi et al., "Hydrogen radical balanced steam oxidation for growing ultra-thin high reliability gate oxide films," *Symp. VLSI Technol. Dig. Tech. Pap.*, p. 109, 1994.

[2.36] H. Nozawa et al., "An EEPROM cell using low barrier height tunnel oxide," *IEEE Trans. Elect. Dev.*, vol. ED-33, p. 275, 1986.

[2.37] A. Gupta et al., "A 5V-only 16K EEPROM utilizing oxynitride dielectric and EPROM redundancy," *ISSCC Dig. Tech. Pap.*, p. 184, 1982.

[2.38] G. Q. Lo et al., "Improved performance and reliability of MOSFETs with ultrathin gate oxides prepared by conventional furnace oxidation of Si in pure N_2O ambient," *Symp. VLSI Technol. Dig. Tech. Pap.*, p. 43, 1991.

[2.39] H. Hwang et al., "Electrical and reliability characteristics of ultrathin oxynitride gate dielectric prepared by rapid thermal processing in N_2O," *IEEE IEDM Tech. Dig.*, p. 421, 1990.

[2.40] H. Fukuda et al., "Novel N_2O—oxynitridation technology for forming highly reliable EEPROM tunnel oxide films," *IEEE Elect. Dev. Lett.*, vol. EDL-12, p. 587, 1991.

[2.41] Y. Okada et al., "Gate oxynitride grown in nitric oxide," *Symp. VLSI Technol. Dig. Tech. Pap.*, p. 105, 1994.

[2.42] M. Bhat et al., "Electrical properties and reliability of MOSFETs with rapid thermal NO-nitrided SiO_2 gate dielectrics," *IEEE Trans. Elect. Dev.*, vol. ED-42, p. 907, 1995.

[2.43] M. Offenberg et al., "Role of surface passivation in the integrated processing of MOS structures," *Symp. VLSI Technol. Dig. Tech. Pap.*, p. 117, 1990.

[2.44] M. Miyashita et al., "Dependence of thin oxide films quality on surface micro-roughness," *Symp. VLSI Technol. Dig. Tech. Pap.*, p. 45, 1991.

[2.45] T. Ohmi et al., "Wafer quality specification for future sub-half micron ULSI devices," *Symp. VLSI Technol. Dig. Tech. Pap.*, p. 24, 1992.

[2.46] M. Gardner et al., "Hydrogen denudation for enhanced thin oxide quality, device performance, and potential epitaxial elimination," *Symp. VLSI Technol. Dig. Tech. Pap.*, p. 111, 1994.

[2.47] K. Lai et al., "Effects of surface preparation on the electrical and reliability properties of ultrathin thermal oxides," *IEEE Elect. Dev. Lett.*, vol. EDL-15, p. 446, 1994.

[2.48] M. Depas et al., "Ultra thin gate oxide yield and reliability," *Symp. VLSI Technol. Dig. Tech. Pap.*, p. 23, 1994.

[2.49] Y. Fong et al., "Oxide grown on textured single crystal Si-dependence on process and application in EEPROMs," *IEEE Trans. Elect. Dev.*, vol. ED-37, p. 583, 1990.

[2.50] C. Y. Kwok et al., "Effects of controlled texturization of the crystalline silicon surface on the SiO_2/Si effective barrier height," *IEEE Elect. Dev. Lett.*, vol. EDL-15, p. 513, 1994.

[2.51] S. L. Wu et al., "Tunnel oxide prepared by thermal oxidation of thin polysilicon film on silicon," *IEEE Elect. Dev. Lett.*, vol. EDL-14, p. 379, 1993.

[2.52] R. B. Sethi et al., "Electron barrier height change and its influence on EEPROM cells," *IEEE Elect. Dev. Lett.*, vol. EDL-13, p. 244, 1992.

[2.53] K. Yoneda et al., "Reliability degradation mechanism of ultrathin tunneling oxide by post-anneal oxidation," *Symp. VLSI Technol. Dig. Tech. Pap.*, p. 121, 1990.

[2.54] S. J. Fonash et al., "A survey of damage effects in plasma etching," *Sol. St. Tech.*, p. 99, July 1994.

[2.55] S. Samukawa and K. Terada, "Pulse time modulated ECR plasma etching for highly selective, highly anisotropic and less-charging poly-Si gate patterning," *Symp. VLSI Technol. Dig. Tech. Pap.*, p. 27, 1994.

[2.56] L. Faraone, "Thermal SiO_2 films on n^+ polycrystalline silicon: Electrical conduction and breakdown," *IEEE Trans. Elect. Dev.*, vol. ED-33, p. 1785, 1986.

[2.57] D. J. DiMaria and D. R. Kerr, "Interface effects and high conductivity in oxides grown from polycrystalline silicon," *Appl. Phys. Lett.*, vol. 27, p. 505, 1975.

[2.58] R. M. Anderson and D. R. Kerr, "Evidence for surface asperity mechanism of conductivity in oxides grown on polycrystalline silicon," *J. Appl. Phys.*, vol. 48, p. 4834, 1977.

[2.59] M. Sternheim et al., "Properties of thermal oxides grown on phosphorous in-situ doped polysilicon," *J. Electrochem. Soc.*, vol. 130, p. 1735, 1983.

[2.60] H. R. Huff et al., "Experimental observations on conduction through polysilicon oxides," *J. Electrochem. Soc.*, vol. 127, p. 2482, 1980.

[2.61] D. K. Brown and C. A. Barile, "Ramp breakdown study of double polysilicon RAMs as a function of fabrication parameters," *J. Electrochem. Soc.*, vol. 130, p. 1597, 1983.

[2.62] L. Faraone et al., "Characterization of thermally oxidized n^+ polycrystalline silicon," *IEEE Trans. Elect. Dev.*, vol. ED-32, p. 577, 1985.

[2.63] C. Y. Wu and C. F. Chen, "Transport properties of thermal oxide films grown on polycrystalline silicon—Modelling and Experiments," *IEEE Trans. Elect. Dev.*, vol. ED-34, p. 1590, 1987.

[2.64] P. J. Wright et al., "Hot electron immunity of SiO_2 dielectrics with fluorine incorporation," *IEEE Elect. Dev. Lett.*, vol. EDL-10, p. 347, 1989.

[2.65] H. N. Chern et al., "Improvement of polysilicon oxide characteristics by fluorine incorporation," *IEEE Elec. Dev. Lett.*, vol. EDL-15, p. 181, 1994.

[2.66] S. Mori et al., "Polyoxide thinning limitation and superior ONO interpoly dielectric for nonvolatile memory devices," *IEEE Trans. Elect. Dev.*, vol. ED-38, p. 270, 1991.

[2.67] M. Aminzadeh et al., "Conduction and charge trapping in polysilicon-silicon nitride-oxide-silicon structures under positive gate bias," *IEEE Trans. Elect. Dev.*, vol. ED-35, p. 459, 1988.

[2.68] H. P. Su et al., "Superthin O/N/O stacked dielectrics formed by oxidizing thin nitrides in low pressure oxygen for high-density memory devices," *IEEE Elect. Dev. Lett.*, vol. EDL-15, p. 440, 1994.

[2.69] T. Ishijima et al., "A deep-submicron isolation technology with T-shaped oxide (TSO) structure," *IEEE IEDM Tech. Dig.*, p. 257, 1990.

[2.70] P. C. Fazan and V. K. Mathews, "A highly manufacturable trench isolation process for deep submicron DRAMs," *IEEE IEDM Tech. Dig.*, p. 57, 1993.

[2.71] T. Hori, "A 0.1 μm CMOS technology with Tilt-Implanted Punchthrough Stopper (TIPS)," *IEEE IEDM Tech. Dig.*, p. 75, 1994.

[2.72] K. Yoshikawa et al., "0.6 μm EPROM cell design based on a new scaling scenario," *IEEE IEDM Tech. Dig.*, p. 587, 1989.

[2.73] K. Yoshikawa et al., "Flash EEPROM cell scaling based on tunnel oxide thinning limitations," Symp. VLSI Technol. Dig. Tech. Pap., p. 79, 1991.

[2.74] Y. Yamaguchi et al., "ONO interpoly dielectric scaling limit for non-volatile memory devices," Symp. VLSI Technol. Dig. Tech. Pap., p. 85, 1993.

[2.75] S. K. Lai et al., "Comparison and trends in EEPROM technologies," *IEEE IEDM Tech. Dig.*, p. 580, 1986.

[2.76] N. Mielke, "New EPROM data loss mechanism," Proc. Int. Rel. Phys. Symp., p. 106, 1983.

[2.77] R. Shiner et al., "Data retention in EPROMs," Proc. Int. Rel. Phys. Symp., p. 238, 1980.

[2.78] B. Euzent et al., "Reliability aspects of a floating gate EEPROM," Proc. Int. Rel. Phys. Symp., p. 11, 1981.

[2.79] P. P. Apte and K. C. Saraswat, "SiO_2 degradation with charge injection polarity," *IEEE Elect. Dev. Lett.*, vol. EDL-14, p. 512, 1993.

[2.80] J. Maserjian and N. Zamani, "Observation of positively charged state generation near the Si/SiO_2 interface during Fowler–Nordheim tunneling," *J. Vac. Sci. Tech.*, vol. 20, p. 743, 1982.

[2.81] K. Naruke et al., "Stress induced leakage current limiting to scale down EEPROM tunnel oxide thickness," *IEEE IEDM Tech. Dig.*, p. 424, 1988.

[2.82] P. Olivio et al., "High field induced degradation in ultra-thin SiO_2 films," *IEEE Trans. Elect. Dev.*, vol. ED-35, p. 2259, 1988.

[2.83] R. Moazzami and C. Hu, "Stress-induced current in thin silicon dioxide films," *IEEE IEDM Tech. Dig.*, p. 139, 1992.

[2.84] F. C. Hsu and K. Y. Chiu, "Hot electron substrate current generation during switching transient," *IEEE Trans. Elect. Dev.*, vol. ED-32, p. 375, 1985.

[2.85] M-S. Liang et al., "Degradation of very thin gate oxide MOS devices under dynamic high field/current stress," *IEEE IEDM Tech. Dig.*, p. 394 (1986).

[2.86] Y. Fong et al., "Dynamic stressing of thin oxides," *IEEE IEDM Tech. Dig.*, p. 664, 1986.

[2.87] S. Yamada et al., "Degradation mechanism of flash EEPROM programming after programming cycles," *IEEE IEDM Tech. Dig.*, p. 23, 1993.

[2.88] T. H. Ning and H. N. Yu, "Optically induced injection of hot electrons into SiO_2," *J. Appl. Phys.*, vol. 45, p. 5373, 1974.
[2.89] M. S. Liang and C. Hu, "Electron trapping in very thin thermal silicon dioxides," *IEEE IEDM Tech. Dig.*, p. 396, 1981.
[2.90] S. Horiguchi et al., "Interface trap generation modelling of Fowler–Nordheim tunnel injection into ultra-thin oxide," *J. Appl. Phys.*, vol. 58, p. 387, 1985.
[2.91] Y. Nissan-Cohen et al., "Dynamic model of trapping-detrapping in SiO_2," *J. Appl. Phys.*, vol. 58, p. 2252, 1985.
[2.92] J. G. Simmons et al., "Thermally stimulated currents in semiconductor and insulators having arbitrary trap distribution," *Phys. Rev.*, vol. B7, p. 3714, 1973.
[2.93] Y. Nissan-Cohen et al., "High field and current induced positive charge in thermal SiO_2 layers," *J. Appl. Phys.*, vol. 57, p. 2830, 1985.
[2.94] M. Knoll et al., "Comparative studies of tunnel injection and irradiation in metal oxide semiconductor structures," *J. Appl. Phys.*, vol. 53, p. 6946, 1982.
[2.95] C. F. Chen and C. Y. Wu, "A characterization model for constant current stressed voltage–time characteristics of thin thermal oxides grown on Si substrates," *J. Appl. Phys.*, vol. 60, p. 3926, 1986.
[2.96] S. Manzini and A. Modelli, "Tunneling discharge of trapped holes in silicon dioxide," *Insulating Films on Semiconductors*, J. F. Verwey and D. R. Wolters (Eds.), Elsevier Science Publishers, p. 112, 1983.
[2.97] E. Harari, "Dielectric breakdown in electrically stressed thin films of thermal SiO_2," *J. Appl. Phys.*, vol. 49, p. 2478, 1978.
[2.98] J. F. Verwey and D. R. Wolters (Eds.), *Insulating Films on Semiconductors*, Elsevier Science Publishers, p. 125, 1986.
[2.99] D. R. Wolters and J. J. van der Schoot, *Insulating Films on Semiconductors*, p. 145, 1986.
[2.100] A. Modelli and B. Ricco,"Electric field and current dependence of SiO_2 intrinsic breakdown," *IEEE IEDM Tech. Dig.*, p. 148, 1984.
[2.101] I. C. Chen et al., "Electrical breakdown in thin gate and tunneling oxides," *IEEE Trans. Elect. Dev.*, vol. ED-32, p. 413, 1985.
[2.102] I. C. Chen et al., "Oxide breakdown dependence on thickness and hole current-enhanced reliability of ultra thin oxides," *IEEE IEDM Tech. Dig.*, p. 660, 1986.
[2.103] T. H. DiStefano and M. Shatzkes, "Impact ionization model for dielectric instability and breakdown," *App. Phys. Lett.*, vol. 25, p. 685, 1974.
[2.104] P. P. Budenstein and P. J. Hayes, "Breakdown conduction in Al–SiO–Al capacitors," *J. Appl. Phys.*, vol. 38, p. 2837, 1967.
[2.105] C.-F. Chen and C.-Y. Wu, "A characterization model for ramp-voltage stressed I-V characteristics of thin thermal oxides grown on silicon substrate," *Sol. St. Elect.*, vol. 29, no. 10, p. 1059, 1986.
[2.106] M.-S. Liang et al., "MOSFET degradation due to stressing of thin oxide," *IEEE Trans. Elec. Dev.*, vol. ED-31, p. 1238, 1984.
[2.107] A. Kolodny et al., "Analysis and modeling of floating gate EEPROM cells," *IEEE Trans. Elect. Dev.*, vol. ED-33, p. 835, 1986.

[2.108] A. Bhattacharyya, "Modelling of write/erase and charge retention characteristics of floating gate EEPROM devices," *Sol. St. Elect.*, vol. 27, p. 899, 1984.

[2.109] S. K. Lai and V. K. Dham, "VLSI EEPROMs," Unpublished report, Intel Corporation, 1986.

[2.110] K. K. Schuegraf et al., "Ultra-thin silicon dioxide leakage current and scaling limit," *Symp. on VLSI Tech. Digest of Technical Papers*, p. 18, 1992.

Chapter 3

H. A. R. Wegener
and W. Owen

Floating Gate Nonplanar Devices

3.0. INTRODUCTION

The "textured poly device" is the first nonvolatile MOS memory cell to use Fowler–Nordheim tunneling from nonplanar surfaces. These surfaces are formed on the edges of polysilicon lines, so that curvature-enhanced tunneling is made possible through oxide layers that are relatively thick. Thicker oxides have an inherently higher reliability. They also have lower capacitances, resulting in smaller memory cell layouts, and greater ease of scaling cells to tighter design rules. Since all nonvolatile writing is done between two polysilicon layers, they are physically and electrically separated from the logic gates, and the two dielectrics can be separately optimized for their specific purposes.

The processes used to form the textured poly are minor variants of procedures used routinely to form oxide layers and to delineate poly lines. Therefore, normal integrated circuit processes and facilities are used to mass produce silicon devices that include textured poly EEPROMs. Historically, the textured poly technology was the first vehicle to use a single 5 V supply for reading, programming, and erasing nonvolatile memory devices. As of this writing (1997), this technology has prospered for two decades, and it has generated a respectable market share. Poly-to-poly tunneling has recently become part of several promising "flash" technologies.

3.1. CELL STRUCTURES AND OPERATION

The main representative of floating gate nonplanar devices is the *textured poly* device which derives its name from the nonplanar polysilicon surface features used to tunnel between two polysilicon layers. The textured poly device has a floating gate as its nonvolatile storage element, but, unlike other silicon-based nonvolatile memory devices, it is charged and discharged from neighboring poly layers instead of a diffused region in a silicon substrate. In the following sections, both the operation and physical structure of the textured poly device will be discussed in detail.

3.1.1 General Principles of Operation

Although the general structure and operation of floating gate nonvolatile memory devices are very similar, this section presents a detailed description of the physical structure of the textured poly device and discussions of the writing and reading operations, biasing of the floating gate by capacitive coupling, and Fowler–Nordheim tunneling of charge between the polysilicon electrodes and the polysilicon floating gate.

3.1.1.1 Description of Device. The textured poly nonvolatile memory device consists of three polysilicon electrodes that partially form MOS gates with the substrate as can be seen with the help of Fig. 3.1. The middle electrode forms a small rectangular poly2 area that is isolated from the other two poly layers forming the neighboring electrodes. This is the floating gate. The electrode on the left side is a line extending a large distance below and above the plane of the paper. It forms the wordline of the memory array and provides the receiving plate of the tunnel device used for erase. The electrode on the right side also extends a large distance beyond a single cell. It forms both the deselect gate of the array and the emitting plate of the tunneling device used for programming.

Another feature that is not obvious from these cross sections is the fact that the oxide layer thicknesses under the gate portions are typically 500 Å, and those between polys are typically in excess of 600 Å. There is no thin oxide region that can translate a moderate potential into the high field necessary for Fowler–Nordheim tunneling. Instead, the nonplanar poly surfaces and edges enhance the

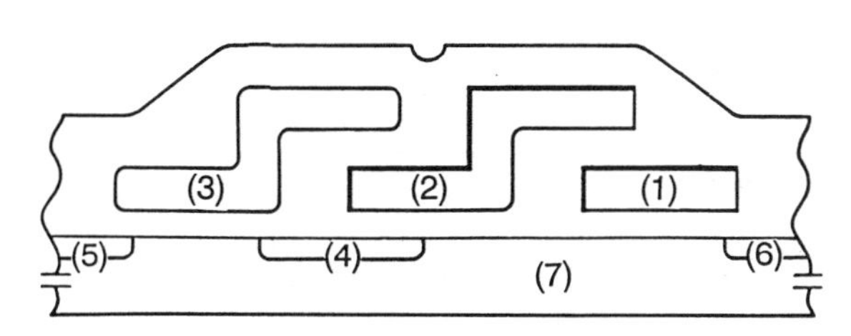

Figure 3.1 Typical textured poly nonvolatile memory cell. (1) is the poly1 deselect transistor and deselect line, (2) is the poly2 floating gate, (3) is the poly3 select transistor and wordline, (4) is the n-doped coupling capacitor plate, (5) is the n-doped connection to the bitline, (6) is the n-doped array ground, and (7) is the p-type substrate.

field from an applied potential by as much as a factor of five. In this way, tunneling is made possible from the right electrode to the floating gate and from the floating gate to the left electrode.

The equivalent circuit of the cell is shown in Fig. 3.2. It consists of three n-channel MOS transistors in series. At the top is the "select transistor" whose drain is connected to the bitline. A large number of cells share the same bitline, and many bitlines form the parallel columns of the memory array. The gate of the "select transistor" is connected to the wordline. Equivalent locations of the separate parallel bitlines share the same wordline, which defines one row of a rectangular memory cell matrix. To read, the wordline is taken to V_{cc}, which couples the cell to the bitline. During writing, the wordline is taken to a high voltage that both turns on the select transistor and provides one of the potentials necessary for tunneling. The tunneling capability between the floating gate and the wordline is depicted by a diode-like capacitor. This is intended to portray the fact that the textured poly features at this site enhance tunneling of electrons only toward the wordline. When no tunneling occurs, the structure behaves like a capacitor.

Below the select transistor on Fig. 3.2 is a coupling capacitor that is charged or discharged by the bitline, when the select transistor is turned on. The coupling capacitor uses part of the floating gate as one of its capacitor plates, and an n-doped region under the floating gate as the other. It functions as a steering element to determine whether the floating gate is coupled high or low, and thus programmed or erased, during a nonvolatile write operation.

The element below the coupling capacitor is the "floating gate transistor." It forms a channel between the n-doped region of the coupling capacitor and the edge of the "deselect transistor." Depending on the charge residing on the floating gate, it either turns on this central series transistor or shuts it off. Clearly, this structure is at the heart of this cell's memory function.

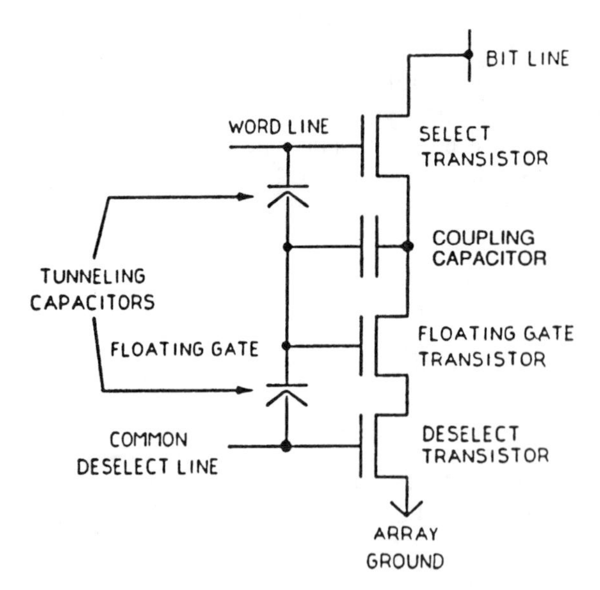

Figure 3.2 Equivalent circuit of the cell.

At the lower end of this series of transistors is the "deselect transistor." Its poly gate is the electrode that permits tunneling of electrons to the floating gate. During writing, it is taken low, which isolates the selected cell from the array ground. During reading, it is turned on and permits discharge of the bitline when sufficient positive charge is stored in the floating gate. The array ground is common to all the cells of the memory array. The tunneling capability from the poly gate of the deselect transistor toward the floating gate is indicated by a second diode-like capacitor structure.

3.1.1.2 Description of Nonvolatile Writing. "Erasing" is initiated by grounding the deselect-line. Next, the bitline is grounded. Then, a positive voltage ramp is initiated on the wordline. Through a small amount of capacitive-coupling, the potential on the floating gate is pulled in a positive direction, but the larger area of the coupling capacitor keeps it more closely coupled to ground.

Therefore, the difference between wordline voltage and floating gate potential increases until a critical voltage, called the "tunnel voltage," is reached. At this point, tunneling electrons from the floating gate to the wordline poly maintain a constant current, and the potential difference between the two electrodes is clamped at a constant value, the tunnel voltage. The constant current is defined by the relationship $I = C\ dV/dt$, where C is the capacitance between the wordline and the floating gate, and dV/dt defines the voltage ramp that is increasing the potential on the wordline. This continues until a maximum wordline voltage is reached.

Then, charge transfer at constant current terminates, although a decreasing amount of charge is transferred during the constant high-voltage plateau, called "flat top." Since all charge transfer is by electrons from the floating gate to the wordline, a net positive charge is formed on the floating gate, and the cell is considered erased. Figure 3.3 describes the potential versus time relationships on different electrodes during "erase." If the tunnel voltage had been much higher, and if, therefore, no tunneling had occurred, the floating gate potential would have followed the broken line of Fig. 3.3.

"Programming" is initiated by again grounding the deselect gate. But now, both the bitline and the wordline are ramped to a high positive voltage. With the select transistor turned on, the bitline voltage is transmitted to the n-doped region, forming one plate of the steering capacitor. The steering capacitor couples the floating gate high, aided by the potential on the wordline.

In this way, the potential difference between the deselect-line and the floating gate increases until a critical voltage is reached. Then, electrons tunnel from the deselect-line poly to the floating gate in a constant current until the two associated ramps reach the flat top level. During the flat top time, an additional small amount of electrons is transferred between the tunneling electrodes. If no tunneling had occurred during the high-voltage ramps, the floating gate potential would have followed the broken line of Fig. 3.4.

Since it is electrons that are transferred from the deselect-line to the floating gate, a net negative charge resides on the floating gate after this process, and the cell is considered "programmed." The details of charge transfer follow those of erasing

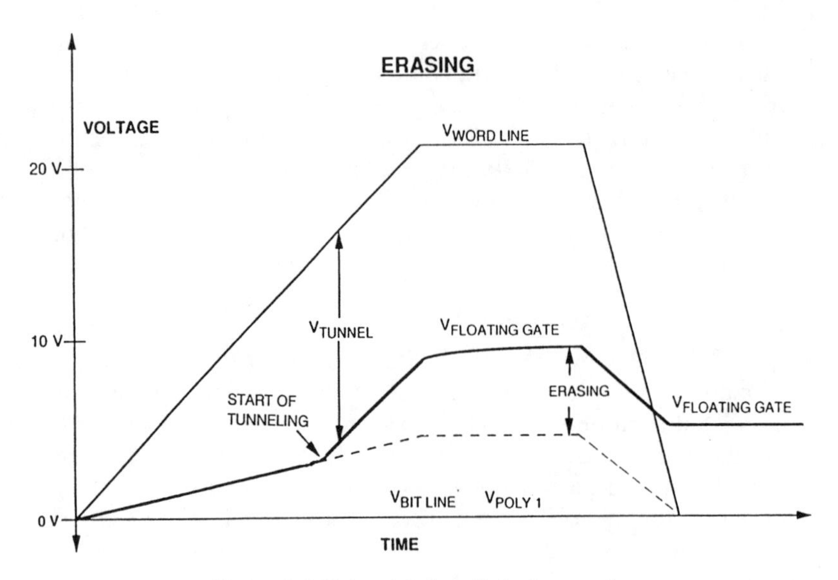

Figure 3.3 Potentials in cell during erasing.

quite closely, including a clamped voltage constant-current phase and a decreasing current during flat top. Figure 3.4 helps identify these stages.

Both erasing and programming require the same potentials on the wordline and deselect-line. The only difference lies in the powerful action of the coupling capacitor. This eliminates many possible write-disturb conditions.

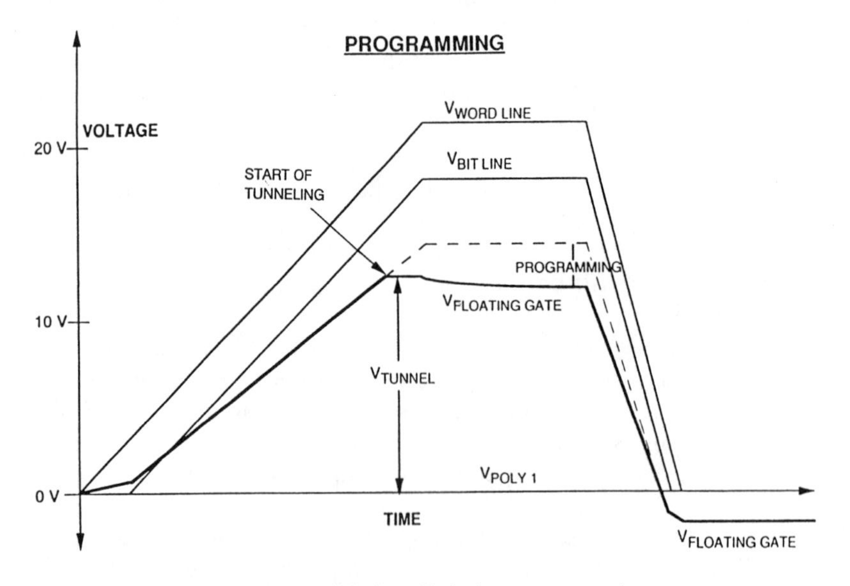

Figure 3.4 Potentials in cell during programming.

During writing, only one selected wordline receives the high-voltage ramp; all other wordlines are held at ground. This means that the coupling capacitor is disconnected from the bitline. No erasing can occur since there is no high voltage on the wordline, and no programming can occur since the bitline voltage cannot be transmitted to the steering capacitor.

All cells on the selected wordline are read first, and latches are set to select the correct voltage for each bitline so that, in the subsequent write operation, programmed cells will be re-programmed and erased cells will be re-erased. The bitlines of any byte or bytes that are to be written with new data from the outside have their latches overridden and set according to the voltages on the individual I/O lines. This means that, on all unchanged bytes, an erased cell will be erased again, and a programmed cell will be programmed again. Only a negligible amount of charge tunnels through the dielectric, and a negligible effect on endurance is incurred. The cells in the directly addressed byte may undergo a complete data change and thereby experience the maximum possible flow of electrons that will ultimately limit their endurance.

3.1.1.3 Description of Reading. The translation of floating gate charge polarity into a current that can be recognized in a sense amplifier is best understood with the help of Fig. 3.5. It shows a plot of the current through the cell as a function of voltage on the floating gate. This current may flow when all three series transistors are turned on.

The deselect transistor is always turned on, except during a write cycle. The select transistor is turned on when that particular wordline is addressed. Simultaneously, a 2 V potential is placed on the addressed bitlines. The amount of current that actually flows depends on the voltage of the floating gate. This voltage depends, in part, on the amount of charge coupled into it by nearby electrodes and, in part, on the amount and polarity of the charge stored on the floating gate. For the part whose cell characteristic is shown in Fig. 3.5, a total of about 1 V is coupled in from the select gate, deselect gate, and steering plate.

After erasing, a +3 V charge results in a +4 V potential on the gate, which gives rise to a current of about 67 microamperes. Similarly, after programming, a charge of –1 V on the gate gives rise to a net potential of 0 V, or a current of 0 microamperes. The sense amplifier and I/O amplifier translate these current levels into a TTL-compatible output signal.

The characteristic in Fig. 3.5 shows the effects of three MOS transistors connected in series. One of these transistors is the floating gate transistor at a low voltage (or a high series resistance), which gives rise to the characteristic below 3.5 V. The threshold voltage and transconductance of the floating gate transistor may be deduced. A change becomes apparent above 3.5 V, where the transconductance is determined by the combination of the three series transistors.

3.1.1.4 Capacitive Coupling of Voltages. As indicated in Section 3.1.1.2 on nonvolatile writing, biasing the floating gate is accomplished by capacitive-coupling. We can understand it by viewing the floating gate poly as one plate of a capacitor

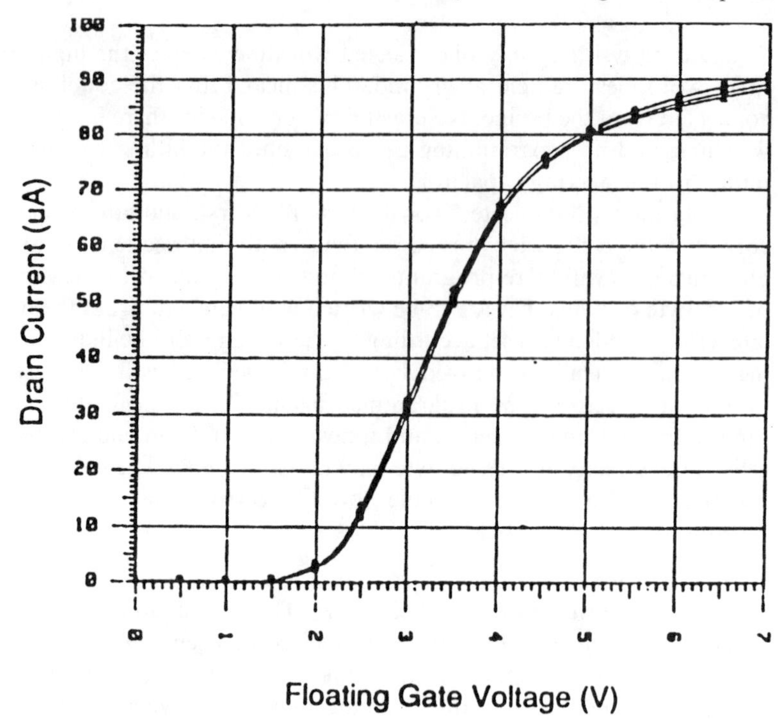

Figure 3.5 Floating gate transfer curve. This is a plot of the current between the bitline and the ground of a single memory cell. Deselect-lines and wordlines were held at 4 V, and the bitline at 2 V.

whose other plate is separated into several elements, each connected to a different bias. This can be represented by

$$Q_{FG} = \sum C_i(V_2 - V_i) \tag{3.1}$$

where Q_{FG} = net charge on floating gate
V_2 = potential on floating gate
C_i = capacitance of element i of other plate
V_i = potential applied to element i

In a typical textured poly cell, these elements are the capacitance between the floating gate and the deselect-line (C_1), the capacitance between the floating gate and the wordline (C_3), the capacitance between the floating gate and the steering plate (C_N), the capacitance between the floating gate and its channel (C_c), the capacitance between the floating gate and the substrate under the field region (C_{bb}), and the capacitance between the floating gate and the metal over it, forming the bitline (C_M).

We can determine the potential on the floating gate as a function of the potentials on the neighboring electrodes by solving Eq. (3.1) for V_2:

$$V_2 = V_{FG} - \sum CR_i V_i \tag{3.2}$$

where $V_{FG} = Q_{FG}/C_T$
$C_T = \sum C_i$ = total cell capacitance
$CR_i = C_i/C_T$ = coupling ratio for capacitance element

If we want to know the amount of charge transferred to the floating gate by the end of the voltage ramp applied during erase (see Fig. 3.3), we can solve Eq. (3.2) for V_{FG}. We know that V_2 is clamped to some fixed difference with V_3 during the constant-current tunneling process. We call this difference the tunnel voltage, V_{TUN3}, which can be measured separately. Therefore, we can replace V_2 by V_3–V_{TUN3}, and solve.

3.1.1.5 Fowler–Nordheim Emission from Textured Surface Features. The currents flowing between the different polysilicon electrodes and the polysilicon floating gate arise from Fowler–Nordheim tunneling [3.1]. As applied to silicon-to-silicon dioxide injection [3.2], the applicable I-V characteristic has the form

$$J_{FN} = a\ E^2 \exp -(b/E) \tag{3.3}$$

where J_{FN} is the current density of the tunneled current and E is the field between the tunneling electrodes, assuming that the spatial arrangement of the emitting electrode and the receiving electrode is plane parallel. For such an arrangement, a = 2E – 6 amps/volts^2 and b = 2.385E + 8 V/cm [3.3]. This expression predicts that, for a fixed current I_{FN}, there exists one voltage V that can sustain it. This is the origin of the constant-voltage drop V_{TUN3} for constant-current tunneling through the tunnel dielectric. It is the exact match of V in the applicable Fowler–Nordheim equation for a constant I_{FN}, which is defined by the constant-current C dV/dt that can be supplied by the external circuitry.

In the usual utilization of Eq. (3.3), the currents required in a cell, within an order of magnitude of 1 nanoampere, can be attained by making the tunneling dielectric near 100 Å thick. This simple approach can be made to work. Consequently, it has given rise to the family of thin oxide EEPROMs described elsewhere in this book.

There is another way of obtaining high fields in silicon/silicon dioxide structures that does not require the use of thin oxide layers. This approach involves use of the enhanced fields that emanate from surface features associated with bumps and edges. Such behavior was first consciously noted in the late 1970s [3.4], when unaccountable "leakage" currents were observed in EPROM structures that had polysilicon gates with some surface features. Such structures were then optimized for deliberate application as tunneling elements in floating gate devices.

As it turns out, the proper tailoring of the radii of curvature of the edges and bumps of emitting poly electrodes results in field enhancement by a factor of about five, without the need for a decrease in oxide thickness. The relative simplicity of formation and the high reliability of the tunneling devices fabricated in this way have made possible the large volume production, the large market share, and the acceptance of products incorporating the "textured poly" features.

An easy way to understand the physical relationships involved is to look at fields in a cylindrical coordinate system (to represent emission from an edge), and in a spherical coordinate system (to represent emission from the tip of a bump). These can be obtained by solving Laplace's equation in the relevant coordinate systems. In the rectangular coordinate system, which applies to plane parallel electrodes, the relationship between electric field, E_x, and applied voltage is

$$E_x = V/t_{ox} \tag{3.4}$$

where t_{ox} is the distance between the electrodes, or the thickness of the dielectric. The field E_r in the cylindrical coordinate system can be represented by

$$E_r = \{V/t_{ox}\}\{t_{ox}/r_c \cdot \ln(1 + t_{ox}/r_c)\} \tag{3.5}$$

where t_{ox} is the radial spacing between concentric cylindrical electrodes and r_c is the radius of curvature of the inner electrode. In the current tunneling scheme, the inner electrode emits the electrons and, therefore, is called the cathode. Moving on to the representation of an electrical field E_R in spherical coordinates,

$$E_R = \{V/t_{ox}\}\{1 + t_{ox}/R_c\} \tag{3.6}$$

where t_{ox} is again the spacing between two concentric spherical shells forming the electrodes and R_c is the radius of curvature of the inner shell.

The expressions in Eqs. (3.5) and (3.6) just give the maximum field at the interface between silicon and silicon dioxide. From expressions derived in the same manner as these equations, the value of the electric potential versus the distance from the interface can be plotted. As shown in Fig. 3.6, the voltage due to a spherical field falls off in a highly nonlinear manner.

Values of r_c and R_c have been obtained with the help of transmission electron micrographs (TEMs). A measured average value for r_c was 35 Å, and a measured average value for R_c was 155 Å [3.5]. When these values are substituted into Eqs. (3.5) and (3.6)—$t_{ox}=350$ Å is used in Eq. (3.5), and $t_{ox}=600$ Å is used in Eq. (3.6)—we find that $E_r/E_x=4.17$ and $E_R/E_x=4.87$. These are the amplification factors of curvature fields over linear fields. If we had chosen t_{ox} for E_x to be 100 Å, then these ratios would be 1.2 and 0.8, showing the importance of radius of curvature relative to the oxide thickness.

It is tempting to substitute these fields directly into the Fowler–Nordheim equation, but the standard derivation assumes the field to be linear and a first derivative of the voltage at the maximum. This would result in a reduced cross section and in a calculated t_{ox} that is much too small. Other first-order approximations have their own shortcomings.

For this reason, the Fowler–Nordheim equation was derived from the beginning for cylindrical and spherical coordinates. Specifically, this involved solving the barrier cross section integral used in the Wentzel–Kramers–Brillouin (WKB) approximation, which determined the field and voltage relationships in the exponent. It turned out that a closed form expression was obtained for the spherical case and a series solution for the cylindrical case. When expressed as a series, truncated after the

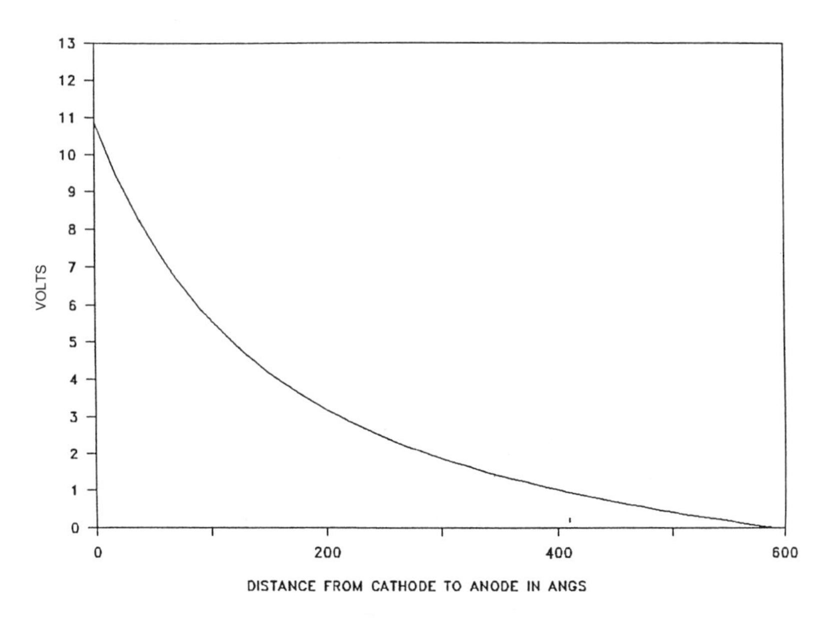

Figure 3.6 Spherical potential versus distance. This is a plot of $V = E_c R_c [(R_c/R) - (R_c/(R_c + t_{ox})]$, where V is the local potential in the dielectric and R varies between R_c and $R_c + t_{ox}$.

second term, and the coefficients are optimized for the truncation, then $E_r(ex)$, the cylindrical field expression in the exponent of the Fowler–Nordheim equation, can be represented by

$$E_r(ex) = \{V - [0.34\,\phi \cdot \ln(1 + t_{ox}/r_c)]\}/r_c \cdot \ln(1 + t_{ox}/r_c) \tag{3.7}$$

Equation (3.7) differs from Eq. (3.5) only by the term in the square brackets. ϕ is the electron barrier height between silicon and silicon dioxide. When $\phi = 3.2$, $r_c = 35$, and $t_{ox} = 350$ are substituted into the expression included in the square brackets, the correction term is 2.6 V. This expression is a good approximation down to V = 3.2 V, when the barrier cross section becomes trapezoidal.

The same approach to the derivation of the field $E_R(ex)$ in the exponential for the spherical case results in

$$E_R(ex) = \{V - [0.86\,\phi \cdot t_{ox}/(R_c + t_{ox})]\}\{(1 + t_{ox}/R_c)/t_{ox}\} \tag{3.8}$$

Again, the difference between Eq. (3.6) and Eq. (3.8) is in the term in the square brackets. Using $\phi = 3.2$, $R_c = 155$, and $t_{ox} = 600$ results in a correction term of 2.2 V. Again, the approximation is good from high potentials down to lower values near 3.2 V, when the geometric construction of the barrier cross section becomes invalid anyway.

The exact expression for the field in the pre-exponential part of the Fowler–Nordheim equation arises from the Taylor series expansion of the barrier integral. For both the cylindrical and spherical cases, the result can be expressed in a form that is very similar to Eq. (3.7) and Eq. (3.8). The pre-exponential field expressions differ from their respective exponential expressions in the numerical constants of the correction terms. These constants are near two-thirds of the constants in Eq. (3.7) and Eq. (3.8).

We can now construct a representation of the Fowler–Nordheim equation for cylindrical and spherical surfaces. It is

$$J_{FN} = a' E'^2 \exp(b/E') \tag{3.9}$$

Here, E' is the field calculated for the exponent and given in Eq. (3.7) and Eq. (3.8), which really has the form $E + dE$. But the exact pre-exponential field expression has been replaced by the exponential field expression. It is expected that this simplification will basically only affect the constant a, turning it into the slightly different constant a'.

In this section, it has been assumed that image forces will not have to be taken into account in the modeling of the tunneling process. This assumption is based on conclusions given in [3.6], [3.7], and [3.8].

3.1.2 Description of Cell Structures

3.1.2.1 Early Cell. An early version of a textured poly EEPROM cell (see Simko [3.9] and Jewell-Larsen et al. [3.10]) is shown in a plan drawing in Fig. 3.7, and as a scanning electron microscopy (SEM) cross section and top view in Fig. 3.8. The floating gate, formed by poly2, has the shape of the letter "L" lying on its side. The narrower arm of the "L" forms the memory transistor. A thin poly1 line runs vertically under the right side of the floating gate. The wide poly3 line runs horizontally and in parallel with the floating gate. It overlaps the floating gate along its lower edge and forms a gate in series with the memory transistor. This poly3 line also functions as the wordline of the array. The silicon substrate under the right side of the "L" and along the top part of the floating gate is made n-type before poly deposition. This large area is connected to the source of the memory transistor on the left, and a metal source-line strap runs vertically on the right. The bitline runs vertically on the left and connects to the memory cell through a contact shared with a neighboring memory cell.

This cell is erased by placing a high-voltage ramp on the poly3 wordline. Simultaneously, the source-line is grounded and poly1 is left to float. The floating gate is coupled toward ground by the n-doped source region underneath. This increases the potential between the poly2 floating gate and poly3. As a result, electrons tunnel from the floating gate to the wordline, and a net positive charge remains on the floating gate. This action occurs simultaneously in all cells sharing the same wordline.

In order to program selected cells of the previously erased wordline, a second high-voltage ramp is now put on the same poly3 wordline. Simultaneously, the source-line is ramped up to a high positive voltage and the poly1 line is held at ground. Now, the potential between poly3 (high positive) and poly2 (coupled to a high positive source) is low. A high potential now exists between the grounded poly1

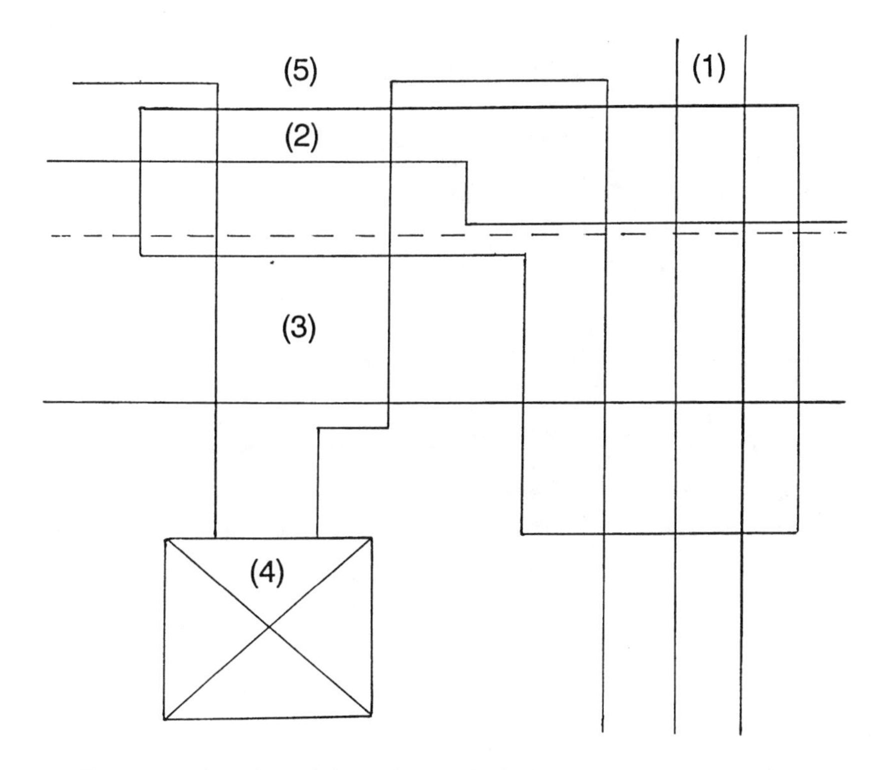

Figure 3.7 Plan view of the early cell. (1) is the poly1 line, (2) is the poly2 floating gate over the memory cell channel, (3) is the poly3 wordline over the memory cell channel, (4) is the memory cell drain contact to the bitline and (5) is the array ground. The broken line indicates the cross section through the cell shown in Fig. 3.8.

line and the floating gate. Electrons tunnel from poly1 to poly2. This results in an increase in negative charge on the floating gate.

Reading the information stored on the floating gate is initiated by placing V_{cc} on the wordline. Simultaneously, the source-line is grounded, and 2 V are placed on the bitline. If a sufficiently high positive charge resides on the floating gate, its channel is on and a current flows from the bitline through the wordline series gate to ground. This current is sensed at the base of the column. Conversely, if the floating gate is charged negatively, its channel is turned off and no current flow to ground is sensed.

3.1.2.2 The Direct Write Cell. A more recent cell structure, in production since 1986, was described by Guterman [3.11]. This cell incorporates several advances. Operationally, it is a "direct write" cell; that is, only specifically

Figure 3.8 SEM cross section through the early cell. This picture shows a combination SEM top view and SEM cross section of the early cell. In the cross section, the three dark poly layers are delimited by white thermal oxides. The memory transistor channel on the left is covered with poly3 and poly2. The triple-poly stack of the cell is seen at the right.

addressed cells have their data changed in one pass. Cells to be programmed and to be erased are written simultaneously in a single high-voltage operation. This entails advantages in write cycle time and complexity. It also increases the endurance of the memory cells, since tunneling occurs only when information in the cell changes.

Yeh [3.12] and Lambertson et al. [3.13] have recently described a two-poly version of the direct write cell. Plan and cross-sectional drawings are shown in Figs. 3.9 and 3.10, respectively. The fabrication technology has been improved by requiring only two polysilicon layers for a textured poly cell. This was accomplished by connecting portions of poly1 and poly2 together by a sublithographic corner contact to form a "bi-level" floating gate. In all operational details, this version functions in exactly the same way as the original direct write cell.

Nonvolatile writing is performed as described earlier in this chapter. Erasing one cell and programming its neighbor are achieved by having the individual bitline carry the signal for either. Writing a "1" requires grounding the bitline of one cell, and writing a "0" requires ramping it to a high voltage for the adjoining cell. Crosstalk is eliminated by having the deselect-line grounded. The deselect-line also functions as a tunneling electrode to emit electrons into the floating gate. The select-line performs a similar double role: it functions as the common wordline of the addressed byte, and it is the receiving electrode for electrons that tunnel from the floating gate. This merging of functions has resulted in the reduction of the memory cell area. The reading of this cell is described in Section 3.1.1.3.

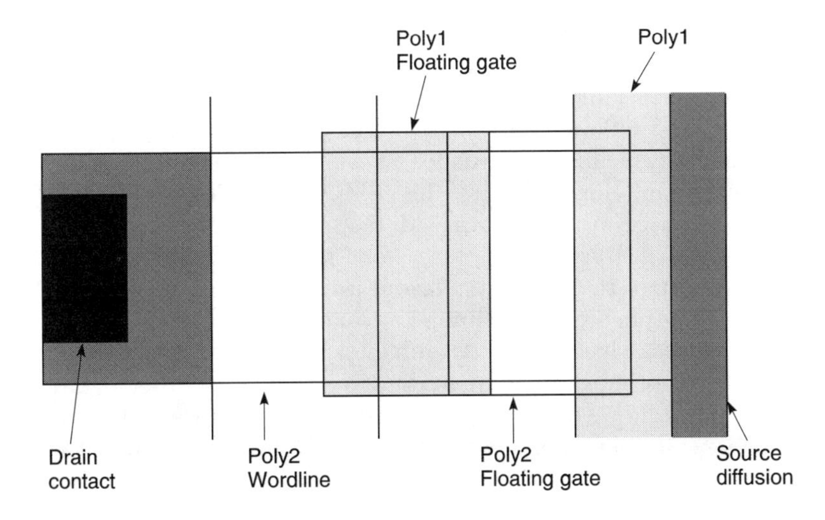

Figure 3.9 Plan view of direct write cell. The contact on the left connects to the bitline. The diffusion on the right is the array ground. The corner contact is hidden under the overlap of poly2 over poly1 at the center of the floating gate. © 1991 IEEE

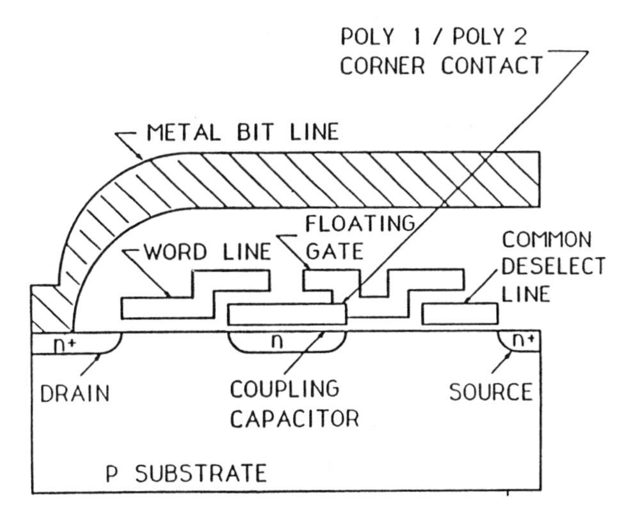

Figure 3.10 Cross section of direct write cell. The array ground is at the right, marked "source." The select transistor gate is formed by the poly2 wordline, and the floating gate transistor is formed by a patch of poly2, which is joined to a patch of poly1 over the coupling capacitor plate, and the deselect transistor gate is formed by the poly1 deselect-line. © 1991 IEEE

3.2. PROCESS TECHNOLOGY

3.2.1 General Description

Aside from the need for the special processing of polysilicon layers and their associated oxides, textured poly EEPROM fabrication technology requires only conventional MOS integrated circuit technology to make it operate satisfactorily.

This has been borne out by the history of this EEPROM technology since 1978. Originally, it utilized a thick field oxide n-MOS process with 4 μm features. This process was sufficient to support the first NOVRAMs and EEPROMs. It also permitted the utilization of high-voltage generating charge pumps, which made this technology the first to introduce 5 V-only external power supply nonvolatile devices. During the following years, this n-MOS technology was refined and feature sizes were reduced to 2 μm. Even with the special requirements for forming the poly-to-poly tunnel structures, the tunnel and logic gate oxides could be made in the same process step. In general, the layout rules did not require minimum state-of-the-art dimensions in order to result in an economic product. In 1987, a CMOS process was introduced, which proved its usefulness for textured poly EEPROM technology, just as it did for most other applications.

Since 1990, the CMOS process has been further refined. This process features two metal layers, only two poly layers, two different gate oxide thicknesses (250 Å and 500 Å), p-channel devices in n-wells, and epitaxial layer substrates. The n-well version of the CMOS technology was a natural extension of the preceding n-MOS process. The EEPROM array, which was based on n-channel transistors, could be used without change, requiring only the addition of isolating wells for the new p-channel devices in the periphery.

The process flow is typical. There is a front section that is devoted to generating n-wells and field oxides. During the midsection, gate oxides, tunnel oxides, poly layers, and substrate junctions are formed. The back end then consists of creating both metal layers, contacts and vias, intermetal oxide layers, and surface passivation.

3.2.2 High-Voltage Circuitry

The present process is designed to support circuitry that operates at voltages as high as 25 V. For this purpose, separate devices are used for normal low-voltage logic gates and for high-voltage circuitry. The low-voltage gates that operate below 7 V have 250 Å oxides. They may be n-channel or p-channel devices. The high-voltage gates have 500 Å oxides. The field regions must withstand the highest voltages without inversion. This is accomplished with 7500 Å oxides, with properly placed field implants and a substrate bias of –3 V.

3.2.3 Cell Processing

In general, every textured poly cell requires two locations with poly-poly capacitors that each permit curvature-enhanced tunneling from the lower poly level to the higher poly level. This structure was initially realized by three poly layers separated by two tunnel oxides (see Fig. 3.1). Tunneling occurred from the lowest poly, the program line poly1, to the second poly, the floating gate poly2, and from poly2 to the erase line poly3. The details of every process step of poly deposition, etching, oxide formation, and annealing are important for the final performance of the cell. Taken separately, these steps are routine recipes that are well within the range of capabilities of the equipment used in manufacturing.

A recent refinement has simplified the processing concerned with tunneling structures. Now, two sets of poly2 over poly1 structures are formed simultaneously, and the poly2 patch of one set is connected to the poly1 patch of the second set. These two connected patches form the floating gate. One poly1 line performs the functions of the older poly1 electrode, and the second poly2 line serves as the older poly3 electrode.

A "corner contact" process provides a sublithographic buried contact between poly1 and poly2. The details of this process are shown in four consecutive stages in Fig. 3.11.

3.2.4 Summary of Production-Level Technology

The most recent process used is an n-well CMOS technology, with two gate oxide thicknesses (150 Å and 450 Å). It utilizes two metal and two poly layers. The typical "L effective" is 0.6 μm, and a cell built with it has an area of 9.3 μm^2.

3.3. MEMORY ARRAY CIRCUITRY

3.3.1 Typical Organization

The block diagram for a typical EEPROM array circuit is shown in Fig. 3.12. The rectangular EEPROM array is in the upper right hand corner. Its rows are accessed by the X-decoder, with input buffers that receive a binary address input, and a decoder that selects one row in the array. The columns of the array represent the Y-direction. They feed into sense amps and, from there, to I/O buffers. The selection of individual columns is achieved with a Y-decoder and its buffers and latches. The operating state of the array is controlled by three logic inputs: chip enable, which powers up and selects the chip; write enable, which makes the chip ready for an input to its memory array; and output enable, which causes the chip to put data on its I/O leads.

In some applications, it is advantageous to address the array in a serial mode. This is shown in Fig. 3.13. Using this approach, 16 KB of memory can be operated with only four pins. This is in contrast to the 24 pins needed for a bytewide 16 KB EEPROM. For a simple two-wire bus, the information from one input is clocked into the control logic, which translates it into its functional read or write state, and extracts the address, which is sent to the X-decoder and Y-decoder of the same rectangular array used in a standard EEPROM. The control logic also operates the high-voltage generation and timing control needed for writing, and the data register and output sequence needed for reading.

3.3.2 Nonvolatile Writing

Originally, EEPROM circuits required an external power supply that provided voltages in the 10 to 20 V range, which were necessary to generate the fields used for

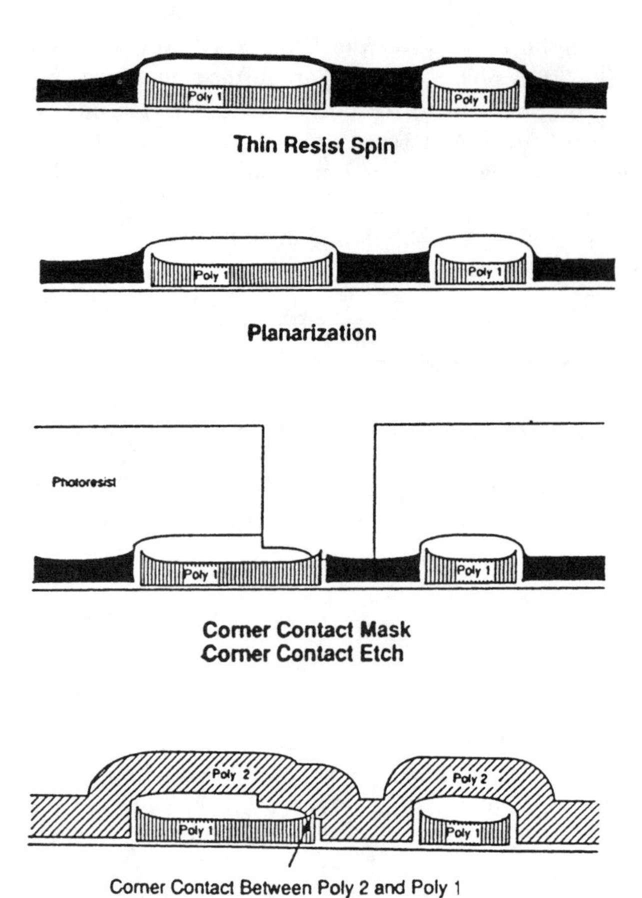

Figure 3.11 Formation of corner contact.

Fowler–Nordheim tunneling. In addition, these external power supplies needed to have circuitry that shaped the rise time characteristic of the high-voltage pulse in order to control the stress on the tunneling element. An innovation, first reduced to practice with a textured poly device [3.14], moved the ability to provide a high voltage of controlled shape inside the chip so that the user only required the normally available 5 V supply to perform the nonvolatile write function. In particular, the high voltage was generated by a charge pump [3.15]. A typical version is shown in Fig. 3.14. Here, alternating capacitors are connected to one of two pulse generators. When A is high, B is low, and vice versa. When B is high, diode D turns on and charges capacitor C_1. Then, B goes low, leaving C_1 isolated. Next, B goes high, turning on D_1 and charging C_2. The amount of charge in C_2 will be proportional to B, plus about half of what had been transferred during A. As the phases alternate, the voltages on the capacitors increase. The increase is greater the further they are to the right. In this way, voltages in the 20 to 30 V range can be attained rapidly.

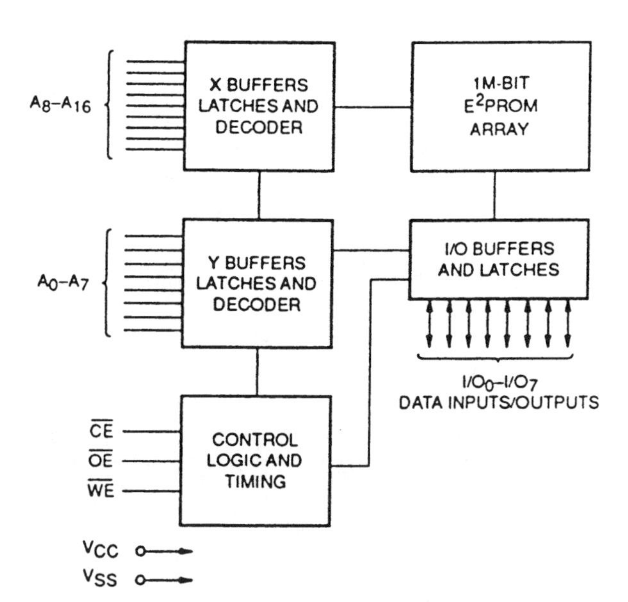

Figure 3.12 Block diagram of a typical bytewide EEPROM.

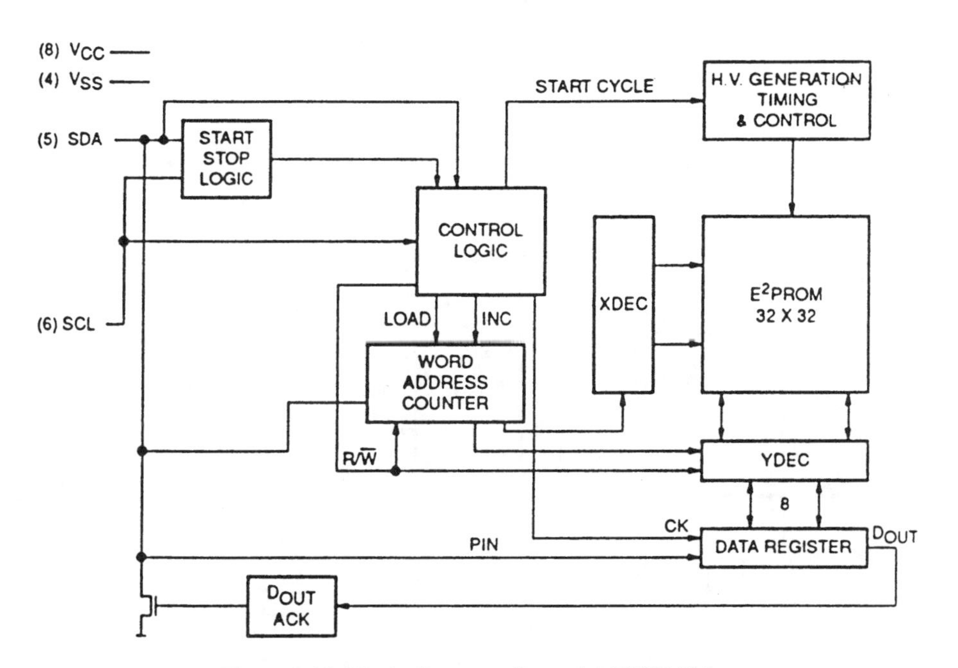

Figure 3.13 Block diagram of a serial EEPROM.

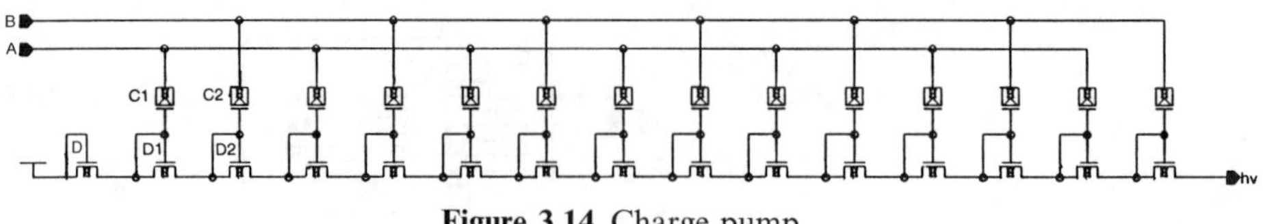

Figure 3.14 Charge pump.

In order to obtain sufficiently high, but minimum stress, tunnel currents, the ramp rate dV/dt must remain at a defined and controlled level. This level is within an order of magnitude of 1E4 V/sec., depending on details of structure and application. Such a ramp rate is about a thousand times slower than could be obtained with typical capacitor and transistor structures. A small area solution to this problem is represented by the ramp rate control circuit described by Simko [3.16] and shown in Fig. 3.15. Here, a leaker transistor T1 will tend to discharge the charge-pump output node V_0. The amount T1 will leak depends on the potential on its gate, which is the instantaneous potential of the output node transmitted by the capacitor, less the amount a second leaker transistor T2 removes from it. The gate of this transistor is biased at a constant voltage by two other transistors acting as a constant current voltage divider. As V_1 decreases because of T_2, discharge through T_1 decreases and ultimately stops.

The maximum voltage is determined by a diode clamp. A corner detector will sense when clamping has begun, which, in turn, starts a timer that discharges V_o when the end of the flat top time has been reached. In this way, the whole trapezoidal shape of the high-voltage pulse needed for tunneling is generated.

3.3.3 Reading

Reading is performed by a tracking sense amplifier that is sensitive to the process variations experienced by the cells in the specific array. The bitline of the

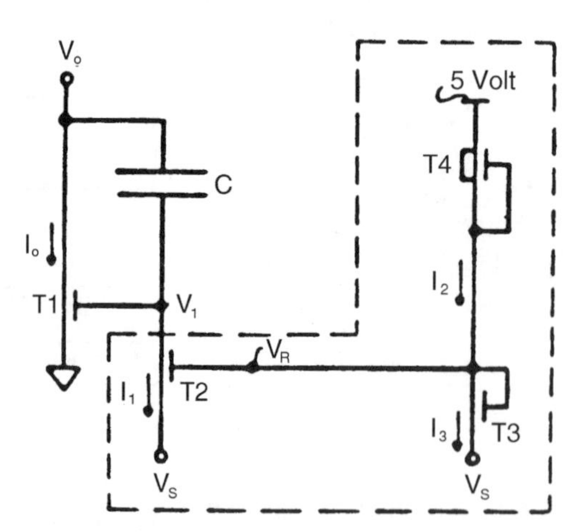

Figure 3.15 Ramp rate control circuit.

addressed cell feeds into one input of an op amp. This bitline is driven by a current derived from a reference cell. The other input to the op amp is controlled by another cell that acts as a bitline reference. The output of the sense amp feeds into a push–pull I/O buffer.

3.3.4 Typical Operation

In standby, Output Enable (OE\) and Chip Enable (CE\) are inactive at a logic high level. Every cell has its wordline at zero, its bitline is precharged, and its deselect-line is turned on. When both OE\ and CE\ are active (set to a logic low), reading is initiated. Inputs at the address pins are latched in, the addressed wordline goes high, and the current through the cell is sensed in the sense amp, which transmits it to the I/O buffers, which, in turn, drive the enabled I/O lines. Output ceases when either CE\ or OE\ returns to high.

Write operations are initiated when both Write Enable (WE\) and CE\ go "Low" and OE\ is "High." The address is latched in by the falling edge of the later one of the first two control signals. The data on the I/O pins are latched into the interior of the chip when either WE\ or CE\ goes "High." This starts the ramp-up of the wordline high voltage by the charge pump, clamped by the ramp reaching the controlled maximum voltage. Now, the flat top timer is started. At its termination, the wordline voltage rapidly discharges to zero. Depending on the latched input signal, a bitline either is grounded or coupled to a bitline high-voltage ramp. The whole process takes several milliseconds and is terminated by the falling wordline high voltage. During this time period, the internal chain of operations cannot be altered by external events.

3.3.5 Special Features

Ever since their existence, the one annoying feature in EEPROMs has been the long time necessary to perform a nonvolatile write operation. It is typically a few milliseconds, or four orders of magnitude slower than comparable volatile operations. A way to gain effective speed is to spread this delay over a large number of devices. A nominal parity in speed could be achieved if 10,000 bytes were written simultaneously, but the software and hardware logistics necessary to do this make this approach cumbersome.

A major improvement in effective write speed is achieved by the ability of an EEPROM to perform "page" writing. With the use of this feature, many bytes can be written during the same write operation. The size of a page is typically determined by the number of cells in one row of the rectangular array of EEPROM cells within one chip. In a megabit part, the page consists of 256 bytes, reducing the effective write time per byte to 19 microseconds.

EEPROMs perform their write operations by making themselves insensitive to external signals, while the nonvolatile write process is underway inside the chip. It can be accessed only after a fixed time period, which is guardbanded to include all possible variations of chip characteristics and environmental operating conditions.

Under most circumstances, this guardband is about two times longer than the actual time used to perform a given write operation. The wait to perform another operation can be minimized by a feature called DATA BAR POLLING, which outputs a signal to an I/O pin when the write operation is concluded internally.

On rare occasions, a combination of noise signals on the external pins of the array may result in an inadvertent write. Such an event can be prevented if the "Software Data Protect" feature is used. It involves a set of specific write operations that have to be performed before data in the array can be changed.

Hardware protection against inadvertent writing includes:

1. A pulse shorter than 10 nanoseconds cannot cause writing.
2. All functions are inhibited when V_{cc}, internally sensed, is lower than 3.6 V.
3. Holding WE\ or CE\ "high,"or OE\ "low," will prevent internal writing.

3.4. DEGRADATION MECHANISMS

3.4.1 Time-Dependent Dielectric Breakdown (TDDBD)

Textured poly devices, in common with other floating gate devices, rely on MOS structures to accomplish the translation of voltage signals into currents and currents into voltages. The main element in MOS structures is the silicon dioxide between silicon electrodes. These electrodes may be single-crystal silicon on one side and polysilicon on the other, or they may be polysilicon on both sides of the dielectric.

These layers fail by dielectric breakdown after a prolonged exposure to an applied electric field. This breakdown is manifested by a sudden onset of high current between the two electrodes, which is followed by a permanent state of low resistance between them. The predictability of this behavior was systematized by Crook [3.17], but it is best explained by Chen, Holland, and Hu's [3.18] model that assumes widespread injection of electrons from one electrode, impact-ionization by the injected electrons, accumulation of some of the generated positive charges in a few isolated locations, and self-regenerative conduction through one of those locations, leading to thermal runaway.

Injection is by a Fowler–Nordheim tunneling mechanism with regions of higher current density provided by submicroscopic surface irregularities on the (intended to be plane) emitting electrode. Differences between the dielectric strength of individual oxides then would be due to silicon surface features and to the propensity of the oxide to form positive charge clusters.

It is useful to examine the oxides that are part of the textured poly cell (Fig. 3.1). Clearly, the gate oxides of the select transistor, the deselect transistor, and the floating gate transistor are standard gate oxides that could ultimately fail by the classic TDDBD mechanism. They share this property with the rest of the chip and with the whole world of MOS technology.

In addition, there are oxides between the flat portions of the poly. As we have seen previously, the surface of the lower poly layer is dotted with conical bumps.

Clearly, emission is favored at their tips. The recessed regions in between, because of a negative radius of curvature and a thicker oxide, will see greatly reduced electrical activity. The high current density at the tips of the cones would seem to invite early breakdown, but positive charge clusters are needed to accomplish this, and the presence of these cones does not appear to especially aid their formation. Experimental data indicate that conditions for breakdown require a prolonged passage of current.

The final regions of interest are the oxides between two poly layers, at the side-walls and the edges of the lower poly layer. At the corners, edge emission is clearly dominant. This again does not appear to accelerate the buildup of the necessary positive charge clusters. Empirically, based on data from test devices maximizing edge emission, breakdown occurs only after much charge has passed. From the columnar nature of the polycrystalline layer, it can be expected that the side-walls of a poly layer will contain vertical ridges, making a kind of edge emission possible. But these same side-walls typically have thicker oxides, reducing the amount of potential emission.

A predictable TDDBD is related to the reproducible aspects of a process, defining the distribution of the details of the emitting surface features of the silicon, and the density and distribution of positive charge cluster generating or storing centers in the oxide. These characteristics have been optimized to such a degree that real-time accelerated data and calculated times to failure indicate, as a minimum, several decades of operating life. The only unmodeled and shorter life aspect of oxide failures are unintended defects incurred during manufacturing, such as inclusions of foreign materials into the dielectric, accidental thin spots, and local changes in the emitting surfaces. These are lumped under the heading "Extrinsic Defects."

3.4.2 Extrinsic Breakdown

Breakdown due to extrinsic defects can be viewed as time-dependent dielectric breakdown of unusual variants of normal structures. The accidental deformations may give rise to material constants and geometric parameters that result in reduced times to failure. Extrinsic oxide breakdown is one of the causes for both infant mortality and long-term reliability failures. As monitored on test structures, there is a small, but finite, chance for such defects. Most of these defects are gross and cause failure and self-elimination during testing. Others fail during burn-in, or as a result of an equivalent or tougher electrical screen. The final hurdle a device must pass is cycling. During this screen, the devices are subjected to several thousand erase/program sequences. Since the number of failures decreases with an increasing number of stress cycles, it can be concluded that these are the last of the infant failures.

3.4.3 Trap-up

As reported by Liang and Hu [3.19], the passage of a current through silicon dioxide is accompanied by the occupation of existing trap sites by electrons and by the generation of new trap sites that also may be negatively charged. Liang and Hu found that the density of the trapped charge was proportional to the current density integrated over time (flux).

This same phenomenon can be observed in textured poly devices. As current tunnels from one poly layer to another, it leaves behind a trail of trapped negative charges. Since these trapped charges have a potential opposite to that which is applied to cause tunneling, every tunneling event causes a need for an increase in applied voltage in order to result in the same tunneling current the next time. Based on the approach that the trapped charge was proportional to the flux, the amount of trap-up as a function of current was modeled and found to agree exactly with measured values [3.20]. This model predicts that, over the life of a cell, its tunnel voltage (the voltage necessary to support a constant cell current [C dV/dt]) will increase by as much as 8 V.

A problem arises when, after many data changes, the tunneling region has trapped up to the level that the tunnel voltage of one cell element (say between the floating gate and the wordline poly) has reached the level of the maximum of the applied high voltage (the flat top of the trapezoidal write pulse). Now, the amount of charge transferred from the floating gate is so small that the correct state cannot be established. The part fails to write correctly. It has reached the end of endurance.

End-of-life endurance by trap-up is the typical failure mode of a textured poly cell. This is because the maximum write voltage has been chosen to be lower than the voltage that would result in intrinsic time-dependent dielectric breakdown. Endurance failure by oxide breakdown is a relatively rare event, indicating that an infant failure site has eluded screening. In general, after many data changes, the part stops writing correctly, which is a relatively benign mode of failure. Even though there is a limit to the end-of-life endurance of a textured poly memory cell, this limit is typically near several million data changes. With all parameters optimized, the end-of-life endurance of a megabit device should be close to the 10 million mark. Optimized serial parts have smaller memory arrays and can be expected to have even higher end-of-life endurances.

3.4.4 Comparison with Thin Oxide

A comparison between textured poly and thin oxide tunneling floating gate devices should start with the details of the charging process. Representative energy band versus oxide thickness plots are shown in Fig. 3.16. They are superimposed at the electron-emitting silicon interface. The thin oxide device has a constant high field through the short length of its thickness. The textured poly device starts at about the same field level, but after about 100 Å, the initially high field has rapidly decreased to a low level and, through most of its 500 Å thickness, it is at a low field (see Fig. 3.6).

Most of the action of the thin oxide structure is at a distance of about 30 – 40 Å. This is where an electron emerges after being emitted from a silicon surface. In addition, the applied potential is reversed during normal operation, and any electron trapped in the region between silicon and 30 Å into the oxide will tunnel back into

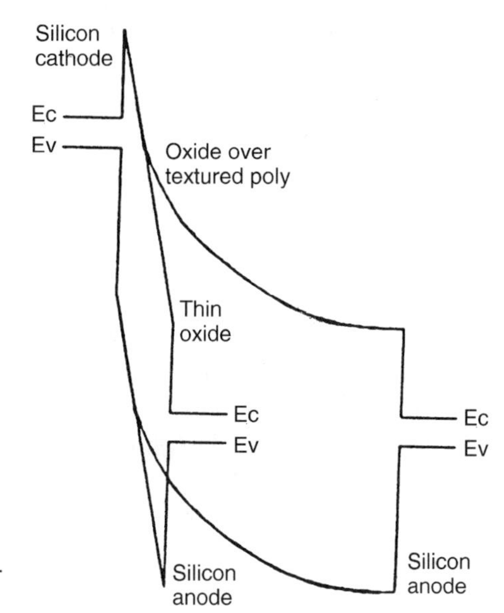

Figure 3.16 Band diagrams of oxide tunneling structures.

the silicon, unless it is in an especially deep trap. This can be expected to be a rare occurrence. Since electron emission may occur from either electrode, the details of the interfaces at both electrodes are very important, and both may contribute to failure with equal probability. This sensitivity to chemical and physical details at both silicon interfaces may contribute to the higher rate of dielectric breakdown.

The textured poly structure shares the same characteristics with the thin oxide structure over the first 30–40 Å. Then, the field rapidly decreases and ends up at a relatively low value for most of the dielectric. The low local fields permit sustained residence of electrons in shallow traps at lower and lower fields. But since the structure is always biased in one direction only, it is relatively insensitive to small details at the electron-collecting electrode (the anode). The low-field region of 450 Å-plus thickness can be seen to act as an effective retardant to regenerative current flow originating near the injecting interface. This should further reduce the probability of dielectric breakdown.

Following Mielke et al. [3.21] and Lai [3.22], the major limit on program/erase cycling in production-level thin oxide devices is oxide breakdown. It is manifested by a single cell failure. This failure mode affects 1 to 5% of the devices of 64 KB EEPROMs during the first 10,000 cycles. It is attributed to latent oxide defects. It could arise at either silicon/silicon dioxide interface, but it could just as easily be associated with oxide defects in the middle of the thin oxide layer, succumbing to time-dependent dielectric breakdown. Device manufacturers never solved this latent defect problem. The solution adopted was to live with the defect by using an error correction code. The approaches used included Hamming code error detection and single-cell-level error correction. The use of such area-intensive approaches makes it possible to achieve an acceptable endurance. The trap-up encountered in these devices may be no more than 2 to 3 V.

In contrast, textured poly devices trap-up at a rate higher by about a factor of three. This is probably because the high-field regions utilized for tunneling amount to a relatively small area whose current density must be high enough to charge up the capacitance of the floating gate to a sufficiently high voltage. But this behavior is predictable. It can be counterbalanced by operating over a larger range of high voltage or over a smaller cell window. Both of these are primarily circuit design challenges. In addition, process improvements have addressed trap-up rate reductions and infant mortality problems. The latent oxide defect is not a limiting factor for textured poly devices. This is probably due to the thick low-field region of the oxide. Internal error correction is, therefore, not necessary for reliability enhancement. Instead, redundancy is used for yield improvement, as is common in most large-area memory devices. The impact on chip area of a few redundant rows is negligible, especially when compared to the yield gained by the ability to repair a few defective bits.

3.5. TYPICAL CHARACTERISTICS

3.5.1 Performance

Textured poly cells have an inherent advantage. The cell size is smaller than a thin oxide cell for the same layout rules. The larger size of the thin oxide cell arises, in part, from the need to be efficient in biasing the floating gate by capacitance coupling, which requires that the coupling ratio between the control-line and floating gate must be high. But the thin oxide tunneling device inherently has a high capacitance. In order to achieve an acceptable capacitive coupling ratio between the floating gate and the wordline, the area of the floating gate has to be made much larger than the minimum lithography limits. Added to this is the need to build in redundancy, or a high level of error correction. This may increase the chip area by as much as a factor of two.

For these reasons, the pressure has been high on thin oxide EEPROM cell technologists to reduce the feature sizes in the layout rules, particularly the area of the thin oxide tunneling region. While this has made life more difficult for manufacturing personnel, it has made it possible to design faster circuits. Conversely, the thick oxide requirements of a textured poly memory circuit have been relatively easy to meet with standard manufacturing technology. However, the thick oxide, which until recently was also used as the gate oxide, limited the high-speed capabilities of the textured poly EEPROM products, compared to those with thin oxide. Table 3.1 gives examples.

This speed refers to parameters related to read access time. The recent introduction of very high-speed products was made possible by separating the thick oxide used for tunneling from the scaled thinner oxide used for the high-speed peripheral circuits. This was done by adding one mask to the process.

TABLE 3.1 REPRESENTATIVE PERFORMANCE OF TEXTURED POLY EEPROMs (AS MEASURED BY ADDRESS ACCESS TIME t_{AA})

Device	Capacity	t_{AA}	Year Introduced	Process Technology
X28C010	1Mb	120 ns	1990	CMOS
X28CT256	256kb	25 ns	1992	CMOS
X28VC256	256kb	45 ns	1992	CMOS
X28HC256	256kb	70 ns	1991	CMOS
X28C256	256kb	150 ns	1988	CMOS
X28256	256kb	250 ns	1986	nMOS
X28HC64	64kb	55 ns	1992	CMOS
X28C64	64kb	150 ns	1988	CMOS
X2864B	64kb	150 ns	1986	nMOS
X2864A	64kb	250 ns	1984	nMOS

The long write cycle time common to all nonvolatile memory technologies currently in production is limited by the need to balance speed with reliability. At present, write speeds are so low, between 0.75 and 3 msec, that improvements of several orders of magnitude would be required before an impact would be felt. There is no obvious approach to achieve this on a physical level. Instead, data architecture and data management approaches have been refined. What is meant here is the full chip writing used in NOVRAMs, page writing, and DATA BAR POLLING.

3.5.2 Endurance

Endurance is a concept important for nonvolatile memories. It is defined by the number of data changes that can be performed on every cell of a given memory chip before one of its cells fails to meet its data sheet specifications. It describes the reliability of a device in terms of the number of write operations that can be performed on it without failure. As with other reliability related characteristics, it varies over the life of a part in accordance with the bathtub curve.

The rapidly decreasing early failure rate is generally caused by manufacturing defects. The fraction of devices affected by such problems is eliminated by electrical screening. The long-term, or steady-state, phase of reliability is caused by long-lived devices with almost negligible defects, or the first failures of normal process-related TDDBD. The final, end-of-life endurance level is determined by physical limits. In textured poly devices, this limit is trap-up. In thin oxide devices, it is dielectric breakdown.

In the design of an application, the maximum number of data changes that a nonvolatile device will perform must be below the "defined" end-of-life endurance. This defined end-of-life endurance may be the standard upper limit specified in the data book of the device. It may also be the experimentally documented end-of-life endurance exemplified by a probability chart (Fig. 3.17). Typical standard upper limits specified for textured poly devices are 10,000 or 100,000 cycles. In contrast, documented end-of-life endurances generally have median values ranging from a few hundred thousand cycles to a few million cycles, depending on specific features of the design.

The endurance probability chart shown in Fig. 3.17 plots the number of data changes of the whole EEPROM array as the logarithm of cycles versus the cumulative probability of failure (i.e., one "cycle" for this graph means that 512 KB cells have been subjected to one data change). A failure event is defined by the first cell of the 512 KB array that failed to write correct data. The sample of nine devices indicated a median end-of-life endurance of 5 million cycles. The cycles were accumulated by a special test mode wherein the data in all cells of the array were changed simultaneously.

In order to predict the reliability of the equipment containing an EEPROM, its endurance failure rate must be known. This long-term failure rate can be estimated on the basis of experimental data. It is typically obtained by subjecting a set of devices to a fixed number of program/erase cycles. Let us assume that three hundred 256 KB EEPROMs were page cycled to 100,000 cycles and that one device failed. The failure rate (FR) in terms of the number of page write operations will be

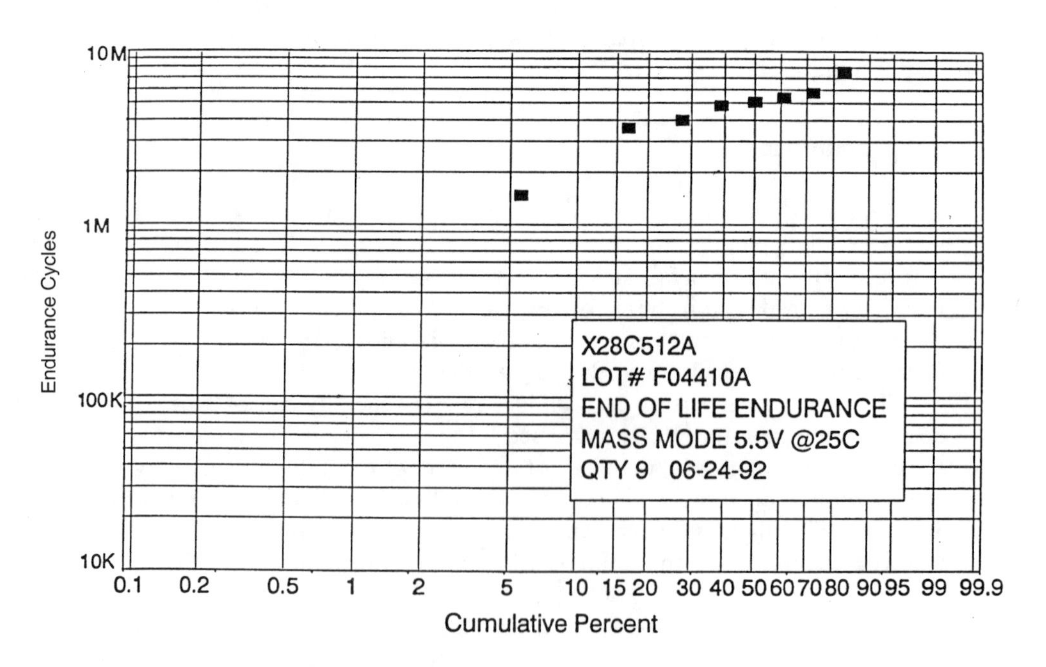

Figure 3.17 Endurance probability chart.

$$\text{FR (per page ops)} = (1+1)/(300)(100,000)(512)$$
$$= 1.3\text{E} - 10 \text{ or } 0.13 \text{ ppb} \qquad (3.10)$$

If this part is used in such a way that 20 pages will see 100,000-page operations each, another 100 pages will see 10,000, and the remaining pages will see 1,000-page cycles, the expected failure rate, E_{FR}, will be

$$E_{FR} = 0.13\text{E} - 9 \cdot (20 \cdot 100{,}000 + 100 \cdot 10{,}000 + 392 \cdot 1{,}000)$$
$$= 0.00044 \text{ or } 0.044\% \qquad (3.11)$$

This parts per billion (ppb) approach permits us to make such estimates by adding the number of data changes for different parts of the chips as long as the defined end-of-life endurance of individual cells is not exceeded. Typical values for textured poly devices are in the 0.1 to 1 ppb range for write operations in the page mode and 0.01 to 0.1 ppb for write operations in the byte mode.

3.5.3 Retention

Ultimately, it is retention that differentiates EEPROMs from other semiconductor memory. It is a measure of the time that a memory cell can retain its information whether it is powered or unpowered. Retention can be quantified by measuring or estimating the time it takes for the floating gate capacitor to discharge when it is intended to keep the information stored. From

$$I_{FN} = C\ dV/dt \qquad (3.12)$$

where I_{FN} is Eq. (3.3) expressed in volts and amperes, we arrive at

$$\exp(B/V_t) - \exp(B/V_o) = t \cdot AB/C \qquad (3.13)$$

by separating variables and integrating. V_t is the floating gate potential at time t, V_o is the floating gate potential at time zero, C is the floating gate capacitance, $B = b \cdot x_o$ is the Fowler–Nordheim exponential constant, and $A = a/x_o^2$ is the pre-exponential constant. Assuming the term containing V_o to be negligible and using Eq. (3.9) as the basis for I_{FN} lead to

$$\ln t = B'/(V_t - \text{delta}) - \ln A'' \qquad (3.14)$$

This equation was fitted to the data obtained from individual EEPROM cells at relatively high constant voltages applied to the select gate. The value of delta was calculated from first principles. The resulting constants and the extrapolated end of retention with 5.5 V applied are shown in Table 3.2. Clearly, the textured poly floating gate retention is of an astronomical time scale. A plot of measured individual cell discharge rates using the constants of Table 3.2 is shown in Fig. 3.18. The experimental data were obtained over the first six decades of seconds.

The curves in Fig. 3.18 were fitted to the data on which Table 3.2 was based. The constants of that table and Eq. (3.14) were utilized in extrapolating the time-to-failure to the 4 V level. From the definition that the end of retention occurs after storage with 5.5 V applied to the wordline, individual cell retention values may be picked from Fig. 3.18. In contrast to the third column in Table 3.2, where the time

TABLE 3.2 DATA LOSS CONSTANTS AND TIME-TO-FAILURE CALCULATED FROM THE EXPERIMENTAL DATA OF 10 MEMORY CELLS UNDER HIGH-VOLTAGE STRESS [SEE ALSO EQ. (3.12) AND FIG. 3.18]

Constant B′ (in volts)	Constant ln A″	Time-to-Failure (at 5.5 volts in years)
359	18.9	3.1E15
346	15.1	1.0E10
346	19.6	1.2E14
449	24.7	5.9E20
352	20.0	7.5E14
428	23.4	3.3E19
344	19.1	1.3E14
395	21.7	2.6E17
394	21.6	2.5E17
406	22.0	1.8E18

scale is in years, the abscissa in Fig. 3.18 is given as the logarithm of time in seconds. For reference, 1E6 seconds is 11.5 days, 1E9 seconds is 30 years, and 1E18 seconds is 30 billion years.

While such a high retention is the property of all floating gate memories, it is particularly high in textured poly devices. This arises from the form of Eqs. (3.7) and (3.8). As the potential across the floating gate gets smaller, the constant increment

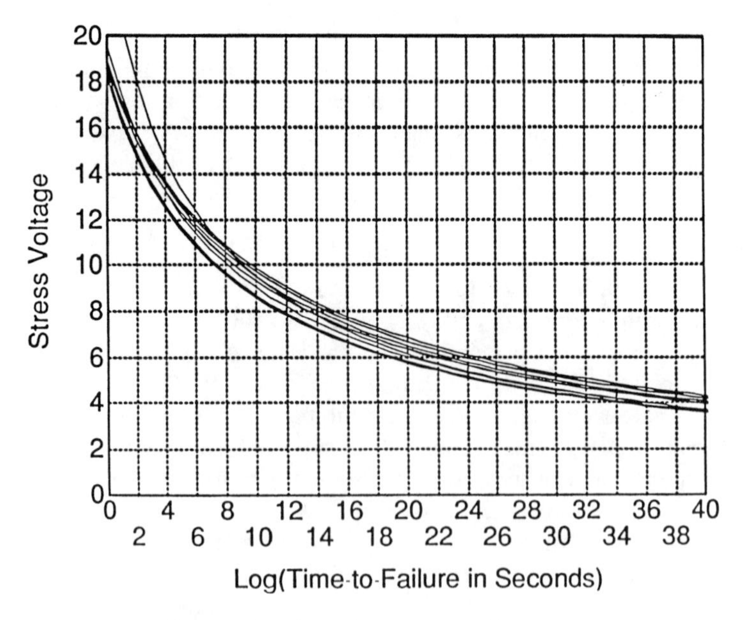

Figure 3.18 Retention versus voltage stress.

subtracted reduces the driving force for leakage disproportionately more strongly. The thick oxide between all poly areas and this special feature combine to ensure good retention even for slightly defective cells. In monitor lots, the failure rates actually found at the accelerated condition of 250°C translate into less than 5 failure units (FITs) at 55°C and a 60% upper confidence limit.

Textured poly memory devices have performed well in applications at 180°C. They show no measurable leakage in nondefective cells at 250°C, and they lose their charge only at temperatures above 300°C.

3.5.4 Technology Features, Cells, and Capacities

Based on the preceding considerations, it is clear that textured poly devices occupy an important position among nonvolatile memories. The formation of the tunneling regions requires both special knowledge and accumulated experience. It does not require special equipment or parameters that are difficult to control.

From the point of view of reliability, textured poly devices have several inherent advantages. The dielectric under high-field conditions for tunneling is electrically in series with a region under a low field. This makes it less sensitive to defects and less susceptible to dielectric breakdown. While the end of endurance is defined by trap-up, this limit represents a benign failure mode in contrast to the unpredictability of dielectric breakdown. It also facilitates the use of electrical screens for the elimination of infant mortality failures and makes error correction unnecessary. Furthermore, retention is fundamentally better in textured poly devices than in thin oxide floating gate devices. Defects that should cause retention failures in other EEPROM technologies will be less effective in doing so in textured poly devices. This is clearly demonstrated by the ability of textured poly devices to withstand very high temperatures without problems, compared to repeated thin oxide failures at temperatures above 125°C.

Another important advantage is the absence of the need to use minimum dimensions to define the area of the thin tunneling oxide. It makes textured poly devices more manufacturable, since small cell areas can be achieved with standard state-of-the-art dimensions. And since the reduced textured polyoxide vulnerability does not necessitate error correction, a large advantage in area, and therefore cost, is achieved. This is demonstrated by Fig. 3.19. This graphic shows the total chip areas of representative 256 K CMOS-EEPROMs, expressed in terms of their area ratio with respect to the most recent textured poly part in production in 1992. It can be seen that the thin oxide parts cluster in the area ratio region of 1.6 to 2.0 at a technology level of about 1.2 μm. The textured poly product with an area ratio of 1.5 can achieve this better ratio even though its representative feature size is as large as 1.8 μm. The second textured poly product is the reference for the other five devices; therefore, its area ratio is 1.0. With a feature size near 1.2 μm, only half the silicon area is required to perform all the functions of the thin oxide 256 K CMOS EEPROMS.

What about the future? Steady progress has been made in the technological development of textured poly devices. This progress has been realized primarily by

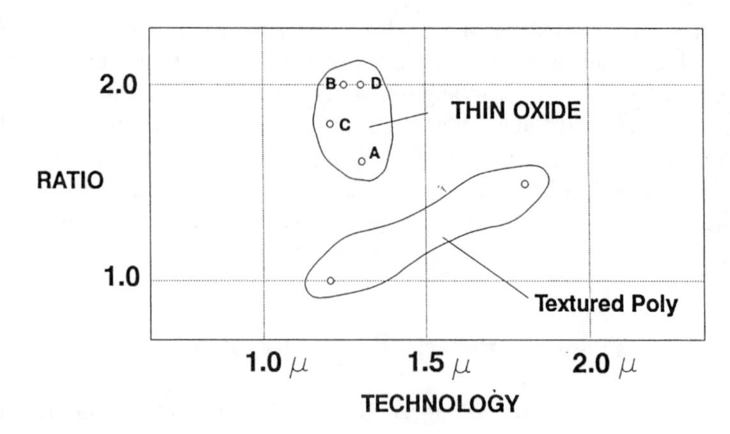

Figure 3.19 Chip area ratio of 256 K CMOS EEPROMs.

the reduction of feature sizes, as exemplified in Table 3.3. Such feature sizes translate into chip capacities, and these are shown in Fig. 3.20 with the die sizes normalized to a 64 K EEPROM made with 0.7 μm lithography. The current technology level permits the manufacture of a 1 Mbit EEPROM chip. From the time scale in Table 3.3, we can expect 4 Mbit textured poly EEPROM chips in 1996 and a 16 Mbit chip in 1998.

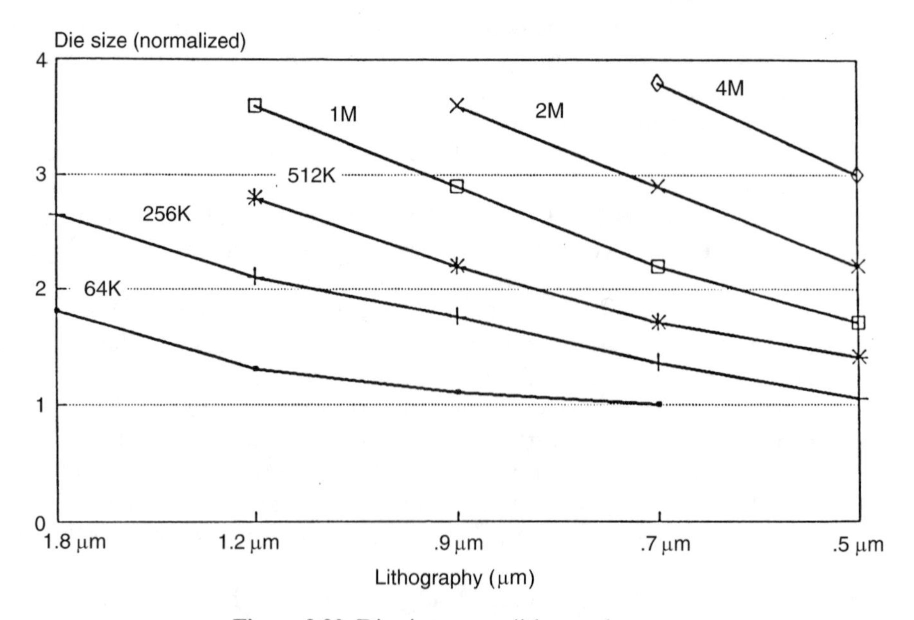

Figure 3.20 Die size versus lithography.

TABLE 3.3 TEXTURED POLY TECHNOLOGY EVOLUTION

Feature	Feature Size				
Lithography	1.8 μm	1.2 μm	0.9 μm	0.6 μm	0.4 μm
EEPROM cell size	72 μm^2	40 μm^2	23 μm^2	9.3 μm^2	6 μm^2
Peripheral gate oxide	500 Å	250 Å	250 Å	150 Å	100 Å
Interpoly tunnel oxide	700 Å	700 Å	700 Å	450 Å	400 Å
Metallization	Single	Dual	Dual	Dual	Dual/triple
Array size	256k	1M	2M	4M	16M
In production	1987	1990	1993	1996	1998

References

[3.1] R. H. Fowler and L. Nordheim, "Electron emission in intense electric fields," *Proc. Royal Soc. London*, 119A, pp. 173–181, 1928.

[3.2] M. Lenzlinger and E. H. Snow, "Fowler–Nordheim tunneling into thermally grown silicon," *J. Appl. Phys.* vol. 40, no. 1, pp. 278–283, 1969.

[3.3] Z. A. Weinberg, "Tunneling of electrons from Si into thermally grown silicon," *Solid State Electronics*, vol. 20, pp. 11–18, 1977.

[3.4] R. M. Anderson and D. R. Kerr, "Evidence for surface asperity mechanism of conductivity in oxide grown on polycrystalline silicon," *J. Appl. Phys.*, vol. 48, no. 11, pp. 4834–4836, 1977.

[3.5] Many instructive TEM cross sections of textured poly tunneling structures were obtained by K. Ritz at Philips Labs.

[3.6] Z. A. Weinberg and A. Hartstein, "Photon-assisted tunneling from aluminum into silicon dioxide," *Solid State Communications*, vol. 20, pp. 179–182, 1976.

[3.7] A. Hartstein and Z. A. Weinberg, "Unified theory of internal photoemission and photon-assisted tunneling," *Physical Review B*, vol. 20, no. 4, pp. 1335–1338, 1979.

[3.8] Z. A. Weinberg, "On tunneling in metal-oxide-silicon structures," *J. Appl. Phys.*, vol. 53, no. 7, pp. 5052–5056, 1982.

[3.9] R. T. Simko, "Substrate coupled floating gate memory cell," U.S. Patent 4,274,012, 1981.

[3.10] S. Jewell-Larsen, I. Nojima, and R. Simko, "5-volt RAM-like triple polysilicon EEPROM," *Conference Proceedings of Second Annual Phoenix Conference*, pp. 508–511, 1983; IEEE Catalog No. 83CH1864-8.

[3.11] D. C. Guterman, "Nonvolatile electrically alterable memory," U.S. Patent 4,599,706, 1986.

[3.12] B. Yeh, "Side wall contact in a nonvolatile electrically alterable memory cell," U.S. Patent 5,023,694, 1991.

[3.13] R. Lambertson, A. Malazgirt, C. Lo, A. Vahidimovlavi, P. Holland, M. Fliesler, and H. Gee, "A high-density dual polysilicon 5 volt only

EEPROM cell," *Technical Digest of the 1991 IEEE IEDM*, Paper 11.2.1, pp. 299–302, 1991. © 1991 IEEE.

[3.14] J. Drori, S. Jewell-Larsen, R. Klein, W. Owen, R. Simko, and W. Tchon, "Single 5V supply nonvolatile static RAM," *Digest of the 1981 Internatl. Solid State Circuits Conf.*, pp. 148–149, 1981.

[3.15] J. Dickson, "On-chip high voltage generation in MNOS integrated circuits using an improved voltage multiplier technique," *IEEE J. Solid-State Circuits*, SC-11, pp. 374–378, 1976.

[3.16] R. T. Simko, "High voltage ramp control systems," U.S. Patent 4,488,060, 1984.

[3.17] D. L. Crook, "Method of determining reliability screens for time dependent dielectric breakdown," *Proceedings of the 1979 International Reliability Physics Symposium*, pp. 1–7, 1979.

[3.18] I. C. Chen, S. Holland, and C. Hu, "A quantitative physical model for time-dependent breakdown in SiO_2," *Proceedings of the 1985 International Reliability Physics Symposium*, pp. 24–31, 1985.

[3.19] M. Liang and C. Hu, "Electron trapping in very thin thermal silicon dioxides," *Technical Digest of the 1981 IEEE IEDM*, pp. 396–399, 1981.

[3.20] H.A.R. Wegener, "Endurance model for textured poly floating gate memories," *Technical Digest of the 1984 IEEE IED*M, Paper 17.7, pp. 480–483, 1984.

[3.21] N. Mielke, A. Fazio, and H.-C. Liou, "Comparison of Flotox and textured-poly EEPROMs," *Proceedings of the 1987 International Reliability Physics Symposium*, pp. 85–92, 1987.

[3.22] S. K. Lai,"Oxide and interface issues in nonvolatile memory," presented at the Santa Clara Valley Section of the IEEE Electron Devices Society 1991 Symposium: Advances in Semiconductor Technologies, pp. 1–19, 1991. © 1991 IEEE.

Chapter 4

Manzur Gill
and Stefan Lai

Floating Gate Flash Memories

4.0. INTRODUCTION

The semiconductor revolution started with the invention of the transistor in 1947. Bipolar transistor technology replaced the vacuum tube which, in turn, was replaced by MOSFET technology in the 1970s, starting the age of LSI/VLSI/ULSI (Fig. 4.1 and Table 4.1). The MOS/VLSI revolution was ushered in by the invention of DRAM, which replaced magnetic memory. In addition, several other semiconductor memories have been in use (Fig. 4.2), the most recent one being Flash memory, which appeared in 1985.

Many predict that Flash memory will be the fuel for the next stage of this revolution [4.1]. This belief is based on the observation that Flash memory has many attributes of an ideal memory (Table 4.2) needed in a computer system (Fig. 4.3). Flash memory's combination of density, rewritability, and nonvolatility (to name a few) makes it unique among semiconductor memories (Fig. 4.2), a superset of these key features.

Among the currently available memory technologies (floppy disk drive, hard disk drive, ROM, EPROM, OTP, EEPROM, DRAM, SRAM, shadow RAM, and Flash), Flash memory satisfies many of the characteristics of an ideal memory (Table 4.3).

Because of the combination of nonvolatility, in-system rewritability, and high density, Flash memory has already found many applications that can broadly be classified [4.2] as:

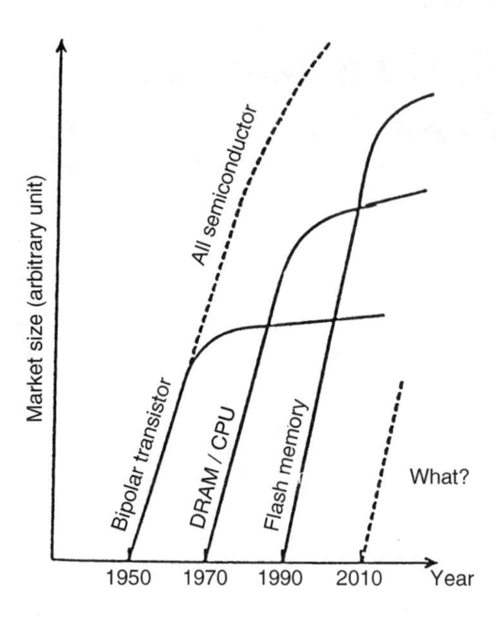

Figure 4.1 History of semiconductor business [4.1].

TABLE 4.1 HISTORY OF SEMICONDUCTOR TECHNOLOGY DRIVERS

Decade	New Technology	Replaced Technology	Market
1950s	Bipolar	Vacuum tubes	Radio/TV
1970s	MOS (DRAM)	Core memory	Computer
1990s	MOS (Flash)	Hard disk/floppy disk	Computer, personal systems

1. *Data Accumulation*—Medical instrumentation, flight recorders, point of sales terminals, handheld instruments, and remote sensing. The benefits of Flash memory in data accumulation applications include density, ruggedness, reliability, and nonvolatility.
2. *Data/Lookup Table Storage*—In this type of application, Flash memory devices store large amounts of infrequently updated system data or look-up tables, for example, PBX switcher and laser printers. Here, the advantages of Flash memory include in-system upgradability, nonvolatility, and density.
3. *Embedded Code Storage*—PC BIOS, digital cellular phones, control software in laser printers, and telecommunication bridges/routers.

Other areas in which Flash memory is finding application are high-end file storage, memory cards, smart cards, PLDs, digital audio machines, satellite systems, and neural nets.

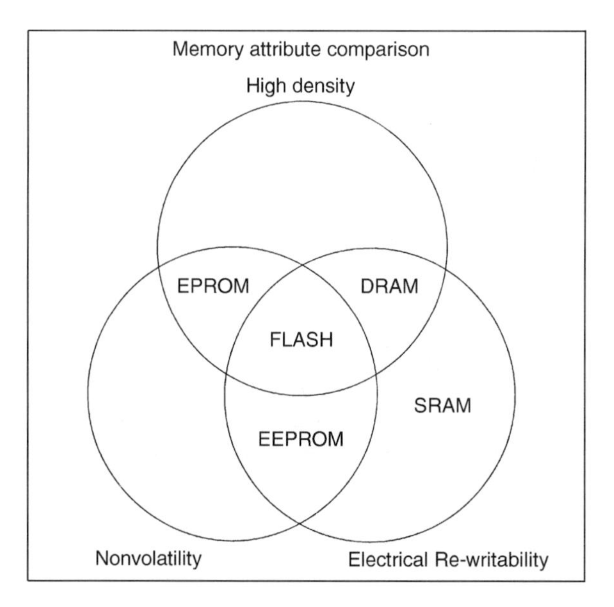

Figure 4.2 Memory attribute comparison. Flash memory is a functional superset [4.2].

TABLE 4.2 THE CHARACTERISTICS OF AN IDEAL MEMORY

1. Nonvolatility
2. Dense [consumes small space/bit]
3. Fast read/write/erase
4. In-system rewritability
5. Bit alterability
6. Endurance [highwrite/erase cycles]
7. Low power consumption
8. High bit count
9. Low cost
10. Highly scalable
11. Single-power supply
12. Ruggedness
13. Portability
14. Small form factor
15. Highly integrable with other system technologies

4.0.1 Why Flash?

Flash memories are designed to be the solution to the scaling problem of EEPROMs by trading off memory cell size with functionality. Standard

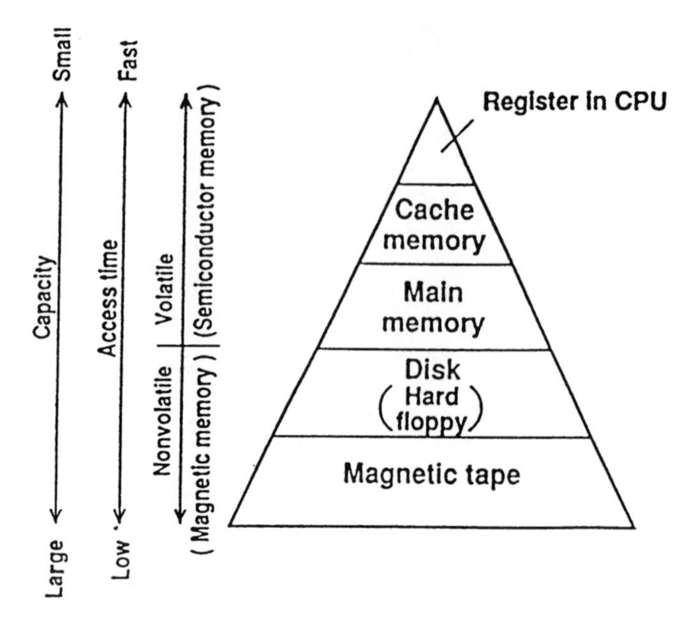

Figure 4.3 Hierarchy of memories in a computer system [4.1].

TABLE 4.3 MEMORY ATTRIBUTES

	Flash	DRAM	SRAM	EEPROM	UV-EPROM	Hard Disk	Floppy Disk
Nonvolatility	O	X	X	O	O	O	O
High density	O	O	X	X	O	O	X
I-T/cell	O	X	X	X	O	X	X
Low power	O	X	X	O	O	X	X
In-system rewritability	O	O	O	O	X	O	O
Bit alterability	X	O	O	O	X	O	O
Fast read	O	O	O	O	O	X	X
Fast write	O	O	O	O	O	O	O
High endurance	O	O	O	O	X	O	O
Low cost	O	O	X	X	O	O	O
Single-power supply	O	O	O	O	X	O	O

EEPROM memory cells have a select transistor that is either separated from or merged with the floating gate memory transistor. The select transistor is required for a number of reasons. The reason why it is needed for FLOTOX EEPROM is described in the following paragraphs. Other EEPROMs have similar limitations.

In a FLOTOX EEPROM, the select transistor minimizes program/erase disturb in an array. In Fig. 4.4, a standard cross-point array without a select transistor is

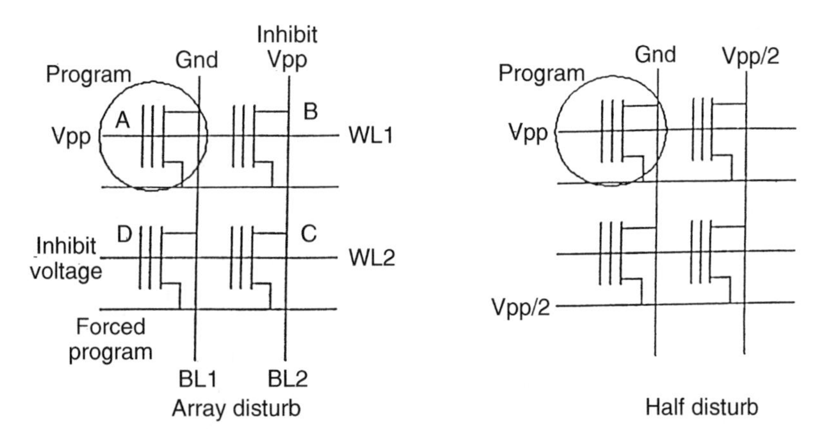

Figure 4.4 Disturb conditions in a typical cross-point single-transistor floating gate array.

shown. If transistor A is to be programmed (defined as electrons into the floating gate), high voltage is applied to wordline 1 and bitline 1 is grounded. If transistor B is not to be programmed, an inhibit voltage of V_{pp} has to be applied to bitline 2 so as not to create a net voltage across the tunnel oxide in transistor B. For the same reason, if transistor C is not to be disturbed, V_{pp} must be applied to the wordline. Given this bias condition, transistor D will be programmed whether or not it is desired to do so. In some designs, $V_{pp}/2$ is applied to the bitline of transistor B and the wordline of transistor C, giving a half-disturb condition where only half of the V_{pp} is applied across the cell. However, because of the logarithmic dependence of tunneling current on voltage, there will still be a small disturb for the half-inhibit condition when exercised over many cycles. Historically, a nondisturb condition has been realized by isolating the floating gate memory transistor from the array using a select transistor.

Another important consideration is that an erased (electrons removed from the floating gate) memory cell can have a net positive charge in the floating gate, giving a depletion state, which, in effect, shorts out the column in the array shown (Fig. 4.4). A select transistor serves to isolate the depletion transistor so as not to give a read error. The drawback of the extra select transistor is that it limits scaling.

4.0.2 A Brief History of Flash EEPROM

Table 4.4 shows the history of Flash EEPROM development . The first modern Flash EEPROM was proposed at the 1984 International Electron Devices Meeting by Masuoka et al. [4.3–4.5]. Figure 4.5 shows the cell structure of this first Flash EEPROM. This EEPROM structure was called a Flash EEPROM because the complete memory array is erased very quickly. At present, the name Flash

EEPROM is used for all EEPROM in which all or a large number of cells, called a block or a page, are erased at the same time.

The Flash EEPROM array reported is based on an EPROM array [4.6]. It has similar high-density, low-cost, and high-reliability advantages of EPROMs. The challenge in Flash memory is to learn how to remove electrons from the floating gate by an electrical process instead of the UV light illumination used for EPROMs. The advantage of Flash EEPROM is that the erase time is less than 1 second, while the erase time for an UV-EPROM is about 10 minutes. Moreover, the UV-EPROM requires an expensive package because of the UV transparent quartz window, and it must be taken out of the system for erasure. Flash EEPROM, however, can be packaged in small, inexpensive plastic packages and reprogrammed in-system (i.e., without removal of the package). The limitation of Flash EEPROM compared to traditional EEPROM is that many bytes are erased simultaneously, instead of a single byte at a time.

In 1985, Mukherjee et al. proposed a source-erase type of Flash memory cell [4.7] called the stacked gate cell (Fig. 4.6). The structure of this cell is the same as that of the stacked gate UV-EPROM, with two modifications: (1) the source junction is graded to support the high voltage during erase, and (2) the gate oxide is considerably thinner than UV-EPROMS to allow F–N (Fowler–Nordheim) carrier tunneling during erase. Since 1985, the stacked gate cell has become the volume shipment leader of Flash EEPROM and, at present, 16 MB Flash memories are in production [4.8]. Several variations of the stacked gate cell have been reported [for example, 4.9–4.12] and will be discussed in detail later in this chapter.

In 1987, Masuoka et al. [4.13] proposed a NAND structured cell. This structure reduces the cell size without scaling of the device dimensions. The NAND structure cell arranges a number of bits in series, as shown in Fig. 4.7. The current EPROM cell has one-half contact per bit. However, for a NAND structure cell, only one contact hole is required per two NAND structures. As a result, the NAND cell can realize a smaller cell area per bit than the current EPROM. At present, 16 MB NAND devices are in production, and 32 MB NAND devices have been discussed [4.15, 4.16].

Finally, in the last few years, several other Flash cell structures and array architectures have been disclosed and are in various stages of development. For example, AND [4.57], DINOR [4.58], and HICR [4.68] employ F–N tunneling for programming in order to achieve low power programming. These cell structures and others are discussed in Section 4.4 of this chapter.

4.1. BASICS OF PROGRAM AND ERASE OPERATIONS

4.1.1 Channel Hot-Electron (CHE) Programming

In this section, we briefly describe the program/erase operations of the majority of Flash memories described in the literature.

TABLE 4.4 HISTORY OF FLASH MEMORY

1978	EAROM	Guterman et al. [4.37]	TI
1984	Flash memory	Masuoka et al. [4.3]	Toshiba
1985	Flash memory (256 KB)	Masuoka et al. [4.4], [4.5]	Toshiba
1985	Source-erase type Flash	Mukherjee et al. [4.7]	Exel
1987	Drain-erase type Flash (128 KB)	Samachisa et al. [4.10], [4.11]	Seeq. UCB
1987	NAND structure EEPROM	Masuoka et al. [4.57]	Toshiba
1987	Source-erase type Flash	Kume et al. [4.57]	Hitachi
1988	ETOX-type Flash (256 KB)	Kynett et al. [4.9], [4.18]	Intel
1988	NAND EEPROM	Shirota et al. [4.19]	Toshiba
1988	ETOX-type Flash	Tam et al. [4.20]	Intel
1988	ETOX-type Flash, reliability	Verma et al. [4.21]	Intel
1988	NAND EEPROM	Momodomi et al. [4.22]	Toshiba
1988	Poly-poly erase Flash	Kazerounian et al. [4.23]	WSI
1988	Contactless Flash	Gill et al. [4.24]	TI
1989	Contactless Flash (256 KB)	D'Arrigo et al. [4.25]	TI
1989	Gate-negative erase	Haddad et al. [4.12]	AMD
1989	NAND EEPROM (4 MB)	Momodomi et al. [4.26]	Toshiba
1989	ETOX-type Flash (1 MB)	Kynett et al. [4.27]	Intel
1989	Side-wall Flash	Naruke et al. [4.28]	Toshiba
1989	Contactless Flash	Gill et al. [4.29]	TI
1989	Punch-through erase	Endoh et al. [4.30]	Toshiba
1990	Well-erase	Aritome et al. [4.31]	Toshiba
1990	NAND EEPROM	Iwata et al. [4.32]	Toshiba
1990	Contactless 1-T Cell Flash, ACEE	Riemenschneider et al. [4.33]	TI
1990	NAND EEPROM, well erase	Kirisawa et al. [4.34]	Toshiba
1990	FACE cell	Woo et al. [4.35]	Intel
1990	Gate-negative erase	Ajika et al. [4.36]	Mitsubishi
1990	Contactless Flash	Gill et al. [4.37]	TI
1990	Bipolarity write/erase	Aritome et al. [4.38]	Toshiba
1990	Negative gate, positive source erase	Miyawaki et al. [4.39]	Mitsubishi
1990	5 V-only NAND (4 Mb)	Tanaka et al. [4.40]	Toshiba
1990	1 Mb ETOX-type Flash	Seki et al. [4.86]	Hitachi
1990	Negative gate, positive source Flash	S. Haddad et al. [4.12]	AMD
1991	Virtual ground array with auxiliary gate	Yamauchi et al. [4.41]	Sharp
1991	5 V-only NAND (4 Mb)	Momodomi et al. [4.42]	Toshiba
1991	1-T Contactless 5 V-only Flash (4M)	McConnell et al. [4.43]	TI
1991	Stack Gate, high source impurity concentration cell (16 M)	Nakayama et al. [4.44]	Mitsubishi
1991	SCSG cell	Kuo et al. [4.45]	Motorola
1991	PB-FACE cell	Woo et al. [4.46]	Intel
1991	Burst-pulse erase	Kodama et al. [4.47]	NEC
1991	Sector-erase	Kume et al. [4.48]	Hitachi
1991	Flash cell, scaling	Yoshikawa et al. [4.49]	Toshiba
1991	Self-convergence erase	Yamada et al. [4.50]	Toshiba
1991	SSW-DSA 64 M Flash	Kodama et al. [4.119]	NEC

TABLE 4.4 HISTORY OF FLASH MEMORY (*Cont'd*)

1992	MEIT	Hori et al. [4.51]	Matsushita
1992	Polysilicon TFT	Koyama [4.52]	NEC
1992	2-Step erase	Oyama et al. [4.53]	NEC
1992	Bipolarity erase	Endoh et al. [4.54]	Toshiba
1992	RT ONO Flash	Fukada et al. [4.55], [4.56]	Oki
1992	FNT prgm/erase Flash	Kume et al. [4.57]	Hitachi
1992	DINOR	Onoda et al. [4.58]	Mitsubishi
1992	HIMOS	Keeney et al. [4.59]	NMRC/ IMEC
1992	SCSG cell	Kuo et al. [4.60]	Motorola
1992	Negative gate, positive source cell with row decoder in triple well (16 M)	Umezawa et al. [4.61]	Toshiba
1992	5 V-only Flash with DSA drain and channel erase	Jinbo et al. [4.62]	NEC
1992	5 V-only sector erase 16 M Flash	T. Jinbo et al. [4.120]	NEC
1992	Serial Flash (9 M)	Mehrotra et al. [4.63]	Sandisk
1993	Virtual ground array DINOR	Ohi et al. [4.64]	Mitsubishi
1993	NOR virtual ground Flash cell	Bergemont et al. [4.65], [4.66]	National
1993	3-D sidewall Flash cell	Pein et al. [4.67]	Stanford/ Phillips
1993	HICR cell	Hisamune et al. [4.68]	NEC
1994	3.3 V ETOX cell (16 M)	Baker et al. [4.8]	Intel
1994	3 V-only DINOR Flash (4 M)	Kobayashi et al. [4.69]	Mitsubishi
1994	Gate negative, source positive, erase Flash (16 M)	Atsumi et al. [4.70]	Toshiba
1994	3.3 V single-power supply HiCR Flash (64 M)	Takeshima et al. [4.71]	NEC
1994	Row redundancy (16 M)	Mihara et al. [4.72]	Mitsubishi
1994	High-speed programming (64 M)	Tanaka et al. [4.73]	Hitachi
1994	Serial Flash 18 MB	Lee et al. [4.74]	Sandisk
1994	Over-erase Detection	Miyawaki et al. [4.75]	Mitsubishi
1994	Quick Boosting Pump	Tanzawa et al. [4.76]	Toshiba
1994	Scaling tunnel oxide	Watanabe et al. [4.77]	Toshiba
1994	Source-side injection 5 V-only	Ma et al. [4.78]	Bright/ Sharp
1994	Subhalf micron NOR Flash cell	Mori et al. [4.79]	Toshiba
1994	Fully depleted floating gate	Sato et al. [4.80]	Sharp
1994	Buried channel Flash memory	Oda et al. [4.81]	Mitsubishi
1994	3 V-only S-injection	Kianian et al. [4.82]	SST
1994	TEFET-Trench Flash EEPROM	Kuo et al. [4.83]	Philips

Channel hot-electron (CHE) programming is achieved by applying high voltage to both drain and gate simultaneously [4.6]. With the control gate high, the high voltage across the drain to source gives a high channel current and channel field which generates hot electrons. The high voltage on the control gate couples a voltage

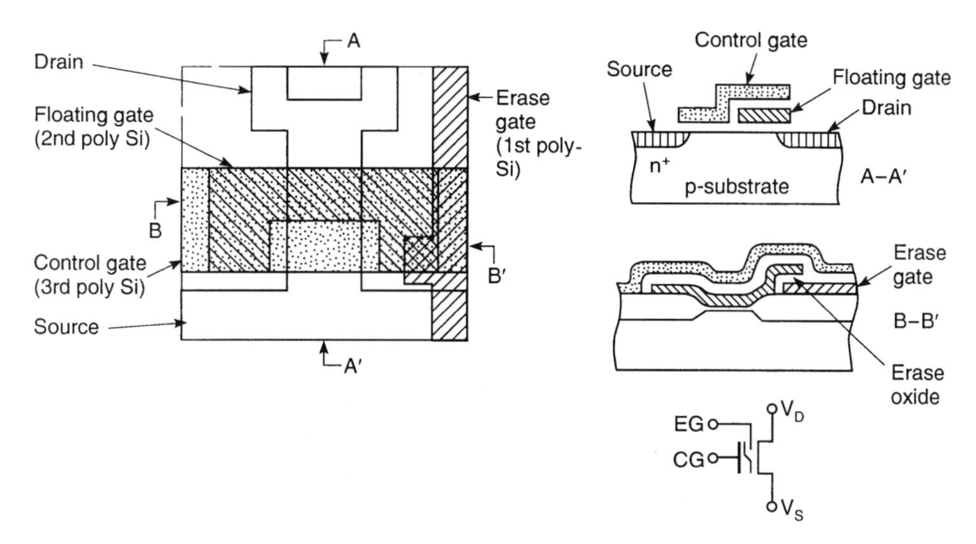

Figure 4.5 Top and cross-sectional view of the Flash EEPROM cell [4.3].

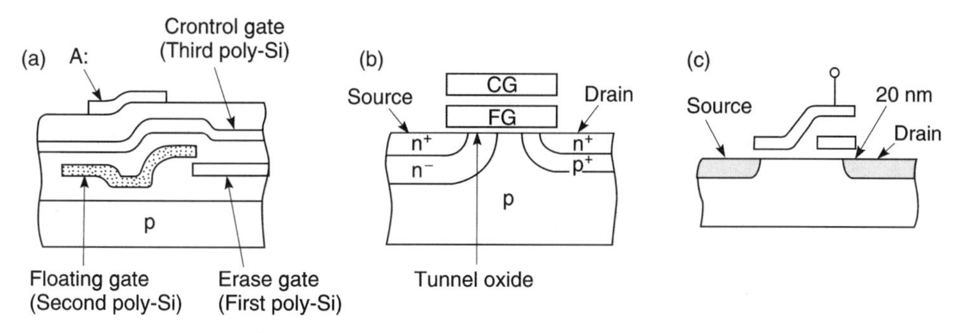

Figure 4.6 Stacked gate Flash EEPROM cell variations: (*a*) [4.3], (*b*) [4.7], and (*c*) [4.10].

to the floating gate and attracts hot electrons to the floating gate. Other cells along the same wordline and bitline are not programmed because no hot electrons are available with only the gate or the drain high. This programming mechanism is well proven, and EPROMs are known to be reliable and manufacturable. Program-disturb conditions in EPROMs are well understood and, in a well-designed EPROM, are insignificant. It is because of these advantages that early Flash memories used an EPROM-like structure [4.3, 4.9].

4.1.2 Source-Side Hot-Electron Programming

Instead of programming on the drain side, it is possible to design a cell that makes use of a narrow gap produced by a stepped gate or poly spacer on the source side to obtain a local high field for programming. For programming, both the floating gate and the stepped gate or poly spacer are biased with positive voltages

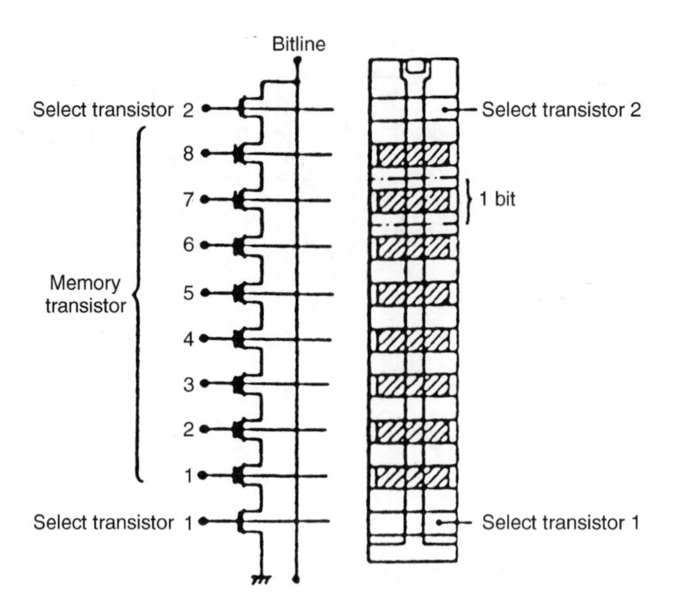

Figure 4.7 Top view and equivalent circuit of the NAND EEPROM [4.13].

to turn on the channel underneath. As a result, most of the source to drain voltage is dropped across the narrow gap. It is important to have the floating gate coupled to a higher positive potential compared to the stepped gate or poly spacer. The hot electrons generated are transported to the floating gate. This hot-electron process is more efficient, compared to channel hot electrons, requiring programming currents an order of magnitude lower. Variations of this technique have been used for a number of different cell structures [4.28, 4.41, 4.78, 4.82, 4.88]; for an example, see Fig. 4.8.

4.1.3 Fowler–Nordheim (F–N) Tunneling

Fowler–Nordheim (F–N) tunneling has been discussed in Chapter 2 with regard to the FLOTOX EEPROM. We will give a brief description here. Tunneling is a process whereby electrons are transported through a barrier. For silicon dioxide, with a 3.2 eV barrier, significant tunnel current can be observed when the oxide thickness is reduced to less than 4 nm. For thicker oxide, it is still possible to have a significant tunneling current when a high field of > 10 MV/cm is applied across the oxide, turning the barrier from a trapezoidal barrier into a triangular barrier of small tunnel distance. The tunneling process in oxide was first reported by Fowler and Nordheim and has the following functional form [4.160]:

$$I = A\,E^2 \exp(-B/E) \tag{4.1}$$

Fowler–Nordheim tunneling is the key transport mechanism for the FLOTOX EEPROM.

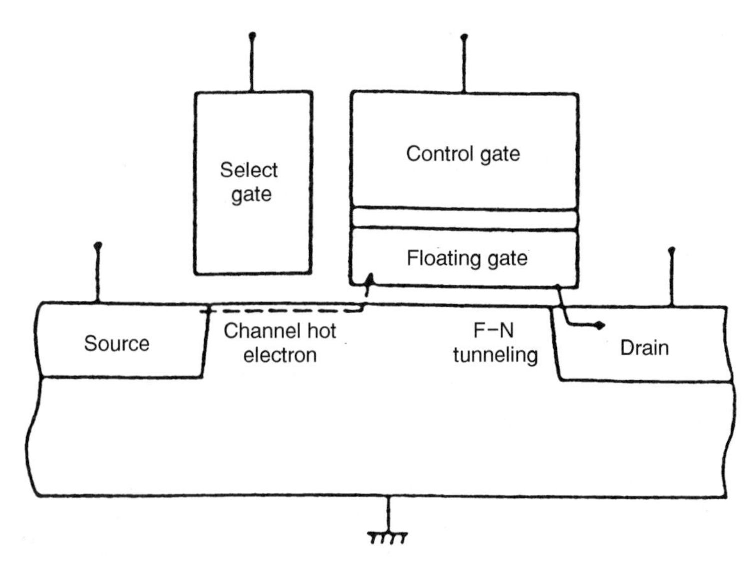

Figure 4.8 Schematic cross-sectional view of SISOS cell [4.28].

4.1.4 Tunnel Programming

Instead of channel hot-electron programming, we can program (electrons into a floating gate) a floating gate cell by Fowler–Nordheim tunneling by applying a high voltage to the top control gate, which couples a high voltage to the floating gate. For a tunnel oxide thickness of about 10 nm, the control gate voltage is 16 to 20 V. The program mechanism is the same as for the FLOTOX EEPROM. Examples of Flash memories based on F–N tunnel programming are: NAND [4.14], DINOR [4.58], AND [4.57], and ACEE [4.25].

As discussed previously, tunnel programming is a half-select process, and historically, a select transistor in series with the floating gate has been used to avoid disturb. More recently, the need for a select transistor has been obviated in tunnel programmable Flash memories by on-chip V_{PP} trimming, array sectoring techniques, and erase control techniques [4.14, 4.57, 4.58, 4.68].

4.1.5 Tunnel Erase Through Thin Oxide

When the first gate oxide in the memory cell is a thin tunnel oxide, it is possible to erase (electrons out of the floating gate) the memory cell by F–N tunneling through the tunnel oxide. The erase process can be accomplished in different ways. With the control gate grounded, erase can be accomplished by applying a high voltage to the source [4.7], to the drain [4.10], or to the substrate in a well [4.14]. It can also be accomplished by applying a negative voltage to the control gate with the other terminals connected to different positive voltages [4.12, 4.65, 4.78].

4.1.6 Tunnel Erase Through Poly-to-Poly Oxide

Instead of erasing to the silicon through the thin tunnel oxide, it is also possible to erase to the poly through the poly-to-poly oxide. Because of texturing in poly-oxide, it is possible to a obtain significant tunnel current even with a thicker oxide [4.3, 4.63]. Although the poly-to-poly erase provides flexibility in the memory cell design, it requires a considerably higher voltage compared to the erase operation through thin oxide.

4.2. FLASH MEMORIES WITH CHANNEL HOT-ELECTRON (CHE) PROGRAM AND TUNNEL OXIDE ERASE

A schematic description of Flash cell programming by CHE is shown in Fig. 4.9. Some of the array architecture and erase conditions of channel hot-electron programmable Flash memories described in the literature will be discussed here. Programming is from the drain in a manner identical to that for an EPROM, and the cell is programmed to a high threshold state as shown in Fig 4.10.

4.2.1 Positive Source Erase

Tunnel erase is achieved by making use of the source to the floating gate overlap area as shown in Fig. 4.11. When the top gate is grounded, a high voltage applied to the source region (with the other junction floating) will pull electrons out of the floating gate by tunneling. The cell is erased to a low-threshold state. A graded source junction allows the application of a high voltage [4.17].

Under a positive source erase condition, the source junction generates significant substrate current. With a sufficiently high voltage, the breakdown process draws significant current and acts as a voltage clamp, limiting any further increase of the junction voltage. When the substrate current is sufficiently high, hot holes generated by the breakdown may start to erase the memory cell. Hot-hole erase is a process that is difficult to control and is avoided in a well-designed memory cell.

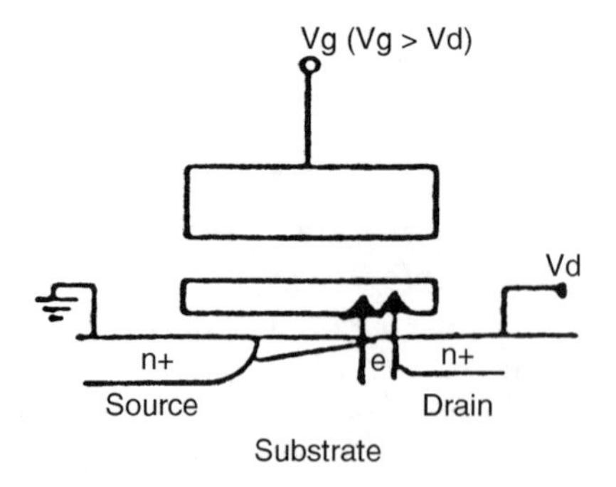

Figure 4.9 Schematic description of Flash EEPROM programming by CHE injection [4.14].

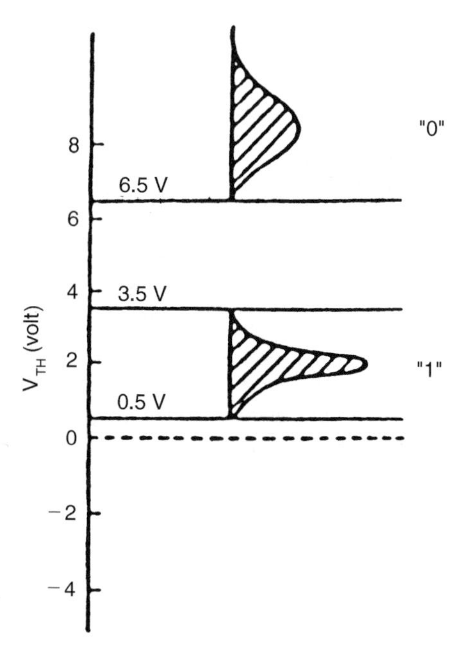

Figure 4.10 Vth distribution in programmed and erased state [4.14].

The source junction is graded by the addition of a phosphorous implant [4.7, 4.9, 4.17, 4.18, 4.20, 4.21, 4.27]. Also, because erase and program occur over different areas of the memory cell, the stresses of program and erase are spread over different regions, resulting in improved reliability. Since the tunnel area is the small source to floating gate overlap determined by lateral dopant diffusion, the coupling ratio for electrical erase is low and there is less need for a high coupling area over the field oxide, favoring a small and compact cell design. Finally, since the source is common to all the memory cells in the memory array, all the cells are erased simultaneously, providing block erase.

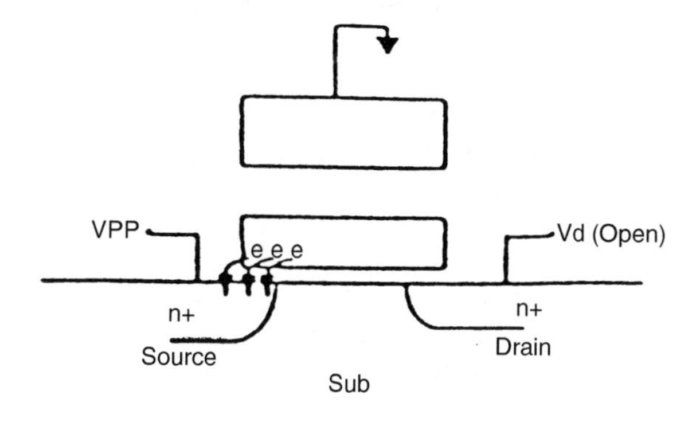

Figure 4.11 Schematic description of Flash EEPROM electrical erase by Fowler–Nordheim tunneling of electrons from the floating gate to the source [4.7, 4.9, 4.14, 4.17, 4.18, 4.20, 4.21, 4.27].

4.2.2 Positive Drain Erase

A split gate, CHE program, drain erase cell is shown in Fig. 4.12 [4.10, 4.11, 4.87].

If it is necessary to erase a smaller number of memory cells using source erase, the sources of the memory cells need to be separated. It is relatively simple to do a small number of sub-blocks by dividing up the array, but it is very area-intensive to do a large number of small blocks. Drains of memory cells are common in a given column, and erasing from the drain (Fig. 4.12) will give natural column blocking. However, since the junction is now used for both program and erase, it can only be optimized for one, and typically, it has to be optimized for erase because of the higher voltage requirement. As a result, programming performance is sacrificed. Second, now the junction is subjected to both program and erase stress so there is also more damage with cycling. Consequently, the cycling performance is limited compared to the case of separate program and erase junctions.

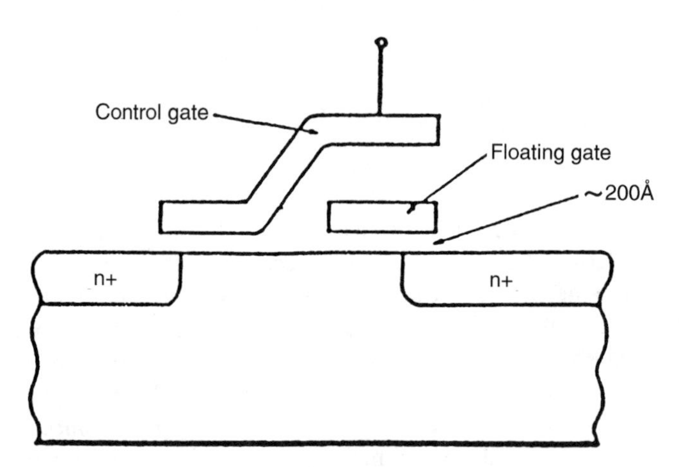

Figure 4.12 Drain erase Flash memory cell [4.10, 4.11, 4.87].

4.2.3 Negative Gate Erase

4.2.3.1 Negative (Control) Gate, Floating Gate to Source Erase. As mentioned previously, both source erase and drain erase suffer from a large substrate current that is not efficient for low power operation. However, the same erase field can be developed across the tunnel oxide by applying a negative voltage to the control gate. The source and drain can either be grounded or kept at a lower voltage, for example, V_{cc}. However, generation and switching of negative voltages increase circuit complexity.

The negative control gate, floating gate to source erase approach with the gate at a high negative voltage and the source typically at V_{cc} is shown in Fig. 4.13 [4.12, 4.48, 4.101, 4.119]. In this approach, the F–N erase current remains constant, and the adverse effects from holes generated by band-to-band tunneling are reduced.

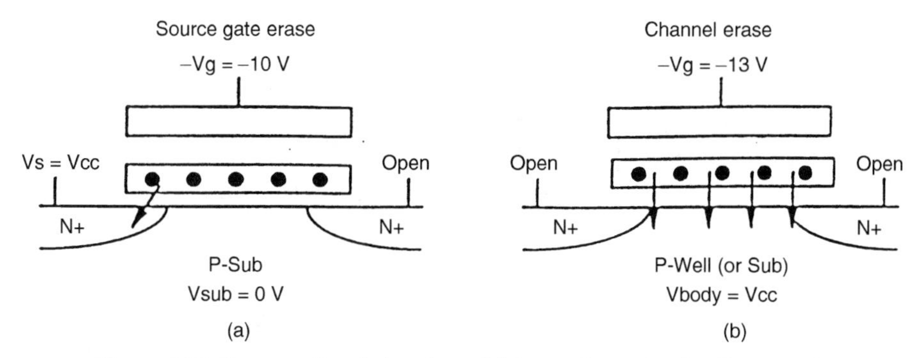

Figure 4.13 Cross-sectional drawing of (*a*) negative gate, floating gate to source erase, (*b*) negative gate, floating gate to channel erase [4.104].

4.2.3.2 Negative (Control) Gate, Floating Gate to Drain Erase. A triple-poly, split-gate Flash memory cell based on source-side injection for programming and negative control gate for erase (electrons tunnel from floating gate through the oxide to the drain) is shown in Fig. 4.14 [4.78]. The cell consists of a pair of stacked gates and a polycide blanket split gate. The select gate controls the weakly-on region during programming and prevents unselected cells from conduction in the program and read modes. The select gate runs along the channel direction. This eliminates the need for below-feature-size poly, reduces select gate delay, and improves the speed over select gate cells of the side-wall type [4.28, 4.88]. The use of a select gate allows the cell to operate in the depletion mode. To achieve a high read current in the erased state, the cell is put into the depletion mode.

The source-side reverse read is used to eliminate read-disturb induced by hot electrons. Figure 4.15 shows the array schematic and its operating conditions. To avoid high-voltage stress in the tunnel oxide during write, the high-voltage control gate (V_{pp}) and bitline (V_{cc}) are arranged to run parallel so that, in the program mode, both drain voltage and control gate voltage are present in the selected column. The self-imposed counterbiasing effect reduces the write-disturb.

The bitlines consist of source-lines and drain-lines arranged in alternating columns. An erase sector is naturally formed along each drain-line without any layout overhead. Each sector consists of two columns of opposing cells that share a common control gate.

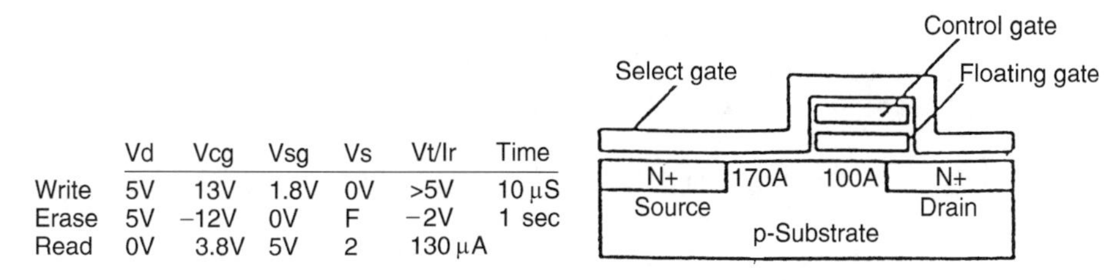

	Vd	Vcg	Vsg	Vs	Vt/Ir	Time
Write	5V	13V	1.8V	0V	>5V	10 µS
Erase	5V	−12V	0V	F	−2V	1 sec
Read	0V	3.8V	5V	2	130 µA	

Figure 4.14 Cell cross-sectional view and operating conditions [4.78].

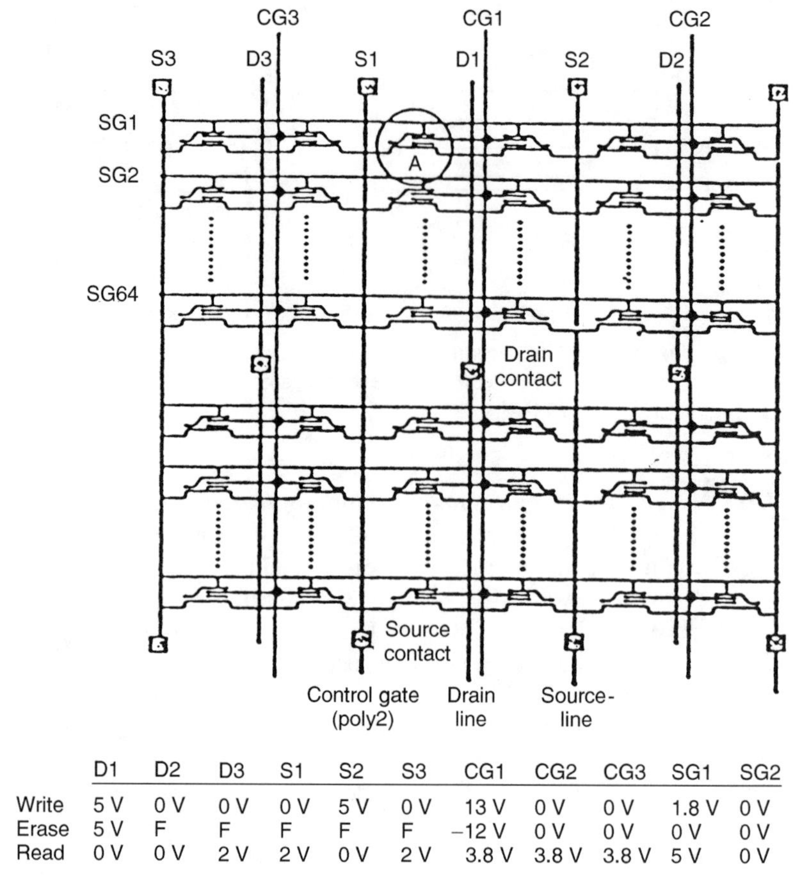

	D1	D2	D3	S1	S2	S3	CG1	CG2	CG3	SG1	SG2
Write	5 V	0 V	0 V	0 V	5 V	0 V	13 V	0 V	0 V	1.8 V	0 V
Erase	5 V	F	F	F	F	F	−12 V	0 V	0 V	0 V	0 V
Read	0 V	0 V	2 V	2 V	0 V	2 V	3.8 V	3.8 V	3.8 V	5 V	0 V

Figure 4.15 Array schematic and selection condition for cell "A" [4.78].

4.2.3.3 Negative (Control) Gate, Floating Gate to Channel Erase. The Nor Virtual Ground (NVG) array concept [4.65, 4.66] combines the Alternate Metal Gate (AMG) programming [4.157] with floating gate to channel erase [4.47]. The floating gate to channel erase avoids the problems associated with source erase discussed previously.

The erase concept operation is shown schematically in Fig. 4.16. The array schematic is shown in Fig. 4.17, and the array operation in Fig. 4.18.

The NVG array concept minimizes the drain turn-on during programming by the presence of a pass gate between the source and the bitline in a block. A column is segmented into blocks by the pass gate.

To achieve erase inhibit, the array well is biased to V_{cc}, necessitating a well-within-well structure. The floating gate to channel erase (F–N tunneling) requires a high electric field between the floating gate and channel. This is provided by a combination of a negative voltage on the control gate and a positive voltage on the array substrate. This necessitates an array well isolated from the substrate.

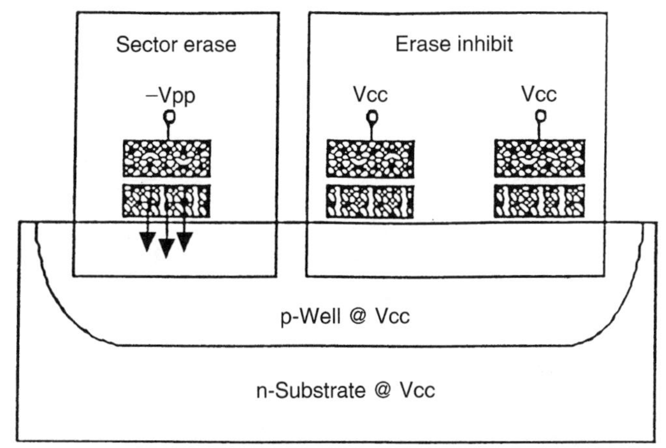

Figure 4.16 Sector erasing scheme [4.65].

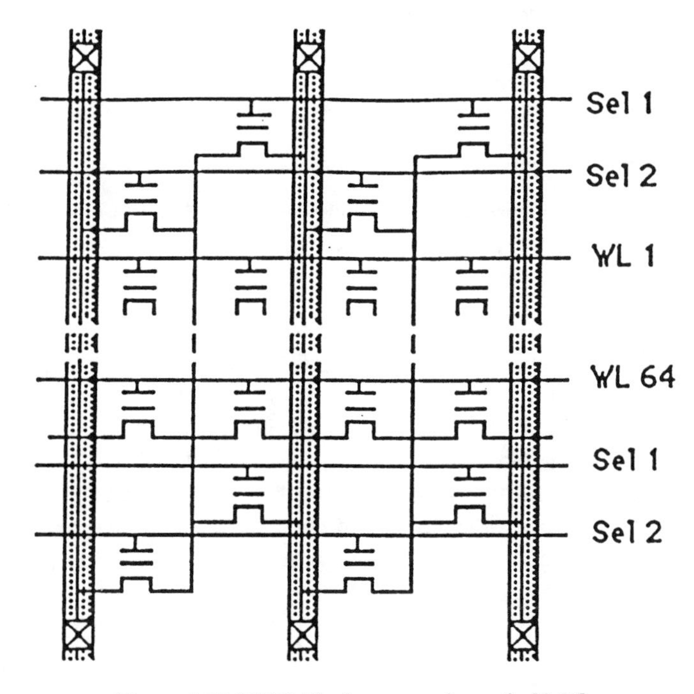

Figure 4.17 NVG Flash array schematic [4.65].

4.2.4 Erase Threshold Control

Electrical erasure presents new challenges compared to UV erase. First, instead of a natural self-limiting UV process, the electrical erase can continue beyond the neutral point to give a net positive charge on the floating gate and a net negative

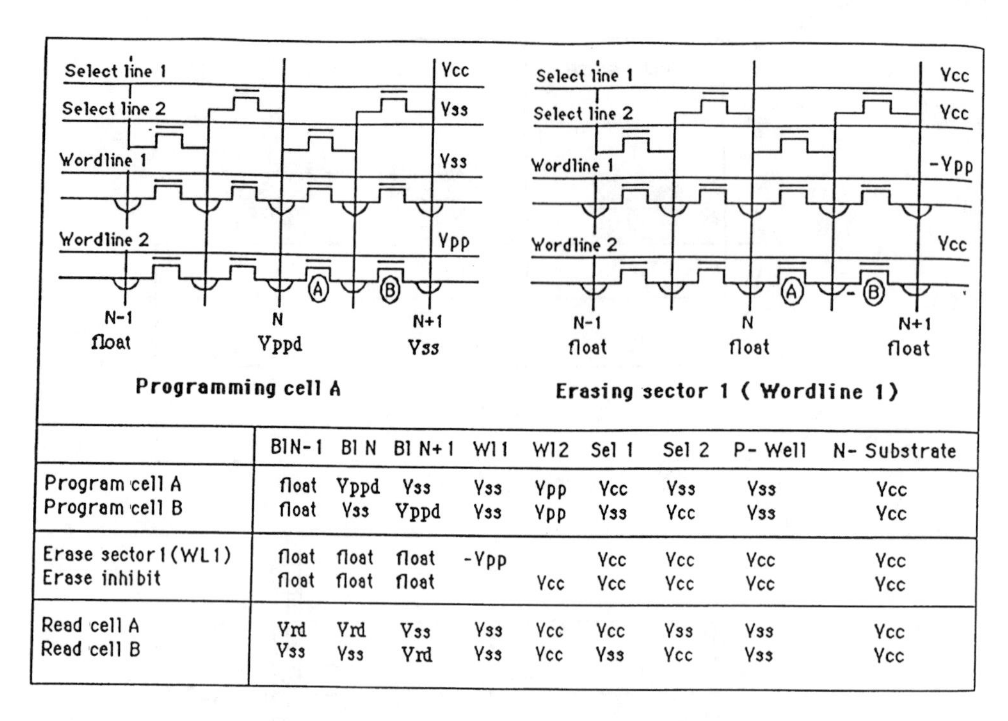

	BlN-1	Bl N	Bl N+1	Wl 1	Wl 2	Sel 1	Sel 2	P-Well	N-Substrate
Program cell A	float	Vppd	Vss	Vss	Vpp	Vcc	Vss	Vss	Vcc
Program cell B	float	Vss	Vppd	Vss	Vpp	Vss	Vcc	Vss	Vcc
Erase sector 1 (WL1)	float	float	float	-Vpp		Vcc	Vcc	Vcc	Vcc
Erase inhibit	float	float	float		Vcc	Vcc	Vcc	Vcc	Vcc
Read cell A	Vrd	Vrd	Vss	Vss	Vcc	Vcc	Vss	Vss	Vcc
Read cell B	Vss	Vss	Vrd	Vss	Vcc	Vss	Vcc	Vss	Vcc

Figure 4.18 NVG Flash operations table [4.65].

threshold voltage. In the simple EPROM-like array structure, a negative threshold can short out columns, giving false ones for any column read. In the worst case, the conducting cell can draw so much current that it is not possible to program other cells in the same column. Second, any cell-to-cell variation, like coupling ratio and tunneling probability, will change the final threshold of the cell. This potentially results in a much wider threshold distribution after electrical erasure unless the process is well controlled. Several approaches have been proposed to solve this over-erase problem [4.9, 4.27, 4.28].

The verified-erase method [4.27] is the key method for controlling placement of erase cells (Fig. 4.19). An erase step is carried out by applying a suitable combination of voltages to source junctions and control gates. Subsequently, a read operation is performed with a read voltage of, for example, 3.2 V, applied to the control gate. This 3.2 V is to be the upper limit for the threshold voltage of a cell in the erased state. It is derived from a worst case read voltage of V_{cc} = 4.5 V and a gate drive of 1.2 V for sufficient read current. If some bits require more time to reach the erased state, erasing is performed again. The erase verify sequence is repeated until all cells in the array have a threshold voltage less than or equal to the 3.2 V maximum. This verified erase method has been shown [4.27] to be very effective in accurately placing (locating) the threshold of erased cells.

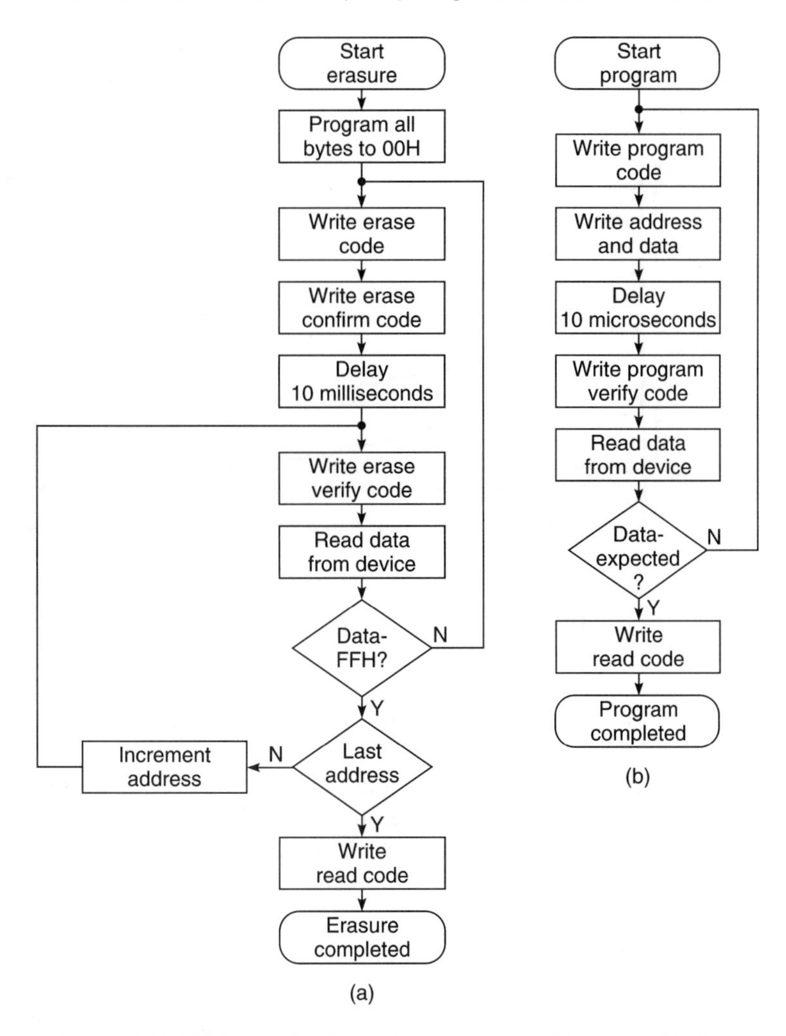

Figure 4.19 (*a*) Erase algorithm for source erase V_t control [4.27]; (*b*) program algorithm [4.27].

On the low-threshold end, the V_t limit is determined by the amount of cell leakage current under worst case conditions so that there won't be a false read on other cells. The low-end threshold is not verified by the algorithm, but is guaranteed by process control and product testing. Any arrays with leaky cells are either corrected by redundancy or discarded as bad. The current process can give erase times as short as 100 msec to as long as 30 secs, with the tunnel oxide thickness being the dominant factor in determining the erase time (Fig. 4.20). A typical erase time is 1 sec.

In the erase algorithm, if there are memory cells that are never programmed, the cells will be erased repeatedly over many cycles. Given enough cycles, these cells will eventually become leaky cells. In order to prevent this problem, all cells need to be pre-programmed before erase. The pre-program resets all cells to a programmed

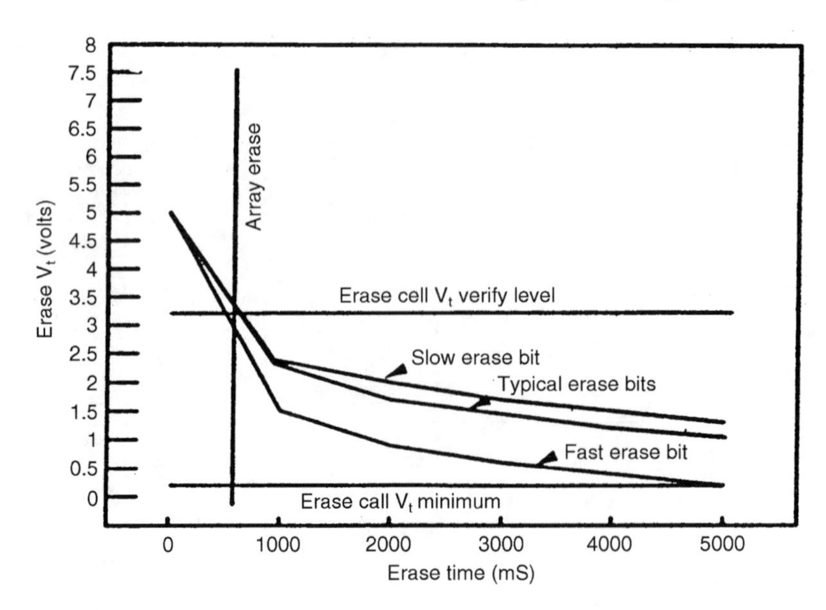

Figure 4.20 Array erase V_t profile as function of erase time [4.27].

charge state and assures good threshold control over a large number of write/erase cycles.

As noted previously, a series enhancement transistor, a merged passed gate—for example, Fig. 4.43 [4.24], Fig. 4.12 [4.10], Fig. 4.40 [4.82], or a side-wall select gate Fig. 4.8 [4.28]—can be used to prevent the leakage current. However, any of these approaches increases the cell size, as well as the process complexity.

4.2.5 Erased V_t Spread Reduction

The erase algorithm described previously controls the placement of the erased cell V_ts. It does not change the spread of the V_ts. The erased V_t spread can be reduced from both process optimization and cell-operating conditions.

4.2.5.1 Post-gate Oxide Poly Process. The fluence, or Q_{bd}, for a tunnel oxide versus annealing time for various annealing temperatures is shown in Fig. 4.21 [4.128]. More recently, careful material analysis studies [4.129, 4.130] have revealed that local oxide thinning (Fig. 4.22) and oxide surface roughness at the poly–oxide interface (Fig. 4.23) are greatly influenced by post gate growth temperature treatment. These are the direct consequence of the grain growth of the polysilicon gate and viscous flow of the oxide, which are enhanced with increasing annealing temperature and time. In addition, poly gate dopant level and species can play a major role in the local variation in tunneling current (Fig. 4.24) [4.130]. Phosphorous atoms segregated at the grain boundaries of the poly gate were found to diffuse to the poly–oxide interface and react with the oxide to form a phosphorous-rich SiO_2, or so-called oxide ridges, as shown in Fig. 4.23. The phosphorous-doped SiO_2, due to the change in stoichiometry, exhibits a barrier height lowering effect for tunneling (Figs. 4.25 and 4.26), thereby causing the F–N current to vary, depending on the amount of dopant incorporation. It was found that, in the sample implanted with phosphorus ($1 \times 16\,cm^{-2}$) and annealed at 950°C, oxide ridges were not observed with a scanning electron microscope (SEM). Thus, post gate oxide process optimization can play a major role in the control of the erased V_t distribution.

4.2.5.2 Two-Step Erase [4.53]. Since the erase current has more variation when electrons are emitted from the poly–SiO_2 interface, a two-step erasing scheme employing wordline post–stress has been proposed [4.53]. In the first step, which is equivalent to the conventional channel erase, a negative high voltage is applied to the control gate of the cell (first F–N erasing step). In the second step, a positive high voltage is applied to program the cell back (the second F–N programming-back step) to decrease the erased V_t distribution. The scheme is based on the observation that the distribution of F–N current emitted from the substrate, or the channel, is narrower than that emitted from polysilicon (Fig. 4.27) for the reasons discussed previously. Regardless of the wide erased V_t distribution after the conventional erase (the first step erase), the F–N tunneling from the channel during the programming-back step (the second step erase) tends to tighten the V_t distribution (Fig. 4.28). It was demonstrated [4.53] that, by applying this two-step erasing scheme in a 16 KB array, the erased V_t distribution could be tightened from 2.0 V to 0.9 V (Fig. 4.29).

4.2.5.3 Self-convergence Erasing Scheme [4.50]. Another technique that employs drain or source post–stress to tighten the erased V_t distribution is called the self-convergence erasing scheme [4.50, 4.134]. It uses drain or source bias with the

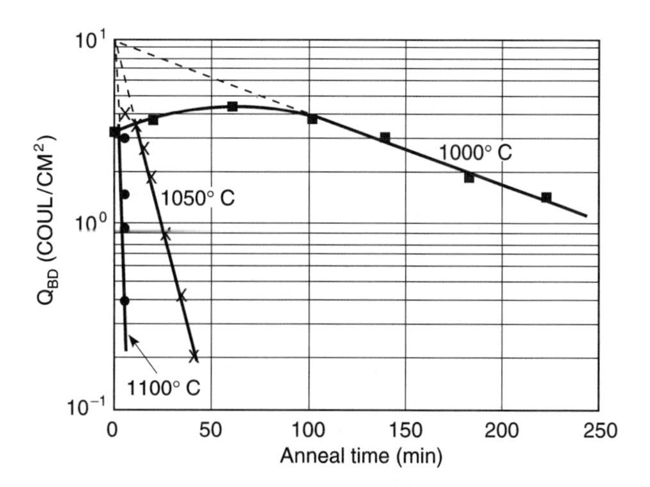

Figure 4.21 Q_{bd} versus annealing time for various annealing temperatures. Q_{bd} was measured using a current density of 50 mA/cm^2 (gate injection). T_{ox} = 80 Å [4.128].

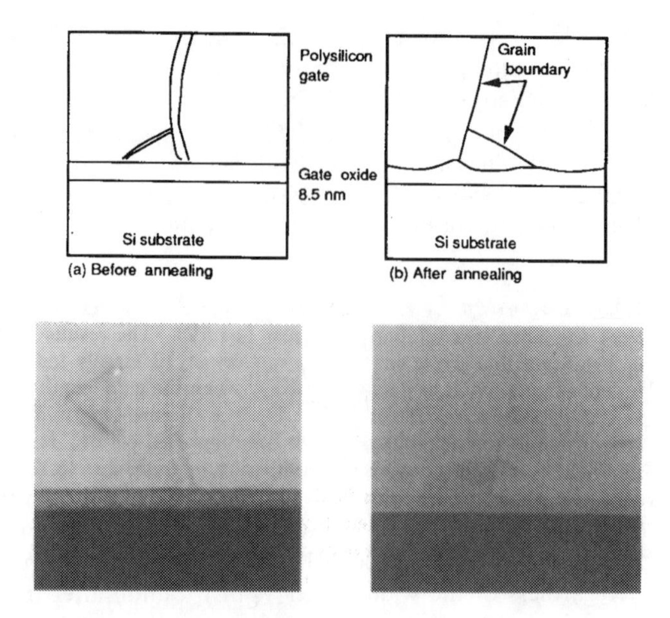

Figure 4.22 Cross-sectional TEM photographs of the polysilicon gate electrode/ultra thin tunneling oxide interface (*a*) before and (*b*) after the post–annealing [4.129].

other terminal grounded to produce channel hot-carrier injection to tighten the V_ts after the conventional erase. Figure 4.30*a* presents typical characteristics of the channel hot-carrier-induced gate current versus V_{fg} of a typical cell in which V_g^* corresponds to the saddle point where channel hot-hole injection is in balance with channel hot-electron injection. Using this device phenomenon, cells starting with

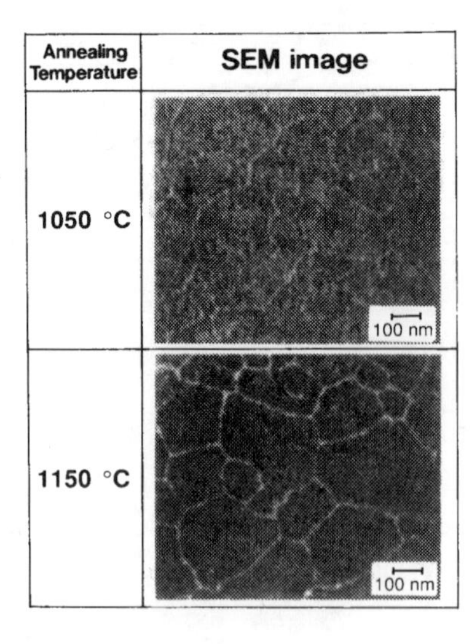

Figure 4.23 Oxide ridge structure dependence on annealing temperature (phosphorus dose: 1E16 cm^{-2}) [4.130].

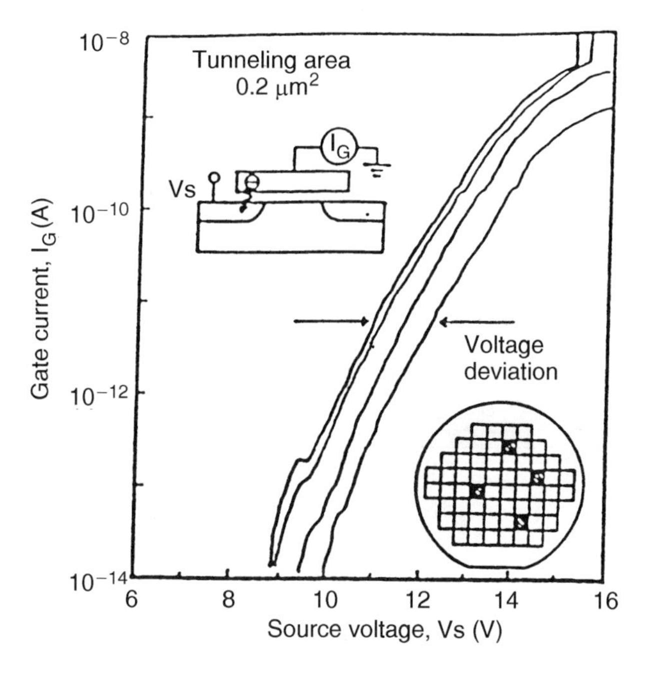

Figure 4.24 Relationship of gate current to source voltage (annealing temperature: 1100°C, phosphorus dose: 1E16 cm^{-2}) [4.130].

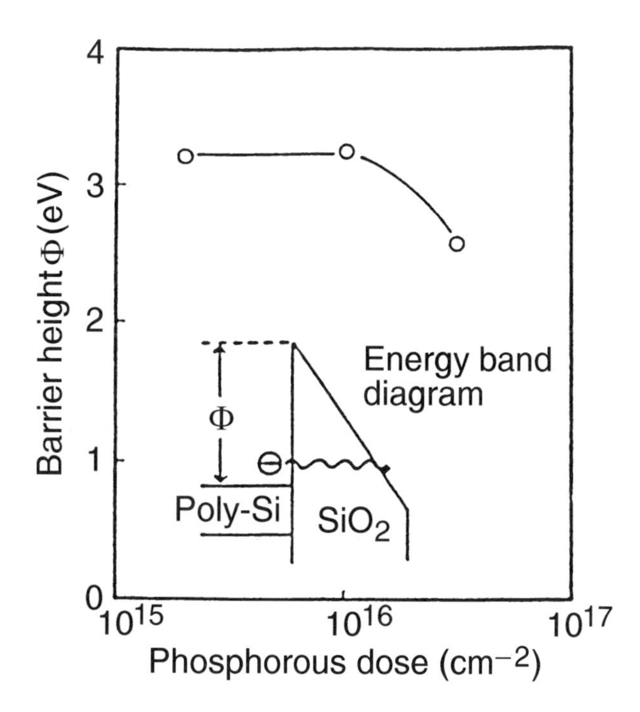

Figure 4.25 Barrier height dependence on phosphorus dose (annealing temperature = 1050°C) [4.130].

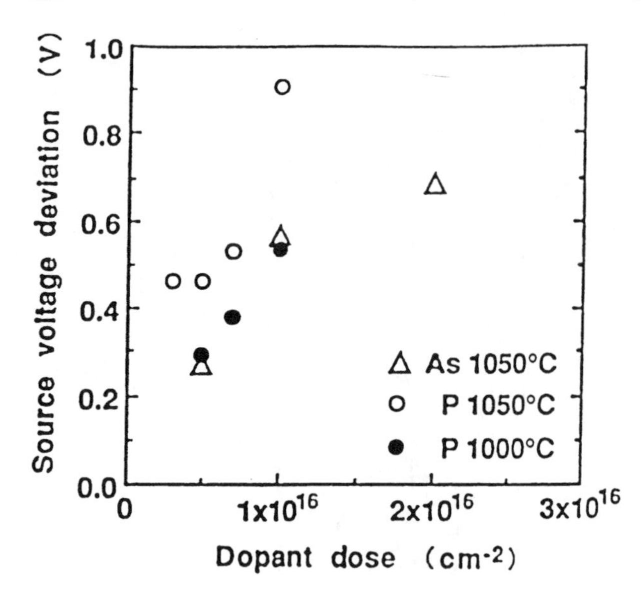

Figure 4.26 Voltage deviation dependence on annealing temperature, dopant dose, and dopant species [4.130].

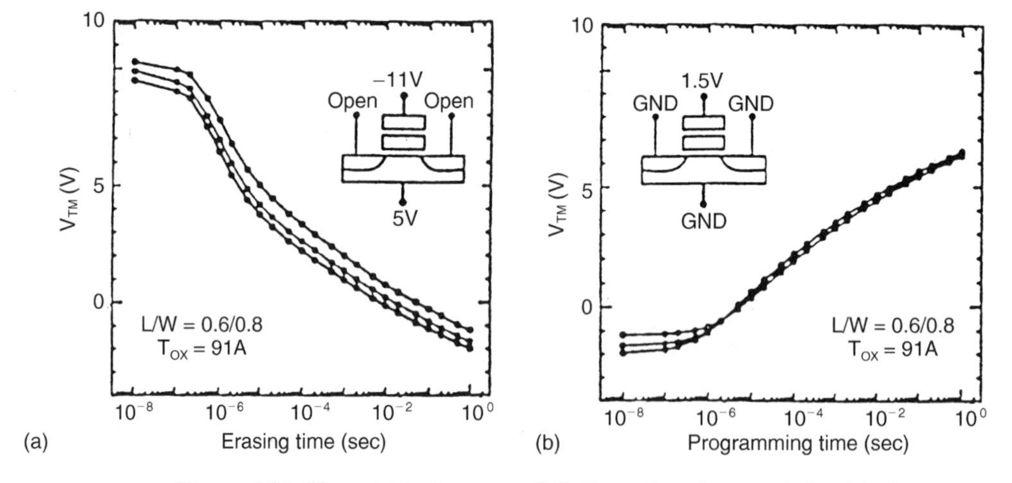

Figure 4.27 Three typical curves of F–N erasing characteristics (*a*) electrons emitted from the poly–SiO_2 interface, (*b*) electrons emitted from the Si/SiO_2 interface [4.53].

different erased V_ts will converge to a common threshold voltage, V_{th}^*, with an appropriate drain bias (Fig. 4.30*b*) [4.50]. The self-convergent V_{th}^* is a function of channel doping, and the total voltage is capacitively coupled to the floating gate. Figure 4.31 demonstrates the tightening of erased V_ts before and after the source voltage is used for convergence [4.50].

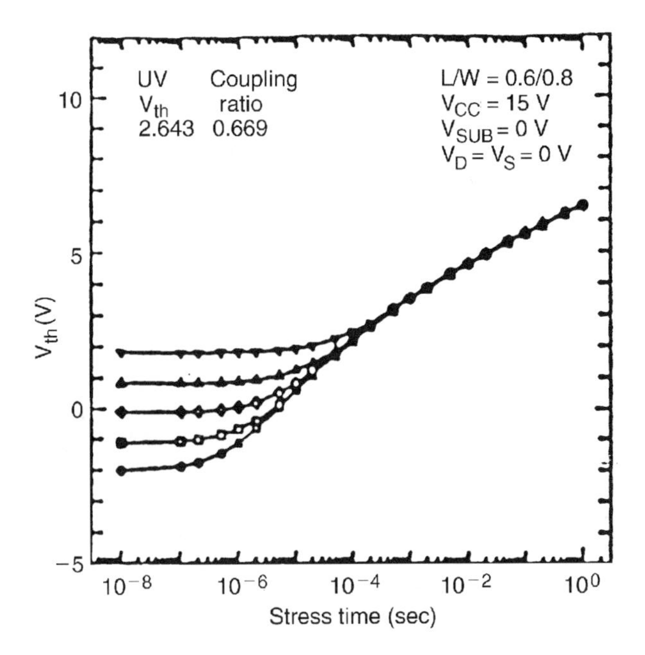

Figure 4.28 F–N programming (electrons emitted from the Si/SiO_2 interface) characteristics of a cell. F–N programming characteristics have good convergence of the programmed V_{tm} regardless of the initial V_{tm} [4.53].

4.2.6 Program-Disturb Mechanisms

The two principal memory cell-disturb mechanisms that can occur during programming of an array are called gate-disturb (DC program) and drain-disturb (program disturb) [4.21]. These mechanisms can occur in memory cells sharing a common wordline (WL) or a common bitline (BL) while one of the cells is being programmed. The effect of these disturbs on the different cells of the memory array is shown in Fig. 4.32.

Gate-disturb occurs in unprogrammed or erased cells that are connected to the same wordline as the cell being programmed. These cells have a low cell threshold voltage. During the programming operation, the common wordline is connected to a high voltage. The electric field across the tunnel oxide becomes high and may cause tunneling of electrons to the floating gate from the substrate. The threshold voltage of the cell will increase and reduce the sense margin. In severe cases, the cell is programmed unintentionally.

Drain-disturb occurs in programmed cells that are on the same bitline as the cell being programmed. These cells will experience a high electric field between the floating gate and the drain. This may cause electrons to tunnel from the floating gate to the drain and lead to a reduced cell threshold voltage.

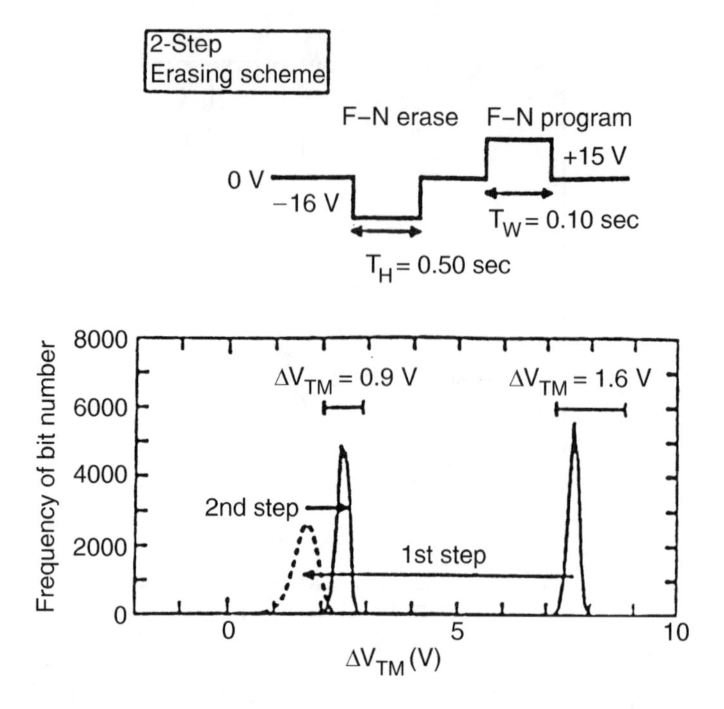

Figure 4.29 The erased-V_{tm} distribution of 16 Kbit cell array using two-step erasing. The erased V_{tm} distribution after two-step erasing operation is suppressed from $\Delta V_{tm} = 2.0$ V to 0.9 V in width [4.53].

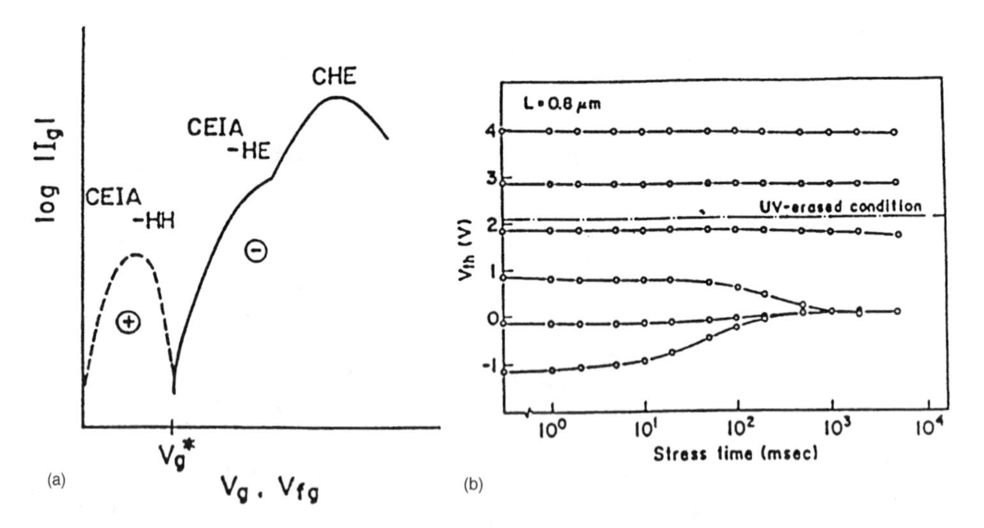

Figure 4.30 (*a*) Characteristics of gate current, I_g, versus gate voltage, V_g, for a NMOSFET, (*b*) threshold voltage, V_{th}, versus drain-stress time with different starting threshold voltages as parameters [4.50].

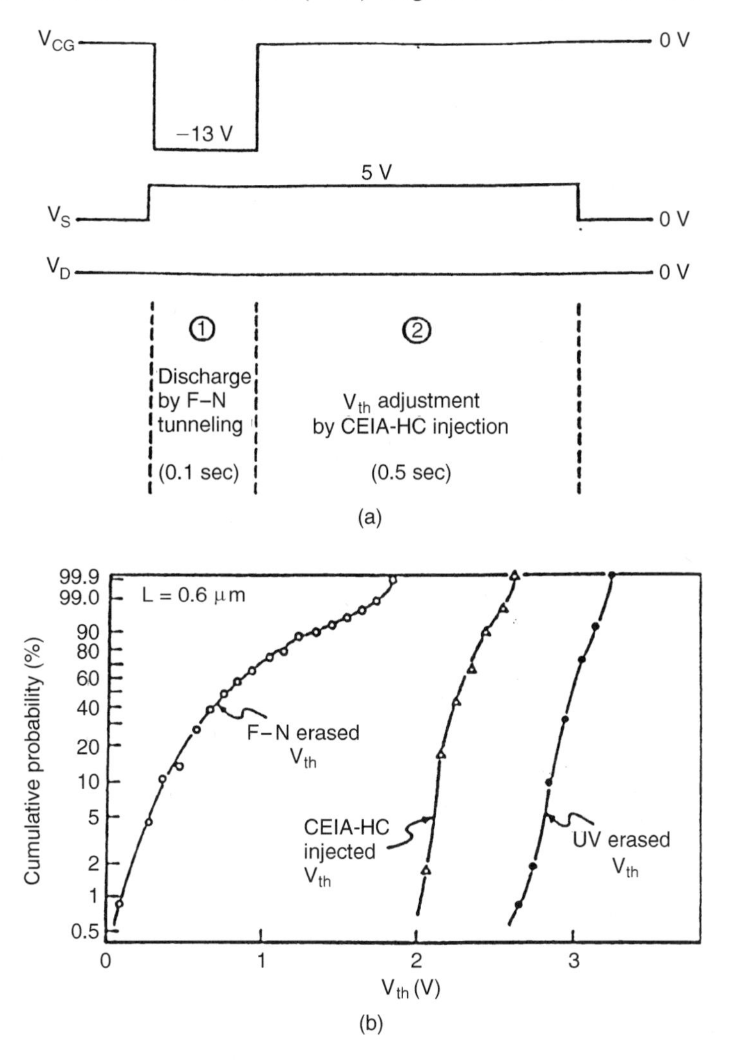

Figure 4.31 (*a*) Timing diagram used in experiments on new erase sequence, (*b*) threshold voltage distribution before and after adjustment by CEIA-IIC injection [4.50].

An important design consideration is the proper selection of the programming voltages to minimize these disturbs. Write/erase cycling of the cell also affects these disturbs. Verma and Mielke [4.21] showed that the drain-disturb characteristics of Flash memory devices are excellent with no measurable change until several thousand cycles. However, write/erase cycling has an influence on the gate-disturb behavior. Figure 4.33 shows the gate-disturb time as a function of cycling. Before cycling, the disturb time is about 100 seconds. After 100 cycles, this margin is decreased about two to three orders of magnitude. This degradation in gate-disturb sensitivity is caused by hole trapping during erase [4.21].

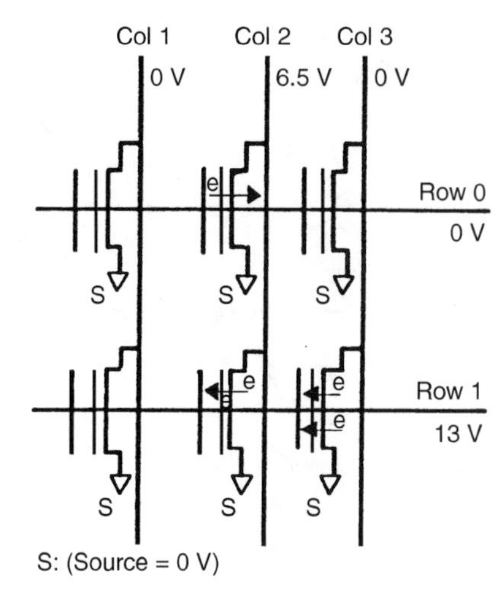

Figure 4.32 Schematic description of disturb during programming [4.21].

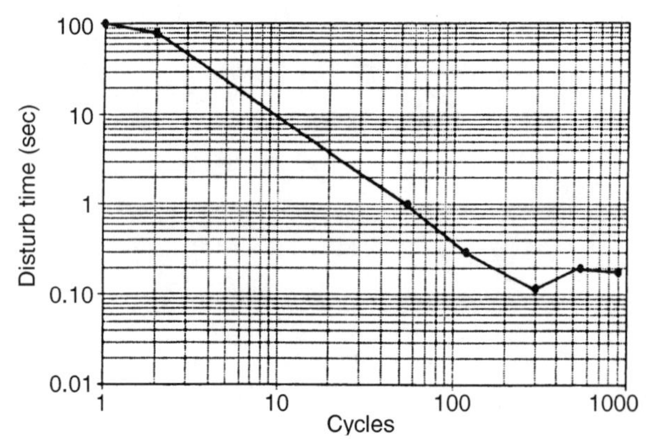

Figure 4.33 Gate-disturb time as a function of cycling [4.21].

4.3. FLASH MEMORIES WITH CHANNEL HOT-ELECTRON PROGRAM AND POLY-TO-POLY ERASE

Even though erasing through the source or drain does not require an extra transistor terminal, thereby giving the most compact cell, the use of junctions does create the problem of breakdown current. Also, a thin oxide is required for the cell gate oxide. Instead of erasing through the silicon, it is also possible to have an extra polysilicon electrode that connects through an oxide to the floating gate. This extra poly electrode results in a slightly larger memory cell or a more complicated process.

The poly-to-poly erase (floating gate to control gate in a double-poly structure, or floating gate to an erase gate in a triple-poly structure) avoids stressing of the gate oxide during erase. The first modern Flash EEPROM [4.3–4.5] from Toshiba used poly-to-poly erase and will be described first, followed by a contactless, virtual ground array version from Sandisk. In both cases, erase occurs from the floating gate to an erase gate, although the locations of the erase gate in the cell are different.

4.3.1 Triple-Poly NOR

Triple polysilicon Flash memory cells [4.3–4.5] are shown in Figs. 4.34, 4.35, and 4.36. The select gate is integrated with the floating gate (merged pass gate). The first polysilicon layer is used as an erase gate and is located between the field oxide and the floating gate. The second polysilicon layer is used as a floating gate in a manner similar to an UV EPROM. The third polysilicon layer is used as a control gate and is, in fact, used as a bit select-line for programming and reading.

The cell is programmed by a channel hot-carrier injection mechanism similar to EPROM. The erasure is accomplished using the first level of polysilicon which serves as an erase electrode causing field emission of electrons from the bottom of the floating gate. In the triple-poly NOR cell, as shown in Fig. 4.34, one part of the channel is controlled by the control gate. The third polysilicon layer is used both as a gate of the selection transistor and a control gate of the triple-poly NOR cell. This selection transistor enables the triple-poly cell to remain in the enhancement mode even if the floating gate transistor becomes a depletion mode device after erasure. The erase gate of the memory cell is supplied with the boosted voltage (V_{EG}), which enables field emission from the floating gate. This boosted voltage (V_{EG}) is produced from the program voltage (V_{PP}). All of the memory cells are erased simultaneously by the erase gate because they are commonly connected to each other.

The read data depend on the stored electric charges on the floating gate similar to the case for EPROM. The erased memory cell (the floating gate not charged with electrons) conducts by applying voltage as shown in Fig. 4.36. The virgin memory cell, prior to being programmed, has the same threshold voltage as the erased memory cell. However, their g_m values are different. The threshold voltage of the F-E2PROM cell is controlled by the selection transistor. However, the g_m depends on the value of the stored electric charges in the floating gate, even if the memory cell is erased.

4.3.2 Triple-Poly, Virtual Ground Contactless [4.63]

The Flash memory cell uses a triple-polysilicon, single-metal, split-channel structure with buried n^+ diffused source-drain, integrated into a contactless, virtual ground array architecture. The buried diffusion bitlines are contacted periodically

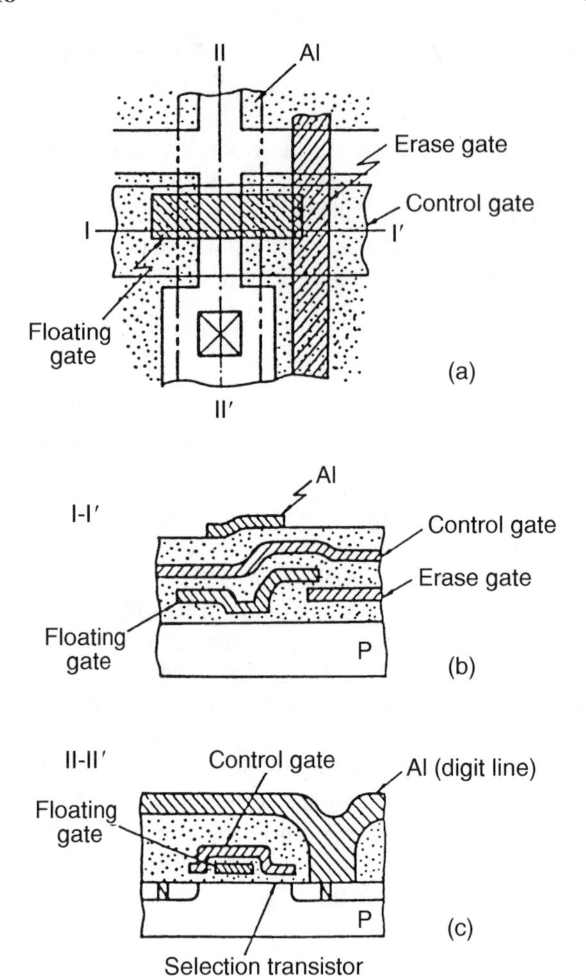

Figure 4.34 (*a*) Top view of F-E2PROM cell, (*b*) cross-sectional view along I–I′ line in (*a*), (*c*) cross-sectional view along II–II′ line in (*a*) [4.3–4.5].

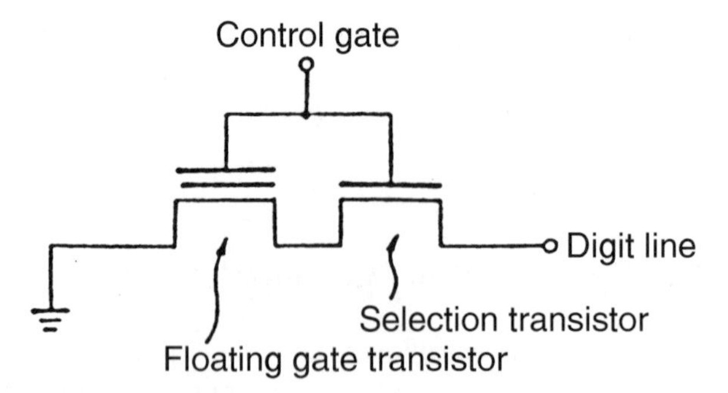

Figure 4.35 Equivalent circuit of Fig. 4.34 [4.3–4.5].

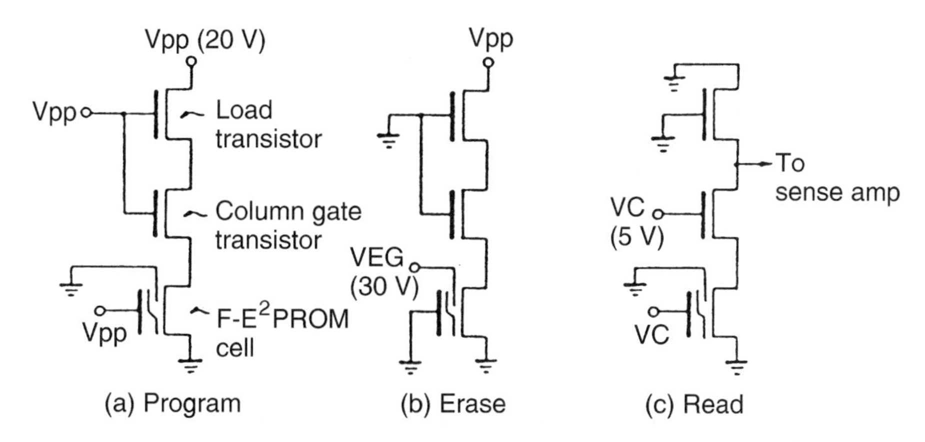

Figure 4.36 Operating conditions (*a*) program, (*b*) erase, (*c*) read [4.3–4.5].

with metal to reduce series resistance. The cell is programmed using channel hot-electron injection and is erased using inter-polysilicon dielectric tunneling. The split-channel memory transistor structure provides an integrated fixed enhancement threshold select transistor in series with the floating gate channel, allowing the floating gate to be erased to negative threshold conditions without introducing channel leakage current when unselected. This eliminates the program-before-erase requirement of 1-T cell Flash memories, greatly simplifies the erase operation, and relieves the very stringent process controls associated with maintaining a tight erase distribution on a large population of bits.

4.3.2.1 Cell/array Architecture. The structure of the cell, represented in Fig. 4.37, consists of three levels of polysilicon that form the floating gate, control gate, and erase gate, and uses conventional isolation. No oxide in the process is thinner than 25 nm. Employing a contactless, virtual ground array and using only a single contact for every 32 cells improves the array density. This reduces the cell size relative to nonvirtual ground cell/arrays, which use one contact for every two cells.

The memory array architecture is shown schematically in Fig. 4.38. In the virtual ground array, the source (S) diffusion of one column of cells is also the drain (D) diffusion of the adjacent column of cells. The memory array is organized into sectors of 576 bytes each—512 user data bytes (compatible with most mass storage operating systems) and 64 bytes containing sector overhead and ECC information. The erase gate (EG) is used for tunnel erase between it and the floating gates of all cells within a sector. The top view and a cross-sectional view of the cell-array are shown in Fig. 4.39.

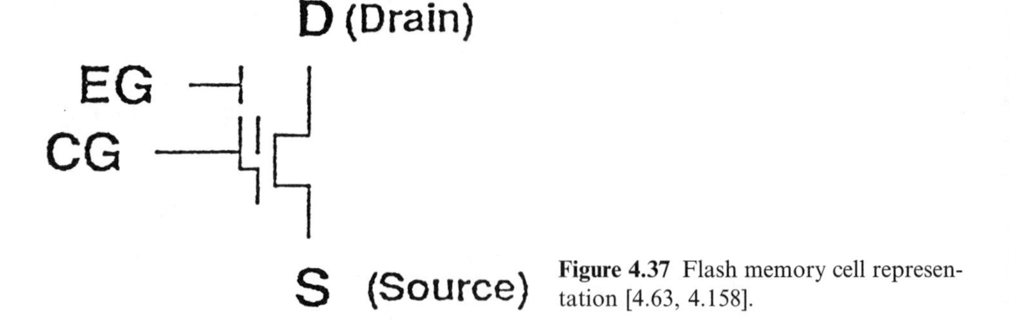

Figure 4.37 Flash memory cell representation [4.63, 4.158].

*4.3.2.2 **Programming.*** The conditions for programming are given in Fig. 4.38. To program, the control gate (CG) of a selected row (the wordline) is raised to 12 V, and the drain bitlines of the selected cells within this row are raised to 7 V (data conditional), while their source bitlines are held at 0 V. For cells whose data require them to be left in the erased state, the associated drain bitlines are held at 0 V, thereby preventing programming. The Flash cell is designed to program at relatively

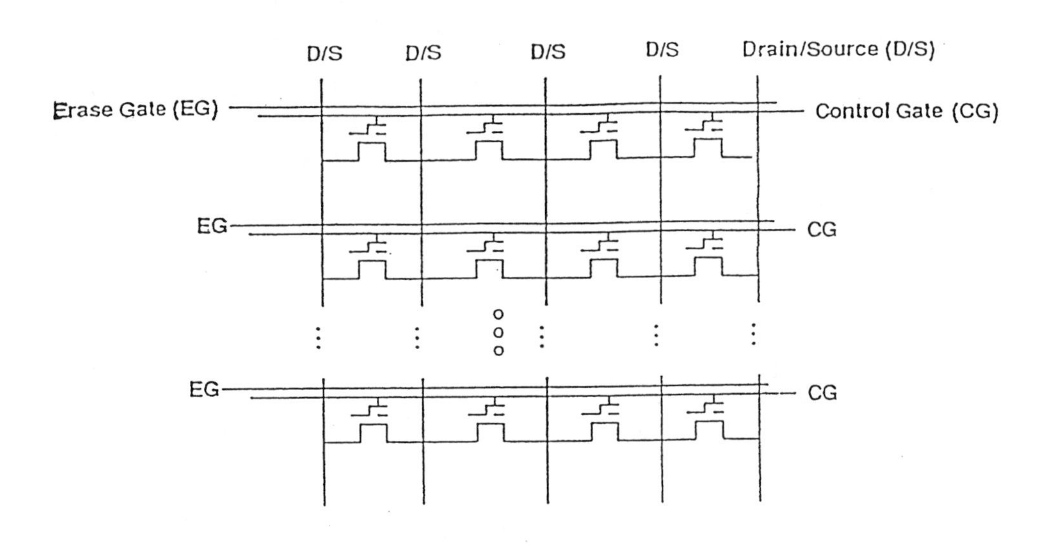

	CG	EG	D	S
Program	12	0V	7V	0V
Erase	0V	12V to 22V	0V	0V
Read	5V	0V	1.5V	0V

Figure 4.38 Flash memory array configuration [4.63, 4.158].

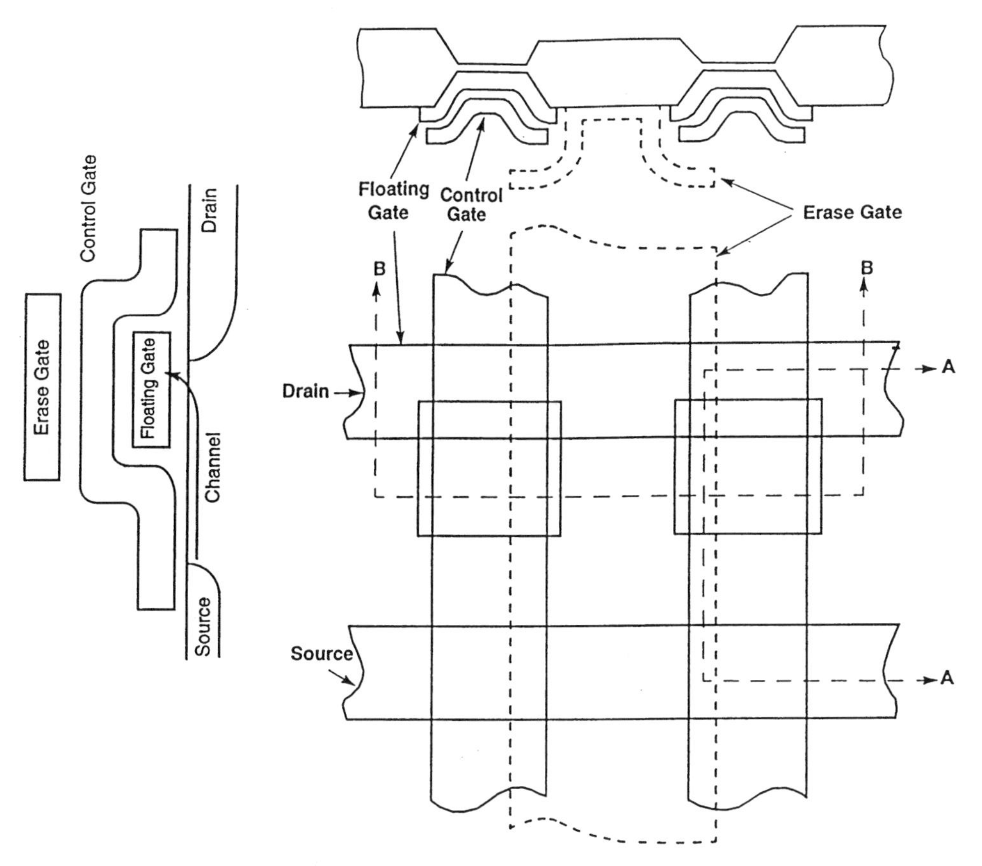

Figure 4.39 Memory array top view and cross-sectional views [4.63, 4.158].

low channel currents, thus allowing for programming 64 bits (termed a *chunk*) in parallel. A typical cell programming time is 5 μs, giving an effective programming time of 0.6 μsec/byte. An on-chip verify operation follows each programming operation to ensure full programming. While the majority of cells program in a single pulse, additional program/verify pulses are applied to each chunk, as needed, to accommodate the full programming of slower bits.

4.3.2.3 Erase. Cell erase uses polysilicon (floating gate side) to polysilicon (erase gate) tunneling through a relatively thick inter-polysilicon oxide to remove electrons from the floating gate. The smallest unit of erase is a sector. Typical sector erase time is 1 msec, giving an effective erase time of 2 μsec per byte. Because of the split-channel feature, there is no cell program-before-erase requirement and, therefore, no associated erase speed loss. The erase voltage, VE, applied to the erase lines,

EG, ranges from 12 V to 22 V, depending on the sector requirements. The VE is automatically increased to compensate for charge trapped in the erase tunnel oxide during extended cycling. As with programming, a verify operation follows each erase operation.

To further reduce the effective erase time, simultaneous, random, multiple-sector erase capability is provided by a sector "tagging" feature that sets erase latches of selected sectors. During multiple sector, parallel erase, sectors verified to be properly erased are "untagged," thereby preventing further erasure, while other sectors in the group, needing additional erasure, continue the erase/verify sequence. In this way, erase oxide stress is minimized, achieving a sector endurance exceeding 100,000 program/erase cycles.

This cell/array approach offers improved erase performance and reliability by confining the relatively high erase voltages within the memory array to the erase polysilicon layer, rather than to diffusion regions within the substrate. This eliminates the relatively high erase currents and device degradation associated with high junction fields and leakages.

4.3.2.4 Read. Cell reading is accomplished by raising the selected wordline to 5 V and biasing the drain of the selected cell to 1.5 V via a current sense amplifier, with the source driven to 0 V. During read, a programmed cell will not conduct, resulting in a "1" state, while an erased cell will sink current, resulting in a "0" state. The 64 cells of the chunk are read simultaneously. The memory incorporates a 20 MHz, double input/double output path, serial interface with a read protocol designed to output data from the device at high rate under continuous read, with no intersector delay. While one chunk's data are being shifted out, the next chunk is addressed and sensed, absorbing all the associated memory access time, to give an effective read time of 200 nsec/byte.

4.3.2.5 Write Inhibit Conditions. Erasing of unselected sectors is inhibited by maintaining their erase-line voltage at 0 V. Programming of unselected sectors/rows is inhibited by maintaining their wordline voltage at 0 V. Programming of unselected cells in selected rows is inhibited by biasing approaches that limit the drain to source voltage differential, thereby eliminating channel hot-electron generation. A further advantage inherent in this split-channel cell over other Flash cells, which have only the floating gate channel, is the attribute of no cell programming when the bias conditions of its drain and source are interchanged. This facilitates integrating the cell into the virtual ground array, without incurring any disturb conditions under reverse drain-source bias conditions.

In all Flash memory arrays, unselected cells sharing wordlines or bitlines of cells being programmed are subjected to program-disturb conditions as their corresponding wordlines or bitlines are biased at programming voltage levels. In this Flash technology, no oxides are thinner than 25 nm, and the oxide fields associated with the disturb conditions are lower than those which may exist in thin tunnel oxide technologies.

4.3.3 The Field-Enhancing Tunneling Injector EEPROM Cell

The field-enhancing tunneling injector EEPROM cell [4.82] is a double-poly, split gate memory cell using poly-to-poly (floating gate to control gate) F–N tunneling for erasing and source-side hot-electron injection for programming. Poly-to-poly tunneling is from a field-enhancing tunneling injector formed on the floating gate. Source-side, hot-electron injection is very efficient, thus allowing the use of small on-chip charge pumps from a single low-voltage power supply, for example, 3.0 or 2.5 V. Cells are normally erased prior to programming. The split gate memory cell size is comparable to traditional stacked gate memory cells because (1) the tunneling injector cell does not need the extra spacing to isolate the higher voltages required for programming the stacked gate array and (2) floating gate extensions are not needed to achieve the required stacked gate coupling ratios.

4.3.3.1 Cell Structure. A top view and cross-sectional view along the wordline and bitline directions are presented in Fig. 4.40. A common source is used for each page; that is, each pair of bits sharing a common source along a row pair (even plus odd row). A single wordline is referred to as a row. The combination of the even and odd rows form a page, which is erased as an entity. Programming may be either byte by byte individually or for all bytes within the same page simultaneously.

The drain region consists of an n^+ S/D diffusion, which is aligned with the edge of the second poly control gate. The source region consists of an n^+ S/D diffusion, which overlaps the floating poly. A cell implant beneath the floating gate is used to control the intrinsic cell threshold (V_t) and the punch-through voltage.

The select gate is separated from the channel by a 40 nm oxide. The floating gate is separated from the channel and source diffusion by a thermally grown 15 nm gate oxide. The floating gate is separated from the control gate by a 40 nm oxide on the side-wall and a 200 nm oxide vertically between the gates. The tunneling injector on the floating gate is formed by oxidation of the polysilicon, similar to the formation of the field oxide "bird's beak" on single-crystal silicon. A silicide or policide can be formed on the control gate to reduce the poly wordline resistance. The cell array schematic is presented in Fig. 4.41. An equivalent circuit representation used to illustrate capacitive-coupling is presented in Fig. 4.42.

Table 4.5 gives the conditions for the memory cell terminals during erase, program, and read operations. V_{cc} is the power supply, for example, a nominal 3 or 2.5 V, V_{ss} is ground, and V_t is the nominal cell threshold. V_{REF} is the reference voltage used to access the memory cell during the read cycle. The high voltages on the wordline during erase and the source-line during programming are generated by on-chip charge pumps.

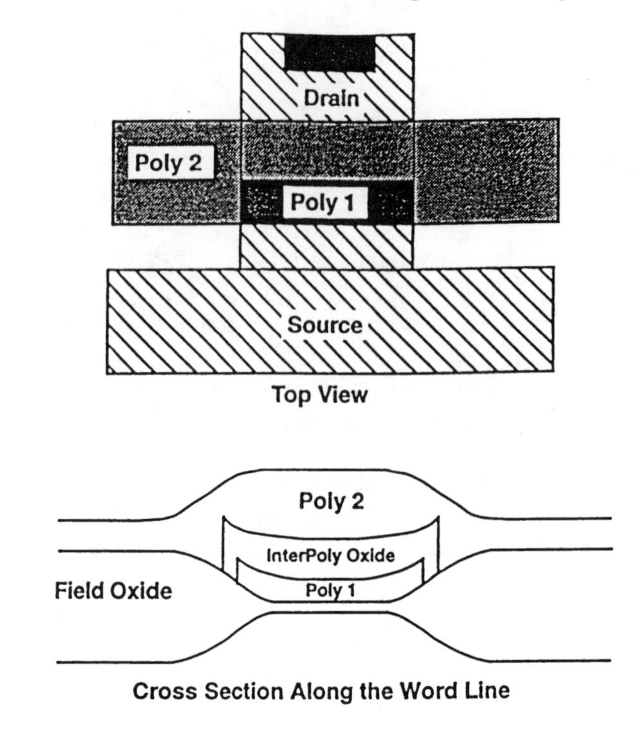

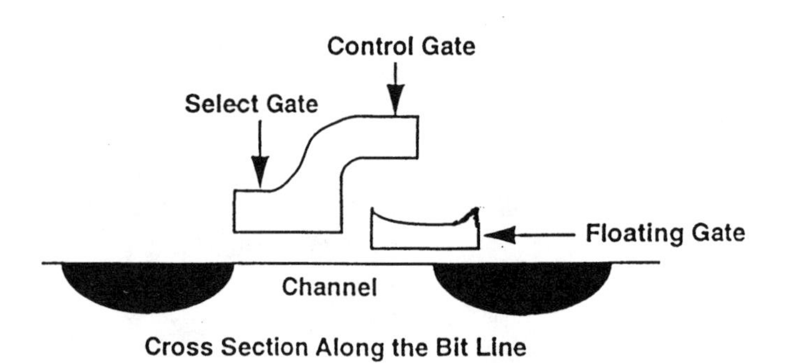

Figure 4.40 Top view, cross section along the wordline, and cross section along the bitline [4.82, 4.158].

During erase, the channel is inverted due to the wordline voltage. This increases the value of C1c. During programming, the channel is in depletion and the value of C1c is negligible. Therefore, the coupling ratios are different during erase and programming. During programming, the capacitive coupling ratio between the source and the floating gate is approximately 80%. This means that approximately 80% of the voltage at the source will be coupled to the floating gate.

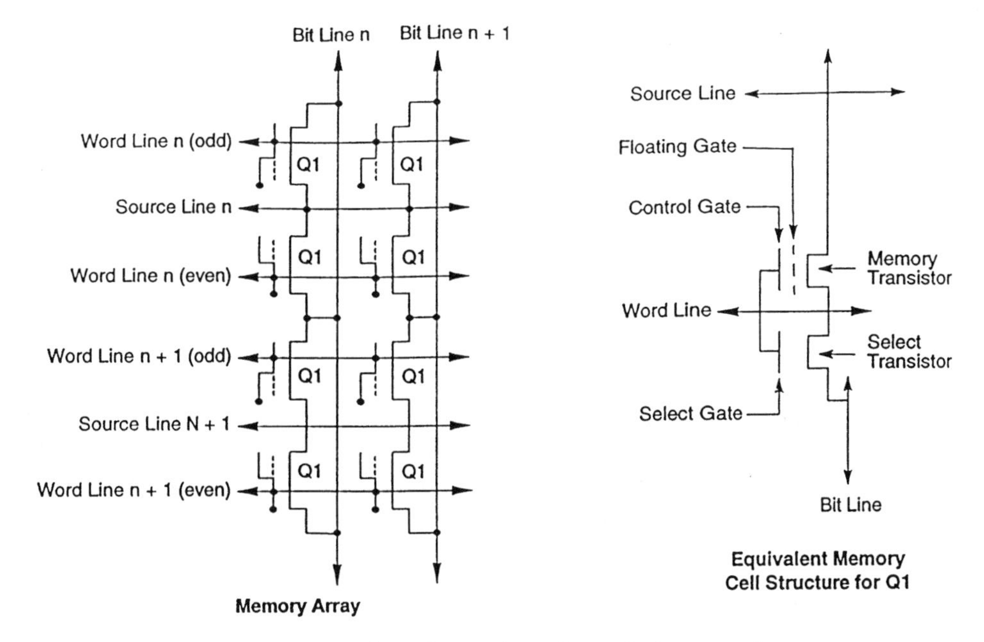

Figure 4.41 Representation of a section of a typical cross-point memory array arranged as eight memory cells in two columns (bitlines), two source-lines, and four wordlines. Note that the wordline is split into an even and odd row, which isolates the source-line from all other source-lines. On the right is an equivalent memory cell showing how the split gate cell provides the logical equivalent of a select transistor and a memory transistor [4.82, 4.158].

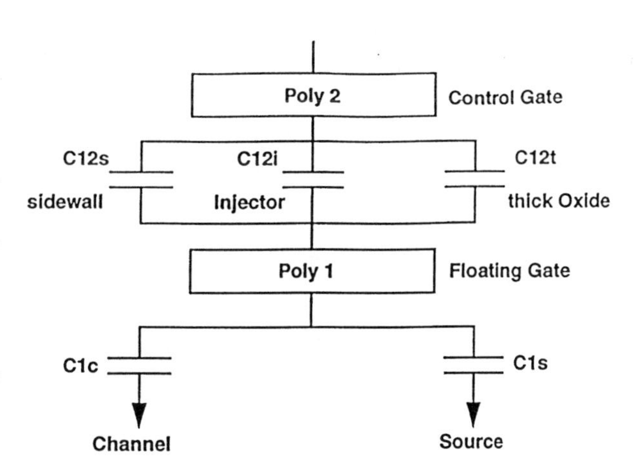

Figure 4.42 Equivalent capacitive-coupling circuit [4.158].
C10 = C1c + C1s; C12 = C12s + C12i + C12t
Coupling Ratios (CR) are defined as:
1. CR10 – CR (poly1 to substrate) – (C1c + C1s)/ (C1c + C1s + C12)
2. CR12 = CR (poly1 to poly2) = C12/ (C12 + C10)
3. CR10 + CR12 = 1

TABLE 4.5 MEMORY CELL OPERATION CONDITIONS

Operation	Wordline	Source-Line	Bitline
Erase	14 V	0 V(V_{ss})	0 V(V_{ss})
Program	V_t	11 V	0 V data "0" V_{cc} data "1"
Read	V_{cc}	0 V(V_{ss})	2 V

*4.3.3.2 **Erasing.*** The cell is erased by Fowler–Nordheim tunneling of electrons from floating gate to control gate through interpoly oxide. During erase, the source and drain are grounded and the wordline is raised to 14 V. The conditions for erase are given in Table 4.5. The low coupling ratio between the control gate and the floating gate provides a significant voltage drop across the interpoly oxide, which is the same everywhere between poly1 and poly2. A high field is generated primarily in the area of the tunneling injector. Charge transfer is very rapid and is eventually limited by the accumulation of positive charge on the floating gate. This positive charge raises the floating gate voltage such that there is insufficient voltage drop across the poly-to-poly dielectric to sustain Fowler–Nordheim tunneling.

The removal of charge can leave a net positive charge on the floating gate. The positive charge on the floating gate reduces the memory cell's threshold voltage to about the select gate V_t. The applied sense voltage is sufficient to turn on both the select transistor and the memory transistor in the addressed memory cell.

Erase can either be by fixed program pulses generated by an internal timer or algorithmically generated by an external controller to optimize erase conditions. Internal verify circuits assure an adequate erase margin.

4.3.3.3 Erase Disturb. Enhanced-field tunneling injector devices are internally organized by pairs (pages) of even and odd rows. Each row pair (page) shares a common source-line and has the wordline at the same voltage potential during erase. Thus, all bytes along the common wordlines are erased simultaneously. All other wordlines (pages) do not receive the erasing high voltage. Therefore, erase-disturb is not possible. The column leakage phenomenon caused by "overerase" in 1-T cells is avoided because the split gate provides an integral select gate to isolate each memory cell from the bitline.

4.3.3.4 Programming. The cell is programmmed using high-efficiency source-side channel hot-electron injection. The conditions for programming are given in Table 4.5. During programming, a voltage of a cell threshold of approximately V_T volts is placed on the control gate via the wordline. This is sufficient to turn on the channel under the select portion of the control gate. The drain is at approximately V_{SS} if the cell is to be programmed. If the drain is at V_{CC}, programming is inhibited. The drain voltage is transferred across the select channel because of the voltage on the control gate. The source is at approximately 12 V. The source to drain voltage differential (i.e., 11 V—approximately V_{SS}) generates channel hot electrons. The source voltage is capacitively coupled to the floating gate. The electric field between the floating gate and the channel sweeps the channel hot electrons that cross the Si–SiO_2 barrier height of approximately 3.2 eV to the floating gate very efficiently.

The programming effect is eventually self-limiting as negative charge accumulates on the floating gate. The programming source-drain current is low. Thus, the source voltage can be generated by an on-chip charge pump. The program time is fast because of the relatively high efficiency of source-side injection. Programming can be by fixed program pulses generated either by an internal timer or by an external controller to optimize program conditions.

4.3.3.5 Program-Disturb. There are two possible types of program-disturb with the field-enhanced tunneling injection cell: reverse-tunnel disturb and punch-through disturb.

Reverse-tunnel disturb can occur for unselected erased cells sharing a common source-line, but on the other row of the selected page to be programmed. Thus, the wordline is grounded. The source voltage is capacitively coupled to the floating gate of the unselected erased cell. If there is a defect in the oxide between the control gate and the floating gate, Fowler–Nordheim tunneling may occur. This could program the unselected erased cell.

Punch-through disturb can occur for selected erased cells, that is, those sharing a common source-line and wordline in an adjacent inhibited bitline. An inhibited bitline is taken high to prevent normal channel hot-electron injection. If there is a defect that reduces the bitline voltage or creates punch-through along the select gate channel, hot electrons could be available to program the inhibited erased cell.

Proper design and processing can prevent both mechanisms. Devices with this memory architecture do not have program-disturb caused by accumulated erase/programming cycles because each page is individually isolated. Each cell is only exposed to high voltage within the selected page along the row- or source-line. There is no high voltage on the bitline.

4.4. FLASH MEMORIES WITH FOWLER–NORDHEIM TUNNEL PROGRAM AND ERASE

Although EPROM-like cell structures offer density and historical learning advantages for Flash memory implementation, cell structures based on Fowler–Nordheim tunneling offer low power and single, low-voltage supply advantages. Examples are DINOR [4.58], AND [4.57], NAND [4.14], and HiCR [4.68]. Based on current understanding, the single-voltage supply Flash memories, below 3 V V_{CC}, will require

Fowler–Nordheim tunneling for programming. The half-select problems associated with F–N tunnel programmming are minimized by small blocks, on-chip V_{PP} trimming, and process control.

4.4.1 Array Contactless EEPROM (ACEE)

In Array Contactless EEPROMs (ACEEs), the program and erase operations occur at the same node. The basic structure, schematic, and operation of the ACEE cell [4.24, 4.25, 4.29, 4.37, 4.159] are shown in Fig. 4.43. The floating gate is formed in first-level polysilicon, and the merged control and select gates are formed in second-level polysilicon. The ACEE cell is a Self-Aligned Thick Oxide (SATO) device, with the source and drain buried under a thick oxide. The oxide between the floating gate and the substrate near the source is 10 nm thick to allow for programming and erasing the cell by Fowler–Nordheim tunneling. A 35 nm gate oxide is employed under the floating gate structure to increase the poly2/poly1 capacitive-coupling ratio as a whole and to increase the gated diode breakdown voltage near the source. The pass gate, with 50 nm gate oxide, prevents the cell from drawing current when the floating gate structure is over-erased into depletion. Thus, the cell current is dominated by the pass gate. The buried source-drain of the

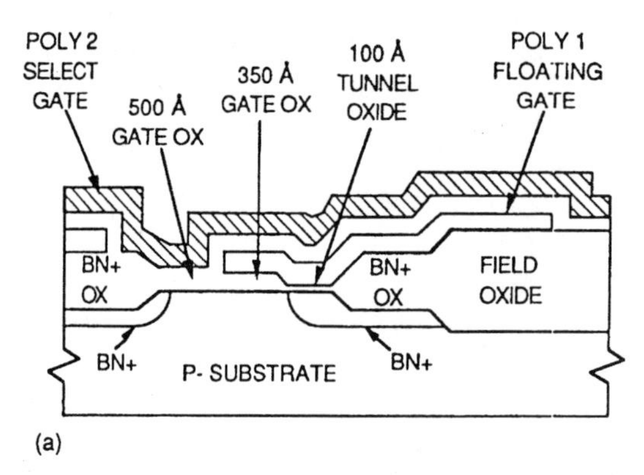

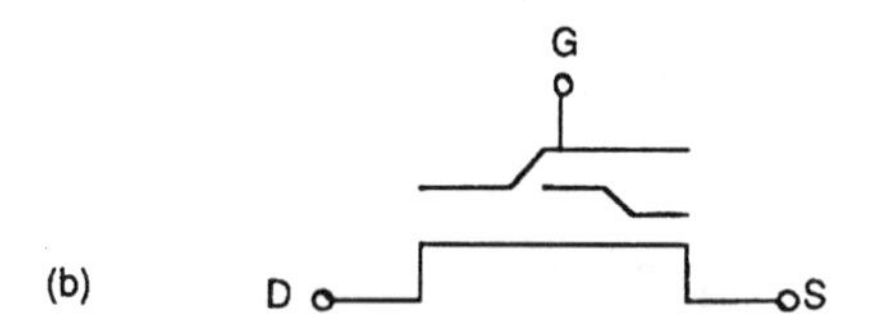

(c)

	D	G	S
ERASE	FLOAT	VEE	5V
PROGRAM	FLOAT	VPP	0
READ	1.5 V	3 V	0

Figure 4.43 (*a*) Cross-sectional view through the ACEE cell, (*b*) schematic drawing of the cell, and (*c*) cell operating conditions [4.159].

cell constitutes a continuous buried bitline. Since negligible supply current is required during program or erase, the voltage drop along the buried bitline in a VLSI memory is small.

4.4.1.1 Charge Transfer. Charge transfer between the floating gate and the source region occurs in a manner similar to the FLOTOX memory cell. A high voltage, on the order of 20 V, is applied to the control gate, the source is held at ground, and the drain is left floating. Because of capacitive-coupling, a voltage of greater than 10 V is coupled to the floating gate. This establishes an electric field greater than 10 MV/cm across the tunnel oxide from the floating gate to the source, and electrons are transferred to the floating gate. Charge transfer to the floating gate raises the threshold voltage of the cell above that of the sense voltage, and the cell stops conducting current. To bring the cell to a conducting state, the charge stored on the floating gate must be removed; that is, a sufficiently high electric field must be applied across the tunnel oxide in the reverse direction.

4.4.1.2 Inhibit Condition. Programming of deselected cells is prevented by increasing the source-line voltage for cells sharing the same wordline and by applying a small positive voltage to unselected wordlines as is illustrated in Fig. 4.44.

4.4.1.3 Read. Reading of the ACEE cell is accomplished by raising the drain (the diffusion near the select gate with 50 nm gate oxide) to about 1.5 V, holding the source at ground potential, and raising the selected wordline to about 3 V. The read operation from the select transistor side provides a margin against read-disturb. The floating gate structure can be erased to a negative threshold voltage since the series select gate is an enhancement device. The threshold voltage of the select gate is lower than the sense voltage applied to the top gate. Thus, during the erased state, the cell current is determined predominately by the select gate and, during the programmed state, by the floating gate structure.

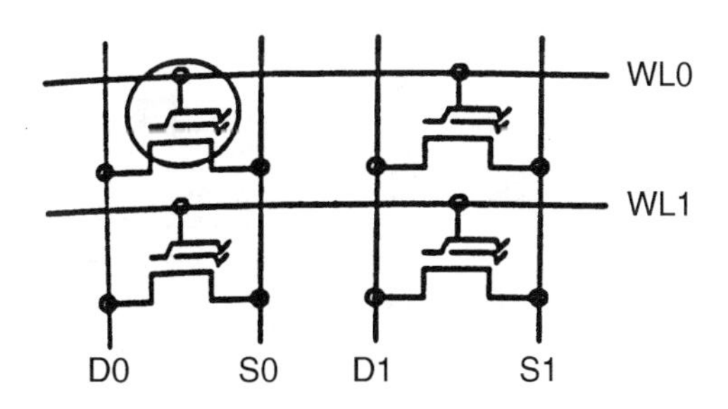

Cell (0,0) Operating conditions

	WL0	WL1	S0	D0	S1	D1
Erase	−11 V	−11 V	5 V	Float	5 V	Float
Program	17 V	7 V	0 V	Float	7 V	Float
Read	3.3 V	0 V	0 V	1.5 V	0 V	1.5 V

Figure 4.44 Schematic drawing and operating condition table showing the inhibit conditions for the ACEE cell array [4.37].

4.4.2 The AND Cell

The AND cell [4.57] program and erase operation is performed by F–N tunneling, as shown in Fig. 4.45 and Table 4.6. The memory cell is programmed through the drain by tunnel injection of electrons and is erased by tunnel injection from the whole channel region. The schematic diagram of this contactless memory array and the program/erase scheme are shown in Fig. 4.46 and Table 4.6, respectively. The memory cells are arranged in parallel between each pair of local data and source-lines, and ST1 and ST2 act as switches for connecting and disconnecting the target block to a global data-line and a common source-line. Both the local data-line and source-line are n^+ diffusions. Internal operating voltages applied to the selected wordline, –9 V for programming and +13 V for erasing, are generated from the single 3 V power supply voltage by on-chip voltage converters, due to the low-current dissipation of tunneling. The rewrite size is as small as 512 bytes (which corresponds to memory cells connected to each wordline), suitable for various silicon file

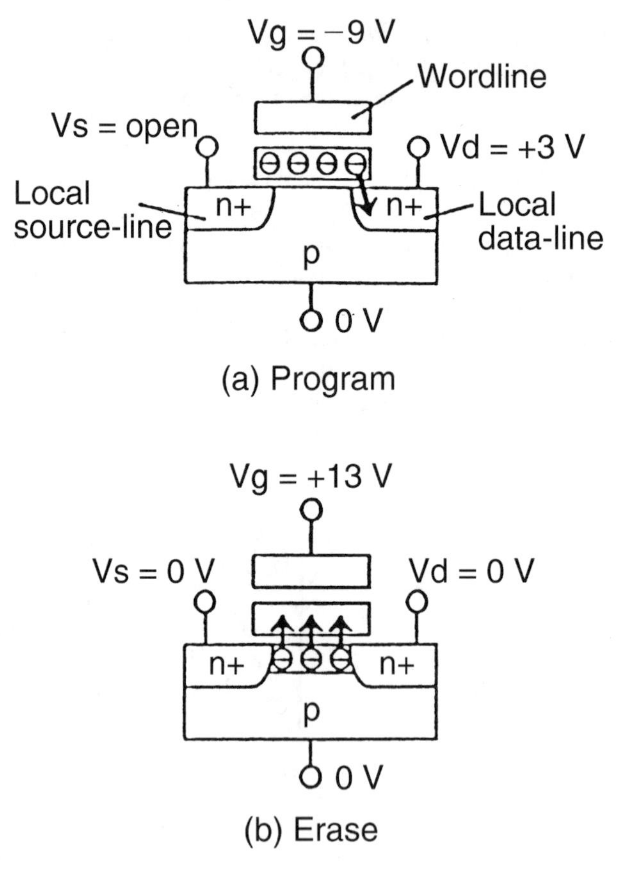

Figure 4.45 Program/erase diagrams and their bias conditions for the AND cell [4.57].

TABLE 4.6 OPERATING CONDITIONS FOR THE AND MEMORY ARRAY

		Erase	Program	Read
Word	Selected	13 V	−9 V	3 V
	Deselected	0 V	3 V	0 V
Local data-line		0 V	3 V (program) 0 V (inhibit)	1 V
Local source-line		0 V	Floating	0 V

Erase: F–N tunneling
Program: F–N tunneling

applications. Each local source-line is disconnected from the common source-line when programming to allow program inhibit for the cells connected to the deselected wordlines. Adjacent bitlines are separated by field oxide isolation.

Scatter in the "low-level" threshold voltage of programmed memory cells can be suppressed below 0.5 V, as shown in Fig. 4.47 by using program and program-verify sequences. In page-mode programming for the selected wordline, the program is controlled bit by bit by setting the potential in each data-line at 3 V for successive program and at 0 V for program termination. Obtaining a tight distribution in the "low-level" threshold voltage would lead to a successful read operation at a 3 V

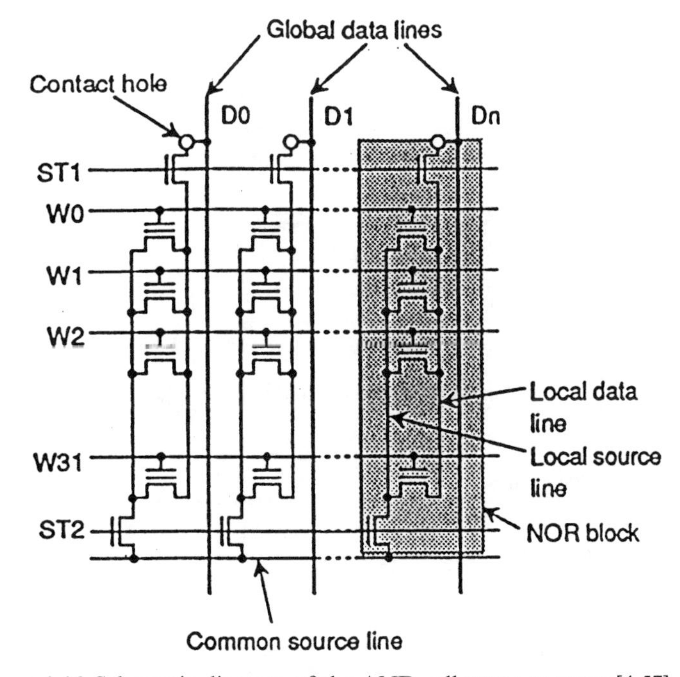

Figure 4.46 Schematic diagram of the AND cell memory array [4.57].

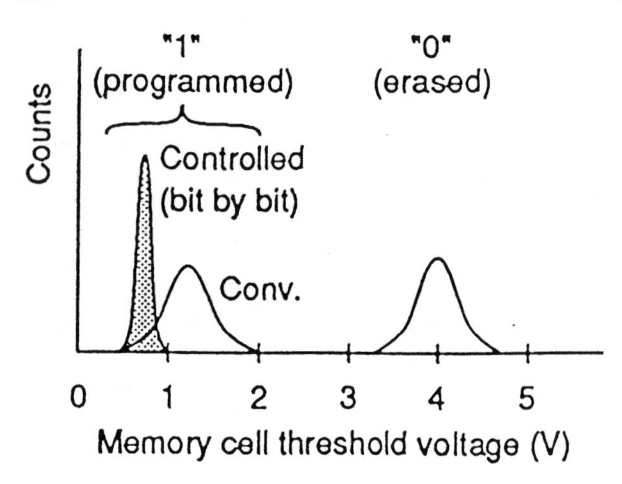

Figure 4.47 Schematic of the low-threshold voltage control for the programmed memory array [4.57].

power supply voltage without introducing a complicated wordline boost scheme [4.39].

Schematic cross sections of the memory array parallel to a wordline and a dataline are shown in Fig. 4.48. A floating gate consists of two layers of poly-Si films electrically combined to each other. The first polysilicon defines a channel length, and the second polysilicon realizes a large capacitive-coupling between the control gate and the floating gate.

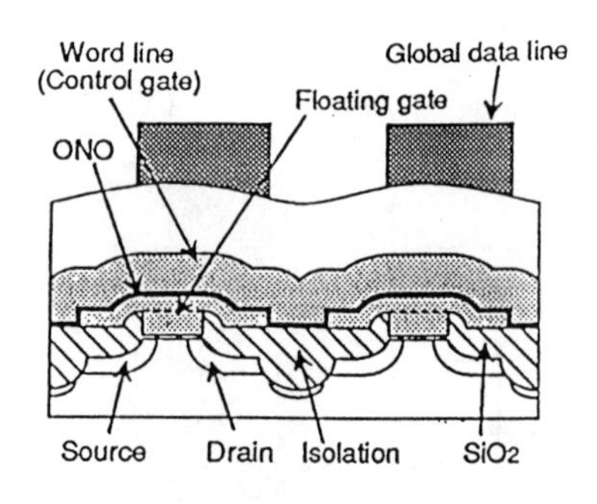

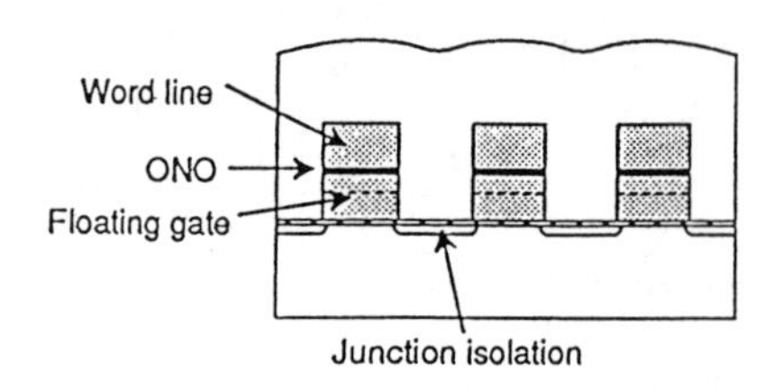

Figure 4.48 Schematic cross sections [4.57].

4.4.3 The DINOR Cell

The **DI**vided bitline **NOR** (DINOR) cell operation is schematically shown in Fig. 4.49. In the case of the DINOR [4.58, 4.69] cell, the "program" operation means to set the V_{th} of the selected cell to the low state, and the "erase" operation means to set the V_{th} of the cells of the selected sector to the high state. That is, the direction of the operation is opposite to the conventional NOR.

The "program" condition uses F–N electron ejection from the floating gate to the drain through the gate-drain overlapped area, while "erase" causes F–N electron injection from the channel area to the floating gate. By using the low V_{th} as the

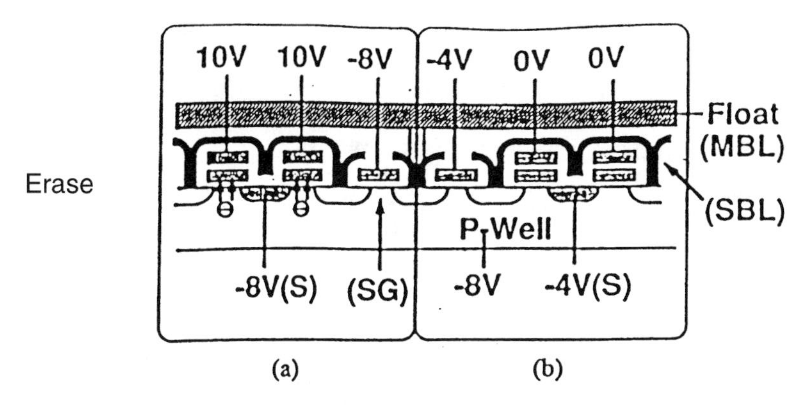

Voltage condition of DINOR cell in the erase operation. (a) Selected sector. (b) Unselected sector.

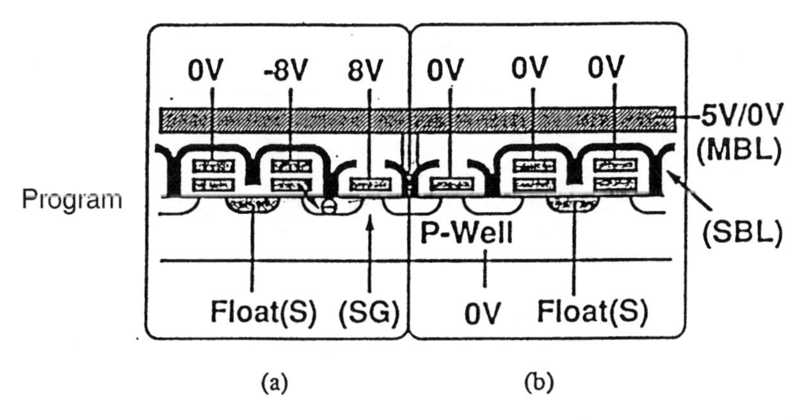

Voltage condition of DINOR cell in the program operation. (a) Selected sector. (b) Unselected sector.

Figure 4.49 Program/erase diagrams and the bias conditions for DINOR cell [4.58].

programmed state, the DINOR cell is able to use the bit-by-bit verify programming sequence. This makes the V_{th} distribution tight and realizes high over-erasure tolerance.

4.4.3.1 Program/Erase Operation. The DINOR cell utilizes the gate-biased F–N write/erase operation, which realizes a maximum internal voltage as low as 10 V. The conditions for various operations are summarized in Table 4.7, and a circuit diagram of the memory cell is shown in Fig. 4.50. Because of a low-current requirement, all the high voltages shown in Table 4.7 can be generated internally from a single external 3 V supply.

The DINOR cell uses a triple-well, triple-level polysilicon and a double layer of metal. The triple well is utilized to apply a negative bias voltage to the substrate in the erase operation. A symmetrical source-drain structure without a p^+ pocket is used for the memory cell.

TABLE 4.7 OPERATIONAL CONDITIONS FOR THE DINOR CELL

	Bitline	Control Gate	Source-Line	Substrate
Write	5 V	−8 V	Floating	0 V
Erase	Floating	10 V	−8 V	−8 V
Read	1 V	3 V	0 V	0 V

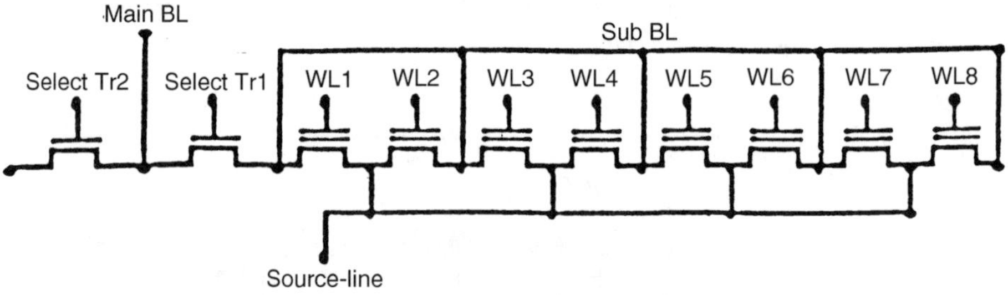

Figure 4.50 Circuit diagram of the DINOR cell [4.58].

4.4.3.2 Cell Structure. The DINOR cell is a unit consisting of one select transistor, eight stacked gate cells, and one sub-bitline connected through the select transistor to a main bitline. Figure 4.51 shows a schematic drawing of the DINOR cell.

The third polysilicon layer, which is used as the sub-bitline, is added to the conventional stacked gate cell structure formed by the first and second polysilicon layers. The low-current requirement of F–N operation makes it possible to use a polysilicon layer as the sub-bitline instead of a metal layer, and this is advantageous

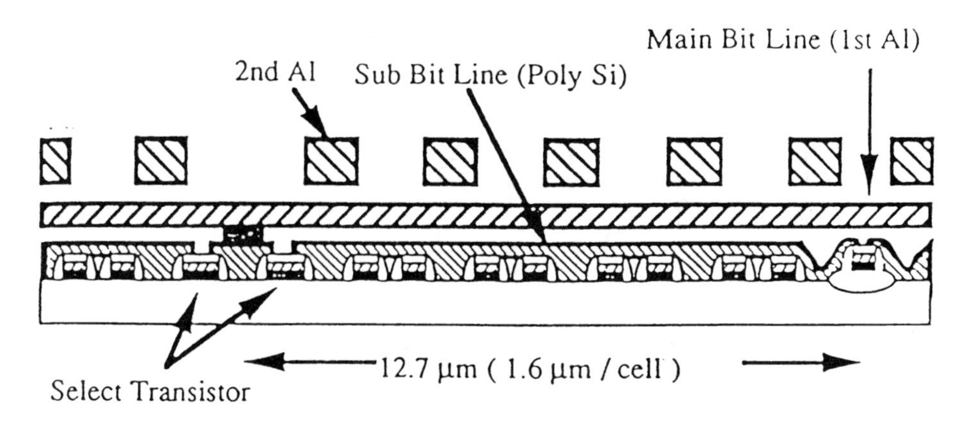

Figure 4.51 Schematic drawing of the DINOR cell [4.58].

in terms of wafer processing simplicity. The first aluminum layer forms the main bitline. The second aluminum layer in the cell array is used to stitch the gates of the select transistors and the main wordlines.

The sector size of the DINOR is 1 KB in which the cells are arrayed in 16 bits (=2 basic units described above) × 512 bits (connected to the same wordline).

4.4.3.3 Disturb and Endurance Characteristics. The DINOR cell solves disturb problems by utilizing the divided bitline and parallel programming. By setting all select transistors off except for the one that is in the programming mode, the maximum drain-disturb is only seven write cycles from the cells connected to the same sub-bitline. The maximum gate-disturb is only one write cycle due to the use of a latch circuit. Characteristics of drain-disturb and gate-disturb, with sufficient margin for device operation, have been demonstrated [4.58].

In the case of the DINOR cell, one more new disturb mode, substrate disturb, exists. Substrate disturb is the undesired erasure caused by V_{sub} (substrate voltage) stress during the erasure of other sectors. The longest time over which substrate-disturb occurs is of the order of one year for actual 16 Mbit devices. By applying an erase inhibit voltage (1/2 of V_{sub}) to the source-lines of the unselected sectors, the effective V_{sub} can be reduced by one-half, making the substrate-disturb immunity longer than 1E10 seconds.

4.4.3.4 Virtual Ground DINOR. A virtual ground DINOR array, utilizing F–N tunneling (for program and erase) has been described by using the asymmetrical offset source-drain structure [4.64] shown in Figs. 4.52 and 4.53. One side of the drain region has an offset against the floating gate, and the other side is overlapped with it. During programming, F–N tunneling current only flows to the overlapped drain region that is selected. Thus, only the selected cell is programmed. The operating conditions for the array are shown in Table 4.8.

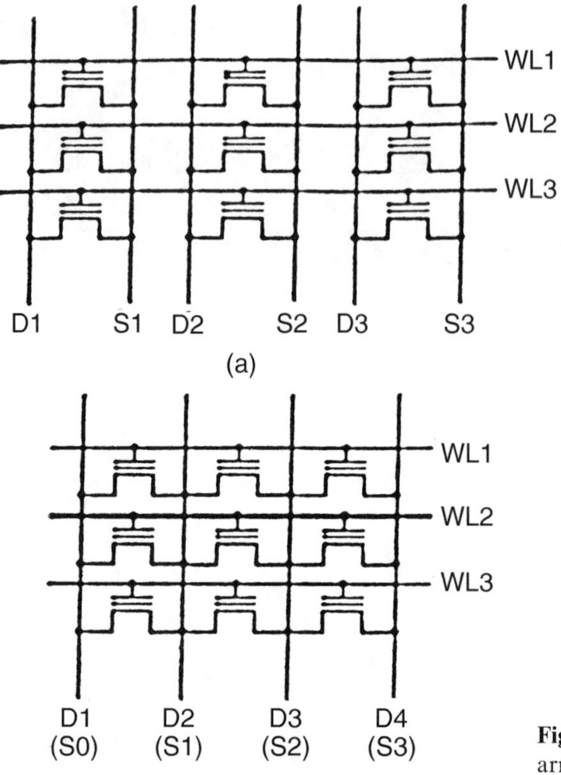

Figure 4.52 (*a*) Conventional contactless array with F–N operation, (*b*) simple virtual ground array [4.64].

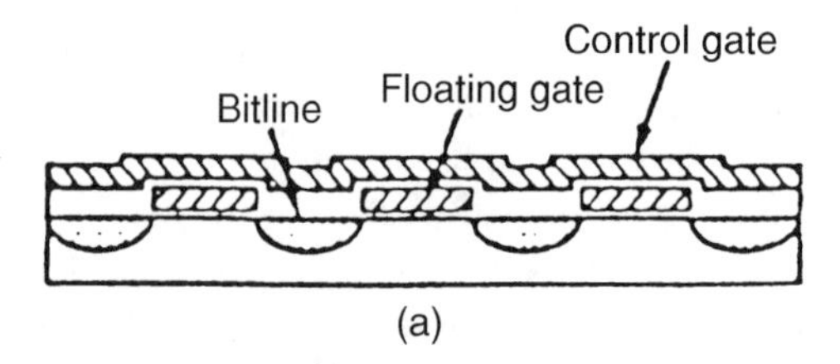

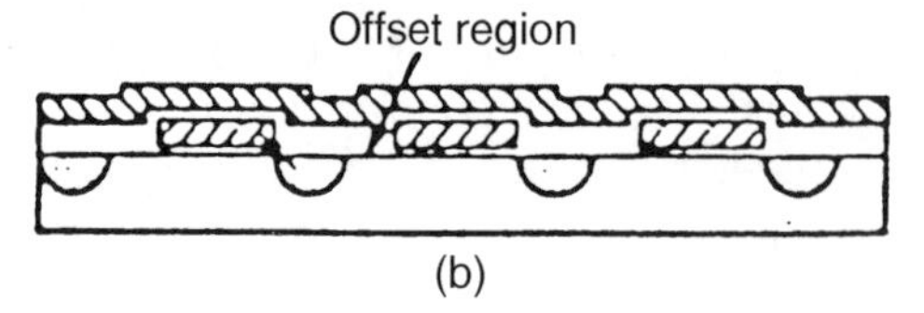

Figure 4.53 (*a*) Conventional virtual ground array that cannot use F–N mechanism, (*b*) virtual ground array using F–N mechanism with offset source/drain structure [4.64].

TABLE 4.8 OPERATING CONDITIONS FOR THE DINOR VIRTUAL GROUND ARRAY

	Bitline	Control Gate	Source-Line	Substrate
Program	5 V	−9 V	Floating	0 V
Erase	Floating	10 V	−9 V	−9 V
Read	1.5 V	3 V	0 V	0 V

4.4.4 HiCR Flash Memory Cell

The High Capacitive-coupling Ratio (HiCR) cell is a floating gate transistor, self-aligned with a buried n^+ source and drain, in a contactless array configuration [4.68]. The cell is programmed by F–N tunneling on the drain side and erased by F–N tunneling from the source-drain junction to the floating gate. The high capacitive-coupling ratio is realized by (1) a small tunneling region between the floating gate and the source-drain junction regions and (2) a 20 nm channel oxide and a 7.5 nm tunneling oxynitride. The floating gate consists of a lower polysilicon layer above the channel and an overlapping upper polysilicon layer above the tunneling oxide. The upper polysilicon layer is connected to the lower poly (Fig. 4.54).

The cell operation is shown in Figs. 4.55 and 4.56. Each pair of bitlines has a common source-line (1 1/2 rail array). Each pair of bitlines is isolated from the adjacent pairs of bitlines by field oxide.

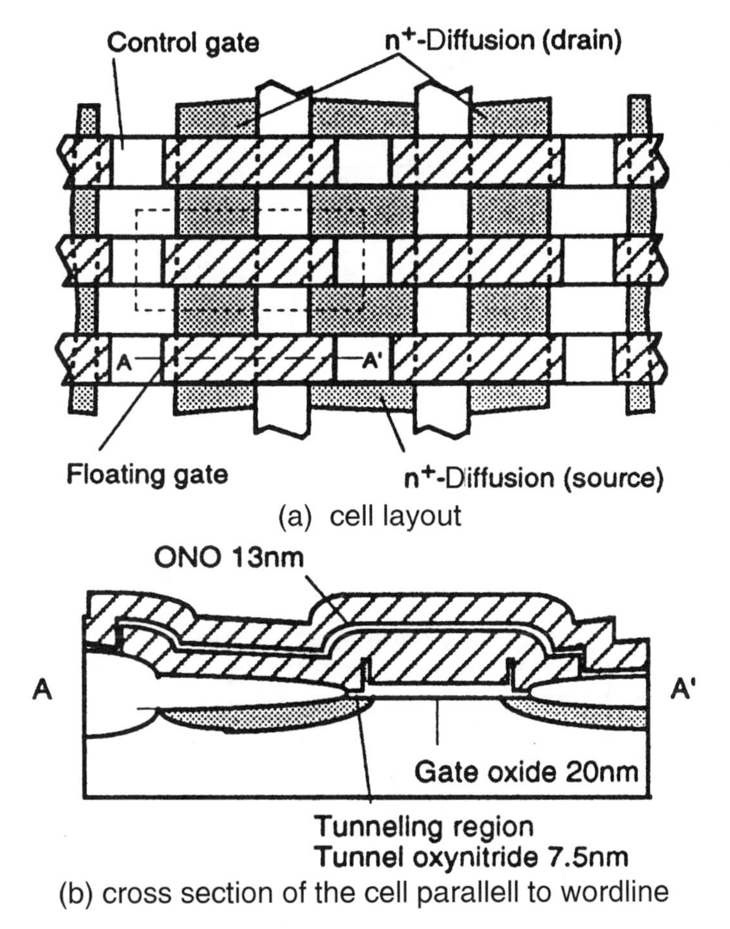

Figure 4.54 HiCR cell structure: (*a*) cell layout, and (*b*) cross section of the cell parallel to wordline [4.68].

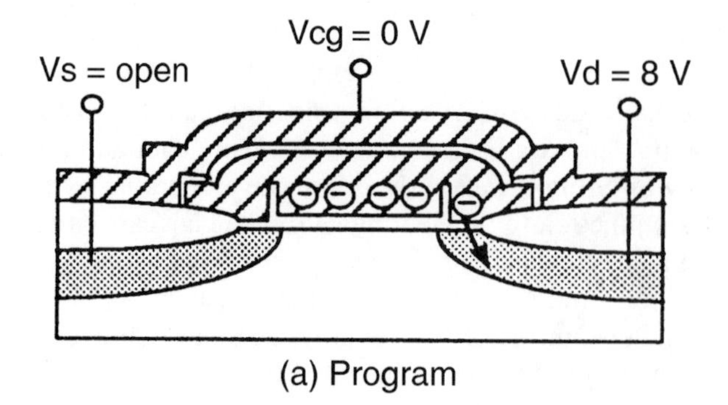

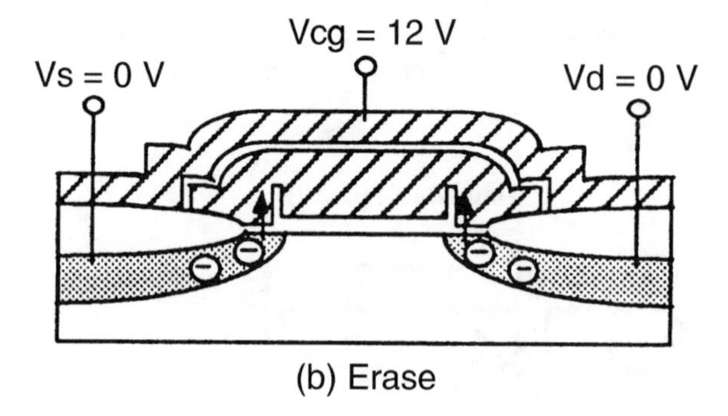

Figure 4.55 Schematic of HiCR cell operation: (*a*) program and (*b*) erase [4.68].

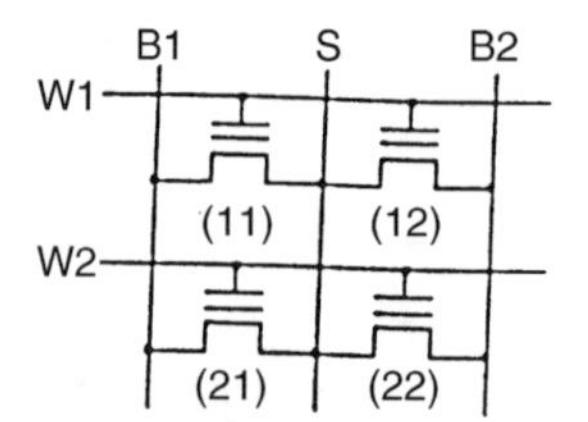

Cell (1,1) operating conditions

(a)

	W1	W2	B1	B2	S
Program	0V	5V	8V	open	open
Erase	12V	0V	0V	0V	0V
Read	3V	0V	1V	0V	0V
Program Verify	2.5V	0V	1V	open	0V
Erase Verify	6V	0V	1V	1V	0V

(b)

Figure 4.56 (*a*) Array schematic, (*b*) array operating condition [4.68].

During programming, the unselected bitline is kept open, and the unselected wordline is at 5 V to minimize program-disturb. During erase, 12 V is applied to the wordlines to be erased, and bitlines and source-lines are grounded. Because of the F–N tunnel program/erase operation, high voltage can be generated on-chip from the supply voltage. The memory cell array (Fig. 4.57) has the NOR structure. The select transistors act as switches for connecting and disconnecting the target block to a main data-line and a main source-line.

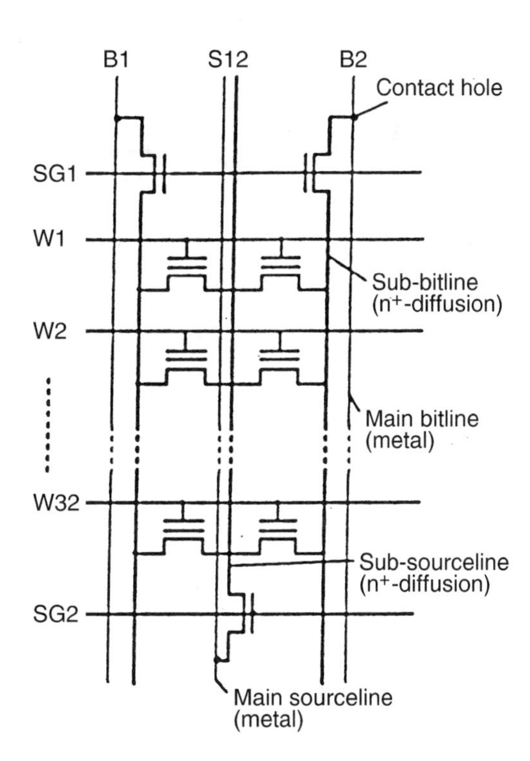

Figure 4.57 Schematic of HiCR memory array [4.68].

4.4.4.1 Program Disturb. Two principal disturbances can occur during programming of an array (Figs. 4.58 and 4.59). The first (I) is unexpected programming of the cell (12) connected to the same source-line (S) and to the selected wordline (W1). It is caused by the source voltage, $V_{CG} - V_{TM}$, of the cell, where V_{CG} is the program inhibit voltage of the unselected wordline (W2) and V_{TM} is the threshold voltage of the erased cells (cell 21 works as a pass transistor when it is erased). The second disturbance (II) is the unexpected programming of cells (21, 22, . . .) connected to the selected bitline (B1).

These problems can be solved by utilizing parallel programming and sub-bitlines. Disturbance I, which has a maximum duration of only one write cycle (1 ms), can be solved by the use of a latch circuit [4.71]. The solution to disturbance II,

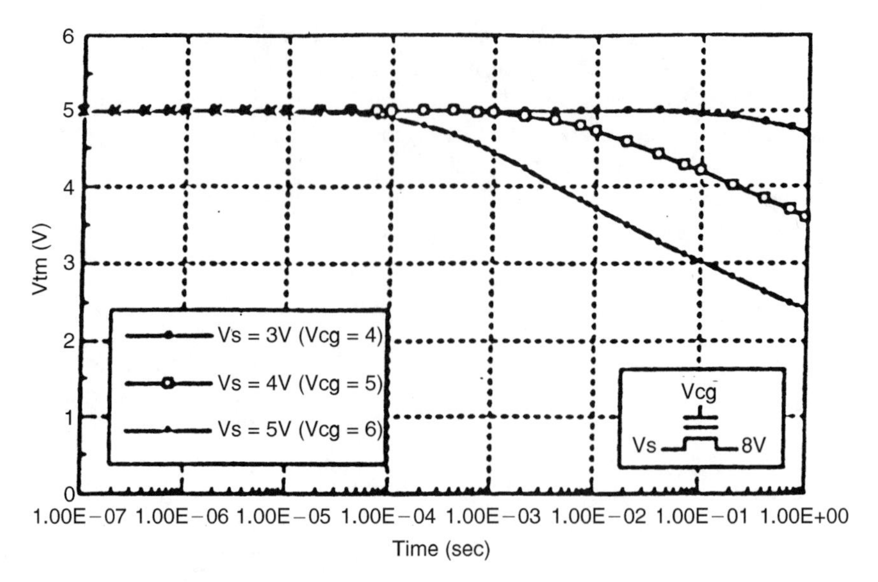

Figure 4.58 Threshold-voltage shifts during disturbance: I. The cell was programmed to a V_t of 5 V before disturbance measurement. Maximum disturbance I is 1 write cycle (1 ms) [4.68].

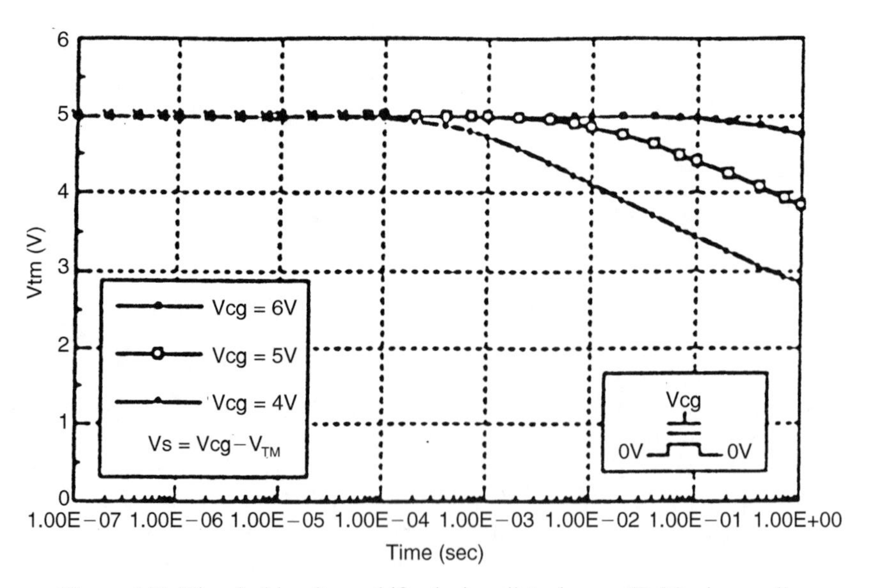

Figure 4.59 Threshold-voltage shifts during disturbance: II. Maximum disturbance II is 31 write cycles (31 ms) [4.68].

which has a maximum duration of 31 write cycles (31 ms), can be realized by setting the select transistors off except in the programming mode. Characteristics of disturbances I and II, which have sufficient margin for device operation, are shown in Figs. 4.58 and 4.59, respectively.

4.4.5 NAND

In the drive to reduce cell size, eliminating the select transistor in a full-feature EEPROM results in a bulk/block erasable Flash EEPROM. The Flash EEPROMs discussed so far are the NOR structure, in which memory cells are connected to a bitline in a parallel manner, and the NAND structure, which reduces the cell size by connecting the cells in series between a bitline and a source-line, thus eliminating the contact hole [4.13, 4.26, 4.32]. The resulting cell structure occupies 85% of the NOR cell area of a stacked gate array.

4.4.5.1 NAND Structure. Figure 4.60 shows the layout and the equivalent circuit of the NAND-structured cell. As shown in the figure, the NAND-structured cell arranges eight or sixteen memory transistors in series, sandwiched between two select gates, select gate 1 (SG1) and select gate 2 (SG2). The first gate (SG1) ensures selectivity, and the second (SG2) prevents the cell current from passing during a programming operation. The floating gates are made of first-level polysilicon. The control gates, which are wordlines in an array, are made of second-level polysilicon. The dielectric between the floating gate and the control gate is an ONO stack.

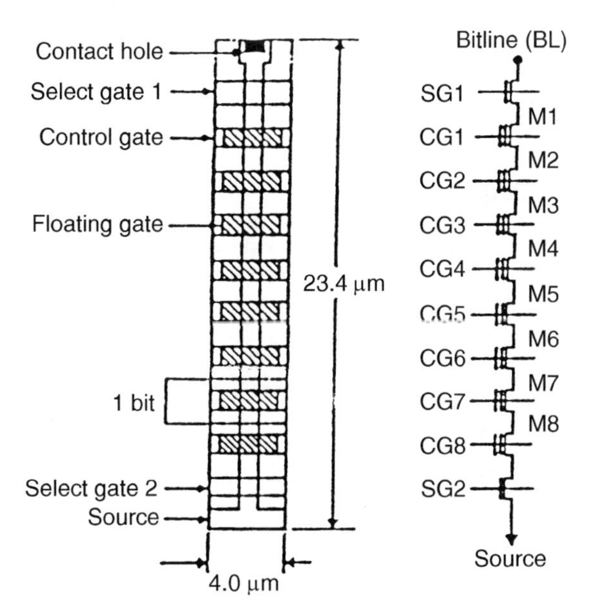

Figure 4.60 Top view and equivalent circuit of NAND-structural cell [4.13].

A NAND-structured cell array is positioned in a p-well region of the surface portion of the n-substrate as shown in Fig. 4.61. The peripheral CMOS circuitry and NAND cell array are located in different p-wells [p-well(1) and p-well(2)], which are electrically isolated from each other. The peripheral circuitry is fabricated in p-well(1). The NAND cell array is located in p-well(2). In an erase operation, p-well(2) is raised to 20 V and p-well(1) is always grounded. Thus, p-well(2) can be biased to V_{PP} during erase, while keeping p-well(1) and the n-well at V_{ss} and V_{cc}, respectively.

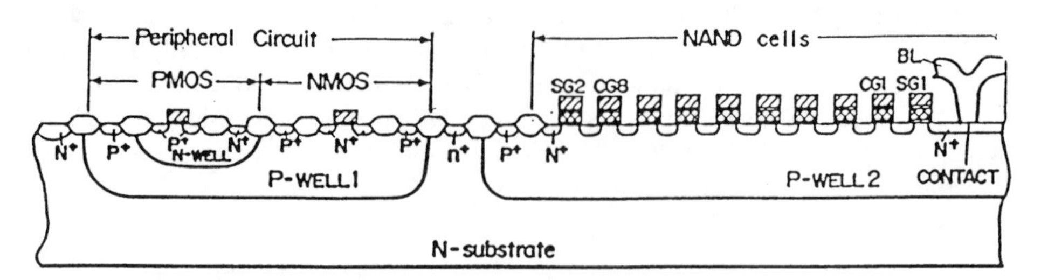

Figure 4.61 Cross-sectional view of NAND EEPROM [4.26].

4.4.5.2 NAND Operation. The operation mechanism of a single-memory transistor in a NAND structure is comparable to that of the conventional EEPROM. However, the programming and reading of the memory cells are more complex. Therefore, additional peripheral circuitry is required, and the reading speed is lower than that of the conventional type because a number of memory cells are connected in series.

The reading method is shown in Fig. 4.62 and is essentially the same as that of the NAND-type MASK ROM. Erased cells ("0") have a negative threshold voltage, and programmed cells ("1") have a positive threshold. Zero volts is applied to the gate of the selected memory cell, while 5 V is applied to the gate of the other cells. Therefore, all of the other memory transistors, except for the selected transistor, serve as transfer gates. As a result, in the case where a "0" is being written, the

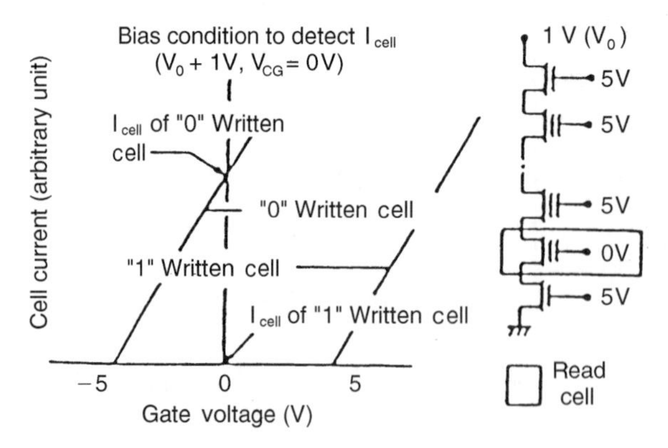

Figure 4.62 Reading of the NAND EEPROM [4.26].

memory transistor is in the depletion mode, and current flows. On the other hand, current does not flow in the case where a "1" is being written because the memory transistor is in the enhancement mode. The state of the cell is detected by a sense amplifier that is connected to the bitline. The difference between a "0" and a "1" depends on whether or not negative charge is stored in the floating gate. If negative charge is stored in the floating gate, the threshold voltage becomes higher and the memory transistor is in the enhancement mode. This is analogous to the conventional EEPROM.

Figure 4.63 shows the equivalent circuit and operating conditions during program/erase and read. During erase, the electrons tunnel from p-well(2) to the floating gate. During program, the electrons tunnel from the floating gate to p-well(2).

In the erase operation, all control gates are grounded, 20 V are applied to the n-substrate and p-well(2), and the source- and bitlines are floating. Erasing can be performed on the whole chip or selected blocks in the Flash manner. Electrons are emitted from the floating gate to p-well(2) by F–N tunneling, and the V_t of the erased cells becomes negative. During the erase operation, there is no voltage difference between the n^+-drain and p-well(2). Therefore, gated-diode breakdown does not occur. The minimum erase block size is equal to the length of a wordline in bytes × number of series of bits between a drain and a source. For a 16 MB NAND, the minimum erase block is 4 KB. The typical erase time is 10 ms (2.5 μs/byte). No preprogramming before erase is required. Hence, the NAND erase time is faster than many other Flash memories.

In the program operation, p-well(2) is grounded, 20 V is applied to the selected control gate, and 10 V is applied to the unselected control gate. The bitline of the

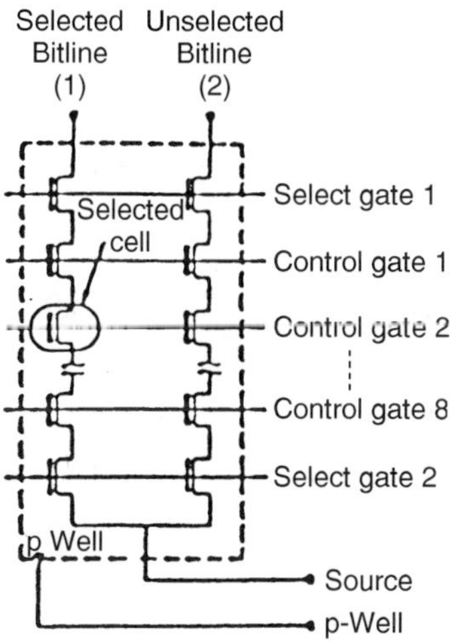

Selected Bitline (1) | Unselected Bitline (2) | Selected cell | p Well

	ERASE		WRITE		READ	
	BL (1)	BL (2)	BL (1)	BL (2)	BL (1)	BL (2)
	OPEN	OPEN	0 V	7 V	1 V	0 V
Select gate 1	20 V		20 V		5 V	
Control gate 1	0 V		7 V		5 V	
Control gate 2	0 V		20 V		0 V	
⋮						
Control gate 8	0 V		7 V		5 V	
Select gate 2	20 V		0 V		5 V	
Source	OPEN		0 V		0 V	
p-Well	20 V		0 V		0 V	
n-Sub	20 V		0 V		0 V	

Figure 4.63 Equivalent circuit and operation voltage of NAND EEPROM [4.34].

ZERO data programming cell is grounded. Electrons are injected from p-well(2) to the floating gate by F–N tunneling. The V_t of the selected cell becomes positive. Ten volts is applied to the bitline of the ONE data programming cell. The V_t of the unselected cell remains negative because the voltage across the tunnel region is inadequate to start the tunneling current. In this case, while 10 V is applied between the drain and p-well(2), gated-diode breakdown does not occur because the control gate is biased at 10 V. In both erase and program operation, the gated-diode breakdown is suppressed entirely. As a result, all high voltages are internally generated easily from a 5 V power supply through charge-pump circuits because there is no breakdown leakage current. Furthermore, oxide degradation [4.9] from hole trapping, due to gated-diode breakdown, is avoided. This results in improved write/erase endurance.

4.4.5.3 Disturb Mechanisms. The operating conditions, including the inhibit conditions, are shown in Fig. 4.63. Erasure of deselected cells in the deselected NAND block is prevented by applying a high voltage of the order of 20 V to the wordlines. Thus, the tunnel dielectric field is zero in the deselected NAND block, and no erase-disturb condition exists.

In order to prevent programming of deselected cells sharing the same bitline in the same NAND block, the control gate voltage is raised to V_m (a medium voltage) to reduce the tunnel dielectric field between the floating gate and the channel. Programming of deselected cells sharing the same control gate is inhibited by applying a medium voltage (V_m) of around 7 V to the deselected bitline to reduce the electric field between the floating gate and the channel.

Programming of cells in each NAND block is performed in a serial order from the source-line side to the bitline side to avoid unintentional charging of deselected cells sharing the same control gate. This ordering prevents the existence of programmed cells on the drain side and avoids the channel potential decrease below 8 V of the deselected bitline voltage to minimize the disturb for deselected cells. Disturbs for deselected erased cells sharing the bitline exist, although the programming (disturb) time is fast enough to prevent the charge gain.

4.4.5.4 Special Features

Advantages

1. Since the number of bitline contacts in the NAND array is reduced by one-eighth to one-sixteenth compared to that of the standard T-cell Flash array, the unit cell size is smaller.
2. A NAND cell with n^+ source and drain junctions is more scalable than the E-Tox cell since the E-Tox cell has a graded-source junction.
3. Programming and erasing are achieved by F–N tunneling and need less power (i.e., low currents), thus allowing a single power supply operation by utilizing internal charge-pump circuits.

4. Uniform F–N tunneling used in the NAND cell does not generate band-to-band tunneling related hot holes, leading to high reliability. Typical endurance is 1 million write/erase cycles.

Disadvantages

1. Since the negative voltage erase is not used in erasing or programming, the required voltage for high-voltage transistors is of the order of 20 V, similar to that of FLOTOX cells. This is higher than that of E-Tox and E-Tox-like cells.
2. The random access time is slow. NAND applications are currently limited to hard disk replacement and to systems that do not require fast random access.

4.5. SPECIAL AND ADVANCED CELL STRUCTURES

This section covers Flash cell structures for embedded applications not covered earlier (SCSG and HIMOS) and advanced cell concepts (three-dimensional structures and multistate cells).

4.5.1 Source-Coupled Split-Gate (SCSG) Flash EEPROM Cell

The SCSG memory cell has a split gate configuration that is formed by merging an enhancement transistor (a split gate transistor) to the source side of a floating gate transistor. The floating gate is formed in the first-level polysilicon, and the merged control and split gates are formed in the second-level polysilicon. This serial configuration allows the memory cell threshold voltage to be determined by the floating gate transistor when the memory cell is programmed (floating gate charged with electrons) and by the split gate transistor, which typically has a threshold voltage of 1.5 V, when the memory cell is erased (floating gate charged with holes). A piece of the floating gate is extended into the source region to allow the erasure of the memory cell to take place at the source. In so doing, both the drain and the source junctions can be separately profiled for optimal performance.

4.5.1.1 Charge Transfer. The basic structure and operation of the SCSG memory cell are shown in Fig. 4.64. For programming, both a high-drain voltage, typically 7 V, and a high control gate voltage, typically 9 V, are applied with the source held at ground. Because of the gate coupling, the actual floating gate potential is less than the drain potential, which is in favor of the generation of hot electrons near the drain junction. A small fraction, typically 1 ppm, of electrons with the proper energy and momentum eventually enter the floating gate. Electrons transferred to the floating gate raise the memory cell threshold voltage, thereby reducing the cell current. For erasure, a high voltage, typically 12 V, is applied to the source, with the control gate grounded and the drain floating. A sufficient electric field is established across the tunnel oxide between the floating gate and the source, and

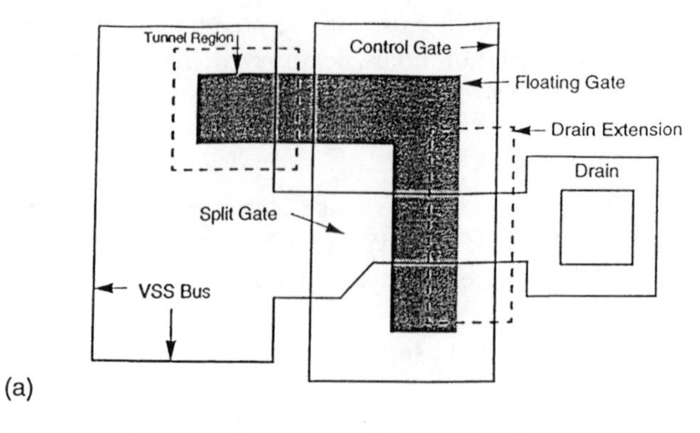

Operation	VD	VCG	VS	VB
Program	7V	9V	0V	0V
Erase	F	0V	12V	0V
Read	1V	VDD	0V	0V

(b)

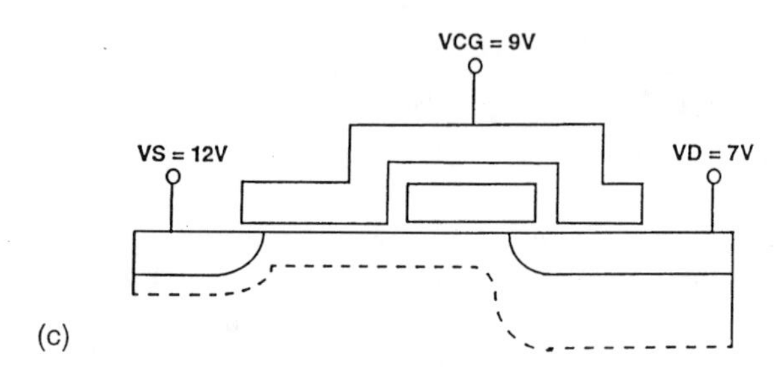

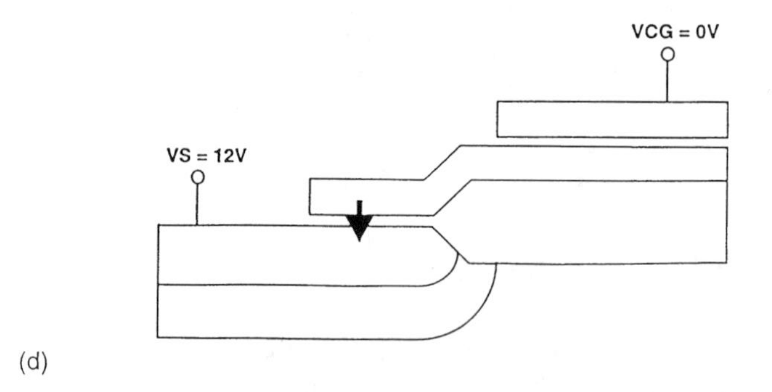

Figure 4.64 Structure and operation of SCSG cell [4.60, 4.158].

electrons are extracted from the floating gate. The extraction of electrons lowers the memory cell threshold voltage, thereby increasing the cell current. The incorporation of the split gate allows the floating gate transistor to be over-erased to depletion without shorting out the column. The region where the floating gate overlaps the source is doped n-type to eliminate the leakage current through the substrate. With the drastically reduced current requirement during erasure, an optional on-chip charge pump can be built to enhance the erase speed to less than 1 msec.

4.5.1.2 Inhibit Conditions. A small section of the memory array during programming is depicted in Fig. 4.65. The selected bit to be programmed is encircled. The bits that share the same wordline will see the wordline voltage repeatedly during programming. The worst case bit will be subjected to this wordline voltage for a duration that equals the total time to program the other bits on the same wordline. The resultant threshold voltage gain is called the gate-disturb. The bits that share the same bitline will see the bitline voltage repeatedly during programming. The worst case bit will be subjected to this bitline voltage for a duration that equals the total time to program the other bits on the same bitline. The resultant threshold voltage loss is called the drain-disturb. As in most Flash EEPROM cells, the tunnel dielectric thickness in the floating gate channel region is around 100 Å. At this thickness and without a select transistor between the drain and the floating gate transistor, suppression of program-disturbs is achieved through the proper balance of the programming voltages and the programming speed. Typically, the duration of a particular half-selected memory bit, exposed to the programming bias conditions, is kept below 1 second.

4.5.1.3 Read. Reading of the SCSG cell is accomplished by raising the drain to about 1 V, grounding the source, and raising the control gate to the supply voltage. The charge state is determined by the relative magnitude of the cell current to a reference current in the sense amplifier. The split gate configuration allows the

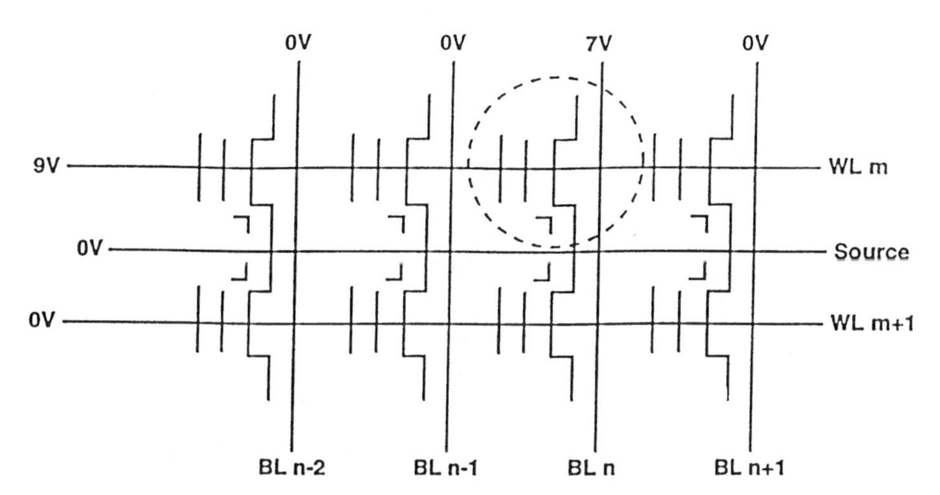

Figure 4.65 Program inhibit conditions [4.99]. Circled cell indicates the selected cell (row#m, column#n). The half-selected bits along the same wordline (WL m) will see the gate disturb, while the half-selected bits along the bitline (BL n) will see the drain disturb. Other bits are unaffected [4.158].

erased threshold voltage distribution of the entire memory array to be targeted and tightly controlled at around 1.5 V, provided that all the floating gate transistors are over-erased to depletion, which is easily achieved with an on-chip charge pump. This, in turn, allows the array to be read at low-supply voltages due to the enhanced gate drive.

4.5.2 High-Injection MOS (HIMOS)

High Injection MOS (HIMOS) [4.99] is a two-poly split gate structure. The presence of a coupling capacitor enhances the injection efficiency through source-side injection, suitable for 5 V-only applications. The erase occurs by electrons tunneling from the floating gate to the drain junction. The programming mechanism is by source-side injection.

Two perpendicular cross sections of the HIMOS device are shown in Fig. 4.66. The device consists of a control gate (CG) channel and a floating gate (FG) channel in series. An additional program gate (PG) is located on the field region in order to couple a voltage onto the FG during programming. During the write operation, the CG voltage is typically 1 to 1.5 V (just above the threshold voltage of the CG channel), and the PG voltage is typically 10 to 12 V, which can be generated on-chip by charge-pumping techniques. The programming efficiency is determined mainly by the interpoly dielectric thickness [4.100]. The interpoly dielectric layer is a thermally grown polyoxide with a thickness of 30 nm.

Moreover, because of this feature, drain engineering is strongly simplified. And 5 V-only (or 3.3 V-only) Flash erase and wordline-oriented sector erase are achieved by Fowler–Nordheim tunneling of electrons from the FG toward the drain junction

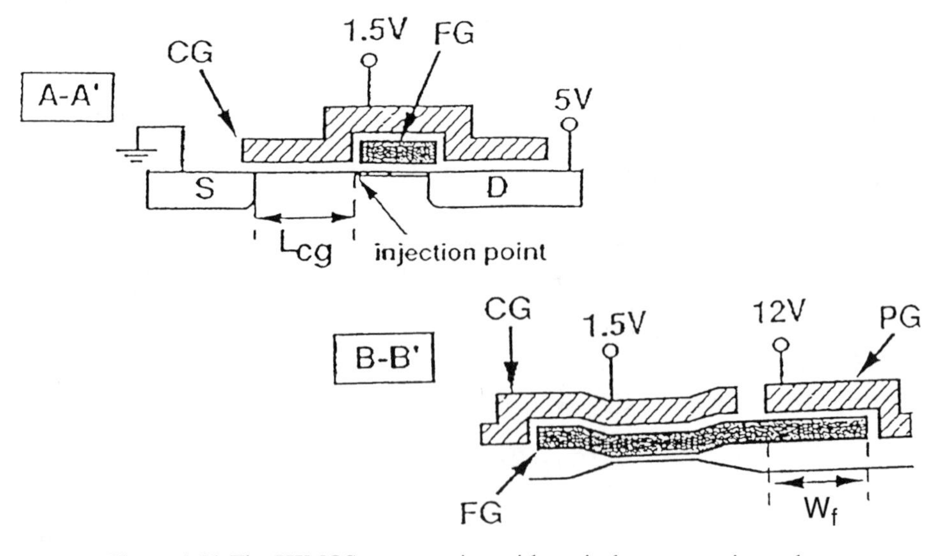

Figure 4.66 The HIMOS cross section with typical programming voltages [4.99].

by applying 5 V (or 3.3 V) to the drain junction and a negative voltage (typically −10 to −12 V) to the PG.

To minimize the additional area that is necessary for the PG, a virtual ground array (VGA) configuration has been chosen. The VGA layout principle and the circuit equivalent are shown in Figs. 4.67 and 4.68, respectively. Because of the strongly asymmetrical programming characteristics of the HIMOS device, the bitlines can be shared between adjacent bits in a row. The additional area for the PG can be reduced by sharing the Program Line between adjacent bytes. This is possible because of the triple gate structure of the device; the CG can be used as the wordline to access every byte separately during the write and read-out operations. The operating conditions of the HIMOS are shown in Table 4.9.

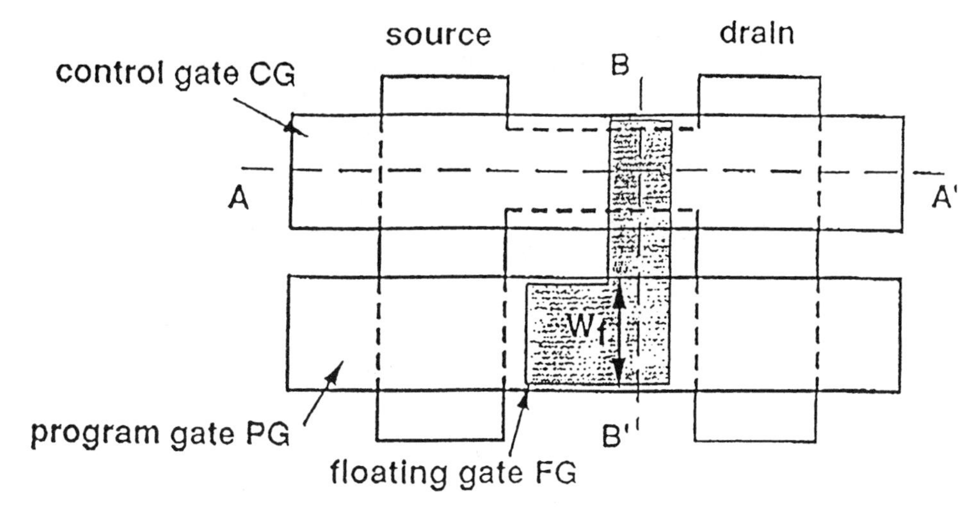

Figure 4.67 Virtual ground array principle [4.99].

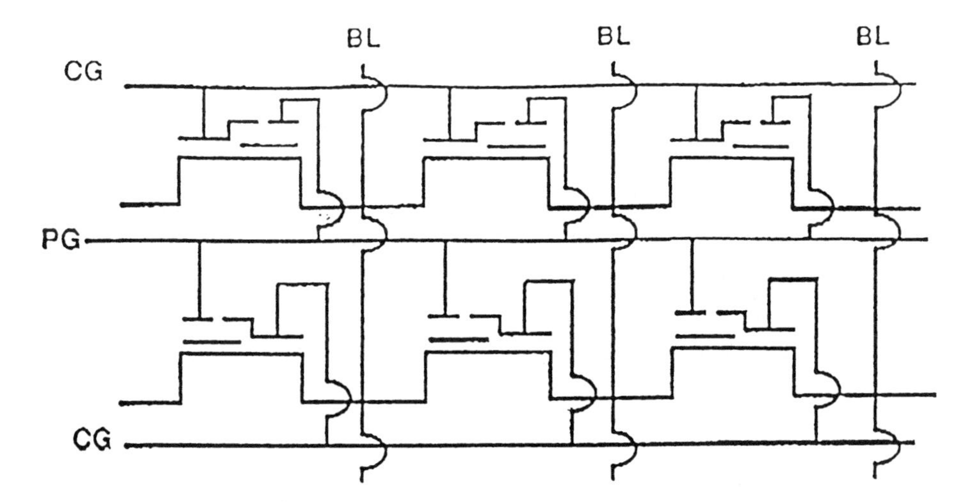

Figure 4.68 Virtual ground array equivalent circuit [4.99].

TABLE 4.9 TYPICAL OPERATING VOLTAGES FOR THE HIMOS DEVICE

	Source	Drain	CG	PG
Write	0	3.3/5	1/1.5	12
Erase	3.3/5	3.3/5	0	−12
Read-out	0	2	2	0

Figure 4.69 shows the programming characteristics of the cell for 5 V and 3.3 V applications. During programming, the CG voltage is very close to the threshold voltage of the built-in select device, and the optimal V_{cg} value is equal to 1 V for a 3.3 V drain voltage and equal to 1.5 V for a 5 V drain voltage. Therefore, the drain current is a constant during the entire programming cycle, which ensures a constant supply of electrons for the hot-carrier generation process. This drain current is also much smaller than in conventional devices where the optimal condition for hot-electron injection occurs at the onset of saturation (V_{fg} @ V_d). The latter feature reduces bitline voltage drops and strongly decreases the necessary programming power, which offers the possibility of page-mode programming and additional charge-pumping techniques for boosting the bitline voltage on-chip, in the case of 3.3 V-only operation.

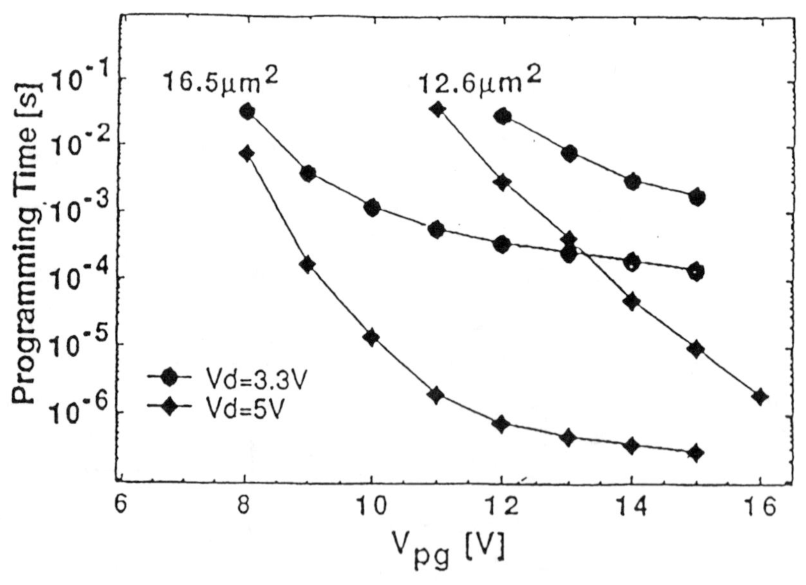

Figure 4.69 The programming time versus V_{pg} for a 12.6 and 16.5 μm^2 cell and for 3.3 V/5 V drain voltages [4.99].

4.5.3 3-D Cell Structures

Recently, three-dimensional (3-D) structures have been reported to realize small cell structures [4.67, 4.83]. Cell size reduction is realized by forming the channel in a vertical direction in contrast to conventional cells where the channel length is lateral, along the surface of the silicon. The major challenge in such structures is the uniformity of the tunnel oxide. Three-dimensional structures offer both promise and challenges.

4.5.3.1 Trench-Embedded Field-Enhanced Tunneling (TEFET). The TEFET cell is a three-dimensional, single-transistor, floating gate memory cell. As in a conventional planar Flash EEPROM cell [4.83], the TEFET cell is programmed by channel hot-electron injection and erased by F–N tunneling. The small cell size is achieved by exploiting trench technology to integrate the NMOS transistor vertically along the side-walls of a trench, as shown in Fig. 4.70. The etched trench extends from the planar surface through the n^+ drain, p-body, and a graded n-type region into the underlying heavily doped n^+ layer that serves as the source. By patterning this underlying n^+ layer, a memory array can be divided into erase sectors. Both the control gate and the floating gate are embedded within the interior of the trench. This vertical structure provides inherently smaller cell size than a conventional planar cell. For example, with 0.7 μm lithography, the cell size can be as small as 2.7 μm^2, which is a factor of two smaller than a conventional planar cell. In addition, the trench EEPROM cell's vertical channel allows the cell pitch to be scaled inde-

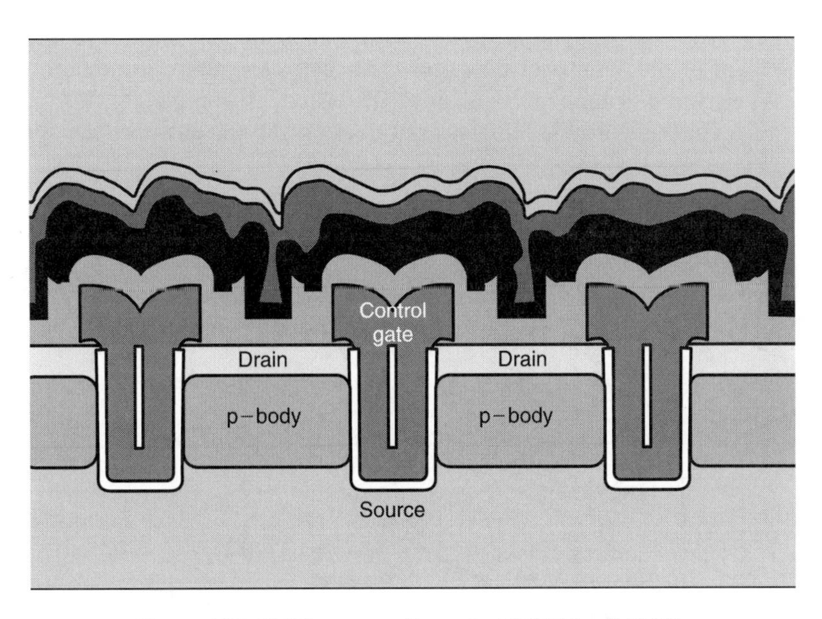

Figure 4.70 SEM cross section of a TEFET cell [4.83].

pendent of the channel length. Thus, the inherent hot-electron effects and punch-through limitation associated with the scaling of planar cells are eliminated, and the freedom of tailoring the channel doping profile for optimal cell performance need not be traded off for scaling. Consequently, the trench cell is suitable for scaling down to the very deep sub-micron regime.

Moreover, the trench cell has a substantially larger channel width than a planar cell of comparable lithography because the trench cell's channel width is equal to the circumference of the trench, whereas the planar cell's channel width is only comparable to one edge of the trench opening. The larger channel width leads to a higher read current and, in turn, a shorter access time. Furthermore, the enhanced electric field at the bottom of the trench allows a low erase voltage. The field enhancement also permits a Fowler–Nordheim (F–N) tunneling erase without inducing band-to-band tunneling and the associated hot-hole injection, which is the major limiting factor for a planar cell's cycling endurance [4.12]. By eliminating hot-hole injection, the trench cell's endurance can be greatly improved.

4.5.3.2 3-D Side-Wall Flash EEPROM Cell. The side-wall Flash EPROM cell is a 3-D, single-transistor, floating gate memory cell that is formed on the side-walls of a narrow silicon pillar (Fig. 4.71) [4.67]. The pillars are approximately 1 μm high, and each pillar forms a separate cell. Current flows vertically along the side-walls of the pillar from the top of the pillar (drain) to the base of the pillar (source) (Fig. 4.72). The floating gate completely encircles each pillar, and similarly, the wordline encircles the floating gate. The cell is programmed by hot-electron injection at the drain end (top) of the silicon pillar, while erase is performed by Fowler–Nordheim tunneling of electrons from the floating gate to the source. The key idea in implementing the cell within an array is to have a smaller space between pillars in the same wordline row than between pillars in adjacent wordline rows. When the wordline polysilicon is deposited, it completely fills the gaps between pillars in the wordline direction but does not fill the gaps between pillars in adjacent

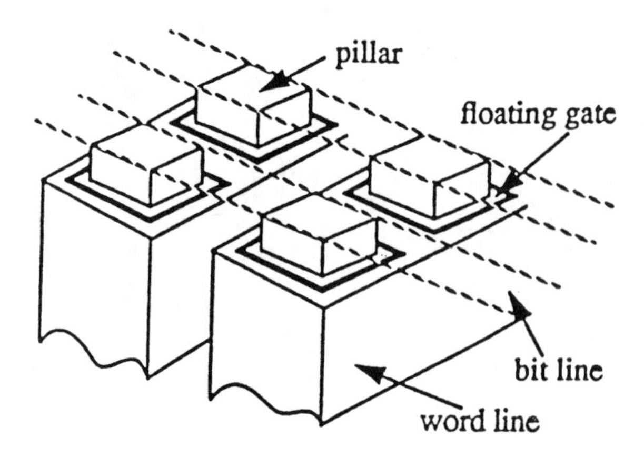

Figure 4.71 Schematic of the side-wall Flash EPROM array [4.67].

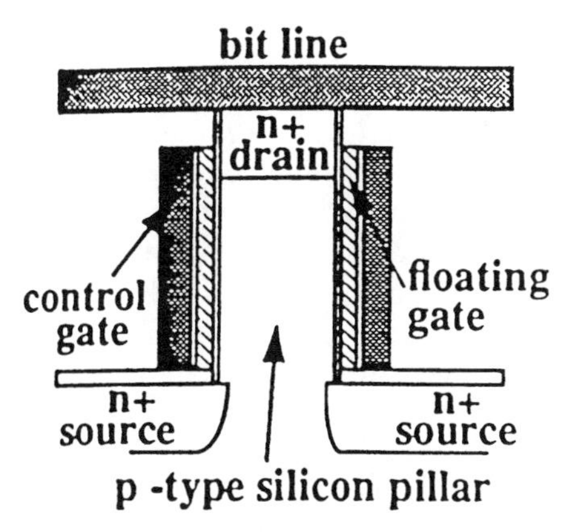

Figure 4.72 Cross-sectional diagram of 3-D cell [4.67].

wordline rows. Therefore, after blanket etching of the wordline polysilicon, the wordlines remain continuous in the wordline direction but are isolated from adjacent wordlines. Bitlines run perpendicular to the wordlines making contact to the top of each pillar. The wordline and floating gate edges are 300 nm below the top of the silicon pillar, which provides the margin that allows the bitline contact to overlap the edges of the pillar without shorting to the gates. Thus, the bitline does not have to have a separate contact opening to the top of every pillar, but can be formed by etching a continuous trench in the planarizing dielectric to expose the top of the pillars along a single bitline. These trenches are then filled with the bitline metal to form a continuous bitline. With this approach, no registration tolerance is required for the bitline contact, and thus, the bitline pitch can be equal to the minimum lithographic pitch.

The key advantages of the cell and array are (1) a cell size within 10% of the square of the minimum pitch with $> 50\%$ reduction when compared to the conventional NOR-type planar cell [4.101]; (2) a channel length independent of the lateral dimensions, thereby allowing aggressive lateral scaling without concern for short-channel effects; (3) inherent complete isolation between cells; (4) read current that is effectively three times that of an equivalent generation planar cell; (5) true cross-point array, thereby simplifying circuitry; (6) metal or silicide low-resistance bitlines; (7) simple optimization of the cell channel doping profile, and (8) low bitline diffusion capacitance.

4.5.4 Multilevel Cell

The primary technique for increasing the memory density is to reduce the size of the memory cell. Another approach to improving memory density is to increase the number of possible states in a cell. The multilevel cell concept has been discussed for volatile and nonvolatile memories [4.92–4.96], but has been difficult to implement on a commercial level. Recently, it has been reported that Intel, Sandisk, AMD, and Toshiba have been working on Flash memories based on multilevel cell technologies [4.97, 4.118, 4.125].

Multi-bit storage takes advantage of the analog nature of the Flash storage element. The conventional 1 bit/cell approach would place the cell in one of two states, a "0" or "1," using a program or erase operation. Erase might be denoted as the absence of charge and program as the presence of charge on the floating gate. Thus, the cell is placed in one of two discrete charge bands. If programming can be done accurately enough, the cell can be placed in one of four discrete charge bands achieving 2 bits/cell storage (Fig. 4.73). An example of state assignment is shown in Table 4.10.

The number of states is determined by

1. Total available charge range (window)
2. Ability to accurately program a state
3. Ability to accurately read a state
4. Disturbance of a state over time

Because of the high accuracy required for programming and sensing, multilevel storage is expected to be somewhat slower than conventional 1 bit/cell Flash for these operations. Multilevel storage also has a lower tolerance for disturbance of the

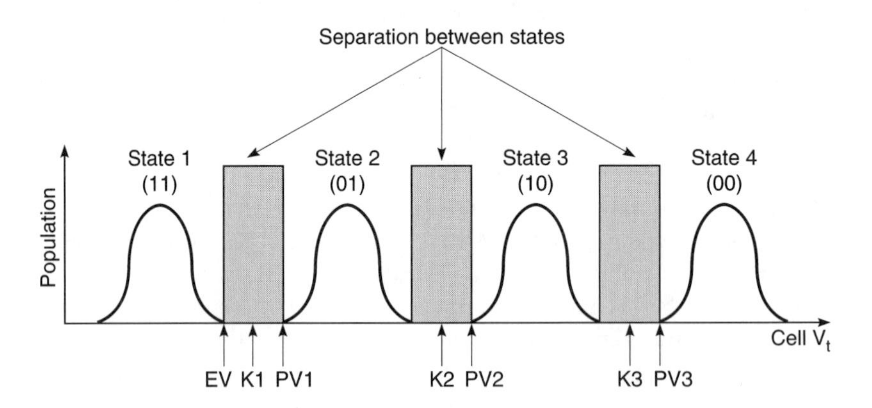

Figure 4.73 Multilevel-cell threshold voltage distribution [4.118].

TABLE 4.10 AN EXAMPLE OF STATE ASSIGNMENT FOR A 2 BIT/CELL

State	V_t Range
0	< 3.0 V
1	3.25 V – 3.75 V
2	4.25 V – 4.75 V
3	> 5.0 V

stored data (e.g., data retention) than conventional Flash, and thus will most likely require some form of error correction in the system or will be used in loss tolerant applications for the higher bit/cell densities. At what bit/cell density error correction is required is not yet clear, but it is expected to be a strong function of the process technology capability and the system requirements.

4.6. FLASH RELIABILITY ISSUES

As with other semiconductor-based memory devices, there are reliability issues with the programming, erasing, and read operation of Flash memory devices. For example, device degradation due to programming with channel hot-electrons and erasing with hot-holes lead to serious changes in the performance of Flash devices. This and other aspects of Flash reliability, such as dielectric breakdown, program/erase endurance, data retention, etc., are addressed in the following sections.

4.6.1 Channel Hot-Electron Programming

The channel hot electron (CHE) has been widely used for programming EPROM's and several Flash memories. This section describes the reliability issues of channel hot electron programming and how to minimize degradation effects associated with CHE programming.

4.6.1.1 Electron Trapping (N_{ox}) in $V_{fg} > V_d$ regime. Under typical programming conditions of $V_{cg} = 12V$, $V_d = 5.5V$, and $V_s = V_{sub} = 0V$, the EPROM cell is biased in the so-called $V_{fg} > V_d$ regime so that the direction of the oxide field near the drain favors electron injection [4.104, 4.121]. In MOS transistor degradation studies under similar stress conditions, Doyle et al. [4.106] found that the V_t shift was caused predominantly by electron charge trapping (N_{ox}) in the nearby oxide. Their experimental results also suggested that the trapped electrons could be subsequently detrapped by channel hot-hole injection when stressed at $V_g = 1/2\ V_d$. In a floating gate structure, the presence of trapped electrons in the oxide near the drain edge can repel the channel hot electrons and reduce electron injection. This effect can be fairly significant in Flash memory cells where the programmability degrades substantially after repeated program/erase (P/E) cycles as seen in Fig. 4.74. That is, the cell program V_t decreases progressively versus P/E cycling when programming time duration is fixed, and the V_t versus cycling window closes (Fig. 4.74).

As a result, electron trapping in the gate oxide, possibly accompanied by electron trap generation, under $V_{fg} > V_d$ bias conditions is of real concern for cell reliability. Liang, et al. [4.107] observed that preexisting electron traps in the oxide would trap-up electrons to a saturation level independent of the oxide field when it is below 5 MV/cm (Fig. 4.75). At oxide fields larger than 5 MV/cm, electron trap generation and the amount of electron trapping can increase dramatically. This suggests that the voltage difference (V_{fg}–V_d) in a memory cell during programming needs to be kept reasonably low in order to avoid excessive electron trapping at the

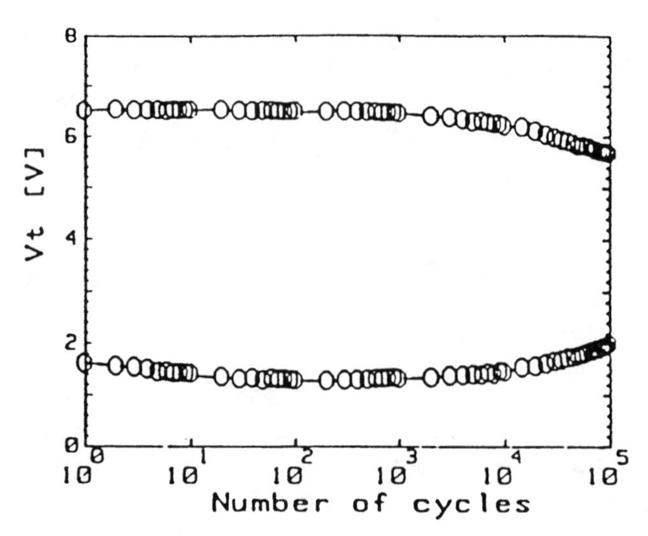

Figure 4.74 Program/erase endurance curve of a typical Flash memory cell [4.104].

drain edge. Nishida and Thompson [4.108] further observed that, for sub-10 nm oxides, there is a reduced trap generation and a lower saturation oxide charge density compared to thicker oxides as shown in Fig. 4.76. For the 10 nm gate oxides being used in Flash cells, electron trapping is reduced over its thick oxide EPROM counterpart.

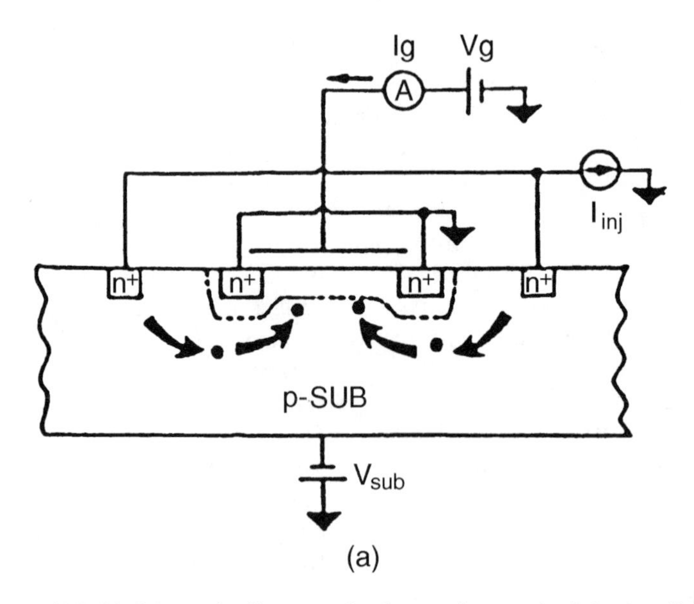

Figure 4.75 (*a*) Schematic diagram of substrate hot-carrier injection [4.107].

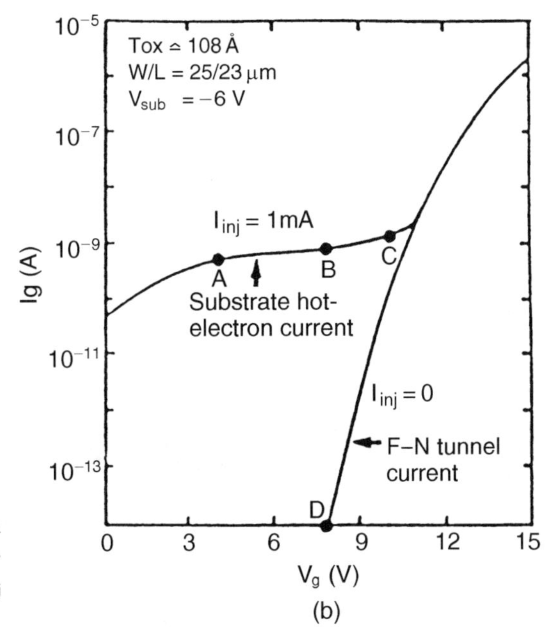

(*b*) I_g versus V_g for F–N tunneling (with $V_{sub} = 0$, $I_{inj} = 0$) and substrate hot-electron injection (with $V_{sub} = -5$ V, $I_{inj} = 1$ mA) [4.107].

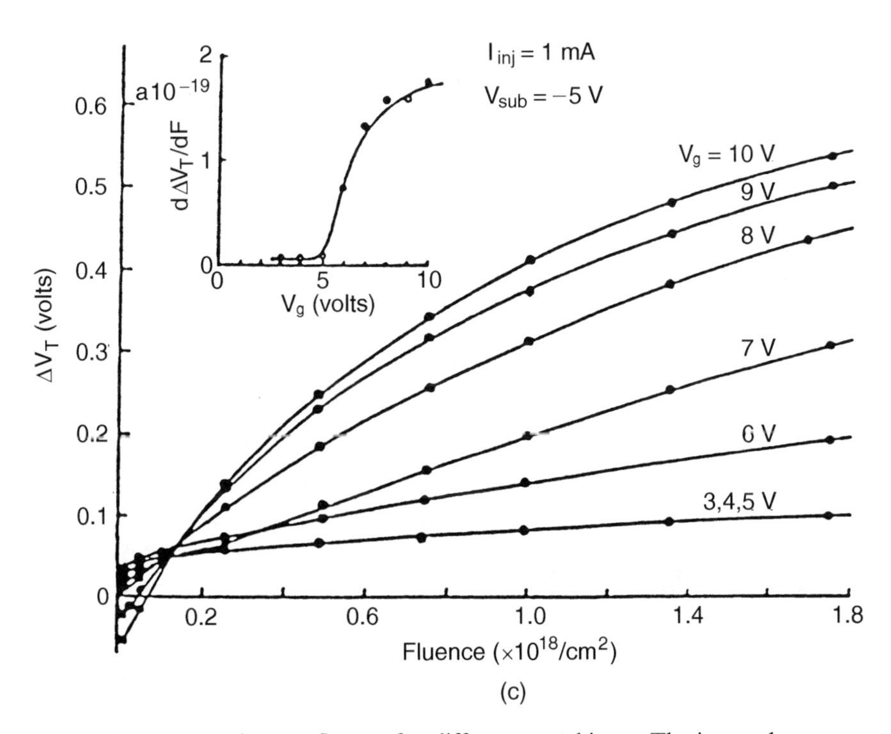

(*c*) ΔV_t versus electron fluence for different gate biases. The insert shows the slope of the curves at high fluence versus V_g [4.107].

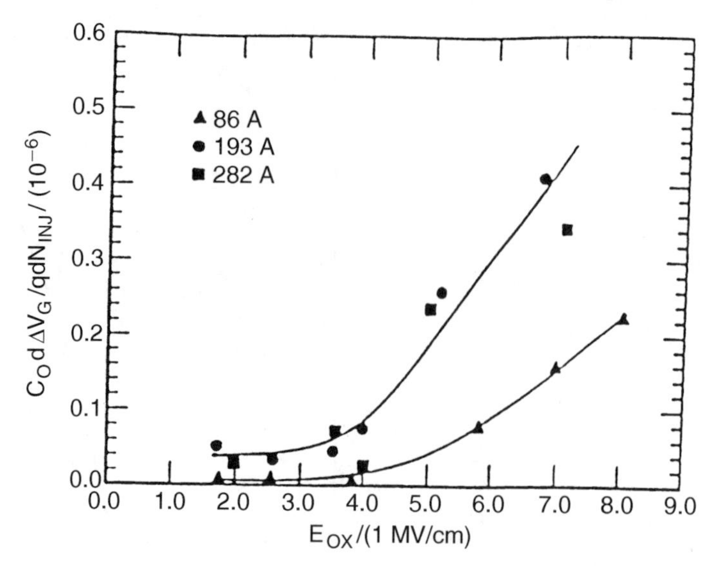

Figure 4.76 Oxide field dependence of slope of gate voltage shift versus electron fluence calculated at constant electron fluence of 6E17 cm^{-2} for 900°C, 0% TCA dry oxide with thicknesses of 86 Å and 282 Å [4.108].

4.6.1.2 Surface State Generation (N_{ss}) in $V_{fg} < V_d$ Regime. As the cell is being programmed, the floating gate potential rapidly collapses. The lateral channel electric field, as well as the average electron energy, rises (Fig. 4.77) [4.104]. Moreover, near the drain end, the oxide field reverses direction and can attract hot holes generated in the channel. Upon co-injection of hot electrons and hot holes, it is well known that interface state (N_{ss}) generation can become important [4.106]. This can lead to device transconductance (G_m) degradation as a result of mobility reduction. Therefore, the ending condition of cell programming operation also needs to be carefully considered [4.104].

4.6.1.3 Trap Reduction Through Nitridation. Besides optimizing programming and erasing conditions, the composition of a gate oxide can also play a part in reducing the trap density in the tunnel oxide. Several studies have shown that nitridation of tunnel oxide makes the tunnel oxide more resistant to electron trap generation [4.131–4.133]. Fukuda et al. [4.56] showed that both electron and hole trap densities are greatly reduced in heavily nitrided tunnel oxide films (RTONO). Figure 4.78 shows trap density generation versus P/E cycles, and Fig. 4.79 presents the improved endurance characteristics of a heavily nitrided, tunnel oxide-based Flash memory cell. Figure 4.80 shows that 1E20 N-atoms/cm^3 are incorporated into the oxide and that the number of H-atoms, which are the origin of electron traps [4.132], decreases in RTONO films.

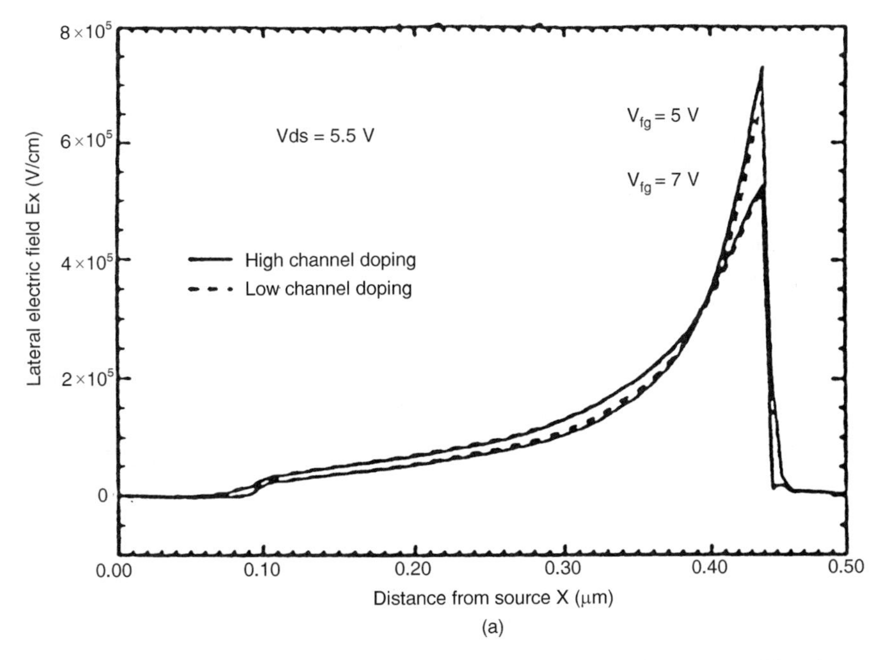

Figure 4.77 (*a*) Lateral electric field across the channel for $V_{fg} = 5\,V$ and $V_{fg} = 7\,V$ for two-channel doping calculated by 2-D device simulator UMDFET2 [4.104].

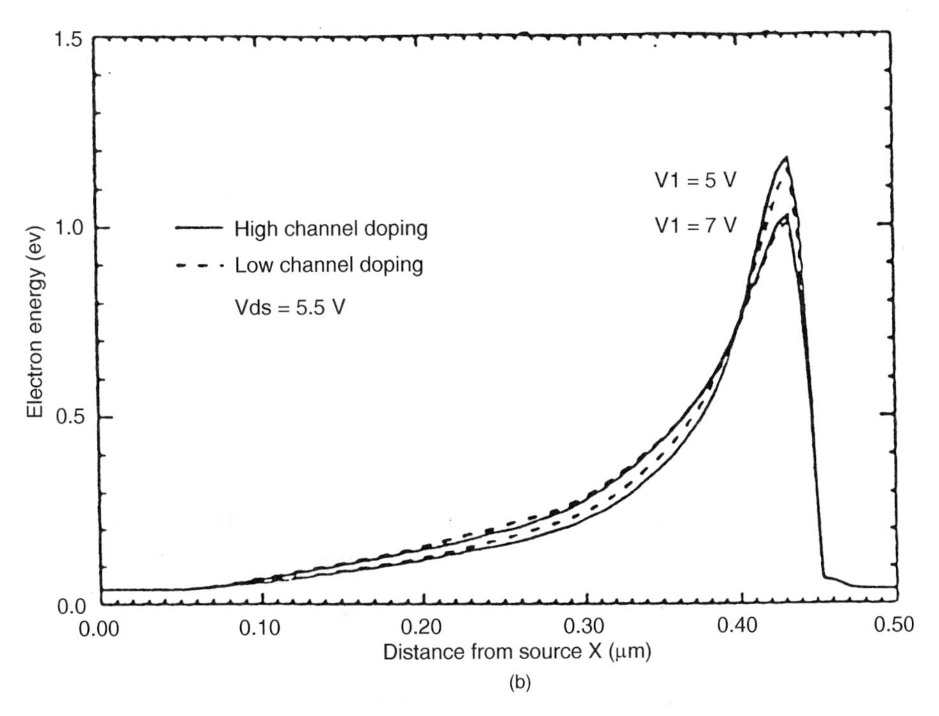

(*b*) Corresponding average electron energy of Fig. 4.77(*a*) [4.104].

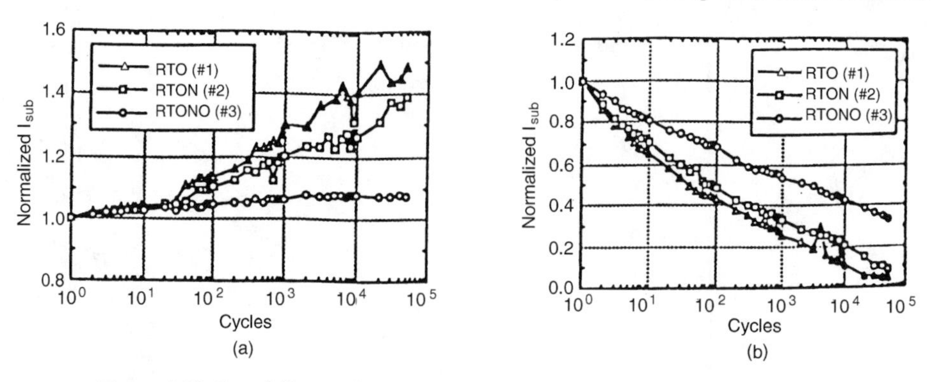

Figure 4.78 I_{sub} shift as a function of the P/E cycles for the (*a*) forward and (*b*) reverse biases [4.56].

Preparation sequences employed

	RTO	RTN (RTON)	RTON
RTO (#1)	O_2,1100 °C , 50 s		
RTON (#2)	O_2,1100 °C , 40 s ⟶	N_2O ,1100 °C, 20 s	
RTONO (#3)	O_2 ,1100 °C , 37 s ⟶	NH_3 , 900 °C , 60 s ⟶	N_2O,1100 °C,30 s

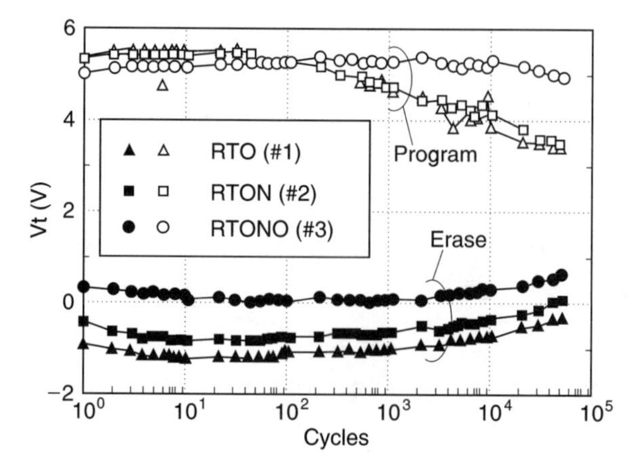

Figure 4.79 Endurance characteristics of the flash cells #1–#3 [4.56].

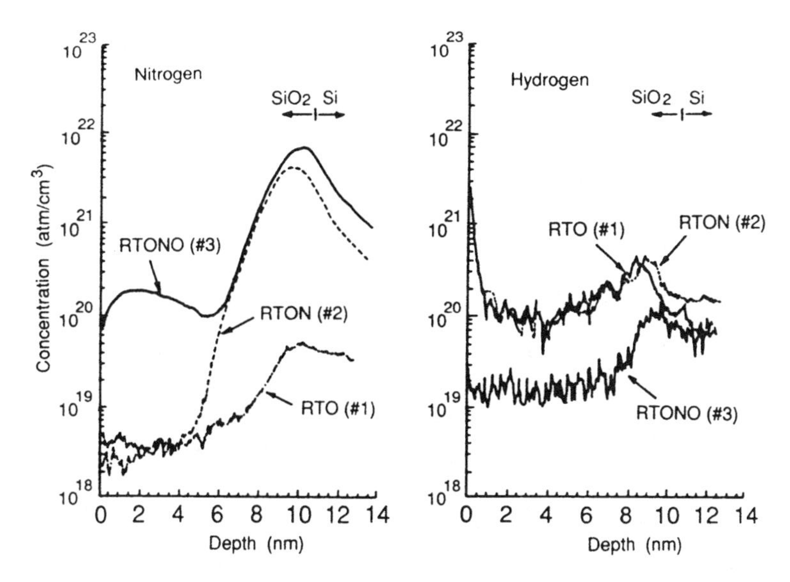

Figure 4.80 SIMS depth profiles of N and H atoms in the oxide films RTO (#1), RTON (#2), and RTONO (#3) [4.56].

4.6.2 Grounded-Gate Source-Erase-Induced Cell Degradation

Unlike the conventional EEPROM cell, the Flash EPROM cell almost always has to contend with the existence of band-to-band tunneling (BBT) current, as shown in Fig. 4.81 [4.113]. The BBT current is due to electron hole pairs generated in the gated-diode source junction during erase (Fig. 4.82). Electron hole pair generation occurs as a result of the presence of a high surface field in the n^+ junction depletion region [4.113–4.115, 4.121–4.123].

When high voltage is applied to the source junction with the control gate grounded, a deep depletion region is formed underneath the gate-to-source overlap region. Electron hole pairs are generated by the tunneling of valence band electrons into the conduction band. The electrons are collected by the source junction, and the holes are collected by the substrate. Since the minority carriers generated thermally or by band-to-band tunneling in the source region flow to the substrate due to the lateral field near the Si–SiO_2 interface, the deep depletion region remains present and the band-to-band tunneling process can continue without creating an inversion layer. The generated holes gain energy because of the electric field in the depletion region. While the majority of these generated holes flow into the substrate, some of them gain sufficient energy to surmount the Si–SiO_2 barrier and are trapped in the oxide [4.113–4.115].

4.6.2.1 Band-to-Band Tunneling Generated Hot-Holes. Band-to-band tunneling generated hot holes have been reported to cause device degradation

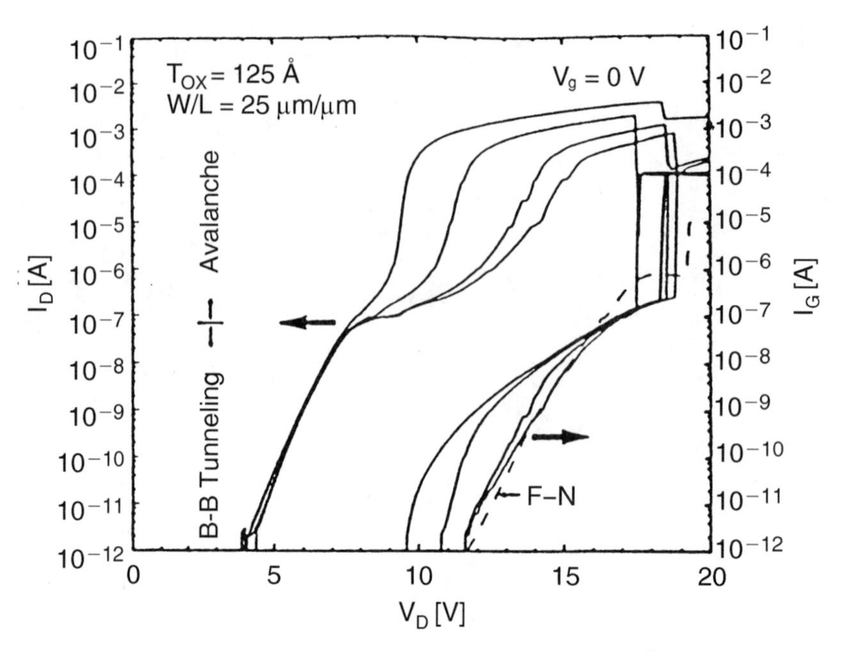

Figure 4.81 Measured drain and gate currents for the 125 Å oxide devices with channel surface concentrations of 10, 5, 2, and 1 E16 cm^{-3} (from left to right). The dashed curve represents the normalized F–N gate current [4.113].

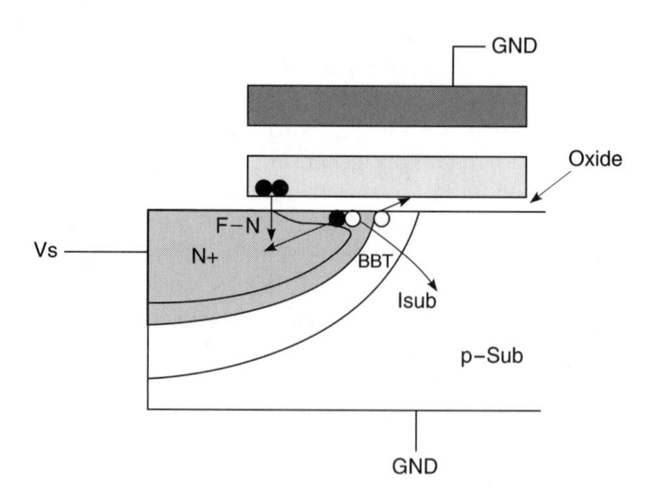

Figure 4.82 Cross-sectional drawing of the cell source diffusion with band-to-band (BBT) current and Fowler–Nordheim (F–N) current [4.104].

[4.113–4.117, 4.123], and hot-hole injection during erase has been reported to cause erased V_t nonuniformity (Fig. 4.83) in the memory array [4.17] and to speed up erase time (Fig. 4.84) versus cycling [4.104]. Trapped holes in the oxide have also been shown to degrade the charge retention (Fig. 4.85) of the memory cell [4.12, 4.21] and to speed up gate-disturb (Fig. 4.86) during the program mode [4.12, 4.21].

Band-to-band tunneling leakage current has also been reported in trench-capacitor DRAM cells [4.124]. This may pose a serious problem in scaling down the trench-capacitor cell since the leakage current from band-to-band tunneling increases drastically with decreasing capacitor oxide thickness [4.84].

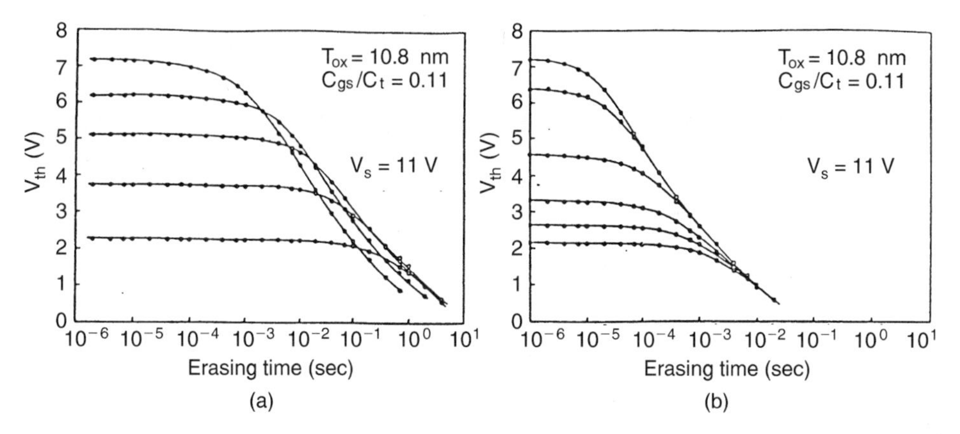

Figure 4.83 Erased V_{th} versus erasing time with programmed V_{th} as a parameter (type-a: n+ doping < 2E15 cm^{-2}, type-b: n+ doping > 5E15 cm^{-2}) [4.17].

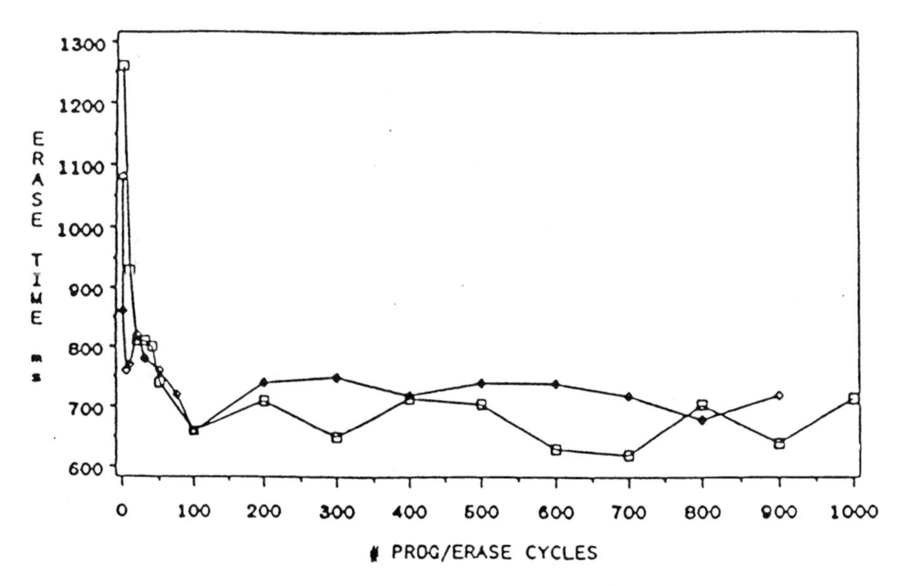

Figure 4.84 Erase time versus P/E cycles. Erase time speedup is due to trapped holes [4.104].

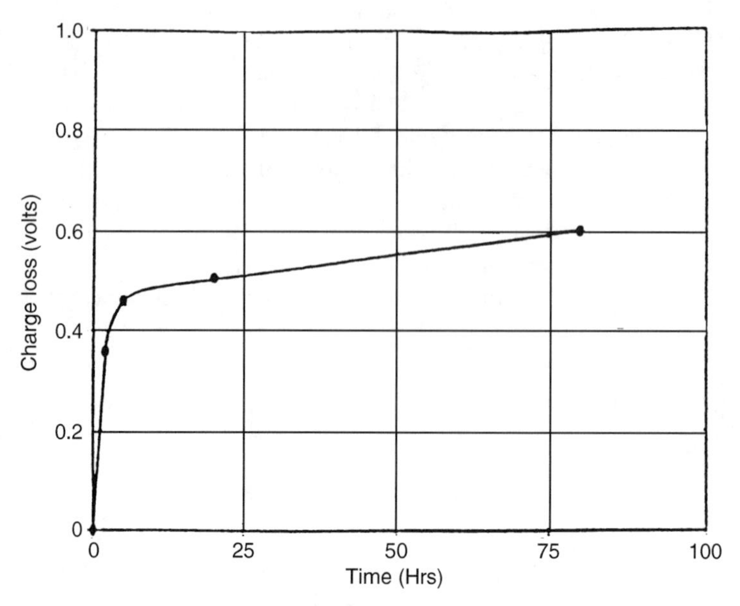

Figure 4.85 Temporal behavior of charge loss seen for system arrays at 250°C. Each point represents the mean worst bit margin shift in the arrays [4.21].

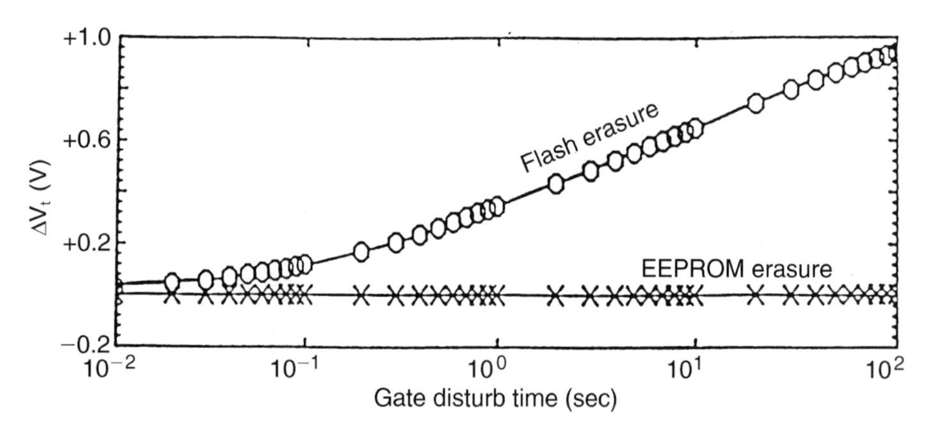

Figure 4.86 Threshold voltage shifts during gate-disturb after the cells having been cycles 20 times, using Flash (O) and EEPROM (X) erasure. The cells were erased to a V_t of 1 V before gate-disturb measurement [4.12].

4.6.2.2 Avalanche Breakdown-Induced Cell Wearout. Aside from BBT, avalanche breakdown can occur during erase if the source junction of the cell is not made robust relative to the erase voltage. In an arsenic n^+ to p-substrate abrupt junction, this event can be particularly detrimental to the cell. The avalanche-generated hot holes can gain sufficient energy to cause severe interface damage (Fig. 4.87) [4.116, 4.117], which can lead to serious degradation of transconductance (G_m) (Fig. 4.88) [4.117].

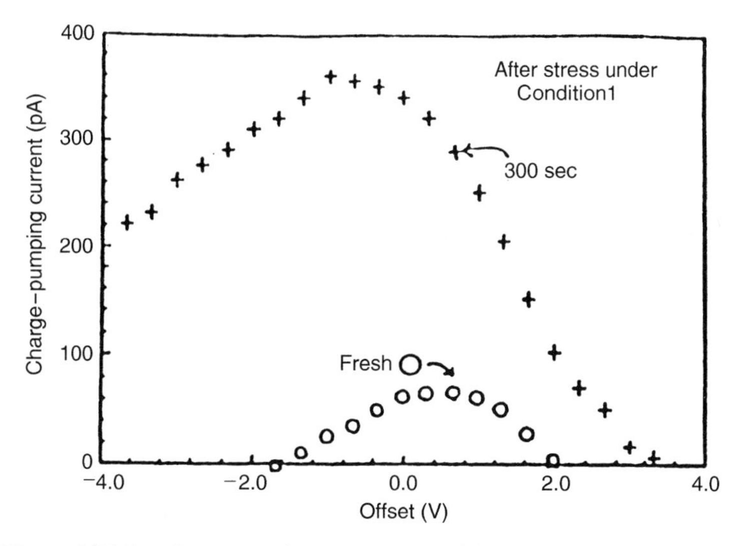

Figure 4.87 Interface states increase measured by charge-pumping current after 300 sec of stress at $V_g = 2.5\,V$ and $V_d = 9.6\,V$ with $V_{sub} = 0$ and V_s floating [4.117].

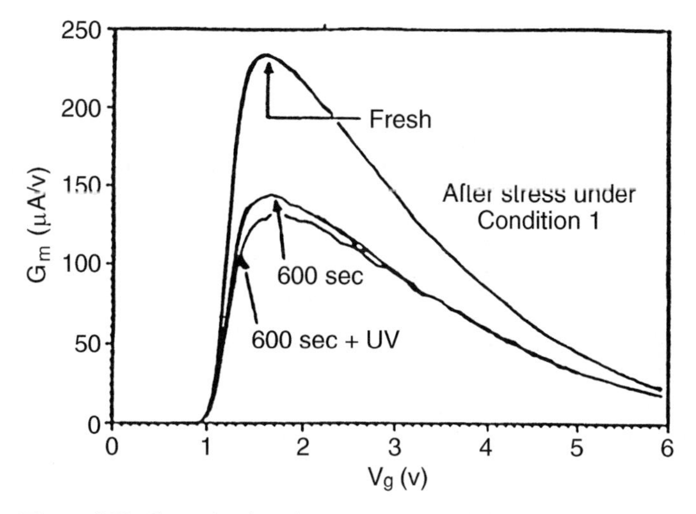

Figure 4.88 G_m reduction due to stress condition in Fig. 4.87 [4.117].

4.6.2.3 Reduction of Hot-Hole Injection. Generally speaking, junction-generated hot holes can be reduced through changes in the cell structure or changes in the erase operation.

4.6.2.3.1 Double-Diffused Junction. Since the heating of BBT-generated holes is greatly affected by the lateral electric field in the source junction, reducing the electric field by a graded source junction would reduce heating of holes. By introducing an additional phosphorous diffusion, not only is the BBT current shifted to a higher n^+ voltage, but the avalanche breakdown is also softened up (Fig. 4.89) [4.89]. This provides some margin against hot-hole injection during a memory cell erase operation. The optimization of the n-region is a tradeoff between the source junction breakdown voltage and the capacitive-coupling ratio between the source junction and the floating gate [4.17].

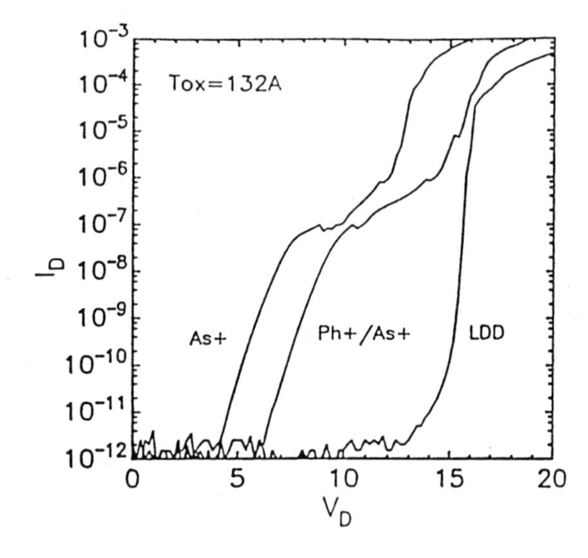

Figure 4.89 Comparison of BBT leakage currents in devices made with As^+, P^+/As^+, and LDD junction [4.89].

4.6.2.3.2 Reduction of Applied Source Voltage. In a grounded-gated, source-erase configuration, one method of reducing hot-hole injection from band-to-band tunneling is to reduce the electric field by reducing the applied voltage to the source.

4.6.2.3.3 Negative Gate, Floating Gate To Source Erase. Erase can be accomplished with the gate at a high negative voltage (–10 V) and the source at V_{cc} (+5 V) as shown in Fig. 4.13*a* [4.12, 4.48, 4.101, 4.119]. In this approach, the vertical field is kept constant, and the F–N erase current is unchanged. However, the lateral field is greatly reduced (only 5 V at the source junction); in that way, the BBT-generated holes are not heated as much. In addition, the "apparent" avalanche breakdown voltage has been shifted up (Fig. 4.90) [4.12] so that the breakdown-related damage to the device is no longer an issue [4.12, 4.101].

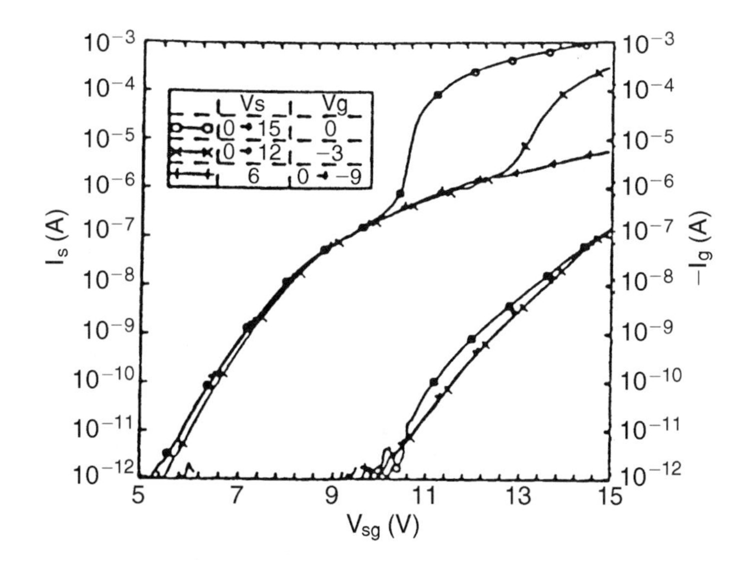

Figure 4.90 Source current, I_s, and gate current, I_g, versus source-to-gate voltage, V_{sg}, for three different bias conditions as defined in the inset. $V_{sg} = V_s–V_g$ defines the oxide field for tunneling [4.12].

4.6.2.3.4 Negative Gate, Floating Gate to Channel Erase. This method, also called channel erase (Fig. 4.13*b* [4.119, 4.120]), is typically accomplished with the gate at a high negative voltage (–13 V) and the channel in the well at V_{cc} (+ 5 V). Since there is no reverse-biased junction in this case, BBT hot holes and junction breakdown are no longer a concern, and the cell is erased with uniform F–N tunneling across the gate oxide.

4.6.2.4 Drain-Leakage-Induced Over-Erase. If bits, which are outside the main V_t distribution, are over-erased and become sufficiently depleted, they can cause program verify failures in the following program cycle. For example, it has been reported that fast bit erase can occur if a low level of bitline leakage is present during high-voltage source erase (Fig. 4.91) [4.126]. This DC leakage path will cause the channel current to flow as the memory cell is sufficiently erased. A large amount of channel hot holes are easily generated under such weak channel turn-on and large V_{ds} conditions. The injection of channel hot holes will accelerate the cell erase and over-erase the fast bit. Such bits, which are outside the main distribution, can be screened out during wafer sort if they are well-behaved or stable.

4.6.2.5 Erratic Bits. It has also been reported [4.126, 4.127] that, in a large memory array, a small number of bits can behave erratically during cycling in a somewhat random fashion, as shown in Fig. 4.92. This phenomenon has been

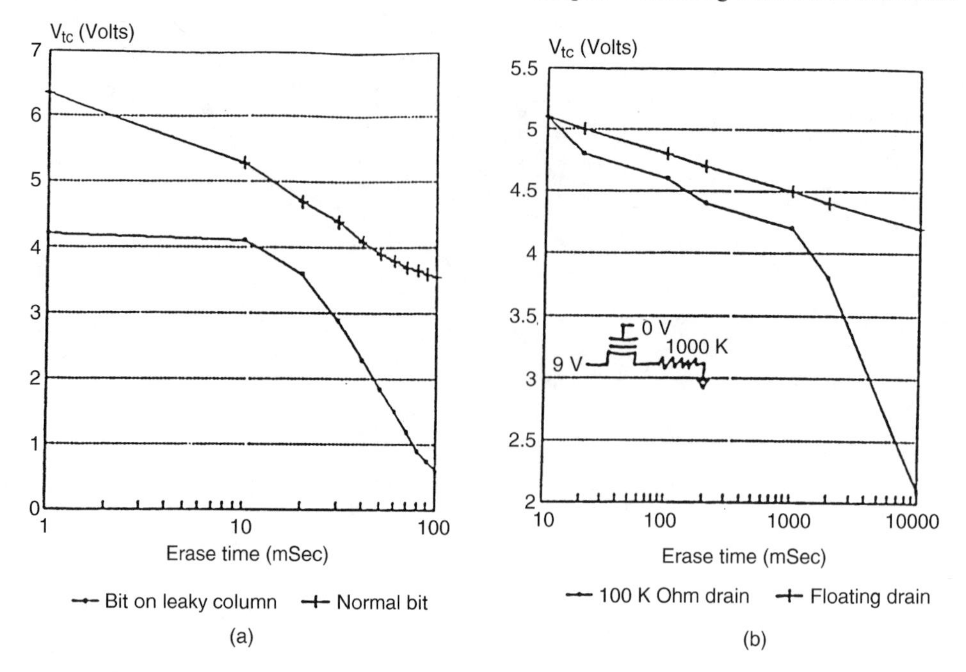

Figure 4.91 (*a*) Single fast erase bit on column with bitline leakage. (*b*) Single-cell erase with and without drain leakage [4.126].

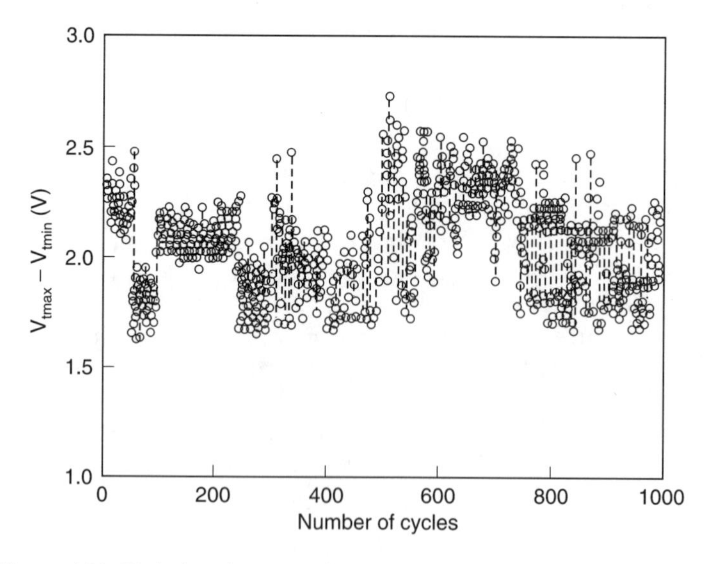

Figure 4.92 Variation in erase distribution (V_t spread) as a function of number of cycles [4.126].

attributed to dynamic hole trapping and detrapping/neutralization in the oxide, which leads to F–N tunneling current fluctuation. Under a worst case scenario, a bit can become so depleted during erase as to cause a soft or temporary failure in cycling. After reconditioning, the problem bit, by some external stimulus such as gate stress or temperature bake, can become a normal bit again [4.127]. This random bit instability effect, if not controlled, can actually lead to early endurance failure.

4.6.3 Source-Side Injection

While source-side injection Flash memory cells have a relatively higher injection efficiency than cells utilizing channel hot-electron injection, the injection is concentrated near the narrow gap. Because of this split gate gap structure, a small amount of electron trap-up in the spacer-like oxide between the floating gate edge and the control gate can change the local electric field and limit the endurance of these structures.

4.6.4 Stress-Induced Oxide Leakage

The reliability of the tunnel oxide is one of the fundamental reliability parameters for Flash memories. These devices are either erased (e.g., stacked gate) or both programmed and erased by high-field (Fowler–Nordheim) injection of electrons into a very thin dielectric film to charge and discharge the floating gate. Unfortunately, this current through the oxide degrades the quality of the oxide and eventually leads to breakdown. The wearout of tunnel oxide films during high-field stress has been correlated with the buildup of both positive or negative trapped charges.

Recently, it has been shown that high-field stress also induces a low-field leakage current in thin oxides, a thickness less than 10 nm [4.136, 4.137, 4.140]. The mechanism of this leakage current has been attributed to the generation of localized defects or weak spots [4.137] and trap states near the injecting interface [4.135, 4.140, 4.141]. These low-field leakage currents degrade the data retention of the EEPROM cell [4.140, 4.1138, 4.38]. Unfortunately, these oxide leakage currents increase with decreasing oxide thickness [4.137, 4.138] and make it difficult to scale down the oxide thickness of the memory cell. The read-disturb observed in Flash memories [4.135], as discussed in Section 4.6.9, is the direct result of high-field stress during P/E cycling.

Aritome et al. [4.38] studied the influence of the waveform of the stress voltage on the degradation of the tunnel oxide. Three types of high-field dynamic stress were used to study the thin oxide leakage currents. Figure 4.93 shows the applied dynamic stress waveforms. In the case of bipolarity stress, positive high voltages were applied to the gate or to the substrate and source-drain (S/D) regions. Fowler–Nordheim tunneling occurs alternately from the substrate to the gate and from the gate to the

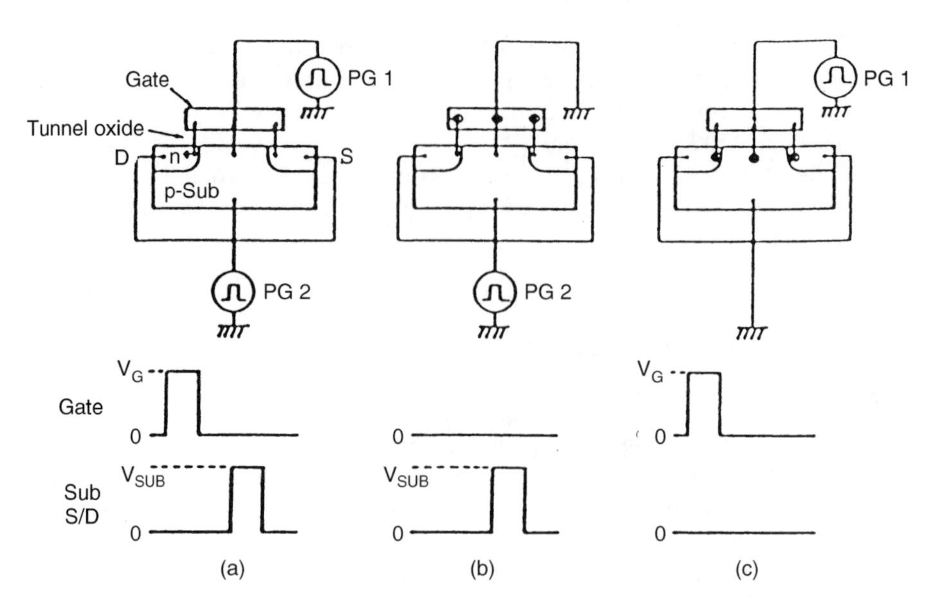

Figure 4.93 Setup and stressing waveform for (*a*) bipolarity stress, (*b*) electron-emitted stress, and (*c*) electron-injected stress [4.38].

substrate. Electron-emitted stress (emission of electrons from the gate) and electron-injected stress (emission of electrons from the substrate) are performed for comparison. The applied gate voltage is comparable to the floating gate voltage of a memory cell during the write operation. The substrate voltage (V_{sub}) is chosen in such a way that the tunnel current during emitted and injected stress is the same.

The $I_g - V_g$ characteristics before and after dynamic stress for a 5.6 nm oxide thickness are shown in Fig. 4.94. The thin oxide leakage currents, as a function of three types of high-field dynamic stressing, are compared. It is observed that the thin oxide leakage current induced by bipolarity stress is about one order of magnitude smaller than that induced by both the electron-emitted and electron-injected stresses. This result shows that the origin of the thin oxide leakage current can be suppressed by a reverse high-field stress.

4.6.4.1 Time-Dependent Dielectric Breakdown (TDDB) under High-Frequency Stress. It has been shown that TDDB characteristics (Fig. 4.95) are dependent on the stress waveform and its frequency [4.142, 4.143]. The stress waveforms are the electron-emitted stress, the electron-injected stress, and the bipolarity stress. The time-to-breakdown during bipolarity stress is longer than that for both electron-emitted stress and electron-injected stress. Moreover, it is shown that the TDDB characteristics depend on the frequency of the stress wave. The breakdown time under unipolarity stress does not depend on frequency. However, the breakdown time to bipolarity stress depends strongly on the frequency. The bipolarity

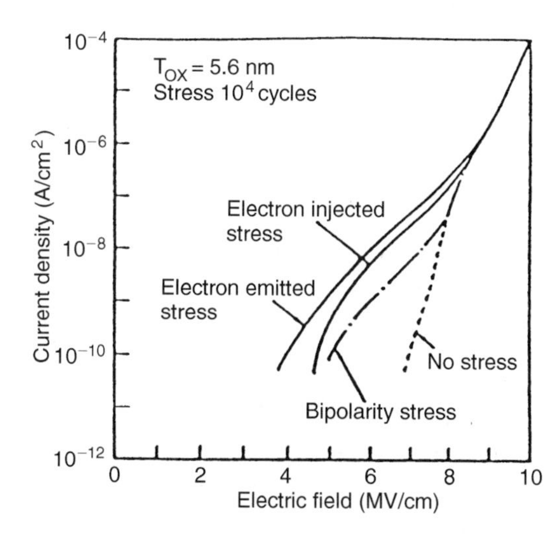

Figure 4.94 Leakage current of the thin oxide at low voltages for 5.6 nm oxide after bipolarity stress, electron-emitted stress, and electron-injected stress [4.38].

stress lifetime in the megahertz range is about 40 times higher than it is at 10 Hz for the same applied voltage. It is suggested that the improved TDDB obtained with bipolarity stress is due to the field-assisted detrapping of holes at the Si–SiO_2 interface [4.143].

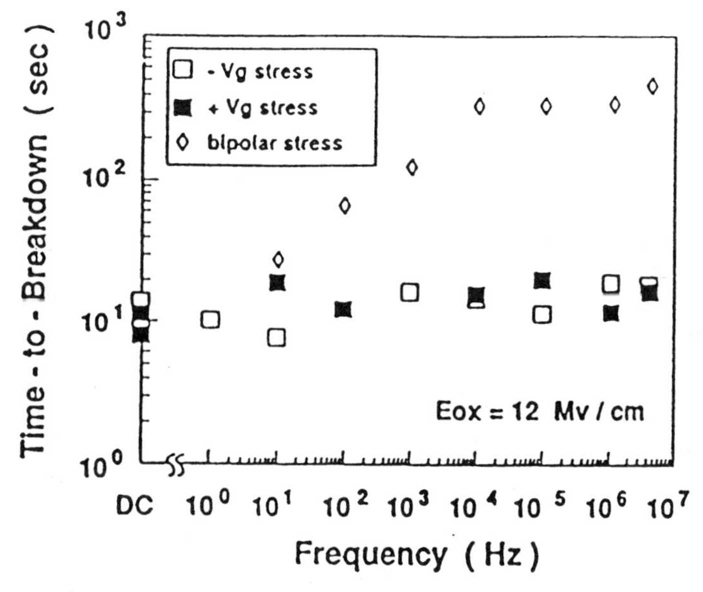

Figure 4.95 Dependence of time-to-breakdown on stress frequency [4.142, 4.143].

4.6.5 Poly-to-Poly Erase

While thicker oxide for poly-to-poly erase does not suffer from stress-induced oxide leakage, it does suffer from increased electron trapping compared to planar tunnel oxide. There are two reasons for this. First, the oxide grown on polysilicon is of lower quality than oxide grown on silicon. Typically, some phosphorus is embedded in the oxide. Second, with enhanced tunneling, the tunnel current flows primarily from the enhanced tunneling points. Electron trapping with the tunnel current tends to reduce the local field. The tunnel current reduction is larger as compared to planar tunneling. Typically, the window closing is about 10 times faster compared to FLOTOX EEPROM.

4.6.6 Reliability of Interpoly Dielectric

ONO (oxide–nitride–oxide) interpoly dielectrics are currently used for nonvolatile memories. In this case, the important issues are not only low-defect density and long mean time to failure, but also charge retention capability. Mori et al. investigated the ONO interpoly dielectric thickness scaling effect on charge retention characteristics in nonvolatile memories [4.144].

For the top oxide layer, a certain thickness (3 nm), which can block hole injection, is required. Si_3N_4 thickness reduction leads to a reduction in initial rapid charge loss. However, too much reduction sometimes leads to enhanced charge loss in a long bake test. Bottom polyoxide thinning, down to 10 nm, does not result in degradation. The NO (nitride–oxide) double-layer structure shows an initially large rapid charge loss, which may be prevented by applying the ONO structure. However, if the bottom oxide layer is thin and its quality is poor, a portion of the bits may suffer from rapid initial charge loss. When scaling down the ONO thickness, one must consider these key factors concerning charge retention, as well as dielectric reliability.

4.6.7 Stacked Gate Write/Erase Endurance

As discussed previously, both channel hot-electron program and tunnel erase are degraded by repeated cycling. The typical result is a decrease in V_t window size when a fixed program and erase voltage and time are used. Figures 4.74 and 4.96 show V_t distribution versus write/erase cycles [4.27 and 4.104].

4.6.7.1 NAND Write/Erase Endurance. Write/erase endurance characteristics are shown in Fig. 4.97. The uniform write and uniform erase technology (charge transfer between floating gate and substrate) guarantees a wide cell threshold window (as large as 4 V), even after 1 million write/erase cycles (Fig. 4.97*a*). However, the threshold window obtained by the uniform erase and nonuniform write technology begins to decrease rapidly at approximately 1E2 write/erase cycles, and fails at 1E5 write/erase cycles (Fig. 4.97*b*) because the Fowler–Nordheim tunneling current during nonuniform writing is confined to a small region at the drain.

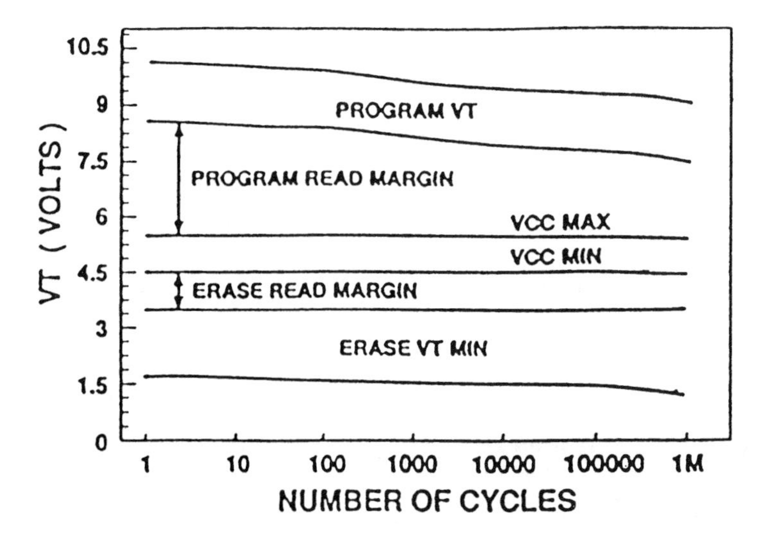

Figure 4.96 Stacked gate FLASH endurance characteristics [4.27].

Therefore, electron traps in the tunnel oxide are generated at a high rate near the drain area, and these electron traps impede electron injection and emission between the floating gate and substrate. This effect may be aggravated by holes that are generated by band-to-band tunneling.

In uniform write and uniform erase technology, the threshold voltage of the erased cell is dependent on the number of write/erase cycles. However, the threshold voltage of the written cell is not dependent on the number of write/erase cycles. This can be explained as follows. Oxide traps and interface traps are generated uniformly over the entire channel area because Fowler–Nordheim tunneling of electrons is

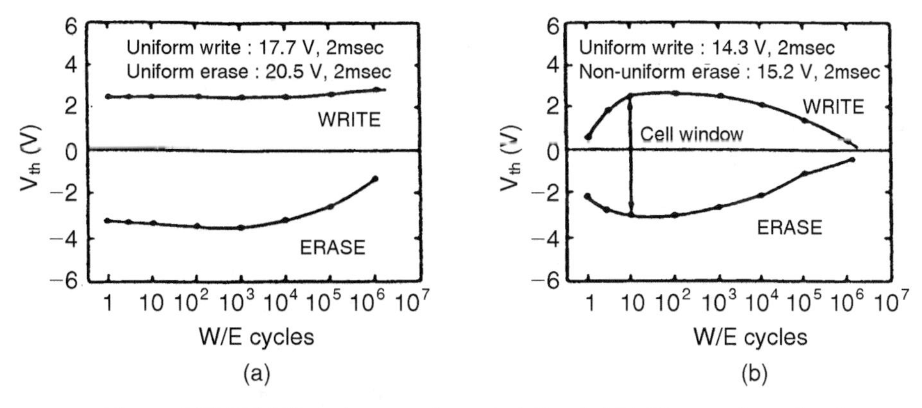

Figure 4.97 NAND endurance characteristics. (*a*) Uniform write and uniform erase technology; (*b*) uniform write and nonuniform erase technology [4.14].

performed uniformly during both the writing and erasing operations. The charges trapped uniformly over the channel area affect not only the electron-tunneling current through the oxide, but also the flatband voltage. The threshold voltage of the erased cell decreases slightly for up to 10^3 cycles due to hole trapping. Hole traps affect the increase of stored positive charge on the floating gate, as well as the decrease in flatband voltage. After 10^3–10^6 cycles, the threshold voltage of the erased cell increases due to electron detrapping. On the other hand, the threshold voltage of the written cell remains almost constant up to 10^6 cycles in spite of charge trapping in the oxide because of the influence of stored charge on the floating gate and trapped oxide charge on both the flatband voltage and injection field, canceling each other [4.103].

4.6.8 Data Retention

In floating gate memories, the stored charge can leak away from the floating gate through the gate oxide or through the interpoly dielectric. This leakage, caused by mobile ions, oxide defects, or other mechanisms, results in a shift of the threshold voltage of the memory cell. The threshold voltage shift can also be caused by detrapping of electrons or holes from oxide traps.

Different charge-loss mechanisms have been described [4.90, 4.91]. These mechanisms are briefly discussed here. The first loss mechanism is thermionic emission over the image force lowered potential barrier. However, this mechanism is not dominant since the barrier height for thermionic emission is considerably higher than the activation energies for the other charge-loss mechanisms. The second mechanism is called electron detrapping. The activation energy for this type of charge loss is about 1.4 eV. Because the intrinsic charge-loss rate decreases with time and stops after a threshold voltage drop of about 0.5 V [4.91], this type of charge loss is associated with the detrapping of electrons that are trapped in the oxide. The third mechanism is related to defects in the oxide and results in both charge loss and gain, depending on the biasing condition. The activation energy for this kind of charge loss is about 0.6 eV [4.91]. The fourth mechanism is related to contamination. Positive ions entering the memory cell may compensate a part of the negative charge stored on the floating gate. The activation energy for charge loss by contamination is about 1.2 eV [4.91].

Kynett et al. [4.27] have shown the results of data retention tests for a 1 MB Flash memory. Accelerated retention bake experiments performed at 250°C for 168 hours indicate that, after 10 K write/erase cycles, Flash memory will exhibit a 0.7 V program V_{th} shift [4.27]. Furthermore, a retention bake of 52 hours produced less than a 0.5 volt V_{th} shift in programmed devices which had been through more than 1 million write/erase cycles.

Table 4.11 shows the data loss of a Flash memory cell after various cycling conditions. The average charge loss is small in the case of a low applied voltage to the source because the hot-hole injection due to band-to-band tunneling is suppressed. The low electric field across the tunnel oxide during erasing also results in excellent write/erase endurance characteristics.

TABLE 4.11 CHARGE LOSS OF ETOX FLASH EEPROM CELL AFTER 100 CYCLES

Cycling Condition	Average Charge Loss
EEPROM program ($V_g > 0$ V) and erase ($V_g < 0$ V)	0.1 V
EEPROM program/Flash erase at $V_{source} = 12.5$ V; $V_{cg} = 0$ V	0.51 V
EEPROM program/Flash erase at $V_{source} = 11.5$ V; $V_{cg} = 1.5$ V	0.33 V
EEPROM program/Flash erase at $V_{source} = 8.5$ V; $V_{cg} = -6$ V	0.24 V
Uncycled	0.12 V

The data retention characteristics of Flash memory cells programmed by two different write/erase (W/E) technologies are compared in [4.38]. Figure 4.98 illustrates these two W/E techniques. The first is the bipolarity W/E technology, which is a uniform write and erase technology. During the write operation, a high voltage (V_{cg}) is applied to the control gate, with the substrate and source-drain regions grounded. Electrons are injected from the substrate to the floating gate over the whole channel area of the memory cell. In the erase operation, a high voltage (V_{sub}) is applied to the substrate and source-drain regions and the control gate grounded. Electrons are then emitted from the floating gate to the substrate. In this write/erase method, the high-field stress of the thin oxide corresponds to bipolarity stress. The other write/erase method utilizes channel hot-electron (CHE) for the write and F–N tunneling for the erase. This technology is a nonuniform write

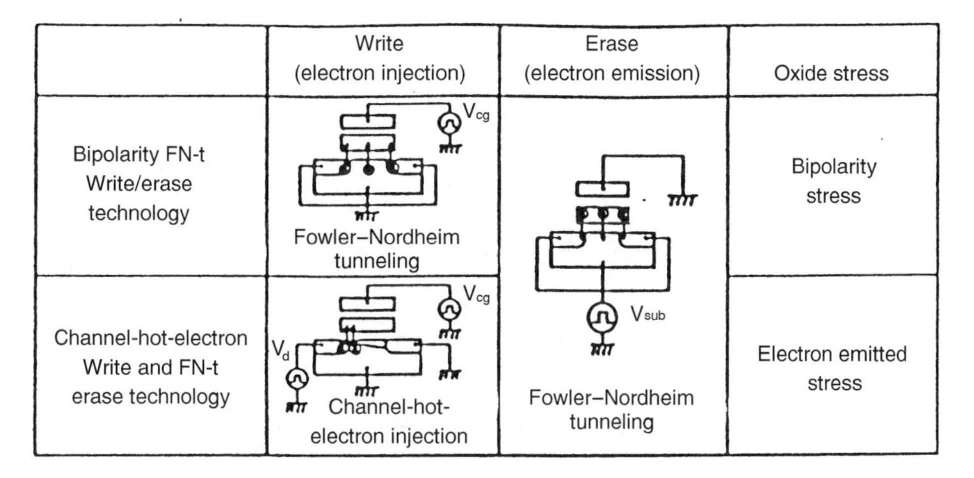

Figure 4.98 Comparison between (*a*) bipolarity F–N tunneling write/erase technology and (*b*) channel hot-electron (CHE) write and F–N tunneling erase technology [4.38].

and uniform erase technology. The erase operation is the same as that in the bipolarity W/E technology. However, during the write operation, high voltages are applied to the control gate and the drain. Thus, channel hot electrons are generated by the lateral electric field, and electrons are injected from the substrate to the floating gate. In this case, the high-field stress of the thin oxide is electron-emitted stress. The data retention will be different for these two W/E technologies because the thin oxide leakage current is different for bipolarity and electron-emitted stresses of thin oxide. Moreover, in the conventional Flash memory erasing method [4.19, 4.135], a high voltage is applied to the source. However, in this experiment [4.38], a high voltage is applied to the substrate, as well as the source-drain regions, in order to prevent the degradation of the thin oxide due to hole injection caused by band-to-band tunneling [4.19, 4.135].

Figure 4.99 shows the write/erase endurance characteristics of both W/E technologies [4.38]. No closure of the cell threshold window occurs up to 1E5 write/erase cycles in either technology. Data retention characteristics are measured under various gate voltage conditions in order to accelerate the retention test. In the case of the CHE write and F–N tunneling erase technology, the stored positive charge rapidly decays as a function of time. As a result, the threshold window decreases (Fig. 4.100). However, in the case of the bipolarity F–N tunneling W/E technology, data loss of the stored positive charge is significantly reduced. This phenomenon can be explained by the fact that the thin oxide leakage current is reduced by the bipolarity F–N tunneling stress. Figure 4.101 shows the data retention time after write/erase cycling as a function of the tunnel oxide thickness. The data retention time is defined by the time it takes V_{th} to reach −1.0 V during the gate voltage stress. In devices with a 7.5 nm tunnel oxide, the data retention time obtained for the bipo-

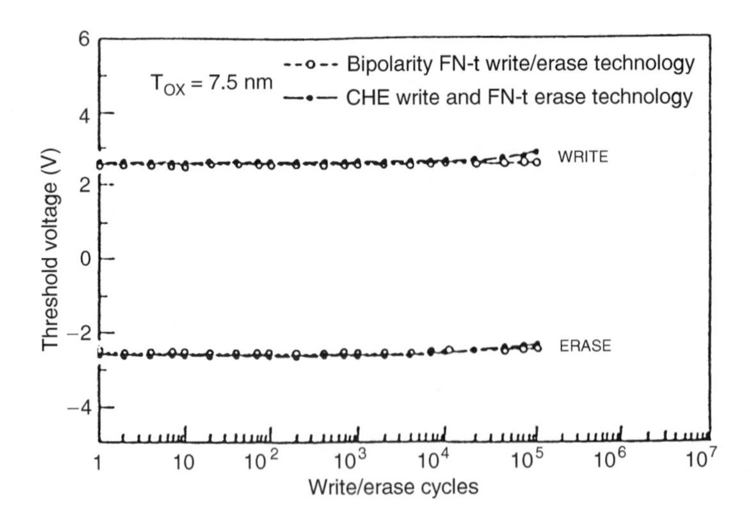

Figure 4.99 Write/erase endurance characteristics of Flash memory cell with 7.5 nm oxide [4.38].

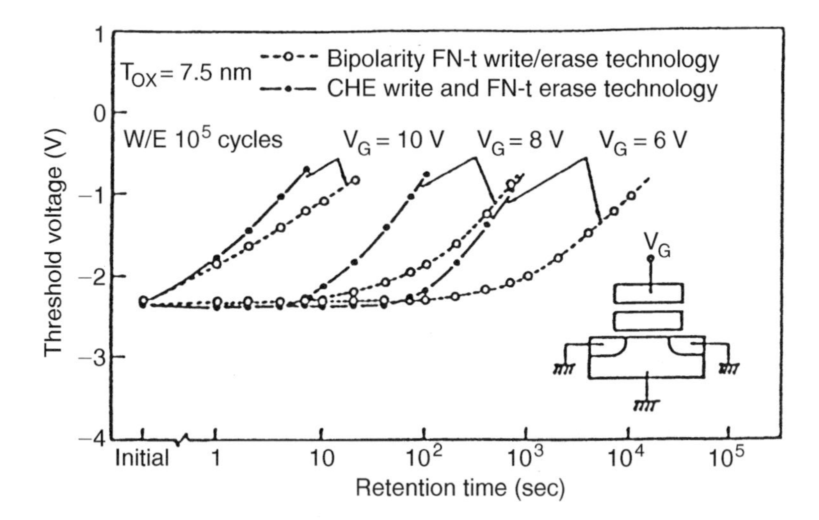

Figure 4.100 Data retention characteristics under gate voltage stress for bipolarity F–N tunneling write/erase technology and channel hot-electron (CHE) write and F–N tunneling erase technology [4.38].

larity F–N tunneling write/erase technology is 50 times longer than that of CHE write and F–N tunneling erase technology, after 1E5 write/erase cycles. However, in devices with a 9 nm tunnel oxide thickness, the data retention time is almost the same for both technologies. For very thin (<9 nm) tunnel oxides, the bipolarity write/erase technology offers improved data retention times in comparison with the CHE write and F–N tunneling erase technology. Therefore, this technology may facilitate the downscaling of tunnel oxides. Reducing the tunnel oxide thickness results in lower programming voltages and in faster read operations because the read current is increased.

NAND data retention characteristics of the programmed memory cell, measured at 300°C after different numbers of write/erase cycles from 10 to 10^6 were carried out, are shown in Fig. 4.102 [4.14]. The stored positive charge effectively remains for up to 100 minutes baking time due to the detrapping of electrons from the gate oxide to the substrate during the retention bake (Fig. 4.102). This retention bake time becomes longer with an increasing number of cycles because the number of trapped negative charges in the thin oxide increases.

The effect of detrapping electrons is equivalent to the effect of trapping holes in the gate oxide. As a result, detrapping of electrons suppresses the data loss of the positively charged cell. Consequently, the stored positive charge, which effectively increases at the beginning of the bake, extends the data retention time of the memory cell programmed by uniform write and erase. Figure 4.103 [4.14] shows the data retention time after write and erase cycling. Thus, the data retention time can be extended by using uniform write and uniform erase technology, especially beyond 10^5 write/erase cycles.

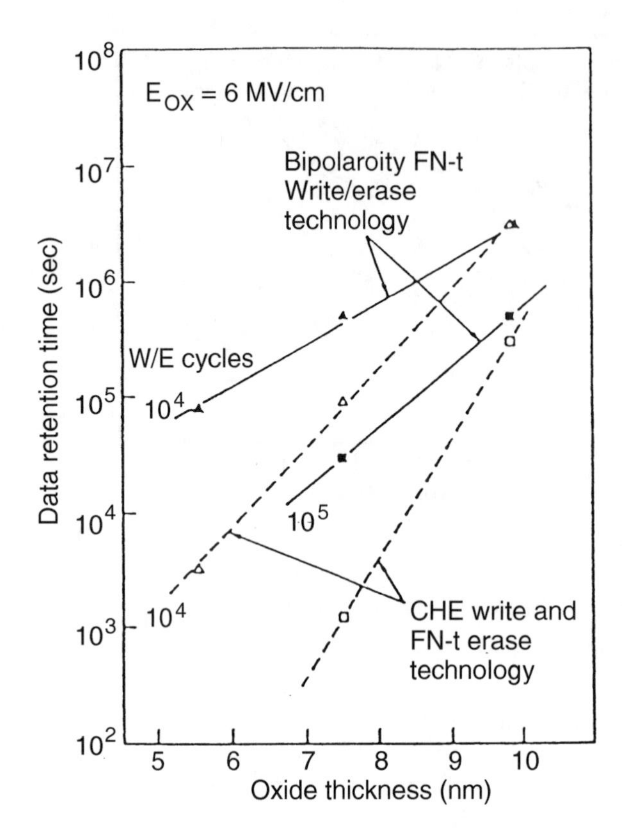

Figure 4.101 Data retention time of Flash memory cell after write and erase cycling as a function of tunnel oxide thickness. The data retention time is defined by the time at which V_{th} reaches –1.0 V during the applied gate voltage stress [4.38].

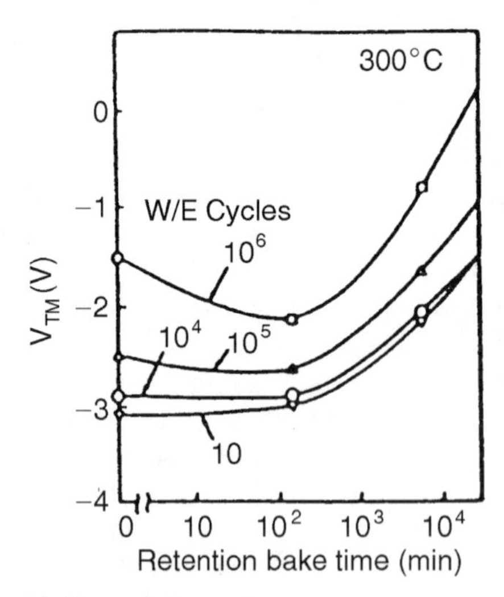

Figure 4.102 NAND data retention characteristics [4.14].

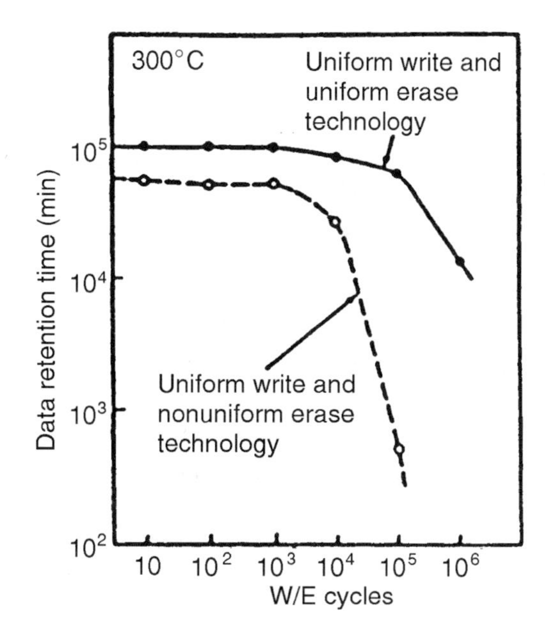

Figure 4.103 NAND data retention time [4.14].

4.6.9 Read-Disturb

The read-disturb stress condition occurs on erased bits that share a wordline with a bit that is being read. The common wordline places the control gate of the erased cells at 5.0 V (V_{cc}). The selected device's drain is driven to about 1.0 V. The unselected bits have their source, drain, and substrate at 0.0 V. In high-density Flash memories, many bits are put into a low-field stress condition when one bit is read. This condition is shown in Fig. 4.104 [4.135].

It has been reported that thin gate oxides exposed to high-field stress and high levels of charge injection can develop a pre-breakdown leakage condition [4.136, 4.137]. As program/erase cycling requirements increase, and the Flash tunnel oxide thickness is decreased, the EEPROM cell becomes more susceptible to tunnel oxide leakage [4.138]. During P/E cycling, some Flash bits can exhibit a leakage condition that permits charge gain on erased bits due to the low electric field present during a read operation. Under prolonged or DC read conditions, the defective cells can appear programmed.

It was shown that the charge gain takes place by electron tunneling through a corrupted oxide barrier [4.135]. The barrier reduction is caused by positive charge trapping at the tunnel oxide to source junction. The charge trapping is due to hole generation during F–N erase. Since the effective barrier takes on various levels, it was proposed that the configuration of trapped charge determines the extent of barrier lowering. The effective barrier of leaking cells was determined by tracking the cell threshold voltage during read-disturb stress conditions.

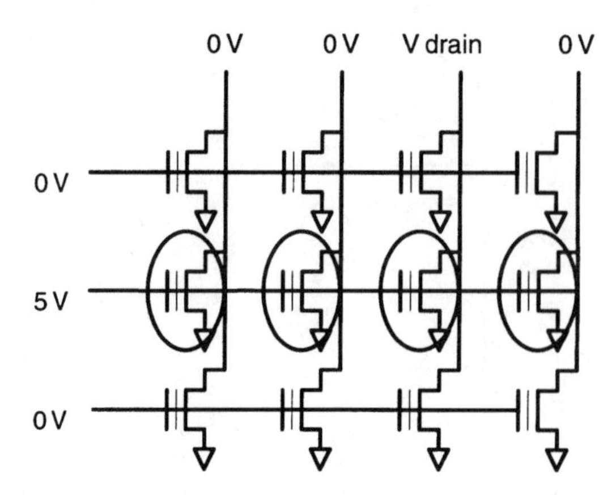

Figure 4.104 The read-disturb stress is applied to all cells sharing the gate select wordline with the cell being read. While the solidly circled cell is read, all of the dotted circled cells are stressed [4.135].

The distribution of calculated effective barrier heights is shown in Fig. 4.105. The distribution of barrier heights suggests that there is variation in the physical configuration of the trapped charge responsible for the read-disturb leakage current.

The leakage current is due to tunneling, and the generation of leakage sites is related to program/erase cycling. It has previously been shown that Flash cycling causes hole trapping [4.20] in the vicinity of the source to tunnel oxide junction. The actual tunneling mechanism could be barrier lowering due to a positive charge cluster located close to the source-oxide junction [4.136] or trap-assisted tunneling [4.139], with traps distributed over the cross section of the oxide. If the mechanism is barrier lowering, the degree of effective barrier reduction depends on the number of trapped charges and their positions. Positive charges, clustered within 2–4 nm of the

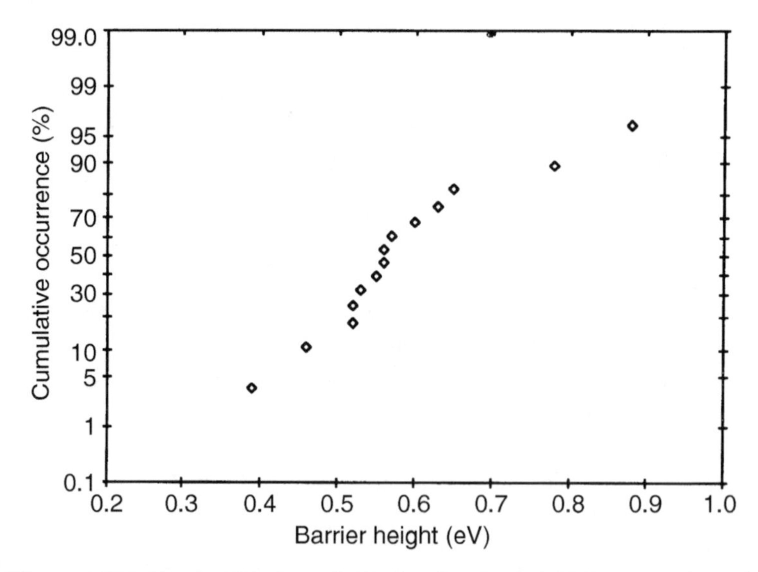

Figure 4.105 The distribution of effective barrier height has a median of 0.56 eV, although there are some higher and lower points [4.135].

source, will have the greatest effect on barrier reduction [4.135], as sketched in Fig. 4.106, by reducing the tunneling distance into the conduction band of the floating gate. It is speculated that some trap configurations are more likely to occur, therefore favoring the appearance of the corresponding barrier height. The effective barrier ranged from 0.39 to 0.88 eV [4.135]. Figure 4.106 shows the conduction band of source, tunnel oxide, and floating gate regions under a read-disturb stress.

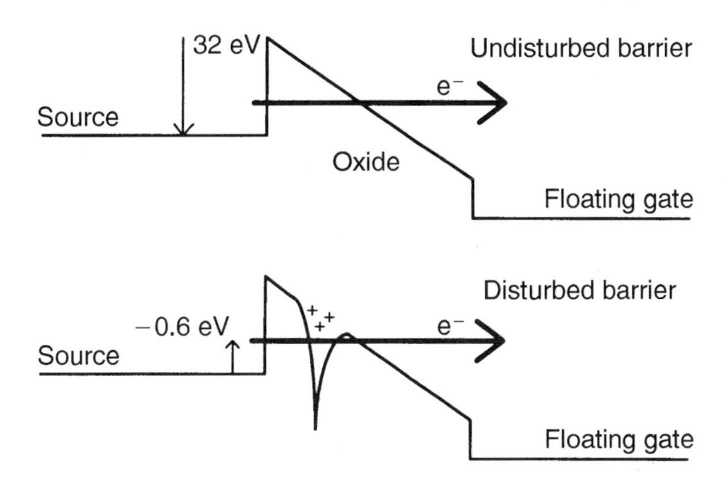

Figure 4.106 Sketch of the conduction band of the Flash source, tunnel oxide, and floating gate region under read-disturb stress. The disturbed barrier has a reduced tunneling distance due to trapped positive charges located near the source region [4.135].

4.7. PROCESS TECHNOLOGY

The process technology required for floating gate devices is derived from standard polysilicon gate technology that has been used so successfully in the manufacturing of SRAMs, DRAMs, and logic devices. The most significant addition is floating gate technology, which requires the highest quality insulating dielectric. Another important consideration is the high voltage that is required for EPROM programming and tunnel erase. It puts special requirements on isolation, junction breakdown, and transistor technology. A further reliability consideration is the quality of the dielectric used in the transistors and in the floating gates.

4.7.1 Floating Gate Technology

In a typical EPROM technology, poly1 is used exclusively for the floating gate. After poly1 is defined, a poly-to-poly dielectric is formed that surrounds the poly1 floating gate completely. In the case of poly-to-poly tunneling, the polyoxide is a specially processed oxide that gives poly-texturing to enhance tunneling probability. If the cell is not designed for poly-to-poly tunneling, then an oxide–nitride–oxide (ONO) composite dielectric is typically used. The composite layer gives a much lower defect density. Much thinner ONO layers can be formed, compared to oxides, with much lower leakage currents. This, in turn, gives a higher coupling-capacitance and smaller cell sizes.

4.7.2 High-Voltage Technology

The high voltage required to operate Flash memories ranges from a low of 12 V for stacked gate Flash to nearly 25 V for poly-to-poly tunnel erase, even though the voltage may be generated internally. Furthermore, the EPROM programming voltage is not expected to be scalable much below 10 V. This is because the gate program voltage is determined by the intrinsic silicon–silicon dioxide barrier (3.2 eV), which is not scalable, as well as the floating gate voltage change which may actually be growing with ever increasing density.

In order to handle these voltages, the field oxide isolation process has to give a sufficiently high-field turn-on voltage, as well as a sufficiently high junction breakdown voltage. The standard approach is to use a much thicker field oxide compared to logic technology of a comparable generation. However, there is a limit on how much a standard field oxide isolation can be scaled. It is believed that eventually more exotic isolation types, like trench isolation, will be used for the smallest geometry.

In addition to isolation technology, the transistor technology also has to handle the high voltage. This is contrary to transistor scaling for performance. Special high-voltage devices are required. In the case of CMOS, for example, it is difficult to have a full CMOS circuit operate above 20 V, and a depletion device is generally used as a load device for higher voltages. A 20 nm transistor can operate reliably at 12 V, but, with further scaling, a 15 nm transistor will not be reliable for 12 V operation. Thus, for scaled technologies or for EEPROM-based Flash technologies, dual gate oxide technologies are used—one to handle high voltages and the other, a high-performance, low-voltage transistor for high-speed logic.

4.8. MEMORY CIRCUITRY

The standard array architecture for Flash memory is just an EPROM array. In a conventional EPROM array, the cells are arranged in an X-Y cross-point array with an EPROM cell at each cross-point. In the horizontal X-direction, control gates of the EPROM cells are connected together by poly to form the wordline. In the vertical Y-direction, the drains of the EPROM cells are connected together by metal through a contact shared by two memory cells. The sources of the cells are connected together horizontally by diffusion. Spaced every 16 or 32 cells, the sources are connected together by a metal line, called a source strap, running in the Y-direction. This is a well-proven EPROM array architecture.

The wordlines are selected by X-decoders, which pass a high V_{pp} to selected wordlines during programming and V_{cc} during read. Nonselected wordlines are held at ground. The columns are selected by Y-select, which passes V_{dd} to a cell being programmed and V_{drain} for read. The array is organized into eight blocks for each bit in a byte. During program or read, one column in each block is activated. If the programmed bits are not separated, but are adjacent to each other, there will be an excess voltage drop when they are programmed at the same time. Finally, there is a sense amplifier for each bit for reading the state of the memory cell.

4.8.1 Row Decoder Circuits

The row decoder circuits are standard circuits derived from EPROM X-decoders. For Flash memories using negative gate voltage erase, the circuit requires modifications in the row decoders. During the erase operation, the negative voltage generated by a negative charge pump is applied to the wordline. The n^+/p^- substrate diode is forward-biased if a negative voltage is applied to the n^+ region with the p-substrate at ground level. P-MOS transfer gates have been used in the first generation of Flash memories using negative gate erase as one way to overcome this problem. In this scheme, the p-MOS transfer gate isolates the negative voltage from the wordline driver n-MOS transistors during the erase operation as shown in Fig. 4.107.

In this row decoder, p-MOS transfer gates are inserted between each wordline and an n-MOS driver, and a gate-drain connected p-MOS transistor is utilized as a diode between each wordline and the negative charge pump. A positive voltage is applied to the p-MOS gate during the erase operation to isolate the negative voltage. The p-MOS gate is negatively biased during other operations.

There are several problems with this scheme. First, a severe stress voltage is applied to the p-MOS gate. The p-MOS gate has to be deeply negatively biased in order to supply zero volt to a nonselected wordline during the read and program

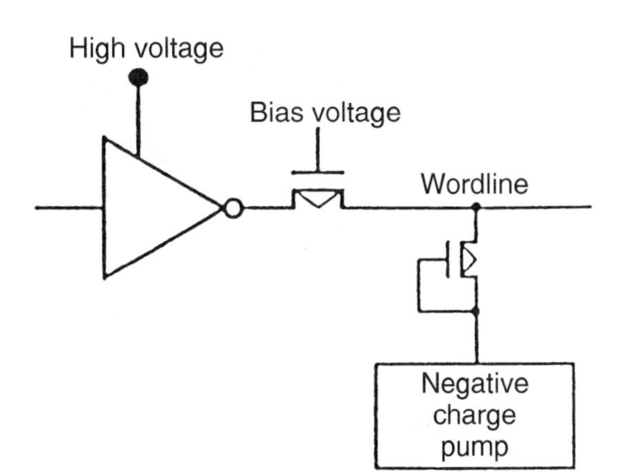

Figure 4.107 Conventional row decoder circuit for negative gate, positive source erase [4.61].

operations. Therefore, a large electric field is applied to the p-MOS transistor gate oxide in the selected wordline. The stress is about 15 V, which is higher by 50% than a normal programming voltage. The high voltage prevents gate oxide scaling. Second, the negative voltage required for the output of the negative charge pump is very large because the negative voltage applied to the wordlines is reduced by the diode-like p-MOS transistor. Finally, this conventional scheme requires a standby current because a negative voltage is needed during standby operation. This is not preferable for portable systems. At the same time, the resistance of the transfer gate increases the wordline delay time.

To overcome these problems, a row decoder utilizing a triple-well process technology was developed [4.61] and is shown in Fig. 4.108. In this circuit, n-MOS transistors used in the row decoder circuit are fabricated in the p-well region,

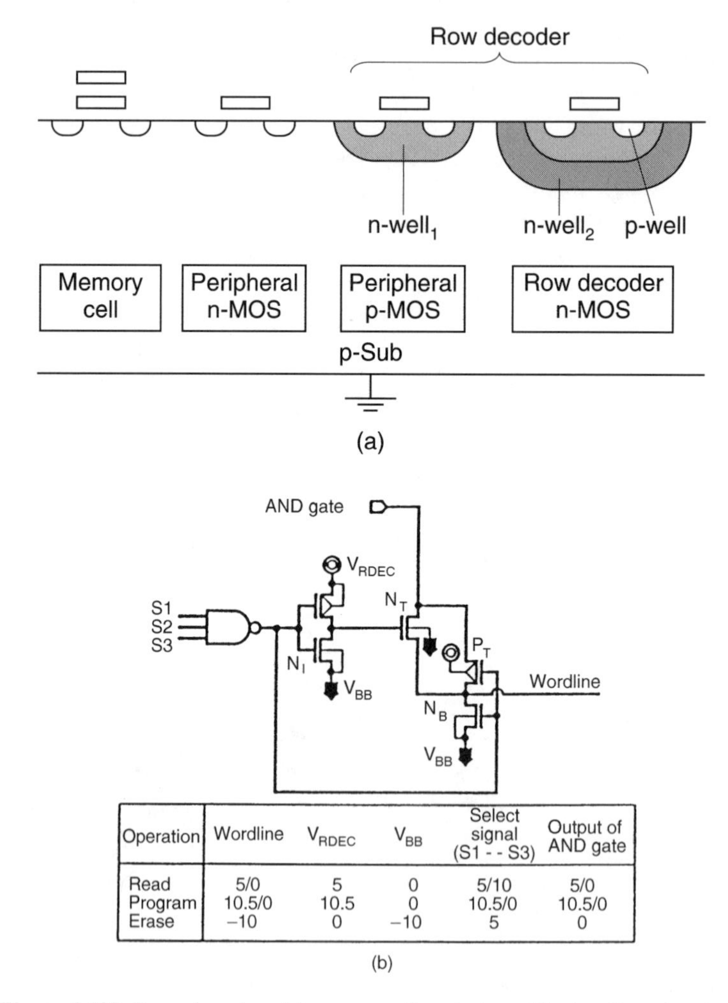

Operation	Wordline	V_{RDEC}	V_{BB}	Select signal (S1 - - S3)	Output of AND gate
Read	5/0	5	0	5/10	5/0
Program	10.5/0	10.5	0	10.5/0	10.5/0
Erase	−10	0	−10	5	0

Figure 4.108 Row decoder: (*a*) cross-sectional view, (*b*) circuit and operation [4.61].

which is isolated from the p-substrate by a deep n-well region. During the erase operation, a negative voltage of −10 V, generated by the negative charge-pump circuit, is applied to the p-well, and all wordlines are set to −10 V by the n-MOS transistors in the p-well. During other operations, the p-well is set to ground level. This scheme prevents the forward-bias problem. As shown in Fig. 4.108, the selected wordline is set to 10.5 V/5 V for programming and reading operations, respectively, and the nonselected wordlines are set to ground level. During the erase operation, all the wordlines are set to the negative voltage. The row decoder circuit consists of a NAND gate, an inverter, and transfer gates. It is designed to minimize the electric field on the gate oxide. During the erase operation, VRDEC is set to ground level, while all the decoder select signals of S1–S3 are set to a high level and the output of an AND gate is set to ground level. Thus, the maximum stress voltage on the gate oxide is reduced to a value as small as 10 V, which is almost the same as the voltage stress during programming. Consequently, a thick gate oxide is no longer needed.

This row decoder is appropriate for a relatively large-size block erasure. A compact row decoder circuit [4.70], in which individual wordlines are selectively biased to a negative voltage during the erase operation, is shown in Fig. 4.109. This circuit consists of a decoder select NAND gate A and wordline select transmission gates Q1–Qn. All the NMOS transistors in this row decoder are formed in the p-well region and biased to an internal voltage V_{BB}. During the erase operation, V_{BB} is

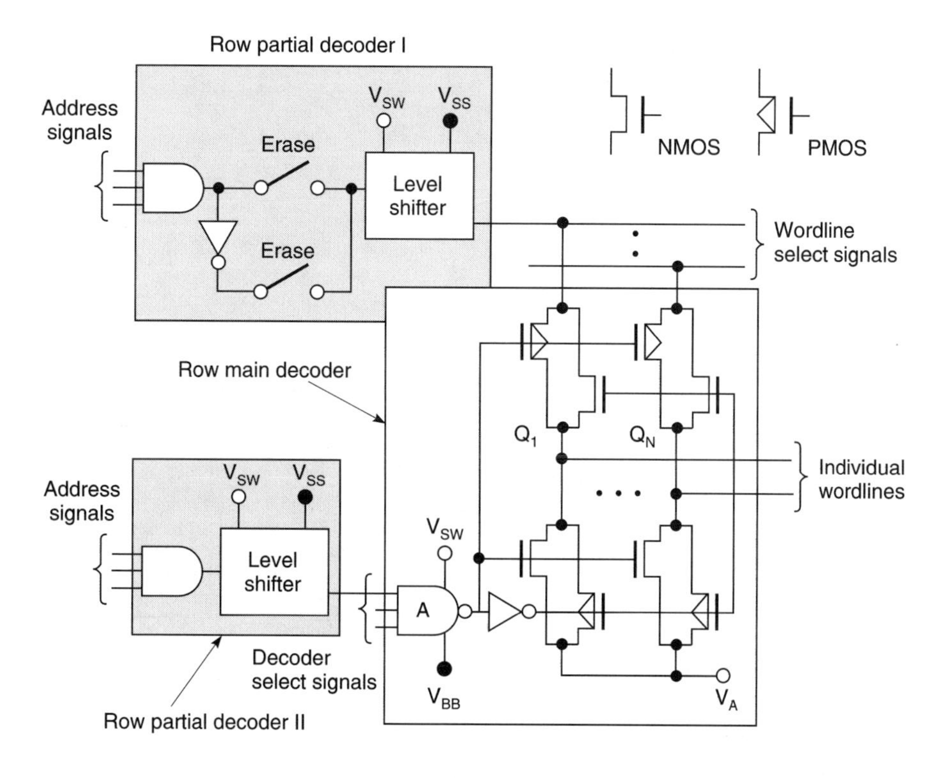

Figure 4.109 Row-decoder circuit in a triple-well technology [4.70].

set to a negative voltage, and during other operations V_{BB} is set to ground. In the selected row decoder, the individual wordlines are connected to the partial decoder's outputs. In the unselected row decoders, the individual wordlines are connected to the signals V_A. The V_A is set to a positive level, V_{SW}, during the erase operation to avoid erase-disturb problems [4.48] and is set to ground during other operations. A parallel-arranged NMOS and PMOS transmission gate scheme enables the application of any voltage, for both selected and unselected wordlines. Both the decoder select signals and wordline select signals are shifted to V_{SW}/V_{BB} in the partial decoders. During the erase operation, the wordline select signals are inverted in the partial decoder to apply a negative voltage to the selected wordline. A minimum sector erase operation, consisting of a single wordline, is possible in principle. A compact row decoder has been obtained by introducing the parallel arranged transmission gate scheme.

During the erase operation, the high-level V_{SW} is set to about 4 V and the low level V_{BB} is set to –7.5 V. The maximum voltage stress applied to the gate oxide during the erase operation is limited to less than that for the programming operation.

In order to accomplish a selective erase in a sector (one wordline) with negative-gate-bias channel erase, a row decoder [4.62], shown in Figs. 4.110 and 4.111, is used. Table 4.12 shows its operating voltages. This row decoder consists of two row-main

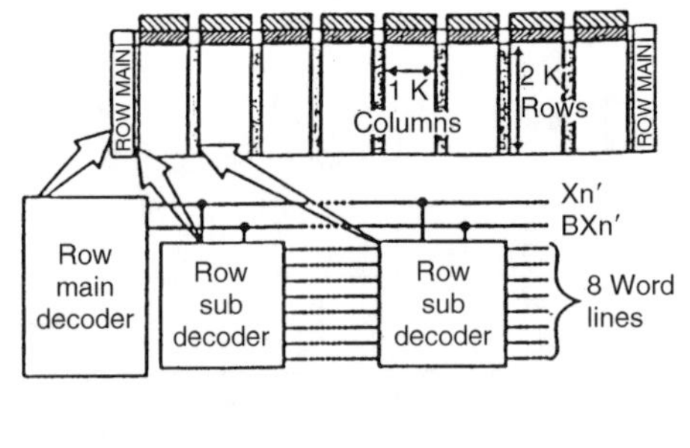

Figure 4.110 Block diagram of row decoder [4.62].

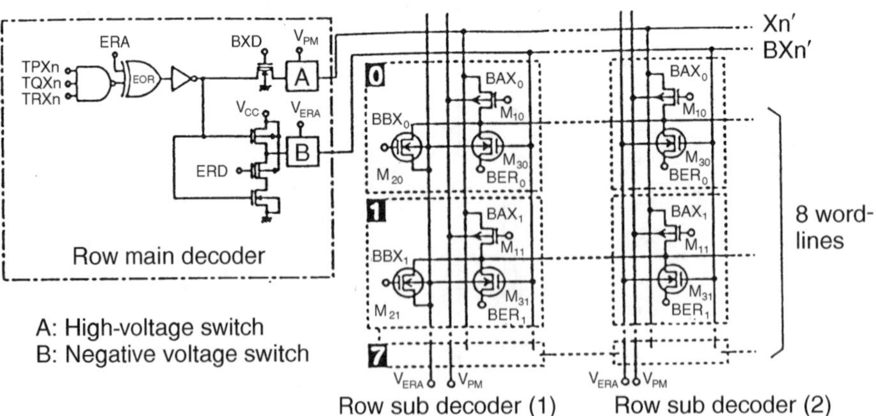

Figure 4.111 Circuit diagram of row decoder [4.62].

TABLE 4.12 ROW-DECODER OPERATING VOLTAGES

Signals	Read		Program		Erase	
	Select	Deselect	Select	Deselect	Select	Deselect
TPXn, TQXn, TRXn	V_{cc}	V_{ss}	V_{cc}	V_{ss}	V_{cc}	V_{ss}
BAXm	V_{ss}	V_{cc}	V_{ss}	12 V	V_{ss}	V_{ss}
BBXM	V_{ss}	V_{cc}	V_{ss}	V_{cc}	−13 V	−13 V
BERm	V_{ss}	V_{ss}	V_{ss}	V_{ss}	−13 V	V_{ss}
Xn^n	V_{cc}	V_{ss}	12 V	V_{ss}	V_{ss}	V_{cc}
BXn^n	V_{ss}	V_{cc}	V_{ss}	V_{cc}	V_{cc}	−13 V
Wordline	V_{cc}	V_{ss}	12 V	V_{SS}	−13 V	V_{cc} OR V_{ss}
ERA, ERD	V_{ss}		V_{ss}		V_{cc}	
BXD	V_{cc}		V_{cc}		V_{cc}	
V_{PM}	V_{cc}		12 V		V_{cc}	
V_{ERA}	V_{ss}		V_{ss}		−13 V	
p-Well for cell	V_{ss}		V_{ss}		V_{cc}	

decoders and nine row-sub decoders. The row-main decoders are placed on both sides of the cell array and select eight of 2,048 wordlines. Its outputs, Xn and BXn, are wired to the row-sub decoders using second metal interconnection. The row-sub decoders are placed on each side of the cell array blocks and select a single wordline of the eight wordlines. This row decoder organization has the advantage of reducing wordline delay and its occupied area.

In a row-main decoder, an NMOS depletion transistor for high-voltage isolation and a high-voltage switch (block A) are the same as those used in conventional EPROM. They are active during the programming mode. The negative voltage switch (Fig. 4.112) in block B becomes effective during the erase mode. The row-sub decoder (1) consists of one PMOS transistor, M1m, and two negative voltage NMOS transistors, M2m and M3m, as shown in Fig. 4.113. In the row-sub decoder (2), the negative voltage NMOS transistor M2m, which is used to discharge the wordline, is omitted in order to reduce the area occupied by the row-sub decoders.

During the reading operation, the high-voltage switch and the negative voltage switch are unavailable. For the selected row-main decoder, the outputs Xn″ (active high) and Bxn" (active low) become 5 and 0 V, respectively. At the same time, one of the PMOS transistors (M1m) in the row-sub decoder turns on by a row pre-decoder; thus, 5 V is applied to the target wordline.

During the programming operation, the gate signal of the NMOS depletion transistor (BXD) is set to 0 V and the high-voltage switch is active. The output Xn′ is converted to 12 V by the high-voltage switch and is applied to the target wordline through one of the PMOS transistors (M1m) in the row-sub decoder.

During the erasing operation, one input signal of the exclusive OR (EOR) circuit (ERA) is set to 5 V, and the negative voltage switch becomes effective. For

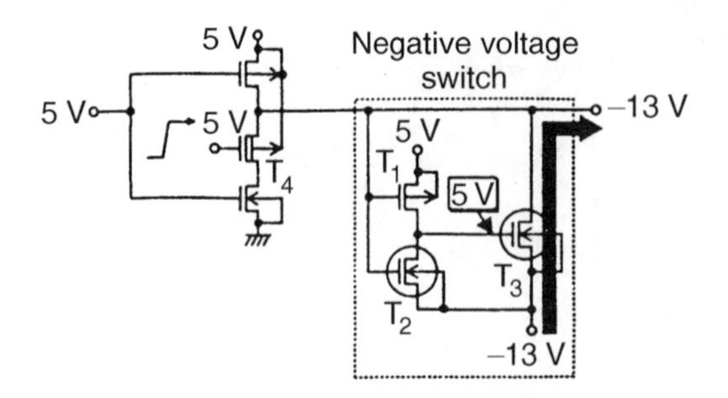

Figure 4.112 Circuit diagram of negative voltage switch [4.62].

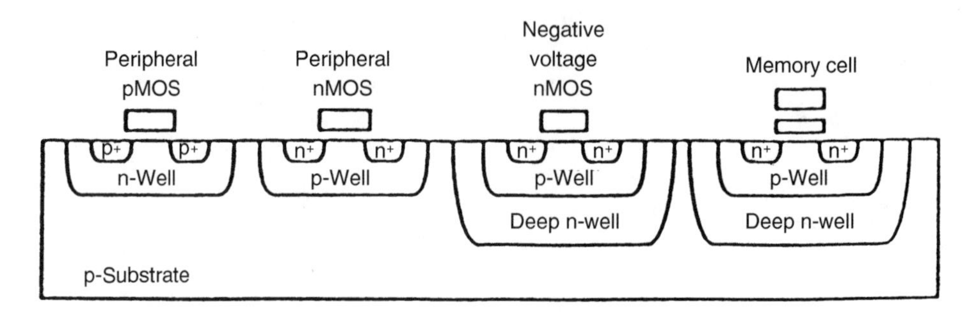

Figure 4.113 Schematic cross-sectional view of the triple-well CMOS structure.

the selected row-main decoder, the output of the EOR circuit becomes 5 V. Then, outputs Xn and BXn become 0 and 5 V, respectively. All of the eight NMOS transistors, M30–M37, in the row-sub decoder turn on because BXn is at 5 V. At the same time, the row pre-decoder supplies –13 V to one of the source signals (BERm) of the NMOS transistors (M3m), and seven others are at 0 V. Therefore, only the target wordline is charged to –13 V, while seven others are at 0 V. On the other hand, for the deselected row-main decoders, the outputs Xn and BXn become 5 V and Z – 13 V, respectively. Thus, the deselected wordlines wired to the deselected row-main decoders are 5 V through the PMOS transistors M10–M17, and any sector erase and block erase are possible by controlling the pre-decoding signals for the row decoder.

4.8.2 Erase Circuits

Electrical erase circuits are the new circuits required for Flash, compared to EPROM. As discussed previously, the exact circuitry depends on the cell type. For source erase, instead of connecting the source straps to ground, these are connected through a switching circuit that switches between ground for normal operation and V_{pp} for erase. For drain erase, the same switch circuits can be used to switch V_{pp}. Finally, for poly-to-poly erase, new circuits are incorporated into the X-decoders or Y-select.

4.8.3 Charge-Pump Circuits

Charge-pump circuits have been discussed in previous chapters. The concept of a quick boosting charge-pump circuit for high-density and low-voltage Flash memories [4.102] is shown in Fig. 4.114.

For the conventional charge-pump circuit, the number of stages connected in series between the power supply and the output of the charge-pump circuit is fixed. This fixed number of stages is designed so as to generate a certain high voltage required for programming. However, the charge-transfer efficiency, I_{PP}/I_{CC}, for the charge-pump circuit is given by $1/(n+1)$, where I_{CC} and I_{PP} are the mean input and output currents for the charge-pump circuit, respectively, and n is the number of stages connected in series between the power supply and the output of the charge-pump circuit. As a result, many stages connected in series are so redundant that the conventional charge-pump circuit not only consumes too much power, but also lowers the charge-transfer efficiency, while the boosted voltage is not much higher than the power supply voltage at the beginning of the operation. Thus, it takes a longer time for the conventional charge-pump circuit to generate a desired voltage for programming when a higher density EEPROM increases the load capacitance for the charge-pump circuit and the power supply voltage is lowered.

The charge-pump scheme conceptually presented in Fig. 4.114 can change the number of stages and the capacitance used for charge pumping. While the output voltage is not much higher than the power supply voltage in the beginning, the number of stages is controlled to be small and the capacitance for charge pumping is controlled to be large in order to increase the charge-transfer efficiency. Accordingly, as the output voltage is boosted, the number of stages is increased and the capacitance for charge pumping is decreased step by step. As a result, the charge-pump scheme shown in Fig. 4.114 offers a high charge-transfer efficiency and reduces the rising time of the output voltage.

The switches SW1 and SW2 are controlled by a set of complementary voltages of a high and low voltage converted by a level shifter circuit (LS).

4.9. FLASH APPLICATIONS

In this section, the current and potential applications of Flash memory are discussed and a brief discussion on DRAM versus Flash is presented.

4.9.1 EPROM Replacement

EPROM was important in the product development phase but was replaced by Mask ROM as the product went into volume production. On the other hand, with Flash, product designers and end-users can continue to update the product more easily and in situ. Another advantage of Flash over EPROM is a smaller footprint, thus saving board space. In embedded microcontroller applications, Flash is replacing EPROM, as well as EEPROM, Boot RAM, and battery RAM. Low-voltage

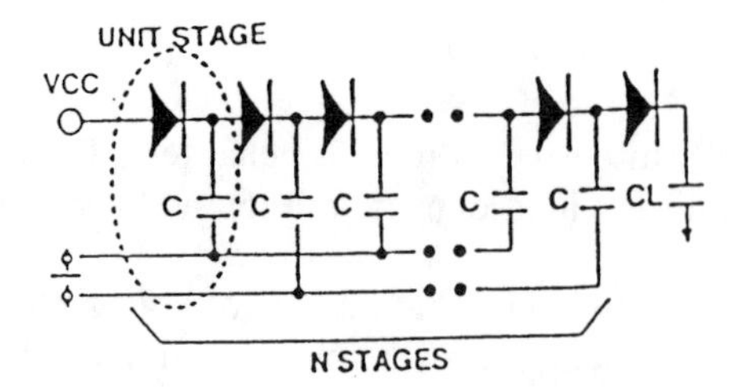

Conventional charge pump circuit.

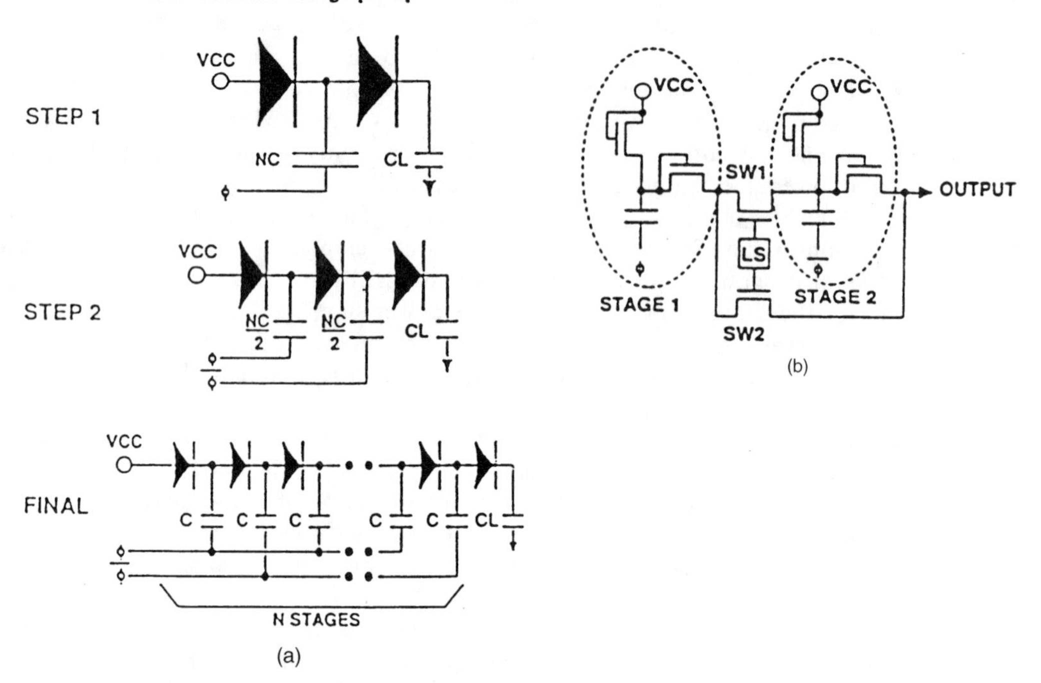

Figure 4.114 (*a*) Charge-pump scheme; (*b*) charge-pump circuit [4.102].

Flash is also expected to replace EPROMs in mobile computer microcontroller applications.

4.9.1.1 Code Storage. Except in cases where disk or tape memory is practical, most embedded computers store code in ROM. But as embedded systems have evolved from 8 bit single-chip controllers to 16 and 32 bit computers, the amount of code in a typical system has increased enormously. Along with the increase in code size, the need to provide code changes has also grown. Bigger programs contain more errors, and systems with more features tend to be upgraded more often.

When code is stored in Mask ROM or UV EPROM, each update can mean that vendors must visit the system, physically take it apart, remove ROMs, and replace them with new ones. Both the cost and the opportunity for damaging the system can be prohibitively large. Boards can be designed so that Flash EPROMs can be erased and reprogrammed from an edge connector, reducing the update problem to opening the cabinet, connecting the programmer, and waiting a few minutes. No physical disassembly of the system is necessary.

Boards can be designed to let the embedded central processing unit (CPU) erase and write the Flash parts in-circuit, under the control of the embedded CPU. The system contacts a remote update source. Then, the local CPU selectively erases blocks of Flash EPROM, writes updated code into them, and continues operating as modified. So, Flash offers a cost advantage over EEPROMs and in-system rewritability over UV EPROMs.

4.9.2 Automotive Applications

In-system reprogrammability makes Flash well suited to many automotive applications.

1. In-system code change for development and optimization without hardware changes.
2. Just-in-time manufacturing flexibility to select appropriate codes and programs for particular modules for vehicles.
3. Inventory control—Minimization of part numbers.
4. In-vehicle selection of features—engine, transmission, ride options.
5. Testability using a test bus and system service diary chips to store service history and codes.
6. Continuous system optimization through the vehicle life cycle by real-time system programmability; in-field repair/alteration, maintainability, and recall avoidance by in-vehicle alteration at the dealer.

4.9.3 Joint Testability Action Group (JTAG)

Another class of system applications for reprogrammable memory elements is storage of system intelligence—typically in a few bytes of EEPROM, distributed in various parts of the system. Using such an approach, for example, we can place system security codes, fault information, system service diaries, manufacturer's system IDs, code IDs for overall system testability, and, so on, throughout the system. The Joint Testability Action Group (JTAG) has, for example, defined a system bus that can include serial Flash EEPROM memory to store information related to board service history. As the electronic content within a future system expands, such testability features will take on additional urgency during development, manufacturing, and maintenance (i.e., the field service environment). The EEPROM element can also be used in applications for system identification and vehicle location (as on toll roads, parking lots, garages, etc.), and eventually will be used as part of intelligent sensors for remote diagnosis and servicing.

4.9.4 Cards

Until recently, Flash memory cards were used primarily in specialized, less cost-sensitive applications in which their ruggedness, removability, small size, low power consumption, and high performance were valued. For example, factory-floor equipment, robots, and industrial control systems use Flash memory cards to increase the flexibility and functionality of precise electromechanical systems. But lower memory media prices and new packaging technology have reduced costs and enhanced the density and functionality of Flash memory cards, making them attractive in mainstream portable equipment and mobile computers. Flash memory cards are now feasible in data-file and application-program storage or transfer.

In mobile designs, the nonvolatility, updateability, and removability of the Flash memory card makes it suitable for file exchange between systems. Solid-state memory access speeds increase system performance over that of rotating disk technology. In addition, the low power consumption of Flash memory enables systems to be developed that can run on common batteries. It is feasible to increase the system battery life of handheld computers using Flash memory storage.

Personal notebook and sub-notebook computers, with standard disk interfaces, can also derive benefit from Flash memory in a hard disk drive configuration. A Flash drive is defined as a mass storage device based on Flash memory technology and integrated with such industry-standard disk interfaces as integrated drive electronics (IDE), small computer systems interface (SCSI), or PCMCIA/AT attachment. A Flash drive differs from a Flash memory card in its system software and hardware requirements, as well as in its relative cost.

4.9.4.1 Smart Cards. Smart cards consist of rewritable, nonvolatile memory and a microprocessor. The microprocessor makes the card "smart." The microprocessor controls the communication between the system and memory, allowing multiple applications, and provides security via interpretation of personal ID number [PIN] codes. Power is supplied to the card externally.

Smart cards are carried in purses or wallets and have many potential applications, such as financial transactions, pay phones, employee ID cards, authentication, and storage of personal information. The smart card concept can utilize high-density Flash in supermarket cards, current credit cards, driver's licenses, access control devices, and car maintenance history, to name a few.

4.9.5 Look-Up Tables/Data Acquisition

Flash can be used in many other applications, including field updateable look-up tables, capture and recording of important data, updateable firmware as BIOS, and the more sophisticated disk drive-like solid-state memory subsystems that have attained sizes up to 200 MB.

Another application, data acquisition, takes full advantage of Flash memory's byte–write capability and full nonvolatility. Any kind of data logger or sensor that depends on some form of battery-backed RAM can benefit from using Flash memory. Flash memory offers cost savings, higher density, and critical data security by eliminating the battery for retaining memory contents.

4.9.6 Personal Systems

Notebooks, laptops, and other portable PCs, such as the luggable, which are light enough to be portable are called portable systems. On the other hand, a mobile computer is a portable computer that can be operated while in motion, for example, a handheld data acquisition computer, a palm-type scheduler, or palm-type and/or pen-based personal computing systems. Flash will be a preferred solid-state disk in mobile applications in the form of PCMCIA cards, for example.

4.9.7 Analog Applications

Flash memories, based on Fowler–Nordheim programming (as in EEPROMs) can be used in mixed-signal chips to store calibration data. Laser trimming for precision can be replaced by Flash trimming. The Flash cells would drive small, precise D/A converters that would control offset, filter parameters, and the like. One application could be in equalization filters for disk-read channels and communication circuits.

Analog memories are also possible candidates for Flash application. Analog memories provide direct audio record/playback without digital conversion. For example, ISD uses 2 bits/cell EEPROM and combines these with audio microphones, speakers, amplifiers, and other components [4.145, 4.161].

4.9.8 Logic

While the advantages of high density Flash in programmable logic are evolving, low to moderate density PLDs have been implemented in Flash. Even PCMCIA designs can almost be debugged in real time using Flash reprogrammability in combination with field programmable gate array (FPGA) technology. PCMCIA I/O cards often use a combination of chipsets, glue-logic, and memory to accomplish their function. In many cases, the glue-logic portion of the design is implemented using programmable logic, such as FPGAs. The FPGA configuration memory can be stored in Flash memory on the card and modified if needed [4.146].

4.9.9 New Architectures

Mobile PC and communications manufacturers are discovering new ways to exploit Flash memory's capabilities in their system designs. Placing applications on the motherboard in a resident Flash array configuration, directly connected to the processor, dramatically speeds up system performance. Frequently used software becomes immediately available, eliminating the lengthy download step required by RAM/disk-based system architectures. A resident Flash disk, acting as a nonvolatile RAM drive, emulates disk drive functionality with no spin-up time or seek/rotation delay. Finally, in systems designed with advanced power management, the contents of system DRAM can be stored in resident Flash memory during the system sleep mode for longer battery life (DRAM refresh can be disabled) and fast system wake-up.

Using Flash memory, a new system architecture is possible (Fig. 4.115). Flash memory can replace system RAM in the RAM's traditional role of direct execute code storage, while some RAM is retained for data manipulation and storage. This allows a reduction in RAM content, with no impact on the total system's memory budget.

For small form-factor systems, converting the application portion of system RAM to nonvolatile memory reduces overall system cost and size, since a disk drive is no longer needed for nonvolatile application storage. The resident application can be upgraded by means of an interface port or Flash memory plug-in cards.

The recent introduction of low-voltage Flash memories with synchronous or DRAM-like host interfaces can replace code-RAMs and nonvolatile code storage in 16 or 32 bit microprocessors or microcontroller-based systems [4.149].

4.9.10 DRAM Versus Flash

Modern DRAMs are based on a 1-transistor MOSFET charge storage cell. The cell consists of a transistor and a capacitor. The information is stored on the capacitor, in the form of charge, and must be refreshed after every read. On the other hand, a Flash memory cell is a single MOS transistor consisting of a double poly-stack structure, and the stored information is nonvolatile, in contrast to DRAM.

The DRAM cell is about one and a half times larger than the 1-T Flash memory cell. Thus, Flash chips are smaller than DRAM memory chips of comparable feature size and density. Furthermore, it is anticipated that, as the manufacturing volume of Flash increases, the cost of Flash memory will become lower than the cost of DRAM (Fig. 4.116).

DRAM and Flash cells have different limitations in scaling (Table 4.13). For DRAM, because of soft errors, the capacitance must remain relatively constant for each successive generation. To maintain the constant capacitance for a small area, 3-D structures, such as trenches or stacked capacitors, are required, complicating the manufacturing process. For Flash, the scaling limitation is the high voltage ($\sim$10 V minimum) that is required to be applied to the gate for channel hot electron or $>$ 10 V for the Fowler–Nordheim tunneling process. This means that the technology has to support a relatively high voltage.

The performance of first-generation 16 Mbit NOR Flash memories is in the same range as 16 Mbit DRAMs. Recent Flash memories, with synchronous or DRAM-like host interfaces, provide DRAM-like data accesses [4.149] and can replace shadow-RAM in 16 and 32 bit microprocessor or microcontroller-based systems. In general, the rate of acceptance of and demand for Flash memory for varied and sundry applications has been much greater than expected, thanks to the marketing efforts of major manufacturers.

At present, most of the major DRAM manufacturers and other silicon suppliers are pursuing the development of Flash memories. According to various projections,

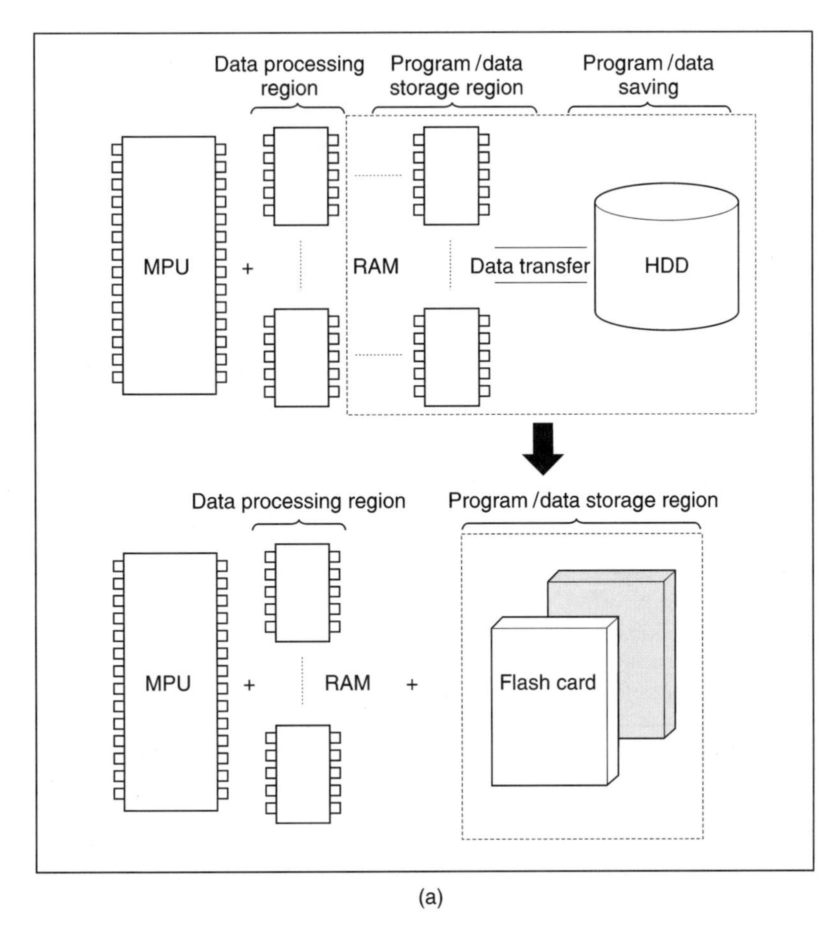

(a)

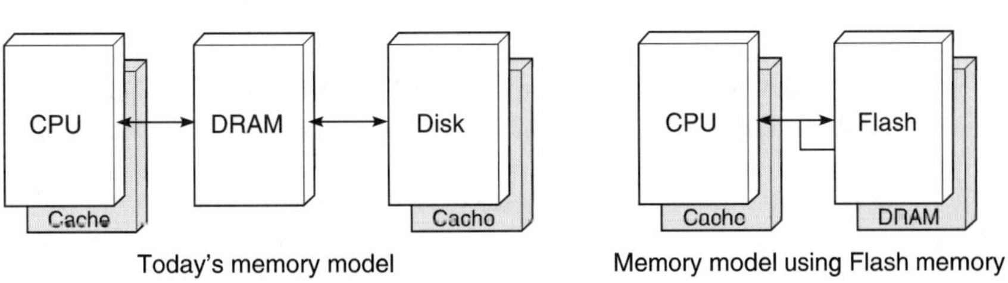

CPU = central processing unit; Dram = dynamic RAM.

(b)

Figure 4.115 Flash memory's nonvolatility and updateability displace the combination of dynamic RAM and hard disk drive used for code storage, creating a new memory architecture (*a*) [4.148] and (*b*) [4.147].

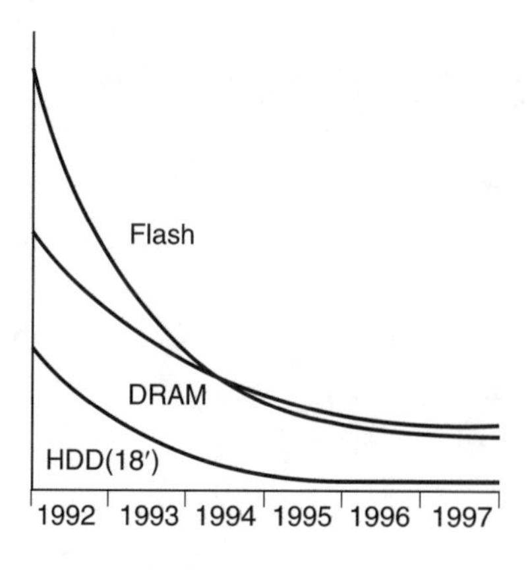

Figure 4.116 Projected prices of Flash, DRAM, and HDD [4.148].

TABLE 4.13 SCALING RULES FOR CONSTANT-FIELD SCALING

	Scaling Factor
Surface dimension	1/K
Vertical dimensions [T_{ox}, X_j]	1/K
Impurity concentration	K
Currents/voltages	1/K
Current density	K
Capacitance/area	K
Transconductance	1
Circuit delay time	1/K
Power dissipation	$1/K^2$
Power density	1
Power-delay product	$1/K^3$

Flash memory will become cheaper than DRAMs sometime between 1995 and 2000. This will further fuel the growth of Flash memory and help make it a commodity memory.

Flash memory cannot compete with DRAMs in applications where bits or bytes must be altered individually or frequently. Flash's slow write speed and limited number of rewrite cycles are serious disadvantages, compared to DRAMs, for many applications.

4.10. CONCLUSIONS AND A LOOK INTO THE FUTURE

Since the emergence of a 256 Kbit Flash chip in 1987 [4.3, 4.9], Flash memory has rapidly evolved into a high-density memory (Table 4.14). Today, several makers offer 16 Mbit Flash. As of 1995, 32 Mbit Flash products and cell technologies up to 256 Mbit had been presented at the IEDM-94 and ISSCC-95.

TABLE 4.14 FLASH MEMORY EVOLUTION AND PROJECTIONS

Year	1987	1991	1993	1994	1998 (Proj.)	2000 (Proj.)
Density (bits)	256 K	1 M	4 M	16 M	64 M	256 M
$V_{CC}(V)//V_{PP}(V)$	5/12	5/12	5/12	3/12	—	—
	—	—	5/5	5/5	5/5	—
	—	—		3/5	3/5	—
	—	—	—	—	3/3	3/3
	—	—	—	—	—	2/3
	—	—	—	—	—	2/2
Access time (ns)	150	150	120	60	50	50
Erase block size	Chip	Chip	16 KB	512 bytes	——	⟶
Erase time (ms)	10^3	10^3	—	10 ms/sector	——	⟶
Program time [μs/byte]	—	25	—	2.5	2.5	0.25
Write/erase cycles	10^2	10^3–10^4	10^4	10^5	10^6	10^7
Major density break-through	—	—	—	—	—	Multistate cell memory
Major application	EROM replacement	EROM replacement	Memory card	PC	Handheld and mobil system	Consumer electronics

Flash memory development work by DRAM manufacturers in the United States, Japan, Korea, and Europe continues. Many alliances have been formed between memory makers, original equipment manufacturer (OEM), and system houses.

Sub-notebook computers, palmtop and pen computers, personal digital assistants (PDAs), cellular phones, and other mobile devices are gaining popularity. It is anticipated that the personal/mobile market will explode as computers become smaller, lighter, more rugged, and less expensive. Flash memory is expected to be the optimum choice for mass storage in personal mobile systems. In mobile and handheld computers, for example, Flash will replace DRAM in order to contain operating systems, BIOS, and applications software.

Because of Flash's attributes, this new market segment will prefer it, while hard disk and floppy disk will remain the medium of choice in most nonportable applications. Although the innovative designs continue to shrink the Flash cell size (e.g., 0.4 μm^2 cell by Hitachi @ 0.25 μm rules (4.151) and 0.67 μm^2 cell by Toshiba (4.152) for 256 Mbit Flash memory), the cost/bit of hard disk continues to improve unabated [4.153]. Therefore, it is becoming clear that Flash cost/bit may overtake hard disk for high-density systems in spite of the many projections over the last few years [4.13, 4.57, 4.58, 4.68, 4.154]. In addition to cost, Flash memory must match the hard disk in reliability (write/erase endurance and data retention) and programming speed to replace it.

At the present time, Flash has already displaced EPROM in many applications, and Flash memory will be the technology of choice in many portable applications in the future. The Flash memory will replace hard disks in low-end applications (1 MB–100 MB), but will not be able to replace hard disks in most applications, primarily because of cost. Solid-state disks and hard disks will be complementary technologies.

DRAM, which has been the fuel for LSI/VLSI/ULSI, is intrinsically more expensive per bit because of its relatively larger cell size and more complex processing, compared to Flash. Flash memory speeds are in the same range as DRAM, and with a DRAM-like interface [4.149], Flash memory can perform many of DRAM's functions [4.145, 4.147, 4.153]. On the other hand, research efforts are underway to improve the cost and nonvolatility of DRAM, for example, NAND-structured DRAM cells to reduce the cell size [4.155] and a nonvolatile RAM (NVRAM) using ferroelectric materials [4.156].

So the semicondutor memory revolution continues unabated.

4.10.1 Flash Market Development Trends

- PC BIOS
- PC operating system and application storage
- Portable computer main memory and mass storage
- Cost/MB crossover with DRAM
- Dominance in mobile disk drive applications
- Faster read/write, lower power
- Application specific Flash
- Commodity memory

4.11. ACKNOWLEDGEMENTS

Flash memory technology and applications continue to explode at an ever increasing rate. Thanks are due to many colleagues in the industry who have contributed the material for this chapter. This list includes, but is not limited to, Dave Sweetman of SST, Kuniyoshi Yoshikawa of Toshiba, Greg Atwood of Intel, Chi Chang of AMD, and Rob Frizzell of National Semiconductor.

References

[4.1] K. Sakui and F. Masouka, "Sub-half micron Flash memory technologies," *IEICE Trans. Electron.*, vol. E-77, no. 8, pp. 1251–1259, 1994.

[4.2] B. Dipert and M. Levy, "Designing with Flash memory," Published by Annabooks, 1993.

[4.3] F. Masuoka, M. Assano, H. Iwahashi, T. Komuro, and S. Tanaka, "A new Flash EEPROM cell using triple polysilicon technology," *IEEE IEDM Tech. Dig.*, pp. 464–467, 1984.

[4.4] F. Masuoka et al., "A 256K Flash EEPROM using triple polysilicon technology," *IEEE ISSCC Dig.* Tech. Pap., pp. 168–169, 1985.

[4.5] F. Masuoka et al., "A 256K Flash EEPROM using triple polysilicon technology," *IEEE J. Sol. St. Cir.*, vol. SC-22, pp. 548–552, 1987.

[4.6] The following is an example from a large literature base on EPROMs:
M. Van Buskirk et al., " EPROMs graduate to 256-k density with scaled n-channel process," *Electronics*, pp. 89–93, February 24, 1983.
J. J. Barnes et al., "Operation and characterization of N-channel EPROM cells," *Solid State Electr.*, vol. 22, pp. 521–529, 1978.
M. Wada et al., "Limiting factors for programming EPROM of reduced dimensions," *IEEE IEDM Tech. Dig.*, pp. 38–41, 1980.
N. R. Mielke, "New EPROM data loss mechanisms," *Proc. IRPS*, pp. 106–113, 1983.

[4.7] S. Mukherjee, T. Chang, R. Pan, M. Knecht, and D. Hu, "A single transistor EEPROM cell and its implementation in a 512K CMOS EEPROM," *IEEE IEDM Tech. Dig.*, pp. 616–619, 1985.

[4.8] A. Baker et al., "A 3.3V 16Mb Flash memory with advanced write automation," *IEEE ISSCC Dig.* Tech. Pap., pp. 146–147, 1994.

[4.9] V. N. Kynett, A. Baker, M. Fandrich, G. Hoekstra, O. Jungroth, J. Kreifels, and S. Wells, "An in-system reprogrammable 256K CMOS Flash memory," Proc. *IEEE ISSCC Dig.* Tech. Pap., pp. 132–133, 1988.

[4.10] G. Samachisa, C. S. Su, Y. S. Kao, G. Smarandoiu, T. Wong, and C. Hu, "A 128K Flash EEPROM using double polysilicon technology," *IEEE ISSCC Dig.* Tech. Pap., pp. 76–77, 1987.

[4.11] G. Samachisa, C. S. Su, Y. S. Kao, G. Smarandoiu, C. Y. M. Wang, T. Wong, and C. Hu, "A 128K Flash EEPROM using double polysilicon technology," *IEEE J. Solid-State Circuits*, vol. SC-22, no. 10, pp. 676–683, 1987.

[4.12] S. Haddad, C. Chang, B. Swaminathan, and J. Lien, "Degradation due to hole trapping in Flash memory cells," *IEEE Elect. Dev. Lett.*, vol. EDL-10, no. 3, pp. 117–119, 1989.
S. Haddad et al., "An investigation of erase-mode dependent hole trapping in Flash EEPROM memory cell," *IEEE Elect. Dev. Lett.*, pp. 514–516, 1990.

[4.13] F. Masuoka, M. Momodomi, Y. Iwata, and R. Shirola, "New ultra high density EPROM and Flash EEPROM cell with NAND structure cell," *IEEE IEDM Tech. Dig.*, pp. 552–555, 1987.

[4.14] S. Aritome et al., "Reliability issues of Flash memory cell," *Proc. IEEE*, vol. 81, no. 5, pp. 776–788, 1993.

[4.15] Y. Iwata et al., "The internal voltage system of the 32 Mbit NAND Flash EEPROM for low-voltage power supply," *Proc. IEEE Nonvolatile Semiconductor Memory Workshop*, Paper #4-2, 1995.

[4.16] K. D. Suh et al., "A 3.3V 32 Mb NAND Flash memory with incremental step pulse programming scheme," *IEEE ISSCC Dig.* Tech. Pap., pp. 128–129, 1995.

[4.17] H. Kume, H. Yamamoto, T. Adachi, T. Hagiwara, K. Komori, T. Nishimoto, A. Koike, S. Meguro, T. Hayashida, and T. Tsukada, "A Flash-erase EEPROM cell with an asymmetric source and drain structure," *IEEE IEDM Tech. Dig.*, pp. 560–563, 1987.

[4.18] V. N. Kynett, A. Baker, M. Fandrich, G. Hoekstra, O. Jungroth, J. Kreifels, S. Wells, and M. Winston, "An in-system reprogrammable 32Kx8 CMOS Flash memory." *IEEE J. Solid-State Circuits*, vol. SC-23, no. 10, pp. 1157–1162, 1988.

[4.19] R. Shirota,Y. Itoh, R. Nakayama, M. Momodomi, S. Inoue, R. Kirisawa, Y. Iwata, M. Chiba, and F. Masuoka, "A new NAND cell for ultra high density 5V-only EEPROMs," Symp. VLSI Tech., pp. 33–34, 1988.

[4.20] S. Tam, S. Sachdev, M. Chi, G. Verma, L. Ziller, G. Tsau, S. Lai, and V. Dham, "A high density CMOS 1-T electrically-erasable non-volatile (Flash) memory technology," Symp. VLSI Tech., pp. 31–32, 1988.

[4.21] G. Verma and N. Mielke, "Reliability performance of ETOX-based Flash memories," *Proc. IRPS*, pp. 158–166, 1988.

[4.22] M. Momodomi, R. Kirisawa, R. Nakayama, S. Aritome, T. Endoh, Y. Itoh, Y. Iwala, H. Oodaira, T. Tanaka, M. Chiba, R. Shirota, and F. Masuoka, "New device technologies for 5V only 4 Mb EEPROM with NAND structure cell," *IEEE IEDM Tech. Dig.*, pp. 412–415, 1988.

[4.23] R. Kazerounian, S. Ali, Y. Ma, and B. Eitan, "A 5 volt high density poly-poly erase Flash EPROM cell," *IEEE IEDM Tech. Dig.*, pp. 436–439, 1988.

[4.24] M. Gill, R. Cleavelin, S. Lin, S. D.'Arrigo, G. Santin, P. Shah, A. Nguyen, J. Esquivel, B. Riemenschneider, and J. Paterson, "A 5-volt contactless array 256K Flash EEPROM technology," *IEEE IEDM Tech. Dig.*, pp. 428–431, 1988.

[4.25] S. D'Arrigo, G. Imondi, G. Santin, M. Gill, R. Cleavelin, S. Spagliccia, E. Tomasetti, S. Lin, A. Nguyen, P. Shah, G. Savarese, and D. McElroy, "A 5V-only 256K bit CMOS Flash EEPROM," *IEEE ISSCC Dig.* Tech. Pap., pp. 132–133, 1989.

[4.26] M. Momodomi, Y. Itoh, R. Shirota, Y. Iwata, R. Nakayama, R. Kirisawa, T. Tanaka, S. Aritome, T. Endoh, K. Ohuchi, and F. Masuoka, "An experimental 4-Mbit CMOS EEPROM with a NAND structure cell," *IEEE J. Solid-State Circuits*, vol. SC-24, no. 10, pp. 1238–1243, 1989.

[4.27] V. N. Kynett, M. Fandrich, J. Anderson, P. Dix, O. Jungroth, J. Kreifels, R. A. Lodenquai, B. Vajdic, S. Wells, M. Winston, and L. Tang, "A 90-ns one-million erase/program cycle 1-Mbit Flash memory," *IEEE J. Solid-State Circuits*, vol. SC-24, no. 10, pp. 1259–1264, 1989.

V. N. Kynett et al., "A 90 ns 100K erase/program cycle mega bit Flash memory," *IEEE ISSCC Dig.* Tech. Pap., pp. 140–141, 1989.

[4.28] K. Naruke, S. Yamada, E. Obi, S. Taguchi, and M. Wada, "A new Flash erase EEPROM cell with sidewall select-gate on its source side," *IEEE IEDM Tech. Dig.*, pp. 603–606, 1989.

[4.29] M. Gill, R. Cleavelin, S. Lin, S. D'Arrigo, G. Santin, P. Shah, A. Nguyen, R. Lahily, P. DeSimone, G. Piva, and J. Paterson, "A 5-volt only Flash EEPROM technology for high density memory and system IC applications," IEEE CICC, pp. 1841–1844, 1989.

[4.30] T. Endoh, R. Shirota, Y. Tanaka, R. Nakayama, R. Kirisawa, S. Aritome, and F. Masuoka, "New design technology for EEPROM memory cells with

10 million write/erase cycling endurance," *IEEE IEDM Tech. Dig.*, pp. 599–602, 1989.

[4.31] S. Aritome, R. Kirisawa, T. Endoh, N. Nakayama, R. Shirota, K. Sakui, K. Ohuchi, and F. Masuoka, "Extended data retention characteristics after more than 10^4 write and erase cycles in EEPROMs," *Proc. IRPS*, pp. 259–264, 1990.

[4.32] Y. Iwata, M. Momodomi, T. Tanaka, H. Oodaira, Y. Itoh, R. Nakayialtla, R. Kirhiawa, S. Aritome, T. Endoh, R. Shirota, K. Ohuchi, and F. Masuoka, "A high-density NAND EEPROM with block-page programming for microcomputer applications," *IEEE J. Solid-State Circuits*, vol. SC-25, no. 4, pp. 417–424, 1990.

[4.33] B. Riemenschneider, A. L. Esquivel, J. Paterson, M. Gill, S. Lin, D. McElroy, P. Truong, R. Bussey, B. Ashmore, M. McConnell, H. Stiegler, and P. Shah, "A process technology for a 5-volt only 4 Mb Flash EEPROM with an 8.6 μm^2 cell," IEEE Symp. VLSI Tech., pp. 125–126, 1990.

[4.34] R. Kirisawa, S. Arilome, R. Nakayama, T. Endoh, R. Shirota, and F. Masuoka, "A NAND structured cell with a new programming technology for highly reliable 5V-only Flash EEPROM," Symp. VLSI Tech., pp. 129–130, 1990.

[4.35] B. J. Woo, T. C. Ong, A. Fazio, C. Park, G. Atwood, M. Holler, S. Tam, and S. Lai, "A novel memory cell using Flash array contactless EPROM (FACE) technology," *IEEE IEDM Tech. Dig.*, pp. 91–94, 1990.

[4.36] N. Ajika, M. Ohi, H. Arima, T. Malsukawa, and N. Tsubouchi, "A 5 volt only 16 Mbit Flash EEPROM cell with a simple stacked gate structure," *IEEE IEDM Tech. Dig.*, pp. 115–118, 1990.

[4.37] M. Gill, R. Cleavelin, S. Lin, M. Middendorf, A. Nguyen, L. Wong, B. Huber, S. D.'Arrigo, P. Shah, E. Kougianos, P. Hefley, G. Satin, and G. Naso, "A novel sublithographic tunnel diode based 5-volt Fash memory," *IEEE IEDM Tech. Dig.*, pp. 119–122, 1990.

[4.38] S. Aritome, S. Shirota, R. Kirisawa, T. Endoh, N. Nakayama, K. Sakui, and F. Masuoka, "A reliable bi-polarity write/erase technology in Flash EEPROMs," *IEEE IEDM Tech. Dig.*, pp. 111–114, 1990.

[4.39] Y. Miyawaki et al., "A new erasing and row decoding scheme for low supply voltage operation 16 Mb/64 Mb Flash EEPROMS," Symp. VLSI Tech., pp. 85–86, 1990.

[4.40] T. Tanaka et al., "A 4-Mbit NAND-EEPROM with tight programmed Vt distribution," Symp. VLSI Circuits, pp. 105–106, 1990.

[4.41] Y. Yamauchi et al., "A 5V-only virtual ground Flash cell with an auxiliary gate for high density and high speed applications," *IEEE IEDM Tech. Dig.*, pp. 319–322, 1991.

[4.42] M. Momodomi et al., "A 4-Mb NAND EEPROM with tight programmed Vt distribution," *IEEE J. Solid-State Circuits*, vol. SC-26, no. 4, pp. 492–496, 1991.

[4.43] M. McConnell et al., "An experimental 4-Mb Flash EEPROM with sector erase," *IEEE J. Solid-State Circuits*, vol. SC-26, no. 4, pp. 484–491, 1991.

[4.44] T. Nakayama et al., "A 60ns 16Mb Flash EEPROM with program and erase sequence controller," *IEEE J. Solid-State Circuits*, vol. SC-26, no. 11, pp. 1600–1604, 1991.

[4.45] C. Kuo et al., "A 512Kb Flash EEPROM for a 32 bit microcontroller," Symp. VLSI Circuits, pp. 87–88, 1991.

[4.46] B. J. Woo, T. C. Ong, and S. Lai, "A poly-buffered FACE technology for high density Flash memories," Symp. VLSI Tech., pp. 73–74, 1991.

[4.47] N. Kodama, K. Saitoh, H. Shirai, T. Okazaki, and Y. Hokari, "A 5V only 16 Mbit Flash EEPROM cell using highly reliable write/erase technologies," Symp. VLSI Tech., pp. 75–76, 1991.

[4.48] H. Kume, T. Tanaka, T. Adachi, N. Miyamoto, S. Saeki, Y. Ohji, M. Ushiyama, T. Kobayashi, T. Nishida, Y. Kawamoto, and K. Seiki, "A 3.42 μm^2 Flash memory cell technology comformable to sector erase," Symp. VLSI Tech., pp. 77–78, 1991.

[4.49] K. Yoshikawa, S. Mori, E. Sakagami, N. Arai, Y. Kaneko, and Y. Ohshima, "Flash EEPROM cell scaling based on tunnel oxide thinning limitations," Symp. VLSI Tech., pp. 79–80, 1991.

[4.50] S. Yamada, T. Suzuki, E. Obi, M. Oshikiri, K. Naruke, and M. Wada, "A self-convergence erasing scheme for a simple stacked gate Flash EEPROM," *IEEE IEDM Tech. Dig.*, pp. 307–310, 1991.

[4.51] T. Hori et al., "A MOSFET with Si-implanted gate-SiO_2 insulator for non-volatile memory applications," *IEEE IEDM Tech. Dig.*, pp. 17.7.1–4, 1992.

[4.52] S. Koyama, "A novel cell structure for giga-bit EPROMs and Flash memories using polysilicon thin film transistors," Symp. VLSI Tech., pp. 44–45, 1992.

[4.53] K. Oyama et al., "A novel erasing technology for 3.3V Flash memory with 64Mb capacity and beyond," *IEEE IEDM Tech. Dig.*, pp. 607–610, 1992.

[4.54] T. Endoh et al., "New write/erase operation technology for Flash EEPROM cells to improve the read disturb characteristics," *IEEE IEDM Tech. Dig.*, pp. 603–606, 1992.

[4.55] H. Fukuda et al., "Heavy oxynitridation technology for forming highly reliable Flash-type EEPROM tunnel oxide films," Electr. Letts., vol. 28, no. 19, pp. 1781–1783, 1992.

[4.56] H. Fukuda et al., "High performance scaled Flash-type EEPROMs with heavily oxynitrided tunnel oxide films," *IEEE IEDM Tech. Dig.*, pp. 465–468, 1992.

[4.57] H. Kume et al., "A 1.28 μm^2 contactless memory cell technology for a 3V only 64Mb EEPROM," *IEEE IEDM Tech. Dig.*, pp. 991–993, 1992.

[4.58] H. Onoda et al., "A novel cell structure suitable for a 3V operation, sector erase Flash memory," *IEEE IEDM Tech. Dig.*, pp. 599–602, 1992.

[4.59] S. Keeney et al., "Simulation of enhanced injection split gate Flash EEPROM device programming," *Microelectronic Engineering*, vol. 18, pp. 253–258, 1992.

[4.60] C. Kuo et al., "A 512Kb Flash EEPROM embedded in a 32b microcontroller," *IEEE J. of Solid State Circuits*, vol. SC-27, pp. 574–582, 1992.

[4.61] A. Umezawa et al., "A 5V only operation 0.6 μm Flash EEPROM with row decoder scheme in triple-well structure," *IEEE J. Solid State Circuits*, vol. SC-27, no. 11, pp. 1540–1546, 1992.

[4.62] T. Jinbo et al., "A 5V only 16Mb Flash memory with sector erase mode," *IEEE J. of Solid State Circuits*, vol. SC-27, no. 11, pp. 1547–1554, 1992.

[4.63] Mehrotra et al., "Serial 9 Mb Flash EEPROM for solid state disk applications," Symp. VLSI Cir., pp. 24–25, 1992.

[4.64] M. Ohi et al., "An asymmetrical offset source/drain structure for virtual ground array Flash memory with DINOR operation," VLSI Technology Symp., pp. 57–58, 1993.

[4.65] A. Bergemont et al., "NOR virtual ground (NVG) — A new scaling concept for very high density Flash EEPROM and its implementation in a 0.5 μm process," *IEEE IEDM Tech. Dig.*, pp. 15–18, 1993.

[4.66] A. Bergemont, "Status, trends, comparison and evolution of FLASH technology for memory cards applications," IC Card Conf. Proc., pp. 63–71, 1993.

[4.67] H. B. Pein and J. D. Plummer, "Performance of the 3-D sidewall Flash EEPROM cell," *IEEE IEDM Tech. Dig.*, pp. 11–14, 1993.

[4.68] Y. S. Hisamune et al., "A high capacitive coupling ratio (HiCR) cell for 3V only 64 Mbit and future Flash memories," *IEEE IEDM, Tech. Dig.*, pp. 19–22, 1993.

[4.69] S. Kobayashi et al., "Memory array architecture and decoding scheme for 3V-only sector erasable DINOR Flash memory," *IEEE J. of Solid State Circuits*, vol. SC-29, no. 4, pp. 454–460, 1994.

[4.70] S. Atsumi et al., "A 16-Mb Flash EEPROM with a new self data refresh scheme for a sector erase operation," *IEEE J. of Solid State Circuits*, vol. SC-29, no. 4, pp. 461–469, 1994.

[4.71] T. Takeshima et al., "A 3.3V single-power-supply 64Mb Flash memory with dynamic bit-line latch (DBL) programming scheme," *IEEE ISSCC Dig.* Tech. Pap., pp. 148–149, 1994.

[4.72] M. Mihara et al., "Row-redundancy scheme for high-density Flash memory," *IEEE ISSCC Dig.* Tech. Pap., pp. 150–151, 1994.

[4.73] T. Tanaka et al., "High speed programming and program-verify methods suitable for low voltage Flash memories," Symp. VLSI Cir., pp. 61–62, 1994.

[4.74] D. J. Lee et al., "An 18 Mb serial Flash EEPROM for solid-state disk applications," Symp. VLSI Cir., pp. 59–60, 1994.

[4.75] Y. Miyawaki et al., "An over-erasure detection technique for tightening Vth distribution for low voltage operation NOR type Flash memory," Symp. VLSI Cir., pp. 63–64, 1994.

[4.76] T. Tanzawa et al., "A quick boosting charge pump circuit for high density and low voltage Flash memories," Symp. VLSI Cir., pp. 65–66, 1994.

[4.77] H. Watanabe et al., "Scaling of tunnel oxide thickness for Flash EEPROMS realizing stress-induced leakage current reduction," Symp. VLSI Tech., pp. 47–48, 1994.

[4.78] Y. Ma et al., "A novel high density contactless Flash memory array using split-gate source-side-injection cell for 5V-only applications," Symp. VLSI Tech., pp. 49–50, 1994.

[4.79] S. Mori et al., "High speed sub-half micron Flash memory technology with simple stacked gate structure cell," Symp. VLSI Tech., pp. 53–54, 1994.

[4.80] S. Sato et al., "An ultra-thin fully depleted floating gate technology for 64 Mb Flash and beyond," Symp. VLSI Tech., pp. 65–66, 1994.

[4.81] H. Oda et al., "New buried channel Flash memory cell with symmetrical source/drain structure for 64-Mbit or beyond," Symp. VLSI Tech., pp. 69–70, 1994.

[4.82] S. Kianian et al., "A novel 3 volts-only, small sector erase, high density Flash E^2PROM," Symp. VLSI Tech., pp. 71–72, 1994.

[4.83] D. Kuo et al., "TEFET—A high density, low erase voltage, trench Flash EEPROM," Symp. VLSI Tech., pp. 51–52, 1994.

[4.84] J. Chen, T. Y. Chan, I. C. Chen, P. K. Ko, and C. Hu, "Subbreakdown drain leakage current in MOSFET," *IEEE Elect. Dev. Lett.*, vol. EDL-8, p. 515, 1987.

[4.85] T. Kamata, K. Tanabashi, and K. Kobayashi, "Substrate current due to impact ionization in MOSFET," *Jpn. J. Appl. Phys.*, vol. 15, p. 1127, 1976.

[4.86] K. Seki, H. Kume, Y. Ohji, T. Tanaka, T. Adachi, M. Ushiyama, K. Shimohigashi, T. Wada, K. Komori, T. Nishimoto, K. Izawa, T. Hagiwara, Y. Kubota, K. Shoji, N. Miyamoto, S. Saeki, and N. Ogawa, "An 80 ns 1Mb Flash memory with on-chip erase/erase-verify controller," *ISSCC Dig.* Tech. Pap., pp. 1147–1151, 1990.

[4.87] R. A. Cernea, G. Samachisa, C. S. Su, H. F. Tsai, Y. S. Kao, C. Y. Wang, Y. S. Chen, A. Renninger, T. Wong, J. Brennan, and J. Haines, "A 1Mb flash EEPROM," *ISSCC Dig.* Tech. Pap., p. 138, 1989.

[4.88] A. T. Wu et al., "A novel high-speed, 5-V programming EPROM structure with source-side injection," *IEEE IEDM Tech. Dig.*, pp. 584–587, 1986.

[4.89] C. Chang and J. Lien, "Corner-field induced drain leakage in thin oxide MOSFETs," *IEEE IEDM Tech. Dig.*, pp. 714–717, 1987.

[4.90] R. E. Shiner et al., "Data retention in EPROMs," Proc. IRPS, pp. 238–243, 1980.

[4.91] R. N. Mielke, "New EPROM data-loss mechanisms," Proc. IRPS, pp. 106–113, 1983.

[4.92] K. C. Smith, "The prospects for multivalued logic: a technology and application view," *IEEE Trans. Computers*, vol. C-30, pp. 619–634, 1981.

[4.93] S. J. Wei and H. C. Lin, "Multivalued SRAM cell using resonant tunnelling diodes," *IEEE J. Solid-State Circuits*, vol. SC-27, no. 2, pp. 212–216, 1992.

[4.94] M. Horiguchi et al., "An experimental large-capacity semiconductor file memory using 16-level/cell storage," *IEEE J. Solid-State Circuits*, vol. SC-23, no. 1, pp. 27–33, 1988.

[4.95] T. Furuyama et al., "An experimental 2-bit/cell storage DRAM for macrocell or memory-on-logic applications," *IEEE J. Solid-State Circuits*, vol. SC-24, no. 2, pp. 388–393, 1989.

[4.96] E. Harari, "Flash EEPROM memory system having multi-level storage cells," U.S. Patent #5,043,940, 1991.

[4.97] "Intel working on multilevel Flash," *EE Times*, p. 1, August 1, 1994.

[4.98] R. M. Anderson and D. K. Ker, "Evidence for surface asperity mechanism of conductivity in oxides grown on polycrystalline silicon," *J. Appl. Phys.*, vol. 48, no. 4, p. 483, 1977.

[4.99] J. Van Houdt et al., "A 5V/3.3V-compatible Flash E^2PROM cell with a 400 ns/70 μs programming time for embedded memory applications," Proc. 5th Biennial Nonvolatile Memory Technology Review, Linthicum Heights, Md., pp. 54–57, 1993.

[4.100] J. Van Houdt et al., "Analysis of the enhanced hot-electron injection in split-gate transistors useful for EEPROM applications," *IEEE Trans. Elect. Dev.*, vol. ED-39, p. 1150, 1992.

[4.101] N. Ajika et al., "A 5V-only 16 Mbit Flash EEPROM cell with a simple stacked gate structure," *IEEE IEDM Tech. Dig.*, pp. 115–118, 1990.

[4.102] T. Tanazawa et al., "A quick boosting charge pump circuit for high density and low voltage Flash memories," Symp. VLSI Cir., pp. 65–66, 1994.

[4.103] J. S. Witters et al., "Degradation of tunnel-oxide floating gate EEPROM device and the correlation with high field current-induced degradation of thin gate oxide," *IEEE Trans. Elect. Dev.*, vol. ED-36, no. 9, pp. 1663–1682, 1989.

[4.104] Chi Chang, "Flash memory reliability," a tutorial at IRPS, 1993.

[4.105] Stefan Lai, "Flash memory reliability," a tutorial at IRPS, 1993.

[4.106] B. Doyle, M. Bourcerie, J. C. Marchetaux, and A. Boudou, "Interface state creation and charge trapping in the medium-to-high gate voltage range ($V_d/2 < Vg < Vd$) during hot-carrier stressing of n-MOS transistors," *IEEE Trans. Elect. Dev.*, vol. ED-37, no. 3, p. 744, 1990.

[4.107] M.-S. Liang, C. Chang, W. Yang, C. Hu, and R. W. Brodersen, "Hot carriers-induced degradation in thin gate oxide MOSFET's," *IEEE IEDM Tech. Dig.*, p. 186, 1983.

[4.108] T. Nishida and S. E. Thompson, "Oxide field and temperature dependent gate oxide degradation by substrate hot electron injection," *Proc. IRPS*, p. 310, 1991.

[4.109] C. S. Jeng, T. R. Ranganath, C. H. Huang, H. S. Jones, and T.T.L. Chang, "High-field generation of electron traps and charge trapping in ultra-thin SiO_2," *IEEE IEDM Tech. Dig.*, p. 388, 1981.

[4.110] M.-S. Liang and C. Hu, "Electron trapping in very thin thermal silicon dioxides," *IEEE IEDM Tech. Dig.*, p. 396, 1981.

[4.111] M.-S. Liang, C. Chang, Y. T. Yeow, C. Hu, and R. W. Brodersen, "MOSFET degradation due to stressing of thin oxide," *IEEE Trans. Elect. Dev.*, vol. ED-31, no. 9, p. 1238, 1984.

[4.112] J. S. Witters, G. Groeseneken, and H. E. Maes, "Programming mode dependent degradation of tunnel oxide floating gate devices," *IEEE IEDM Tech. Dig.*, p. 544, 1987.

[4.113] C. Chang, S. Haddad, B. Swaminathan, and J. Lien, "Drain-avalanche and hole trapping induced gate leakage in thin-oxide MOS devices," *IEEE Elect. Dev. Lett.*, vol. EDL-9, no. ll, p. 588, 1988.

[4.114] Y. Igura, H. Matsuoka, and E. Takeda, "New device degradation due to 'cold' carriers created by band-to-band tunneling," *IEEE Electron Device Letters*, vol. 10, no. 5, p. 227, May 1989.

[4.115] K. Yoshikawa, S. Mori, E. Sakagami, Y. Ohshima, Y. Kaneko, and N. Arai, "Lucky-hole injection induced by band-to-band tunneling leakage in stacked gate transistors," *IEEE IEDM Tech. Dig.*, p. 577, 1990.

[4.116] R. Rakkhit, S. Haddad, C. Chang, and J. Yue, "Drain-avalanche induced hole injection and generation of interface traps in thin oxide MOS devices," *Proc. IRPS*, p. 150, 1990.

[4.117] Y. Tang, C. Chang, S. Haddad, A. Wang, and J. Lien, "Differentiating impacts of hole trapping vs. interface states on TDDB reduction in MOS transistors," Proc. SRC Topical Res. Conf. on Floating Gate Non-Volatile Memory Research, Berkeley, Calif., 1992.

[4.118] M. Baur et al.,"A multilevel cell 32 Mb Flash memory," *IEEE ISSCC Tech. Dig.* Pap., pp. 132–133, 1995.

[4.119] N. Kodama, K. Oyama, H. Shirai, K. Saitoh, T. Okazawa, and Y. Hokari, "A symmetrical side wall (SSW)-DSA cell for a 64Mbit Flash memory," *IEEE IEDM Tech. Dig.*, p. 303, 1991.

[4.120] T. Jinbo, H. Nakata, K. Hashimoto, T. Watanabe, K. Ninomiya, T. Urai, M. Koike, T. Sato, N. Kodama, K. Oyama, and T. Okazawa, "A 5V-only 16 Mb Flash memory with sector-erase anode," *IEEE ISSCC. Dig.* Tech. Pap., p. 154, 1992.

[4.121] E. Takeda, N. Suzuki, and T Hagiwara, "Device performance degradation due to hot-carrier injection at energies below the Si-SiO_2 energy barrier," *IEEE IEDM Tech. Dig.*, p. 396, 1983.

[4.122] T. Endoh, R. Shirota, M. Momodomi, and F. Masuoka, "An accurate model of subbreakdown due to band-to-band tunnelling and some applications," *IEEE Trans. Elect. Dev.*, vol. ED-37, no. 1, pp. 290–296, 1990.

[4.123] H. Matsuoka, Y. Igura, and E. Takeda, "Device degradation due to band-to-band tunnelling," Intl. Conf. Sol. State Dev. and Mat., Tokyo, pp. 589–592, 1988.

[4.124] T. Ozaki, A. Nitayama, T. Hamamoto, K. Sunouchi, and F. Horiguchi, "Analysis of band-to-band tunnelling leakage current in trench-capacitor DRAM cells," *IEEE Elect. Dev. Lett.*, vol. EDL-12, no. 3, pp. 95–97, 1991.

[4.125] R. Shirota et al., "A new programming method and cell architecture for multilevel NAND Flash memories," IEEE Nonvolatile Semiconductor Memory Workshop, Paper #2.7, 1995.

[4.126] S. K. Lai, "Oxide/silicon interface effects in E^2PROMs and ETOX™ Flash," Proc. SRC Topical Res. Conf. on Floating Gate Non-Volatile Memory Research, Berkeley, Calif., 1992.

[4.127] T. C. Ong, A. Fazio, N. Mielke, S. Pan, G. Atwood, and S. K. Lai, "Instability of erase threshold voltage in ETOX™ Flash memory array,"

Proc. SRC Topical Res. Conf. on Floating Gate Non-Volatile Memory Research, Berkeley, Calif., 1992.

[4.128] S. Lassig and M.-S. Liang, "Time-dependent degradation of thin gate oxide under post-oxidation high-temperature anneal," *IEEE Elect. Dev. Lett.*, vol. EDL-8, no. 4, p. 160, 1987.

[4.129] K. Yoneda, Y. Fukuzaki, K. Satoh, Y. Hata, Y. Todokoro, and M. Inoue, "Reliability degradation mechanism of the ultra-thin tunneling oxide by post-annealing," *Dig. Symp. VLSI Tech.*, p. 121, 1990.

[4.130] M. Ushiyama, Y. Ohji, T. Nishimoto, K. Komori, H. Murakoshi, H. Kume, and S. Tachi, "Two dimensionally inhomogeneous structure at gate electrode/gate insulator interface causing Fowler–Nordheim current deviation in nonvolatile memory," *Proc. IRPS*, p. 331, 1991.

[4.131] H. Fukuda, M Yasuda, T. Iwabuchi, and S. Ohno, "Novel N_2O-oxynitridation technology for forming highly reliable EEPROM tunnel oxide films," *IEEE Elect. Dev. Lett.*, vol. EDL-12, no. 11, pp. 587–589, 1991.

[4.132] S. K. Lai, J. Lee, and V. K. Dham, "Electrical properties of nitride-oxide systems for use in gate dielectric and EEPROM," *IEEE IEDM Tech. Dig.*, pp. 190–193, 1983.

[4.133] H. Fukuda, T. Arakawa, and S. Ohno, "Thin-gate SiO_2 films formed by in-situ multiple rapid thermal processing," *IEEE Trans. Elect. Dev.*, vol. ED-39, no. 1, pp. 127–133, 1992.

[4.134] K. Yoshikawa, S. Yamada, J. Miyamoto, T. Suzuki, M. Oshikiri, E. Obi, Y. Hiura, K. Yamada, Y. Ohshima, and S. Atsumi, "Comparison of current Flash EEPROM erasing methods: stability and how to control," *IEEE IEDM Tech. Dig.*, p. 595, 1992.

[4.135] A. Brand, K. Wu, S. Pan, and D. Chin, "Novel read disturb failure mechanism induced by Flash cycling," *Proc. IRPS*, pp. 127–132, 1993.

[4.136] J. Maserjian and N. Zamani, "Behavior of the Si/SiO_2 interface observed by Fowler–Nordheim tunneling," JAP, pp. 559–567, January 1982.

[4.137] P. Olivio, T. N. Nguyen, and B. Ricco, "High-field-induced degradation in ultra-thin SiO_2 films," *Trans. Elect. Dev.*, vol. 35, no. 12, pp. 2259–2267, December 1988.

[4.138] K. Naruke, S. Taguchi, and M. Wada, "Stress-induced leakage current limiting to scale down EEPROM tunnel oxide thickness," *IEEE IEDM Tech. Dig.*, pp. 424–427, 1988.

[4.139] R. Moazzami and C. Hu, "Stress-induced current in thin silicon dioxide films," *IEEE IEDM Tech. Dig.*, pp. 139–142, 1992.

[4.140] D. A. Baglee and M. C. Smayling, "The effects of write/erase cycling on data loss in EEPROMs," *IEEE IEDM Tech. Dig.*, pp. 624–626, 1985.

[4.141] R. Rofan and C. Hu, "Stress-induced oxide leakage," *IEEE Elect. Dev. Lett.*, vol. EDL-12, pp. 632–634, 1991.

[4.142] E. Rosenbaum and C. Hu, "High-frequency time-dependent breakdown of SiO_2," *IEEE Elect. Dev. Lett.*, vol. EDL-12, pp. 267–269, 1991.

[4.143] E. Rosenbaum, Z. Liu, and C. Hu, "The effect of oxide stress waveform on MOSFET performance," *IEEE IEDM Tech. Dig.*, pp. 719–722, 1991.

[4.144] S. Mori, E. Sakagami, H. Araki, Y. Kaneko, K. Narita, Y. Ohshima, N. Arai, and K. Yoshikawa, "ONO inter-poly dielectric scaling for nonvolatile memory applications," *IEEE Trans. Elect. Dev.*, vol. ED-38, no. 2, pp. 386–391, 1991.
[4.145] R. D. Hoffman, "Speciality memories gain currency," EBN, p. 25, April 18, 1994.
[4.146] C. Westmont and B. Wall, "Reprogrammability eases PCMCIA designs," *Electronic Design*, pp. 110–113, April 18, 1994.
[4.147] B. Bipert and L. Hebert, "Flash memory goes mainstream," *IEEE Spectrum*, pp. 48–52, October 1993.
[4.148] Suguru Kuki, "Speed of Flash memories sparks maker interest," JEE, pp. 40–43, November 1992.
[4.149] D. Bursky, "Flash memory interfaces simplify systems," EDN, pp. 168–171, November 7, 1994.
[4.150] R. H. Dennard et al., "Design of Ion-Implanted MOSFET's With Very Small Physical Dimensions," *IEEE J. Solid-State Circuits*, vol. SC-9, no. 10, pp. 256–268, 1974.
[4.151] M. Kato et al., "A 0.4-μm^2 self-aligned contactless memory cell technology suitable for 256-Mbit Flash memories," *IEEE IEDM Tech. Dig.*, pp. 921–923, 1994.
[4.152] S. Aritome et al., "A 0.67 μm^2 self-aligned shallow trench isolation cell, [SA-STI cell] for 3V-only 256 Mbit NAND EPROMs," *IEEE IEDM Tech. Dig.*, pp. 61–64, 1994.
[4.153] Data storage report, Jonas Press Publishing Company, 1994.
[4.154] S. Aritome et al., "Advanced NAND-structure cell technology for reliable 3.3V 64 Mb electrically erasable and programmable read only memories (EEPROM)," *Jap. J. Appl. Phys.*, vol. 33, no. 1B, pp. 524–528, 1994.
[4.155] M. Aoki et al.," Triple density DRAM cell with silicon selective growth channel and NAND structure," *IEEE IEDM Tech. Dig.*, pp. 631–634, 1994.
[4.156] S. Onishi, "A half-micron ferroelectric memory cell technology with stacked capacitor structure," *IEEE IEDM Tech. Dig.*, pp. 843–846, 1994.
[4.157] B. Eitan et al., "Alternate metal virtual ground (AMG)—a new scaling concept for very high density EPROMS," *IEEE Elect. Dev. Lett.*, vol. EDL-12, no. 8, pp. 450–452, 1991.
[4.158] *IEEE Floating Gate Standards – to be published in 1997.*
[4.159] *IEEE Floating Gate Standards, IEEE Std.–1005, IEEE Standard Definitions and Characterization of Floating Gate Semiconductor Arrays, 1991.*
[4.160] M. Lenzlinger and E. H. Snow, "Fowler–Nordheim tunneling in thermally grown SiO_2," JAP, vol. 40, p. 278, 1969.
[4.161] F. Goodenough, "IC holds 16 seconds of audio without power," *Electronic Design*, pp. 39–44, January 31, 1991.

Chapter 5

Frank R. Libsch
and Marvin H. White

SONOS Nonvolatile Semiconductor Memories

5.0. INTRODUCTION

The state-of-the-art techniques for nonvolatile semiconductor memories (NVSMs) are the SONOS (silicon–oxide–nitride–oxide–silicon) and the floating gate-type memories. These memories are electrically erasable programmable read-only memories (EEPROMs) and generally require a two-transistor cell (2TC) layout—a NVSM memory transistor and a "select" transistor. The result is higher chip cost and a density for conventional (full-feature) EEPROMs at least one generation behind single-transistor cell EPROMs. EEPROMs of the Flash floating gate variety have been specified at 100 K erase/write cycles and find applications in areas that do not require extensive erase/write cycling but necessitate memory management techniques to extend their usefulness to a semiconductor disk application. SONOS EEPROMs have extended erase/write cycling to 10 M cycles and, combined with 0.25 μm design rules and a so-called split gate (1.5TC) design, can provide a 1.0 μm^2 device suitable for a 256 MB memory array and PCMCIA cards. In order for EEPROMs to be more competitive with EPROMs, industry needs a manufacturable single-transistor cell (1TC) EEPROM to exploit the possibility of a semiconductor disk.

In the NVRAM (nonvolatile RAM), the nonvolatility of the EEPROM is combined with the ease of use and fast programming of the static RAM. The content in the static RAM section of the memory is transferred in parallel to the nonvolatile EEPROM portion at power down or power failure. Sometimes this type of memory is referred to as a shadow RAM. The requirements for a NVRAM are generally more stringent in terms of endurance or erase/write cycling since a 10 M cycle NVRAM refreshing every minute will have a 20-year life cycle and offers replacement for DRAMs in low power applications such as portable computers and other consumer products. The advantages of SONOS NVSMs lie in the low programming voltages, endurance to extended erase/write cycling, inherent resistance to radiation, and compatibility with high-density scaled CMOS for low power, portable electronics.

5.1. THE SONOS NONVOLATILE MEMORY TRANSISTOR

The SONOS is a multi-dielectric device consisting of an oxide–nitride–oxide (ONO) sandwich in which charge storage takes place in discrete traps in the silicon nitride layer. A net positive or negative charge can be stored in almost equal amounts in Low-Pressure Chemical Vapor Deposition (LPCVD) nitrides. One of the oxide layers adjacent to the silicon semiconductor, called the tunnel oxide, permits the transfer of signal charge from the silicon substrate to the discrete traps in the silicon nitride by a process called *tunneling*. The process of tunneling can take several forms. However, the tunnel oxide must be quite thin (i.e., less than 2.0 nm) to provide efficient high-speed transfer of signal charge. The other oxide layer adjacent to the gate polysilicon electrode, called the blocking oxide, separates the control gate from the silicon nitride storage region. The function of this oxide layer is to isolate or "block" the transfer of charge to and from the nitride storage region. This oxide permits the scaling of the nitride film in the ONO sandwich to realize NVSMs with low programming voltages (5 to 10 V) [5.1]. This structure is discussed in detail later in this chapter, but first we review the historical development of NVSMs based on the use of silicon nitride as the storage medium.

Figure 5.1 illustrates the development of the multi-dielectric nonvolatile memory device over three decades of research, development, and manufacturing of NVSMs. The early MNOS devices of the 1960s and mid-1970s were p-channel devices and used an aluminum gate technology, relatively thick nitrides (45 nm), and thin tunnel oxides (2 nm). The programming voltages were typically 25 to 30 V, and the technology was p-channel, diffusion-isolated, epitaxial layers for separation of the high-voltage MNOS memory array from the logic and control circuitry. In the latter 1970s and early 1980s, the technology moved into polysilicon gates with n-channel SNOS devices with reduced nitride thicknesses (25 nm) and corresponding lower programming voltages (12 to 16 V). In the mid-1980s and 1990, the technology moved in the direction of adding another dielectric, called the blocking oxide, to create a so-called SONOS device with a "scaled" nitride dielectric (5 to 15 nm) so that even lower programming voltages of 5 to 10 V could be achieved [5.2]. The sections to follow detail the development of the basic memory device, the memory cells that exploit the advantages of charge storage in multi-dielectric films, the electrical characteristics and technology of these cells, and the memory arrays constructed with these cells.

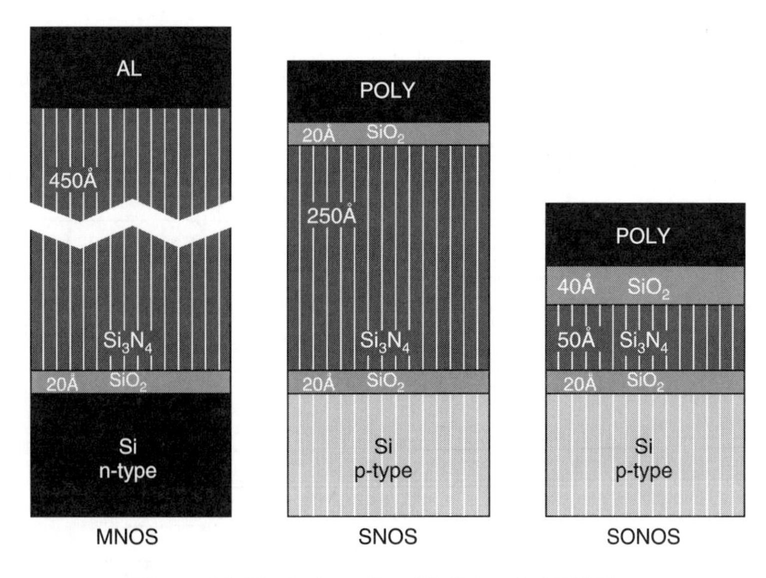

Figure 5.1 Evolution of multi-dielectric NVSM.

5.1.1 Trigate MNOS Memory Cell (3TC)

Over 25 years ago, the first nitride storage device, in the form of a MNOS capacitor, was introduced which demonstrated charge injection and storage in the nitride region. Subsequently, the MNOS memory transistor was developed, together with the first electrically alterable read-only memories (EAROMs) which were NVSMs. In 1975, MNOS EAROMs were fabricated with p-channel MNOS memory transistors and diffusion-isolated, epitaxial technology with programming voltages of ± 25 V. These early devices had silicon nitride thicknesses of 450 Å (45 nm) and aluminum gates, and used a so-called trigate transistor known as a drain-source protected (DSP) geometry. Figure 5.2 shows a cross section and equivalent circuit diagram of a DSP trigate MNOS EEPROM memory transistor which was employed in early NVSMs [5.14]. The gate structure is composed of an electrically reprogrammable MNOS memory gate structure in the middle, with a thin tunnel oxide sandwiched between two nonmemory MNOS gate structures with thicker (nontunneling) oxides. These nonmemory MNOS gate structures served as access gates to isolate MNOS memory cells which shared a common read-out column. The presence of thick oxide regions over the source and drain junctions of the device also served to protect these regions from high electric fields during programming.

In operation, the nonmemory sections are enhancement-mode devices in series with the memory section, and control the drain current that flows during the read operation when the memory section is in the erase (high-conductance) state. In the write (low-conductance) state, the memory section controls the current flow during the read operation. These early MNOS memories lacked a developed technology base with a major problem in the control of the silicon nitride thickness under

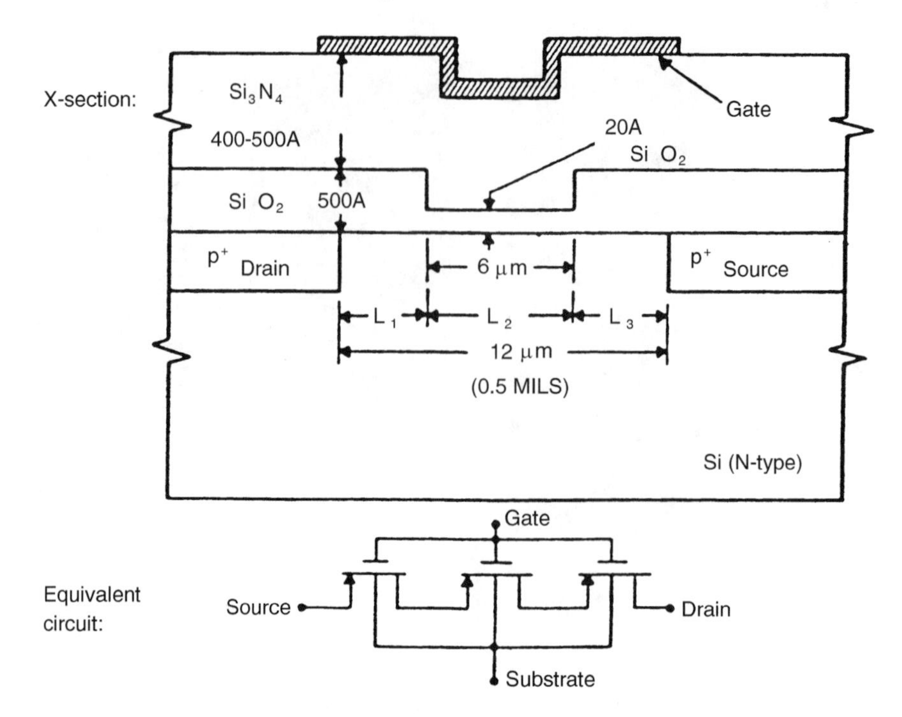

Figure 5.2 Cross section and equivalent circuit of the drain-source protected (DSP) MNOS memory device [5.14].

atmospheric pressure chemical vapor deposition (APCVD). Process control over the nitride thickness was difficult, and nitride thickness variations translated into variations in electric fields that deleteriously impacted cell endurance across the memory array. An aluminum gate technology, coupled with the lack of suitable isolation (i.e., LOCOS process), penalized the reliability, density, and speed. Thus, the limitations of the early p-channel Al-gate process were (1) relatively low read speeds (approximately 850 ns), (2) nonstandard power supplies, including a non-5 V supply in the read mode, (3) nonstandard pin configurations, and (4) read disturb effects since the access and memory gates were common [5.4].

Figure 5.3 illustrates a more recent version of the trigate SNOS memory cell which uses a polysilicon gate electrode (poly2) sandwiched between two conventional MOS transistors [5.18]. The transistors have a polysilicon gate electrode (poly1) for independent control over the isolate (access) and select functions. The metal gate (M) has been replaced by a polysilicon gate electrode (S), although the literature still refers to SNOS devices as MNOS devices. This review of multidielectric NVSMs, makes a distinction between the gate electrodes and the physical construction of the device. The SNOS device in Fig. 5.3 is fabricated in a p-well with a 1.6 nm SiO_2 tunnel oxide and a 28 nm Si_3N_4 charge storage layer as shown in the transmission electron microscope (TEM) image of a MNOS device cross section in Fig. 5.4 [5.49]. The tunnel oxide layer exhibits stable morphologic and stoichiometric characteristics.

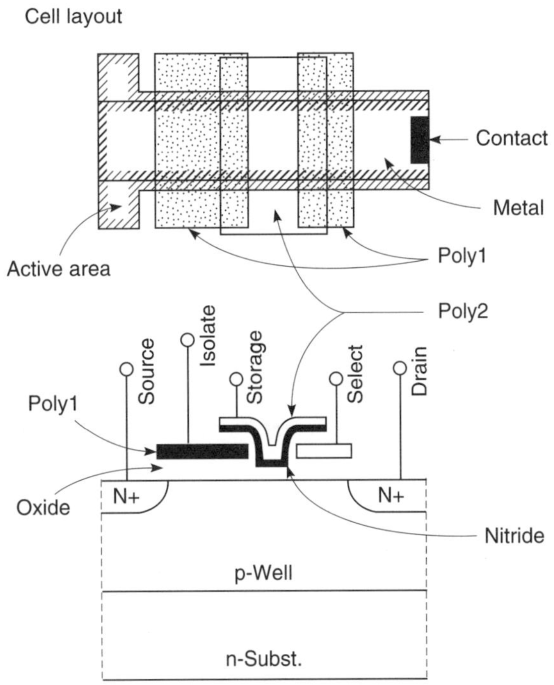

Cell layout
Contact
Metal
Poly1
Active area
Poly2
Source
Isolate
Storage
Select
Drain
Poly1
Oxide
N+
N+
Nitride
p-Well
n-Subst.

Figure 5.3 Cross section and layout of SNOS trigate memory cell [5.18].

Poly Si
Si3N4
SiO2
Si
50 nm

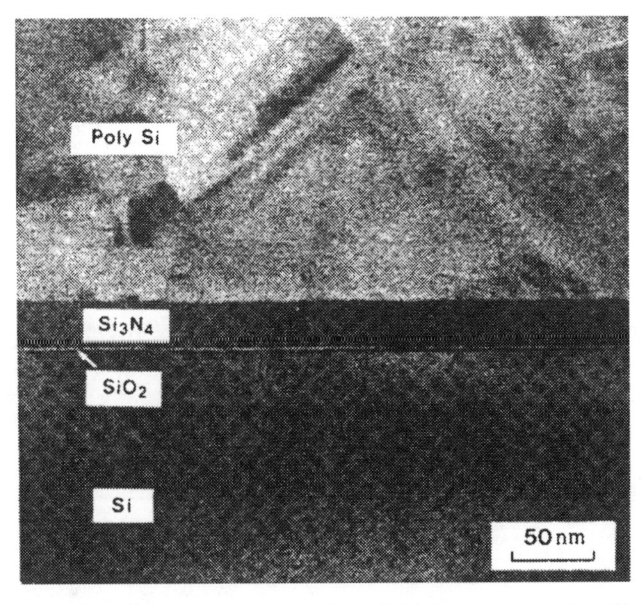

Figure 5.4 TEM observation image of an SNOS memory device. SNOS is fabricated on a p-well having 1.6 nm SiO_2 and 28 nm Si_3N_4 [5.49].

5.1.1.1 Trigate Memory Cell Operation. Figure 5.5 illustrates the basic operating conditions for the trigate, n-channel memory cell [5.17]. The programming high voltage is 15 V and is composed of $V_{cc} = 5\,V$ and $V'_{pp} = -10\,V$, where the latter is supplied by an on-chip negative high-voltage generator and clamped by the avalanche breakdown voltage of a reverse-biased p–n junction. In the write or programming operational mode, V'_{pp} is applied to the p-well, while V_{cc} is applied to the gate of the selected memory cell. In this write mode, electrons are injected (tunnel) from the channel through the thin tunnel oxide to store in "deep" electron traps within the Si_3N_4 layer. The resultant electron charge storage shifts the threshold voltage of the MNOS memory cell in a positive direction; the exact amount depends on the programming time, t_p, which is generally less than 10 ms and is generated by an on-chip timer.

In the programming inhibit operational mode, 5 V is applied to the drain of the deselected cell, which reduces the voltage between the gate and the channel, and the effective electric field across the tunnel oxide. Thus, no tunneling or injection of electrons into the nitride takes place, and the programming operation is effectively inhibited. The channel is shielded from the substrate voltage, V'_{pp}, with the application of 5 V to the drain terminal.

In the erase mode of operation, the large negative voltage V'_{pp} is applied to the gate of the memory cell, while V_{cc} is applied to the p-well. Under these conditions, holes are injected or tunneled from the substrate to the deep traps in the nitride. In one model of the device operation, the holes recombine with trapped electrons and,

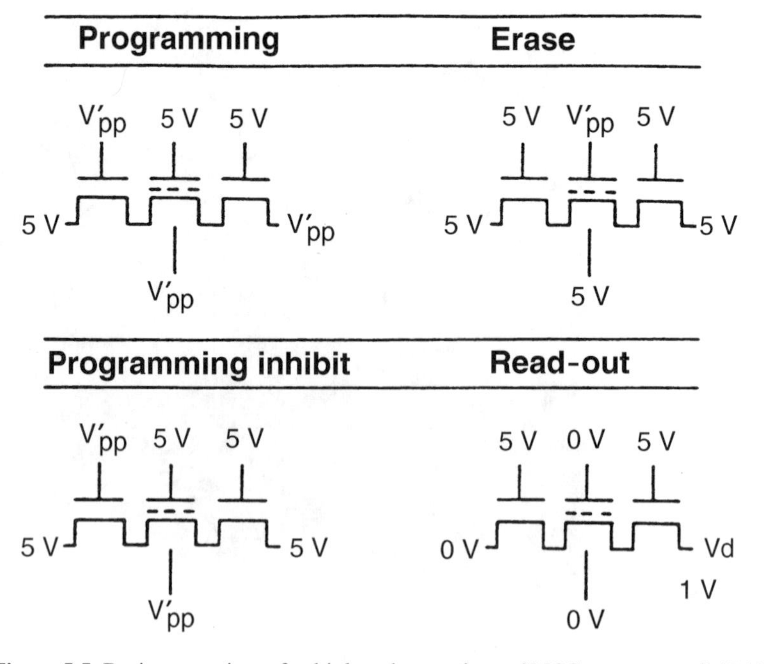

Figure 5.5 Basic operation of a high-voltage trigate SNOS memory cell [5.17].

through a mechanism of multiple occupancy, create a state of positive stored charge in the nitride. These nitride traps are called *amphoteric traps*, and they are pictured to be associated with excess silicon dangling bonds attached to a tetrahedral base of nitrogen atoms [5.2, 5.44]. In another model of device operation, there is a picture of two distinct traps of almost equal densities—one for the storage of electrons and the other for the storage of holes.

In the read mode of operation, the memory gate is grounded (0 V) and the write state (0s) and erase state (1s) are read out depending on the channel current that flows in the memory cell. In the erase state, a channel current flows in the memory cell in response to an applied drain voltage $V_d \approx 1$ V.

5.1.1.2 Trigate Memory Cell Electrical Characteristics. The most important characteristics of a NVSM cell are (1) programming speed (erase/ write), (2) data retention (memory), and (3) endurance (erase/write cycling). These characteristics are determined by measuring the change in threshold voltage, ΔVth, of the memory transistor section of the trigate memory cell. The threshold voltage must be measured as a function of time with very little read-disturb effect to obtain an accurate picture of the memory cell operation. One method of measuring the V_{th} is to employ a source follower technique, while another approach compares the threshold voltage measured at two different current levels. Figure 5.6 illustrates the data obtained on a SNOS memory transistor at 85°C with the cross section shown in Fig. 5.3 [5.17]. The so-called virgin threshold, V_{thi}, is the "unprogrammed" state of the SNOS memory transistor. The erase/write, or programming characteristics, are illustrated in Fig. 5.6*a*, and a 2 V shift in V_{th} is obtained with a minimum programming voltage of 14V in 3 ms.

The memory, or data retention, characteristics of the SNOS memory transistor portion of the trigate memory cell are illustrated in Fig. 5.6*b*, and it can be seen that V_{th} decreases logarithmically with time. The write state decays about 0.3 V/decade, while the erase state decay is about 0.15 V/decade for these particular devices. These characteristics vary depending on the location of the memory window center, the dielectric thicknesses, and the amount of stored charge in the nitride layer. The window between erase and write states, combined with the nonmemory threshold voltage in the trigate memory cell, influences the design of the sense amplifier. For the trigate memory cell described in Figs. 5.5 and 5.6, cell retention will be determined by the intersection of the write data retention curves with the V_{th} = 0 V point, which sets the lower limit of V_p programming to 14 V and 10-year data retention.

The influence of erase/write cycling on the electrical characteristics of the cell is called the endurance (see Fig. 5.6*c*). After 100 K cycles and a typical programming voltage of 15 V, the memory window remains relatively unchanged, and 10-year data retention is maintained at 85°C. The electric fields in the SNOS device influence the speed, retention, and endurance of the memory cell. This will be discussed in Section 5.4.

5.1.1.3 Trigate Memory Cell Technology. The early block-oriented random access memory (BORAM) chips, which were designed with the DSP geometry

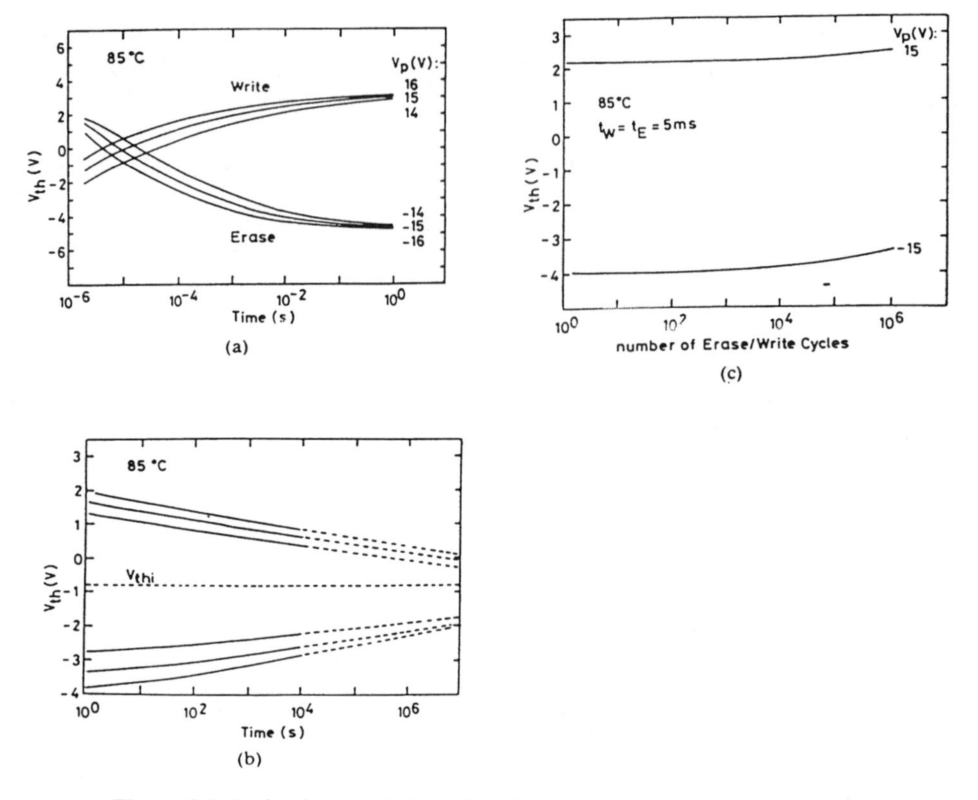

Figure 5.6 Basic characteristics of a SNOS nonvolatile memory device which are applied to 64 Kbit EEPROM; (*a*) erasure and write speed, (*b*) data retention characteristics, and (*c*) erase/write-cycle endurance [5.17].

trigate structure, were fabricated with a p^+ diffusion isolation, n-type epitaxial layer process. The trigate p-channel, aluminum-gate MNOS memory cells were placed in an isolated n-pocket where high voltages could be applied without affecting the address-decoder, sense-amplifier, shift-register, and read-out circuitry. Typical programming voltages for the MNOS memory array were 25 to 30 V (nitride thicknesses of 45 nm), while the low-level logic operated at 5 V TTL levels. This technology was a "carryover" from the bipolar era where collector-to-collector isolation was obtained with separate epitaxial pockets.

With the advent of the local oxidation of silicon (LOCOS) process to isolate active and passive areas on a chip, the technology moved away from the epitaxial approach to the use of p-wells, followed by n-wells, and finally to the twin-tub approach where both p- and n-wells are employed. The limitation of the early trigate memory cells lies in the use of a dual-dielectric MNOS or SNOS memory cell where the nitride layers were typically 25 to 45 nm in thickness and required programming voltages of 15 to 30 V. Since high-density and decreased access time were needed, the memory transistors and the peripheral support circuitry transistors required scaling to geometries of 2 μm, with programming voltages of 14 to 16 V. The solution to the high-voltage MNOS or SNOS memory transistors was to employ a lightly doped drain (LDD) type of structure where the drain was "engineered" to have an n^+ diffusion to minimize the electric fields. The tunnel oxide thickness of these memory cells was typically 1.6 nm, with an accuracy of less than ± 0.05 nm. The tunnel oxide formation is a carefully controlled process with variations among different manufacturers. One growth procedure uses a temperature of 850°C and an oxygen-to-nitrogen ratio of 0.1%. The growth rate under these conditions has been reported to be on the order of 0.01 nm/min after the oxide film has grown to about 1 nm thickness.

The rapid initial growth of the so-called native oxide on silicon (i.e., 0.6 to 1.0 nm) has been the subject of many studies over the years and remains one of the most important areas in scaling these multi-dielectric nonvolatile memory devices. The silicon nitride layer is deposited at about 790°C with a LPCVD process using a mixture of dichlorosilane (SiH_2Cl_2) and ammonia (NH_3) gases. The polysilicon gate electrode is deposited and doped with phosphorus, followed by a phosphosilicate glass deposition and a high-temperature (900°C) hydrogen anneal at the contact window step to improve the memory retention.

5.1.2 Pass Gate Memory Cell (2TC)

A popular memory cell design in present-day SNOS and SONOS memory arrays is the so-called pass gate or 2 transistor cell (2TC). A MOS transistor serves as a select device in series with the nonvolatile SNOS or SONOS memory transistor to minimize any column disturbance during a read operation of the addressed SNOS or SONOS memory transistor. Fig. 5.7 illustrates an early pass gate memory cell cross section for an n-channel SNOS memory transistor. This structure shows an n-n^+ offset drain (source) and a thick oxide layer (40 nm) between the n-layer and the gate electrode. Such a layer reduces the electric field near the drain and permits higher programming voltages to be employed in the cell operation. This feature is referred to in CMOS technology as a lightly doped drain (LDD). The pass gate memory cell has also been used in a p-channel SONOS memory cell with a 1.6 nm

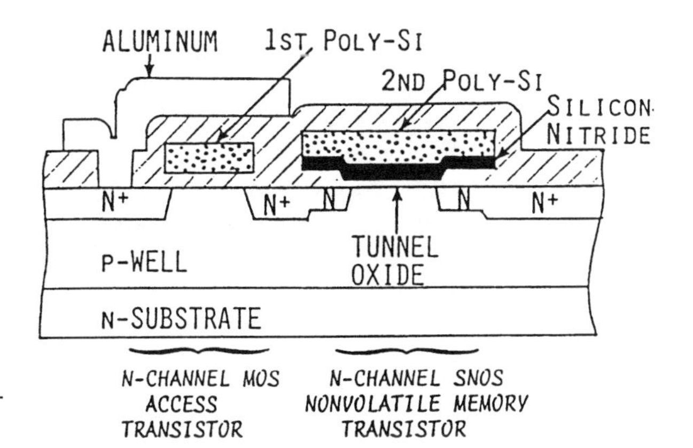

Figure 5.7 Cross section of SNOS memory cell used in 256 Kbit EEPROM.

tunnel oxide, a 15 nm silicon nitride storage layer, and a 4 nm capping or blocking oxide between the nitride and the polysilicon gate electrode, as shown in Fig. 5.8. The limitation of previous SNOS memory cells, caused by the penetration of stored charge into the nitride and subsequent loss to the gate electrode, has been removed by the addition of the so-called blocking oxide between the silicon nitride and the polysilicon gate electrode. The resulting structure is called a SONOS nonvolatile memory device where the silicon nitride may be scaled to dimensions of 15 nm and the 10 V programming voltage is less than the junction breakdown voltage of the peripheral CMOS transistors. Silicon nitride layers as thin as 5 nm have been reported to achieve true 5 V programming and compatibility with peripheral, low-power, scaled CMOS circuits.

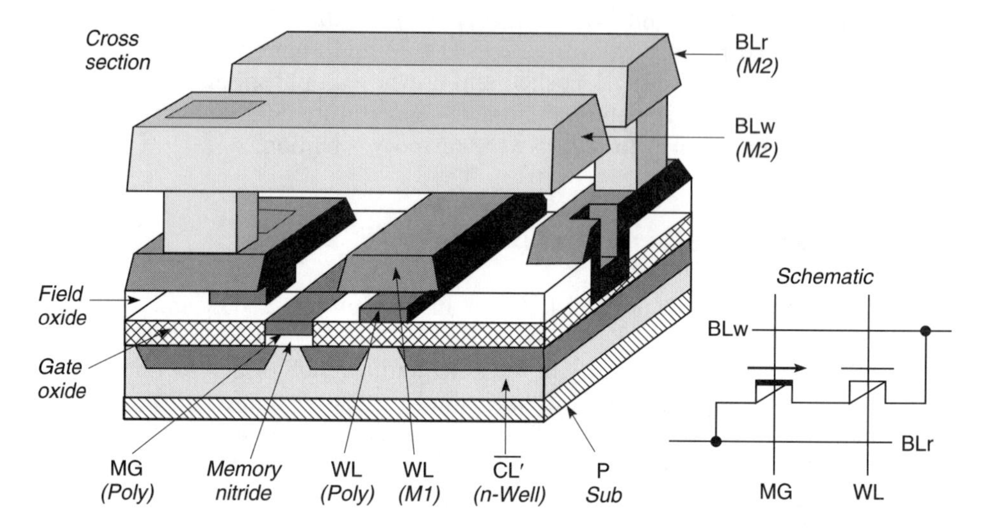

Figure 5.8 2TC p-channel SONOS memory cell.

5.1.2.1 Pass Gate Memory Cell Operation. Figure 5.9 illustrates a memory cell architecture for n-channel SNOS pass gate memory cells where byte erase/write is achieved by placing 8 bits in the direction of the wordline in each n-well corresponding to 1 byte (eight sense amplifiers and eight input/output buffers) [5.5]. An external $V_{cc} = 5\,V$ is supplied to the chip and boosted by an on-chip high-voltage generator to $V_p = 16\,V$. Figure 5.9*a* illustrates the bias conditions for byte write, and Fig. 5.9*b* illustrates the bias conditions for byte erase. Special precautions are required in the erase-inhibit and write-inhibit states of the memory cell. $V_p = V_i$ is selected to avoid repeated exposure for a bit in the erase-inhibit state to a voltage $V_p - V_i$. In addition, the breakdown voltage of the drain $n^+ - p$ junction must be larger than V_p to ensure that the high-voltage generator does not have to supply a large total leakage current to the memory array. A similar problem arises in floating gate Flash EEPROMs where a "hot-carrier" channel current flows in each programmed cell, but this is a consequence of the write mode of operation. In SNOS/SONOS

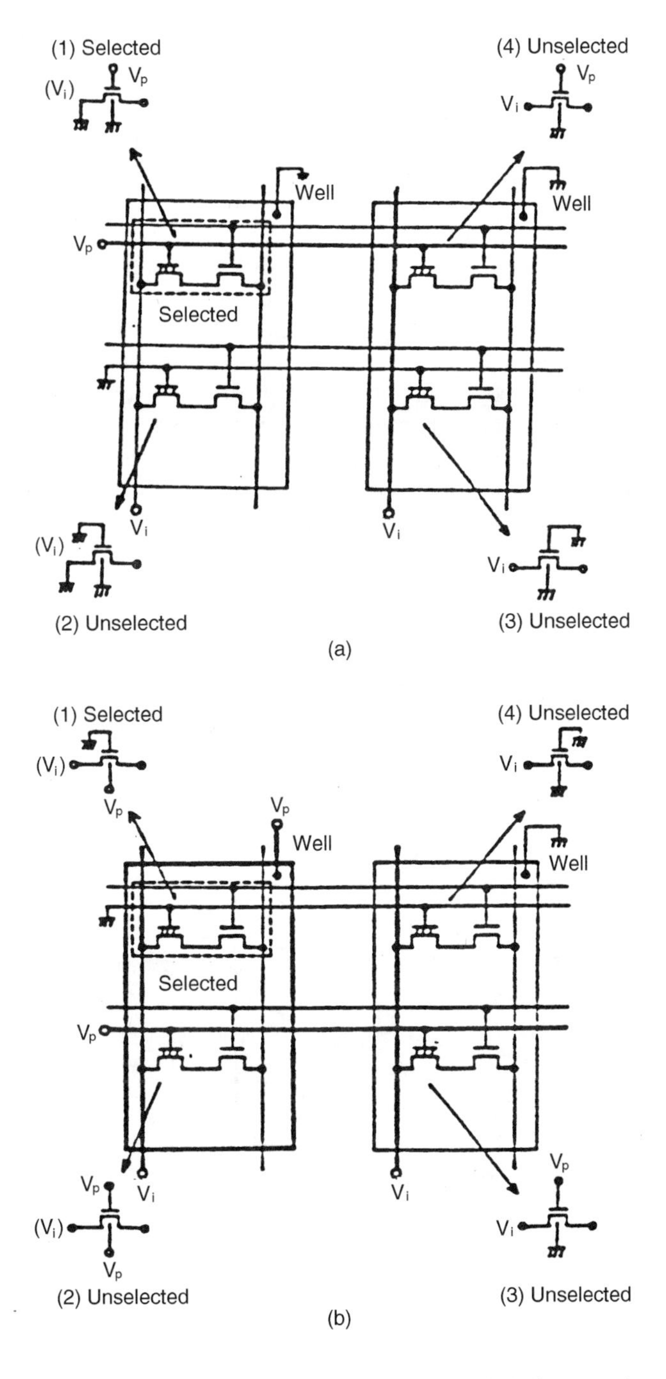

Figure 5.9 Bias condition for (*a*) byte writing and (*b*) byte erasing a 2TC SNOS memory cell [5.5].

memory cells, there is no need for current to flow in the cell during the write mode of operation since a channel need only be formed for tunneling of carriers.

5.1.2.2 Pass Gate Memory Cell Electrical Characteristics. For the n-channel SNOS memory device in the p-well CMOS technology of Fig. 5.9, the tunnel oxide is 1.6 nm and the nitride thickness is 30 nm, with a programming voltage of 15 V. Typical write and erase times are about 10 and 200 μs, respectively, for an applied voltage of 15 V. The corresponding erase/write characteristics of the SNOS memory device are illustrated in Fig. 5.10 for 16 K and 64 K EEPROMs [5.5]. The erase time is shorter by 1.5 orders of magnitude for the 64 K EEPROM due to the reduction in tunnel oxide thickness from 2.1 nm to 1.6 nm. This is a direct consequence of the fact that the erase operation is governed by direct tunneling, while the write operation is controlled by modified Fowler–Nordheim tunneling. Figure 5.11 illustrates the retention characteristics of the 1.6 nm and 2.1 nm tunnel oxides, and 10-year nonvolatility is confirmed at 125°C [5.5]. P-channel SONOS memory devices in an n-well CMOS technology have been reported with 1.6 nm tunnel oxides, 15 nm oxynitride layers, and 4 nm capping or blocking oxides in radiation-hard 64 KB and 256 KB (1 MB EEPROMs under development) NVSMs and ASIC chips operating under 10 V programming voltages (7.5 ms erase, 2.5 ms write), with 10-year retention at 80°C and 10 K erase/write cycles.

5.1.2.3 Pass Gate Memory Cell Technology. The SNOS memory transistor characteristics given in Figs. 5.10 and 5.11 were obtained from devices with tunnel oxides formed in a nitrogen-diluted oxygen ambient (0.1% oxygen) at 850°C as described in the discussion of the trigate memory cell [5.5]. The technology is

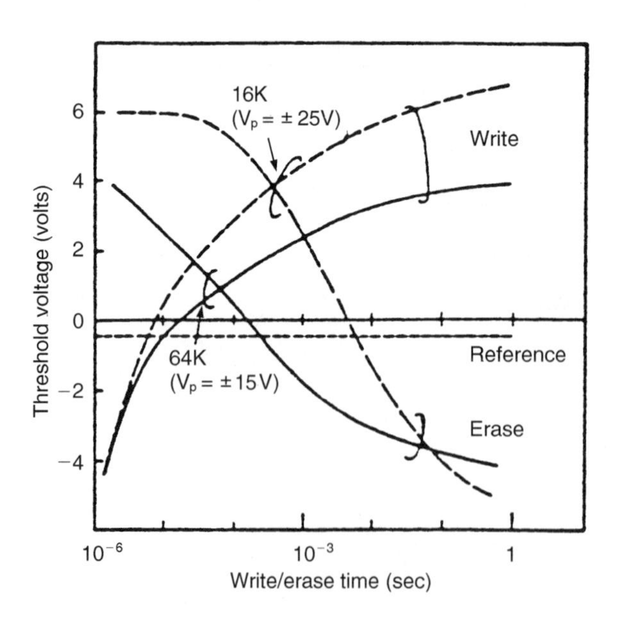

Figure 5.10 Write/erase characteristics of SNOS memory devices [5.5].

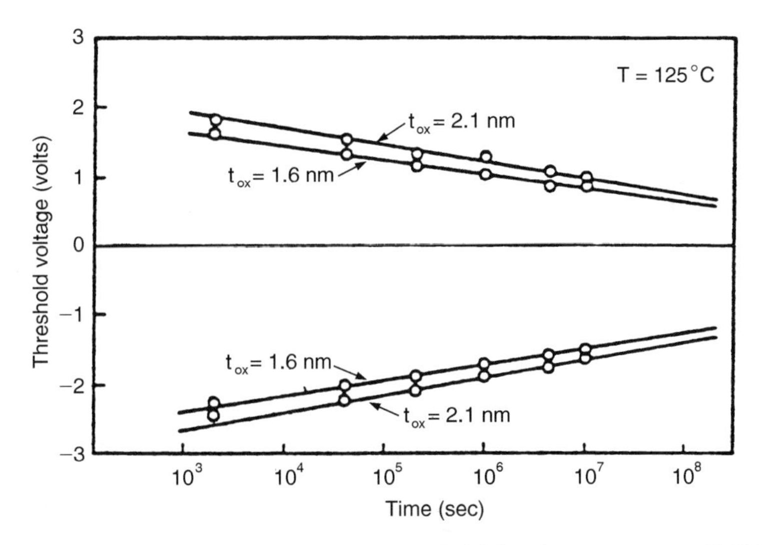

Figure 5.11 Retention characteristics of high-voltage-structure SNOS devices having 1.6 nm and 2.1 nm thick tunnel oxide [5.5].

very similar to that for silicon nitride formed with a mixture of ammonia and dicholorosilane at 790°C. A high-temperature hydrogen anneal is performed at 900°C. The reproducibility of tunnel oxide formation is estimated to be about (± 0.1 nm [3σ]) based on ellipsometric data on thicker (5 nm) oxide samples. Figure 5.12 illustrates the influence of tunnel oxide variations on the erase/write electrical

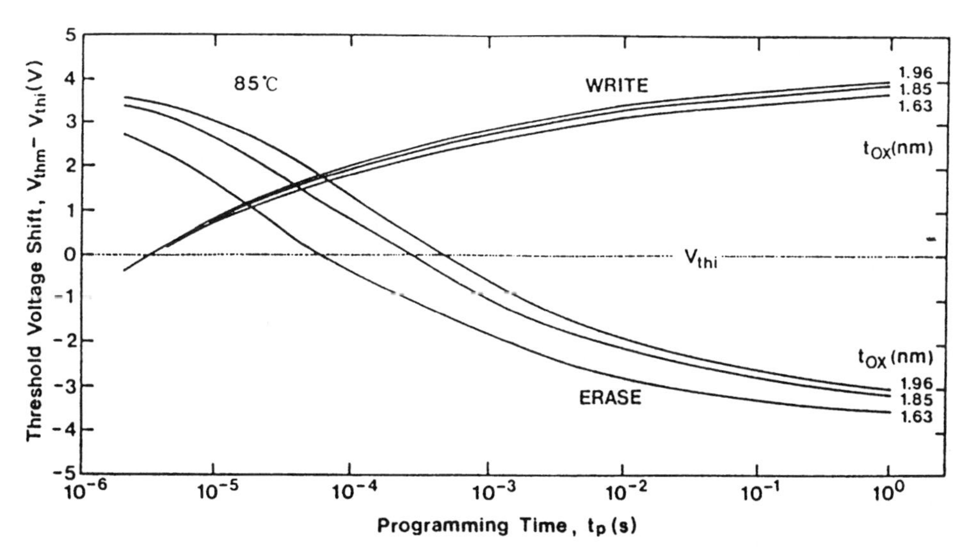

Figure 5.12 Tunnel oxide thickness dependence of programming (erase/write) characteristics of a SNOS memory transistor. V_{thi} is the virgin level (never programmed) [5.49].

characteristics of a SNOS memory transistor with a 26.3 nm silicon nitride film and programmed at approximately 14 V [5.49]. It can be seen that the write operation is not influenced to a large extent by variations in the tunnel oxide since electron injection occurs by modified Fowler–Nordheim tunneling where nitride thickness variations are more important in setting the programming electric field.

The threshold voltage shift ($V_{thm} - V_{thi}$) is slightly more positive for the thicker tunnel oxides, which indicates a reduction in electron backtunneling (i.e., an increase in retention). The erase operation, however, is influenced by the variation in tunnel oxide thickness since hole injection is by direct tunneling, which is highly dependent on the thickness of the tunnel barrier (i.e., the tunnel oxide thickness). Figure 5.13 illustrates the dependence of erase time on tunnel oxide thickness, with a delay of one decade in time for each 0.3 nm change in tunnel oxide thickness; Fig. 5.14 illustrates the influence of tunnel oxide variations on data retention characteristics at 85°C, exhibiting 10 year retention; and Fig. 5.15 illustrates the effect of 100 K erase/write cycles on data retention at 85°C, where the so-called insensible level corresponds to the sense amplifier window in the detection circuitry [5.49]. In the case of oxynitride films, for improved retention, a mixture of $NH_3 : SiH_2Cl_2 : N_2O$ is employed, together with a high-temperature hydrogen anneal, to provide improved data retention.

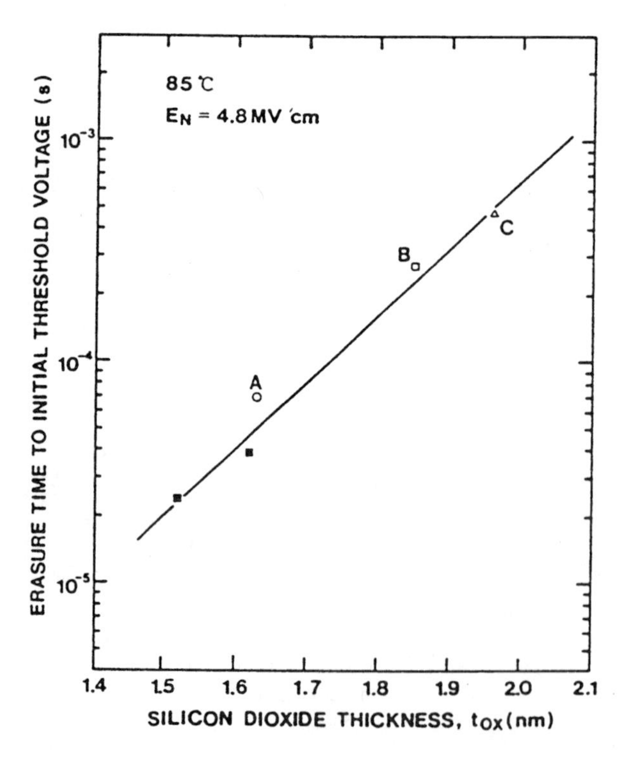

Figure 5.13 Erase time versus silicon dioxide thickness, t_{ox}. The logarithmic erasure time is proportional to t_{ox} according to the direct tunneling mechanism. Erase time is delayed one decade for each 0.3 nm of tunnel oxide thickness [5.49].

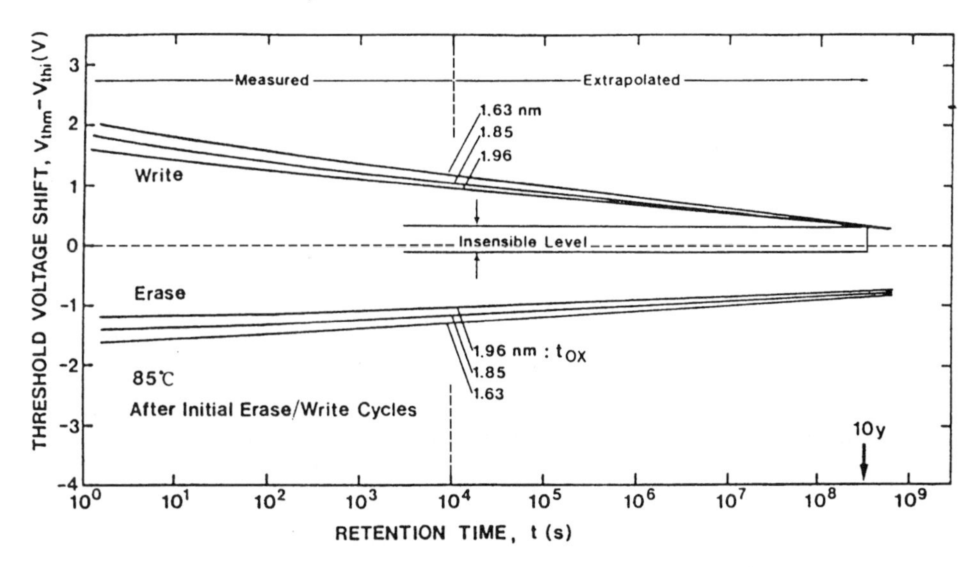

Figure 5.14 Tunnel oxide thickness dependence of data retention characteristics. The programmed threshold voltage, V_{thm}, varies with logarithmic decay after 1000 s, and as a result, retention time is predictable [5.49].

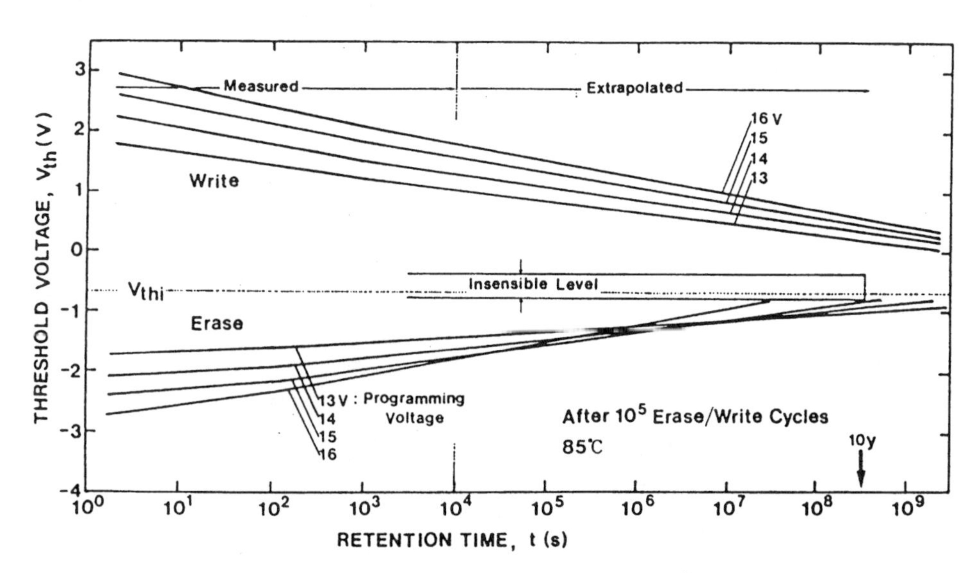

Figure 5.15 The data retention characteristics after 10^5 erase/write cycles with various programming voltages. The total erase and write pulse width is 10 ms, which is the worst case condition for endurance [5.49].

5.1.3 Split (Merged) Gate Cell (1-1/2TC)

The split gate or merged gate SONOS cell is a compaction of the pass gate cell into a single transistor with overlying polysilicon gates, as shown in Fig. 5.16 for an n-channel SONOS memory cell [5.6]. The address gate partially overlaps the memory gate. The SONOS memory device has a 2 nm tunnel oxide, a 5 nm silicon nitride layer, and a 4 nm top capping or blocking oxide. A conventional LDD structure is used in the memory cell, as described previously, to reduce the field in the vicinity of the drain. Figure 5.17 shows a cell layout and the equivalent circuit for 4 bit cells where a design rule of 1.2 μm for the address gate and a memory gate length of 1.5 μm translates into a cell size of 4.0 μm × 4.6 μm (18.4 μm^2), which is comparable to the cell sizes of UV EPROMs and Flash EEPROMs [5.6]. Figure 5.18 illustrates a cell layout for a split gate memory cell with 0.25 μm design rules to achieve a 1.0 μm^2 cell area.

5.1.3.1 Split Gate Memory Cell Operation. The electrical operation of the split gate memory cell shown in Figs. 5.16 and 5.17 is described with the schematic of Fig. 5.19 and the accompanying table [5.6]. In this configuration, all of the memory sources and p-wells are connected in common to reduce the memory area. However, this requires a –4 V (V_{pp}) to be applied to all of the address gates (X1, X2) to disconnect the common sources from each memory region. In the program (write) mode of operation for the SONOS memory cell, all of the cells on the same wordline are first erased and the new data are selectively written into the array. A + 5 V is applied to the memory gate of the SONOS transistor, and –4 V is applied to

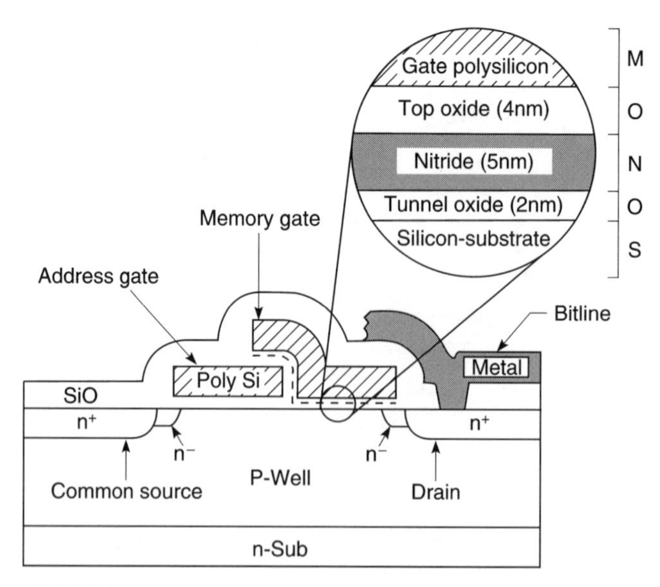

Figure 5.16 Schematic cross section of the dual gate SONOS memory cell. The inset shows the SONOS structure [5.6].

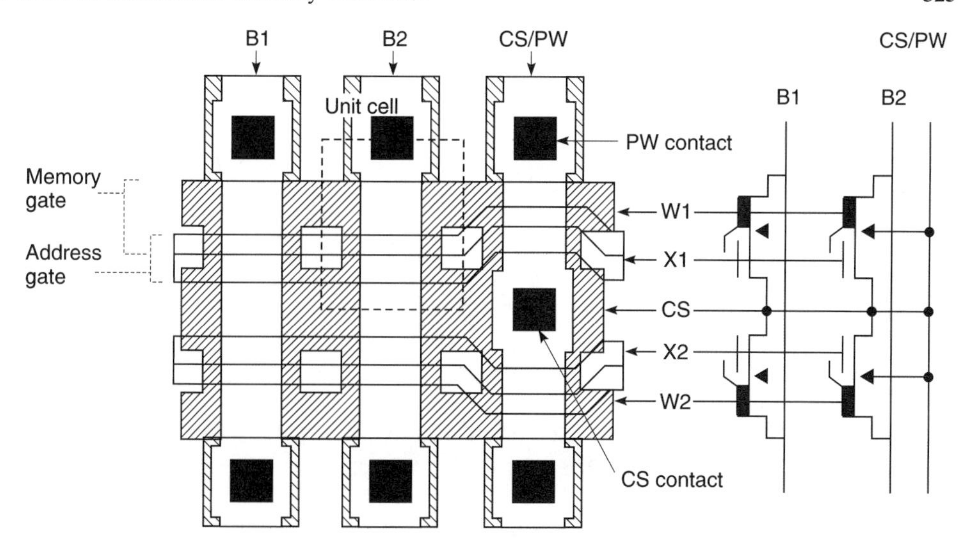

Figure 5.17 Design and equivalent circuit of 4 bit cells. The dotted squares signify unit cell areas [5.6].

the p-well and the selected drain. Thus, a +9 V is applied to the memory gate of cell A relative to the p-well and the selected drain, which causes electrons in the channel to be injected (tunneled) into the ONO structure with subsequent trapping and storage in the thin silicon nitride layer.

The write-inhibit of cell B is achieved by the +5 V bitline (B2). In the erase mode, –4 V is applied to the memory gate of cell A, while the p-well receives +5 V. Thus, –9 V is applied to the memory gate relative to the p-well, which causes holes to accumulate at the silicon surface and be injected (tunneled) into the ONO structure with subsequent trapping and storage in the thin silicon nitride layer. All of the cells on the same wordline are erased simultaneously, and since +5 V is applied to the

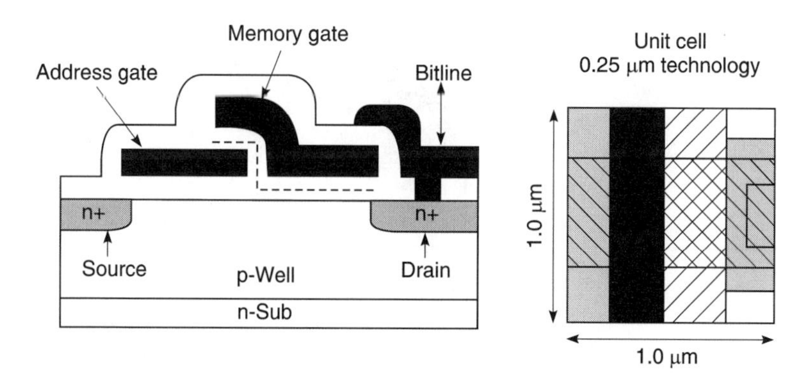

Figure 5.18 0.25 μm CMOS technology with scaled dual gate SONOS memory cell.

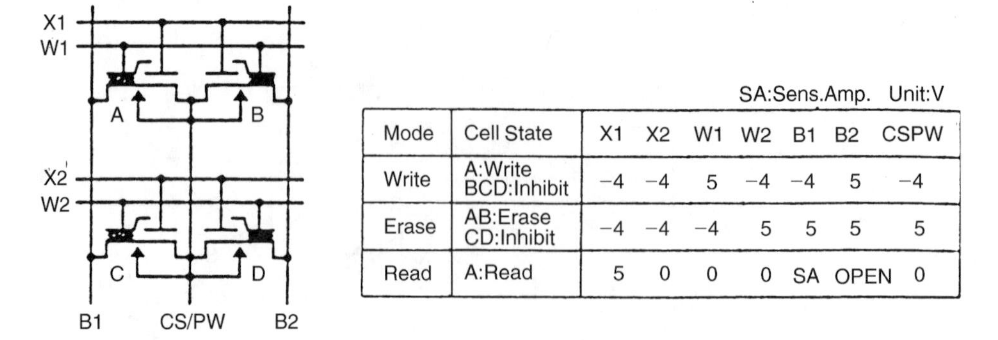

Mode	Cell State	X1	X2	W1	W2	B1	B2	CSPW
Write	A:Write BCD:Inhibit	−4	−4	5	−4	−4	5	−4
Erase	AB:Erase CD:Inhibit	−4	−4	−4	5	5	5	5
Read	A:Read	5	0	0	0	SA	OPEN	0

Figure 5.19 Scheme of applied voltages in memory write, erase, and read modes [5.6].

n-substrate and p-well, there is no leakage current to the substrate. In the read mode, 0 V is applied to all of the memory gates, while +5 V is applied to the selected address gate. The cell current is detected by a current-sensing amplifier, while the memory gate and source are maintained at 0 V and the drain is limited to less than 1.5 V to minimize read-disturb, which can cause a soft-write. The split gate memory cell has been proposed to operate from a single power supply with an appropriate selection of the erase/write window.

5.1.3.2 Split Gate Memory Cell Electrical Characteristics. Figure 5.20 shows the erase/write characteristics of the split gate memory cell shown in Figs. 5.16, 5.17, and 5.19 [5.6]. The write and erase voltages are ± 9 V, with a write time of 100 μs and an erase time of 10 ms, which is suitable for a multi-block-erase/page write programming scheme. The erase time may be improved to 1 ms by reducing the tunnel oxide thickness, as shown in Fig. 5.21, where comparable nitride and blocking

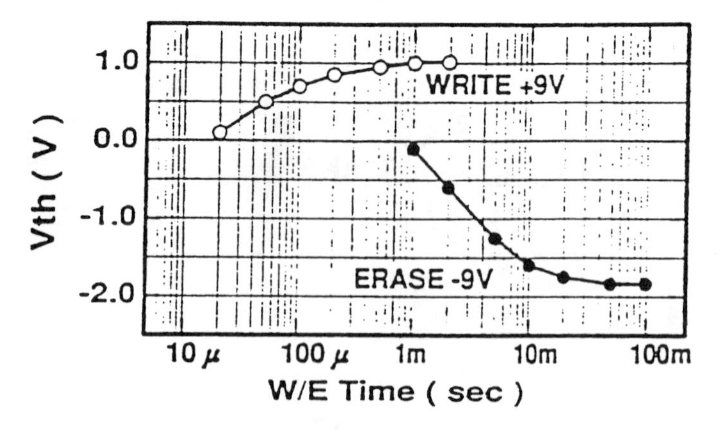

Figure 5.20 Threshold voltage versus write/erase time [5.6].

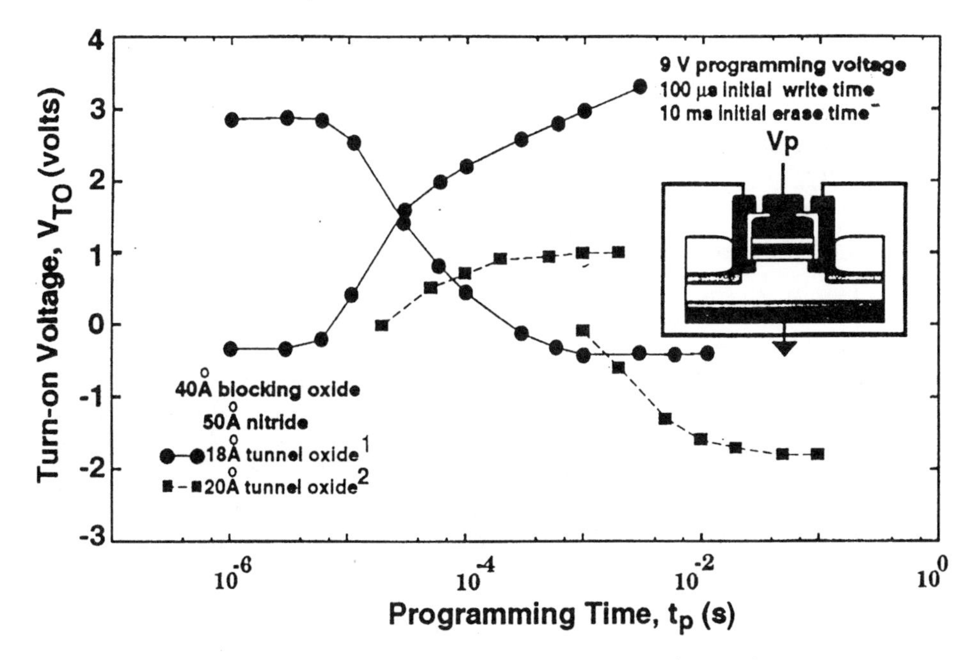

Figure 5.21 SONOS erase/write characteristics [5.36].

oxides were employed except for the reduction in tunnel oxide thickness [5.21]. The endurance characteristics of the SONOS memory transistor are shown in Fig. 5.22, where an erase gate voltage of –9 V for 10 ms and a write gate voltage of +9 V for 100 μs was employed [5.6]. The endurance level is at least 10 M cycles.

5.1.3.3 Split Gate Memory Cell Technology. The top oxide in the SONOS memory transistor may be formed by either a thermal oxidation procedure in dry oxygen or steam to provide a high density of trapping sites near the top oxide/nitride interface or a combination of deposited oxide (i.e., NH_3 and SiH_2Cl_2 mixture) and thermal oxide for densification of the blocking oxide. A conventional LDD structure may also be employed to reduce the electric fields near the drain junction as shown in

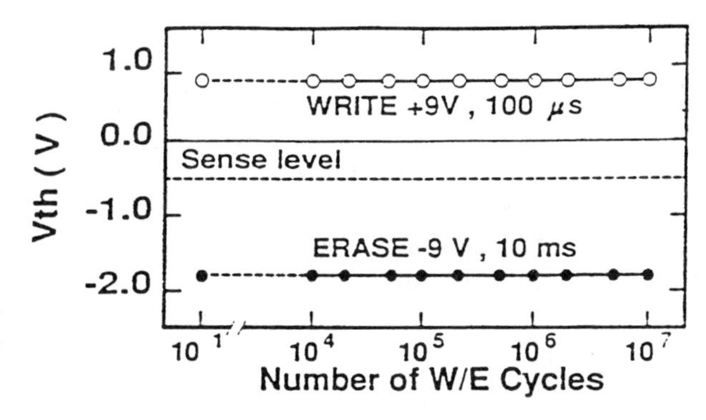

Figure 5.22 Endurance characteristics of threshold voltage versus the number of write/erase cycles [5.6].

Fig. 5.16. The top oxide prevents hole transport to the gate electrode. However, it does not effectively inhibit electron injection both to and from the gate electrode. The trap density in the silicon nitride layer may be increased by adjusting the ratio of ammonia to dichlorosilane, which will increase the memory window by providing more charge storage sites.

5.1.4 Single-Transistor Memory Cell (1TC)

A goal of nonvolatile semiconductor memory (NVSM) technologists has been to realize a true single-transistor memory cell (1TC) for high-packing density and manufacturability. To date, however, there has not been such a memory cell. The basic limitation of the 1TC approach is the isolation of the erase state memory cells (normally on) from the column sense circuitry, while the desired memory cell is addressed during the read operation. One approach shifts the erase/write memory window with selective ion implantation, whereas another approach shifts the memory window with the use of a source–substrate bias. Still another technique employs localized hot-electron injection near the drain junction to program the SONOS device over a limited region, thereby leaving the rest of the channel free to serve as an access transistor. All of these methods have limitations with respect to integration into commerical processes for EEPROM and NVRAM products.

Another application for the 1TC approach is for a DRAM concept of a smaller cell size, simpler cell structure, and simpler operation than the conventional DRAM structure. Since the SONOS 1TC stores the information, or charge, in the insulator of the transistor, no additional storage capacitor is needed, thus minimizing the cell size and increasing the single-event upset immunity. In this instance, long-term retention is not as important as in EEPROM and NVRAM applications. A recent thin tunnel oxide (12 Å) approach has shown that data retention can be as long as 1,000 seconds at 80°C after a 10E11 cycle endurance test with +7 V/–7 V, 500 nsec write/erase pulses [5.53]. Other work has shown low programming voltages (5 V) and high endurance (greater than 10E7 cycles), and has proposed a SONOS 1TC memory cell in a NOR architecture with a cell area of $6F^2$, where F is the technology feature size [5.55]. A 0.20 μm feature size permits a 1TC area of 0.24 μm^2 for advanced 1-Gb memory chips.

5.1.5 NVRAM Transistor Cell (SRAM + EEPROM)

One of the early examples of the use of MNOS memory technology was in an automatic meter reader that used p-channel MNOS memory transistors in a memory counter stage in series with static p-channel load elements. The idea of NVSM transistors as load elements to store the metered count was a forerunner of the concept of a shadow RAM or nonvolatile RAM (NVRAM). NVRAMs are high-performance memories where fast read/write access is assured by the SRAM part of the memory cell, while the nonvolatility feature is provided by the EEPROM device. An important function of the NVRAM is to transfer data between the SRAM and the EEPROM in a parallel or block mode during power down or power failure. Thus, data are saved during a power outage, thereby eliminating the need for battery backup. In general, the cell for conventional NVRAMs consists of a 6TC SRAM cell and a floating gate EEPROM with its associated high-voltage circuitry, although recent progress has been made on low-voltage SONOS EEPROMs.

5.1.5.1 NVRAM Cell Operation. Fig. 5.23 illustrates one of the early versions of the NVRAM with p-channel conventional MOS transistors in the static memory cell of Fig. 5.23*a*, and a dual MOS-MNOS memory cell as shown in Fig. 5.23*b* with depletion-mode load transistors T_3 and T_4 [5.7, 5.47, 5.48]. The gates of transistors T_5 and T_6 are controlled by the wordline (WL), while the gates of transistors T_7 and T_8 are controlled with the MG (memory gate) signal voltage. Conventional SRAM operation is achieved with information exchange between the digit lines (D) and the bistable nodes (Q) by maintaining MG at a low level. The nonvolatile memory operation is achieved with a sequential +28 V erase voltage pulse followed by a −28 V write voltage to the MG control signal line. The erase pulse shifts the threshold voltages of the MNOS memory transistors MT_1 and MT_2 to a high level (+2 V), and then the application of the write pulse shifts the threshold voltage of the memory transistor, with its source near V_{ss}, to a low level (−6 V). The threshold voltage of the other memory transistor, which is near V_{dd}, remains at the high level (+2 V) (i.e., the write-inhibit). Thus, the content of the MOS SRAM is written into the paired MNOS memory transistors. This information may be retained without power supplies for a long period of time. The recall function provides retrieval at the power turn-on stage, although retrieval may also be activated with a stable power supply. Both operations require the deactivation of transistors T_7 and T_8 with a high-level MG in the early stages and then a low-level MG for MOS SRAM operation.

In a more recent development of the NVRAM cell, a 3.3 V 6T CMOS SRAM latch has been combined with 2 p-channel thin film transistors (TFTs), S1 and S2,

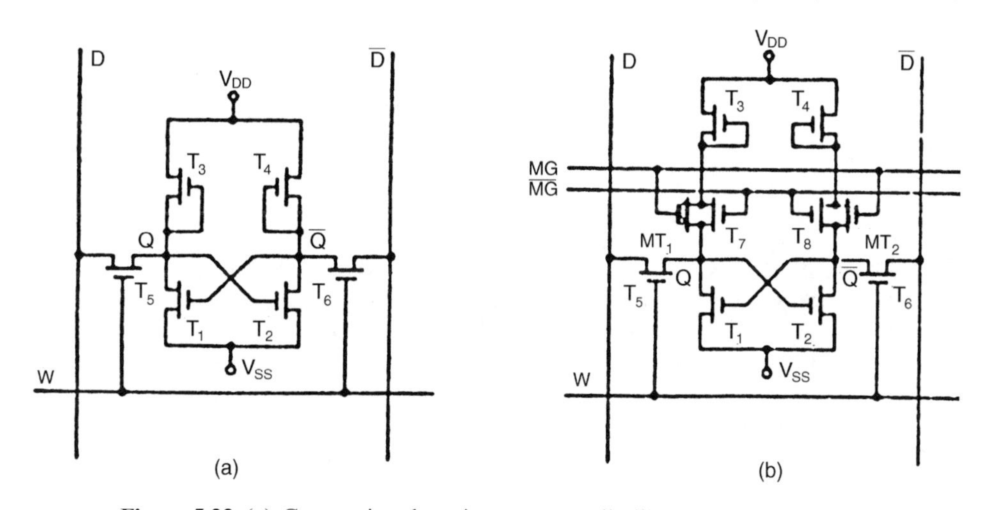

Figure 5.23 (*a*) Conventional static memory cell. (*b*) Read/write memory cell [5.7].

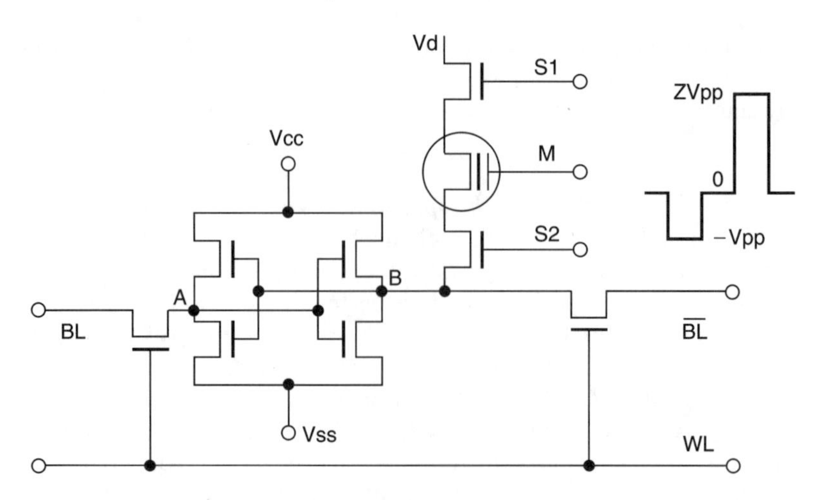

Figure 5.24 The proposed NVRAM bit cell. M is the nonvolatile memory element [5.46].

and a NVSM p-channel SONOS element (M) as illustrated in Fig. 5.24 [5.46]. In the power-up condition, node B always comes up in the low state. There are three operational modes for the NVRAM cell: NVSTORE, NVRECALL, and READ. In the read mode, transistors S1 and S2 are off, which isolates the latch from the NVSM device and the cell operates as a conventional SRAM. In the NVSTORE operation, the contents of the SRAM are transferred to the NVSM as S2 is turned on, and the pulse sequence shown in Fig. 5.24 is applied to the gate of M, where M will be either erased or written depending on the voltage at node B. In the NVRECALL mode, the data are transferred from M to the SRAM cell since both S1 and S2 are turned on during the power-up sequence. If M is in the erase state, then a current flows into node B and pulls the latter into a high state, whereas, if M is written, then node B automatically goes into the low state. The operational voltages for the SRAM are 3.3 V and the NVSM is programmable with $V_{pp} = \pm 3 - 4$ V (i.e., twice the supply voltage) in a 0.4 μm CMOS technology. The advantage of this NVRAM cell is the use of NVSM p-channel load elements, which is compatible with the design of high-density BiCMOS SRAMs and p-channel TFT load resistors.

5.1.5.2 NVRAM Cell Electrical Characteristic. The electrical characteristics of the NVRAM cell shown in Fig. 5.23 are described in this section. Figure 5.25 illustrates the erase/write and retention characteristics of the p-channel SONOS memory transistors constructed on polycrystalline silicon [5.46]. The SONOS device provides a saturated 2.5 V window for a program voltage of 8 V with a crossover point of 20 ms at ± 6 V programming. A usable 0.6 V window exists at the end of 10 years. The decay characteristics are logarithmic as carriers detrap from the tunnel oxide through the tunnel oxide in a manner similar to the performance of SONOS transistors constructed on bulk silicon. An important feature of this device is the use of relatively low voltages (< 8 V) during program operations. These voltages may be derived by "on-chip" doubling of the supply voltage.

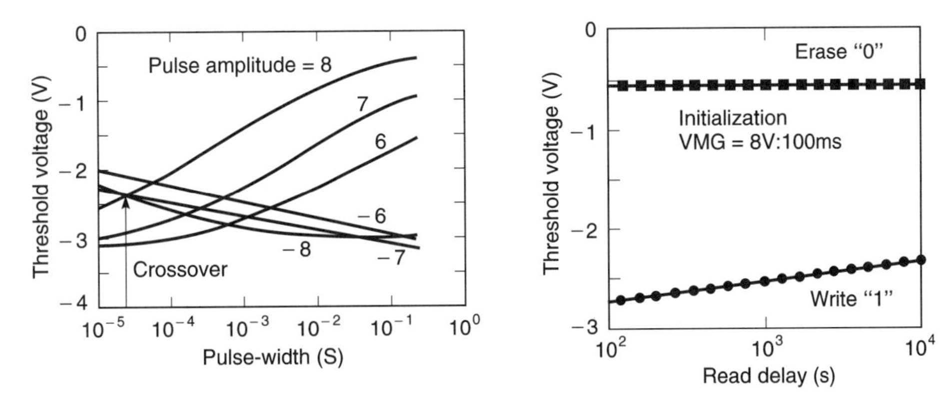

Figure 5.25 Electric characteristics of the p-channel TFT SONOS nonvolatile memory cell (substrate floating) [5.46].

5.1.5.3 NVRAM Cell Technology.

The technology of the p-channel SONOS TFT transistors described in Section 5.1.5.2 is described in this section. The process of forming the SONOS device is similar to an off-set, bottom gate TFT device employed in thin film panel displays, with the exception of the formation of the ONO gate insulator. This is illustrated in Fig. 5.26 where a schematic of the cross section, as well as a high-resolution TEM photograph, are given [5.46]. In addition, a SEM cross section showing the quad-poly process and "stacked" NVSM is illustrated. The inverted SONOS device is formed by first depositing, on the polysilicon gate, a LPCVD blocking oxide of 5 nm followed by a 10 nm silicon nitride layer. The top surface of the nitride is subsequently thermally oxidized to produce a 2 nm tunnel oxide, which receives an 80 nm polysilicon deposition to form the substrate region. There is a 0.25 μm gate-to-drain off-set in the formation of the SONOS device.

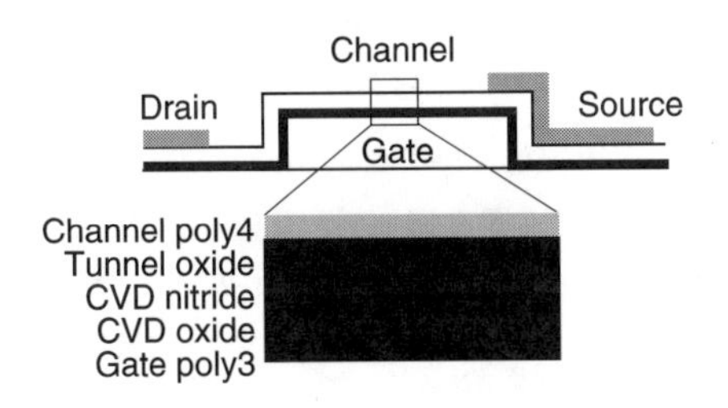

Schematic of the NVM device and the high-resolution TEM through the boxed region

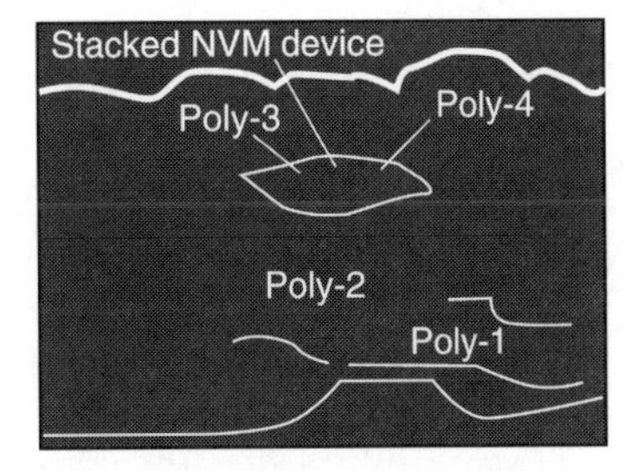

SEM cross section showing the quad-poly process and the stacked nonvolatile memory device

Figure 5.26 P-channel SONOS TFT nonvolatile memory device cross section and quad polysilicon process for NVRAM [5.46].

5.2. MEMORY ARRAY CIRCUITRY

The sections that follow discuss progress made on the various SONOS memory cells, namely, the (1) trigate cell (3TC), (2) the pass gate cell (2TC), and (3) the split gate (merged) cell (1.5TC). Limited progress has been made toward the realization of a manufacturable 1TC. However, comments here are made on several approaches. The progress toward an NVRAM is noted, and a discussion of several important features of SONOS NVSM arrays concludes this section.

5.2.1 Trigate Transistor Array

The trigate transistor cell was first introduced in 1973 as an aluminum gate, p-channel, drain-source protected, MNOS device to reduce the high electric fields near the drain-source junctions during erase/write operations, minimize the stress on the tunnel oxide crossing the junction region, and provide row isolation in a column read-out scheme [5.14]. Figure 5.2 is a schematic cross section and equivalent circuit of the drain-source protected (DSP) MNOS memory device. The gate structure is composed of an MNOS gate in the middle with a MOS gate of thicker gate oxide on either side, overlapping the source and drain. With this integrated split gate geometry, the memory transistor becomes an enhancement device with high-drain and source breakdown voltages. The DSP MNOS memory cell drain current is controlled by the nonmemory section (thicker gate oxide of the MOS transistor) during the erase (high-conductance) operation, and by the center memory section (thin tunnel oxide MNOS transistor) during the write (low-conductance) operation.

P-channel DSP MNOS BORAM (block-oriented random access memory) chips have been produced for high-density military recorder and secondary memory applications from the 8 KB densities introduced in 1975 to today's 64 KB/128 KB densities [5.15, 5.16, 5.21]. Write and read operations were performed over a full –55°C to 125°C temperature range, and data were retained through brief exposures to temperatures in excess of 250°C. The BORAM chips operated properly after an accumulated ionizing radiation dose of more than 1E5 rads(Si). The product of erase/write cycles and minimum expected retention is greater than 1E12 cycle-hours, or

$$\text{retention} > (\text{1E12 cycle-hours})/(\text{accumulated E/W cycles}) \tag{5.1}$$

A novel feature of the 64 KB/128 KB BORAM chip was the ability to select two one-transistor cells/bit (64 KB/chip) versus one one-transistor cell/bit (128 KB/chip) through a high-level address bit. Figure 5.27 gives the layout of a two transistors/bit cell [5.21]. Figure 5.28 shows the BORAM architecture [5.21]. Twelve address inputs are multiplexed through four address pins to control word selection, with bit A12 designating the type of memory storage (one versus two transistors/bit) architecture configuration. The one and two one-transistor cell(s)/bit array is organized as 1024 by 128 bits or 512 by 128 bits, respectively, with all I/O occurring serially via a 129 bit dynamic shift register. There are four control signals (CSbar, AEbar, CL, and MWbar) to access read, write, row erase, and chip erase functions. Reading and writing is accomplished in blocks of 128 bits. Row clear is performed in one

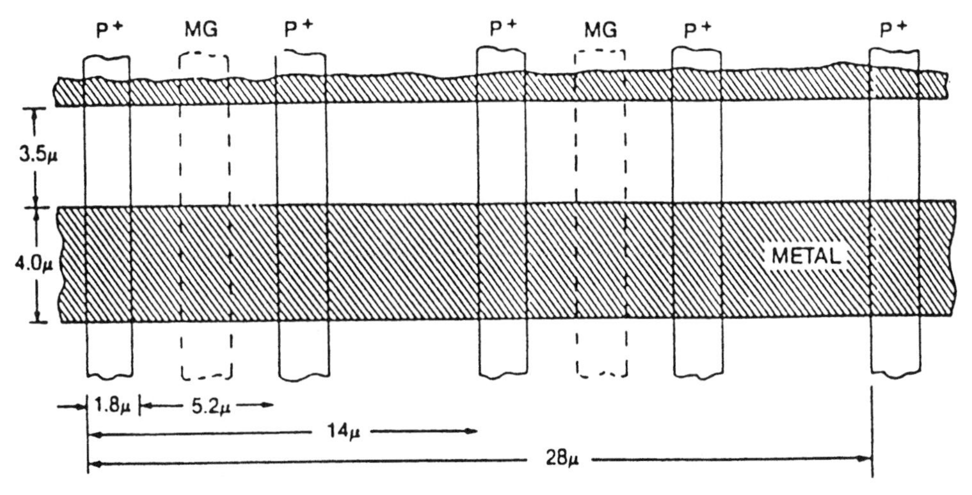

Figure 5.27 Layout of the two-transistor cell [5.21].

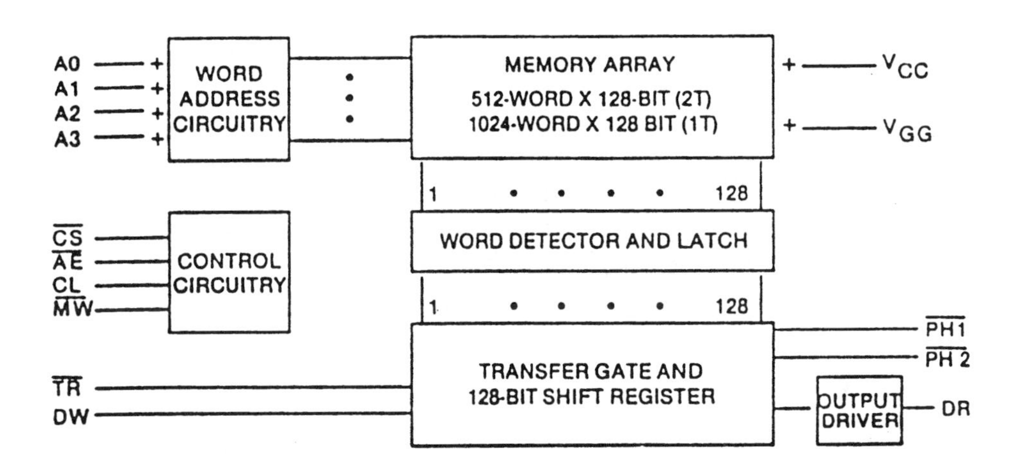

Figure 5.28 Dual mode 64 Kbit/128 Kbit BORAM block diagram [5.21].

(128 KB/chip) or two (64 KB/chip) 128 KB word blocks. The signals TRbar, PH1bar, and PH2bar control data transfer.

An n-channel polysilicon gate SNOS process, introduced in 1979, delivered the first 16 KB SNOS EEPROM, followed by the 32 KB and 64 KB in 1983 and the introduction of the 64 KB CMOS SNOS trigate memory in 1986 [5.4, 5.5, 5.19, and references therein]. The limitations of the previous p-channel Al-gate process, namely, (1) relatively low read speeds ($\approx$ 850 nsec), (2) nonstandard power supplies (3 supplies, including a non-5 V supply in the read mode), (3) nonstandard pin configurations, and (4) read cycle limitations, were eliminated by this n-channel, poly-Si gate SNOS process. The introduction of a high-temperature (800–900°C) H_2 anneal was integrated to prevent degradation of the memory nitride nonvolatility properties caused by the necessary high-temperature treatments of the diffused source-drain junctions in poly-Si technology. The high-temperature H_2 anneal ensured 10-year nonvolatility in SNOS devices. In addition, the introduction of a larger trap density in the LPCVD nitride layer, through the introduction of a larger SiH_4/NH_3 gas ratio, was responsible for the faster program speed without the trade-off of data loss during the read mode. The read reference voltage to the gates was eliminated, thus eliminating the read-disturb condition by essentially employing a static read-out with the SNOS gates grounded. The evolution of the n-channel trigate is based on two levels of poly-Si gates, with two or three gate lines/cell. A typical n-channel trigate memory cell, incorporating two MOS transistors with the SNOS memory transistor, is shown in Fig. 5.3. Bias conditions for each of the memory cell operation modes are shown in Fig. 5.5, with the chip block diagram presented in Fig. 5.29 [5.17]. Today's n-channel trigate cell is implemented in a 1.5 μm, double-poly, CMOS 256 KB EEPROM technology.

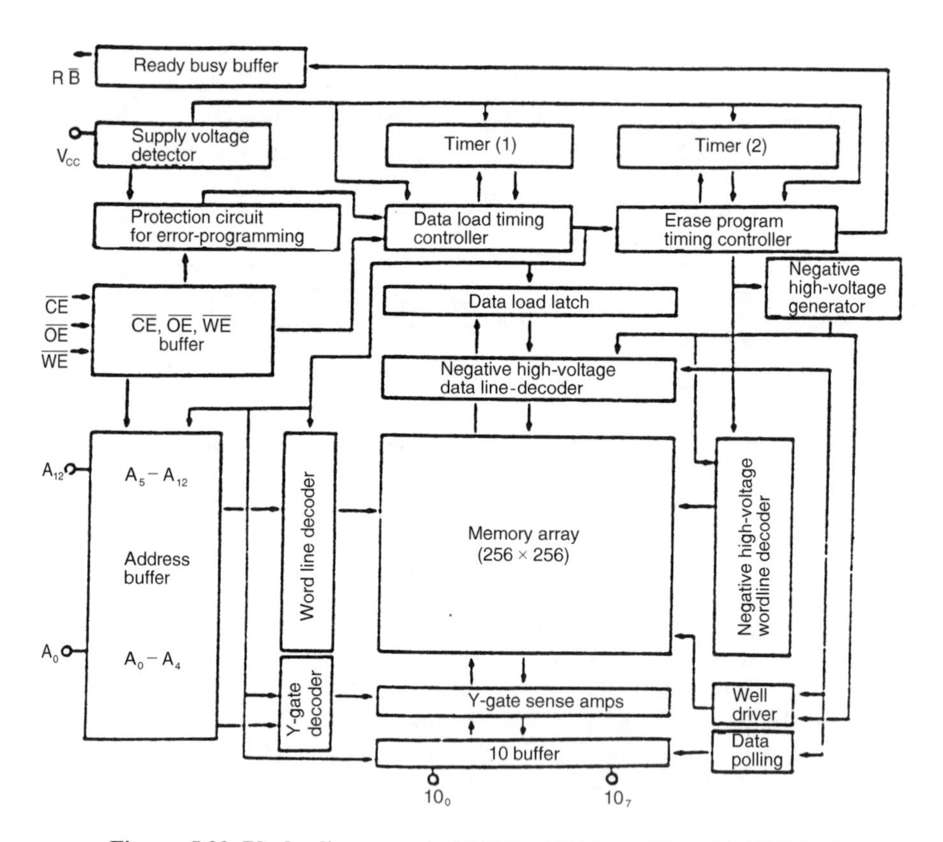

Figure 5.29 Block diagram of CMOS 64 Kbit EEPROM HN58C65. Negative high voltage is used for programming. In the write operation, the negative voltage is delivered to the well and data-line decoder and in the erase operation. The wordline decoder is supplied with negative high voltage [5.17].

5.2.2 Pass Gate Transistor Array

Today's 64 KB and 256 KB SNOS and SONOS EEPROMS use a two-transistor (one MOS access transistor and one SNOS transistor) per bit configuration as shown in Fig. 5.7, implemented in a 1.25 μm to 1.3 μm CMOS technology, with on-chip generated programming voltages ranging from 10 to 13 V [5.29]. The MOS transistor acts as the select device to eliminate the problem of accidentally disturbing the unaddressed SNOS/SONOS cells during a read operation of an addressed SNOS/SONOS cell. A comparison scaling between each generation is shown in Table 5.1 [5.37]. Other features include separate wells within the memory array to allow full byte function and a differential memory cell design (two SNOS memory transistors per bit) resulting in improved memory retention. Figure 5.30 shows an electrical schematic of the basic operation of memory programming [5.30]. For chip density, the two SNOS memory transistors per bit may be replaced by a one SNOS transistor cell and a common reference SNOS cell, as shown in Fig. 5.31 [5.30].

TABLE 5.1 SCALING SCHEME FOR SNOS EEPROMS [5.37]

Item	16 k (k = 1)	64 k (k = 1.5)	256 k (k = 2.5)	SF
Technology (μm)	3	2	1.2	k^{-1}
Cell area (μm^2)	400	180	60	k^{-2}
MOS : t_{ox} (nm)	75	50	30	k^{-1}
SNOS : t_n (nm)	50	32	20	$k^{-1.1}$
: t_{ox} (nm)	2.1	2.0	1.9	$k^{-0.1}$
Program voltage (V)	25	16	10	$k^{-1.1}$

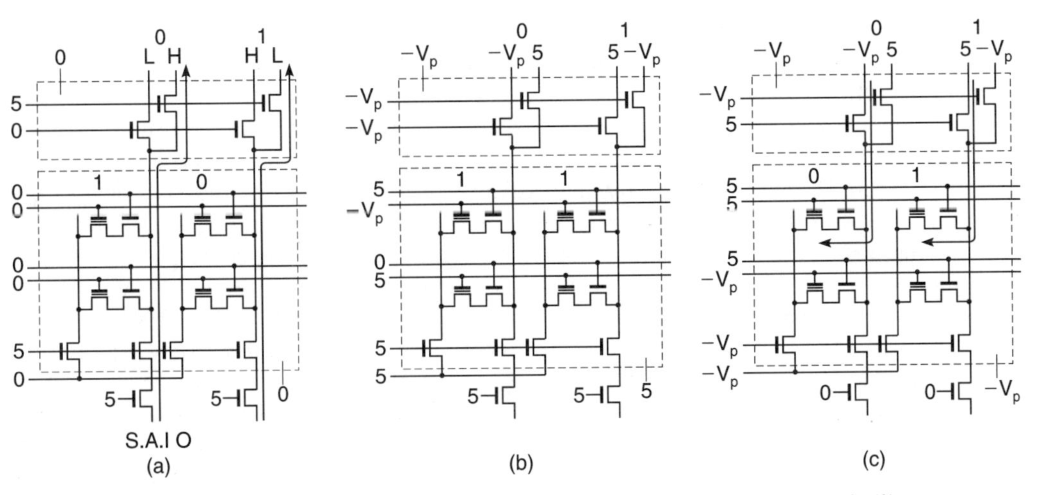

Figure 5.30 Basic operation of memory programming. (*a*) Data load; (*b*) erase; (*c*) write [5.30].

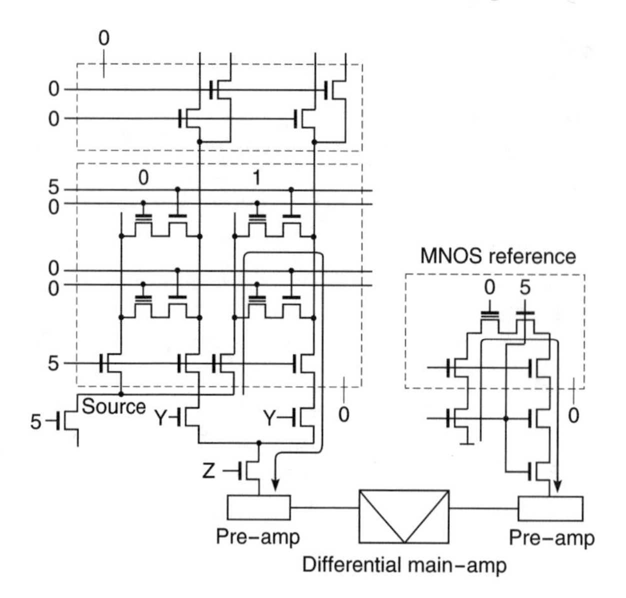

Figure 5.31 Read-out of memory data and differential sense amp [5.30].

5.2.3 Split (Merged) Gate Transistor Array

By 1991, further lateral scaling of the pass gate cell had been developed for realizing a small-area n-channel transistor cell, targeting high-density markets such as semiconductor disk applications [5.6]. The cell consists of an effective 1.2 μm channel length address gate partly overlapping a 1.5 μm effective channel length SONOS memory transistor, as shown in Fig. 5.16. The incorporation of a scaled SONOS gate dielectric limits the erase and write voltage to a 9 V differential. A conventional LDD structure is used, with no extra structure needed for high voltage in the cell or the periphery circuits, thus proving very advantageous in a high-density integration demonstration of 1 MB EEPROM. Process parameters and EEPROM characteristics are listed in Tables 5.2 and 5.3, respectively [5.6]. A single 5 V cell programming voltage scheme using p-channel SONOS in an n-well has been proposed and is accomplished by aggressive vertical scaling of the SONOS device to a 1.1 nm tunnel oxide, a 5.0 nm nitride, and a 4.0 nm blocking oxide. Figure 5.19 summarizes the applied voltages for the write, erase, read, and corresponding inhibit modes for the neighboring cells. Fig. 5.32 shows the erase/write timing diagram and Fig. 5.33 illustrates the peripheral circuitry for the 1 MB SONOS EEPROM [5.6].

Block (one block consists of four pages, or 512 bytes total) erase is divided into a block-address-load cycle and an erase cycle. Blocks to be erased are selected by address latches controlled by A9–A16, where inverters I1 and I2 compose the block address latches. One- to four-page erase is selected by A7–A8. A 1 MB array has 256 blocks. Next, the erase cycle is implemented by bringing PGM and OE low for 10 ms. Thus, in the selected latches, V_{pp} = –4 V is applied through the selected decoder/driver circuits to the wordlines using p-channel transistors MW3 and MW4. Internal

TABLE 5.2 PROCESS PARAMETERS [5.6]

Process	1.2-μm lithography n-sub twin-tub CMOS Double polysilicon
Peripheral transistor	
Gate oxide thickness	35 nm
Gate length	NMOS 1.6 μm PMOS 2.0 μm
Cell transistor	
Gate insulators thickness	
Top oxide	4.0 nm
Memory nitride	5.0 nm
Tunnel oxide	2.0 nm
Memory channel length	1.5 μm
Address channel length	1.2 μm

TABLE 5.3 SONOS EEPROM CHARACTERISTICS [5.6]

Organization	Erase: 256 block × 512 byte Write: 1024 page × 128 byte
Power supply	V_{cc} = 4.5–5.5 V
Standby current	20 μA
Operation current	7.0 mA (1 MHz)
Cell size	4.0 μm × 4.6 μm
Die size	5.3 mm × 6.3 mm
Equivalent prog. time	1.1 ~ 20.5 μs/byte
Erase time	10 ms (1 ~ 256 block)
Write time	100 μs (1 page)

negative voltage generators and zener diodes are used to generate V_{PP} = –4 V. A +5 V is applied to both the bit lines and the common-source/p-well lines. Inverters I3 and I4 are the data latches.

Page-write is divided into a byte load and a write cycle. In the byte-load cycle, a series of 128 byte data are loaded by address A0–A6 to the data latches with valid data on D0–D6. Next, in the write cycle, OE is high, with PGM brought low for 100 μsec. All the data are written simultaneously. A7 and A8 control the selection of one of four pages through the block-address latch controls and place +5 V on the high-voltage wordline via MW2 and MW4. Write and Write–inhibit are executed by the application of –4 V or +5 V, respectively, to the SONOS drain. In the read mode, current sense type amplifiers in the I/O buffer are used to detect the cell current: 0 V is applied to all memory gates, the drain is limited to less than 1.5 V,

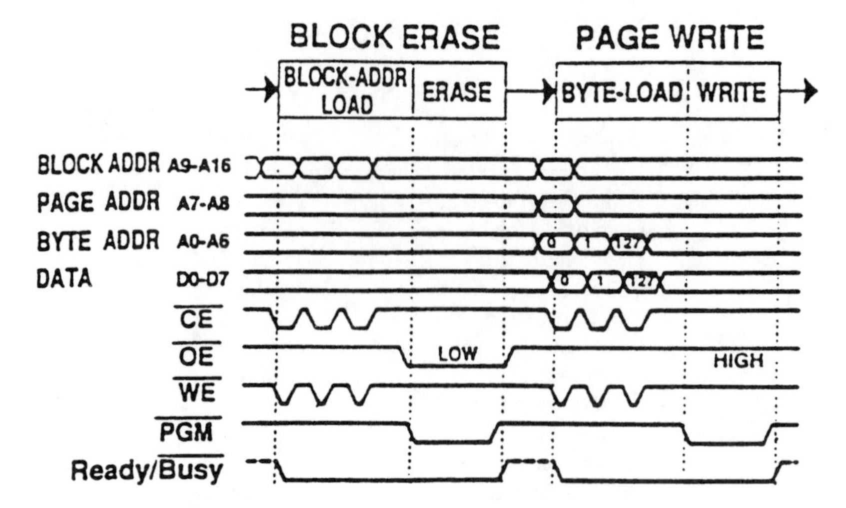

Figure 5.32 Erase/write timing chart of the 1 MB SONOS EEPROM [5.6].

and +5 V is applied to the selected address gate. Note that, since +9 V/100 μs is sufficient for writing, a conventional page-write method was implemented. On the other hand, –9V/10 ms is required for erasing, so the method of multiblock erase was developed, which enables erasing from a minimum of 1 to 256 blocks.

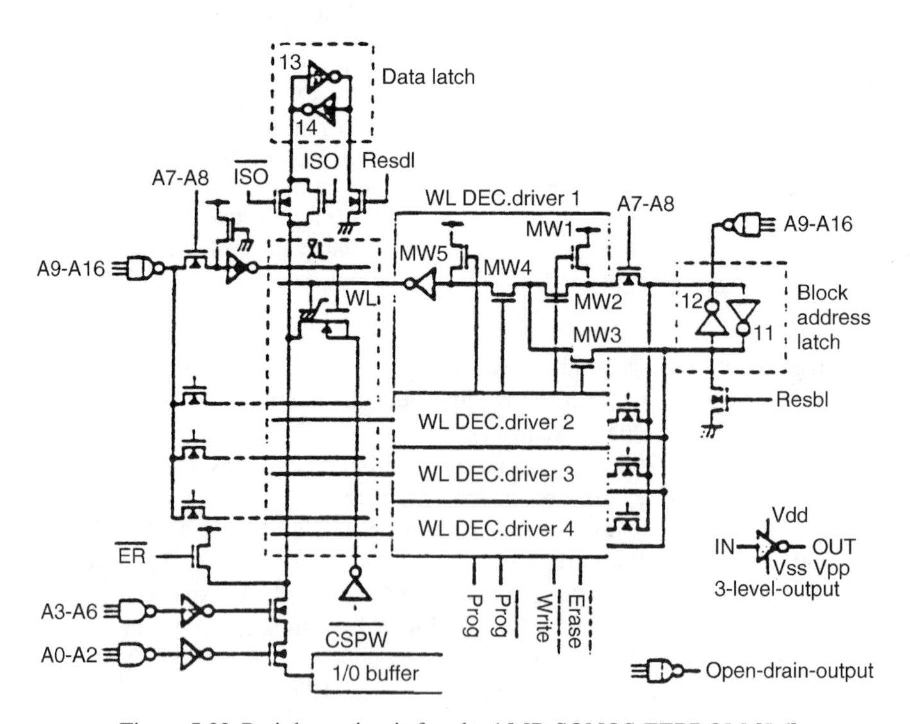

Figure 5.33 Periphery circuit for the 1 MB SONOS EEPROM [5.6].

5.2.4 Single-Transistor Array

The goal of lateral cell scaling, as has been demonstrated, started with a trigate transistor cell (3TC) and progressed to a two-transistor pass gate cell (2TC), a 1.5-transistor merged-gate cell (1.5TC), and the goal of a one-transistor (or device) cell density. Figure 5.34 shows the layout and cross section of an experimental, one-transistor, MNOS memory cell [5.50]. Compared to the floating gate EEPROM, the one MNOS transistor has three advantages: (1) the memory array requires only one level of polysilicon, (2) the memory array requires a small number of contacts periodically on the n+ lines, and (3) the cell permits further scaling (cell size can be $2F^2$, where F is the minimum feature size). The challenges of the cell are: (1) the threshold voltage rises with increasing numbers of write and erase cycles (the end of life is approximately 10^4 cycles) and (2) the threshold window is much smaller (approximately 2 to 3 V) than that of traditional MNOS structures. Specifications for the 256 KB and 1 MB EEPROM are summarized in Table 5.4 [5.50]. Clearly, the incorporation of the SONOS transistor will go a long way toward meeting these challenges.

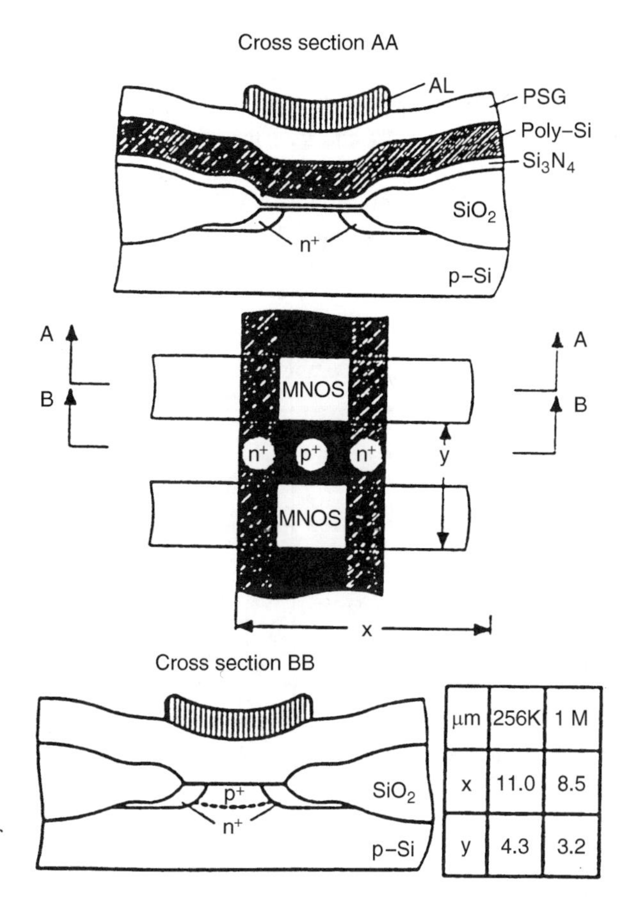

µm	256K	1 M
x	11.0	8.5
y	4.3	3.2

Figure 5.34 Layout and cross section of a MNOS memory cell (1TC) [5.50].

TABLE 5.4 FEATURES OF 256 KB AND 1 MB SNOS EEPROMS WITH 1TC DESIGN

Parameter	256 KB EEPROM	1 MB EEPROM
Organization	32 KB × 8	128 KB × 8
Chip size	8.0 to 5.6 mm	8.9 to 6.5 mm
Cell size	4.3 to 11 μm	3.2 to 8.5 μm
Power (read)	150 mW	150 mW
Power (standby)	50 mW	30 mW
Access time	350 ns	700 ns
Endurance	1000	10000
Retention time	> 3 years	> 3 years
Minimum channel length	2.2	2.2
Page programming time	5 ms	5 ms
Erase time	10 s	10 s

5.2.5 NVRAM Transistor Array

The purpose of the NVRAM is to preserve data when a disturb event occurs. The NVRAM must be capable of accomplishing the NV store operation while the external power supply is being shut down [5.8, 5.9, 5.10, 5.12]. This implies that the only source of energy for the store operation is the charge stored in the decoupling capacitors.

A present 256 KB NVRAM memory cell schematic is shown in Fig. 5.35 [5.11]. The lower portion of the cell is a typical four-transistor SRAM that may be read and

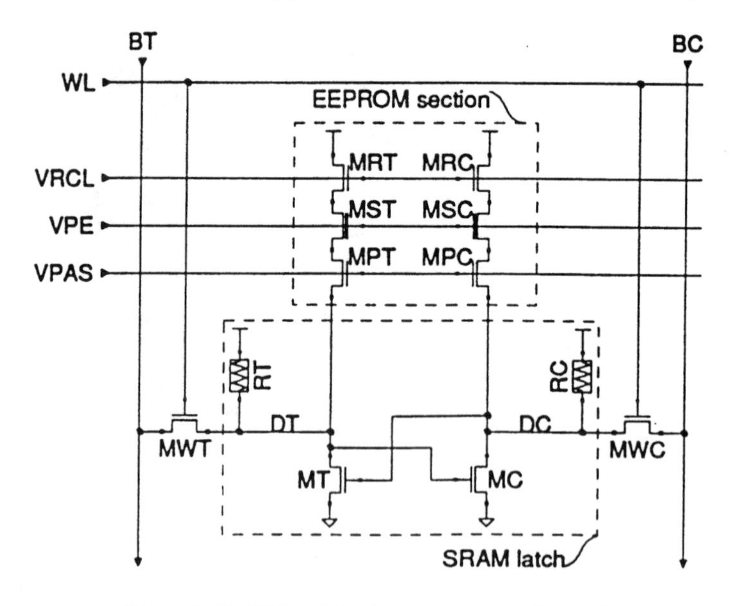

Figure 5.35 NVRAM trigate memory cell [5.11].

written to at SRAM speeds and for an unlimited number of cycles. The upper portion of the cell consists of the EEPROM, entailing two SNOS transistors and four isolation MOS transistors.

The store cycle is accomplished with the wordline (WL) and VRCL lines off. The nonvolatile data are first erased by pumping the VPE line to a negative supervoltage (–10 to –15 V), and the SNOS transistors (MST and MSC) are positively charged to depletion-mode thresholds. The VPE line is returned to ground. Assume that the SRAM node DT is high (5 V) and that DC is low (ground). The VPE line is pumped up to a positive supervoltage (10 to 15 V). With VPAS high and DC low, the channel of SNOS transistor MSC is held at ground. The entire supervoltage appears across the SNOS transistor, and it is negatively charged to an enhancement-mode threshold. Both DT and VPAS are high, so that n-channel transistor MPT is cut off. The surface potential below the SNOS gate on the high side of the cell will couple up with VPE, and, assuming that little potential is dropped across the SNOS dielectric, the MST threshold remains unchanged due to deep depletion channel shielding. This program inhibit scheme confines the supervoltage to VPE only, thus allowing the remaining lower voltage devices to be packed tightly without the need for additional isolation and leading to a higher density layout.

A recall operation begins by returning the volatile data through the bitlines, leaving both DT and DC low due to the very high impedance of resistors RT and RC. The wordline is returned low, VPAS is held high, and VPE is low. Line VRCL is then raised to 5 V, producing a differential current through the enhancement and depletion SNOS transistors. If SNOS transistors MST and MSC have a negative and a positive threshold voltage, respectively, a larger current will flow into node DT versus DC. DT will charge up and turn on transistor MC when its threshold voltage is reached. With MC on, node DC is clamped to ground and DT continues to charge to a stable high voltage. The original nonvolatile data are now restored into the SRAM without any data inversion.

A simplified block diagram of the 256 K NVRAM is shown in Fig. 5.36 and implemented in an 0.8 μm technology [5.11]. The hardware store and recall can be commanded through SEbar and REbar, respectively. Both commands may be software clocked with software control pin SSbar or one of the chip enables. The store busy pin, SBbar, indicates when hardware store is completed.

5.3. DEGRADATION MECHANISMS

A SONOS memory device has certain characteristics that are unique to its nonvolatile memory application [5.20, 5.22, 5.23, 5.24, 5.25, 5.34, 5.35, 5.38, 5.39]. These characteristics have to do with the process by which charge is injected and removed from the Si_3N_4 memory charge storage layer (erase/write), whether the device characteristics are degraded as a function of erase/write cycles (endurance), whether the charge can be reliably held within the nitride for long periods of time (retention), and how these characteristics are altered by temperature and, in some instances, radiation.

Erase/write characteristics depend on the nitride thickness. For a given amount of trapped charge in the nitride insulator, thinning the nitride thickness produces a smaller memory window because the capacitance between the gate and the stored memory charge in the nitride increases. The choice of a blocking oxide in scaling MNOS technology helps preserve the memory window by minimizing the charge injection from and to the gate electrode, thereby improving the charge-trapping efficiency in the nitride.

Retention is the ability of the Si_3N_4 to retain charge when the writing (or erasing) pulse returns to the bulk potential. Usually, it is measured as the change

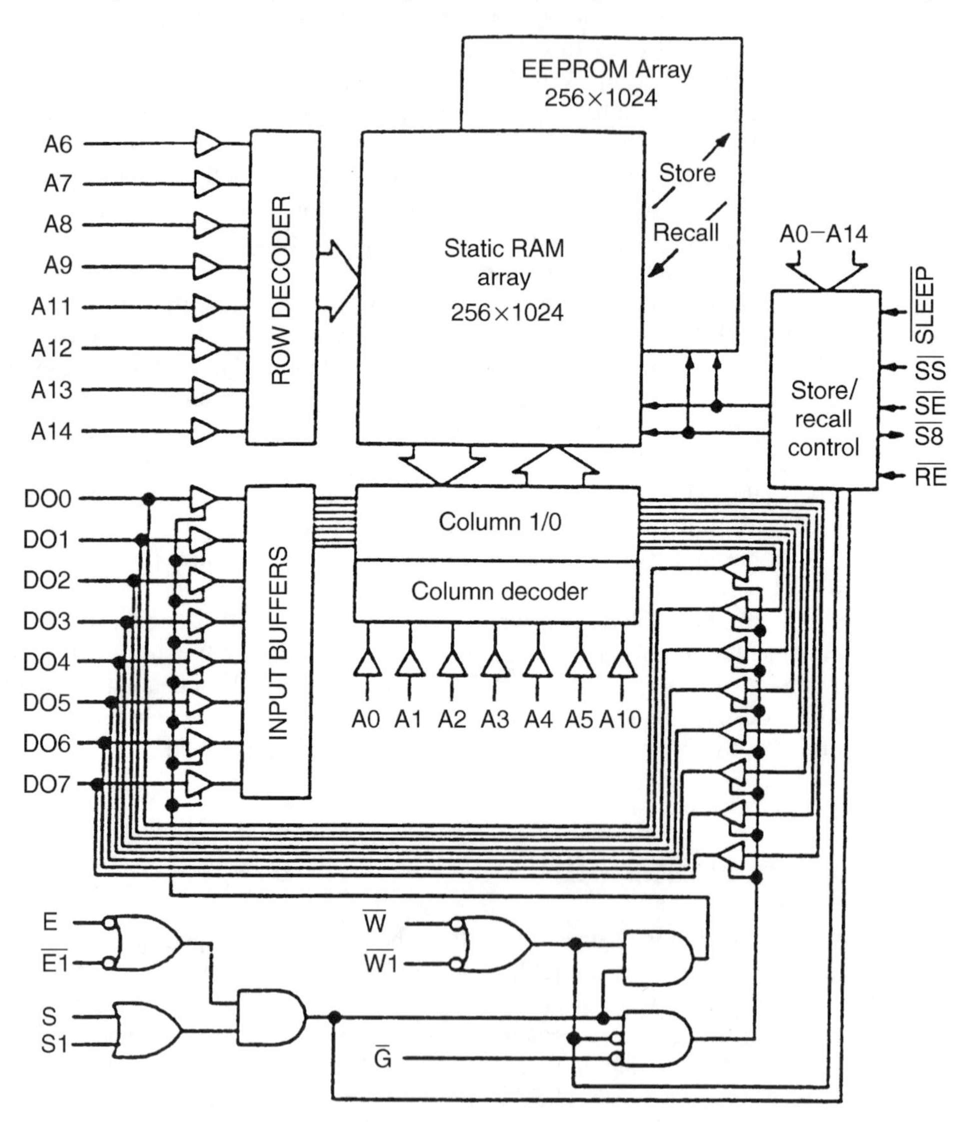

Figure 5.36 256 K SONOS NVRAM block diagram [5.11].

in the threshold voltage as a function of time after the voltage pulse returns to the bulk potential. In conventional MNOS devices, short-term decay and long-term decay rates can be identified, with the threshold voltage changing logarithmically with time. The break point from short-term decay to long-term decay is a function of device cycling, with the breakpoint moving to shorter times as the device is cycled. The mechanism for short-term decay is thought to be backtunneling of stored nitride charge assisted by interface traps, while long-term decay is due to the rearrangement of charge within the nitride. Since data retention characteristics depend on the programming depth of the stored charge, data retention time decreases with decreasing programming voltage, or increasing temperature after programming. Therefore, two acceleration tests are usually implemented, one at high temperatures and the other at a low programming voltage to ensure a 100% screen for 10-year retention. In addition, the penetration depth in nitrides is larger for holes than electrons. When scaling the nitride thickness, holes will, therefore, be trapped closer to the gate and will, in fact, be lost primarily through the gate electrode, thereby degrading the erase state. Thus, the introduction of a blocking oxide layer in more recent SONOS scaling has helped counter this charge loss to the gate electrode.

A MNOS memory transistor may be pictured as possessing an inherent endurance characteristic (or ability to withstand repeated erase/write cycling), which is related mainly to the change in the nitride conductivity, exacerbated by the very high electric fields employed to inject charge into the nitride. However, by keeping the programming voltage low so that the equivalent electrostatic field in the nitride is below 5 MV/cm, the degradation is minimized and the device performance is extended to beyond 10 M erase/write cycles.

5.4. TYPICAL CHARACTERISTICS

Characteristics shown in the following sections represent typical present-day device characteristics of commercially available memory arrays. The list of papers in the reference section of this chapter and that of Chapter 9 is the best source of characteristics for the many different device dimensions and operating conditions available.

5.4.1 Erase/Write

A SONOS device allows the insertion and removal of charge from the nitride layer by application of electrical signals to the device terminals. Typical erase/write characteristics of an n-channel MNOS memory device, with the corresponding energy band diagrams during programming, are shown in Fig. 5.37 [5.40]. As has been discussed in detail in Section 5.1, in the write mode, a positive gate to substrate bias is applied, which causes electrons to be injected by modified Fowler–Nordheim tunneling from the device channel, through the tunnel oxide layer, and into traps in the Si_3N_4 layer. The net negative stored charge causes a positive shift in threshold voltage and places the n-channel SONOS transistor in the low-conduction state.

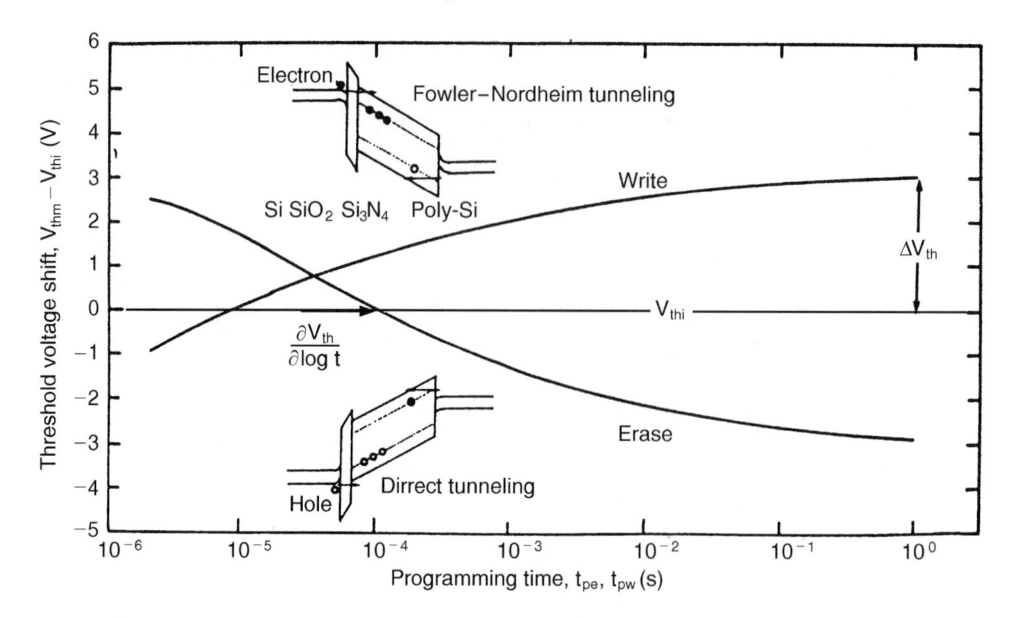

Figure 5.37 Erase/write characteristics and energy-band diagram during programming of an n-channel SNOS memory device [5.40].

Conversely, in the erase mode, a negative gate to substrate bias is applied, which causes holes to be injected by direct tunneling from the device channel through the tunnel oxide layer and into traps located in the Si_3N_4 layer. The resulting net positive charge causes a negative shift in threshold voltage and places the n-channel SONOS transistor in the high-conduction state.

5.4.2 Retention

Retention is defined as the time between the storage of data (i.e., nitride charge) and the time at which it can no longer be read out correctly. In the past, MNOS memory device retention characteristics have been a center of focus. Retention data, after various erase/write cycles at 85°C, are shown in Fig. 5.6. A good quality of scaling is that the retention characteristic may be improved as the Si_3N_4 memory storage layer thickness is reduced. Retention time may be estimated from backtunneling of trapped nitride charge, which is nitride thickness dependent and may be expressed as

$$\tau = \tau_o \exp(\alpha_{ot} X_{ot} + \alpha_n x) \tag{5.2}$$

where the backtunneled charge depends exponentially on X_{ot}, the tunnel oxide thickness [5.54]. The distance of the trapped charge in the silicon nitride from the nitride/tunnel oxide interface, x, for a 10-year storage (memory) time is estimated to be 4.4 nm, where τ_o is the semiconductor conduction band to trap tunneling time constant of 10^{-13} sec, $\alpha_{ot} = 1.07 \times 10^{10}\,m^{-1}$, and X_{ot}, the tunnel oxide thickness, is 1.9 nm. Note that, in scaled SONOS structures, the tunneling of trapped nitride charge to the gate electrode is also possible through the blocking oxide.

5.4.3 Endurance

Endurance of a SONOS device is a measure of the device's ability to meet a specified retention time as a function of accumulated erase/write cycles. The degradation of the SONOS structure comes from two effects; (1) deterioration of the Si–SiO_2 interface in the form of the creation of interface states (traps) with cycling, and (2) deterioration of the bulk Si_3N_4 layer trap density, which manifests itself in the form of increased charge centroid penetration into the nitride film as a function of write/erase cycling. Both mechanisms are linked to the breaking of bonds, which are believed to be Si–H bonds. Figure 5.38 shows the relation between erase/write cycles and the cumulative failure rate of 64 KB EEPROM products.

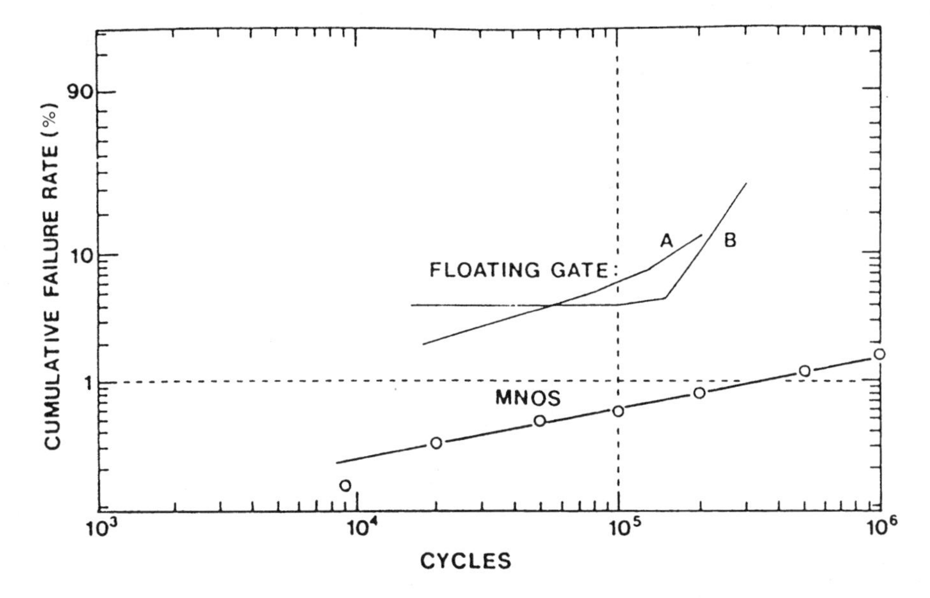

Figure 5.38 Relation between the cumulative failure rate and erase/write cycles. The failure rate of MONOS (SNOS) products is less than 1% at 10^5 cycles. The intrinsic failure mode is at more than 10^6 cycles [5.17].

5.4.4 Radiation Hardness

In general, solid-state devices may be susceptible to gamma, neutron, electron, proton, and alpha source radiation. The difference in radiation response arises primarily from the nature of the incident radiation and its interaction with critical layers of devices. For gamma ray photons, or charged electron or proton particles, most of the energy transfer occurs through a single intermediate process involving the production of plasmons and their decay within a picosecond to produce electron hole pairs. The amount of damage due to ionization is related directly to the charge yield per unit dose (i.e., the number of electron hole pairs created per rad) and the field-dependent recombination. For a specific dose and a specific electric field, the recombination rate of electron hole pairs is significantly different for gamma rays and electron radiation. At a fixed electric field, the recombination rate is determined by the line density of electron hole pairs created by the incident radiation. The line density, which is determined by the linear energy transfer (LET), is inversely proportional to the average separation distance between the created electron hole pairs. Obviously, the closer the spacing of the pairs, the larger the probability that recombination occurs, and the lower the final yield.

Low LET particles, such as high energy electrons and gamma ray photons, create a sparse density of charge pairs along their tracks. (For a 1 MeV electron, the average pair separation is approximately 50 nm.) High LET particles, such as protons, alpha particles, and heavier ions, generate a large line density of pairs along their tracks. (For a 1 MeV proton, the average separation distance is approximately 0.3 nm.) In addition to ionization damage, permanent damage due to the rupture of chemical bonds and the displacement of atoms in irradiated materials can result from larger mass particles such as neutrons.

Several radiation effects on electrical behavior are deleterious to the operation of semiconductor memories in general. The first effect, a transient effect or "photoeffect" from a high-intensity burst of radiation, is brought about by the separation of electron hole pairs generated by the incident radiation in reverse-biased junctions. All charges stored dynamically on junctions, for example, biases in flip-flops found in SRAMs and DRAMs, are reduced until lost by this effect. The SONOS transistor is not affected by this condition since the stored information is dependent on the threshold voltage shift brought about by stored charge in the nitride layer and is not dependent on the maintenance of external power. However, transient radiation during a write or read operation of a memory may produce a reduced window size, may result in the storage of false data, or may cause the improper detection of a stored memory state. Therefore, the transient hardness of the control logic and the details of the memory circuit design are the primary considerations for transient radiation hardness. Transient radiation testing with 50 ns electron beam pulses indicates a read upset threshold of 1×10^8 rad(Si)/sec and a write upset level of 3×10^9 rad(Si)/sec for a 10 V, 64 K SONOS technology [5.31, 5.33]. The read upset threshold can be increased to 5×10^8 rad(Si)/sec by increasing the size of an NMOS switch in the memory array. The memory did not lose data or latchup when exposed to the maximum available transient radiation level of 1.4×10^{11} rad(Si)/sec.

The other radiation effect, total dose radiation, is due to the occurrence of ionization and charge separation in the insulating layers of devices. Generally, ionization radiation causes an increase in the density of fast interface traps, and positive charges accumulate in the silicon dioxide layer, which, for example, reduces the current gain in bipolar transistors and increases the threshold voltage in MOS transistors. In addition, the increase in the density of interfacial defects changes the channel charge, and mobility in inversion is significantly reduced by the increase in coulombic scattering from the generated defects. Overall, the MOS transistor's response is also slower. SONOS devices are inherently hard because, unlike silicon dioxide, the mobilities of electrons and holes are not much different in nitrides. Thus, when exposed to ionizing radiation, both generated carriers can be swept out of the nitride, resulting in a negligible amount of trapped charge.

In the thin tunnel and blocking oxide layer, the density of electron hole pairs generated upon irradiation is quite small because of the small generation volume. Even if a few Si–H bonds are broken at the interface by electron hole pairs that have escaped recombination, the formation of a stable dangling bond (interface trap) requires that hydrogen, released in the process, diffuse away from the generation site. The low ion diffusivity of the nitride layer prevents the hydrogen, released as a byproduct of the interface reaction, from diffusing away from the interface. Thus, the hydrogen, which is confined to the thin oxide region (tunnel oxide < 2.0 nm), recombines with the dangling bond to regenerate the Si–H center, and a negligible change in interface trap density is measured upon irradiation of SONOS devices. Because of these properties, it is expected that further scaling of the SONOS device can only help render a more radiation-hardened device. Note that floating gate nonvolatile memory devices inherently contain thicker oxide layers than SONOS devices and are, therefore, less radiation hard since the shift in threshold voltage produced in thick oxide MOS-like structures is proportional to the square of the oxide thickness.

The reduction of the programming voltages and the absence of thick oxide in present-day, scaled SONOS devices have improved their radiation hardness. Radiation-hardened fabrication processing has been integrated into SNOS/CMOS memory technology. Acceptable shifts for total radiation doses up to Mrad levels at 77 K have been obtained for SNOS structures. Results of total dose radiation testing of the 64 K SONOS EEPROMs mentioned above, using a Co^{60} source at a dose rate of 100 rad(Si)/sec and a Cs^{137} source at a dose rate of 0.2 rad(Si)/sec, are shown in Fig. 5.39 [5.32]. All parts were fully functional at 1 Mrad(Si).

5.4.5 Low-Voltage Operation

An important scaling issue relates to the voltages that can be tolerated by the peripheral circuitry. For a given fabrication technology, 1.25 μm CMOS for example, a maximum breakdown voltage ceiling exists, usually in the range of 12 to 14 V. In spite of external 5 V off-chip compatibility, until recently, EEPROMs were programmed at higher voltages generated on-chip with charge-pump and high-voltage regulation circuitry. While a higher programming voltage is desirable for a fast programming time and reduced noise margin, a lower programming voltage is necessary to ensure device reliability and fabrication technology compatibility with peripheral circuits. For conventional MNOS devices, the most practical "scaling" is to reduce the thickness of the nitride layer, and hence, reduce the voltage drop across the nitride layer. However, charge centroid considerations dictate a lower nitride thickness limit of about 12.5 nm and a minimum programming voltage of 8 V.

Table 5.5 illustrates commercial production margins of programming voltages and insulator film thicknesses in scaled-down MNOS memory devices [5.40]. Lower

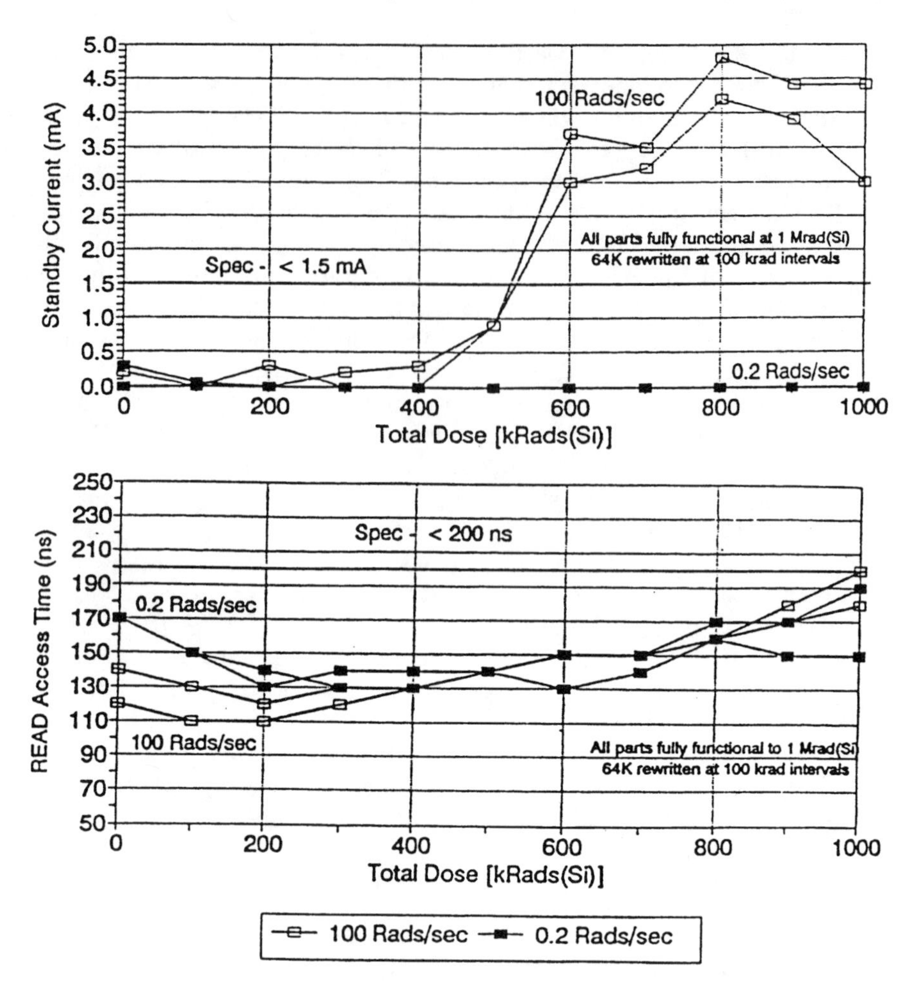

Figure 5.39 Standby current and read access time versus total dose [no standby current increases at 1 Mrad (Si)]. 64 KB SONOS EEPROM data [5.32].

TABLE 5.5 PARAMETERS FOR THE SCALING OF SNOS MEMORY DEVICES [5.40]

Bit Size (bits)	Design Rule (μm)	t_{ox} (nm)	t_N (nm)	V_p (V)
64 K	2	1.8±0.1	28±1.0	14.5±0.9
256 K	1.3	1.8±0.1	24±0.9	12.6±0.8
1 M	0.8	1.8±0.1	20±0.8	10.8±0.7
4 M	0.5	1.8±0.1	17±0.7	9.4±0.6
16 M	0.3	1.8±0.1	14±0.5	8.0±0.5

programming voltages are possible with the SONOS device since the blocking action of the added blocking oxide removes any limitation on reduction of the nitride thickness. The addition of the blocking oxide layer shows the additional advantages of memory window enlargement by (1) reducing unwanted injection from the gate that reduces the memory window charge in the nitride, and (2) possibly larger oxygen-related electron trap densities at the nitride–top oxide interface due to the oxidation of the nitride. Programming voltages down to 5 V on experimental SONOS devices with 5.0 nm of memory nitride and tunnel oxides of 1.1 and 1.8 nm are illustrated in Fig. 5.40 [5.36].

5.4.6 CMOS Compatibility

The SONOS fabrication process is based on standard polysilicon gate technology and is CMOS compatible. The ONO triple dielectric film of the SONOS device has recently been adapted into submicron CMOS processes for the manufacturing of high-density DRAMs and SRAMs because of increased reliability and yield realized by the inherent ability of the top (blocking) oxide film process to fill any underlying pinholes in the nitride. When fabricated on normal CMOS production lines, the SNOS-type EEPROM technology yield is higher than that of SRAM. This high yield is attributed to the low defect density and ion-barrier properties of the Si_3N_4 film, as well as to the insensitivity of the MNOS EEPROM's device characteristics to SiO_2 failures.

5.4.7 Scaling Issues

Since the erase or write voltage pulse is not applied between the source and drain of the SONOS, a shorter channel length can be used than is possible in conventional floating gate type memory. To guarantee reliability, conventional MNOS device scaling must be implemented in parallel with the scaling down of the periph-

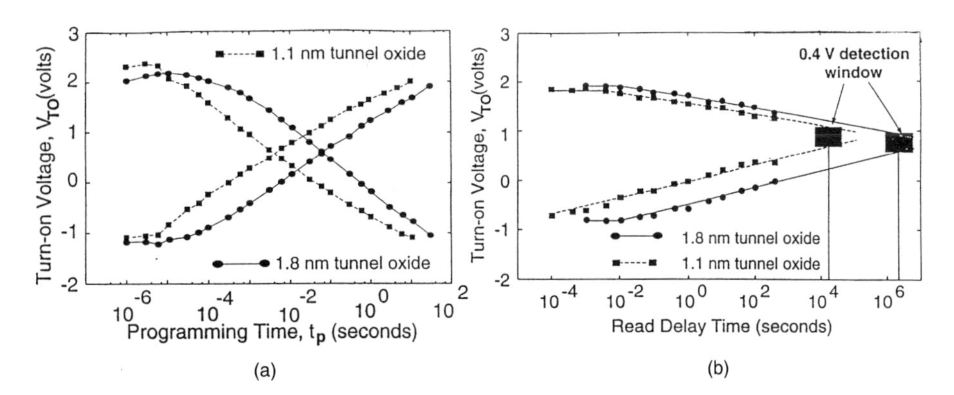

Figure 5.40 Electrical characteristics of a scaled SONOS memory. (*a*) Erase/write characteristics; (*b*) Retention characteristics [5.36].

eral CMOS transistors. Fig. 5.41 illustrates the allowable nitride thickness variation tolerance for a given programming voltage and chip integration level [5.42]. Retention characteristics (> 10 years with sufficient margin at 85°C) and electric field (6 MV/cm across the nitride film, assuming no stored charge) dictate the maximum and minimum nitride thickness, respectively, for a given programming voltage. Table 5.6 shows a SNOS scaling method based on the following relationships:

$$X_N \propto V_p \propto X_{ot} \tag{5.3}$$

and

$$\tau_e \propto \exp(X_{ot}/\lambda_{ox}) \tag{5.4}$$

where V_p is the programming pulse voltage amplitude, X_N is the nitride film thickness, τ_e is the erase time, and λ_{ox} is the de Broglie wavelength in the oxide (approximately 0.1 nm) [5.41]. As can be seen in these equations, the programming voltage is reduced by decreasing the nitride thickness, and the erase time is shortened by decreasing the tunnel oxide thickness. These observations have been confirmed experimentally, as shown in Fig. 5.42 [5.51].

For thinner nitrides, gate injection can seriously limit the size of the memory window. A reduction in the saturation of the written state (approximately 2 V) may

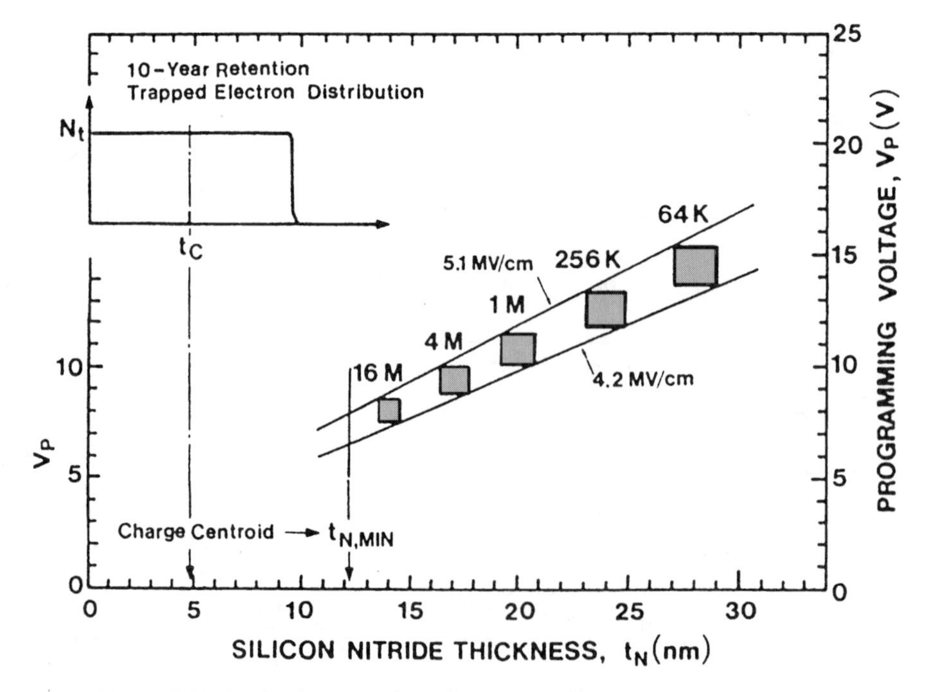

Figure 5.41 Production margins of programming voltage and Si_3N_4 film thickness in scaled-down SNOS memory devices. 4 and 16 MB EEPROMs are predictable from the trend of actual results obtained from 64 KB, 256 KB, and 1 MB EEPROMs [5.42].

TABLE 5.6 HIGH-DENSITY MEMORY SCALING APPROACH OF (1) CONVENTIONAL SNOS, (2) SONOS1 WITH A CONSTANT $X_{ob} = 2.5$ NM, AND (3) SONOS2 WITH A THICKER X_{ob}

SNOS/SONOS High-Density Memory Scaling Approach						
1st Generation	—	—	1 M	4 M	16 M	—
Aggressive Scaling	16 K	64 K	256 K	1 M	4 M	—
Property	(k = 1)	(k = 1.5)	(k = 2.5)	(k = 3.75)	(k = 6)	SF
Technology (μm)	3	2	1.2	0.8	0.5	k^{-1}
Cell area (μm^2)	400	180	60	28	5	k^{-2}
Program voltage (V)	25	16	10	6	4	$k^{-1.1}$
ΔV_{fb}(V)			6	3.5–4	2.5	
MOS: X_{ox} (nm)	75	50	30	20	12.5	k^{-1}
SNOS: X_{OT} (nm)	2.1	2.0	1.9	N.A.	N.A.	$k^{-0.1}$
X_N (nm)	50	32	20	N.A.	N.A.	$k^{-1.1}$
SONOS1: X_{OT} (nm)	—	—	1.9	1.9	1.9	k^0
X_N (nm)	—	—	15.5	7.8	3.2	$k^{-1.8}$
X_{OB} (nm)	—	—	2.5	2.5	2.5	k^0
SONOS2: X_{OT} (nm)	—	—	1.9	1.9	1.9	k^0
X_N (nm)	—	—	11.8	4.5	2.5	—
X_{OB} (nm)	—	—	4.5	3.8	3.0	$k^{-0.5}$

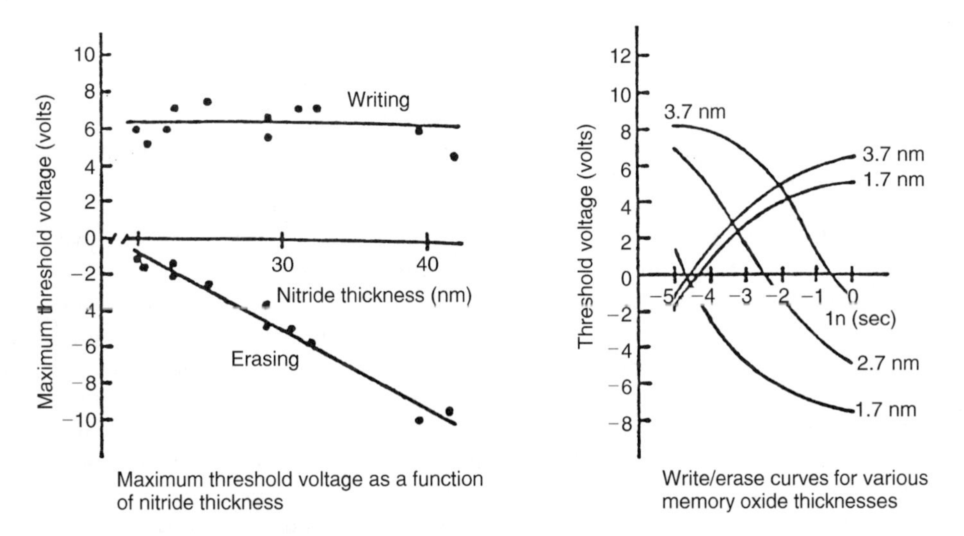

Figure 5.42 V_{th}(max) versus X_N and write/erase curves for conventional MNOS n-channel transistors of various tunnel oxide (1.7 nm, 2.7 nm, and 3.7 nm) and nitride (20–40 nm) film thicknesses [5.51].

be achieved with the addition of a thin (approximately 2.5 nm) top blocking oxide to form a SONOS structure. The literature suggests that there is a larger penetration length in nitrides for holes (typically 15 to 20 nm) than for electrons (typically 5 to 10 nm). In reducing the nitride thickness, the holes will, therefore, be trapped closer to the gate. The thin blocking oxide of 2.5 nm may not be thick enough to prevent hole loss from the nitride to the gate.

A second SONOS scaling approach has been described where a thinner nitride thickness (< 10 nm) and a thicker top oxide (> 3 nm) are employed in the SONOS device. The goal is to inhibit gate injection as well as to stop charges injected from the silicon at the nitride-blocking oxide interface, resulting in a higher trapping efficiency, thus preserving the memory window as the nitride thickness is reduced. The process of steaming the nitride to form a blocking oxide has previously been expected to result in a high density of oxygen-related electron traps at the blocking oxide–nitride interface. The beneficial effects are (1) better retention, (2) a larger memory window in spite of the decreasing nitride thickness, and (3) a write independent threshold state if the blocking oxide were constant with nitride scaling. Retention would improve exponentially with the increase in backtunneling distance for the charge at the nitride–blocking oxide interface.

A third scaling approach has the same scaling factor (SF = $K^{-0.5}$; see SONOS2 in Table 5.6) for the blocking oxide thickness, but stems from the knowledge that the majority of traps responsible for the memory window are located in the nitride bulk rather than at the blocking oxide–nitride interface. A minimum nitride trap density is thus necessary for maximizing the memory window that is independent of threshold voltage saturation due to nitride trap saturation. Blocking oxide versus nitride film thickness scaling as a function of minimum nitride trap density is shown in Fig. 5.43. To ensure optimum reliable scaled-down SONOS, the minimum desirables are an ample memory window at the maximum needed retention time for the maximum number of erase/write cycles. To this end, the scaled SONOS devices can find applications ranging from EEPROM functions (very slow programming of 1 to 100 msec with long nonvolatile retention of years) to NVRAM functions (fast programming of 1 to 10 μs for limited nonvolatile retention).

Estimations for a 1 MB, 5 V-only (± 9 V erase/write voltage), SONOS EEPROM with 1.2 μm design rules, intended for semiconductor disk application, with a die size of 5.3 mm × 6.3 mm and a high programming speed of 1.1 μs/byte shows a device density of 19.3 MB/in^3 and 15.9 MB/W (operating) [5.55]. The thicknesses of the top oxide, the silicon nitride, and the tunnel oxide layers are 4, 5, and 2 nm, respectively. Further reduction of the tunnel oxide from 2 to 1.8 nm decreases the erase time from 10 ms to 1 ms as seen in Fig. 5.21, and still further reduction to 1.1 nm reduces the erase time to 100 μs, as seen in Fig. 5.40. A promising SONOS device structure is the so-called buried-channel device which has extended endurance and improved retention characteristics, as shown in Fig. 5.44 [5.26]. The saturation in the SONOS transistor threshold voltage for extended pulse widths has been eliminated due to an increase in the nitride trap density. These thin ONO gate insulators may be used with 0.25 μm design rules to realize a 256 MB NVSM compatible with low power CMOS and suitable for high-density (semiconductor disk) PCMCIA cards in portable electronics and high-density NVRAMs for eventual DRAM replacement.

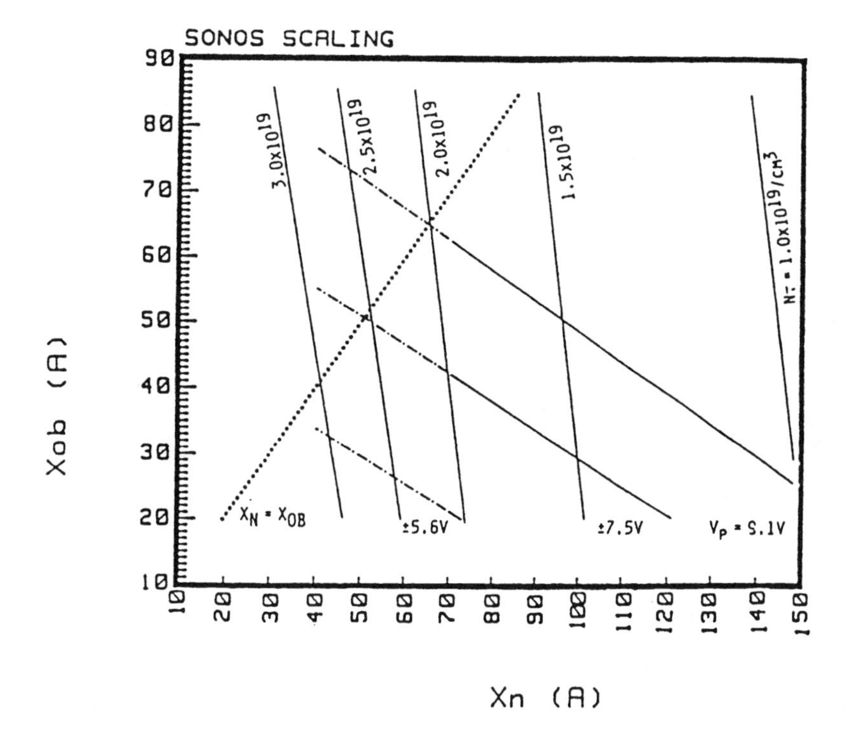

Figure 5.43 X_{ob} versus X_N scaling reliability requirements for different programming biases not exceeding 10 MV/cm across the tunnel oxide of 1.9 nm. A minimum N_T must be present to maximize the memory window.

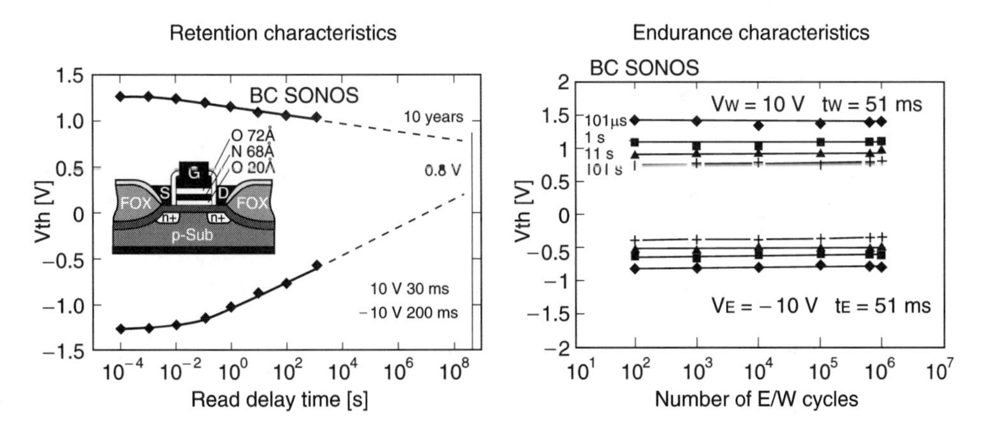

Figure 5.44 Buried channel scaled SONOS device electrical characteristics (As implant, 40 KeV, $1 \times 10^{12}/cm^2$) X_{ot} = 2.0 nm, X_n = 6.8 nm, X_{ob} = 7.2 nm [5.26].

ACKNOWLEDGMENTS

The authors would like to express their appreciation to the National Science Foundation's Microstructure Program, the Office of Naval Research under the SDIO/BMDO sponsorship, the Naval Research Laboratories through the DNA, and the Westinghouse Electric Corporation's Advanced Technology Laboratories for their continued support of graduate and undergraduate research in areas relevant to nonvolatile semiconductor memories. In addition, we would like to express our appreciation to thoughtful discussions with many of our colleagues over the years, particularly with Yoshiaki Kamagaki and Shin-ichi Minami (Hitachi), Ted Dellin and Herman Stein (Sandia Labs), Dennis Adams, Anant Agarwal, Mark Jacunski, Joe Brewer and Ronald Cricchi (Westinghouse), Umesh Sharma (Motorola), Margaret French (Lehigh), Yin Hu (Texas Instruments), Anirban Roy (WSI), Bill Brown (University of Arkansas), Richard Wiker (Honeywll), Roman Fedorak (NADC), Herb Mette (ERADCOM), Chen-Chung Chao (Intel), Loren Lancaster (NVX), Al Goodman (ONR), and Nelson Saks (NRL).

References

[5.1] F. R. Libsch, A. Roy, and M. H. White, "A true 5V EEPROM cell for high density NVSM," IEEE 45th Annual Dev. Res. Conf., Santa Barbara, Calif., *IEEE Trans. Elect. Dev.*, ED-34 (11), p. 2371, 1987.

[5.2] F. R. Libsch, A. Roy, and M. H. White, "Charge transport and storage of low programming voltage SONOS/MONOS memory devices," *Sol. State Elec.*, vol. 33, no. 1, p. 105, 1990.

[5.3] A. Roy, F. R. Libsch, and M. H. White, "Electron tunneling from polysilicon asperities into poly-oxides," *Sol. State Elec.*, vol. 32, no. 8, p. 655, 1989.

[5.4] Y. Yatsuda et al., "An advanced MNOS memory device for highly-integrated byte-erasable 5V-only EEPROMs," IEEE Int. Elec. Dev. Meeting, p. 733, 1982.

[5.5] Y. Yatsuda et al., "Hi-MNOS II technology for a 64-kbit byte-erasable 5-V-only EEPROM," *IEEE Trans. Elec. Dev.*, ED-32(2), p. 224, 1985.

[5.6] T. Nozaki et al., "A 1-Mb EEPROM with MONOS memory cell for semiconductor disk application," *IEEE J. Sol. State Cir.*, SC-26(4), p. 497, 1991.

[5.7] Y. Uchida et al., "1K-bit nonvolatile semiconductor read/write RAM," *IEEE Trans. Elec. Dev.*, ED-25(8), p. 1066, 1978.

[5.8] D. D. Donaldson, M. D. Eby, R. Fahrenbruck, and E. H. Honningford, "SNOS 1K $\times$ 8 static nonvolatile RAM," *IEEE J. Sol. State. Cir.*, SC-17(5), p. 847, 1982.

[5.9] R. A. Haken et al., "An 18V double-level poly CMOS technology for nonvolatile memory and linear applications," IEEE Int. Sol. State Cir. Conf., p. 90, 1983.

[5.10] J. R. Cricchi, M. D. Fitzpatrick, F. C. Blaha, and B. T. Ahlport, "Hardened MNOS/SOS electrically reprogrammable nonvolatile memory," *IEEE Trans. Nucl. Sci.*, NS-24(6), p. 2185, 1977.

[5.11] C. E. Herdt, "Nonvolatile SRAM—The next generation," 5th Nonvolatile Memory Technology Review, Linthicum Heights, Md., p. 28, 1993.

[5.12] C. E. Herdt, "Analysis, measurement, and simulation of dynamic write inhibit in an nvSRAM cell," *IEEE Trans. Elec. Dev.*, ED-39(5), p. 1191, May 1992.

[5.13] V. Tjulkin and V. A. Miloshevsky, "A 4kb nMOS static NVRAM with extended 16kb non-volatile memory," IEEE ISSCC, p. 226, 1991.

[5.14] J. R. Cricchi, F. C. Blaha, and M. D. Fitzpatrick, "The drain-source protected MNOS memory device and memory endurance," *IEEE IEDM Tech. Digest*, p. 126, 1973.

[5.15] J. R. Cricchi et al., "Nonvolatile block-oriented RAM," IEEE ISSCC, p. 204, 1974.

[5.16] J. E. Brewer et al., "Army/Navy MNOS BORAM," GOMAC 76.

[5.17] Y. Kamigaki et al., "Yield and reliability of MNOS EEPROM products," *IEEE J. Sol. State Cir.*, SC-24(9), p. 1714, 1989.

[5.18] A. Lancaster et al., "A 5V-only EEPROM with internal program/erase control," IEEE ISSCC, p. 164, 1983.

[5.19] D. D. Donaldson, E. H. Honnigford, and L.J. Toth, "+5V-only 32K EEPROM," IEEE ISSCC, p. 168, 1983.

[5.20] Y. Uchida et al., "A 1024-bit MNOS RAM using avalanche-tunnel injection," *IEEE J. Sol. State Cir.*, SC-10(5), p. 288, October 1975.

[5.21] T. B. Smith, P. C. Smith, and J. E. Brewer, "Dual mode 64k-bit/128k-bit MNOS BORAM chip," product literature.

[5.22] K. Dimmler et al., "50ns 256K SNOS EEPROM," 10th IEEE NVSM, Vail, Colo., 1989.

[5.23] S. Minami et al., "Improvement of written-state retentivity by scaling down MNOS memory devices," *Jap. J. Appl. Phys.*, vol. 27, no. 11, p. L2169, 1988.

[5.24] E. Suzuki, Y. Hayashi, and H. Yanai, "Degradation properties in metal–nitride–oxide–semiconductor structures," *J. Appl. Phys.*, vol. 52, no. 10, p. 6377, 1981.

[5.25] M. H. White, J. W. Dzimianski, and M. C. Peckerar, "Endurance of thin-oxide nonvolatile MNOS memory transistors," *IEEE Trans. Elec. Dev.*, vol. 24, no. 5, p. 577, 1977.

[5.26] A. K. Banerjee, Y. Hu, M. G. Martin, and M. H. White, "An automated SONOS NVSM dynamic characterization system," 5th Nonvolatile Memory Technology Review, Linthicum Heights, Md., p. 78, June 22–24, 1993.

[5.27] R. E. Paulsen, R. R. Siergiej, M. L. French, and M. H. White, "Observation of near-interface oxide traps with the charge-pumping technique," *IEEE Trans. Elec. Dev.*, vol. 13, no. 12, p. 627, 1992.

[5.28] S. Minami and Y. Kamigaki, "New scaling guidelines for MNOS nonvolatile memory devices," *Proc. Solid-State Devices and Materials*, Japan, p. 9, 1990.

[5.29] Hitachi, Ltd, Technical Marketing Dept., Semi. and Integr. Cir. Div., Musashi Works, Japan, product literature.

[5.30] Y. Kamigaki et al., "Highly reliable 256-Kb MNOS EEPROM," 10th IEEE NVSM Workshop, Vail, Colo., p. 1, 1989.

[5.31] U. Sharma and M. H. White, "Ionization radiation induced degradation of MOSFET channel frequency response," *IEEE Trans. Elect. Dev.*, ED-36, p. 1359, 1989.
[5.32] D. Adams et al., "SONOS technology for commercial and military nonvolatile memory applications," 5th Nonvolatile Memory Technology Review, Linthicum Heights, Md., p. 96, June 22–24, 1993.
[5.33] M. Jacunski et al., "Radiation hardened nonvolatile memory for space and strategic applications," GOMAC Conference, p. 449, 1992.
[5.34] F. R. Libsch, A. Roy, and M. H. White, "Comparison of technologies for high density NVSM," 1st Nonvolatile Memory Technology Review, Linthicum Heights, Md., June 1987.
[5.35] R. L. Wiker, "High density memory technology in advanced signal, data, control processors and other memory systems," 5th Nonvolatile Memory Technology Review, Linthicum Heights, Md., p. 38, June 22–24, 1993.
[5.36] M. French, H. Sathianathan, and M. White, "A SONOS nonvolatile memory cell for semiconductor disk application," 5th Nonvolatile Memory Technology Review, Linthicum Heights, Md., p. 70, June 22–24, 1993.
[5.37] Y. Yatsuda et al., "Scaling down MNOS nonvolatile memory devices," Proceedings of the 13th Conf. on Solid State Devices, Tokyo, 1981; *Jap. J. Appl. Phys.*, vol. 21, Suppl. 21–1, p. 85, 1982.
[5.38] E. Suzuki, H. Hiraishi, K. Ishii, and Y. Hayashi, "A low voltage alternable EEPROM with metal–oxide–nitride–semiconductor (MONOS) structures," *IEEE Trans. Elec. Dev.*, vol. 30, no. 2, p. 122, 1983.
[5.39] T. A. Dellin and P. J. McWhorter, "Scaling of MONOS nonvolatile memory transistors," Electrochem. Soc. Meeting, San Diego, Calif., October 1986.
[5.40] S. Minami and Y. Kamigaki, "New scaling guidelines for MNOS nonvolatile memory devices," *IEEE Trans. Elect. Dev.*, ED-38(11), p. 2519, November 1991.
[5.41] F. R. Libsch, A. Roy, and M. H. White, "Scaling reliability requirements in SONOS/MONOS EEPROM devices," 10th IEEE NVSM Workshop, Vail, Colo., p. 45, August 1989.
[5.42] Y. Kamigaki et al., "New scaling guidelines for MNOS memory devices," 10th IEEE NVSM Workshop, Vail, Colo., p. 42, August 1989.
[5.43] A. Roy, F. R. Libsch, and M. H. White, "Investigations on ultra-thin silicon nitride and silicon dioxide films in nonvolatile semiconductor memory (NVSM) transistors," Proc. Sym. Silicon Nitride and Silicon Dioxide Insulating Films, 170th Electrochem. Soc. Meeting, San Diego, Calif., October 1986.
[5.44] F. R. Libsch, A. Roy, and M. H. White, "Amphoteric trap modeling of multidielectric scaled SONOS nonvolatile memory structures," INFOS 87 Conf., Applied Solid-State Science Series of Elsevier Science, Publ. B.V., Eindhoven, The Netherlands, October 1987.
[5.45] S. L. Miller, "Effect of temperature on data retention of SNOS transistors," 8th IEEE NVSM Workshop, Vail, Colo., August 1986.
[5.46] U. Sharma et al., "A novel technology for megabit density, low power, high speed NVRAMs," VLSI Symposium, Japan, p. 53, 1993.

[5.47] S. Saito et al., "A 256 bit nonvolatile static random access memory with MNOS memory transistors," Proc. 7th Conf. Sol. State Devices, Tokyo, 1995, *Suppl. Jap. J. Appl. Phys.*, vol. 15, p. 185, 1976.

[5.48] Y. Nishi and H. Iizuka, "Invited: Nonvolatile semiconductor memory," *Suppl. Jap. J. Appl. Phys.*, vol. 16, p. 191, 1977.

[5.49] S. Minami and Y. Kamigaki, "Tunnel oxide thickness optimization for high-performance MNOS nonvolatile memory devices," *IEICE Trans.*, vol. E74, no. 4, p. 875, 1991.

[5.50] A. Nughin, A. Multsev, and A. Milosheysky, "N-channel 256 Kb and 1Mb EEPROMs," IEEE Intl. Sol. State Cir. Conf., p. 228, 1991.

[5.51] J. A. Topich, "Charge storage model for SNOS memory devices," Abstract 268, *Proc. Electrochem. Soc.*, 1978.

[5.52] M. L. French and M. H White, "Scaling of multidielectric nonvolatile SONOS memory structures," *Sol. St. Elect.*, vol. 37, p. 1913, 1994.

[5.53] C. Wann and C. Hu, "High endurance ultra-thin tunnel oxide for dynamic memory application," *IEDM Tech. Dig.*, p. 867, 1995.

[5.54] F. R. Libsch and M. H. White, "Charge transport and storage of low programming voltage SONOS/MONOS memory devices," *Sol. St. Elect.*, vol. 33, p. 105, 1990.

[5.55] M. White et al., "A low voltage SONOS nonvolatile semiconductor memory technology," *Proc. of the Sixth Biennial IEEE International Nonvolatile Memory Technology Review*, p. 52, 1996.

Chapter 6

Yukun Hsia
and Vance C. Tyree

Reliability and NVSM Reliability

6.0. INTRODUCTION

Reliability is a measure of a product's performance consistency. A reasonable quantitative definition of reliability is the probability that the product item remains effective in performing its designated function as required in an operating environment for a specific period of time. However, at the component level, in view of the necessary high measure of reliability to support system product reliability, the reliability metric is often given by the device's rate of failure to meet performance requirements within the expected (or specified) operating environment. For example, to specify component reliability, we may define a failure rate unit (FIT), in which

$$\mathit{1}\ \mathit{F}\text{ailure Un}\mathit{IT} \equiv 1\ \text{FIT} \equiv 1\ \text{failure}/10^9\ \text{device-hours}$$

A commonly accepted set of assumptions, generally found to be representative of the failure behavior of electronic components and other complex systems, has been systematized as the reliability "bathtub" curve shown in Fig. 6.1.

The early failure or the so-called infant mortality period represents high failure rates attributable to manufacturing defects. Screening, testing, and burn-in procedures implemented by the manufacturer and user alike are traditional primary means of quality control and assurance with which enhancement of reliability is attempted. Fig. 6.2 illustrates the expected impact of burn-in on the failure rate during infant mortality.

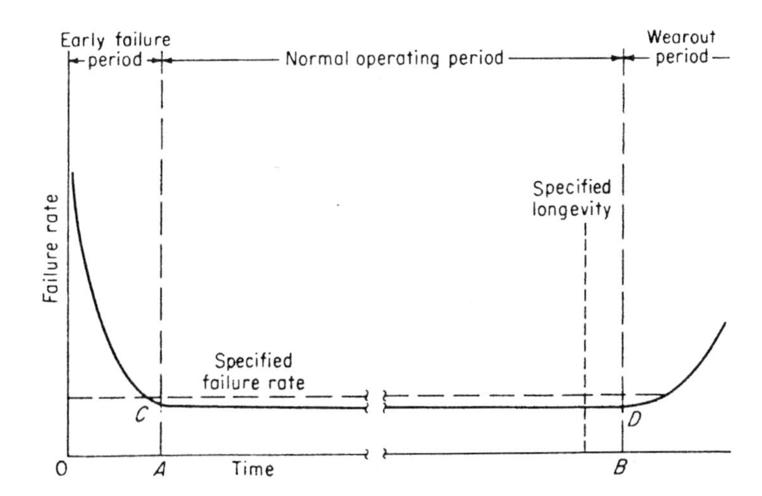

Figure 6.1 The reliability lifetime "bathtub" curve (after Grant and Leavenworth [6.1] from "Reliability of Military Electronics Equipment [6.2]).

The failure rate attained during the normal operating period defines quantitatively the quality of the component. Proper design, with sufficient design margins to exclude failure modes from the operating regime of the component, ensures a low failure rate during the product lifetime. Similarly, good engineering design practices increase the longevity of the component and the postponement of the onset of the wearout period or end-of-life of the product.

An alternative view of reliability is presented in Fig. 6.3 in which the utility of reliability, effected through design and manufacturing, is emphasized. Reliability is designed into a product through the elimination of failure modes in the use of the component, the incorporation of margins in the component circuits, the application of redundancy, fault tolerance, and other reliability enhancement techniques, and the implementation and monitoring of the manufacturing processes to minimize the introduction of latent defects to the manufactured component. Thus, reliability is incorporated at all levels of design, in the manufacturing process, in the device, in the circuit, and in the system component design. Reliability is assured through screen

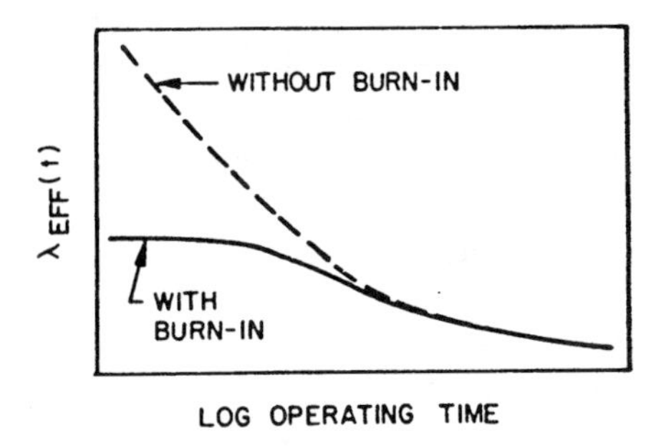

Figure 6.2 Failure rate versus time in use for devices with and without burn-in (after Sze [6.3]).

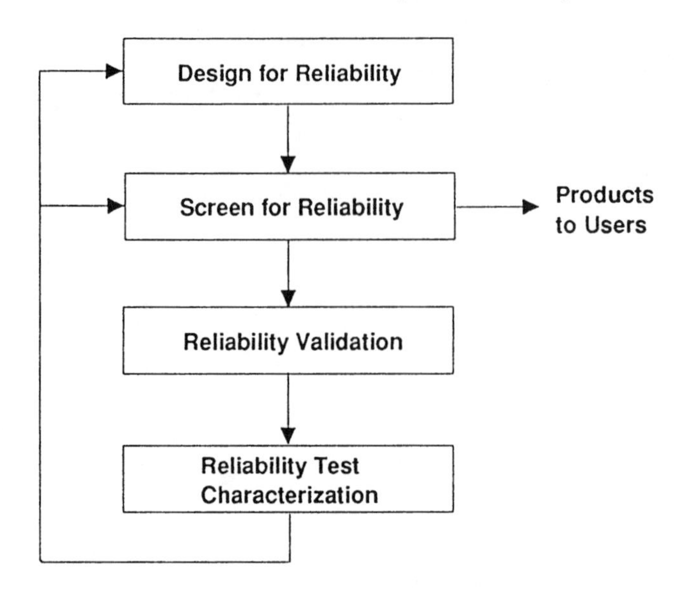

Figure 6.3 An alternative view of product reliability.

testing, sampling for failures, life testing, and application of field use data. In addition, reliability test characterization is performed to provide for design and screening methodology.

NVSM is a subset of MOS/semiconductor integrated circuit technology. Therefore, NVSM reliability failure considerations include all those of MOS devices. It should also incorporate the NVSM unique definition of reliability failure: "the retrieval of incorrect data from memory as it is accessed within the specified operating conditions" [6.4]. This definition of reliability is inclusive of NVSM failures, such as retention failures, because of reduced operating margins, as well as failures due to catastrophic endurance wearout. This chapter discusses unique NVSM reliability issues. For completeness, Table 6.1 is provided to summarize some of the known semiconductor integrated circuit (IC) reliability failure mechanisms that are not unique to NVSM [6.5]. As a result of NVSM's inherently higher operating electric fields, if not properly considered, some of the known MOS failure mechanisms can unnecessarily become prominent contributors to device failures. Examples of some applicable failure modes include contaminants in dielectrics, electromigration of interconnect metallization, as well as dielectric breakdowns from high-field effects.

Not unlike other VLSI semiconductor technologies, NVSM reliability has a significant impact on yield, and consequently, on cost. A high infant mortality failure rate often manifests itself as a low yield in the process line. Worse yet, a high infant mortality failure rate leads to a high return rate as failures occurring within the product warranty period. A more complex relationship between reliability, yield, and cost can be exemplified by the consideration of incorporating testability and redundancy to improve reliability, which may increase circuit com-

TABLE 6.1 TIME-DEPENDENT FAILURE MECHANISMS IN SILICON SEMICONDUCTOR DEVICES (AFTER SZE AND PECK [6.5] PER [6.3]).

Device association	Failure mechanism	Relevant factors	Accelerating factors	Acceleration (E_a = apparent activation energy for temp.)
Silicon oxide and silicon–silicon oxide interface	Surface charge accumulation	Mobile ions, V, T	T	E_a = 1.0–1.05 eV (depends upon ion density)
	Dielectric breakdown	$\mathcal{E}$, T	$\mathcal{E}$, T	E_a = 0.2–1.0 eV $\mathcal{E}^{\gamma}$, $\gamma(T)$ = 1–4.4
	Charge injection	$\mathcal{E}$, T, Q_f	$\mathcal{E}$, T	E_a = 1.3 eV (slow trapping) $E_a \approx -1$ eV (hot electron injection)
Metallization	Electromigration	T, J, A, gradients of T and J, grain size	T, J	E_a = 0.5–1.2 eV J^{γ}, $\gamma(T)$ = 1–4
	Corrosion (chemical, galvanic, electrolytic)	Contamination, H, V, T	H, V, T	Strong H effect $E_a \approx$ 0.3–1.1 eV (for electrolysis) V may have thresholds
	Contact degradation	T, metals, impurities	Varied	
Bonds and other mechanical interfaces	Intermetallic growth	T, impurities, bond strength	T	Al-Au: E_a = 1.0–1.05 eV
	Fatigue	Bond strength, temperature cycling	Temp. extremes in cycling	
Hermeticity	Seal leaks	Pressure differential, atmosphere	Pressure	

Note: V—voltage; T—temperature; $\mathcal{E}$—electric field; J—current density; A—area; H—humidity.

plexity and die size. Both of these factors determine the yield of a semiconductor process, as well as being direct contributors to product cost. Yet, as has been verified by many effective early product introductions, use of redundancy and fault-tolerant techniques can affect yield and cost performance competitiveness in an otherwise noneconomical process. This important complex relationship between reliability and yield is noted herein, for the purpose of encouraging future research studies and analyses.

6.1. PRIMARY RELIABILITY CONCERNS

Both floating gate and floating trap NVSM devices are dependent on the integrity of the memory device dielectrics for reliable device performance. Failures in the gate dielectrics contribute to both retention and endurance failures. In addition, since NVSM devices are MOS integrated circuit devices, they suffer reliability failures in a similar fashion as MOS devices. Furthermore, because of the high electric field needed to write or erase NVSM devices, higher than typical MOS integrated circuit V_{DD} voltages are usually routed through some of the integrated circuit elements. Therefore, NVSM IC devices typically can be expected to experience higher electrical field stresses under normal operating conditions. These devices would exhibit higher than typical failure rates if the design was not appropriately executed with those considerations carefully incorporated in the manufacturing process, device structure, or circuit configuration.

Table 6.2 lists the common EPROM failure mechanisms that have been observed [6.6]. Of the many failure modes listed, a significant number are related to typical MOS IC device failure mechanisms. The table also shows the region of the reliability lifetime "bathtub" curve affected by the failure mechanism, as well as the corresponding thermal activation energy and the primary detection methods for failure. In Sections 6.1.1 and 6.1.2, detailed discussions on failures uniquely related

TABLE 6.2 EPROM FAILURE MECHANISMS (AFTER ROSENBERG [6.6]).

Mode	Lifetime Region Affected	Thermal Activation Energy (eV)	Primary Detection Method
Slow trapping	Wearout	1.0	High-temp bias
Surface charge	Wearout	0.5–1.0	High-temp bias
Contamination	Infant/wearout	1.0–1.4	High-temp bias
Polarization	Wearout	1.0	High-temp bias
Electromigration	Wearout	1.0	High-temp operating life
Microcracks	Random	—	Temperature cycling
Contacts	Wearout/infant	—	High-temp operating life
Silicon defects	Infant/random	0.3	High-temp operating life
Oxide breakdown/leakage	Infant/random	0.3	High-temp operating life
Hot-electron injection	Wearout	—	Low-temp operating life
Fabrication defects	Infant	—	High-temp burn-in
Charge loss	Infant/random/wearout	1.4	High-temp storage
Oxide-hopping conduction	Infant/random	0.6	High-temp storage/burn-in

to NVSM are presented. It suffices here to note that, in the case of the EPROM, for example, charge loss, oxide trapping, and hot-electron injection are primarily responsible for EPROM device failures. Charge traps typically result in EPROM wearout failures. For EEPROMs (E^2PROMs) and flash E^2PROMs, the buildup leading to end-of-lifetime dependent dielectric breakdown and dielectric charge trapping contribute similarly to retention and endurance reliability failures. For floating trap NVSM devices such as the MNOS, SONOS, and other equivalent memory devices, the behavior of traps in the gate dielectric is the primary contributor to the reliability performance of NVSM devices vis-à-vis their retention and endurance features.

Not unnoticed by many device technologists is a possibly significant difference in reliability failure related to pinhole defects between floating gate and floating trap NVSM devices. A conductive pinhole defect in the thin oxide dielectric of the floating gate NVSM device destroys the charge storage capability of the floating electrode. On the other hand, such a defect may not result in catastrophic failure of the floating trap NVSM device. Therefore, it is reasonable to assume, given that other factors are equivalent, that the reliability of floating trap NVSM devices can be expected to be superior to floating gate NVSM devices.

6.1.1 Physics of Traps

In the context of NVSM, we will review the physics of traps. Traps are the storage sites for MNOS-type floating trap NVSM devices. They constitute the very means that lead to reliability failures in EEPROM-type floating gate NVSM devices. Therefore, the fundamental understanding of traps is very important to the NVSM technology.

6.1.1.1 Charge and Traps in SiO_2. Charge and traps in SiO_2 have been studied extensively because of their technological importance in MOS technology. Following Deal and the recommendations of an IEEE-sponsored committee [6.7], the charges in the Si–SiO_2 and SiO_2 systems, applicable also to the floating gate NVSM technology, are categorized below and illustrated in Fig. 6.4. They are:

1. Interface trapped charge, located at the Si–SiO_2 interface, in electrical communication with the silicon, readily charged or discharged dependent on the surface potential. They are attributed to intrinsic dangling bonds at the interface, other structural or oxidation-induced defects, metal impurities, and other physically induced defects, such as by radiation. In the literature, this category of charge is also known as surface states, fast states, interface states, Q_{SS}, and so on.
2. Fixed oxide charge, located in the oxide in close proximity to the interface, not in direct electrical communication with the silicon, and primarily positively charged. The charge originates with the oxidation process in which ionized silicon is not satisfied with an oxygen bond.
3. Oxide trapped charge, in the bulk of the oxide, with either negative or positive polarity, dependent on the nature of the charge trapped. Particularly in the case

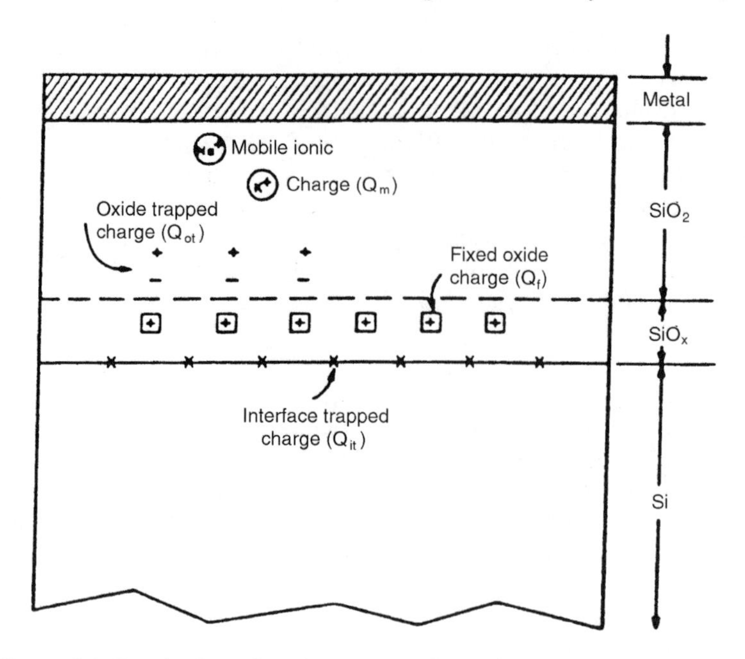

Figure 6.4 Terminology for charges associated with thermally oxidized silicon (after Deal [6.7]).

of floating gate NVSM, these trapped charges are dependent on the endurance cycling of the device, with a reliability failure mode dependent on the fluence or the cumulative transport of charge through the oxide.

4. Mobile ionic charge, primarily due to ionic contaminants, such as Na^+, and to a lesser degree, ions such as Li^+ and K^+.

Table 6.3 summarizes the recommended symbols for these categories' charge. The interface trap density further reflects the quantification of the interface trapped charges both spatially (per unit area) and energetically (per eV in the energy band gap of SiO_2).

The primary failure mechanism of the oxide pertains to dielectric breakdown due to high field/current stressing, modifying the physical properties dominated by the traps, and trapped charge in the oxide. Conceptually, the breakdown process can be viewed in stages: (1) localized high-field regions form as a result of charge trapping, (2) these localized regions cause positive feedback effects and trapping of more charge, and (3) eventually, the local high-field region generates sufficient current density to result in dielectric failure due to electrical runaway processes. A physical model that clearly describes all the various known factors of breakdown in SiO_2 is still being pursued by researchers in the field. The following discussion presents a generic model suitable for elaboration on some of the physical concepts for a better understanding of NVSM reliability [6.8].

TABLE 6.3 OXIDE CHARGES (AFTER DEAL [6.7]).

Interface trapped charge	Q_{it}, N_{it}
Fixed oxide charge	Q_f, N_f
Oxide trapped charge	Q_{ot}, N_{ot}
Mobile ionic charge	Q_m, N_m
Interface trap density	D_{it} (number/$cm^2 \cdot$ eV)

Notes: Q = Net effective charge per unit area
N = Net number of charges per unit area

Under high-field bias, electrons are injected from the substrate silicon into the gate dielectric, SiO_2, via Fowler–Nordheim tunneling. (Hot holes injected from the gate into the oxide are also possible.) In the oxide, some of the electrons can gain sufficient energy to cause impact-ionization, creating hole-electron pairs. Some of the holes, instead of being driven by the channel field to be collected by the substrate silicon, are trapped in localized weak spots, resulting in localized field enhancement and further increasing the Fowler–Nordheim tunnel current. A positive regenerative process is initiated, leading to dielectric breakdown or endurance failure. Figure 6.5 illustrates the process just described with energy band diagrams.

In the case of thin gate MOS and tunnel oxide NVSM devices, it has been shown experimentally that the endurance of the oxide is dependent on the time required for the total hole fluence, Qp, to reach some critical value, where

$$Qp \propto J\alpha t \tag{6.1}$$

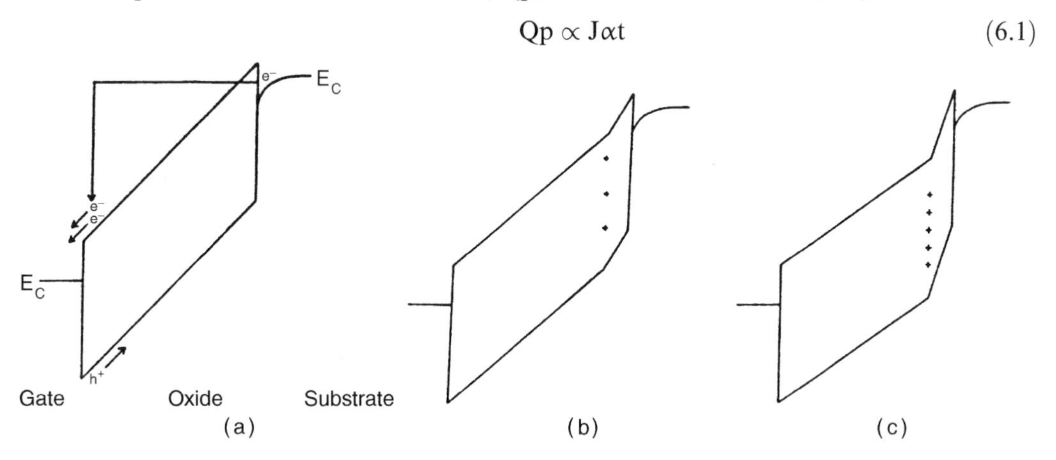

Figure 6.5 Breakdown in SiO_2 due to high-field current injection. (*a*) Electron injection from the cathode results in some of the electrons gaining sufficient energy to create hole-electron pairs in the SiO_2. Some of the generated holes are driven to the cathode. (*b*) Hole trapping in local areas at the cathode increases the cathode field. (*c*) Localized high densities of trapped holes result in the formation of localized high-field/current density regions, causing positive feedback eventually resulting in breakdown (after Chen et al. [6.8]).

in which

$(J \propto \exp -(B/E_{ox}) =$ the Fowler–Nordheim tunnel current density

$\alpha \propto \exp -(H/E_{ox}) =$ the hole-generation coefficient

and thus, t_{BD}, time to endurance breakdown, can be given as

$$t_{BD} \propto \exp(B + H)/E_{ox} \tag{6.2}$$

As seen from this equation, a linear relationship exists between log t_{BD} and $1/E_{ox}$, as is illustrated in Fig. 6.6, which compares the model with experimental data.

From this equation, the "electric field acceleration factor," β, can be derived as

$$\beta \equiv \frac{d[\log(t_{BD})]}{d\ E_{ox}} = \frac{-(B + H)}{2.30\ E_{ox}^2}\left[\frac{\text{decade}}{\text{mv/cm}}\right] \tag{6.3}$$

which predicts, from the model, that β is proportional to $1/E^2_{ox}$. Figure 6.7 compares the model-based β factor with results reported in the literature, showing the agreement between experiment and theory.

The β factor, though widely used to extrapolate oxide lifetime from accelerated stress testing, exhibits a wide spread in measured values and thus requires more accurate modeling to take into consideration the defect-related, statistical features of failure. A simple concept of "effective oxide thinning," ΔX_{ox}, describing a defect density as D (ΔX_{ox}), has been introduced [6.9], which yields

$$t_{BD} \propto e^{(B+H)\ (X_{ox}-\Delta X_{ox})/V_{ox}} \tag{6.4}$$

where the ΔX_{ox} term, with a corresponding defect density D (ΔX_{ox}), represents effects such as localized weak spots, surface aspirates (regions of effective high E_{ox}), and localized defects where increased trap generation occurs. Figure 6.8 illustrates the effect of ΔX_{ox} on β and concurrently references the calculated β factor to experimental results.

The temperature dependency of both the acceleration factor and the effective activation energy for oxide breakdown can be obtained from Eq. (6.2) by substituting

$$B + H = (Bo + Ho)(c + E_a/kT) \tag{6.5}$$

from which

$$\beta \propto 1/T \tag{6.6}$$

can be derived.

The inverse β and temperature relationship was found to be representative of oxides of different thicknesses, as seen in Fig. 6.9.

The temperature dependency of the effective activation energy for oxide failure can be derived from Eqs. (6.2) and (6.5) as

$$\begin{aligned} E_{aeff} &\equiv k\frac{d\ln(t_{BD})}{d\ (1/T)} \\ &= \frac{(B_0 + H_0)\ (X_{ox} - \Delta X_{ox})E_a}{V_{ox}} - E_b \end{aligned} \tag{6.7}$$

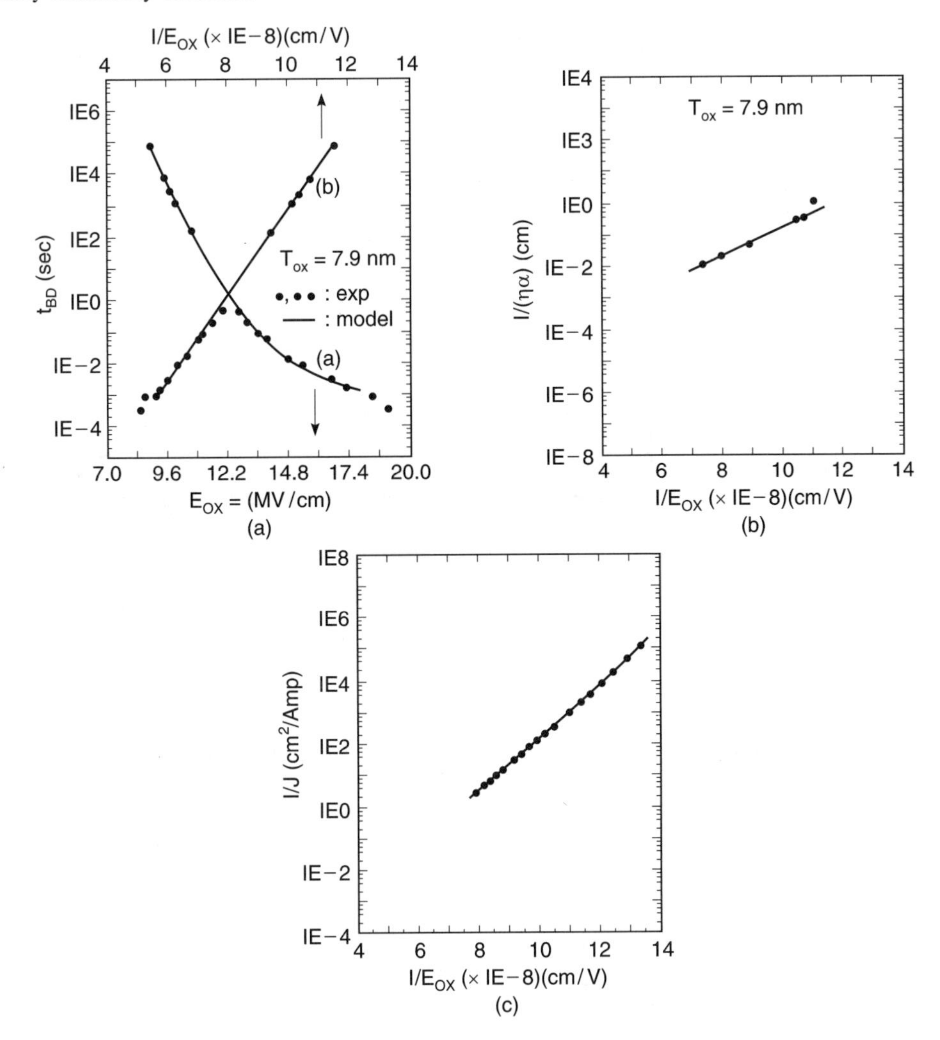

Figure 6.6 Experimental data on dielectric breakdown. (*a*) Log t_{BD} versus E_{ox} and $1/E_{ox}$ for the 7.9 nm tunneling oxide. Log t_{BD} follows a linear relationship with $1/E_{ox}$, not E_{ox} as hitherto assumed. (*b*) Log $1/\alpha$ versus $1/E_{ox}$ for the 7.9 nm tunneling oxide. The slope corresponds to the H factor of α ($\alpha = \alpha_o\, e^{-H/E}$). (*c*) Log $1/J_{total}$ versus $1/E_{ox}$ for the same tunneling oxide. The slope in the figure is approximately equal to the B factor of the Fowler–Nordheim current ($J = AE^2e^{-B/E}$) (after Chen et al. [6.8]).

An alternative and frequently used model for time-dependent dielectric breakdown of silicon dioxide is given as follows [6.11–6.13]:

$$t_f(F) = A \exp\left\{\beta[V_{BO}(F) - V] + \frac{\Delta H}{kT}\right\} \tag{6.8}$$

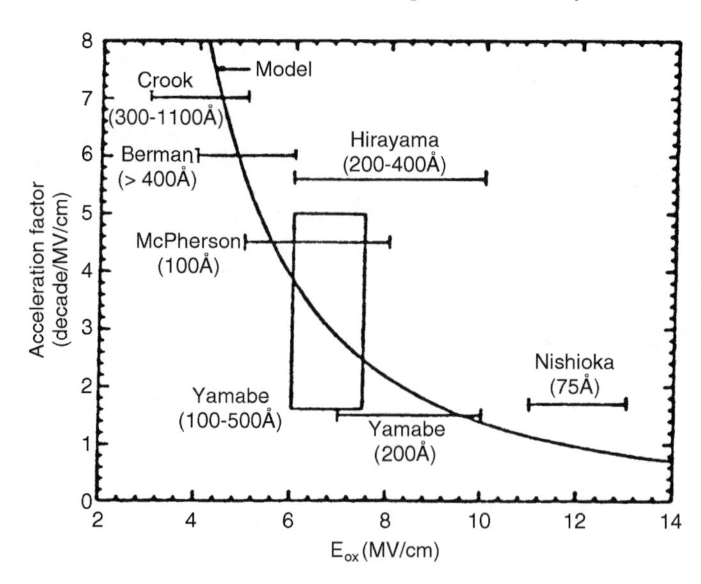

Figure 6.7 Theoretical field acceleration factor compared with reported results. The solid curve is derived from log $(t_{BD})(B+H)/E_{ox}$ (after Lee et al. [6.9]).

where $t_f(F)$ = time-to-fail associated with the population percent F
β = a voltage parameter
$V_{BO}(F)$ = a limiting breakdown voltage
ΔH = the activation energy of the failure mechanism
A = a pre-exponential term that has the units of time

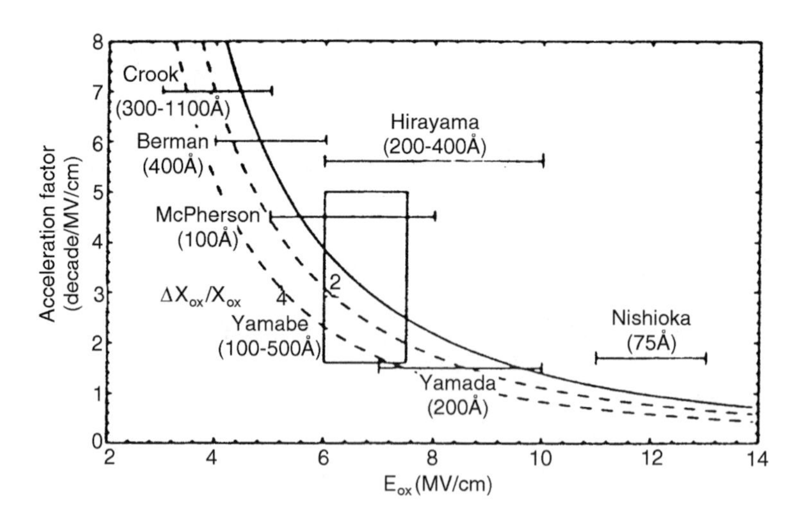

Figure 6.8 Field acceleration factor β versus E_{ox} as a function of $\Delta X_{ox}/X_{ox}$. β decreases with increasing ΔX_{ox} (after Lee et al. [6.9]).

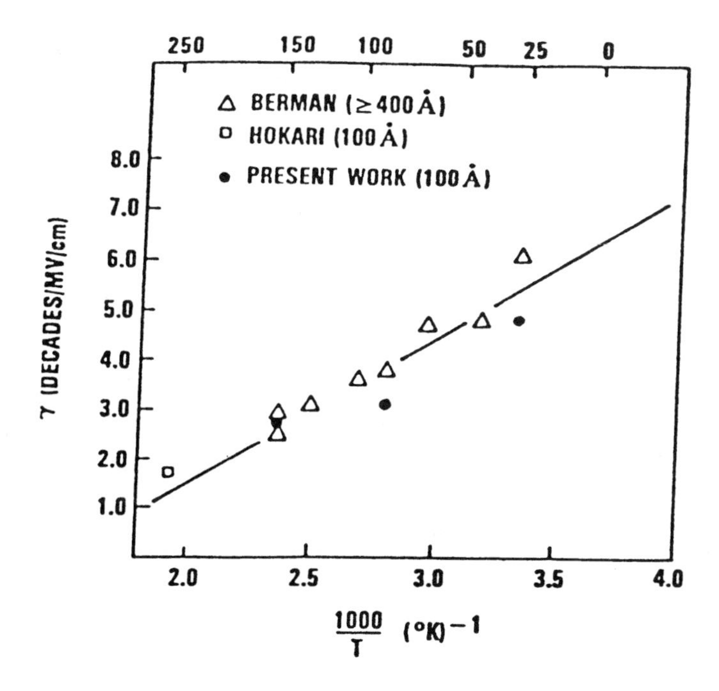

Figure 6.9 Dependence of the electric-field acceleration parameter on temperature (after McPherson and Baglee [6.10]).

This model is primarily derived empirically and predicts a linear dependency on the electric field, even though it could be argued that, for high electric fields typically experienced by EEPROM erase/write operations, both experimental data and theory point to the reciprocal field model. Contradictory data found in recent literature [6.14] indicate the need for caution in the use of one generic model for failure prediction. Table 6.4 illustrates the orders of magnitude difference in calculated reliability parameters in time-dependent dielectric breakdown data for 210 Å oxide films, based on differences of the two models in question.

6.1.1.2 Charge and Traps in Si_3N_4. In floating trap NVSM devices such as MNOS and its various derivatives, including SNOS, MONOS, and SONOS, the spatially distributed traps in the silicon nitride replace the floating gate as the storage element, or more correctly, medium. Unlike the study of traps in SiO_2, there has been much less in the literature regarding the nature of traps in Si_3N_4.

A microscopic physical model of Si_3N_4 traps has been proposed [6.15] based on known attributes of traps in silicon nitride, including:

1. Electron and hole traps as being equal in number
2. Traps spatially distributed in the entire dielectric

TABLE 6.4 LINEAR AND RECIPROCAL MODEL ACCELERATION PARAMETERS AND FIT RATES FOR HIGH- AND LOW-FIELD REGIONS (AFTER BOYKO AND GERLACH [6.14].

Temp. (C)	Model, Regime	Acceleration Parameter*			FIT Rate		
		Median	Lower 95%	Upper 95%	Median	Lower 95%	Upper 95%
150	Linear, high-E	1.7	1.3	2.2	140,000	63,000	320,000
	Linear, low-E	5.4	4.4	6.4	250	140	430
	Recip., high-E	84	62	105	.006	.00006	.5
	Recip., low-E	83	68	99	.04	.02	.09
60	Linear, high-E	3.2	2.5	3.8	1300	610	2900
	Linear, low-E	4.7	3.0	6.4	100	43	240
	Recip., high-E	150	82	180	$<$ le-6		
	Recip., low-E	94	61	130	.0006	.0001	.004

*Units Linear = decades/MV/cm Reciprocal = decades/cm/MV

3. A large electron capture cross section for electron traps, and similarly, a large capture cross section for hole traps, indicating that both types of traps are charged
4. Spectroscopic evidence of excess silicon, equivalently dangling Si bonds, as well as Si–H and N–H bonds in the microscopic model of silicon nitride

In the model, the memory traps are attributed to unsatisfied silicon bonds, or dangling bonds, originating from silicon nitride film formation (see Fig. 6.10). A silicon dangling bond that has an electron in it is electrically neutral. The energy level for this electron is inside the energy gap of Si, at an energy D_o, slightly below the midgap position, as illustrated by the electron energy diagram of a MNOS structure in Fig. 6.11. However, this neutral configuration of the dangling bond is not the lowest energy state. Instead, either of the two charge configurations—one with the dangling bond possessing two electrons, or the other, possessing none (i.e., two holes)—is lower in energy. Hence, the system of Si dangling bonds cooperatively

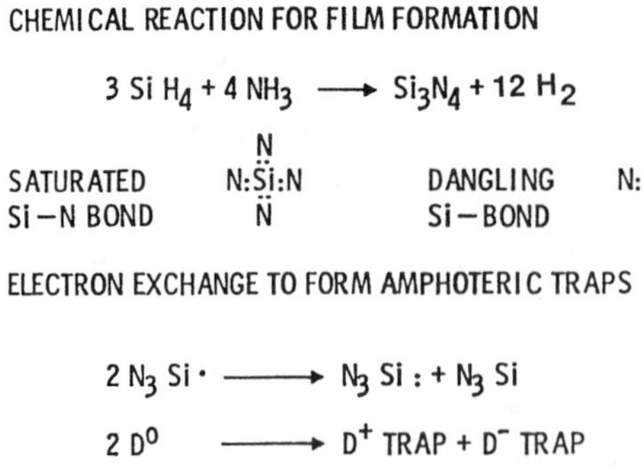

CHEMICAL REACTION FOR FILM FORMATION

$$3\,SiH_4 + 4\,NH_3 \longrightarrow Si_3N_4 + 12\,H_2$$

SATURATED Si–N BOND N:Si:N (N above and below)

DANGLING Si–BOND N:Si (N above and below)

ELECTRON EXCHANGE TO FORM AMPHOTERIC TRAPS

$$2\,N_3\,Si\cdot \longrightarrow N_3\,Si: + N_3\,Si$$

$$2\,D^0 \longrightarrow D^+\ TRAP + D^-\ TRAP$$

Figure 6.10 Microscopic model of memory traps (after Hsia and Ngai [6.15]).

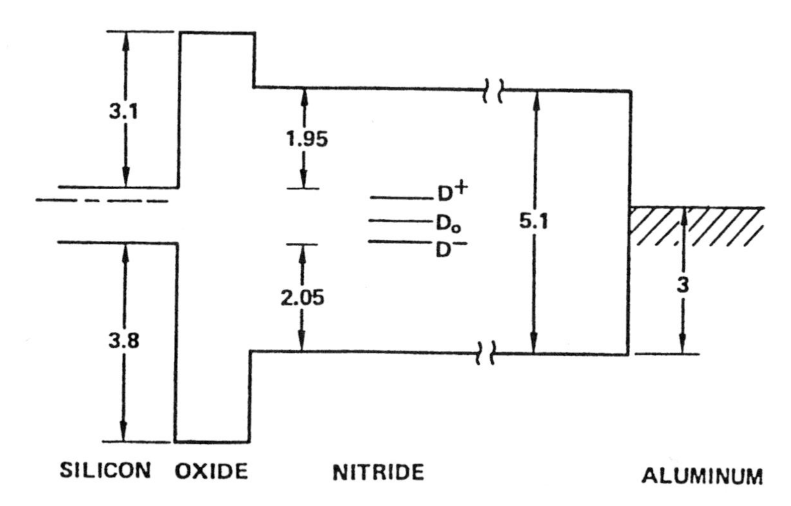

Figure 6.11 Electron-energy diagram of MNOS structure (all energy values in eV) (after Hsia and Ngai [6.15]).

lowers its total energy by electron charge transfer from one dangling bond to the other. Such charge transfers are accompanied by bond distortions. The result is the formation of positively and negatively charged dangling bonds replacing the original neutral ones. The negatively charged dangling bond state D^-, with two electrons, lies in energy below D_o, near the Si valence band edge. D^- is a hole memory trap. The positively charged dangling bond state D^+, with two holes, lies in energy above D_o, near the Si conduction band edge. D^+ is an electron memory trap. If the initial charge state of the silicon nitride film is neutral, then we must have equal numbers of D^+ and D^- states compensating each other, consistent with empirical results [6.16, 6.17]. Since these memory traps, D^+ and D^-, are charged, the large observed capture cross section of 5×10^{-13} cm^2 for both holes and electrons is congruous with the model [6.16, 6.17]. The trap center is amphoteric; that is, the dangling bond can possess either of the two charge states, D^+ or D^-.

Subsequent empirical evaluation of electron and hole injection and trapping in MONOS NVSM devices has provided evidence of the amphoteric nature of silicon nitride charge traps [6.18].

Figure 6.12 depicts n-channel MNOS memory operations as represented by the energy diagram of the nitride–oxide–silicon system. In the write operation, an electron, available at the Si–SiO_2 interface, tunnels through the SiO_2 barrier to the nitride conduction band. The electron then drops to the D^+ level, converting D^+ to D^o. By a similar process, D^o is converted to D^- in accordance with

$$\begin{aligned} e^- + D^+ &\rightarrow D^o \\ e^- + D^o &\rightarrow D^- \end{aligned} \tag{6.9}$$

To erase, one electron in the D^- state tunnels across to the silicon conduction band via the large tunnel barrier comprised of the spatial silicon nitride region and the silicon dioxide barrier. Then, the remaining electron in the D^o state similarly

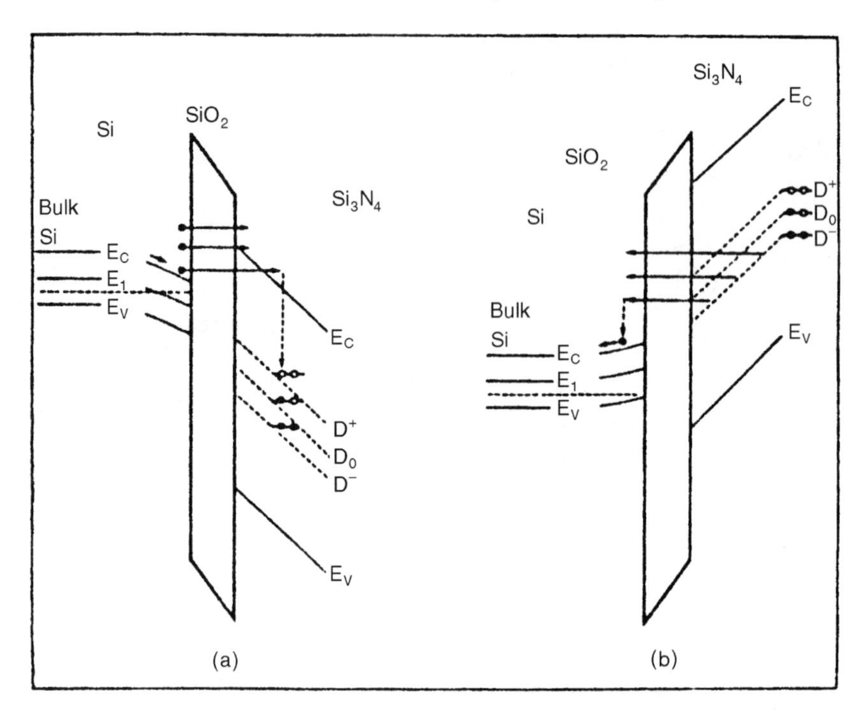

Figure 6.12 Energy-level diagrams describing MNOS memory operations: (*a*) write; (*b*) erase (after Hsia and Ngai [6.19]).

tunnels across the barrier. The asymmetry of the tunnel barrier leads to asymmetry in the memory erase write operation.

For p-channel MONOS NVSM devices, hole injection from the semiconductor has been demonstrated as the dominant mechanism for writing, and the traps in the nitride can undergo a similar reaction:

$$\begin{aligned} h^+ + D^- &\rightarrow D^o \\ h^+ + D^o &\rightarrow D^+ \end{aligned} \tag{6.10}$$

For a given injection condition, one of the capture processes involves a charged state and is characterized by a larger capture cross section compared with the capture process by the neutral state, D^o.

Silicon nitride, prepared by chemical vapor deposition from ammonia–silane mixtures at elevated temperatures, has a significant concentration of hydrogen. The hydrogen contained in the film is bonded to either silicon or nitrogen. The hydrogen content increases with decreasing deposition temperature, indicating more complete Si–N bonding in the nitride at higher deposition temperatures [6.17, 6.20]. Results [6.21] from combining nuclear reaction and infrared absorption studies have indicated hydrogen concentrations of 8.1 and 6.5% for deposition temperatures of 750°C (with NH_3/SiH_4 ratio of 200:1) and 900°C (with NH_3/SiH_4 ratio of 300:1), respectively.

It is expected that there will be more complete Si–N bonding at higher growth temperatures and larger NH_3/SiH_4 ratios. Indeed, for films deposited at 900°C and 950°C, no Si–H bonds are observed [6.21] and H is bonded only to N. A chemical vapor deposition (CVD) nitride grown at a relatively low temperature (around 600°C) will have more incomplete silicon–nitrogen bonding. In addition to existing silicon dangling bonds D^+ and D^-, there are also the hydrogen-decorated silicon dangling bonds, that is, Si–H bonds. Si–H is not a strong bond. Upon annealing [6.17,6.21] at a higher temperature, say 950°C, the Si-H bonds in the low-temperature film will break and H can diffuse out of the film. Indeed, it has been shown that annealing [6.21] a 750°C film for five hours at 950°C removed 50% of H bonded to N and 65% of H bonded to Si. The process of high-temperature annealing of a film grown at a low temperature has the effect of increasing the Si and N dangling bonds, the strained Si–N and Si–Si bonds, as well as D^+ and D^-. All these species can be candidates for memory traps in the silicon nitride bulk. Si dangling bonds and strained Si–Si bonds have been identified [6.22] as the origins of interface states and charges at the Si–SiO_2 interface of Si MOS devices.

On erase/write cycling the MNOS device, electrons or holes of large energy are injected into the nitride. For example, the injected hole makes the following electromechanical reactions possible:

$$\equiv \mathrm{SiH} + \mathrm{h}^+ \rightarrow \equiv \mathrm{Si} - + \mathrm{H}^+ \tag{6.11}$$

and

$$\begin{aligned} &\equiv \mathrm{Si} - \mathrm{H} + \mathrm{h}^+ + \equiv \mathrm{Si} - \mathrm{N} - \mathrm{Si} \equiv \rightarrow \\ &\equiv \mathrm{Si} - + \equiv \mathrm{Si} - \mathrm{N} - \mathrm{H} + -\mathrm{Si} \equiv \end{aligned} \tag{6.12}$$

Here, $\equiv$ Si denotes the Si bonded to three N atoms, and $\equiv$ Si– denotes a Si dangling bond, D^o, which will eventually lower its energy by conversion to D^+ and D^-. Thus, memory traps are created by Si–H bond breaking, and in addition, cause the movement of stored charges away from the silicon.

The microscopic model of memory traps can be considered in connection with a macroscopic model to predict the logarithmic retention behavior of the MNOS devices, as well as increased dielectric conduction after endurance exercise [6.23].

The spatially distributed traps can be viewed as localized regions of the memory dielectric. With high-field stress, delocalization increases, with structural adjustment shifting the local energy levels in the regions of highest applied field, increasing the partial coherence of the nearby site waveforms, and thus enhancing site-to-site (as trap-to-trap) carrier mobility between these sites. With continuing erase/write or endurance cycling, intersections of delocalized states can result in the formation of a "weak link" network in the dielectric, which anticipates the device failing in the end-of-life "conductive" mode.

The macroscopic model can be considered as the generalization of the Boltzmann Principle via-à-vis the localized trap sites. The number of sites, N, available for activation in the range E to E + dE can be written as

$$\mathrm{N} = \mathrm{f(E)dE} \tag{6.13}$$

If we assume that the activation of N for a generalized parameter ξ is proportional to N, then,

$$\frac{dN}{d\xi} = \lambda N \tag{6.14}$$

Now, if we also assume

$$\frac{d\lambda}{dE} = \frac{\lambda}{E_o} \tag{6.15}$$

Then,

$$\begin{aligned} N &= N_o \exp[-\lambda\xi] \\ &= f_o(E)dE \exp[-\lambda\xi] \end{aligned} \tag{6.16}$$

and

$$\lambda = C \exp[-E)_o/E] \tag{6.17}$$

If q is defined to be the average charge of the parameter of interest Q per activation, then

$$\begin{aligned} \frac{dQ}{d\xi} &= \int_{E_1}^{E_2} q \frac{dN}{d\xi} \\ &= q\frac{E_o}{\xi} \int_{E_1}^{E_2} f_o(E) \exp[-\lambda\xi] \exp[-E/E_o] dE \end{aligned} \tag{6.18}$$

and, assuming $f_o(E)$ = constant = r, then

$$\frac{dQ}{d\xi} = \frac{qrE_o}{\xi} * [\exp(-\lambda\xi)] \Big|_{\lambda = C\exp[-E_2/E_o]}^{\lambda = C\exp[-E_1/E_o]}$$

or

$$\frac{dQ}{d\xi} = \frac{qrE_o}{\xi} * [\exp(-C\xi \exp(-E_1/E_o)\{1 - \exp(-\Delta E/E_o)\}] \tag{6.19}$$

where E_1 = lowest site energy
E_o = highest site energy
ΔE = $E_2 - E_1$,
r = $N/\Delta E$

for

$$\Delta E >> E_o \text{ and } E_2 < E_o$$

Therefore, the generic form of the macroscopic behavior is

$$\Delta Q = \frac{qNE_o}{\Delta E} * \log \xi + \text{constant} \tag{6.20}$$

predicting a logarithmic behavior for MNOS retention and endurance.

For example, in the case of a thermally activated system, such as retention and endurance, $E_o = kT$ and the generalized parameter can be taken to be time, that is, $\xi = t$ (time). Then, the generic form is given as Eq. (6.21), representing MNOS retention behavior over extended time.

$$\Delta Q = \frac{qNkT}{\Delta E} * \log t + \text{constant} \tag{6.21}$$

6.1.2 NVSM failure modes

Both types of NVSM memory devices have been considered together as similar technologies having very similar operational properties and limitations; however, their failure modes are quite dissimilar. This is due largely to the significantly different roles the traps played in each corresponding device types. Thus, the failure modes of the floating gate NVSM devices are studied separately from that of the floating trap NVSM devices.

6.1.2.1 Floating Gate NVSM Devices Two distinct contributors to EPROM retention, with different activation energies, have been determined [6.24]. (1) Charge leakage due to oxide-hopping conduction between the floating gate and the silicon substrate is dependent on the oxide dielectric quality. Manufacturing and substrate defects are random and have low activation energies, 0.3 eV for catastrophic defect failures and 0.6 eV for oxide-hopping leakage failures. Both failures manifest themselves as infant mortality and random lifetime failures. (2) Charge loss due to thermal emission has a thermal activation energy of 1.4 eV. This high-activation energy contributes to the nonvolatile retention properties of the EPROM. A plot of initial discharge rate versus storage temperature for the worst case bit of a typical programmed EPROM part is shown in Fig. 6.13.

This figure can be compared with a corresponding plot, Fig. 6.14, of a defective bit of another EPROM. The difference in activation energy of the two should be noted. The screening of the defective EPROM for reliability can be effectively implemented based on an understanding of the failure mechanism of the device, taking advantage of the activation energy differences.

Repeated stress of erase-program operations of the EPROM leads to another major failure mode of the EPROM, namely, the endurance limits and failures. Two types of failure phenomena have been reported—loss of memory window and shift of operating margin. The window collapse can be attributed to field or charge current fluence generation of oxide traps leading to increased leakage discharge of the floating gate. During cycling, the electrons can also become trapped in the oxide. The resulting negative oxide charge can be observed as a shift of threshold window center with endurance. The trapped charge also inhibits further injection charging of the floating gate, causing the memory cell to require a higher program voltage or a

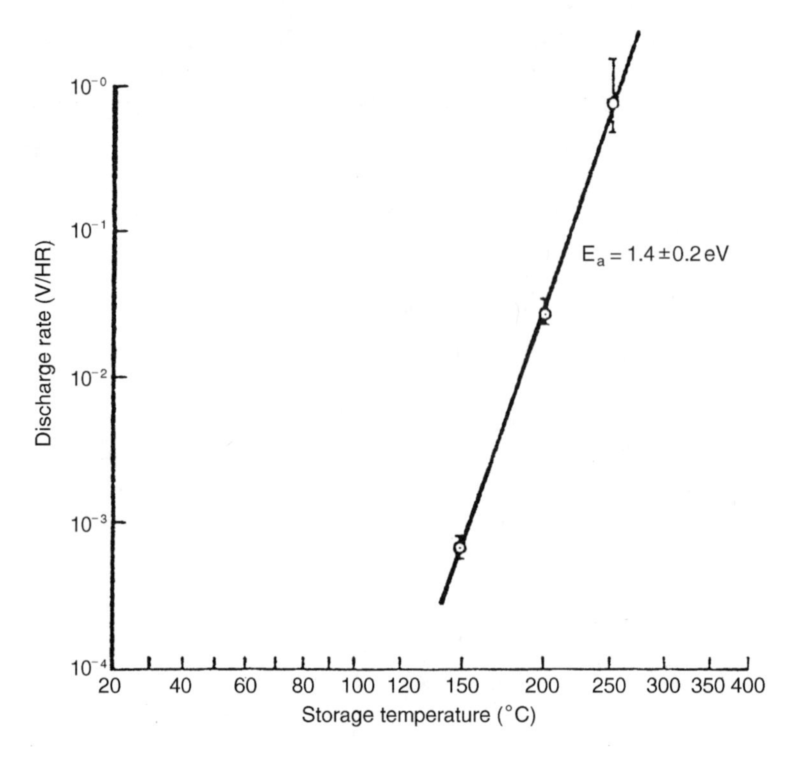

Figure 6.13 Plot of initial discharge rate versus storage temperature for the worst case bit on typical programmed parts (after Shiner et al. [6.24]).

longer program time. For the same program pulse, the memory cell exhibits a narrowing of the threshold window.

There are several approaches for Flash EEPROM. They are generically derived from EPROM technology. The programming, identical to EPROM, is accomplished with channel hot-electron injection. Different from UV erase in EPROM, the Flash EEPROM erase is accomplished with Fowler–Nordheim tunneling through the oxide gate dielectric. Therefore, bipolarity electric stress is experienced by the oxide gate dielectric in the case of Flash EEPROM endurance.

Even as it is generally acceptable to assume that Flash EEPROMs, being based on EPROM technology, have similar reliability characteristics, reliability performance is different. The thin gate oxide in use in EEPROMs is often exposed to a significantly increased level of operating stress since supply voltage scaling is not practiced as other aspects of the device process technology are scaled. Table 6.5 illustrates the relationship between process/device technology and the time-dependent oxide failure acceleration factor that must be used to accommodate the differences in oxide thickness.

The reliability characteristics of the floating gate tunnel oxide EEPROM (FLOTOX), with its thin tunnel oxide of less than 200 Å (typically 100 Å), are expected to be dominated by the properties of the tunnel oxide. In tunnel oxide

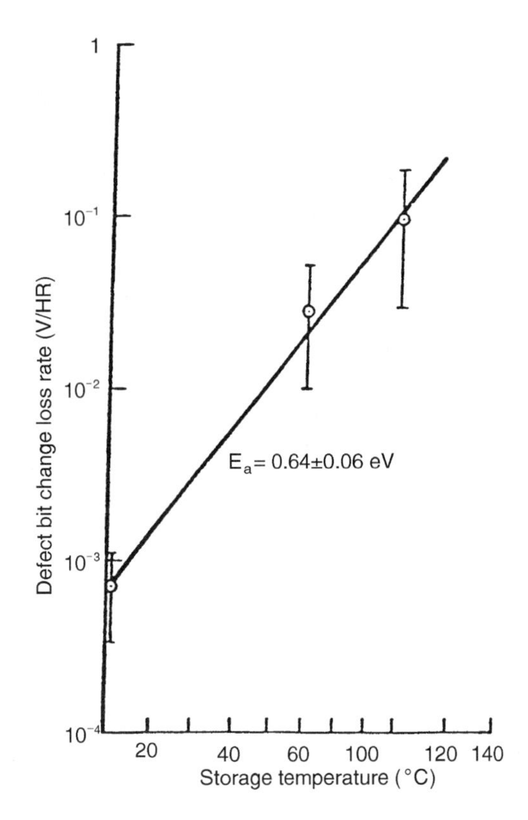

Figure 6.14 Plot of discharge rate versus storage temperature for erased defect bits in the programmed state (after Shiner et al. [6.24]).

TABLE 6.5 TIME-DEPENDENT OXIDE FAILURE ACCELERATION (PER INTEL REPORT [6.25]). [REPRINTED BY PERMISSION OF INTEL CORPORATION, COPYRIGHT 1989, INTEL CORPORATION]

Type	Supply Voltage (Volts)	Oxide Thickness (Å)	Operating Stress (MV/CM)	Acceleration Factor at _% Over Stress				
				10%	20%	30%	50%	100%
HMOS E	5	700	0.714	3.2	10	32	100	1.0E + 5
HMOS II E	5	400	1.25	7.5	55	422	3162	5.6E + 8
ETOX^{TM}	5	400	1.25	7.5	13.4	422	3162	5.6E + 8
ETOX 11	5	235	2.13	3.7	13.4	49.1	658	4.3E + 5

Assumes:
1. No bias generators.
2. Depletion loads.
3. Failure rate calculations use the appropriate acceleration factor for stress voltage and maximum operating voltage (conservative).

EEPROMs (some EEPROMs use nitride as the tunnel dielectric structure, making them similar in properties to the generic MNOS NVSM devices to be discussed later in this section), typically the oxide dielectric is thermally grown. The tunnel oxide is expected to be limited in endurance performance, with electron and hole traps dynamically modified, filled and emptied, and new traps generated, by the tunneling current during erase and write cycling. Figure 6.15 shows the typical endurance plots of a FLOTOX EEPROM illustrating some aspects of the device's charge-trapping properties.

Initially, with erase write cycling, the threshold window is widened. The generation of additional positive charges in the silicon dioxide near either interface of the metal-insulator semiconductor (MIS) structure with each corresponding erase or write cycle, during endurance exercising, results in an additional electric field at the injection interface for electrons to increase charge storage at the electrode and a corresponding field at the anode, that is, the poly–SiO_2 interface, for write threshold increase. Figure 6.16 illustrates the influence of positive charges in the SiO_2 on the EEPROM threshold window. The dominant endurance characteristic of the EEPROM is the collapse of the threshold window with increasing erase/write cycles. This phenomenon has been referred to as trap-up. The tunneling electrons, or fluence through the oxide, can be captured by electron traps. In addition, electron traps are generated by the tunneling current.

The trapped charge in the oxide reduces the effective injection field at the tunneling interface, decreasing the tunneling current and thus narrowing the threshold window of the device until it becomes nonoperative as a circuit element. A single-cell endurance is compared against measured array endurance features in Fig. 6.15. It dramatizes the importance and necessity of designing for worst case margins to ensure the performance of the product which, in this example, is an array of 16,384

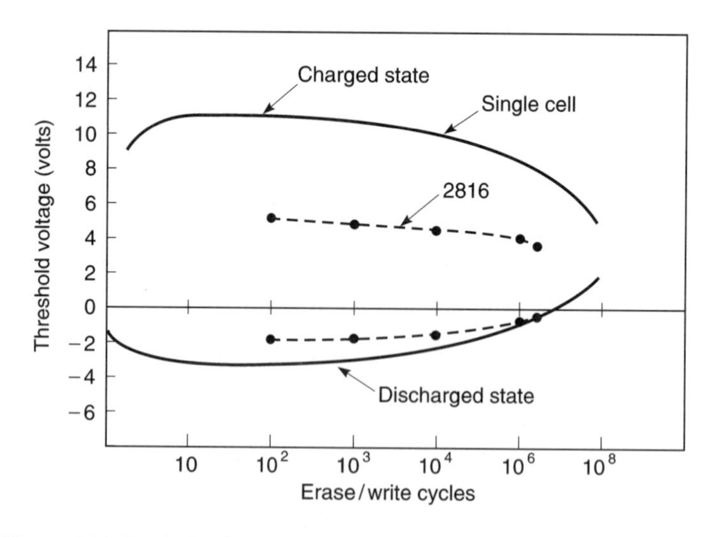

Figure 6.15 Typical cell and device window versus log cycles (after Euzent et al. [6.26]).

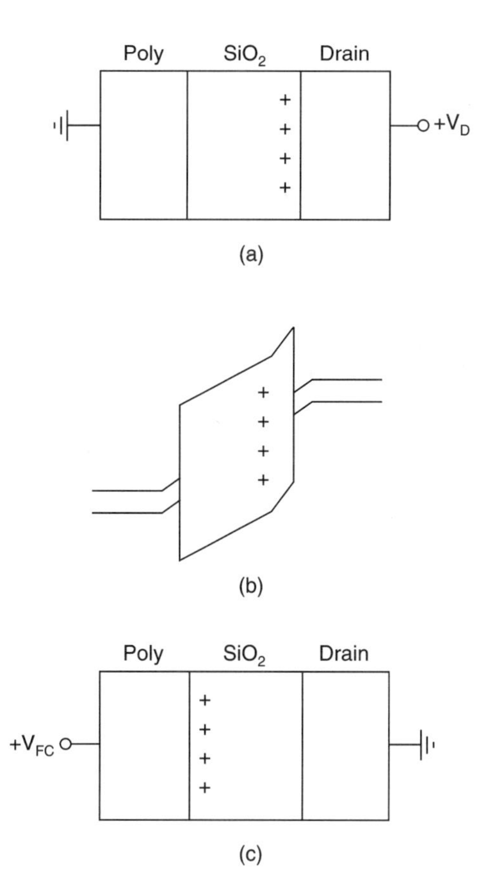

Figure 6.16 Threshold window widening (after Euzent et al. [6.26]). (*a*) Positive charge included at the SiO_2-Si interface at the end of the operation. (*b*) Band diagram of subsequent erase showing lowering of the tunneling barrier by the trapped positive charge. (*c*) Positive charge near polysilicon-SiO_2 interface at the end of the erase operation.

individual EEPROM cells. Figure 6.17 is a threshold window endurance plot of a 1Mbit Flash EEPROM memory illustrating not only the design margins needed to ensure product quality, but also excellent endurance properties of $> 10^6$ erase/write cycles, reflecting the improvement made in the gate dielectric endurance properties through the 1980s. Intrinsic charge loss of EEPROMs is expected to be equivalent to that of EPROMs. However, EEPROMs can be assumed to exhibit a higher rate of data retention failure than EPROMs due to defect-related failures of the oxide dielectric. The EEPROM tunnel oxide typically is thinner, and in addition, the number of erase/write cycles, which is usually specified several orders of magnitude higher for recent EEPROMs, contribute to higher defect charge losses than EPROM devices of the equivalent technology stage (i.e., defect levels of the process as well as lithographic dimensions). The activation energy of this EEPROM failure mechanism has been measured [6.26] at 0.62 ± 0.4 eV.

Other than tunnel oxide floating gate EEPROMs, the other floating gate EEPROM technology is the textured-poly gate cell. Based on much thicker interpoly tunnel oxides than the tunnel oxide typical of FLOTOX EEPROMs, textured-poly devices exhibit excellent reliability in failures due to random defects that contribute

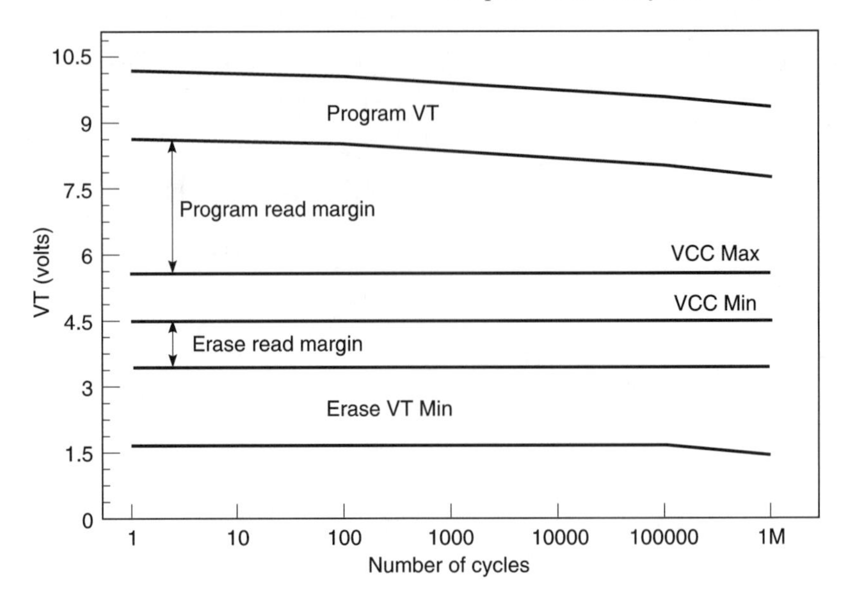

Figure 6.17 Array V_t versus cycles (after Kynett et al. [6.27]).

to intrinsic charge loss causing retention failures with endurance cycling. On the other hand, the textured-poly cells typically suffer from charge-trapping endurance failure at much lower levels of endurance cycling than FLOTOX devices [6.28]. Figure 6.18 compares the trap-up properties of the two floating gate EEPROM devices.

6.1.2.2 Floating Trap NVSM Devices. With spatially distributed traps in the dielectric as the storage means, MNOS retention differs markedly from that of the floating gate NVSM. MNOS thresholds continuously decay as trapped charges in the dielectric are detrapped. A logarithmic threshold decay was established and reported in the literature [6.29]. The end of data retention, typically represented by the closure of the threshold memory window, has also been investigated in terms of temperature; that is, activation energy, derived from Arrhenius plots, ranges from $\approx 0.6\,\text{eV}$ to $1.8\,\text{eV}$ without considering endurance effects. The deleterious effect of endurance exercising on the MNOS device can be presented as a continuous reduction of retention lifetime activation energy with erase/write cycling of the device. Table 6.6 and Fig. 6.19 present a set of such data on MNOS retention and endurance reliability.

6.2. NVSM RELIABILITY AND APPLICATIONS

Current trends in microcomputer chips will lead to chips that have more than 100 million transistors and a required failure rate of 10 FIT [6.30]. Data from Intel Corporation (see Fig. 6.20) clearly support this FIT rate trend along with the asso-

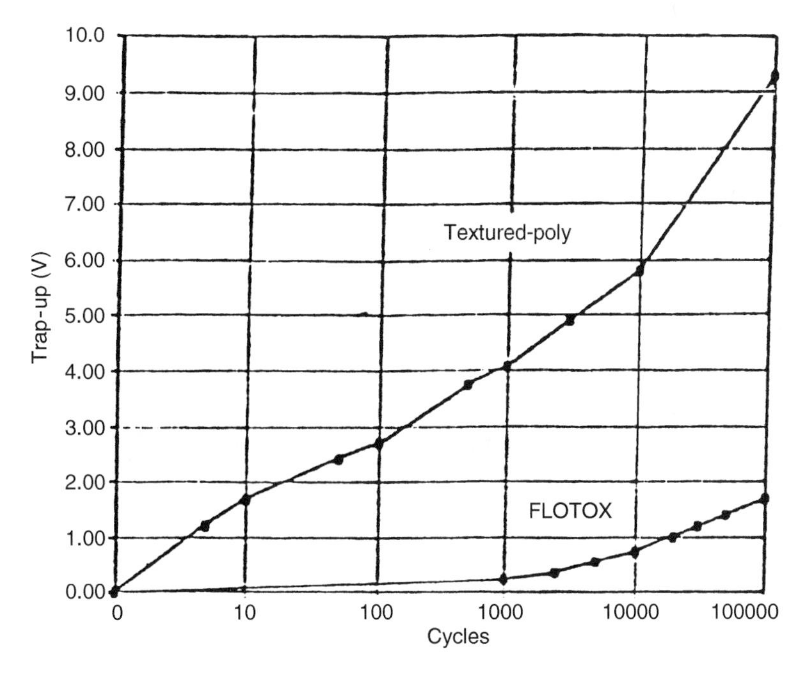

Figure 6.18 Trap-up comparison. Trap-up is measured as the increase in programming voltage required to program the cell as a result of cycling (after Mielke et al. [6.28]).

ciated improved infant mortality goals targeted for the end of this century. This trend will carry over into memory chips as well.

In order to reach these transistor count goals, the feature sizes of CMOS technology will be scaled to a 0.2 μm channel length, a 50 Å gate oxide thickness, and a metal pitch of 0.8 μm as indicated by Intel trends in Figs. 6.21 and 6.22.

These dimensional trends have a profound effect on the requirements for reliability assessment and process control. Chips will be operating with many orders of magnitude less intrinsic reliability margin than was available in the early 1970s. Assuming that existing relationships between time-to-fail and device parameters are valid for electromigration, hot-carrier degradation, and oxide wearout, it can

TABLE 6.6 ACTIVATION ENERGIES AT VARIOUS ERASE/WRITE CYCLES (AFTER AJIKI ET AL. [6.29]).

Erase/Write Cycles	Activation Energy (eV)
10	0.67
10^3	0.60
10^4	0.50
10^5	0.25

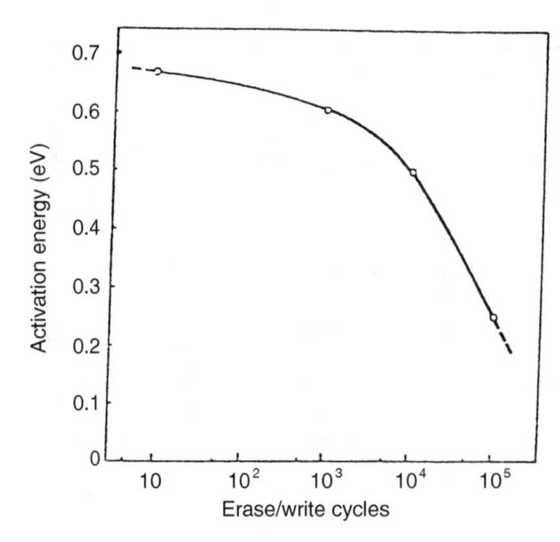

Figure 6.19 Activation energies at various erase/write cycles (after Ajiki et al. [6.29]).

be shown in Fig. 6.23 that relative median time-to-fail has been reduced by the scaling process.

This reduced reliability margin means that the processing line monitoring requirements must grow to offset the smaller latitude. Material quality variations that were acceptable in the past can no longer be tolerated in the future. As new processes and new materials are introduced in submicron fabrication, it will be

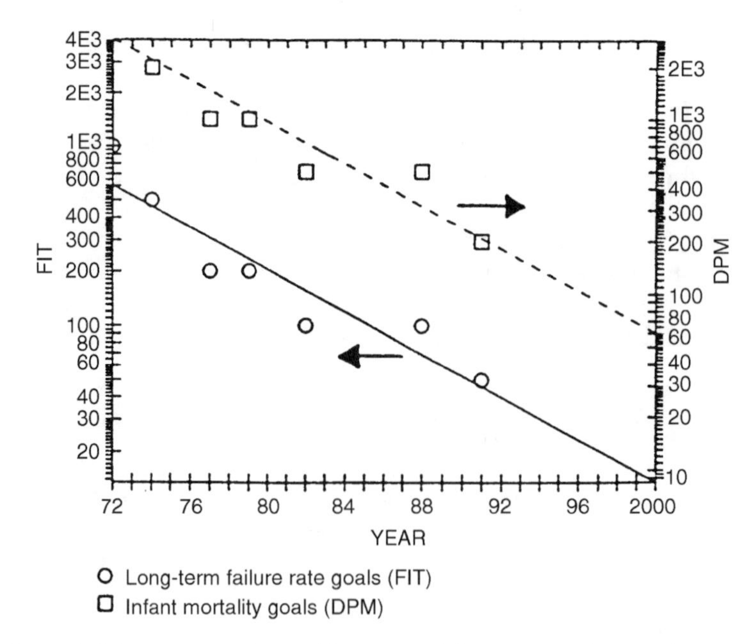

Figure 6.20 Infant mortality and long-term failure rate goals at Intel versus time (after Crook [6.30]).

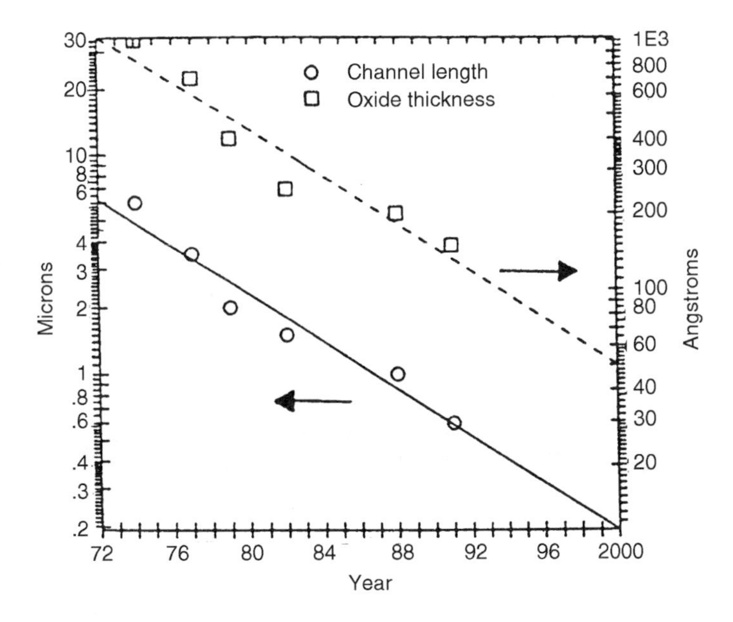

Figure 6.21 VLSI channel length and gate oxide thickness trends (after Crook [6.30]).

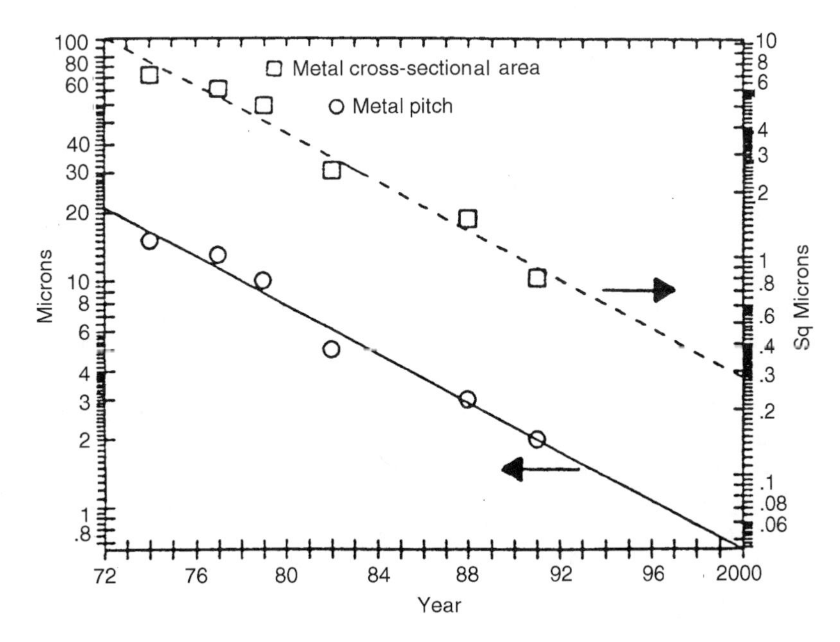

Figure 6.22 VLSI metal pitch and minimum line cross-sectional trends after Crook [6.30]).

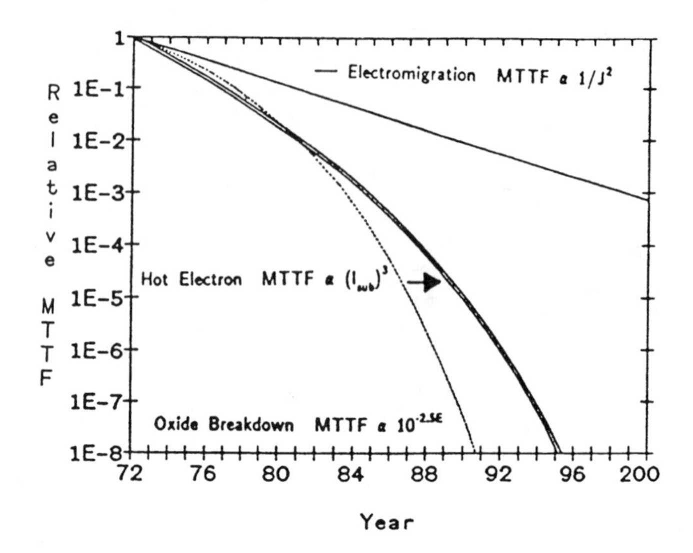

Figure 6.23 Over the years, impact of the effects of device and metallization scaling on time-to-failure margins for electromigration, hot electrons, and oxide wearout failure mechanisms (after Crook [6.30]).

necessary to update or completely change reliability models to accommodate new mechanisms. Test structures must be developed to characterize the reliability limiting processes and provide monitors for process control. It is necessary to monitor the individual mechanisms at the wafer level, at the end of processing, and also to extend wafer-level monitors further back into the processing line. Tighter process controls within an aggressive statistical process control (SPC) regime is critical.

Among the main considerations in reliability of NVSM devices is the fact that normal operation of the devices requires high operating voltages in order to erase or write data. The high voltages result in high-stress fields across the gate oxides of not only the storage devices, but also the MOS circuitry that supports the operation of the storage device. The insulator quality becomes the most serious concern for the fabricator of NVSM devices. Even relatively small insulator defects, which may be of little concern to standard MOS circuits, are of major concern in NVSM devices because of the higher operating fields. This is compounded by the fact that floating gate devices must use ultra-thin insulators to be able to program the storage devices quickly.

In addition to the high-field stress in insulators, another stress factor is caused by the high-peak current required to charge and discharge circuit capacitances to the elevated voltages required in the chip. This places interconnect metalization in jeopardy of early failure due to electromigration. The high-peak current densities result in high-peak temperature and can place more stringent requirements on the current density limit rules that designers must use in designs. Because one of the design goals is to produce the highest circuit density possible, the current density design rules are commonly pushed to the limit permitted by the fabrication process.

Beyond the hard fail aspects of the materials are the issues of retention and endurance failure modes in NVSM devices as outlined in Section 6.1. These mechanisms cause slow degradation in performance. Retention failure is caused by repeated interrogation of the device, resulting in a small amount of charge being changed, which limits the time period within which data can be accessed without error to a time less than is acceptable for an application. Endurance failure is a result of changes in the characteristics of the storage transistor so that the device can no longer meet performance criteria (usually retention time). Changes in device characteristics occur with accumulated erase/write cycles.

The above reliability limiting mechanisms must be understood individually, modeled, and monitored in order to maintain a prescribed level of NVSM reliability. The following discussion addresses the failure mechanisms in terms of intrinsic and extrinsic, or defect-related causes. In order to be able to monitor the reliability of NVSM devices (or any other MOS device), models must be created for the rates of failure due to each mechanism. This enables the development of well-characterized test structures and will lead to more accurate statistical process control in manufacturing.

With appropriate models in place that allow the extrapolation of individual mechanism-accelerated test data to use conditions, additional effort is needed to understand the point in the fabrication process where the reliability hazard is introduced. It is also necessary to be aware that fabrication testing is not, by itself, sufficient to guarantee a flow of highest quality product.

6.2.1 Reliability Modeling

The authors endorse a global perspective of NVSM reliability—the reliability of the NVSM represents the quality of the product and its improvement that can be addressed in all stages of product design and manufacturing, ranging from processing to reliability testing. However, it must be understood that existing insulator reliability models, as described in Section 6.1, are empirical and may be changed as more research is done to understand the physical processes of stress-induced insulator damages. At this time, there are two widely debated models for insulator reliability. One is described in [6.31] and is referred to as the "E" model, indicating the electric field dependency of acceleration factor. The other is included in the paper by Chen et al. [6.8], which has fit data from numerous papers to an equation for time-to-fail that varies as 1/E. Since both models appear to fit available data, the actual physical model will likely have other variables involved in a more complex relationship with electric field and temperature. For purposes of assessing the oxide reliability of NVSM devices, the 1/E model fits experimental results best when the oxide is exposed to high electric fields in the range used to write data into a NVSM chip.

In addition to the insulator failure models, there are models for the other mechanisms that determine chip reliability. This section discusses a number of other models for mechanisms that are of interest to NVSM reliability. A top-down approach is appropriate to gain perspective in the problem of creating models to measure and track reliability on a fabrication line. The familiar "bathtub" failure rate curve has three regions of interest: early life (often called infant mortality), useful life, and wearout. The early-life hazard rate has always been a major concern for manufacturers because it directly impacts product warranties and company reputation.

There is a simplified method for expressing the overall early-life hazard rate of a chip by defining an expression that contains the aggregate effects of all mechanisms involved in limiting the lifetime of a NVSM chip. It is most common to choose an exponential distribution for early failure rates because it generally fits measured early failure data very well, and it also has the property of being mathematically tractable. This expression takes the classical time-dependent hazard rate form:

$$\lambda(t) = \lambda_b * \exp(-\tau * t) \tag{6.22}$$

where λ_b = base failure rate
τ = rate time constant
t = time

The equation assumes that all temperature-dependent mechanisms are referenced to 25°C (room temperature). The failure rate obtained from samples life-tested at other temperatures is obtained by multiplying each time-to-fail by the Arrhenius relationship to correct it to the reference temperature.

$$A(T) = \exp\{[-E_a/k] * [(1/T) - (1/T_o)]\} \tag{6.23}$$

where A(T) = temperature acceleration factor
T = temperature (Kelvin)
T_o = reference temperature (Kelvin)
k = Boltzmann's constant = 8.62310^{-5} (eV/K)
E_a = activation energy (eV)

The failure rate at the new temperature then becomes

$$\lambda(t) = \lambda_b * A(T) * \exp[-\tau * A(T) * t] \tag{6.24}$$

which is the basic temperature acceleration model for failure rates of any CMOS device by any known mechanism. Tables 6.1 and 6.2 in Sections 6.0 and 6.1, respectively, contain activation energies that have been measured for many mechanisms of concern to semiconductor device manufacturers. The actual value of activation energy for early-life failures may vary from the values stated in the tables because the tabulated activation energies are measured for the wearout process. Early failures may contain multiple contributing mechanisms that lead to activation energy values that differ significantly from the tabulated values given here. Table 6.7 summarizes typical values for coefficients for Eq. 6.24 for a range of mechanisms that govern failure rates in IC devices.

The failure rate during the early failure period is also governed by other acceleration factors that relate to the specific mechanism. In this case, it is necessary to understand the physics of the mechanism well enough to at least derive an empirical

TABLE 6.7 EARLY-LIFE FAILURE RATE PARAMETERS

Mechanism	Lambda_b (F/10^6 hrs.)	Tau (1/hrs.)	Ea (eV/deg K)
Metal	0.00102	1.18	0.55
Oxide	0.0788	7.70	0.30
Contamination	0.000022	0.0028	1.0
Miscellaneous	0.010	2.2	0.43

model for the mechanism. Even though the previous discussion relates to early failure, it must be understood that these physical mechanisms are generally responsible for all aspects of the life curve because the presence or absence of small defects in the materials results in local conditions that manifest themselves as early failures (e.g., Eq. 6.4, Section 6.1). For the mechanisms of greatest interest to NVSM designers and users (i.e., time-dependent dielectric breakdown, hot-carrier damage, and electromigration), acceptable empirical models already exist and are a reasonably good fit to a large amount of experimental data.

6.2.1.1 Oxide Failure Models. Section 6.1 discussed one of two models that are commonly found in the literature. In particular, one model discussed is referred to as the 1/E model, as is clear from the form of the field acceleration factor equation (Eq. 6.3 in Section 6.1). There is still considerable discussion in the reliability research community over whether the 1/E model applies over the entire field stress range to which oxides are commonly exposed. As extensive data sets show, the 1/E model fits well for high-field cases, which is precisely the case for NVSM applications. However, existing data at the lower fields experienced by support circuitry on NVSM chips suggest that the earlier field acceleration factor model by Crook [6.12] fits data at more moderate field levels. This is referred to as the "E" model:

$$\text{Aef:oxide} = \exp[(E_{ref} - E_s)/E_o] \tag{6.25}$$

where E_{ref} = reference electric field
E_s = stress electric field (operating field)
E_o = normalizing electric field

Extensive experimental data collected by Domangue et al. [6.33] yielded a value for E_o of 0.134 MV/cm.

In the case of the 1/E model, the field acceleration factor experimentally determined by Lee, Chen, and Hu [6.9] for 1% failure was found to be 192 MV/cm. Based on this observation, the field acceleration factor most appropriate for NVSM devices is given as

$$\text{Aef:oxide} = \exp[-192(1/E_s - 1/E_{ref})] \tag{6.26}$$

Field acceleration factors are significantly influenced by the assumptions made relative to the defect population that is of interest. In the case of early failure, time to

1% fail, the slope of the $1/E_s$ versus time-to-fail curve, is significantly less than is observed at around 50% failures, as shown in Fig. 6.24. This graph clearly shows the degradation effects of oxide defects on the useful life of MOS devices. NVSM devices are particularly susceptible to random defects because of the necessarily thin oxides and high operating fields during write. As was indicated in Section 6.1, the development of oxide lifetime models is based primarily on experimental data and curve fitting to intuitively reasonable model equations. There is not yet a fully physical model that describes all aspects of TDDB life-limiting characteristics that account for the highly complex effects of various defects (including contamination). It is possible that, when routine oxide growth techniques approach defect-free quality, a more refined theoretical model will be published and experimentally verified.

In addition to a field acceleration factor for oxides, experimental results have suggested that an acceleration factor exists for scaling oxide thickness. Yamabe and Taniguchi [6.34] have described experimental data that suggest the existence of a relationship between oxide thickness and oxide thickness at constant electric field. It is apparent that this relationship is second order to the effects of electric field. Also, it is not clear whether this relationship is more a function of the significance of defect size compared to the oxide thickness. If this observation proves to be related to a mechanism other than defects, then it will have a significant bearing on NVSM reliability because the oxide thickness is so small in storage devices.

The models discussed previously for field-accelerated oxide failure assume that the oxide is exposed to a uniform stress level resulting from a roughly equipotential surface on both sides of the oxide. In NVSM devices, stress conditions are far from

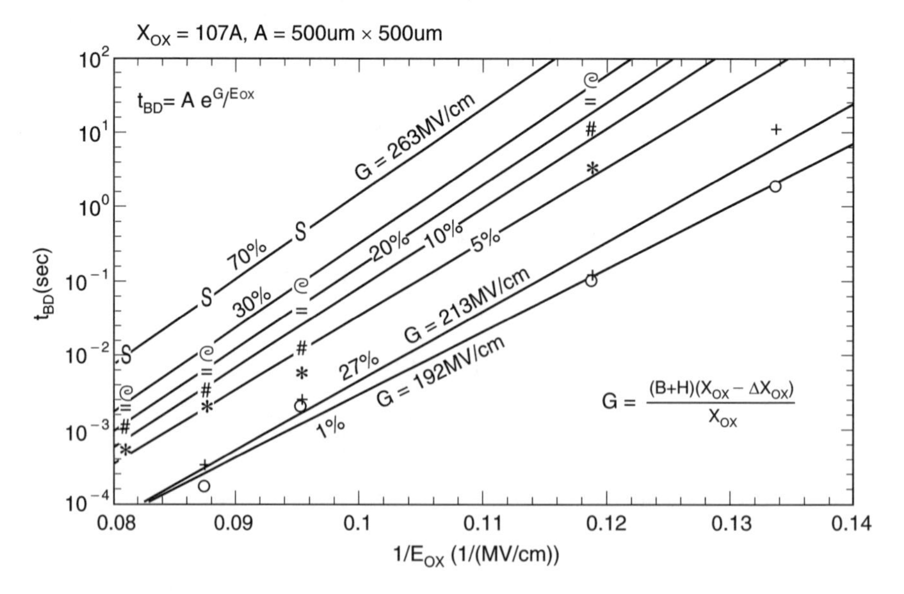

Figure 6.24 t_{BD} versus $1/E_s$, for different percentage failure levels for TDDB data (after Lee et al. [6.35]).

uniform because the method for writing and erasing the contents of the memory involves the use of hot-carrier injection along with Fowler–Nordheim tunneling current. In addition, high-drain voltages can be present along with low gate voltages in NVSM devices and bias transistors close to the snapback region, which is favorable to hot-hole injection into the oxide. Several studies [6.35–6.37] have shown that these bias conditions lead to nonuniform stress conditions in the drain region. In particular, Chen et al. [6.35] have shown a sharp drop in Q_{bd} in transistors that have been biased to inject significant hole current into the oxide. At present, there is no model for predicting oxide failure under these conditions against a general background of uniform field stress. The main concern in this situation is that existing uniform field stress models may underestimate the oxide lifetime.

6.2.1.2 Hot-Carrier Damage Models. Hot-carrier damage to transistors is of considerable concern to circuit designers of MOS circuits. As hot-carrier-induced damage occurs, the transistor transconductance and threshold voltage shift, causing circuitry eventually to fail to perform to the level required for proper operation of the circuit. It is ironic to be concerned about hot-carrier damage in a treatment on NVSM reliability because hot-carrier injection is one of the important mechanisms used for writing into NVSM cells. However, the concern applies to the MOS support circuitry present on the same chip that contains the nonvolatile memory array. If the support circuitry fails to perform its intended task, then the entire chip is considered useless even if the storage array is fully functional.

For short-channel MOS transistors, hot-carrier degradation can be expressed empirically in a simple parameter shift relationship [6.32]:

$$\text{Transistor parameter shift} = \text{Phc} * t^n \tag{6.27}$$

where t = stress time
n = power factor, which changes with injection mode
Phc = magnitude of degradation

Phc varies with bias conditions and has been experimentally shown to be related to drain voltage. This constant has an exponential relationship with drain voltage (V_d) [6.32]:

$$\text{Phc} = \text{beta} * \exp(-\text{alpha}/V_d) \tag{6.28}$$

The proportionality constants, alpha and beta, are found experimentally for transistors produced on a specific processing line. These constants are extremely sensitive to processing variations and also depend on manufacturing methods that have been devised to reduce hot-carrier damage rates on short-channel transistors. In particular, various lightly doped drain (LDD) structures are implemented by several manufacturing techniques that have a profound effect on the parameters in these equations.

Combining Eqs. 6.27 and 6.28, we can solve for the amount of time required for a specified shift in a parameter. This is useful for cases where a designer intends to implement circuits that are tolerant of selected parameter shifts caused by hot-carrier damage. From a manufacturing point of view, it is valuable to have a method that can be used to relate easily measured electrical quantities to expected time to unacceptable parameter change. One easily measured electrical quantity is the amount of substrate current that flows at specific bias conditions known to cause high rates of parameter shift. As experimentally determined, the substrate current is strongly correlated with the transistor parameter degradation rates [6.45, 6.46]. Substrate current generation is modeled by [6.46]:

$$I_{sub} = C * I_d * \exp(-\phi(i)/q * \lambda * E_m) \tag{6.29}$$

where C = process determined constant
I_d = drain current
$\phi(i)$ = impact-ionization energy
λ = mean-free path of electrons in the channel
q = electron charge
E_m = maximum electric field in the channel

An expression that relates substrate current and drain current to transistor lifetime, t, was developed [6.47] and requires relatively simple calibration procedures for parameter shift:

$$t = (A/I_d) * (I_{sub}/I_d)^{-m} \tag{6.30}$$

where A and m (typically m is about 2.5 to 2.9) are determined experimentally for a particular wafer fabrication process. This relationship assumes no temperature-related factors. Since it is widely recognized that a thermal effect is to be accounted for in hot-carrier damage, the following thermal acceleration relationship was defined [6.50], which corrects the constant A for temperature:

$$A_{hc}(T) = A_o * \exp(-0.039/kT)$$

which leads to the following revision of Eq. 6.30:

$$t = (C/A_{hc}(T) * I_d) * (I_{sub}/I_d)^{-m} \tag{6.31}$$

where C is an experimentally determined constant.

Conspicuously absent in the above discussion is an early-life failure rate relationship. This is because hot-carrier damage is strictly a wearout mechanism that affects all transistors at roughly the same rate for the same level of stress. There may, however, be some local structural differences in LDD technology transistors, introduced during manufacture, that would affect the damage rate at different locations on a wafer, which would be treated as a process control issue.

6.2.1.3 Electromigration Models. Electromigration of interconnects is one of the most common life-limiting mechanisms found in VLSI devices. There have been many examples of documented field failures in which this mechanism was found. Its presence as a lifetime limiter is not confined to any one technology, but is most commonly observed whenever there is either a requirement for high DC current in a chip or there is a persistent high-peak pulse current of the type commonly found in memory devices. NVSM chips are among the most vulnerable to pulse-current-driven electromigration failure because high-peak current flows whenever write or erase operations occur.

The early-life failure model for electromigration assumes a form described in [6.32]:

$$\lambda(t) = 1.02E - 3 * A(T) * [\exp(-1.18 * t_o)] * \exp[-1.18 * A(T) * t] \quad (6.32)$$

where $A(T)$ = temperature acceleration factor
t_o, t in hours

This empirically derived failure rate was extracted from a large database described in reference [6.32] and tries to deal with all factors that cause early failure in metal interconnects. This expression, along with an expression describing the wearout aspects of interconnects, provides NVSM designers with a means of analyzing life-limiting factors in a chip.

Given a manufacturing facility that is well controlled with respect to random point defects in the metallization, the lifetime of the metal interconnects is well described by the classic "Black's Model" that relates the combined effects of current density and temperature on the lifetime of the metal [6.91]. Cumulative failures are lognormally distributed.

$$t_{50} = A * ((1/J)^n) * \exp(E_a/kT) \quad (6.33)$$

where A = pre-exponential constant
J = current density (A/cm^2)
E_a = metal thermal activation energy (eV)
n = current density exponent
k = Boltzmann's constant = 8.62 E-5 eV/K
T = metal temperature (K)

Constants E_a and n are determined experimentally in accelerated tests involving ASTM standard test methods [6.38–6.40], and generally result in constants that span a relatively narrow range depending on the type of metal system in use for interconnects. The pre-exponential constant contains the effects of the metal deposition environment and alloy composition, metal thickness, width, and length. The constant A (and for that matter, all of the other constants) must be determined experi mentally on a specific test structure (e.g., the ASTM structure) in order to have a reference point for metal quality that can be extrapolated to normal use conditions.

Activation energy, E_a, ranges from about 0.43 eV to about 0.8 eV, with a typical value of about 0.55 eV. Factors controlling the actual value of the activation energy include alloy composition, annealing conditions, contamination, and the presence or absence of passivation. In addition, special metal systems that utilize composite structuring with a refractory barrier material below (or above and below) the aluminum alloy layer will exhibit different activation energies depending on the criteria defined as failure. These composite, or layered, metal systems typically do not fail in the open-circuit mode observed with aluminum alloys. They experience a resistance increase with time as the aluminum alloy layer migrates, leaving only the barrier material to conduct current. Electromigration testing of these composite materials uses a fractional resistance increase (e.g., 1.30 times initial resistance at the test temperature) as a failure criterion. The choice of magnitude of this fractional change of resistance will affect the value of the activation energy extracted.

The current density exponent, n, has a typical value of 2.0 for metal subjected to nominal design limit current densities. Other values of n are reported in [6.32] for some alloys and composites, depending on the current density used in the electromigration test. It is particularly important that the material being tested be carefully characterized at various ranges of current density in order to be able to extract a value of n that is consistent with the test conditions and that will make extrapolation to at-use conditions meaningful. Extraction of n for extreme acceleration factor conditions is also desirable, as will be discussed later in this chapter.

Black's Model for electromigration assumes that a direct current flows in the interconnect. As a result, there will be an equilibrium mass flow condition which eventually results in open failure. If the current flow is pulsed or bipolar, there may not be an equilibrium mass flow condition in the interconnect. Towner et al. [6.41] have empirically derived a duty cycle correction relationship that translates DC t_{50} test results to pulsed DC operating conditions. Generally, a frequency dependency does not exist as long as the pulse period is shorter than the vacancy relaxation time of the interconnect metal. Suehle [6.42] and others [6.43, 6.44] have shown experimentally that, when the pulse frequency is greater than about 200 Hz, the relationship between t_{50}:pulse and t_{50}:DC is

$$(t_{50}\text{:pulse})/(t_{50}\text{:DC}) = 1/r^2 \tag{6.34}$$

where

$$r = \text{pulse duty cycle.}$$

In the case of complete symmetry of current flow direction (fully bipolar, assuming short periods) [6.43], the electromigration lifetime is greatly increased, yielding an MTF:AC/MTF:DC ratio of about 1000. This ratio was experimentally shown to be independent of current density and is of considerable importance in the case where a metal interconnect is supplying current to charge and discharge a large capacitive load.

6.2.2 NVSM Application — Device Design Examples

An accurate understanding of device performance limitations and reliability failure mechanisms points to device design innovations that can enhance product manufacturability and reliability. As devices are scaled for storage density, drain-source leakages and gate dielectric defects become significant performance, reliability, and manufacturability issues. Recent EPROMs and Flash EEPROMs have incorporated changes in the memory cell to minimize the impact of these failure modes [6.49, 6.50, 6.51]. A series-enhancement MOS transistor is added to the floating gate structure to form a split gate memory cell, as seen in Fig. 6.25. The series transistor eliminates the depletion leakage current of electrically erased memory

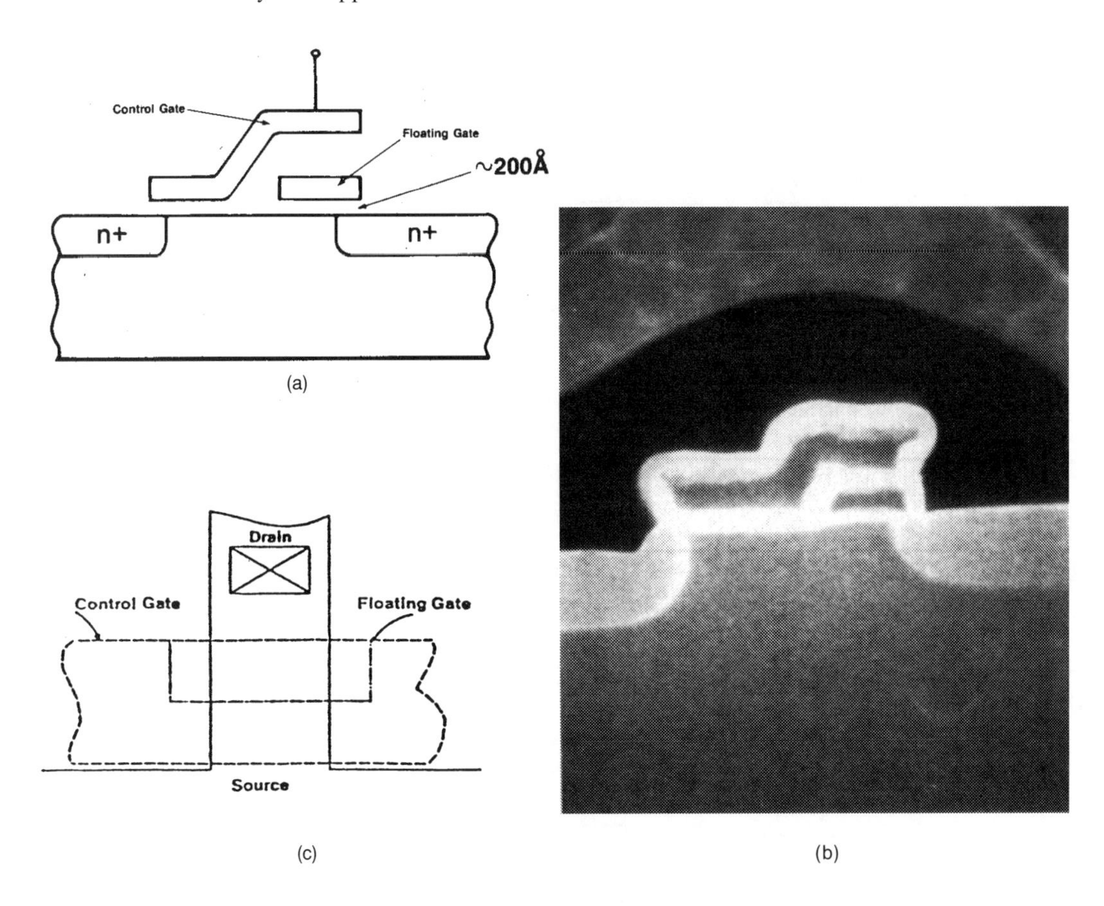

Figure 6.25 (*a*) Crosssection of the Flash EEPROM cell, (*b*) SEM cross section, and (*c*) layout of the Flash EEPROM cell (after Samachisa [6.49]).

devices in the EEPROM array. It also provides immunity to drain-source punchthrough, regardless of the effective channel length of the floating gate memory region. Therefore, with the series transistor as part of the memory cell structure, the yield-limiting memory channel area can be minimized to improve manufacturability. The split gate EEPROM cell size is larger ($\simeq 20\%$) than a similar cell using the same set of design rules, but without the series transistor. The increase in area must be evaluated against the smaller memory channel length that can be utilized and the reduction of erase and read circuit complexity because of the improved memory cell characteristics. To reduce the gate dielectric defect level and, concurrently, to enhance capacitive-coupling of the control gate and the floating gate, an interpoly dielectric sandwich of a $SiO_2/Si_3N_4/SiO_2$ stacked structure [6.51] has been used (see Fig. 6.26). A higher breakdown voltage and lower defect density than a SiO_2 monolayer dielectric for this structure have been reported in the literature [6.53, 6.54]. To obtain better

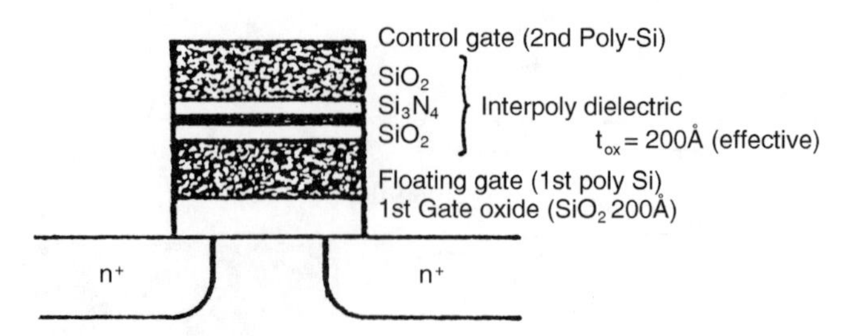

Figure 6.26 Cross-sectional view of the memory cell transistor (after Ohtsuka [6.51]).

endurance characteristics, the FLOTOX EEPROM tunnel insulator utilizing nitrided oxide thin films has been suggested [6.55, 6.56].

The importance of the role of reliability physics and engineering is evident in the device design of floating trap NVSM cells. To eliminate charge injection from the polysilicon gate electrode into the nitride, a blocking oxide is introduced, forming the silicon–oxynitride–nitride–oxide–silicon (SONOS) memory gate structure. Typical cross-sectional views of the SNOS and SONOS are compared in Fig. 6.27. In Fig. 6.28, typical capacitance-voltage (C-V) plots of these structures are illustrated showing the "0" and "1" states of the devices, set/programmed with ± 25 V, 5-second programming pulses. The location of the "0" and "1" states, vis-à-vis the initial C-V plots in SONOS structures, indicates that charge injection is from the substrate, validating the successful blocking of the undesirable gate electrode injection [6.59].

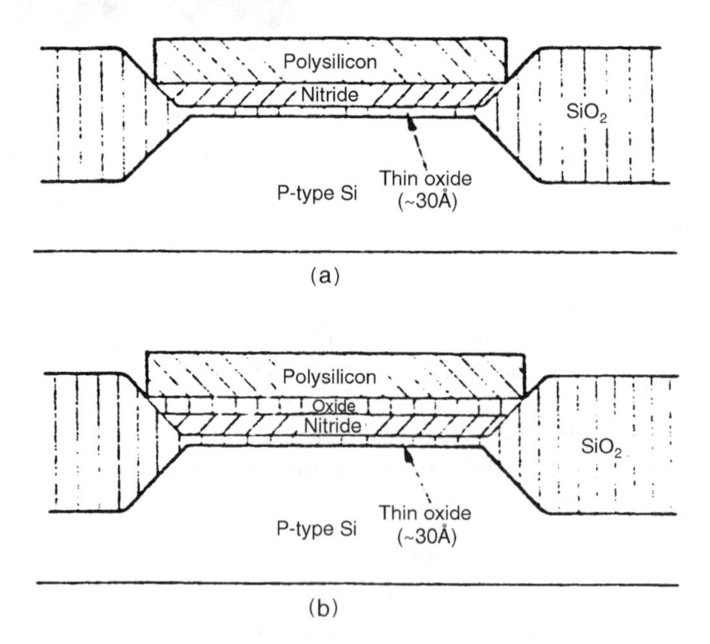

Figure 6.27 Cross-sectional views of (*a*) SNOS and (*b*) SONOS capacitors (after Chen [6.57]).

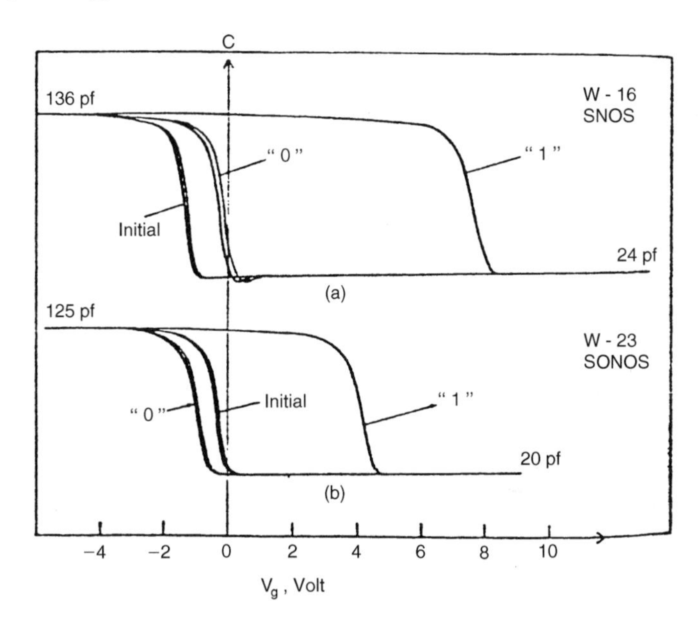

Figure 6.28 C-V plots of SNOS and SONOS capacitors after application of negative- and positive-voltage pulses (after Chen [6.57]).

The nitride trap density, conductivity, and composition can also be tailored vertically in the MNOS memory gate to improve NVSM retention and endurance. By adjusting fabrication process parameters, a two-layer silicon nitride memory dielectric can be formed in which the layer immediate to the tunnel oxide is to be maximized for memory charge traps, and the second layer formed, minimized for trap defects. The charge traps are thus restricted to a region that is most effective for memory threshold shifts. Retention and endurance of the device are improved as charge leakage and charge centroid migration are bounded by the "trap-free" second nitride layer preceding the gate electrode [6.58].

Series MOS structures merged with the MNOS memory transistor (see Fig. 6.29) are used to eliminate depletion-mode operation of the MNOS, as well as to reduce yield and reliability failures due to memory dielectric pinholes over heavily doped source and drain diffusions [6.59]. Such structures also have increased drain-source breakdown voltages, as well as greatly reduced soft turn-on leakages. To eliminate "sidewalk" parasitic leakage currents, cross gate MNOS memory structures have been effective [6.60]. Figure 6.30 shows the center cross-sectional view along the width direction in the gate region of a typical step gate MNOS transistor. The transistor region between the thick gate oxide and thin memory oxide region functions as an MNOS device with gradually increasing oxide thickness. This region has been termed the sidewalk and causes parasitic leakage currents, as is illustrated in Fig. 6.31. Figure 6.32 is a composite schematic of the cross gate MNOS memory structure. Along the width of the gate channel, the sidewalks are removed, with the gate extending beyond the metal electrode. The field oxide at the edge of the MNOS gate width prevents the formation of any parasitic conductive path due to fringing

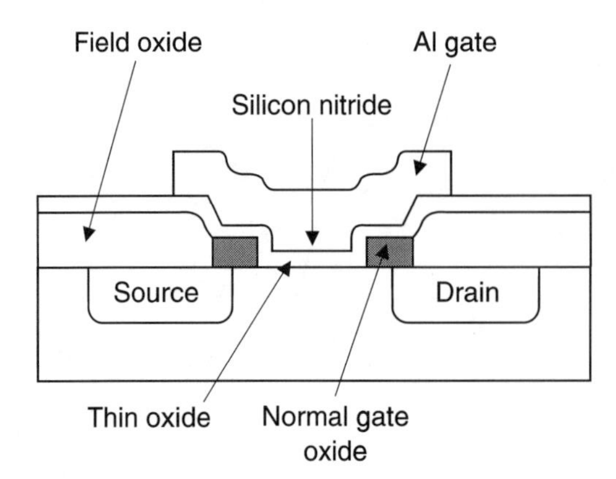

Figure 6.29 Schematic showing representative section through MNOS memory transistor (after Hsia [6.59]).

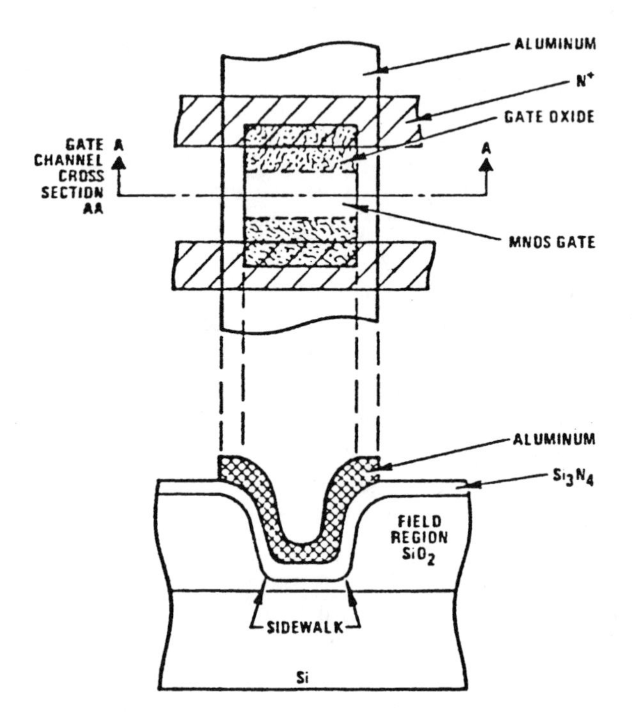

Figure 6.30 Typical MNOS device structure with cross section AA showing the "sidewalk" region (after Hsia [6.60]).

fields. Figure 6.33 shows the transfer characteristics of the cross gate MNOS memory device without the parasitic sidewalk leakage current. The improved margins that resulted from eliminating the "sidewalk" parasitic leakage currents permit single-transistor memory arrays and improved MNOS memory array density, since the more conservative two elements per lot storage/sense techniques are no longer needed, even though the cross gate memory cell is larger in area.

The use of an on-chip program voltage so that a single low-level V_{DD} supply (i.e., 5 V) can be used for EEPROM operation imposes additional design considerations on the memory cell [6.61]. In a conventional, fully aligned MNOS (SNOS) memory transistor, a large program current flows as a result of drain breakdown due

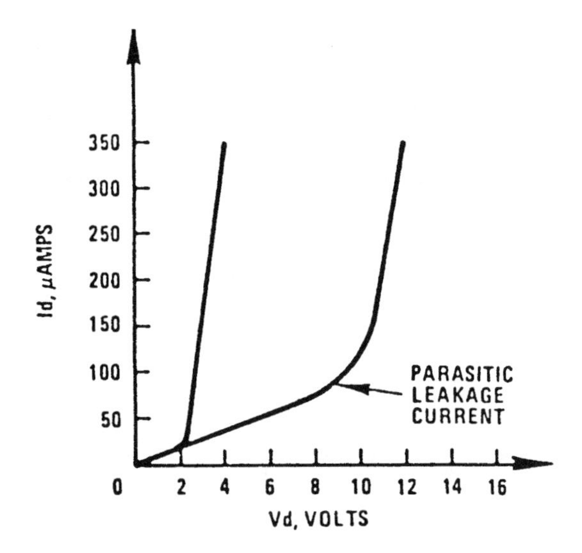

Figure 6.31 Typical MNOS transistor characteristics with parasitic leakage current problems (after Hsia [6.60]).

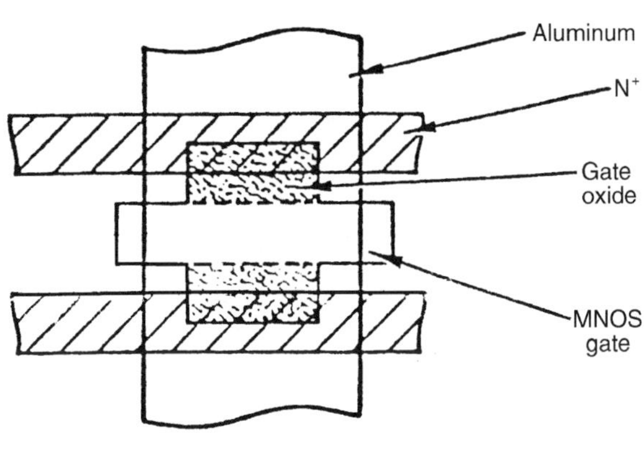

Figure 6.32 The cross-gate MNOS memory device (after Hsia [6.60]).

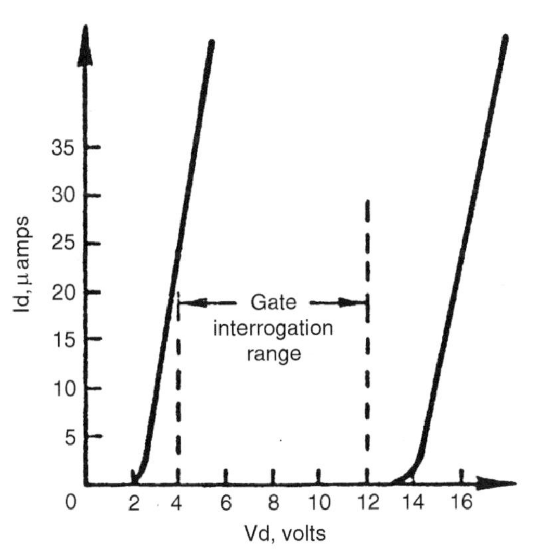

Figure 6.33 Typical device transistor characteristics obtained with the cross gate MNOS memory device (after Hsia [6.60]).

to hard ground biases on gate and substrate. The on-chip program voltage supply current cannot readily sustain such a large programming current to reduce the breakdown voltage. Therefore, an off-set MNOS memory cell structure is utilized, as shown in Fig. 6.34. The $N - N^+$ off-set drain (and source) structure, together with the step polysilicon gate, results in two field-strength regions under the gate electrode (see Fig. 6.35). With the off-set, the breakdown voltage is dependent on impurity concentration. In addition, optimization of the breakdown voltage must be determined in connection with the relationship of concentration and voltage drop through the channel region. Figure 6.36 shows these two relationships and the selection of an N implant dosage of $3 \times 10^{12}\ cm^{-2}$ to $10 \times 10^{12}\ cm^{-2}$ for optimal results.

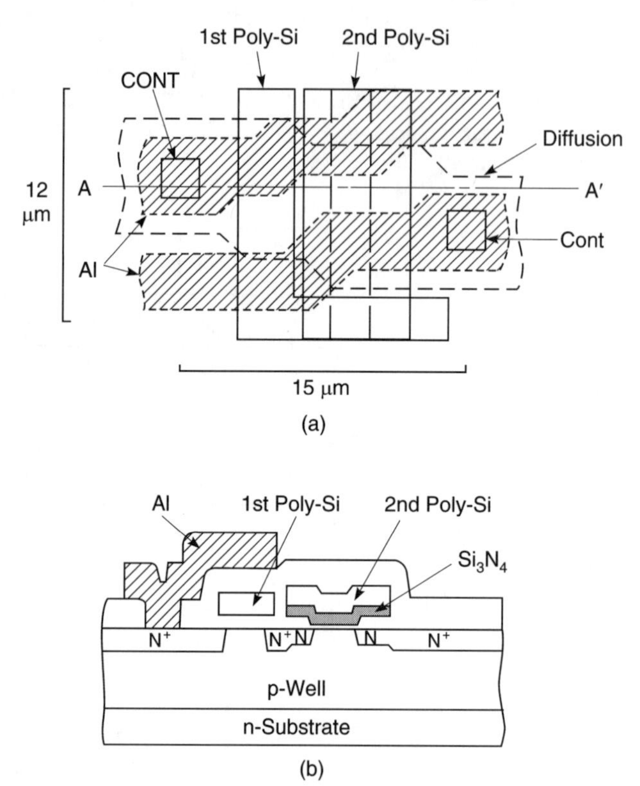

Figure 6.34 Memory cell (*a*) layout pattern and (*b*) cross section for 64Kbit EEPROM (after Yasude [6.61]).

6.2.3 NVSM Application — Circuit Design Examples

Attention to design margins through circuit design enhances circuit performance and improves reliability. First-generation MNOS NVSM products, for example, used two-element-per-bit storage and differential sensing to provide for the necessary margins to achieve performance and reliability [6.62]. In Fig. 6.37, note

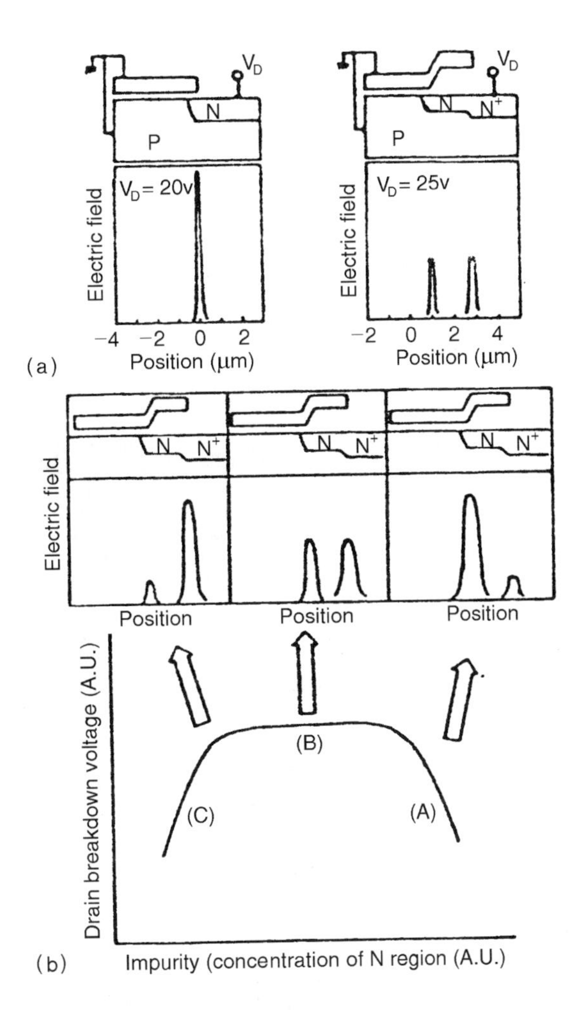

Figure 6.35 Electric-field strength under gate electrode. (*a*) Comparison between conventional MNOS device and high-voltage-structure MNOS device. (*b*) Drain breakdown voltage versus impurity concentration of N-region, in conjunction with electric-field strength (after Yasuda [6.61]).

that the memory data bit is stored as the difference in threshold voltage between the two MNOS transistors in the two legs of the flip-flop load. The load control transistors are bypass elements which ensure that the flip-flop latching current is determined by the high-conductance state memory transistor. The latching sense flip-flop is the difference sensing circuit for data output. Improvement in device technology and the continuing pressure to achieve high-density storage per unit area in silicon provided the incentive for single-element storage. That is accomplished with the use of a reference storage element plus a balanced bitline (see Fig. 6.38), in which a reference bitline is used for each of the two MNOS storage arrays shared by the differential sense flip-flop circuit in common [6.64]. Figures 6.39 and 6.40 are similar sensing schemes using high noise-immunity circuit design techniques to achieve reliable high-speed sensing of floating gate NVSM devices. In Fig. 6.40, the reference voltage for the flip-flop sense circuit is derived in the "dummy" memory cells, which are erased and programmed in every Flash erase cycle of the multi-megabit Flash E^2PROM [6.65]. The characteristic of the "dummy" cell is the same as that of the

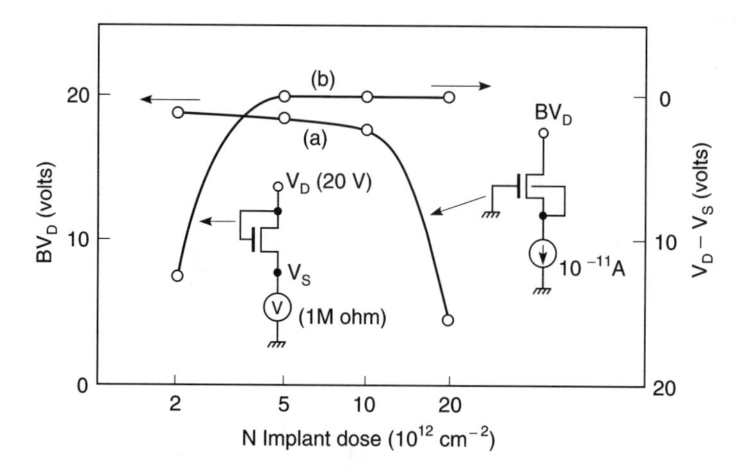

Figure 6.36 N-implant dose dependence of (*a*) breakdown voltage and (*b*) voltage drop through MNOS device (after Yasuda [6.61]).

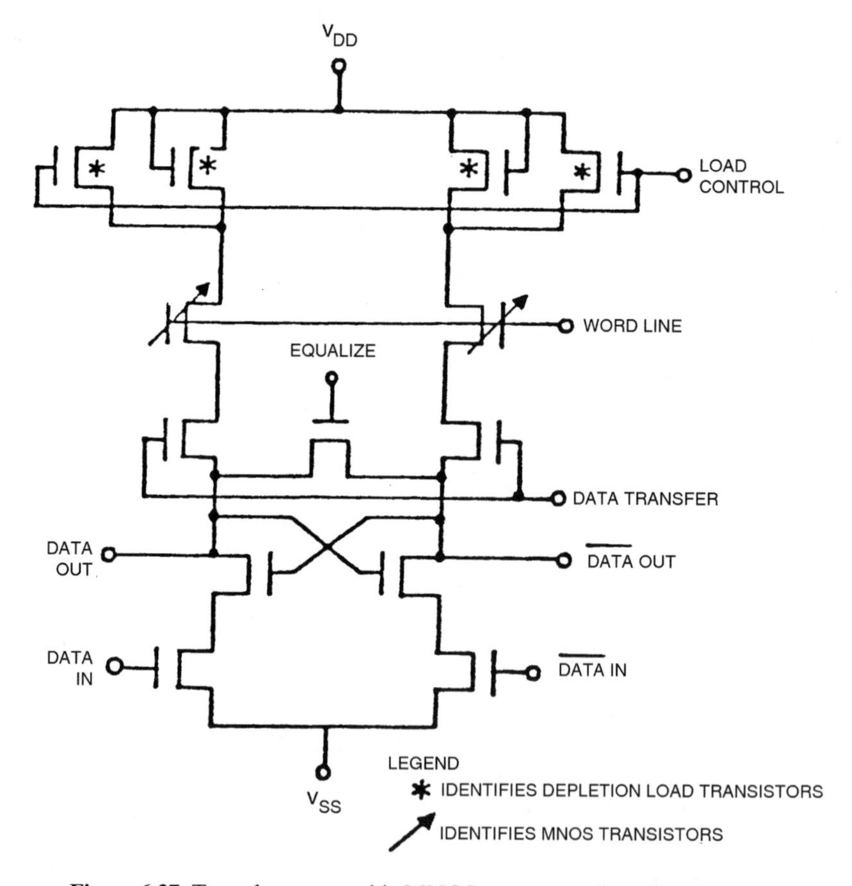

Figure 6.37 Two-element-per-bit MNOS memory cell (after Hsia [6.62]).

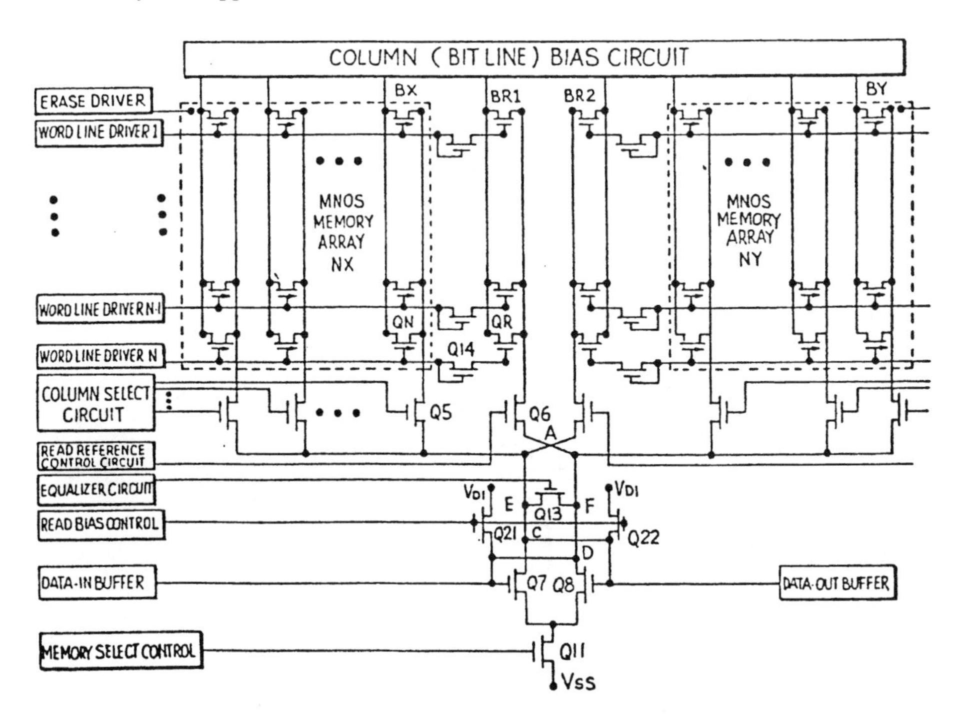

Figure 6.38 Referenced threshold, balanced bitline sensing of MNOS data storage (after Hsia [6.63]).

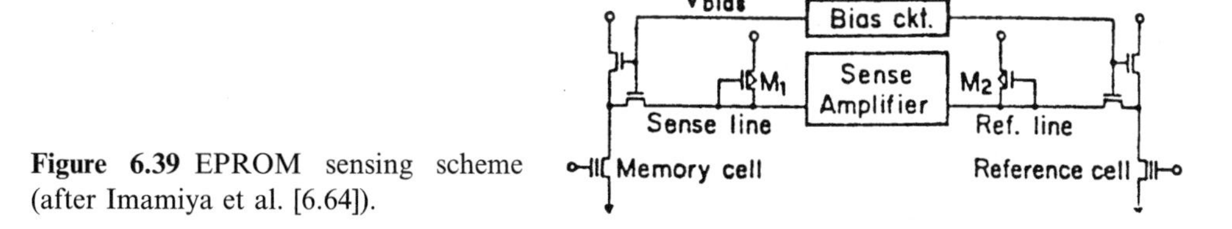

Figure 6.39 EPROM sensing scheme (after Imamiya et al. [6.64]).

data cell. However, the capacitances in the array are such that the reference level is loaded so that it is at a value between a "zero" (programmed state) and a "one" (erased state).

For many EEPROM applications, single-bit endurance failures have been the limiting reliability constraint. To improve endurance reliability, many circuit techniques are utilized, such as redundancy of data storage so that the stored data can be retrieved correctly even if one of the memory devices in a two-element-per-bit storage scheme fails. Figure 6.41 illustrates the schematic of a redundant storage cell concept designated as the "Q" cell [6.66]. The cell is composed of two identical EEPROM memory transistors, which are programmed identically with the same data through adjacent bitlines BL_1 and BL_2. The memory read operation is accomplished separately from both transistors with separate sense amplifiers. The data are, in turn, combined through a logic NOR operation. The correct datum is read out as

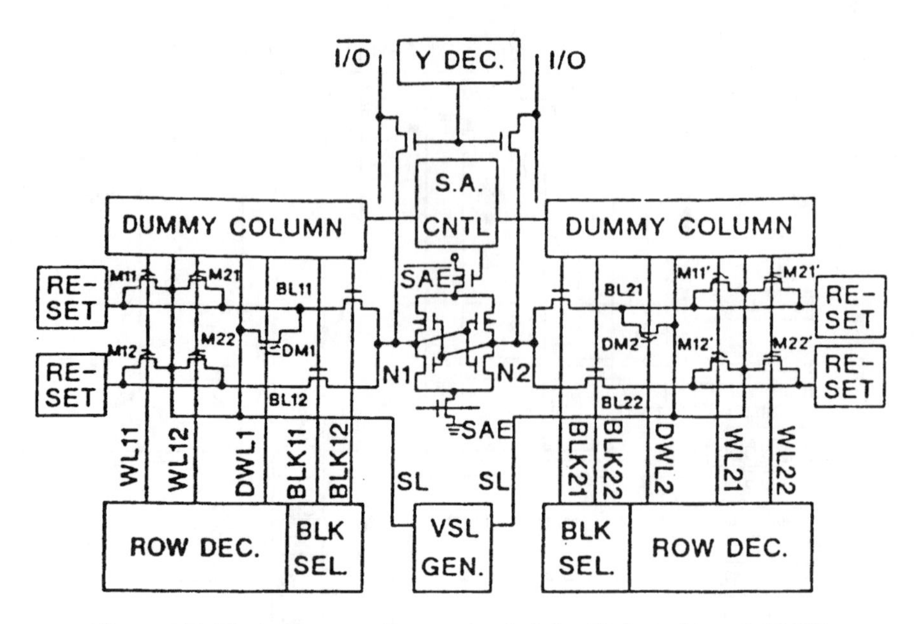

Figure 6.40 Block diagram of sense circuit (after Kobayaski et al. [6.65]).

long as one of the two EEPROM transistors and its corresponding data path is not defective. Assuming an initial 100% goodness screening for parts acceptance, the reliability of parts in use is greatly enhanced by the application of the redundant storage, because the probability of an adjacent cell wearing out (or failing endurance) at the same time is exceedingly low.

To realize high-density EEPROMs with high reliability, error correction circuits (ECCs) are often utilized [6.67–6.70]. ECCs are effective in improving EEPROM reliability failures due to random bit failure caused by tunnel oxide breakdown in erase/write endurance. We can derive the first-order improvement in product reliability as follows. The chip failure probability without ECC, P_o, is simply

$$P_o = 1 - (1 - q)^{1M} \tag{6.35}$$

in which q is the bit failure probability.

In the case of the use of 4 parity bits generated from 8 data bits, that is, ECC (8,4), the byte failure probability, q_A, is given as the probability of more than 2 bits out of 12 bits failing, since ECC corrects for single-bit errors,

$$q_A = 1 - (1 - q)^{12} - {}_{12}C_2 q(1 - q)^{11} \tag{6.36}$$

and the megabit (or 128 KB) chip failure probability, P_A, is given as

$$P_A = 1 - (1 - q_A)^{128K} \tag{6.37}$$

Figure 6.42 illustrates the calculated result of the use of EEC techniques to improve product reliability, in the case of a megabit EEPROM, using three alternative EEC choices [6.69]. The orders of magnitude improvement in reliability with ECC is very impressive.

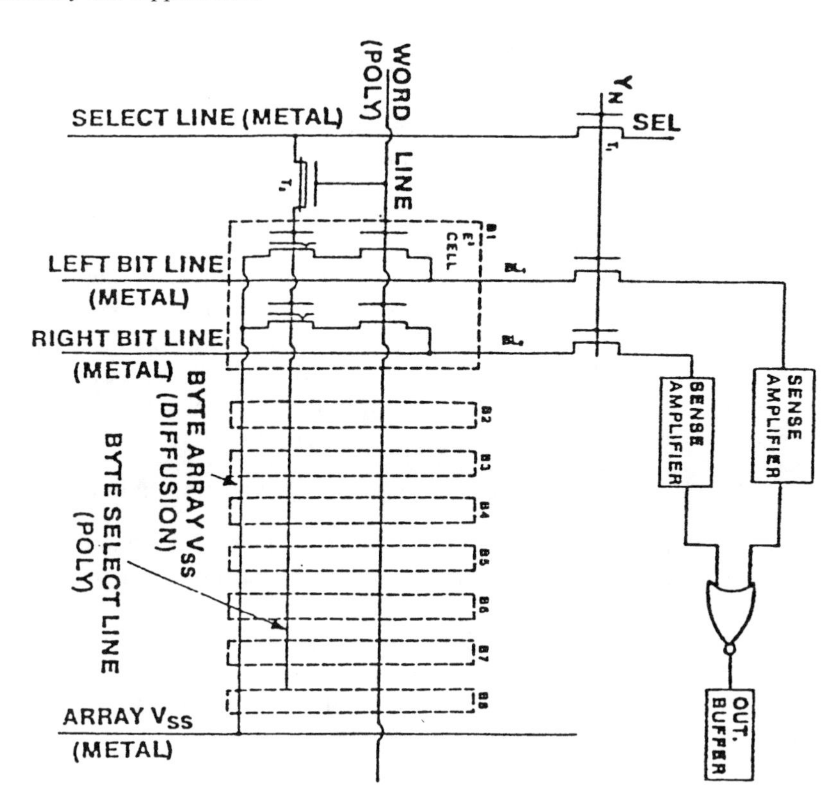

Figure 6.41 Schematic of the byte organization using the Q-cell concept (after Cioaca et al. [6.66]).

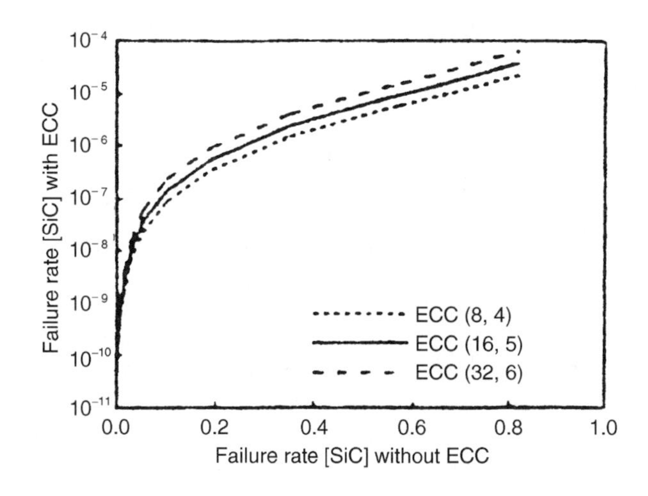

Figure 6.42 Calculated failure rate of 1Mbit EEPROM caused by random bit failure (after Terada et al. [6.69]).

6.3. RELIABILITY TESTING AND YIELD

All semiconductor devices are manufactured with numerous controls in place to guarantee that the manufacturing parameters are stable and are on the process design target. In general, these controls are re-checked by measuring test structures on the final product wafer plus various performance tests conducted on the product itself. Parametric test devices provide information about how well the process achieved the goal of meeting parametric specifications. Assuming that parametric targets are achieved, functional tests conducted on the product will supply data that relate to the yield of the product, which, in turn, reveal faulty parts due to design-related sensitivities to process variations along with faulty parts caused by random defects. Yield is defined as the ratio of functional parts to the total number of parts manufactured in a wafer lot or specified period of time.

$$Y = \text{Number of good chips/Total chips fabricated} \tag{6.38}$$

A basic process yield can be computed based on the number of initially functional parts from the wafer lot. This yield number rarely is representative of the final yield because further parts fallout after burn-in.

Before IC devices can be installed into systems, additional testing is usually performed after the parts are subjected to burn-in in an attempt to remove marginally reliable devices that would fail in a short period of time after being placed into service. Burn-in testing is a mildly accelerated life test that is intended to cause devices with reliability defects to fail, effectively removing most of the devices that fail in the infant mortality (early-life) failure region of the "bathtub" curve." The number of infant mortality failures is generally kept in the production line database along with the number of initially nonfunctional parts detected at initial functional test. A yield figure that accounts for all functional parts after burn-in is the true production yield.

Reliability testing beyond burn-in is typically performed on individual part types using procedures prescribed by either the manufacturer's standard procedures or by military procedures. In general, life testing involves two broad categories of data gathering: failed part field return and accelerated life testing. Gathering reliability data by tracking the failures of parts in the field is an effective, though very expensive, method, even if it is possible to have a complete environmental history of the part while in service. The most common method of assessing parts reliability is to run accelerated life tests on a randomly selected set of parts from one or more wafer lots. Unfortunately, trends in failure rate requirements are becoming more demanding as VLSI technology progresses. As Crook[1] [6.30] has noted, trends in infant mortality and long-term life goals have been projected through the year 2000. Based on this projection, it will be necessary to achieve a long-term life failure rate of 10 FIT by the end of the century.

The reduction in failure rate goals has profound implications for life testing as is commonly described in various military and commercial testing procedures. As

[1]This paper by D. L. Crook is recommended to the reader for projecting testing into a framework of future technology.

mentioned above, life testing is performed on a sample of parts from the production line. The sample size is dictated by requirements for FIT rate and confidence limits needed to have an acceptable assurance of reliability. Typical life tests are performed at 125°C for 1,000 hours. If we assume that the failure distribution is Poisson statistics, then the number of samples required for a life test is easily calculated. Table 6.8 illustrates the sample-size requirements based on some example failure rates and confidence limits. Clearly, in order to achieve the failure rate goals projected for the end of the century, sample size for life testing will have to be very large.

We may add to this the fact that chip complexity is increasing at a rate that would result in 100 million transistor chips by the end of the century as seen in Fig. 6.43. Unfortunately, as chip complexity increases over the next decade, the cost of each chip will be on the order of several hundred to a few thousand dollars. At this cost level, life testing of the scale dictated by these projections will not be cost-

TABLE 6.8 THE EFFECT OF FAILURE RATE GOAL ON THE MINIMUM SAMPLE SIZE REQUIREMENTS, ASSUMING POISSON STATISTICS (AFTER CROOK [6.30]).

Failure Rate Goal (FIT)	No. of Devices 90% UCL		No. of Devices 60% UCL	
	0 Fails	2 Fails	0 Fails	2 Fails
1000	355	835	143	463
100	3550	8350	1437	4630
10	35,500	83,500	14,370	46,300

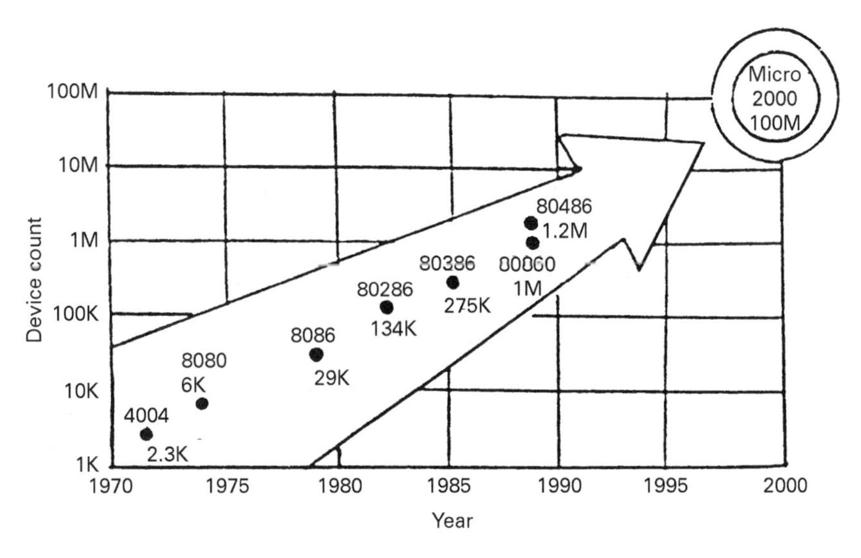

Figure 6.43 Microcomputer transistor count per chip versus time (after Crook [6.30]).

effective. It is very clear that some method other than product life testing is required to assess chip reliability.

During the past 15 years or so, a number of research efforts have been initiated to provide a means by which manufacturers can effectively predict the wearout lifetime of IC devices by using information about physical failure mechanisms. This research effort has produced the well-understood physical failure models that were discussed in Section 6.2.1. Given these physical models, we can construct test structures and test procedures that will enable routine wafer-level monitoring of lifetime limiting mechanisms. Wafer-level reliability testing is a highly effective tool for both in-line and end-of-line reliability monitoring on current technologies. JEDEC Standards are being defined for the three reliability limiting failure modes (electromigration, hot-carrier damage, and oxide wearout) and will soon be in widespread use.

Section 6.3.2 describes the proposed standard reliability test methods in sufficient detail that they can be implemented on a wafer fabrication line. It must be understood that the monitoring and process control required for production of high-reliability integrated circuits is very complex and is becoming more so as the minimum fabricated geometry decreases in size. No single method of monitoring will completely guarantee reliable chips. However, a combination of in-line monitoring, along with end-of-line wafer-level reliability monitoring, will provide both the necessary feedback needed for good statistical process control and valuable "disaster alarms" that will minimize the chances of installing chips containing known global reliability hazards into critical systems.

Before discussing the wafer-level reliability testing procedures mentioned above, it is necessary to point out the limitations that future chip complexity will place on the use of wafer-level reliability testing as a tool for process control. Current technologies are at a level of complexity that permit effective use of wafer-level reliability testing as an end-of-line reliability monitor. However, based on projections of future technologies [6.30], end-of-line wafer-level testing will not be sufficient to address the 10 FIT failure rate requirements of the future.

First, we must realize that, as device geometries scale, the intrinsic reliability margin becomes progressively smaller. This point was discussed earlier and illustrated in Fig. 6.23. Scaling to current technologies necessitated major improvements in contamination control to compensate for the shrinking intrinsic reliability margin.

The continual improvement in processing technology, which has reduced the number of defect-related failures (infant mortality), has a dramatic effect on the ability to detect any remaining defects. Crook uses oxide reliability as an example of a traditional reliability limiter in MOS technology and points out that statistics on oxide failure (Fig. 6.44) indicate that the defect tail of the cumulative failure distribution is becoming a significantly smaller percentage of the total population. This leaves a cumulative failure distribution that is controlled primarily by intrinsic behavior. Monitoring the defect tail with test structures will become a statistical problem similar to the problem of life testing an entire functional device. By the end of the century, it will at best be possible for test structures to monitor one-twentieth of the active gate area of the chip if the test capacitors are confined to the scribe lines. The only way to be able to monitor the defect failure of the oxide is to assign 100% more

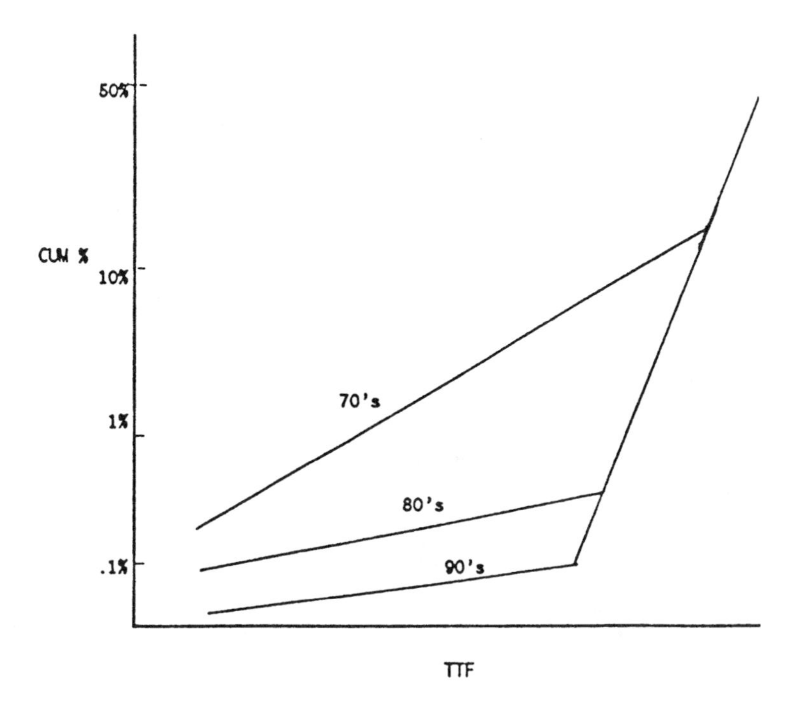

Figure 6.44 Examples of typical distributions for oxide defects in the past, present, and future (after Crook [6.30]).

area to the test structures. This option is surely unacceptable to any manufacturer who produces chips to sell for profit.

A similar argument involving test structure scale could be made for the other failure mechanisms, which leads to the conclusion that wafer-level reliability monitors are not useful as the primary process control monitor. Wafer-level reliability testing must be augmented with monitors that are closer to the processes that control the production of reliable materials. Crook suggests that the most promising place to install reliability monitors is far up the processing fishbone, as illustrated in Fig. 6.45. Each step in the process must have monitors in order to ensure that each step contributes its part to the overall reliability of the product. This, in turn, implies that reliability assurance must depend on a high level of confidence in these processing step monitors. End-of-line wafer-level reliability testing will continue to provide a check for reliability disasters, much as parametric tests on a small number of transistors within the PCM serve to provide an assurance that all of the transistors in a wafer are performing within the designed limits.

The testing technology required for monitoring well into the fishbone of each processing sequence is beyond the scope of this chapter. We will supply some guidelines for end-of-line and in-line wafer-level reliability testing that will serve existing technology well, while providing a model for methods that will be useful in the future.

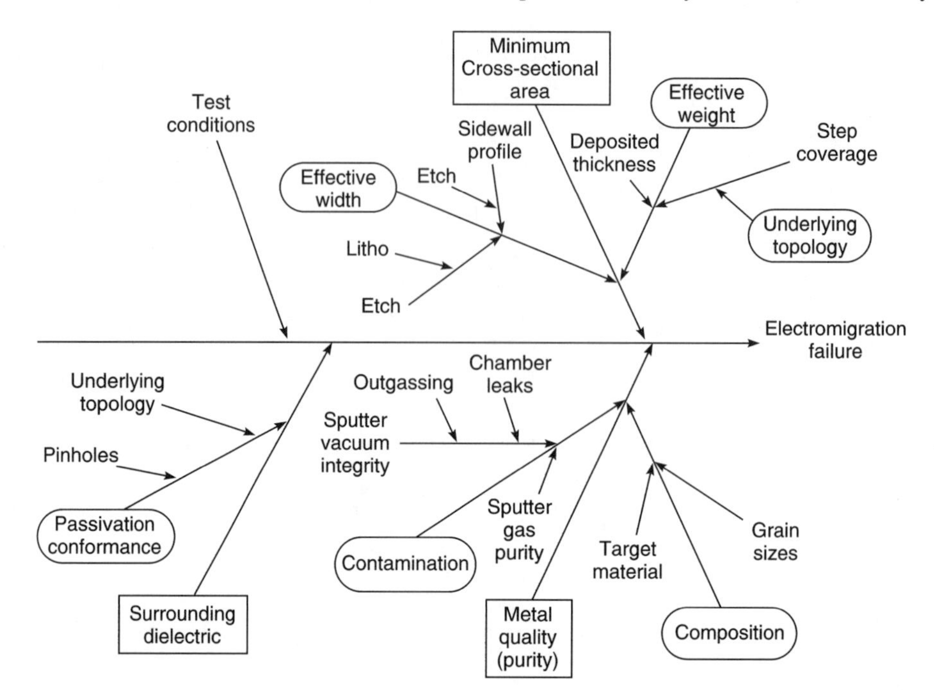

Figure 6.45 Fishbone diagram showing input variables that affect electromigration time-to-fail (after Crook [6.30]).

6.3.1 Validation Testing: Reliability Assurance in Manufacturing[2]

Life test reliability data validate accelerated tests used by manufacturers and consumers alike. Additional reliability data on the breakdown of defect mechanisms and failure modes provide the feedback data necessary not only to improve the manufacturing process, but also to direct future efforts in the areas of process, design, and manufacturing.

Reliability testing must utilize test procedures, sampling plans, and data analysis procedures that are specified in JEDEC standard methods. As of 1992, no official standard was set out in the JEDEC files. The methods described in this section provide a snapshot of the preliminary standards at this time. It is likely that only small changes will be made in these methods when the official JEDEC Standard is published.

The lifetime models used for all the methods included herein are empirical and represent a best fit to existing data. As dimensions scale downward, there is likely to

[2]This section contains paraphrased and extracted material from: FINAL REPORT—Develop Quality Assurance Procedures Based on Wafer Level Testing, by Vance C. Tyree, written for Defense Advanced Research Projects Agency under Contract No. N00140-87-9263, April 10, 1991.

be some deviation in predicted lifetime if the mechanism responsible for failure changes. In this section, all efforts are made to point out the known weaknesses in the existing models.

Validation testing involves two inseparable parts which establish that the fabrication line first is not building in some fundamental reliability hazards, and second, is able to assess the degree to which the design of the part itself contributes to any excessive stress that could lead to early failure.

Wafer-level reliability testing is a powerful tool for measuring the generic quality of the fabrication process itself. Generic quality is absolutely essential to being able to produce reliable parts of any type. In the case of NVSM, in which intentional "overstress" is applied to write data into the device, a tight statistical quality control with rapid, accurate feedback of insulator quality will form the basis of long-life NVSM parts.

Waferlevel reliability testing permits monitoring reliability limiting mechanisms at the wafer level before the parts are packaged and placed into service. This test method must be performed very rapidly in order not to delay the wafer production flow. Highly accelerated stress is applied to a test structure that is designed to represent some worst case at risk structure permitted in a design. Time-to-fail data are gathered and analyzed to assess the relative quality of the wafer. These data are compared with similar data taken from a known good wafer lot and used to decide whether built-in reliability hazards are produced in the wafer lot under test. Since the basis of the quality assessment is the use of test data from highly accelerated tests, it is important to recognize the limitations of such testing relative to the extrapolated life for the product.

6.3.1.1 Accelerated Testing and Its Limitations. Accelerated testing, by definition, enhances the rate of progress of failure mechanisms in order to study their relative importance in a product. The enhancement process involves increases in temperature, voltage, current, and so on, to levels that will result in acceleration factors large enough to reduce lifetime to a manageable value for test purposes. The difficulty is that, as acceleration factors become large, reliability limiting processes that are of secondary interest interfere with the mechanism of interest. If the mechanism is isolated in a test structure, moderate to very high acceleration factors can be used to study the mechanism.

When extreme acceleration factors are used, it is dangerous to extrapolate the results of tests to at-use conditions because either secondary mechanisms may interfere with projection or statistical scatter is so great that extrapolation over many decades yields ambiguous results. Figure 6.46 illustrates an attempted extrapolation of hot-electron degradation test results to at-use conditions. A well-tested and proven model for the mechanism must exist if such extrapolations are even possible. With such a model, experimental errors and defects occur, causing data scattering that leads to inaccurate extrapolation. The scatter boxes and associated median values of test data in Fig. 6.46 illustrate the uncertainty in extrapolation of accelerated test results over several decades. Accelerated life test data must be carefully examined for potentially ambiguous circumstances that can invalidate an extrapolation to at-use

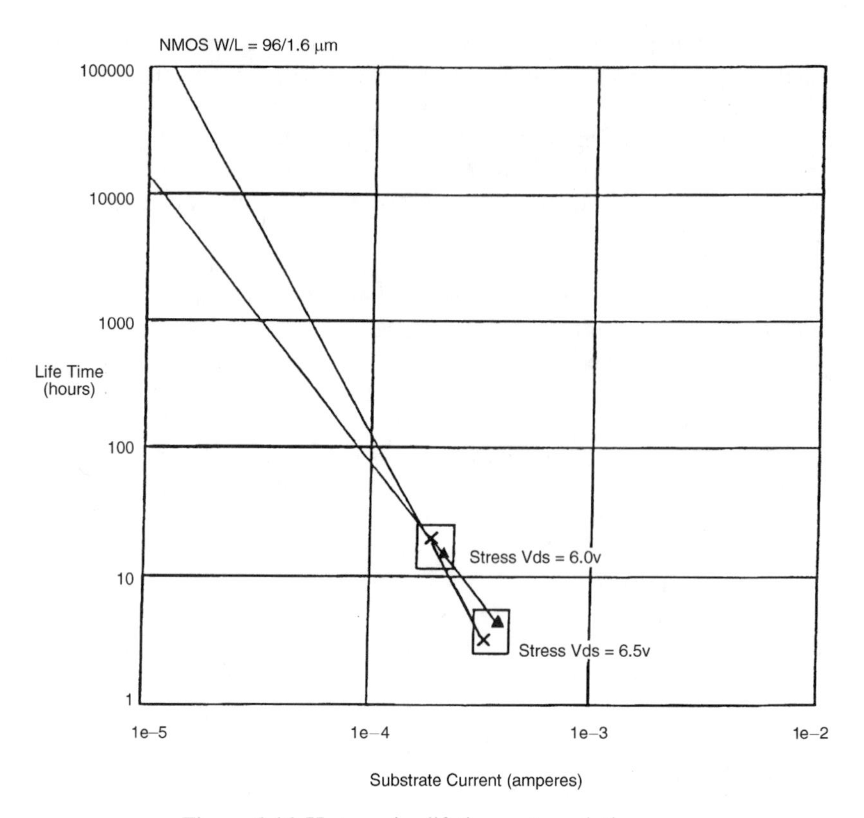

Figure 6.46 Hot-carrier lifetime extrapolation error.

conditions. The most common complicating factors in extrapolation of measured time-to-fail data are the presence of bimodal or multimodal distributions in the data and large standard deviations in the statistics. There is also considerable confusion about what statistical distributions to assume.

6.3.1.2 The Impact of Choice of Distribution. The two most common distributions in use in reliability testing are lognormal and Weibull. Many researchers in reliability have found that both distributions are useful, depending on the mechanism that is monitored or on stress conditions used in the life tests. When sample sizes are large, it may be possible to distinguish between the two distributions if the data contain relatively little scatter. However, in many instances small sample sizes are dictated, as in the case of wafer-level reliability testing. Wafer-level tests are usually confined to the space in the scribe lanes (or kerf) between chips, which represents a relatively small wafer area that must be shared with parametric monitoring test structures.

As shown in [6.71], when there is a difference in assumed distribution (e.g., lognormal versus Weibull) for a Standard Wafer-Level Electromigration Acceleration Test (SWEAT) structure unit [6.83], the ability to distinguish between the two distributions is strongly influenced by the sample size. In the example in the reference [6.71], Fig. 6.47, a "perfect" data set shows no curvature when the SWEAT unit is assumed to be lognormally distributed and plotted on a lognormal graph. However, when the SWEAT unit is iterated, the distribution is no longer lognormal for the string of SWEAT units and shows slight curvature with a "perfect" data set. This curvature is so slight that, if the data sample size is small and contains scatter, there is no way to distinguish the deviation from lognormal. Evidently, attempting to extrapolate what may be a quasi-lognormal distribution to at-use conditions from a small sample size is going to yield a significant error in hazard rate. This error can become large when we are attempting to extrapolate highly accelerated test data to at-use conditions that are several decades longer in time than the actual test time. Noise in the data compounds an already difficult extrapolation of a distribution that is not a straight line on a lognormal graph, as was the case with the SWEAT example.

The choice of an appropriate distribution is significantly influenced by the acceleration factors used in the life testing. The distribution for failure in at-use conditions is very likely different from the distribution that appears to match data from accelerated tests. Life testing strategy must account for the fact that highly accelerated tests may not permit extrapolation to at-use conditions because multiple failure mechanisms may significantly influence the outcome.

6.3.1.3 Recognizing Bimodal Distributions. Bimodal, and in some cases multimodal, distributions in life test data are a frequent source of error in at-use condition extrapolations from accelerated test data. The best way to deal with this

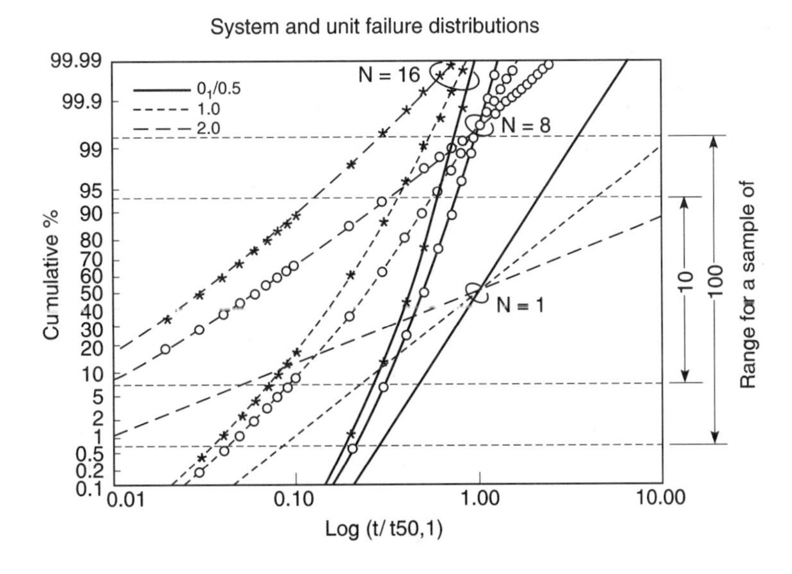

Figure 6.47 Distributions of individual SWEAT units compared with distributions of iterated SWEAT units (after Crowell et al. [6.71]).

problem is to create test conditions that do not stimulate competing reliability-limiting mechanisms. Unfortunately, only rarely can a practical test environment be created that will permit such simplification. Testing time considerations will force higher stress levels that, in turn, force the data analyst to deal with bimodal distributions.

Assuming that each mechanism has its own sigma and mean, we can express the cumulative distribution function (CDF) as

$$F(t) = \sum_{i=1}^{k} q_i * F_i(t) \tag{6.39}$$

where q_i is the fraction of the population that belongs to mechanism i.

Note that

$$\sum_{i=1}^{k} q_i = 1 \tag{6.40}$$

If we have a bimodal distribution (k = 2), then we can express the CDF as

$$F(t) = q_1 * F_1(t) + (1 - q_1) * F_2(t) \tag{6.41}$$

where q_1 would be an early fail population and q_2 would be the main population expressed in terms of the early population: $q_2 = 1 - q_1$. F(t) could equally well be expressed in terms of the main population.

Figure 6.48 shows the general form of lognormal plots of bimodal distributions in which each population has two well-defined means and various sigmas. In this set of examples, $\ln(t_{50}) = 0$ for the freak (early) population, and $\ln(t_{50}) = 10$ for the main population. Also assume that $q_e = 0.1$ and $q_m = 0.9$. The various sigmas of both populations include 1/2, 2.0, and 5.0. The vertical scale is the normalized cumulative fail percentage, and the horizontal scale is the natural log of the lifetime. The first thing to note is that, as long as the sigmas of the two populations are relatively small, it is easy to recognize the presence of two populations simply by the fact that the lognormal plots are not linear. Also, note that the upper vertical bars indicate that 90% of the freak (early) population lies to the left of the vertical bar, and the lower vertical bars indicate that 90% of the main population lies to the right of the vertical bar. Finally, observe that it is intuitively possible to separate the mean and sigma from the plot as long as the sigmas of the two populations are not too large.

Theoretically, given knowledge of the form of the CDFs of the two populations, their mean and sigma can be extracted as long as the two populations do not overlap very much. Figure 6.49 illustrates a typical bimodal lognormal plot of life test data indicated by the circles on the graph. The squares show the lognormal plot of the extracted early population, where $q_e = 0.2$, and the triangles show the lognormal plot of the extracted main population, where $q_m = 0.8$. Clearly, any attempt to fit the set of data points to a line on the lognormal graph will lead to a grossly distorted mean and sigma.

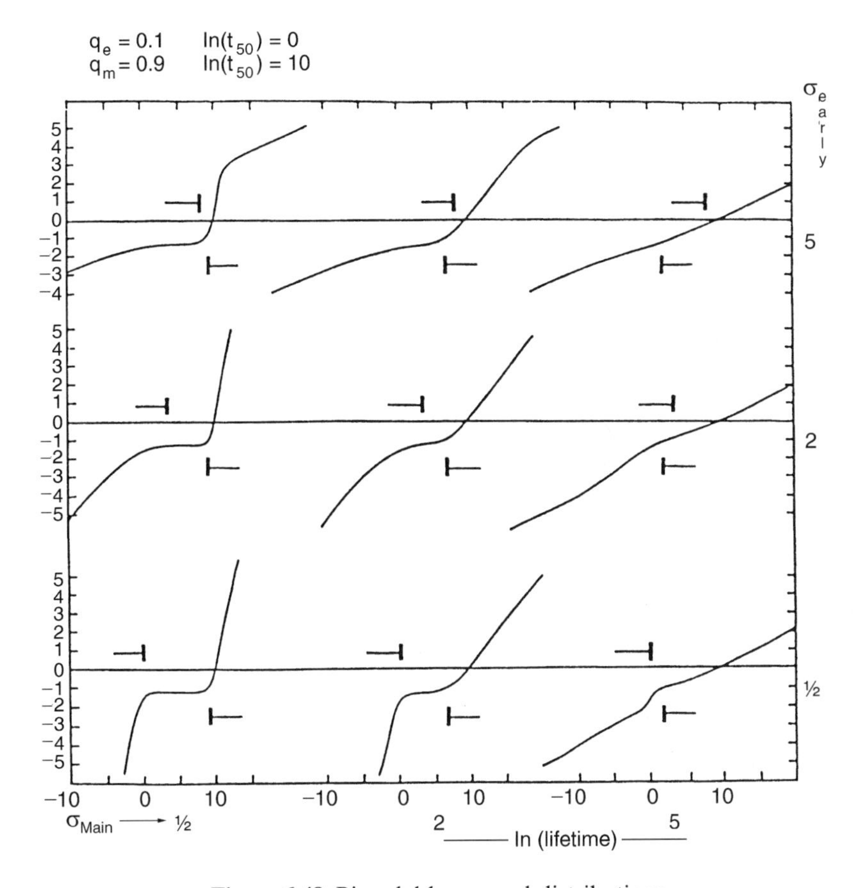

Figure 6.48 Bimodal lognormal distributions.

Mathematically, the CDF of the entire data population is represented as

$$F(t) = q_e * F_e(t) + q_m * F_m(t) \tag{6.42}$$

Solving for $F_e(t)$ yields

$$F_e(t) = \frac{F(t) - q_m * F_m(t)}{q_e} \tag{6.43}$$

Assuming that the early and main populations do not overlap much, the early part of the population can be expressed, assuming that $F_e(t)$ is approximately zero, as

$$F_e(t) = \frac{F(t)}{q_e} \tag{6.44}$$

Similarly, assuming that $F_e(t)$ is approximately one, the main population can be expressed as

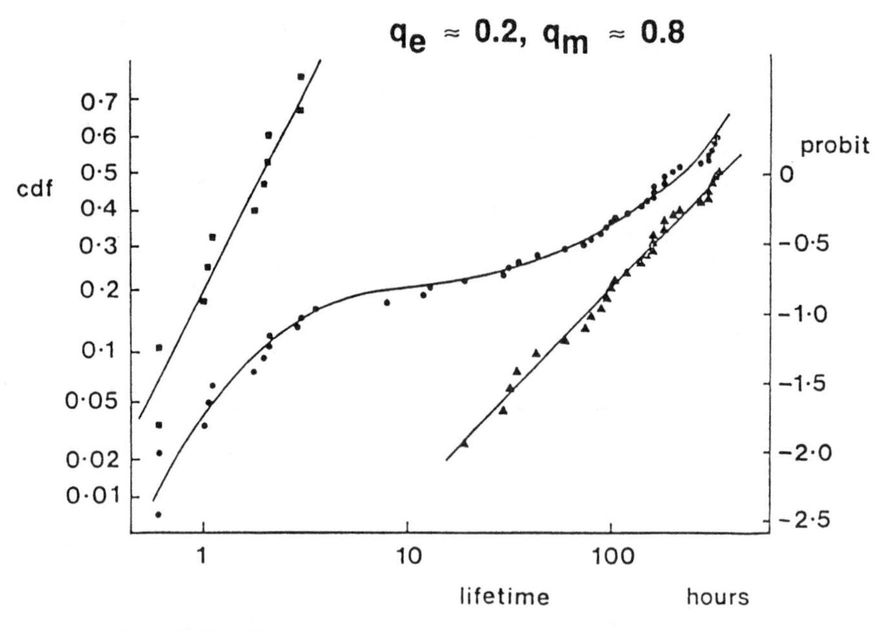

Figure 6.49 Life test data containing early and main populations.

$$F_m(t) = \frac{F(t) - q_e}{1 - q_e} \tag{6.45}$$

The early populations are usually a result of the presence of "weak" structures caused by random defects. As has been indicated, these early failures can be separated from the main population, and permit the evaluation of the global quality of the test structures. Unfortunately, when conducting wafer-level tests that are highly accelerated, the distributions become multimodal because other less dominant mechanisms become activated. The separation process becomes impossible, and some other means must be devised to relate wafer-level reliability test results to at-use conditions. A method for dealing with this additional complication is to create a test structure and test method calibration hierarchy.

6.3.1.4 Hierarchical Calibration: Dealing with Nonideal Data Sets. By recognizing the limitations that highly accelerated testing impose, it is possible to develop wafer-level test structures, test methods, and data analysis procedures that provide valuable tracking of product reliability. Highly accelerated test structures can be defined to minimize the deviation from the life models empirically derived from large sample sizes at more moderate accelerated test conditions. This allows simple representation of measured data from the highly accelerated tests for the purpose of comparison of one wafer lot with another, even though extrapolation to at-use conditions is invalid. Traceability to at-use conditions is achieved by using a database of moderate acceleration factor tests performed on test structures and product or product-like functional devices from calibration wafers that also contain the wafer-level reliability test structures.

This hierarchy of testing creates a well-controlled reliability tracking environment for wafer fabrication. It is important to realize that this hierarchy has interpretation limits. As long as wafer-level test results remain within prescribed bounds, there is no reason to be concerned that the product life expectancy has changed from the reference lot. However, if data from the wafer-level tests indicate either an increase or a decrease in accelerated test lifetime, then interpretation of the test results becomes more complex.

The simplest result to deal with is a consistent increase in wafer-level test lifetime. This indicates that the material under test is becoming more robust for some reason that hopefully is known to the fabrication manager. If this is the case, then it is necessary to re-calibrate the hierarchy and continue monitoring.

Reliability tracking to maintain quality becomes more difficult when there is a reduction in wafer-level test lifetime. If the reduction is relatively small, then the temptation is to ignore it as a statistical glitch caused by relatively small sample sizes. This can be dealt with by having a clear idea of the relationship of acceptable confidence limits (say 90% or 95%) to the calibration point in the hierarchy of test structures. It is clearly dangerous to try to extrapolate the highly accelerated test results to at-use conditions in order to assess the reliability degradation that is represented in the wafer-level test data. The correct procedure is to consult the cumulative database that exists on the wafer fabrication line in order to match the observed deviation with past experience with wafer-level tests, which have corresponding moderate acceleration factor tests, that are extrapolatable, and conclude its impact based on this comparison. In the absence of such a database, as would be the case at the start-up of wafer-level reliability testing, a database must be created.

Creation of a hierarchical reliability database involves periodically conducting moderate acceleration factor life tests (including correlation with a product or an equivalent Standard Evaluation Circuit) alongside wafer-level reliability test structures. Mechanism-specific reliability test structures on a test chip that contains wafer-level test structures for the same mechanism provide the means by which the wafer-level structures can be correlated with test results that can be extrapolated from moderate acceleration factor tests. Where possible, these test structures should be included along with every wafer lot that is fabricated. In any case, calibration wafers should be fabricated periodically to permit this tracking.

To complete the reliability correlation database, a functional device that is part of the product line should be life-tested from wafers from the same lot as the calibration wafers described above. In the absence of a large-volume product, as would be the case for a manufacturer who is involved primarily in ASIC fabrication, a Standard Evaluation Circuit (SEC) designed in the same gate array, or a standard cell library, could be used as a functional device to correlate moderate acceleration factor test structure results with a product or product-like device. An SEC should be designed to reflect the "typical" design style used for the majority of the devices produced on the fabrication line. For a fabrication line that is producing NVSM devices, a proper SEC would be a typical product chip. SEC life testing would be conducted at typical 1,000-hour life test conditions recommended for MIL testing, or at test conditions specified for the most critical application.

As the volume of reliability test data accumulates in the database, the uncertainty bounds for wafer-level test data shrink and permit definition of less ambiguous accept–reject criteria for wafer lots. This hierarchical calibration procedure is the most viable method for implementing a practical wafer-level reliability screening program on a fabrication line. The need for the calibration will remain as long as physical models cannot predict the lifetime of a material in the presence of multiple life-limiting mechanisms under highly accelerated stress testing.

6.3.1.5 Isolating Mechanisms—Models and Testing: TDDB. Time-dependent dielectric breakdown (TDDB) has gained considerable attention in recent years because, even though feature sizes of MOS devices have been steadily shrinking, the supply voltage has remained at 5.0 V. This trend is unabated to the submicron channel-length regime. As a result, with the thinner gate oxides, the in-use electric field has increased from 1 MV/cm to 2 MV/cm, which increases the rate of failure of gate oxides due to stress field.

The result of this scaling is that the quality of the oxide must be as close to theoretically perfect as possible. Thus, screens must be in place to catch global fabrication problems that could lead to early failure in the oxide. Stressing at moderate fields (3 to 5 MV/cm) has been used [6.12, 6.72] to study TDDB and to develop models for extrapolation to at-use conditions. Unfortunately, these tests require weeks of stressing to produce useful life data on oxides. If wafer screens are to be practical, stressing at very high fields (6 to 8 MV/cm) is necessary [6.31] to obtain results in a shorter time. Or even stressing to breakdown (8 to 10 MV/cm) must be used [6.77, 6.78] to gather data in minutes or seconds.

Test structures for oxide testing are simply capacitors constructed in various convenient configurations for either wafer probing or packaged testing. A test structure suitable for gathering statistical TDDB data on a fabrication process for purposes of calibration is given in Fig. 6.50 and fills a rather large die with a large array of capacitors in both the n-substrate and the p-well (or the reverse). This layout of capacitors allows wafer probe access to 38 capacitors in parallel for simultaneous stressing. The pad structure can be turned inside-out and converted to wire bonding pads to allow testing as packaged devices in order to eliminate the need for a probe station and to allow implementation of a large test fixture to stress large numbers of capacitor arrays simultaneously at different stress conditions. A total of 5 to 10 sites would be needed to gather data on 200 to 300 capacitors per wafer. The progress of data gathering is enhanced in such a structure by using step stressing [6.13] or a modified version of this method.

The test configuration illustrated in Fig. 6.51 allows measurement of capacitance, leakage testing to assess the condition of each capacitor, and a stress mode in which all capacitors are stressed in parallel through individual resistors to prevent a failure in one capacitor from modifying the stress to the others.

Moderate acceleration factor testing can be done using a modified-step stress testing procedure. A step stress (voltage step) time-to-fail equation [6.13] will help gather calibration data in a relatively short time. The step stress method described by Anolick [6.13] has been modified to permit convenient, moderate speed testing of

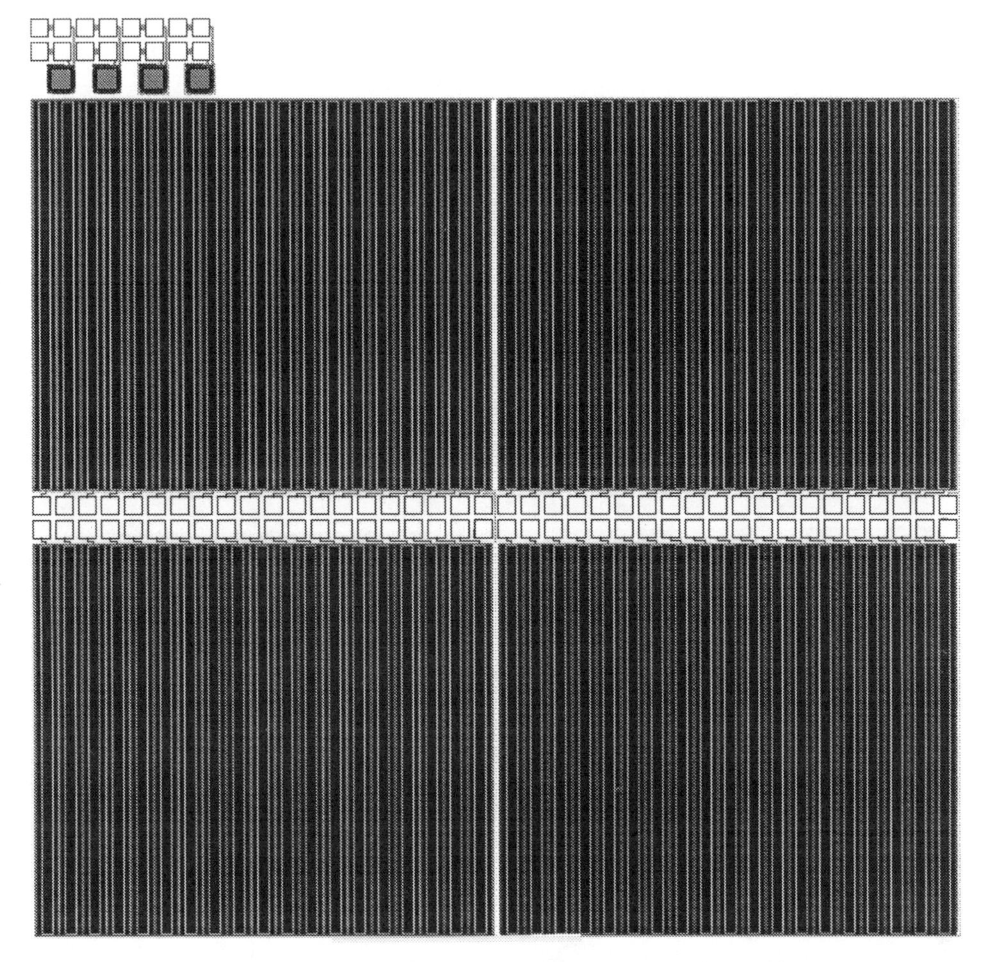

Figure 6.50 Wafer-probed TDDB test capacitor array.

large numbers of capacitors at various fields. The test method diagrammed in Fig. 6.51 utilizes an exponential progression of stress time followed by a test for leakage at the end of each time interval. Time intervals (in seconds) are 0.01, 0.1, 1.0, 10.0, 100, and 1,000 at each stress field level.

Each sweep through the stress time intervals is done at a fixed stress field. After each sweep, the number of failures at the end of each time interval is also saved. The next sweep over the same capacitor array is performed at a higher field, and the data for that time/field are saved. Each surviving capacitor is exposed to progressively higher fields, while accumulating stress from all past stress sweeps. Test data generally must be treated as censored data because each chip may not have all capacitors failing at a particular stress condition. Extrapolation of test results to at-use conditions is done by the method given in Anolick [6.13].

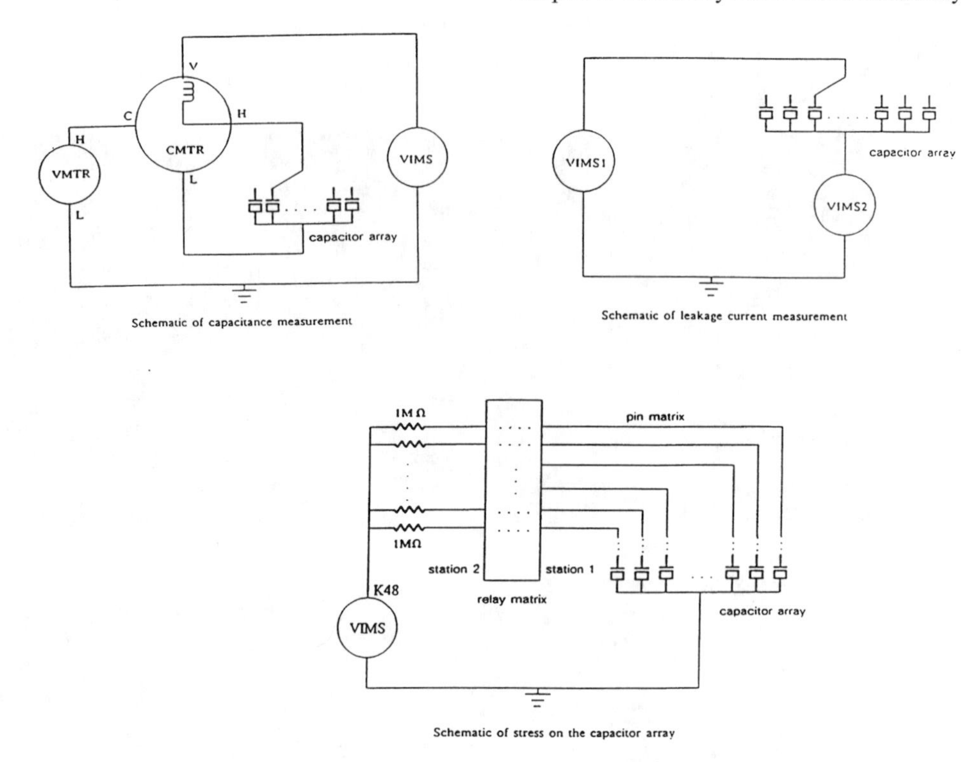

Figure 6.51 Test configuration for capacitor arrays.

This step stress method is more rapid than testing at a fixed field and can be done at wafer-level. Test times are still much longer than can be considered practical for true wafer-level testing. It can also be argued that testing is too rapid for unambiguous extrapolation to at-use conditions. It is, therefore, advisable to include package-level fixed field stress testing in the calibration procedure in order to build a solid confidence bridge with the rapid wafer-level testing used in wafer screening.

Testing at the wafer level typically implies that the testing time will be very short to avoid reducing wafer fabrication line throughput. Rapid wafer-level testing for TDDB is most easily achieved by using a ramped stress procedure. A ramped stress procedure depends on the assumption that oxide failure occurs when a critical level of total charge fluence has been reached. It is further assumed that the time span over which this total charge fluence is reached does not change the magnitude of the total fluence to failure. This leads to two rapid testing strategies that are suitable for wafer screening. Voltage ramp testing increases the stress field in prescribed steps while monitoring the capacitor for signs of failure. Voltage ramp testing starts at stress levels near normal operating fields and progressively increases the stress to failure. The second method is current ramp testing, which forces measured quantities of charge through the capacitor. The charge magnitude is progressively increased by increasing the force current at regular time intervals until the capacitor fails.

A JEDEC Task Group, JC-14.2, whose charter enables it to generate wafer reliability testing standards, has defined a standard test method for TDDB and was in final voting at the time this text was being written. The current version of the document, JEDEC Document No. JESD-35, "A Procedure for the Wafer-Level Testing of Thin Dielectrics," reflects the current status of the TDDB test method standards.

Voltage ramp and current ramp testing provide two valuable quality parameters. A relatively small sample size of test capacitors can distinguish between good quality oxide and lesser quality oxide, as shown in Fig. 6.52. The histogram with single lines is from a high-quality oxide, while the double bar histogram is from one with lower quality. The test can clearly indicate a global problem that will uniformly degrade the expected lifetime of transistors.

The same test methods used on large sample sizes will permit extraction of defect density-limited oxide problems. It can also reveal differences in oxide edge effects involving field edge or poly edge. Figure 6.53 shows a comparison of oxides from two different fabrication lines. Fabricator A has no distinguishable defect tail in the cumulative distribution function (CDF) and unmeasurable differences in edge effects in the gate oxide. Fabricator B is relatively dominated by defect-limited oxide breakdown. This oxide is nominally acceptable, considering that operating fields are around 1.5 MV/cm. However, this oxide would be less desirable for IC devices to be used in a system that must operate for long periods of time without any opportunity for repair. In addition, the fabricator B oxide has no statistically distinguishable edge

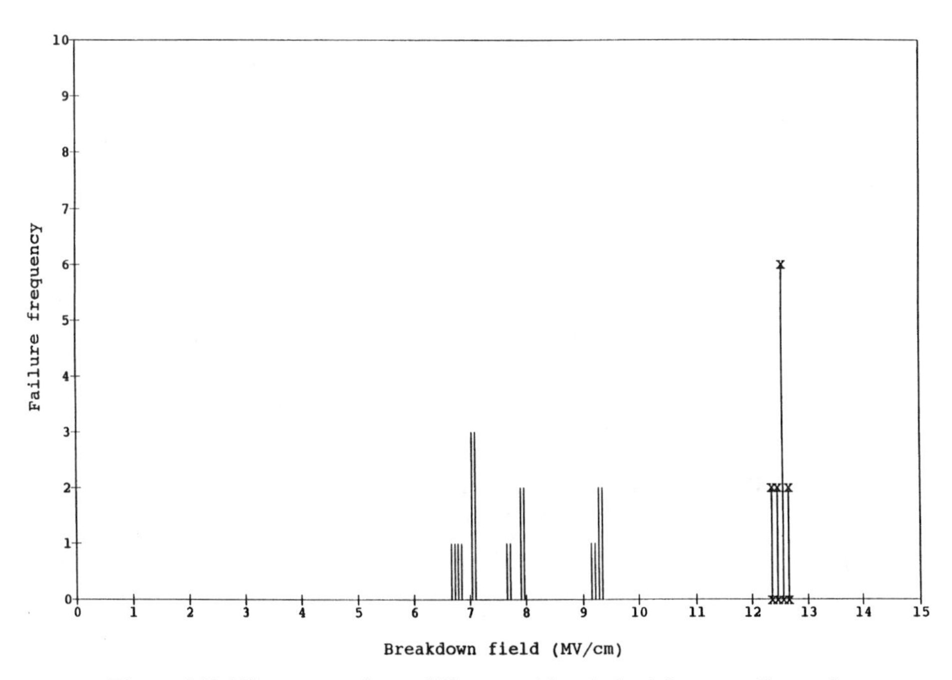

Figure 6.52 Histogram of two different oxides derived from small sample sizes.

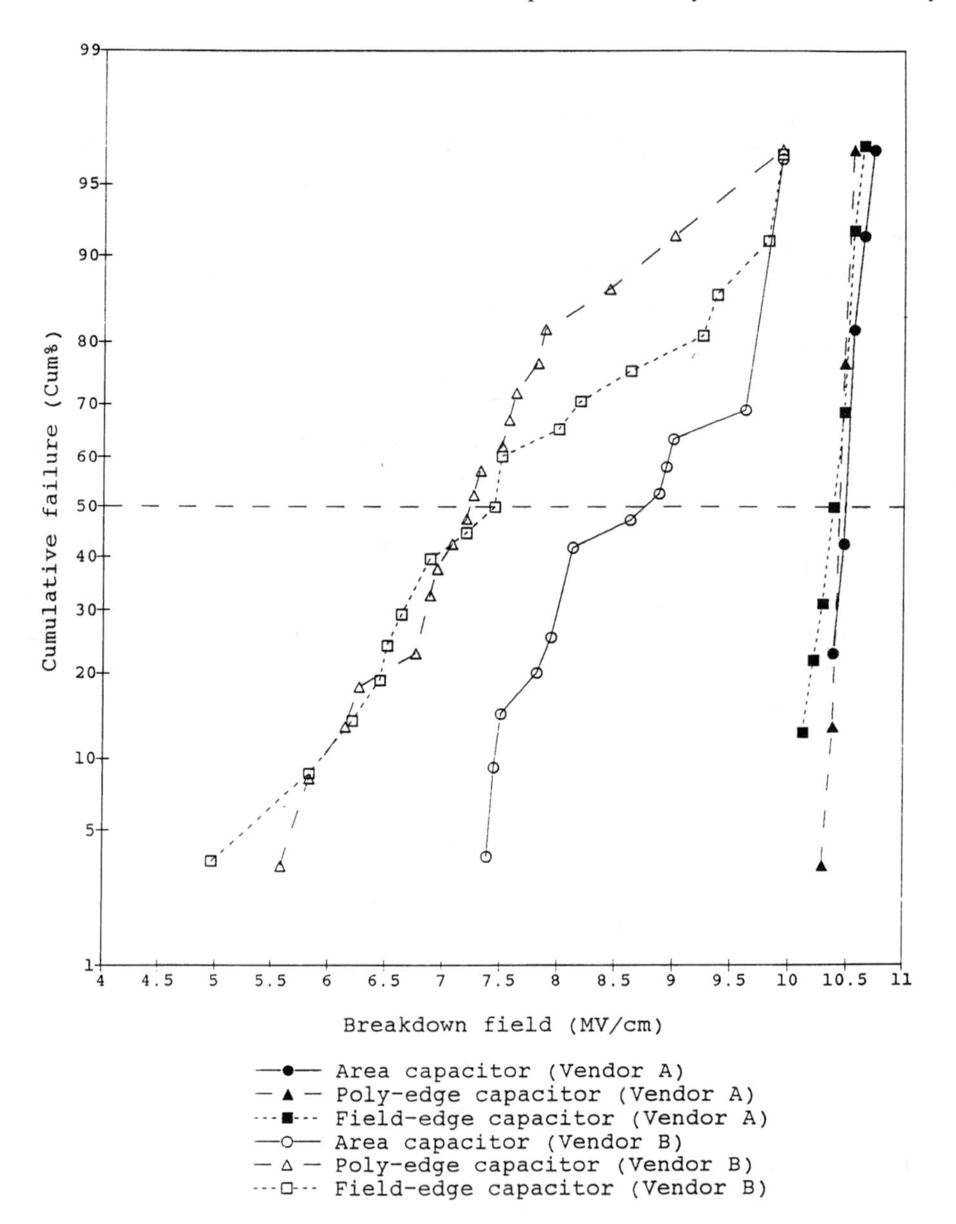

Figure 6.53 CDF plots of voltage ramp TDDB test results from two different fabricators.

factors, largely because of the much larger sigma for this defect-dominated CDF tail. However, there is a suggestion of better performance on area capacitors, with edge factors minimized in the test structure. The JEDEC test procedure includes a description of a defect density extraction process, as well as a suggested test structure design.

6.3.1.6 Isolating Mechanisms—Models and Testing: EM. In discussing electromigration (EM), most of the literature deals with aluminum and its alloys (e.g., Al/Si and Al/Si/Cu). The models that have been developed for electromigration apply fairly well to these metal systems. However, we must not forget that most of the fine-line technologies (e.g., 1.2 μm CMOS) are using barrier metal (e.g., Al/W, Al/Ti-W, etc.) to retard metal and contact electromigration. The behavior of these metals in electromigration tests [6.76] will require some strategy changes that are yet to be defined.

Normal operating current density, by design rule, is typically 1×10^5 A/cm^2, which generally guarantees a 20+ year lifetime for the metal. The current density ranges commonly used in accelerated testing are 0.5×10^6 to 2.5×10^6 A/cm^2 for a moderate acceleration factor and 3.0×10^6 to 1.0×10^7 A/cm^2 for an extreme acceleration factor. Both ranges of acceleration factor are needed to establish a hierarchical electromigration test program with reasonable correlation to at-use conditions.

A widely accepted time-to-fail model for an aluminum alloy interconnect is

$$t_{50} = A * W^p * L^q * t * J^{-n} * \exp\left(\frac{E_a}{k * T}\right) \quad (6.46)$$

where E_a = activation energy, about 0.4 eV to 0.9 eV, and length and width compensation is included to handle grain mechanical stress effects [6.78–6.80]. However, since we are concerned primarily with fine-line technologies, the length, thickness, and width dependencies are negligible and are typically absorbed into the constant A. The activation energy is a function of alloy composition and thermal processes during wafer fabrication. The time-to-fail relationship that is most commonly quoted is

$$t_{50} = A * J^{-n} * \exp\left(\frac{E_a}{k * T}\right) \quad (6.47)$$

The exponent in the current density is of concern because the array of test structures that are being considered for wafer-level reliability (WLR) electromigration testing will use widely different current densities. The value of n is determined by the current density [6.77] and must be known in order to extrapolate test results to at-use conditions. Fortunately, extrapolation from extreme current densities, where n is changing rapidly, will not be required with careful calibration to test results at moderate current densities.

Electromigration in contacts is becoming a much more serious problem than it has been in the past because the junction depths and contact sizes are smaller in fine-line technologies. Testing for contact electromigration is more difficult at wafer level because forcing a larger current through contacts to obtain a large acceleration factor can cause large-voltage gradients in the junction region, which can provide alternate current paths.

Two failure modes should be considered in contacts: high resistance caused by migration of silicon into the contact region, and junction leakage caused by aluminum spikes growing into pits in the contact. A model for leakage and for resistance increase [6.81, 6.82] is useful for extrapolation to at-use conditions. Separate models are needed to deal with the two contact failure mechanisms. Both models are discussed in detail in the referenced papers. The activation energy and magnitude of n are a function of the contact metallurgy and the magnitude of current flowing, all of which make highly accelerated testing difficult.

Fortunately, the use of barrier metals in contacts moderates the contact electromigration problem enough that tests at extreme acceleration factors to screen the wafers are not necessary. Barrier materials have essentially eliminated the classic contact EM problem and have extended contact lifetime to well beyond the expected lifetime for interconnect metal. However, in the spirit of checking, which has been the intent of WLR, some method for determining correct barrier metal deposition will enable an indirect contact screening procedure.

6.3.1.7 Electromigration Testing Sandards: Contact Electromigration. In the case of contact electromigration, there is no standard procedure other than using small contact strings [6.81] and testing them individually at elevated temperatures to obtain a reasonable acceleration factor. A reasonable test is to place the contact test structure at high temperature (135 to 200°C) and to apply a stress current (monitoring contact voltage drop), interrupted periodically to measure junction leakage. Failure criteria must be set for junction leakage and contact resistance.

It is doubtful that much effort will ever be needed to define wafer-level tests for contact electromigration because fabrication technology is evolving in a direction that eventually will completely eliminate any direct contact of aluminium with silicon either at junctions or to polysilicon. Currently, developments are favoring the use of barrier materials or even the use of selectively deposited refractory material within contacts.

6.3.1.8 Electromigration Testing Standards: Interconnect Electromigration. A considerable amount of effort has been directed toward creating wafer-level metal electromigration screens. The first step is to define an electromigration test that can be used for calibration of wafer-level electromigration tests. At this point in time, a rather well-exercised test structure and test method for oven-based electromigration testing has been developed at the National Institute of Standards and Technology, which is currently an ASTM standard [6.38, 6.41, 6.42]. Consistent results were obtained [6.77] when test stripes were tested in packages and at wafer probe. This test is run at a moderate current density (typically 2.5×10^6 A/cm^2) and yields activation energy values that are observed by others.

With an accurate electromigration test method in place for calibration of metal systems, it is possible to pursue wafer-level tests. The structure that has received the greatest attention is the Standard Wafer-Level Electromigration Acceleration Test (SWEAT) [6.83], which can detect metalization problems with amazing accuracy. It can be tested in less than one minute (typically targeted for 30 seconds) per test structure. Test current density is in the range of 1×10^7 A/cm^2, which causes significant Joule heating (self-heating to 300–400°C). These two acceleration factors working together are the reason why the test is so rapid. Temperature and current density are controlled to achieve a target failure time for "normal" metal.

Even though some concern has been expressed that such extreme acceleration factors result in the domination of electromigration by some other mechanism of failure, much experimental evidence [6.92] reveals that the failures are dominated by electromigration. However, without calibration with the NIST structure, it is difficult to extrapolate results to use conditions. It is extremely useful in detecting poor step coverage when several SWEAT structures are placed over various steps.

Two other rapid electromigration test methods have been considered: the isothermal and the breakdown energy of metal (BEM) [6.84] tests. The isothermal test involves using Joule heating to raise the temperature of a metal stripe to a very high temperature (near 400°C) and maintaining it there by controlling the current. This is done in a controlled manner by monitoring the resistance of the metal stripe while under stress. The BEM test is a current ramp test to failure in which the total integrated energy dissipated in the test stripe to cause open-circuit failure is a measure of the metal resistance to electromigration. The median energy to fail in the stripe (in units of energy per unit length) becomes the measure of metal quality.

The main focus of this section is on the SWEAT structure because the SWEAT method is rapidly becoming a de facto industry standard. A layout for the standard SWEAT structure unit is presented in Fig. 6.54, with several candidate voltage tap configurations indicated. The configuration labeled "C" is the proposed standard configuration. The most pressing concern about the SWEAT test procedure is the lack of knowledge of the thermal environment along the test structure. The approximate temperature assessment described in the Root and Turner paper [6.83] used geometric arguments that were at best one-dimensional approximations.

A test method suggested in Root [6.92] uses a rather simple geometrical argument for translating average temperature measured, employing resistometric methods to peak temperature of the test link. However, a fairly broad distribution

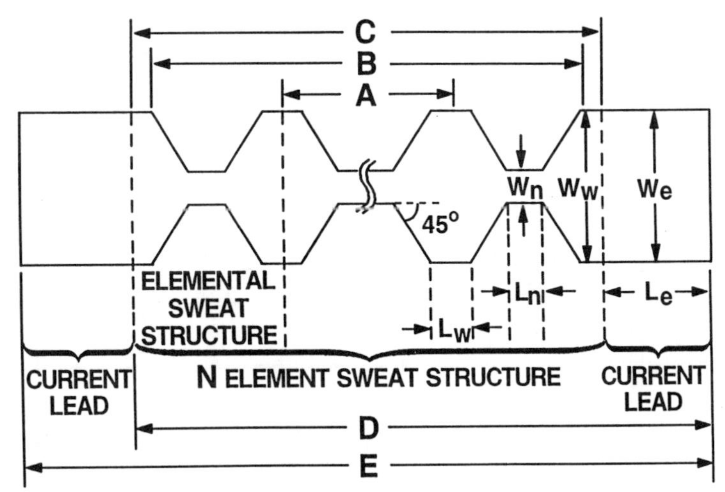

Figure 6.54 SWEAT structure unit and various voltage tap configurations (after Crowell et al. [6.85]).

in the data suggests that the temperature was not very well controlled or even known (see Fig. 6.55). Even though the method implemented was being considered for use as the JEDEC standard SWEAT procedure, a mathematical model of the SWEAT thermal environment (TEARS model) [6.81, 6.85] greatly improves the estimate of the peak temperature, with an observable improvement in test stability.

Evidence of the lower accuracy of temperature assessment is indicated in Fig. 6.55, which shows three different implementations of temperature computation. Figures 6.55*a*,*b* show the two most common methods of temperature calculation. Figure 6.55*a*, labeled "Historical," shows controller stability when the system uses a geometrically calculated temperature, which is the basic method described in the Root and Turner paper. An improvement is achieved by incorporating a local heat-sinking concept in the heat flow through the SWEAT structure. Both methods show evidence of temperature drift that the system must periodically correct. This correction can be seen as a "sawtooth" correction to the calculated time-to-fail graph in Figures 6.55*a*,*b* during the early part of the test.

The TEARS model is sufficiently general that it can be used for other test structure geometries that are tested by using at least partial Joule heating as part of the thermal acceleration process. In addition, it can account for thermal gradient effects in the SWEAT structure that are responsible for causing thermoelectric potentials within the test structure, which, in turn, cause a modification of the temperature profile symmetry (Fig. 6.56).

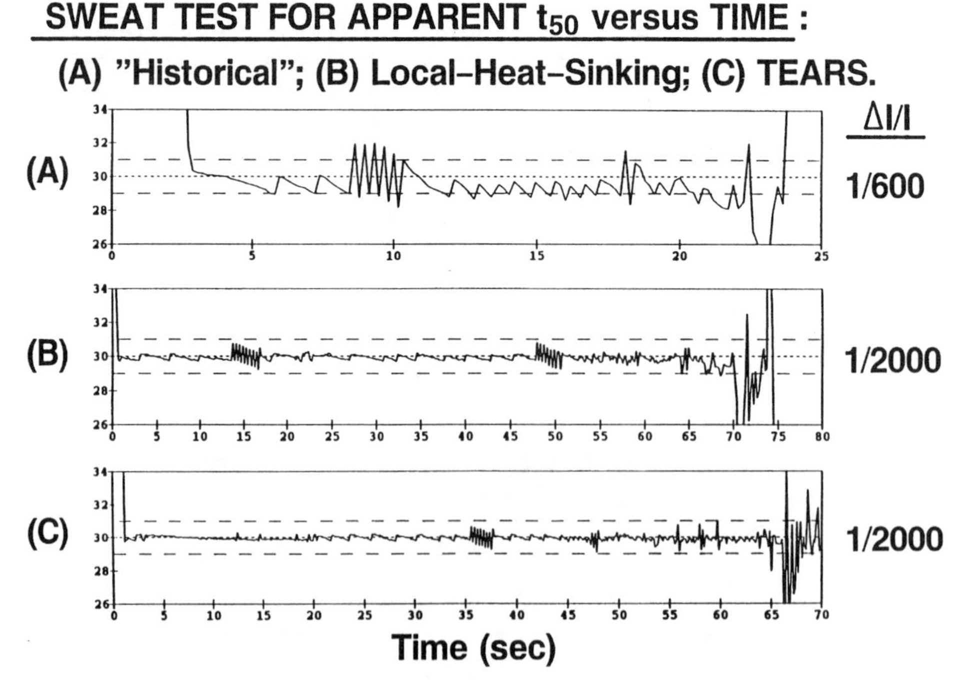

Figure 6.55 Test data showing computed time-to-fail control for three thermal models.

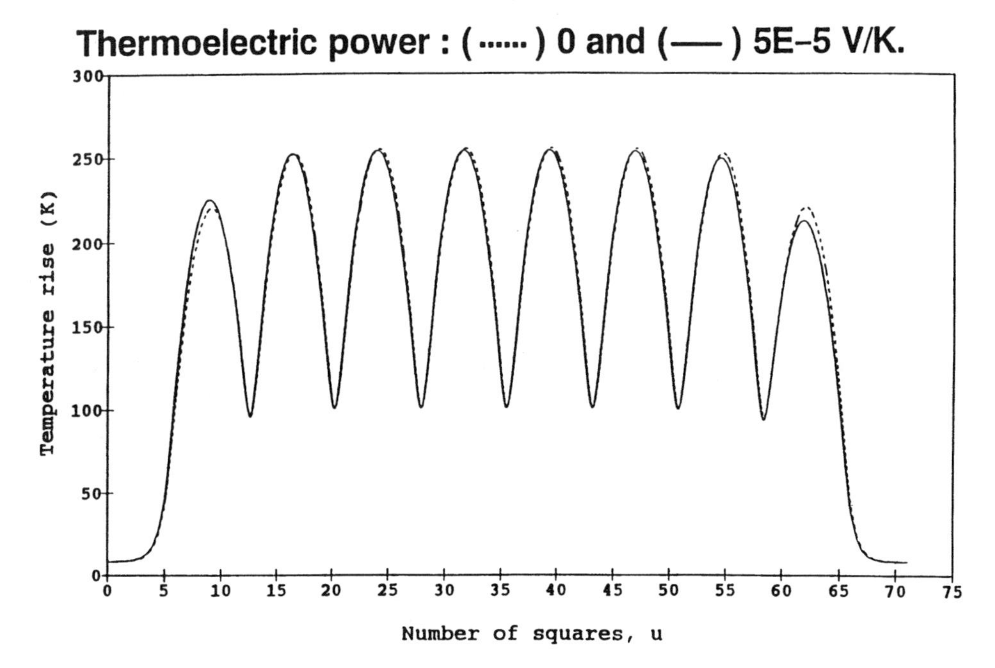

Figure 6.56 SWEAT thermal profile: with and without the thermoelectric term.

Programs for wafer-level monitoring of electromigration in metal interconnects can be implemented with existing test structures and time-to-fail models, but a good thermal model for estimating temperature during testing with substantial Joule heating is essential.

6.3.1.9 Isolating Mechanisms—Models and Testing: HCD. Hot-carrier-induced MOS transistor degradation (HCD) becomes a concern as channel lengths become less than about 2.0 μm. Since normal operating voltages have not been scaling with channel length, this problem will become increasingly important to monitor. Methods for reducing the rate of degradation through fabrication process changes (LDD, etc.) involve delicate fabrication procedures that can get out of control and must be monitored. Accelerated testing is certainly in order, but the danger of extrapolation error is great because most simple testing methods are indirect and extrapolation models are empirically derived.

Generally, stress conditions use an elevated drain voltage to enhance the rate of degradation. Tests generally monitor changes in drain current, in threshold, and in transconductance, G_m. A model for time-to-fail based on threshold change [6.48] is a typical example of an empirical model that uses the ratio of drain current to substrate current as an indicator of damage rate. Thus,

$$t_{50} = H * W * \frac{I_{ds}^{1.9}}{I_{b}^{2.9}} * \Delta V_{th}^{1.5} \qquad (6.48)$$

where H = dielectric dependent constant and W = channel length. However, the observed ratio of changes in drain current versus initial drain current suggests a more complex model [6.86]. Much more experimentation is needed to define a good extrapolation model that will relate the measurements at wafer-level to the shift in parameters that concern the circuit designer.

Currently, hot-carrier testing is typically done at high stress levels to cause changes in threshold and transconductance (hence, drain saturation current) and requires many hours of stress time to detect significant changes in transistor characteristics. These tests are frequently done at wafer level, but the length of stress time required makes them unsuitable for wafer screening. Since substrate current is directly related to the electron hole pair generation rate, monitoring substrate current is a practical method of estimating the hot-carrier damage rate, as was assumed in the above extrapolation model.

The most direct wafer-level test method to use is a four-terminal field-effect transistor (FET) connected to measure drain current and substrate current. A stress drain voltage greater than at-use conditions is applied, and the gate voltage is swept to find the peak in the substrate current. The measured values of drain current and substrate current are used to assess the hot-carrier sensitivity of the transistor. Test transistors should be minimum design rule channel length by a width of at least 20 times the channel length. A pass/fail criterion is based on the ratio of the drain current at the peak substrate current to the peak substrate current. The exact pass/fail threshold is experimentally developed by calibration with long-term stress tests that compare changes in threshold, transconductance, and saturation current with initial tested drain current to substrate current ratio.

Clearly, new wafer-level hot-carrier tests are needed that can measure the actual damage process at the early stages of hot-carrier injection before damage reaches a level that begins to change transconductance. One promising method uses charge pumping to measure the increase in interface states caused by hot-carrier injection [6.87, 6.88]. Figure 6.57 shows the test configuration used to gather data for correlation with long-term test data. Data gathered from this system are illustrated in Fig. 6.58 and show large shifts in charge-pumping current after only a moderate amount of conventional stress time of three hours.

Charge-pumping current tests are compared with test results of transconductance changes in Table 6.9. These data show how many seconds of stressing are required to cause a 10% shift in charge-pumping current and a 10% shift in transconductance at various stress bias conditions. The stress conditions range from moderate stress bias ($V_{ds} = 6.5\,V$, $V_{gs} = 3.0\,V$) to just above at-use bias ($V_{ds} = 5.5\,V$, $V_{gs} = 1.5\,V$). In all cases, the reduction in test time is dramatic. Typically, charge-pumping tests show a 10% shift in 1/1,000 the stress time. The only smaller ratio is at $V_{ds} = 5.5\,V$, which required less than six minutes to change the charge-pumping current by 10%, while available data for transconductance indicate more than 28 hours for an estimated 10% change in transconductance. The trans-

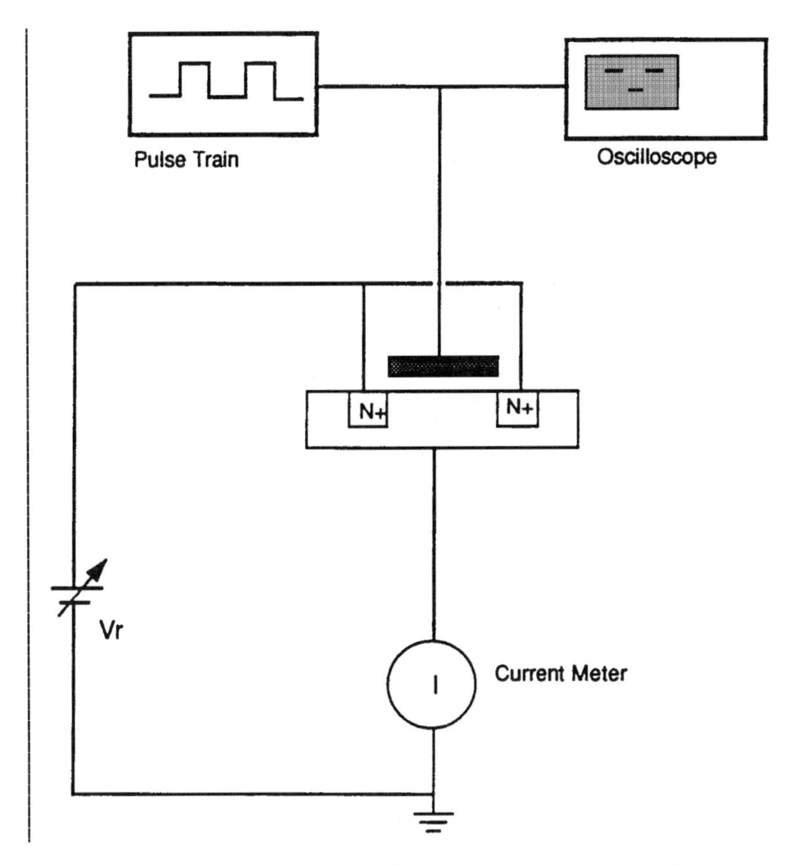

Figure 6.57 Charge-pumping measurement configuration.

conductance test at a drain stress of $V_{ds} = 5.5\,V$ was actually terminated after 28 hours before a transconductance change of 10% was logged.

Charge-pumping measurements are clearly capable of rapid assessment of hot-carrier damage rate at the wafer level. It not only is rapid, but it actually measures the transconductance degradation mechanism directly by measuring the accumulation of surface states. One limitation imposed by charge pumping is that, with the very low current measured (picoampere range), it is almost impossible to conduct the tests in the automated wafer probe environment found in a wafer fabrication line. More research is needed to solve this problem.

6.3.2 Typical Reliability Data

With the diversity of NVSM technology, and the progress and maturity of technology through the years, in addition to the different metrics used by various manufacturers and users to quantify the several aspects of NVSM reliability, a meaningful comparative study of reliability is not practical or desirable. Nevertheless, a quantitative feel for NVSM reliability can be obtained with some

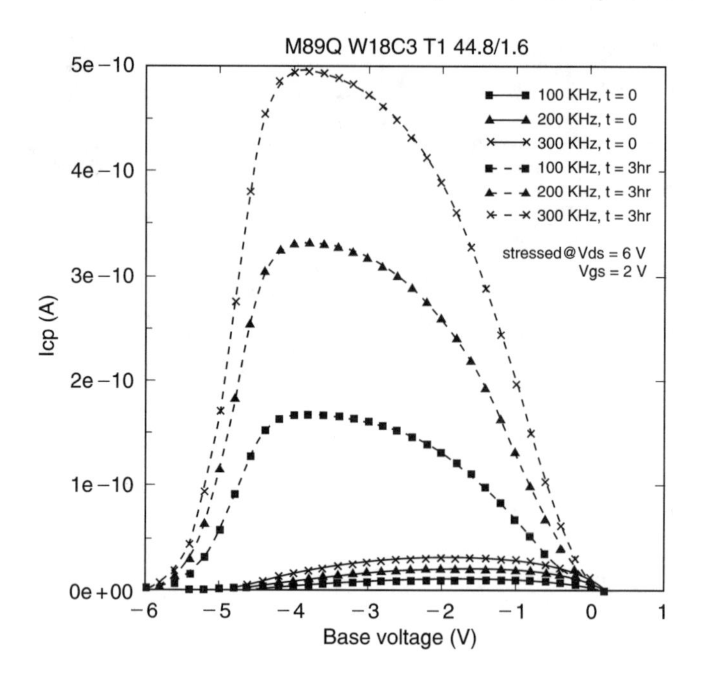

Figure 6.58 Charge-pumping current before and after stress.

reliability data from the literature. Readers are advised (1) to note the date of publication of data compiled in this section, (2) to consult the cited literature for measurement methods and techniques used, and (3) to develop their own data through their supplier or their own reliability test programs.

Table 6.10 compares the defect-limited memory data retention of several NVSM devices, published in 1987 [6.28]. These failure rates can be compared with typical < 100 FIT data established for SRAM and DRAM of that time. For memory applications, data-loss failures (i.e., soft errors from alpha particle, etc.) should be included as overall retention failure considerations, if ECC is not used in the application. Table 6.11 compares retention data-loss failure rates of floating gate EPROM and E^2PROM technology against the other semiconductor memory technologies of

TABLE 6.9 COMPARISON OF TESTING TIME FOR DIFFERENT EVALUATION TECHNIQUES IN HCD ASSESSMENT

Method / HCD indicator / Stress condition	Time-to-10% change (sec)					
	$V_{ds} = 6.5$ $V_{gs} = 2.0$	$V_{ds} = 6.5$ $V_{gs} = 3.0$	$V_{ds} = 6.25$ $V_{gs} = 2.0$	$V_{ds} = 6.26$ $V_{gs} = 3.0$	$V_{ds} = 6.0$ $V_{gs} = 2.0$	$V_{ds} = 5.5$ $V_{gs} = 1.5$
Charge-pumping technique	2.0	6.3	4.5	7.9	25	316
DC transistor parameter characterization technique (gm in linear region)	1920	12,600	7311	15,300	18,026	> 100,800

TABLE 6.10 NONVOLATILE MEMORY DATA RETENTION (AFTER MIELKE ET AL. [6.28]).

Sample				%Fail vs. # Hours				
Product	Type	Size	Temp.	48	168	500	1000	FITS
64 KB	EPROM	1800	250°C	0.9%	1.1%	2.0%	n.a.	15
16 KB	FLOTOX	550	150°C	0.0%	0.0%	0/2%	0.2%	31
64 KB	Tex-Poly	350	250°C	0.0%	0.0%	0.0%	0.3%	1

TABLE 6.11 TYPICAL RETENTION FAILURE RATES (FIT) (AFTER MIELKE ET AL. [6.28]).

Technology	Soft Errors	Total
EPROM/E^2PROM	< 1	1–50
DRAM	50–1000	50–1000
Poly-load SRAM	20–1000	20–1000
Full-CMOS SRAM	10–50	10–50

interest. Table 6.12 further compares these technologies in terms of lifetime catastrophic failure mechanisms, such as oxide breakdowns and data loss from gate leakage. From Table 6.12, it can be noted that floating gate NVSM devices have superior oxide breakdown properties—primarily a consequence of attention to silicon dioxide dielectric integrity necessitated by high-voltage requirements of the floating gate operations. Table 6.13 summarizes some representative values for endurance failure rates of Flash EEPROM and other floating gate NVSM devices. The limited erase/write endurance performance of floating gate devices must be carefully evaluated in terms of end-use memory system erase/write cycle requirements.

Floating trap NVSM devices, such as MNOS memories, have been reported to have superior data retention failure rates as compared to floating gate NVSM devices, such as FLOTOX and textured-poly EEPROMs [6.89]. The spatially distributed traps in MNOS memories are less subject to data loss from electrical charge

TABLE 6.12 LIFE TEST FAILURE RATES (FIT) (AFTER MIELKE ET AL. [6.28]).

Product(s)	Technology	Oxide Breakdown	Data Loss	All Causes
64 KB	Tex-Poly	< 48	< 15	129
16 KB	FLOTOX	< 21	17	43
64 KB–512 KB	EPROM	11	52	100
General Logic	Non-FG	91	0	120

TABLE 6.13 FLASH MEMORY RELIABILITY DATA SUMMARY (PER INTEL DATA BOOK [6.89]). [REPRINTED BY PERMISSION OF INTEL CORPORATION, COPYRIGHT 1988, INTEL CORPORATION]

- Flash memory life test reliability is equivalent to EPROM (64 vs. 103 FITS)
- Flash memory reprogrammability failure rate lower than EPROM/EEPROM

Reprogrammability failure specifications

Flash memory	<.1% AT 10 K-cycles (goal)
EPROM	2%
EEPROM	5% AT 10 K-cycles

- Flash memory cycling test results to date

Product	Cycles	Fail/unit
27F64	20,000	0/1000
28F256	60,000	0/2

- Data retention equivalent to EPROM (~100 years)

leaks through latent failures in the memory oxide film. Figure 6.59 shows the retention failure characteristics of MNOS (measured at accumulated 1% failure) against temperature (Arrhenius plot) and against floating gate NVSM devices A, B, and C. Accumulated failures from erase/write endurance cycling typical of a MNOS EEPROM product is given in Fig. 6.60. The failure rate was within the 10^5 erase/write cycle endurance specification of the product as of November 1987. Other data included in the figure are for floating gate NVSM devices measured in a similar manner.

As has been discussed, the limited endurance of NVSM products precludes their application as random access memories in traditional design or architecture. On the other hand, for applications or designs in which the device is accessed infrequently, such as N times per day (where N is less than 100, for example), the endurance reliability of the NVSM can be calculated and can be shown to be similar to that of other semiconductor IC memory devices such as DRAMs and SRAMs.

From Fig. 6.60, after 10^5 erase/write cycles, the cumulative chip failure rate is about 0.7%. This particular device is organized architecturally as 256 pages of 32 bytes so that each erase/write access is, in effect, an erase/write cycle for 32 bytes. Therefore, failure rate/page is $0.7 \times (1/256)\%$, and the failure rate dependence on N is given as follows:

$$\text{Failure Rate} = (0.7\%/256\ \text{pages}/10^5) \ast (\text{N times}/24\ \text{hours})$$
$$\ast\ \text{N} \times 10^{-11}/\text{hours}$$
$$\ast\ \text{N} \times 10^{-2}\ \text{FIT}$$

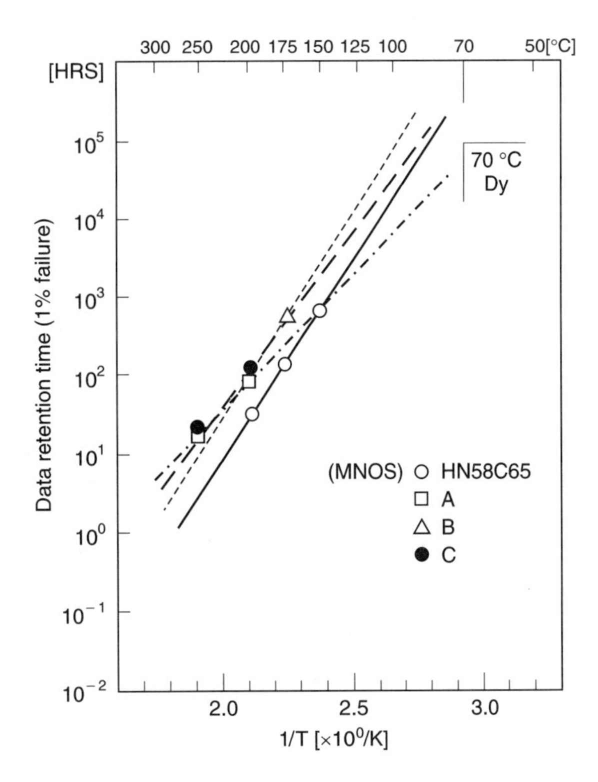

Figure 6.59 Data retention characteristics (Arrhenius plot) (after Hagiwara et al. [6.90]).

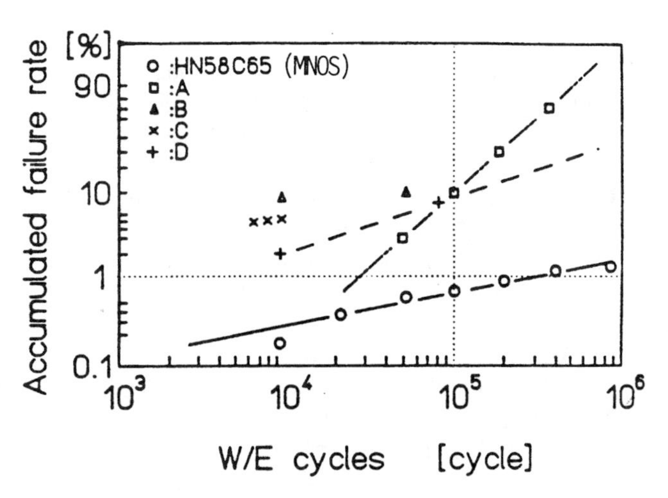

Figure 6.60 Reliability on write/erase cycles (accumulated failure rate) (after Hagiwara et al. [6.90]).

Typical DRAM and SRAM ICs have failure rates in the range of 10 to 100 FIT. Thus, the NVSM endurance reliability for some applications can be treated much like other IC components in the system. In fact, in many applications where $N \simeq 1$, the endurance reliability of NVSM products does not contribute to system reliability concerns.

References

[6.1] E. I. Grant and R. S. Leavenworth, "Statistical quality control," Chapter 18, *Some Aspects of Life Testing and Reliability*, McGraw-Hill, New York, 4th ed., 1974.

[6.2] "Reliability of military electronics equipment," Report by Advisory Group on Reliability of Electronic Equipment, Office of the Assistant Secretary of Defense (Research and Engineering), Superintendent of Documents, Government Printing Office, Washington, D.C., 1957.

[6.3] S. M. Sze, ed., "VLSI technology," Chapter 14, *Yield and Reliability*, by W. J. Bertram, McGraw-Hill, New York, 1983.

[6.4] "IEEE standard definitions and characterization of metal nitride oxide semiconductor arrays," ANSI/IEEE Std 641-1987, New York, 1988.

[6.5] S. M. Sze, from D. S. Peck, "Practical applications of accelerated testing—introduction," Reliability Physics, 13th Annual Proceedings, pp. 253–254, 1975.

[6.6] S. Rosenberg, "E-PROM reliability: Part 2—Tests and screens weed out failures, project rates of reliability," *Electronics Magazine*, pp. 17–22, August 14, 1980.

[6.7] B. E. Deal, "Standardized terminology for oxide charges associated with thermally oxidized silicon," *IEEE Trans. on Electron Devices*, vol. ED-27, pp. 606–608, 1980.

[6.8] I.-C. Chen et al., "Electrical breakdown in thin gate and tunneling oxides," *IEEE Trans. on Electron Devices*, vol. ED-32, no. 2, pp. 413–422, 1985.

[6.9] J. Lee, Chen, and Hu, "Statistical modeling of silicon dioxide reliability," Proc. IRPS, pp. 131–138, 1988.

[6.10] J. W. McPherson and D. A. Baglee, "Acceleration factors for thin gate oxide stressing," *Proc. IRPS*, pp. 1–5, 1985.

[6.11] E. S. Anolick and G. Nelson, "Low field time dependent dielectric integrity," *Proc. IRPS*, pp. 8–12, 1979.

[6.12] D. L. Crook, "Method of determining reliability screens for time dependent dielectric breakdown," *Proc. IRPS*, pp. 1–7, 1979.

[6.13] E. S. Anolick and L.-Y. Chen, "Application of step stress to time dependent breakdown," *Proc. IRPS*, pp. 23–27, 1981.

[6.14] K.C. Boyko and D. L. Gerlach, "Time dependent dielectric breakdown of 210 Å oxides," *Proc. IRPS*, pp. 1–8, 1989.

[6.15] Y. Hsia and K. L. Ngai, "MNOS traps and tailored trap distribution gate dielectric MNOS," Proceedings of the 11th Conference on Solid State Devices, Tokyo, 1979; *Japan. J. of Appl. Phys.*, vol. 19, Supplement 191, pp. 245–248, 1980.

[6.16] B. H. Yun, "Measurements of charge propagation in Si_3N_4 films," Appl. *Phys. Letts.*, vol. 25, pp. 340–342, 1974; P. C. Arnett and B. H. Yun, "Silicon–nitride trap properties as revealed by charge-centroid measurements on MNOS devices," *Appl. Phys. Letts.*, vol. 26, pp. 94–96, 1975.

[6.17] Y. Hsia, W. W. Y. Lee, Y. T. Chen, and K. L. Ngai, "Experiments on graded nitride MNOS," 36th Annual Device Research Conf., Santa Barbara, Calif., 1978.

[6.18] A. K. Agarwal and M. W. White, "New results on electron injection, hole injection, and trapping in MNOS nonvolatile memory devices," *IEEE Trans. on Electron Devices*, vol. ED-32, No. 5, pp. 941–951, 1985.

[6.19] Y. Hsia and K. L. Ngai, "Empirical study of the metal–nitride–oxide semiconductor device characteristics deduced from a microscopic model of memory traps," *Appl. Phys. Letts.*, vol. 41, no. 2, pp. 159–161, 1982.

[6.20] H. J. Stein, "Hydrogen content and annealing of memory quality silicon-oxynitride films," *J. of Electr. Maters.*, vol. 5, pp. 161–177, 1976; H. J. Stein and W. A. R. Wegener, "Chemically bound hydrogen in CVD Si_3N_4: dependence on NH_3/SiH_4 ratio and on annealing," *J. of Electrochem. Soc.*, vol. 124, p. 908, 1977.

[6.21] P. S. Peercy et al., "Hydrogen concentration profiles and chemical bonding in silicon nitride," *J. of Electr. Maters.*, vol. 8, 1979, pp. 11–29.

[6.22] K. L. Ngai and C. T. White, "Reconstructing states at the Si–SiO_2 interface," *J. of Vac. Sci. Technol.*, vol. 16, pp. 1412–1417, July 1979.

[6.23] R. W. Pryor, "A mechanism for endurance failure in metal–nitride–oxide–semiconductor device structures," *J. of Appl. Phys.*, vol. 52, no. 5, pp. 3702–3704, 1981.

[6.24] R. E. Shiner et al., "Data retention in EPROMS," *Proc. IRPS*, pp. 238–243, 1980.

[6.25] Intel Reliability Report, RR-60, "ETOX^{TM} Flash memory reliability data summary," Order No. 293002-006, October 1989.

[6.26] B. Euzent et al., "Reliability aspects of a floating gate E^2PROM," *Proc. IRPS*, pp. 11–16, 1981.

[6.27] V. N. Kynett et al., "A 90-ns one-million erase/program cycle 1-Mbit Flash memory," *J. of Solid-State Circuits*, vol. 24, no. 5, pp. 1259–1264, 1989.

[6.28] N. Mielke et al, "Reliability comparison of FLOTOX and textured-polysilicon E^2PROMs," *Proc. IRPS*, pp. 85–92, 1987.

[6.29] T. Ajiki et al, "Temperature accelerated estimation of MNOS memory reliability," *Proc. IRPS*, pp. 17–22, 1981.

[6.30] D. L. Crook, "Evolution of VLSI reliability engineering," *Proc. IRPS*, pp. 2–11, 1990.

[6.31] D. A. Baglee, "Characteristics and reliability of 100 Å oxides," *Proc. IRPS*, pp. 152–155, 1984.

[6.32] VHSIC/VHSIC-Liek Reliability Prediction modeling, RADC-TR-89-177, Final Technical Report, October 1989.

[6.33] E. R. Domangue, R. Rivers, and C. Shepard, "Reliability predicting using large MOS capacitors," *Proc. IRPS*, pp. 140–145, 1984.

[6.34] K. Yamabe and K. Taniguchi, "Time dependent dielectric breakdown of thin thermally grown SiO_2 films," *IEEE Trans. on Electron Devices*, vol. ED-32, No. 2, 1985.

[6.35] I.-C. Chen et al., "The effect of channel hot carrier stressing on gate oxide integrity in MOSFET," *Proc. IRPS*, pp. 1–7, 1988.

[6.36] B. J. Fishbein and D. B. Jackson, "Performance degradation of N-channel MOS transistors during DC and pulsed Fowler–Nordheim stress," *Proc. IRPS*, pp. 159–163, 1990.

[6.37] R. Rakkhit et al., "Drain-avalanche induced hole injection and generation of interface traps in thin oxide MOS devices," *Proc. IRPS*, pp. 150–153, 1990.

[6.38] ASTM Designation: F 1261–89, Standard Test Method for Determining the Average Width and Cross-Sectional Area of a Straight, Thin-Film Metal Line.

[6.39] ASTM Designation: F 1260-89, Standard Test Method for Estimating Electromigration Median Time-to-Fail and Sigma of Integrated Circuit Metallizations.

[6.40] ASTM Designation: F 1259-89, Standard Guide for Design of Flat, Straight-Line Test Structures for Detecting Metallization Open-Circuit or Resistance–Increase Failure Due to Electromigration.

[6.41] J. M. Towner et al., "Aluminum electromigration under pulsed D. C. conditions," *Proc. IRPS*, pp. 36–39, 1983.

[6.42] J. S. Suehle and H. A. Schafft, "Current density dependence of electromigration t_{50} enhancement due to pulsed operation," *Proc. IRPS*, pp. 106–110, 1990.

[6.43] B. K. Liew et al., "Electromigration interconnect lifetime under AC and pulsed DC stress," *Proc. IRPS*, pp. 215–219, 1989.

[6.44] J. A. Maiz, "Characterization of electromigration under bidirectional (BC) and pulsed unidirectional (PDC) currents," *Proc. IRPS*, pp. 220–228, 1989.

[6.45] E. Takeda and N. Suzuki, "An empirical model for device degradation due to hot-carrier injection," *IEEE Electr. Dev. Letts.*, vol. EDL-4, no. 4, 1983.

[6.46] Chenming Hu et al., "Hot electron induced MOSFET degradation—model, monitor and improvement," *IEEE Transactions on Electron Devices*, vol. ED-32, no. 2, 1985.

[6.47] Y. Chingchi et al., "Structure and frequency dependence of hot-carrier-induced degradation in CMOS VLSI," *Proc. IRPS*, pp. 195–200, 1987.

[6.48] J. J. Tzou et al., "Hot-electron-induced MOSFET degradation at low temperatures," *IEEE Electr. Dev. Letts.*, EDL-6, no. 9, 1985.

[6.49] G. Samachisa et al., "A 128K Flash EEPROM using double-polysilicon technology," *IEEE J. of Solid-State Circuits*, vol. 22, no. 5, pp. 676–683, 1987.

[6.50] S. Ali et al., "A 50-ns 256K CMOS split-gate EPROM," *IEEE J. of Solid-State Circuits*, vol. 23, No. 1, pp. 79–85, 1988.

[6.51] N. Ohtsuka et al., "A 4-Mbit CMOS EPROM," *IEEE J. of Solid-State Circuits*, vol. 22, no. 5, pp. 669-675, 1987.

[6.52] K. Yoshikawa et al., "Technology requirements for mega bit CMOS EPROMS," *IEDM Tech. Dig.*, pp. 456–459, 1984.

[6.53] S. Mori et al., "Polyoxide/nitride/oxide structure for highly reliable EPROM cells," *Symp. VLSI Technology Digest of Technical Papers*, pp. 38–39, 1984.

[6.54] S. Mori et al., "Reliable CVD inter-poly dielectrics for advanced E & EEPROM," *Symp. VLSI Technology Digest of Technical Papers*, pp. 16–17, 1985.

[6.55] S. Mehrotra et al., "A 64Kb CMOS EEPROM with on-chip ECC," *ISSCC Digest of Technical Papers*, pp. 142–143, 1984.

[6.56] A. Gupta et al., "A 5V-only 16K EEPROM utilizing oxynitride dielectrics and EPROM redundancy," *ISSCC Digest of Technical Papers*, pp. 184–185, 1982.

[6.57] P. Chen, "Threshold-alterable Si-Gate MOS devices," *IEEE Trans. on Electron Devices*, vol. ED-24, no. 5, pp. 584–586, 1977.

[6.58] Y. Hsia, E. Mei, and K. L. Ngai, "MNOS retention-endurance characteristics enhancement using graded nitride dielectric," Proceedings of the 14th Conference on Solid State Devices, Tokyo, 1982; *Japan. J. of Appl. Phys.*, vol. 22, Supplement 22-1, pp. 89–93, 1983.

[6.59] Y. Hsia, "MNOS LSI memory device data retention measurements and projections," *IEEE Trans. on Electron Devices*, vol. ED-24, no. 5, pp. 568–577, 1977.

[6.60] Y. Hsia, "Cross gate MNOS memory device," *IEEE Trans. on Electron Devices*, vol. ED-25, no. 8, pp. 1071–1072, 1978.

[6.61] Y. Yatsuda et al., "Hi-MNOS II technology for a 64-Kbit byte-erasable 5-V-only EEPROM", *IEEE J. of Solid-State Circuits*, vol. 20, no. 1, pp. 144–151, 1985.

[6.62] Y. Hsia, "MNOS LSI memory device data retention measurements and projections," *IEEE Trans. on Electron Devices*, vol. ED-24, no. 5, pp. 568–577, 1977.

[6.63] Y. Hsia, "Threshold referenced MNOS sense amplifier," U.S. Patent No. 4,376, 987, March 15, 1983.

[6.64] K. Imamiya et al., "A 68-ns 4 Mbit CMOS EPROM with high-noise-immunity design," *IEEE J. of Solid-State Circuits*, vol. 25, no. 1, pp. 72–78, 1990.

[6.65] K. Kobayashi et al., "A high-speed parallel sensing architecture for multi-megabit Flash E^2PROM's," *IEEE J. of Solid-State Circuits*, vol. 25, no. 1, pp. 79–83, 1990.

[6.66] D. Cioaca et al., "A million-cycle CMOS 256K EEPROM," *IEEE J. of Solid-State Circuits*, vol. 22, no. 5, pp. 684–692, 1987.

[6.67] S. Mehrotra et al., "A 64Kb CMOS EEPROM with on-chip ECC," *ISSCC Dig. Tech. Papers*, pp. 142–143, 1984.

[6.68] T-K. J. Ting et al., "A 50-ns CMOS 256K EEPROM," *IEEE J. of Solid-State Circuits*, vol. 23, pp. 1164–1170, 1988.

[6.69] Y. Terada et al., "120-ns 128K × 8-bit/64K × 16-bit CMOS EEPROM's," *IEEE J. of Solid-State Circuits*, vol. 24, no. 5, pp. 1244–1249, 1989.

[6.70] T. Nakayama et al., "A 5-V-only one-transistor 256K EEPROM with page-mode erase," *IEEE J. of Solid-State Circuits*, vol. 24, no. 4, pp. 911–915, 1989.

[6.71] C. R. Crowell, C.-C. Shih, and V. C. Tyree, "SWEAT structure design and test procedure criteria based upon TEARS characterization and spacial distribution in iterated structures," *Proc. IRPS*, pp. 277–286, 1991.

[6.72] E. S. Anolick and L.-Y. Chen, "Screening of time-dependent dielectric breakdown," *Proc. IRPS*, pp. 23–27, 1982.

[6.73] O. Hallberg, "NMOS voltage breakdown characteristics compared with accelerated life tests and field use data," *Proc. IRPS*, pp. 28–33, 1981.

[6.74] M. Schatzkes and M. Av-Ron, "Statistics of defect related breakdown," *Proc. IRPS*, pp. 167–175, 1981.

[6.75] J. C. Lee, I.-C. Chen, and C. Hu, "Modeling and characterization of gate oxide," *IEEE Transactions on Electron Devices*, vol. ED-35, no. 12, pp. 131–138, 1988.

[6.76] S. S. Iyer and C.-Y. Ting, "Electromigration study of the Al-Cu/Ti/Al-Cu system," *Proc. IRPS*, 1984.

[6.77] H. A. Schafft and T. C. Grant, "Electromigration and the current density dependence," *Proc. IRPS*, pp. 93–99, 1985.

[6.78] B. N. Agrawala et al., "Dependence of electromigration induced failure time on length and width of aluminum thin-film conductors," *J. Appl. Phys.*, 1970.

[6.79] G. A. Scoggan et al., "Width dependence of electromigration life," *Proc. IRPS*, 1975.

[6.80] J. S. Arzigian, "Aluminum electromigration lifetime variations with linewidth: the effects of changing stress conditions," *Proc. IRPS*, pp. 32–39, 1983.

[6.81] S. Vaidya, "Electromigration induced leakage at shallow junction contacts metallized with aluminum/poly-silicon," *Proc. IRPS*, pp. 50–54, 1982.

[6.82] P. A. Gargini et al., "Elimination of silicon electromigration in contacts by the use of an interposed barrier metal," *Proc. IRPS*, 1982.

[6.83] B. J. Root and T. Turner, "Wafer level electromigration tests for production monitoring," *Proc. IRPS*, pp. 100–107, 1985.

[6.84] C. C. Hong and D. L. Crook, "Breakdown energy of metal (BEM)—a new technique for monitoring metallization," *Proc. IRPS*, pp. 108–114, 1985.

[6.85] C. R. Crowell, C.-C. Shih, and V. C. Tyree, "Simulation and testing of temperature distribution and resistance versus power for SWEAT and related joule-heated metal-on-insulator structures," *Proc. IRPS*, pp. 37–44, 1990.

[6.86] K. M. Cham et al., "Self-limiting behavior of hot carrier degradation and its implication on the viability of lifetime extraction by accelerated stress," *Proc. IRPS*, pp. 191–194, 1987.

[6.87] G. Groesenken, H. E. Maes, N. Beltran, and R. F. DeKeersmaecker, "A reliable approach to charge-pumping measurements in MOS transistors," *IEEE Trans. Electron Devices*, vol. ED-31, no. 1, pp. 42–53, 1984.

[6.88] P. Heremans, J. Witters, G. Groeseneken, and H. E. Maes, "Analysis of the charge pumping technique and its application for the evaluation of MOSFET degradation," *IEEE Trans. Electron Devices*, vol. ED-36, no. 7, 1989.

[6.89] "ETOX Flash memory: The cost effective and reliable firmware management solution," Intel Corp., Order No. 296294-001, March 1988.
[6.90] T. Hagiwara et al., "Yield and reliability of MNOS EEPROM products," 9th IEEE NVSM Workshop, February 1988.
[6.91] J. R. Black, "Current limitations of thin film conductors," *Proceedings IRPS*, 1982.

Chapter 7

George Messenger

Radiation Tolerance

7.0. INTRODUCTION

This chapter considers the very important topic of radiation response of nonvolatile semiconductor memories. Modern electronic systems must perform properly in hostile radiation environments created by nuclear explosions for most Department of Defense (DOD) applications. In addition, all space systems are exposed to electrons and protons in or near the Van Allen belts and to cosmic rays [7.1].

7.1. BASIC RADIATION CONSIDERATIONS

First, a brief summary of basic radiation effects is presented as a foundation for considering the radiation response of floating gate and SNOS memory devices.

7.1.1 Radiation Environments

It is important to establish quantitatively the types of radiation environments and the appropriate range of intensities that are of interest to system designers contemplating the use of SNOS and floating gate memories.

7.1.1.1 Nuclear Weapon Environments. Nuclear explosions generate neutrons, gamma rays, and X rays [7.2]. Endo-atmospheric explosions create a radiation threat primarily from neutrons and gamma rays. The X rays are converted to a blast wave and thermal radiation from the fireball by interaction with the atmosphere in the immediate vicinity of the explosion. Balanced hardness considerations dictate the nuclear radiation specification. If the equipment is manned, human survival (approximately 500 Rads), or near instantaneous human incapacitation (approximately 3,000 Rads), defines the ionizing dose specification level [7.3]. The ionizing dose specification for these applications is, therefore, usually less than 3,000 Rads(Si). The accompanying neutron specification, determined in the same manner, is usually less than 10^{12} n/cm^2 (1 Mev Equivalent). This explanation is greatly oversimplified for purposes of clarity and brevity. Factors involved in this determination are the relative biological effectiveness (RBE) to relate Rads(Si) for neutrons and gamma rays to radiation equivalent man (REMs). The actual size of the nuclear weapon is also an extremely important factor. For weapons larger than 20 kilotons (KTs), blast and thermal effects are usually dominant. It is only for scenarios that involve nuclear weapons smaller than 20 KTs that neutron and gamma specifications are normally encountered.

For unmanned electronic equipment, the radiation specification is normally balanced to the capability of the system to survive the nuclear blast. A blast survival capability of approximately 2 psi (pounds per square inch) is typical for most electronic equipment. Determining the neutron and gamma ray exposure coradius with 2 psi coincidentally usually results in gamma specifications less than 3,000 Rads(Si) and neutron levels less than 10^{12} n/cm^2 (1 Mev Equivalent). Finally, fallout is an important factor. The radioactive fission fragments are thrust high into the atmosphere and fall back to earth in a complex pattern dominated by wind and weather. It should be recognized that some special military requirements may be encountered with higher radiation specifications. It is frequently required that components meet a higher radiation specification than the electronic system. This is done to provide a hardness design margin; such design margins usually range from 3 to 10.

Exo-atmospheric systems such as satellites and ballistic missiles operate in the hard vacuum of space. Here, the X rays travel unattenuated and interact directly with the system. The neutrons and gamma rays also arrive at the system with no attenuation. The radiation requirements are established for balanced hardness by determining the system survivability to the thermomechanical effects of X rays. This limit is approximately 1 cal/cm^2 for many satellites. The most vulnerable components tend to be polished optical lenses or sensors that are, by necessity, directly exposed to the unattenuated X-ray flux. The neutron, X-ray, and gamma ray intensities at the 1 cal/cm^2 radius are easily determined for typical nuclear weapons. A neutron fluence of 3×10^{11} n/cm^2 is typical, and a gamma dose of 50 Rads(Si) is usually encountered. However, the total ionizing dose is usually dominated by shine-through X rays and electrons that have been attenuated by the satellite skin, the electronic box, and the component package. The shine-through dose is usually between 500 and 50,000 Rads(Si). Often, high atomic number shielding materials, such as lead, are used to attenuate the X rays. In the practical limit, X-ray shine-through can be reduced to the same level as the gamma exposure, or about 50 Rads(Si).

Tactical and avionic specification considerations are summarized in Table 7.1. The lower limits are levels below which semiconductor devices show no significant degradation in nearly all applications. The upper limits are radiation levels that result in nearly instantaneous human incapacitation and severe blast damage.

A design margin at the component level is used to ensure high system survival probability without the necessity of component radiation lot sample testing. Some military requirements, such as strategic weapon systems, require much higher radiation specifications. These requirements are system specific and are always classified; therefore, they cannot be discussed here.

TABLE 7.1 TACTICAL AND AVIONIC RADIATION REQUIREMENTS

		Range		
	Typical	Lower Limit	Upper Limit	Components with Design Margin
Neutron fluence n/cm**2 1 Mev Equivalent	10exp12	10exp11	2*10exp12	5*10exp12
Ionizing dose Rads (Si)	3,000	50	5,000	10,000
Dose rate Rads (Si)/s	10exp8	10exp6	10exp9	3*10exp9
Overpressure Psi	2	—	5	—

7.1.1.2 The Natural Space Radiation Environment. The natural space radiation environment is dominated by protons and electrons from the Van Allen belts. These belts contain high levels of ionizing radiation, and therefore, nearly all space satellites are established in orbits either below or above the belts. Low earth orbits (LEOs) are substantially below the inner belt as exemplified by the various shuttle orbits. Synchronous altitude satellites are in an orbit substantially above the outer belt. The National Aeronautics and Space Administration (NASA) has done extensive mapping of electron and proton fluxes in the entire operational space around the earth, and computer programs are available to define proton and electron fluences and spectra of any desired orbit for any mission life. However, these mappings still contain considerable uncertainies. A typical worst case mission electron fluence is 10^{15} e/cm^2 with an average energy of 1 MeV.

For all but a few orbits close to the inner belt, the proton fluence contributes a negligible amount of ionizing dose compared to electrons. However, proton fluences are defined in the same manner as electron fluences for desired orbits and mission life. The range of protons is relatively small, and the satellite skin, the box thickness, and the component package combine to reduce the ionizing dose from protons. When a nuclear weapon explodes in space, it produces a very high density of electrons that are trapped by the earth's magnetic field in a manner similar to the natural belts. These are referred to as pumped-up belts, and the electrons are known as Argus electrons after the first test blast that led to their discovery. These electrons decay as a function of time, and the decay time depends strongly on the location of the burst. A large number of bursts can build the electron density up to a saturation value in the pumped-up belts. Worst case mission electron fluences can be an order of magnitude higher than the natural fluence under these circumstances.

The ionizing radiation reaching electronic components is attenuated by the satellite skin, the walls of the electronic box, the walls of the package, and any other structures through which the radiation must pass. This typically amounts to 1 to 3 g/cm^2 and provides attenuation of two to three orders of magnitude. This results in ionizing dose levels in electronics varying from 10^3 to 10^5 Rads(Si) for most applications. Figure 7.1 shows the attenuation characteristics and can be used to calculate the dose reaching electronic devices.

Cosmic radiation is the final major constituent of the space environment that can damage electronic components. The contribution to total ionizing dose is negligible. The satellite and other material surrounding the electronic components provide no significant shielding due to the extremely high energy of cosmic ray components. Some

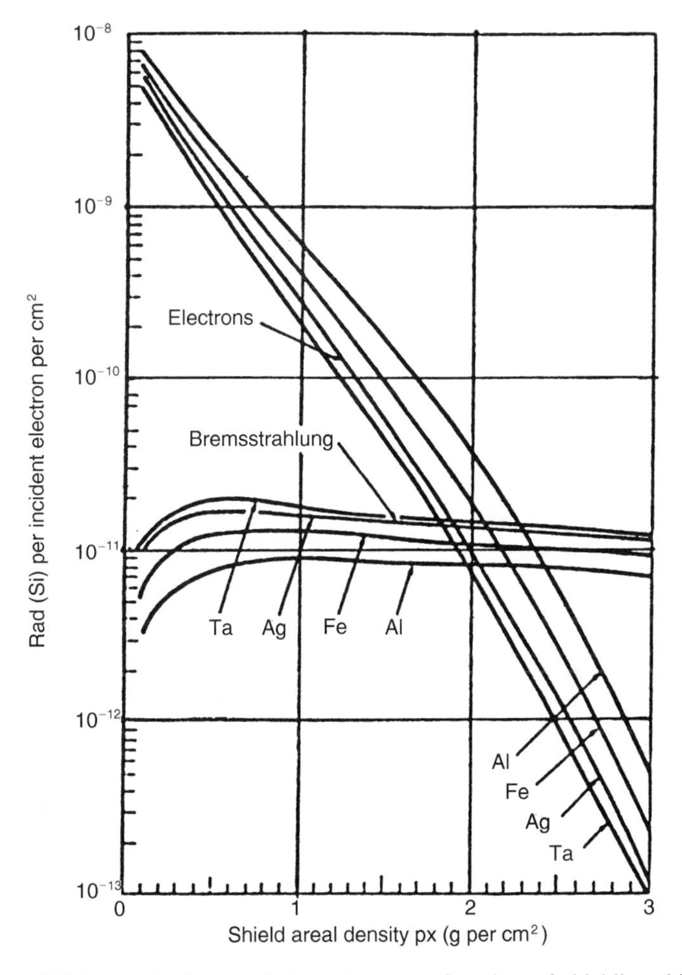

Figure 7.1 Isotropic fission electron dose as a function of shielding thickness for a family of shielding materials.

cosmic rays originate in the sun, and their fluxes are tied into the solar cycle. Again, NASA has mapped cosmic ray behavior in the space around the earth, and estimates of cosmic ray activity can be obtained for anticipated orbits and mission duration. Cosmic rays cause single-event phenomena (SEP) in semiconductor components, which primarily cause only a change of state in a logic cell. Subsequently, the entire component functions properly. This is referred to as single-event upset (SEU). Occasionally, however, the passage of a cosmic ray through the active volume of a semiconductor component can trigger latchup or burnout; these incidents usually result in catastrophic failure. The cosmic ray environment is described by the Heinrich curves shown in Fig. 7.2, which are plots of particle flux as a function of linear energy transfer (LET) [7.5]. These curves reflect various amounts of solar activity, allowing the designer to select his own worst case condition.

7.1.1.3 Nuclear Power Reactors. There is a small usage of semiconductor components in commercial power reactors. For example, pressure and temperature

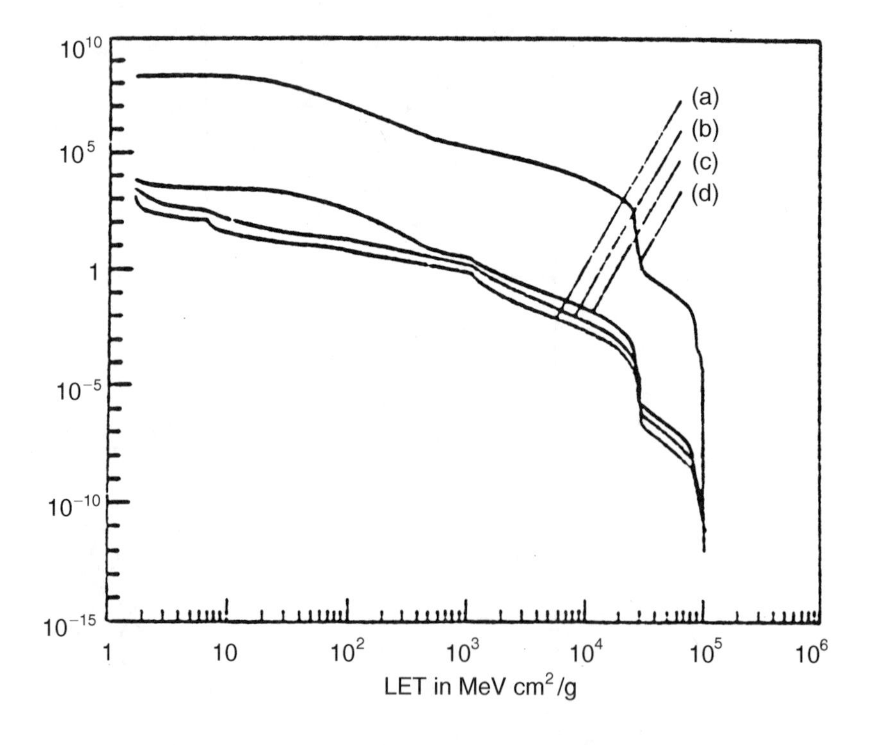

Figure 7.2 Integral LET spectra in the interplanetary medium, inside a spacecraft with 0.064 cm aluminum walls. These spectra include the contribution of all the elements. (*a*) Pure galactic cosmic rays at solar maximum; (*b*) pure galactic cosmic rays at solar minimum; (*c*) an environment so severe that it is exceeded only 10% of the time; (*d*) an extremely large solar flare at its peak.

transducers are used to monitor the reactor core. These components are specified to survive 2×10^8 Rads(Si), reflecting a loss of coolant accident (LOCA) [7.6].

7.1.1.4 Radiation Simulations. Radiation testing of semiconductor components is routinely accomplished in a variety of simulators [7.1]. The most commonly used simulators include: (1) fast burst reactors and training-research-isotopes-general-atomic (TRIGA) reactors for neutron exposure, (2) Cobalt-60 cells for ionizing dose exposure, (3) linear accelerators (LINACs) and Flash X rays for dose–rate testing, (4) Aracor machines for in-line ionizing dose hardness assurance testing of semiconductor components, and (5) cyclotrons for simulating high-energy cosmic rays. It should be noted that simulators do not accurately reproduce the environments for which the equipment is being designed. Extrapolation or interpolation is usually required.

For example, the ionizing dose environment in space is usually caused by shine-through electrons delivered at a very low rate (0.0001 to 1 Rad(Si)/s), and the standard test is in a Co^{60} cell at 300 Rad(Si)/s. This can be very misleading, for ionizing dose effects are strongly dependent on dose-rate. On the other hand, the ionizing dose from a nuclear weapon is usually delivered very rapidly and is strongly scenario dependent. The prompt dose is delivered at rates ranging from 10^6 to 10^{12} Rads(Si)/s, with a pulse width of approximately 10 ns. Almost all of the dose is accumulated within one minute. The ionizing dose simulation in the Co^{60} cell is thus at too low a rate to properly simulate nuclear weapon effects. Most LINACs and Flash X rays cannot exactly simulate the very short nuclear prompt pulse. This is not a serious problem since obtainable simulator pulse widths vary from approximately 10 ns to 100 ns.

Dosimetry can also be a problem, but with good technique, it can be accomplished with 25% repeatability and 50% absolute accuracy [7.1,7.3]. Radiation testing should be carefully planned in advance, with close cooperation between test engineers, facility engineers who usually take responsibility for dosimetry, and health service personnel who ensure that the testing being carried out is consistent with safety and minimum radiation exposure to personnel.

Ionizing dose is usually monitored with thermoluminescent dosimeters (TLDs), and, if the pulse shape requires monitoring, it is often done with a personal identification number (PIN) diode. Neutron dose is usually monitored with sulfur pellets. The facility personnel then convert the sulfur reading to a 1 Mev Equivalent neutron fluence using a factor obtained by using a complex unfolding procedure generic to their facility in its present test configuration. All dosimetry calibrations should be traceable to national standards maintained by the National Institute of Standards and Technology (NIST).

7.1.2 Displacement Damage

Displacement damage is caused by all high-energy radiation and describes the disorder in the single-crystal silicon lattice resulting from collisions between the radiation particles and individual silicon atoms. However, neutron fluence is the primary cause of displacement damage in the environments discussed previously. The predominant source is nuclear explosions.

Occasionally, it is necessary to evaluate displacement damage from protons or electrons. Since charged particles transfer nearly all of their energy to the silicon lattice by ionizing interactions, ionizing dose damage usually exceeds displacement damage by a large factor. When a neutron collides with a silicon atom, the silicon atom is highly energized. It becomes a recoil primary and subsequently dislodges many more silicon atoms before it loses all of its energy and comes to rest in the lattice. The immediate consequence of this process is the formation of a large number of Frenkl defects (vacancy interstitial pairs). The vacancy is an effective recombination and trapping center. Both the vacancy and interstitial are mobile at room temperature. The vacancy diffuses until it forms a more complex defect, which is stable. The vacancy might encounter an interstitial atom, in which case an immediate repair of the lattice occurs at that location. Other, less benign possibilities include: (1) formation of an E center by complexing with a donor atom, (2) formation of an A center by binding to an oxygen impurity atom, or (3) formation of a divacancy by tying up another vacancy. These defects are all effective recombination and trapping centers. They cause a reduction in minority carrier lifetime and an increase in silicon resistivity. The increase in resistivity is a result of majority carrier trapping and mobility reduction of the remaining majority carriers.

7.1.2.1 Lifetime Degradation. The amount of lattice damage is directly proportional to neutron fluence. The recombination rate of minority carriers is directly proportional to the amount of lattice damage, and finally, the minority carrier lifetime is inversely proportional to the recombination rate. Thus,

$$1/\tau = 1/\tau_0 + K\phi \tag{7.1}$$

where τ is the minority carrier lifetime, τ_0 is the initial unirradiated lifetime, K is a proportionality constant, and ϕ is the 1 Mev Equivalent neutron fluence. The derivation of this equation is beyond the scope of this book. The equation holds for values of neutron fluence up to $10^{16}\ \mathrm{n/cm^2}$. Minority-carrier lifetime is a function of injection level that translates to a dependence on bias currents in semiconductor devices. The Shockley–Read formalism can be used to introduce this dependence into Eq. (7.1). The major effect of lifetime degradation is a reduction in current gain in bipolar devices. Lifetime degradation also causes an increase in junction leakage currents. Ionizing dose also increases leakage currents, and these effects are usually dominant in MOS devices. Figure 7.3 shows the dependence of lifetime damage constant on doping density and injection level for silicon [7.1].

7.1.2.2 Carrier Removal. The trapping of majority carriers in silicon devices is directly proportional to neutron fluence. However, as the trapping continues, each carrier removed is a bigger proportion of the remaining carriers. This results in the following expression for carrier density as a function of neutron fluence.

$$n = N_{di}(1 - k\phi)/(1 + k\phi) \tag{7.2}$$

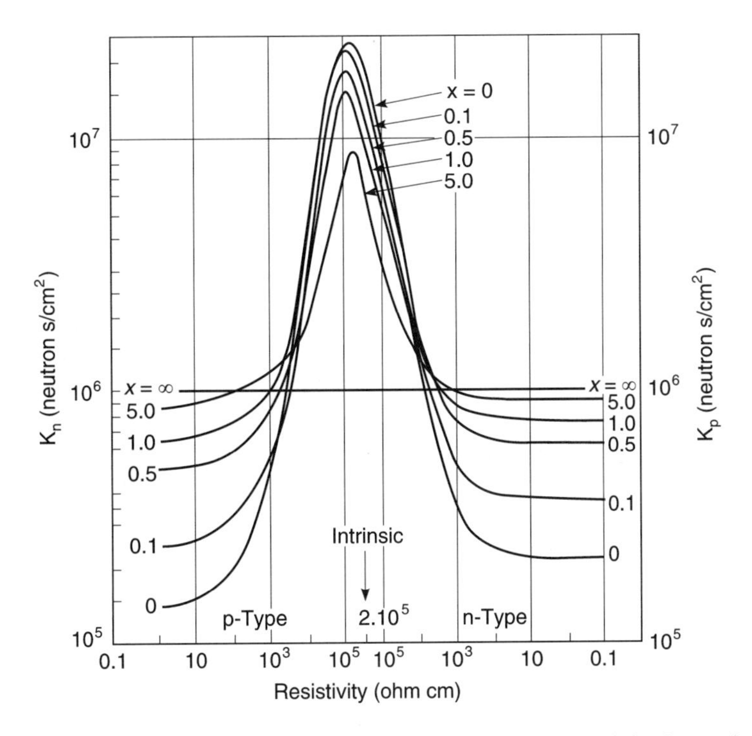

Figure 7.3 Damage constants versus resistivity for various injection ratios (proportional to bias currents).

where N_{di} is the initial donor density assumed to be equal to the initial electron carrier density, k is a constant, and ϕ is the neutron fluence. This approximation assumes that the removal process is dominated by the donor vacancy trapping centers, and it neglects small contributions from oxygen vacancy centers and divacancies. When $k\phi$ reaches one, the material becomes intrinsic. The divacancy has trapping levels at the middle of the band gap, and all silicon material asymptotically approaches intrinsic at high levels of neutron irradiation. The expression for carrier density is often approximated by the exponential relationship

$$n = N_{di} \exp(-k_1 \phi) \tag{7.3}$$

where k_1 is a constant. The carrier removal rate is then

$$dn/d\phi = -N_{di} k_1 \qquad k_1 \phi \ll 1 \tag{7.4}$$

Figure 7.4 shows carrier removal data in N- and P-type silicon for a 10% change in resistivity [7.7]. Figure 7.5 shows the change in resistivity as a function of neutron fluence [7.7].

7.1.2.3 Mobility Degradation. Each carrier removed is trapped at a defect center, becomes charged, and is therefore an effective scattering center for the remaining mobile carriers. The Conwell–Weisskopf formula relates carrier mobility

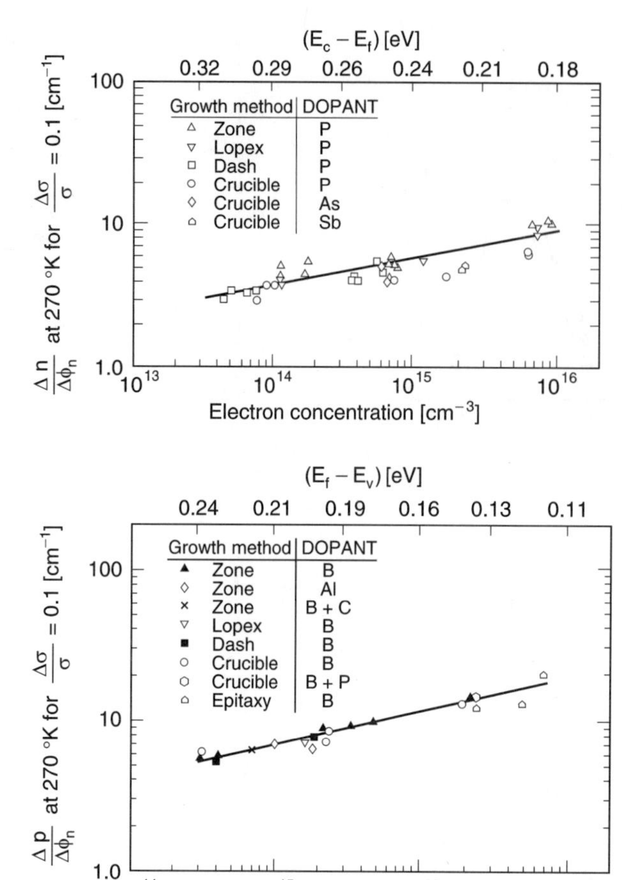

Figure 7.4 Neutron-induced carrier removal rates at 270 K for silicon, evaluated for a 10% shift in resistivity.

to the density of charged scattering centers. Therefore, this formula can be modified by adding $N_{di}k_1\phi$ directly to the density of scattering centers. The resulting mobility decreases monotonically with neutron fluence in a power law fashion [7.1] as $\mu_n = 65 + 1{,}265\{1 + 6.47 \cdot 10\exp(-13) \cdot N_{di}(1 + k_1\phi)^{0.72}\}^{-1}$. A similar expression is available for hole mobility.

7.1.2.4 Particle Equivalence. All high-energy nuclear radiation produces displacement damage. The majority of the radiation damage database has been obtained with reactor neutrons and normalized to 1 Mev Equivalent n/cm^2. Therefore, it is useful to relate displacement damage from other high-energy radiation to neutron damage. It has been conclusively shown that displacement damage is proportional to the energy that goes into atomic displacements, regardless of the type of irradiation. Thus, it is possible to relate the displacement damage for any given type of high-energy radiation to 1 Mev Equivalent neutron damage by a simple proportionality constant.

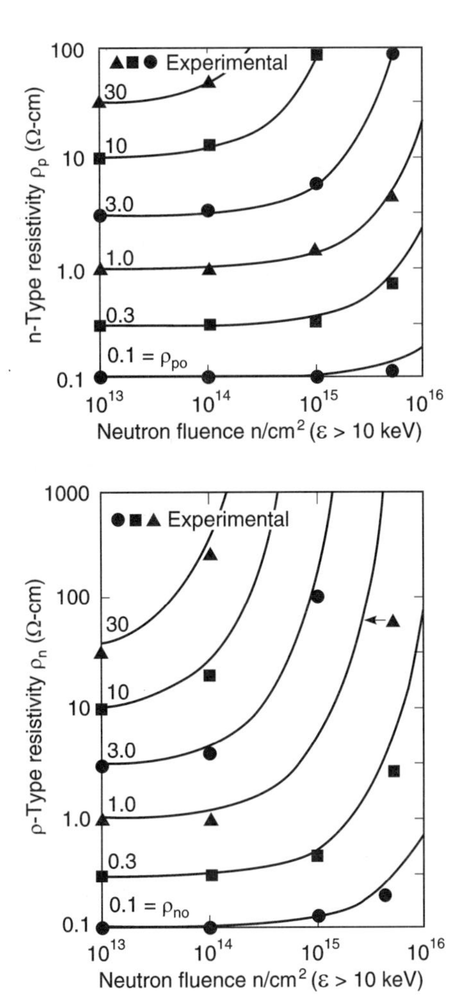

Figure 7.5 Dependence of resistivity on neutron fluence for silicon.

The radiation types of primary interest include protons, electrons, and gammas. Protons and electrons are charged particles that convert most of their energy to ionization in semiconductor materials. Therefore, damage to electron devices is usually dominated by ionizing dose effects. The production of displacements by gamma radiation is a two-stage process: first, a Compton electron is produced, and then the Compton electron can produce displacement damage. Consequently, device damage from gamma radiation is almost always dominated by ionizing dose effects.

7.1.2.5 Device Effects. The dominant effect of displacement damage is the reduction of common emitter current gain in bipolar devices, which is a direct result of the reduction in minority carrier lifetime [7.1]. This effect is proportional to radiation fluence and is given by the Messenger-Spratt equation,

$$\Delta 1/\beta = \phi/2\pi f_t K \quad (7.5)$$

where β is the common emitter current gain, ϕ is neutron fluence, f_t is the common emitter gain-bandwidth product, and K is a constant (approximately 1.6×10^6 for silicon). For bipolar ICs, this shows up as a reduction in parameters such as fanout in digital circuits and open loop gain in op-amps. Second-order effects include increases in resistance, leakage current, and breakdown voltage. Figure 7.6 shows typical current gain data as a function of bias current and neutron fluence.

The dominant effect of displacement damage in MOS devices is an increase in channel resistance and a reduction of transconductance. These effects are a direct consequence of carrier trapping, which reduces carrier density and mobility, and increases device resistivity. Neutron irradiation produces a relatively small ionizing dose due to ionization effects associated with the stopping of recoil silicon atoms. This ionizing dose is sufficient to cause failure in unhardened MOS devices at moderate to high levels of neutron exposure. The ionizing dose depends strongly on the nature of the materials that are shielding the silicon device from the neutrons and the materials in the immediate vicinity of the device.

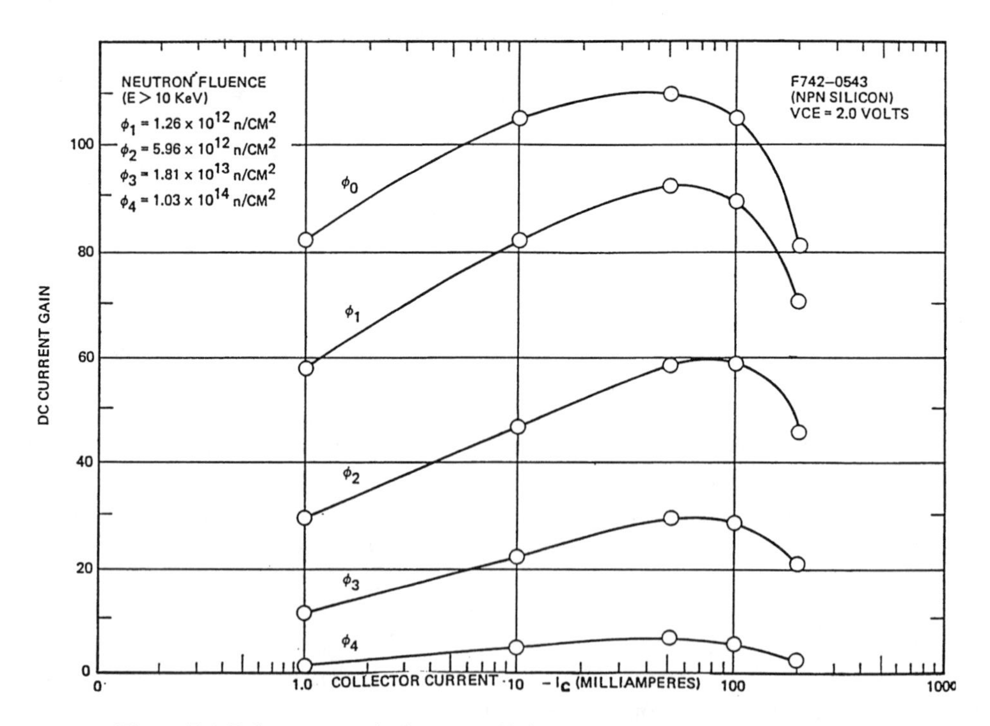

Figure 7.6 DC current gain for a small-signal NPN silicon transistor as a function of bias current for various values of neutron fluence.

7.1.3 Ionizing Dose Damage

Ionizing radiation produces hole-electron pairs in semiconductors and insulators. When these carriers recombine or are swept out of the semiconductor, the effect is transient and no permanent damage can result. However, if some of these carriers are trapped, they can produce significant changes in electrical properties and result in permanent damage. The energy levels of the trapping centers vary from shallow to deep with respect to conduction and valence bands. This leads to a wide variation in annealing properties.

The trapping states at the silicon–silicon dioxide interface exhibit a complex time-dependent buildup. The interplay between time-dependent buildup and annealing processes makes ionizing dose damage in semiconductor devices a complex phenomenon [7.8]. There is no such thing as a unique "total dose" response for semiconductor devices. Ionizing dose quasi-permanent damage is a function of the dose-rate time history, the total dose, and the time after irradiation at which the device is measured. Other important variables include the bias voltage applied and the temperature. The existing ionizing dose database must be used very cautiously. It can only be legitimately extrapolated to environments that closely approximate the dose rate at which the data were taken, as well as the total dose. Often, the dose-rate information is not available; most of the available data are from Co^{60} irradiations done at approximately 300 Rads(Si)/s. The accumulation rate of ionizing dose in space is roughly four orders of magnitude lower, making it virtually impossible to extrapolate the existing data and necessary to obtain additional test data applicable to space dose rates.

A similar problem exists for extrapolation to weapon environments where the prompt dose is delivered in microseconds and the accumulation rate is six to twelve orders of magnitude greater than typical Co^{60} test data. This problem is further exacerbated since some semiconductor devices damage more at Co^{60} rates than at lower dose rates, whereas others damage less. A similar statement holds for higher dose rates compared to Co^{60} rates. A considerable research effort is currently ongoing to improve this situation, with support from the major using agencies, including DOD, DOE, and NASA. At this time, a revision of Mil. Std. 883, Method 1019.4 is planned, which will contain an added anneal step to ensure that the Co^{60} test will yield data that are always more conservative than exposure to either much lower or much higher dose rates.

7.1.3.1 Trapping in Bulk SiO_2. Most silicon dioxide layers encountered in modern silicon semiconductor devices trap substantially all of the photo generated holes, whereas, nearly all the photogenerated electrons are swept out by the field in the oxide. The net result is an accumulation of trapped holes. In N-channel transistors, this accumulation occurs close to the silicon–silicon dioxide interface where it causes a substantial negative shift in threshold voltage and an increase in leakage current. In p-channel devices, the accumulation of trapped holes occurs close to the gate electrode–silicon dioxide interface where it produces a much smaller negative shift in threshold voltage.

Hole trapping in field oxides produces large voltage shifts that cause leakage current increases. The substrate oxide in silicon-on-insulator (SOI) material is believed to trap both holes and electrons. The majority of hole traps are shallow and are characterized by a distribution of energy levels. As a result, substantial annealing occurs even at room temperature and can be accelerated at elevated temperatures. The accumulation of holes is proportional to the ionizing dose, and the annealing rate is proportional to the density of trapped holes. Therefore, the density of holes at any time after the start of irradiation can be obtained as the linear convolution of the dose rate and the unit impulse response. The unit impulse for threshold voltage shift, which is proportional to trapped hole density, is

$$-\gamma_o \Delta V_{to} = -C_1 \ln\ t/t_a + C_2 \tag{7.6}$$

where γ_o is the impulse dose, ΔV_{to} is the threshold voltage shift, t is time in minutes, t_a is one minute, and C_1 and C_2 are constants that depend on the manufacturing process [7.1].

7.1.3.2 Trapping at the SiO_2–Si Interface. Many modern MOS devices also exhibit a buildup of electrons trapped in interface states. This buildup is a complex two-stage phenomenon that leads to a time-dependent buildup, as well as time-dependent annealing behavior. The electron traps are significantly deeper than the hole traps, so that annealing effects at ordinary device application temperatures are dominated by annealing of trapped holes. The trapped holes and electrons tend to cancel each other in terms of their effect on threshold voltage shift.

Interface states tend to be amphoteric, and p-channel devices usually exhibit a buildup of trapped holes. These traps add to the effect of the holes trapped in the silicon dioxide.

Interface state charge, either holes or electrons, is effective in reducing channel mobility, transconductance, switching speed, and increasing channel resistance [7.8].

7.1.3.3 Device Effects. The degrading effects of ionizing dose on semiconductor devices are extremely complex and poorly understood. The radiation effects community continues to utilize a "total dose" terminology that is almost completely incorrect for describing ionizing dose damage. The majority of ionizing dose damage measurements have been made in Co^{60} at approximately 300 Rads(Si)/s. These data can be used to make relative hardness comparisons between devices where the damage is caused primarily by hole storage in the oxide. Linear convolution can usually be relied on for extrapolation to much higher and much lower dose rates for devices where hole storage is the dominant failure mechanism. For devices where hole storage in the oxide and interface charge storage make significant contributions, both effects must be measured and the Co^{60} data cannot be used either to make relative hardness assessments or to extrapolate to other dose rates.

Hole storage and interface charge effects can be measured using special techniques. The damaging effects of hole storage on devices can be determined by a technique that rapidly irradiates the device and then measures the affected parameters before the interface charge buildup occurs. The damage caused by interface states can be determined by irradiating the device and then storing the device at elevated temperatures. The storage temperature and time are selected to anneal substantially all of the oxide charge and allow the buildup of interface states to go to completion. This selection is tricky because too long a time or too high a temperature can also anneal part of the interface states. A storage of 24 hours at 125°C has been used successfully. The best high temperature storage procedure is probably different for different manufacturing processes.

Ionizing radiation effects are poorly understood. There is still no combination of measurements (electrical, chemical, or physical) which can be made on a semiconductor device and used to predict its response in an ionizing radiation environment. This is in marked contrast to neutron, SEU, and dose-rate effects, where our understanding allows reasonable predictions based on device measurements. The microscopic nature of the hole traps and the interface traps is unknown, although a number of models have been proposed in the literature. A quantitative relationship between the density of traps and the manufacturing process is needed, and this must be supplemented by a quantitative relationship between the ionizing dose, density of traps, and trapped charge. The totality of data available suggests that "pure silicon oxide" contains a negligible trap density and that the traps actually observed are crystalline defects, which are strongly affected by the high-temperature processing steps, and chemical defects, which can be inadvertently introduced at many of the process steps. Today, radiation hard processes combine reduced high-temperature processing steps with ultra-clean manufacturing processes to reduce trap densities at the interface and in the oxide. Some success has also been reported by introducing dopants to neutralize interface states.

The effects of ionizing radiation on devices must be obtained by lot sample testing units from a production process and the use of statistical techniques to extrapolate the behavior to be expected from the rest of the lot.

Bipolar devices are relatively unaffected by ionizing radiation. At high levels, however, the current gain is reduced because charge trapping increases the surface recombination velocity at the emitter periphery. Leakage current increases are also experienced in some bipolar structures. Finally, the introduction of oxides as isolating regions in modern bipolar ICs introduces leakage problems; SOI devices are a good example.

MOS devices suffer increased leakage, negative threshold voltage shifts, decreased transconductance, reduced speed, and increased channel resistance. These are the most important failure mechanisms, but it should be remembered that all of the specified electrical parameters will show changes. Figure 7.7 shows threshold voltage changes as a function of ionizing dose for various bias conditions [7.8].

Two insidious results of the complex damage phenomena will now be highlighted. First, the dose-rate dependence makes the total dose failure threshold a strong function of the dose-rate used to test the device. Second, some devices show a failure window. That is, they do not fail until a critical dose level is reached; then they remain in a failed state until a second higher critical dose level is reached; then they again become operative; and finally, they fail again at a third critical dose level. Ionizing dose tests must consider both of these problems to ensure that the test

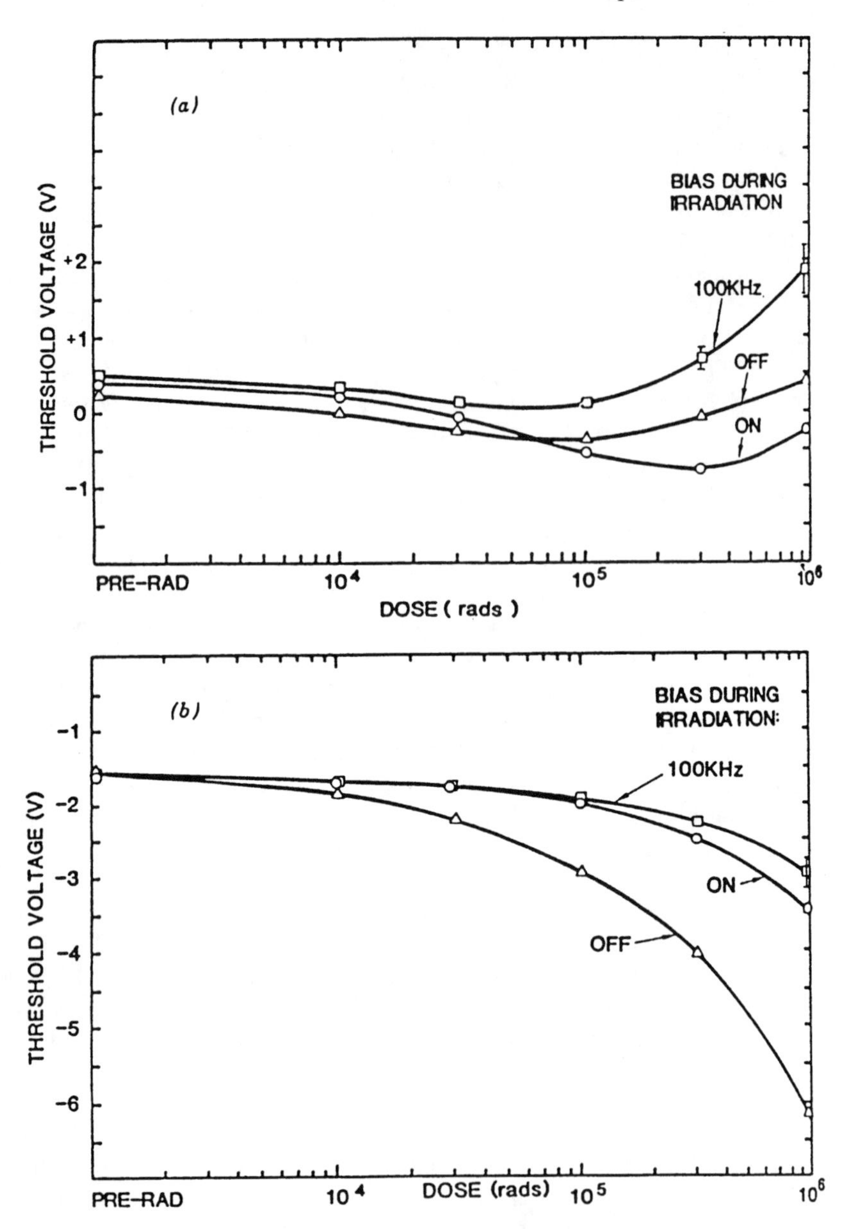

Figure 7.7 Threshold voltage shifts as a function of ionizing dose for transistors biased on and off and switched at a 100 kHz rate.

results are truly applicable to the specified scenario. Figure 7.8 shows ionizing dose failure threshold as a function of exposure rate for a typical unhardened silicon IC [7.8].

Unhardened MOS devices have failure thresholds that range from 500 Rads(Si) to about 50,000 Rads(Si), with PMOS devices about 10 times as hard as comparable

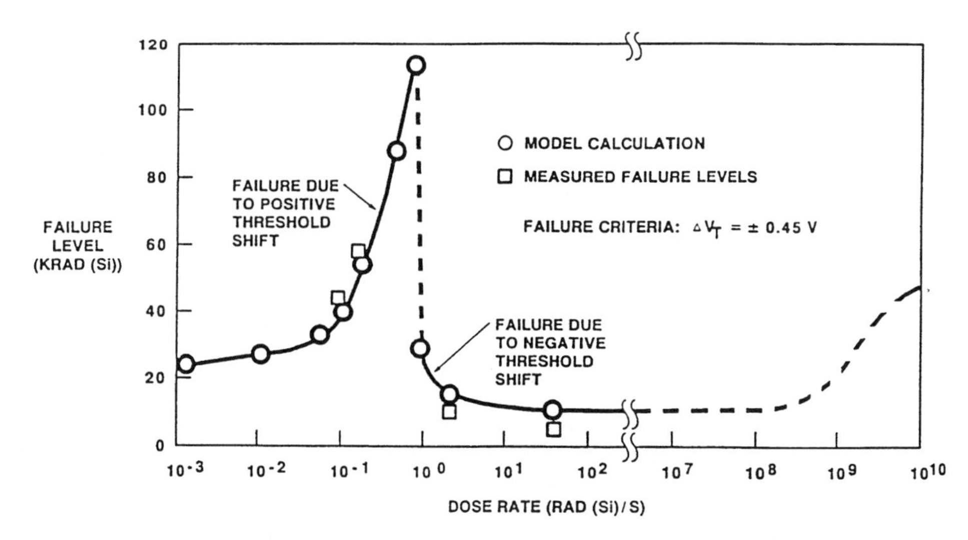

Figure 7.8 Ionizing dose required to produce failure as a function of the dose rate at which the radiation is delivered.

NMOS devices. Most unhardened MOS devices can be used in tactical and avionic applications, although they often require hardness assurance tests. They can also be used in moderate space environments, especially when sufficient shielding is available and hardness assurance tests are incorporated.

Hardened MOS devices are available at ionizing dose levels up to about 1×10^6 Rads(Si). Research programs have yielded operable devices up to 100×10^6 Rads(Si).

7.1.4 Dose-Rate Effects

Dose-rate effects describe the transients excited in devices by bursts of ionizing radiation from nuclear explosions or from simulators such as LINACs and Flash X rays. Photocarriers (hole-electron pairs) are generated in semiconductors and insulators by the ionizing radiation. These carriers are immediately accelerated by the built-in fields and the applied fields resulting in photocurrents. Usually, the magnitude of these photocurrents is well within safe operating limits for semiconductor devices, and no permanent damage results. Frequently, however, these currents exceed the levels required to switch logic devices, and logic errors result in digital circuits. At very high dose-rate levels, for devices biased at high-voltage levels, burnout can occur. This is usually not a problem at tactical and avionic levels but can be a very important problem for space applications with a nuclear weapon specification. At dose-rate levels above 1×10^8 Rads(Si)/s, ICs with parasitic NPNP or PNPN parasitic paths can be caused to latch up. Latchup can be catastrophic if the product of the holding current and the applied voltage exceeds the device power-handling capability. In any case, the device, when in the latched condition, will not respond to input signals.

7.1.4.1 Generation Rates. The generation energies for photocarriers in silicon, silicon dioxide, and silicon nitride are 3.6 eV, 18 eV, and 10.8 eV, respectively. The generation constants are silicon—4.2×10^{13}, silicon dioxide—8.4×10^{12}, and silicon nitride—1.4×10^{13} photocarriers per cm^3 per Rad. When photocarriers are created in the vicinity of a P–N junction, a photocurrent is generated, which behaves in the same manner as junction leakage current. The equilibrium photocurrents are proportional to the dose rate, the generation constant, and the effective volume [7.1]. The effective volume is the junction area times the sum of the junction width and the effective diffusion lengths on both sides of the junction. Thus,

$$\mathrm{Ipp} = \mathrm{qg}\gamma\mathrm{A}(\mathrm{w} + \mathrm{Lp} + \mathrm{Ln}) \tag{7.7}$$

where q is electron charge, g is the generation constant, γ is the dose-rate, w is the junction width, Lp is the hole diffusion length on the N side of the junction, and Ln is the electron diffusion length on the P side of the junction. For short pulses of radiation (pulse width, t_p, less than the minority carrier lifetime, τ), the photocurrent is given by

$$\mathrm{Ip} = \mathrm{Ipp}(t_p/\tau)^{1/2} \tag{7.8}$$

where Ip is the prompt photocurrent [7.1].

7.1.4.2 IEMP. Internal electromagnetic pulse (IEMP) is created by radiation pulses when they penetrate the walls of an electronic system and the package surrounding an electronic device [7.1]. The EMP is created by Compton electrons and photoelectrons accelerated across cavities in the electronics. The magnitude of the effect depends on the energy, magnitude, and type of incident radiation flux, the thickness and composition of the walls and packaging materials traversed, and the dimensions of the cavity. A thorough discussion of these effects is beyond the scope of this work and can be found in referenced literature. IEMP effects are usually insignificant in tactical and avionic applications (although they cannot be completely neglected), but they are extremely important in space applications that have a nuclear weapons survivability requirement. IEMP effects can be minimized by emphasizing good electromagnetic design practices such as minimizing ground loops, keeping all components close to ground planes, and using dense layout designs to minimize volume and void regions.

7.1.4.3 Burnout. Radiation pulses can cause burnout in electronic devices when they are in a powered-up or biased condition. If the pulse energy is high enough, burnout can occur in unpowered electronics. However, the required levels are above those normally encountered in nearly all applications. Radiation pulses of interest are so short that the energy has a very limited time to be diffused by thermal conduction. In the adiabatic limit, all of the energy goes into a temperature rise in the affected volume. Failure usually occurs in metallization runs, vias, or p–n junctions. The energy dissipated during the pulse can be estimated as the product of the resistance, pulse width, and photocurrent squared for resistive elements such as metallization stripes and vias. The resulting temperature rise is the product of the energy dissipated and the mass involved, divided by the specific heat. For p–n junctions, the energy dissipated is the product of the bias voltage, pulse width, and junction photocurrent. If the bias voltage is not "stiff," it may sag or collapse during the pulse, and under these circumstances, the dissipated energy is reduced. The temperature rise is again approximated as the product of the energy dissipated and the mass of the junction, divided by the specific heat. Device failure threshold can be estimated by comparing the temperature rise to maximum device temperature specifications, the intrinsic temperature for the semiconductor resistivities involved, or the melting temperature for the regions involved.

Many modern semiconductor devices exhibit a photoresponse that follows the prompt pulse, whereas others produce photocurrents that persist long after the pulse is over. The extended response comes from the lifetime of photocarriers in extended bulk silicon or substrate regions. Devices that show this response, such as monolithic CMOS ICs, are much more susceptible to burnout than are devices such as CMOS-SOI or CMOS-SOS ICs. The pulse widths encountered are often sufficiently long, especially considering the extended response, so that thermal conduction is possible in one dimension. In this case, a solution to the thermal continuity equation results in a temperature rise proportional to the product of the power per unit area (P/A) and the square root of the pulse width, t_p [7.1]. This result has been applied to semiconductor devices, and measured values of the Wunsch–Bell constant are available in several industry data bases. Thus,

$$\Delta T_f = K t_p^{1/2} P_f / A \tag{7.9}$$

where ΔT_f is the temperature rise at failure, K is the Wunsch–Bell constant, P_f is the power at failure, and A is the junction area. Values of ΔT which result in device failure range from about 100°C to 1400°C. The values of K tabulated in databases are usually worst case measurements and can be used to estimate survivable levels of P/A.

7.1.4.4 Latchup. Some modern IC families are subject to latchup. Latchup is caused when a radiation pulse or an electrical transient turns on a PNPN or an NPNP parasitic path within the monolithic structure [7.1, 7.8]. Bulk CMOS ICs are particularly susceptible, whereas CMOS-SOS, CMOS-SOI, PMOS and NMOS ICs cannot latchup. Latchup can be suppressed on bulk CMOS ICs by utilizing lightly doped epitaxial silicon on heavily doped starting wafers. This approach is coupled with frequent body and substrate ties to reduce parasitic resistance enough to eliminate latchup. Other logic families, such as low power Schottky TTL ICs, have a low probability of latchup. Latchup can be catastrophic if the power dissipated in the latch up path is sufficient to cause failure. Otherwise, the applied bias can be removed, thus breaking the latchup path. Subsequently, the bias can be reapplied and the device will function normally. While a device is latched, it will not respond to input signals to the affected regions.

The electrical requirement for latchup is that the product of the common emitter current gains in the interconnected parasitic PNP and NPN transistors be greater than unity. Current gain increases with bias current, so that any transient, electrical or radiation, that increases the current through the parasitic latchup path may cause latchup. Current gain also increases with temperature, so that it is possible for a device which will not latch at room temperature to exhibit latchup at elevated temperatures. The internal biases on the latchup path are affected by series and parallel resistances that can shift as a function of radiation and temperature. Thus, devices may exhibit latchup windows as the dose rate increases. These devices show no latchup up to a threshold value; then they exhibit latchup up to a higher value of dose-rate, and they do not latch up at even higher dose rates. These complexities make it virtually impossible to design a cost-effective, but highly reliable, hardness assurance screen.

The latchup problem can be most effectively resolved by using a combination of the following proven techniques: (1) use only ICs from families that cannot latch up, (2) use a power-sequencing scheme to immediately remove bias after a prompt radiation pulse and restore the bias no sooner than 100 microseconds later, and (3) place a resistor or inductor in series with the IC to ensure that the current is restricted to a value significantly less than the measured holding current of the latchup path.

7.1.4.5 Upset. Upset usually refers to a logic change produced in a digital device. Frequently, the term is extended to include an output transient that exceeds the device noise margin. There is no general definition of upset for analog circuits, although, for unique applications, an output transient that is large enough to cause functional degradation is often referred to as upset. For the great majority of cases, the transient produced in analog circuits by a short burst of radiation does not cause significant functional degradation and acts much like a large noise pulse. Industry data banks contain upset thresholds for nearly all IC families and a large amount of data on individual devices. The usual hardening techniques are to design the circuitry so that it is fail safe when upset and to go through an appropriate reset procedure to mitigate the effects of the data loss caused by the upset. For serious problems, such as autonomous computer systems in ICBMs, a circumvention subsystem is used. A radiation detector disables the computer logic and clamps the I/Os, and a software program reinitializes the computer to the last valid data set stored in hard memory. The normal computer program then takes over. Upset thresholds vary from approximately 10^7 Rads(Si)/s in unhardened logic families to 10^{11} Rads(Si)/s in hardened CMOS-SOS and CMOS-SOI families [7.1].

7.1.5 Single-Event Phenomena

The trend to higher and higher densities in modern ICs has reached the point at which the ionization produced by the passage of a single cosmic ray, or α particle, through the active volume of a logic device can deposit a charge significantly greater than the charge that maintains the logic state. Under these conditions, the device can be upset. In some cases, the resulting photocurrents can trigger latchup or second breakdown, leading to catastrophic failure [7.1,7.8].

7.1.5.1 Single-Event Latchup/Burnout. Latchup has been observed in some monolithic CMOS ICs after the passage of a high-energy ion through the device. This would obviously be catastrophic in the space environment. Therefore, susceptible devices must be eliminated in the design phase, or the system level fixes discussed in Section 7.1.4.4 must be employed [7.8]. Burnout has been observed in MNOS and SNOS memory devices from a high-energy ion when in a relatively high-gate bias condition typical of the write cycle. The burnout path has been identified through the gate dielectric by subsequent failure analysis. Burnout has also been observed in n-channel power MOS transistors. The high-energy ion triggers a second breakdown, or possibly, a parasitic latchup path leading to burnout. Since n-channel devices are significantly more susceptible than p-channel devices, the second breakdown explanation is probably correct. Power MOS transistors such as HEXFETs are actually composed of thousands of transistors in parallel. The ion track from a cosmic ray particle is capable of triggering the burnout in a single transistor, which then spreads to produce a large burnout area involving many of the neighboring transistors. Hardened n-channel power transistors have been designed which virtually eliminate this failure mode.

7.1.5.2 Single-Event Upset. Single-event upset is of concern in microprocessors and memories. Upsets in a microprocessor invalidate the calculation currently underway, and the only effective mitigating techniques are to use redundancy with recalculation when indicated, or triple redundancy and voting. Upsets in a memory can be detected and corrected using error detection and correction (EDAC) software, with only a small penalty in overhead. The basic technique requires that the bits in a word be separated so that the probability of the same cosmic ray causing multiple bit errors in the same word is vanishingly small. This is readily accomplished, for example, by using sixteen $256\,\text{K} \times 1$ memory chips in parallel to form a $256\,\text{K} \times 16$ memory. Each 16-bit word then has one bit in each chip. The simplest EDAC technique detects two bit errors and corrects one. Because the SEU rate in space is low for even the most sensitive memory chips, presently this technique is completely adequate.

CMOS-SOS and CMOS-SOI memories show much lower SEU rates than monolithic CMOS ICs. An effective hardening technique has been developed which consists of adding cross-coupled resistors to SRAM cells. These resistors, in combination with the cell capacitance, act as a low-pass filter and prevent the high-frequency transient caused by the cosmic ray particle from upsetting the cell. Cross-coupled capacitors can also be used to lower SEU error rates. The Miller effect multiplies the effective capacitance by the voltage gain of the cell, and the enhanced capacitance filters the cosmic ray transient, preventing cell upset. Transistor feature size must be below 10 microns before SEU rates become significant. Therefore, many older generation IC memories are intrinsically hard with respect to SEU. This is an attractive hardening possibility if the amount of data that must be protected is relatively small. The transient photocurrents generated by a cosmic ray track are essentially over in less than a nanosecond for small particles, such as α particles, and can persist for several nanoseconds for larger particles, such as iron ions.

7.2. FLOATING GATE RADIATION EFFECTS

Several significant hardening efforts have been directed toward EPROMs or EEPROMs, and hardened devices may soon become commercially available [7.10, 7.11]. EPROMs are normally erased by exposure to ultraviolet light, whereas EEPROMs can be erased electrically. However, both device types use a similar gate structure so that their radiation response is similar. The failure threshold in an ionizing dose environment is approximately 5 KRads(Si) and may be extended up to approximately 100 KRads(Si) by use of special circuit design and device-hardening approaches. The ionizing dose environment is the most critical problem for floating gate devices, and so, it will not be necessary to evaluate neutron, dose rate, and SEU response in detail. Unhardened floating gate devices are suitable for use in most tactical and avionic applications, as long as a hardness assurance program is in place to ensure adequate survivability. It should also be possible to use them in low to moderate ionizing dose environments in space by using appropriate shielding in combination with a hardness assurance program. These devices are unsuitable for use in severe radiation environments.

The floating gate contains a polysilicon layer insulated from both the gate and the silicon channel by silicon dioxide layers. A "one" is stored in the memory cell by increasing the electron charge in the polysilicon layer, thereby establishing a relatively high positive threshold voltage for the n-channel transistor. A "zero" is stored by removing the electrons or adding holes, thereby establishing a negative threshold voltage for the transistor. Ionizing radiation tends to reduce the electron charge on the floating gate when a one is stored and to increase the electron charge when a zero is stored. This drives the transistor to an intermediate value of threshold voltage and destroys the stored information.

7.2.1 Basic Effects

MOS transistor threshold voltage is a function of the charge on the floating gate and is given by

$$V_t = V_{to} + \sigma d/\varepsilon \qquad (7.10)$$

where V_t is threshold voltage, V_{to} is the initial threshold of the transistor, σ is the charge density on the gate, d is the oxide thickness between the control gate and the floating gate, and ϵ is the oxide dielectric constant. Three processes contribute to the net loss in electronic charge produced by ionizing radiation on the floating gate; (1) hole injection from the oxides on both sides of the floating gate, (2) hole trapping in the oxides, and (3) electron emission from the floating gate. For the relatively low values of field normally encountered in floating gate memory devices, hole injection and hole trapping are proportional to the field, E.

$$f(E) = 1 - \exp(-kE) \approx kE \text{ for } E \leq 0.5 \text{ MV/cm} \qquad (7.11)$$

At high levels of ionizing radiation, the three processes reach an equilibrium characterized by an equilibrium threshold voltage, V_{te}. The continuity equation can be solved for the threshold voltage as a function of ionizing radiation for these conditions [7.12]. The combined effects of endurance and radiation have also been analyzed and measured [7.13].

7.2.2 Response to Ionizing Radiation

Ionizing radiation reduces the threshold voltage of the transistor in the "one" state toward the equilibrium state value and increases the value of the threshold voltage in the "zero" state toward the equilibrium state value. Radiation therefore closes the window between the two states of the memory. Thus,

$$V_t(D) = V_{te} + (V_{to} - V_{te})\exp(-AD) \tag{7.12}$$

where D is the ionizing dose in Rads(SiO_2) and A is a constant of proportionality that depends on the specific manufacturing process. This relatively simple result gives excellent agreement between measured and predicted results using only the fitting parameter A, as is shown in Fig. 7.9. The basic process at work is simply the change in charge on the floating gate which can be accurately modeled as a one-dimensional capacitor. The charge is proportional to radiation dose, and its subsequent motion is closely approximated as being proportional to the field. However, the process-dependent constant A reflects the type, quantity, and energy levels of the traps in the oxide regions. Very little is known about these traps, and it is their mitigation that is essential to hardening floating gate memory devices. It is noted in passing that the ultraviolet radiation used to erase EPROMs is really a massive dose of ionizing radiation that creates the equilibrium threshold voltage, V_{te}, without damaging the device in any permanent fashion.

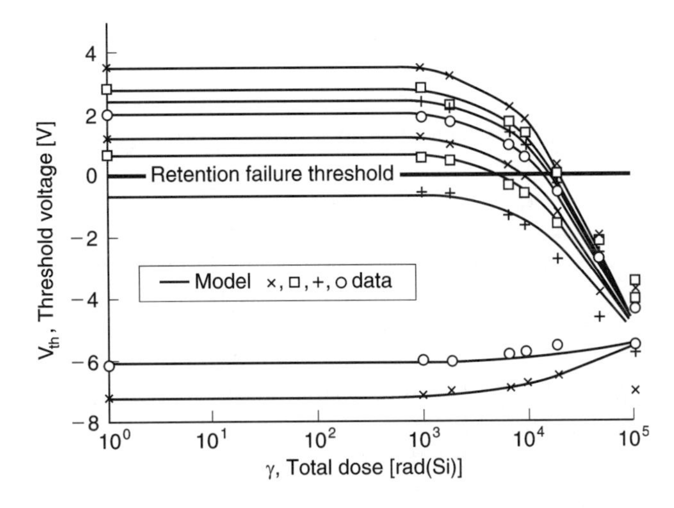

Figure 7.9 Agreement between predicted and measured shifts in threshold voltage as a function of ionizing dose for a floating gate memory device.

7.2.3 Hardening of Floating Gate Devices

Normally, reading the floating gate memory devices is accomplished by comparison with a reference level. Above this level, read "one"; below this level, read "zero." Obviously, a differential sensing scheme could be used to read the memories, and the radiation failure threshold could be significantly increased. However, this would substantially complicate the memory design and the read circuitry. The memories could be set to a significantly higher "one" threshold by increasing the write cycle drive, thus widening the initial window. This could produce a substantial improvement in radiation failure threshold. Perhaps hardened oxide technology can be applied to the oxides on both sides of the floating gate, but this would require major process modifications. For the foreseeable future, it may be possible to increase the radiation failure threshold by a factor of ten.

7.2.4 Dose-Rate Response

The radiation-induced changes in charge on the floating gate occur extremely fast and will, therefore, follow the radiation pulses normally encountered. The change in threshold voltage will therefore depend on the total ionizing dose in the pulse and will be independent of the dose rate. The dose per pulse is approximately equal to the product of the maximum dose rate and the pulse width (full width-half maximum). For nearly all applications, floating gate memory devices will be limited by the ionizing dose failure threshold and not by dose-rate effects. In most applications, the logic device composing the rest of the digital system will upset at relatively low dose-rates. Therefore, from a system standpoint, dose-rate effects will be very important, even though they will not directly erase the floating gate memories.

7.2.5 Other Radiation Effects

The present relatively low ionizing dose failure threshold for floating gate devices makes it unnecessary to evaluate the neutron failure threshold in detail. For any practical exposure scenario, the ionizing dose failure threshold would be reached while the neutron fluence was below the level that would produce any measurable changes. A semiquantitative neutron failure threshold can be estimated as approximately 10^{14} n/cm^2, at which level the accompanying ionizing dose would be approaching the failure threshold. At 10^{15} n/cm^2, the transconductance and channel resistance would also show significant degradation. Floating gate memories are not directly susceptible to SEU since the ionizing charge released by even the highest energy cosmic rays in the vicinity of the floating gate will not cause a significant change in threshold voltage. Again, however, it must be noted that SEU in the controlling logic could result in an erroneous write.

7.2.6 System Considerations

Floating gate memories have a long-term retention problem in that the charge tends to leak off the floating gate under normal operating conditions. This is not a problem for most applications. Radiation also causes charge to leak off the gate. Both of these problems can be mitigated by using a refresh cycle similar to that used for dynamic memories. In fact, it is probably necessary for space applications in which the ionizing radiation accumulates over time. Ionizing radiation effects are known to be strong functions of temperature. The radiation analysis and characterization of floating gate devices have primarily been considered at room temperature. This is probably adequate for many applications, but it should be noted that elevated temperatures will significantly enhance both normal and radiation-induced leakage.

7.3. SNOS RADIATION EFFECTS

Silicon–nitride–oxide–silicon (SNOS) memories have been the traditional devices used where a programmable radiation-hardened nonvolatile memory is required. The silicon–nitride–silicon dioxide gate dielectric sandwich has been effectively used to harden some MOS devices. Adjustment of the oxide and nitride region thicknesses makes the device an effective EPROM. The device is similar to a floating gate device. However, the charge that controls the memory state is a sheet at the nitride–oxide interface. SNOS memories have achieved ionizing dose failure thresholds in excess of 1 MRad(Si).

7.3.1 Basic Effects

Again, the effect of ionizing radiation is to create hole-electron pairs in the gate oxide and nitride. This charge drifts to the oxide–nitride interface where it reduces the threshold voltage if the gate is initially in the "one" state and increases the threshold voltage if the gate is initially in the "zero" state, thus reducing the window. At very large values of ionizing radiation, the threshold voltage reaches an equilibrium value and the stored information is lost. Charge storage must occur in a sheet at the interface between two insulators of different dielectric constants to satisfy the fundamental laws of electrostatics. The problem of dominant radiation effects comes from ionizing dose, and so a detailed discussion of dose-rate, neutron, and SEU effects will not be necessary. Write voltages of 10 to 20 V are used in SNOS memories. Memory gate dielectric burnout from some high-energy ions (typically ions with LETs of Argon or greater) will occur for SNOS memories when a write voltage is applied to the gate. This limits the safe memory cycling specification for SNOS memories in space. For typical applications with less than 10^4 cycles, failure probabilities are very low, with expected error rates less than 10^{-13} errors per bit day.

7.3.2 Response to Ionizing Radiation

The continuity equation can be solved in the gate region (see Fig. 7.10), as a function of ionizing dose and leads to a relationship similar to Eq. (7.12). However, for SNOS devices, a composite solution that combines the effects of normal charge leakage and the effects of radiation-induced leakage has been obtained [7.14] and is given as

$$V_t(D,t) = (V_{to} - V_v)(t/t_o)^a \exp(-AD) - Bt^a \exp(-AD) \int_{t_o}^{t} D(t)t^a \exp(AD)dt + V_v \tag{7.13}$$

where a is a tunneling constant, V_v is the threshold voltage before the device has been written, erased, or irradiated, V_{to} is the initial threshold voltage, t_o is the time immediately after a write pulse when V_{to} is measured (typically several seconds), A is a process-dependent constant that reflects the type and density of the trapping states in the gate, and B is a constant reflecting the normal charge leakage from the gate. In the absence of radiation, Eq. (7.13) simplifies to

$$a = \ln\{[V(t) - V_v]/[V_{to} - V_v]\}/\ln(t/t_o) \tag{7.14}$$

from which the value of a can be determined by measuring V(t) at two different times, for example, 5 and 50 seconds. A simple expression for a device irradiated in the virgin state is

$$V_t(D) = B\{\exp(-AD) - 1\} + V_v \tag{7.15}$$

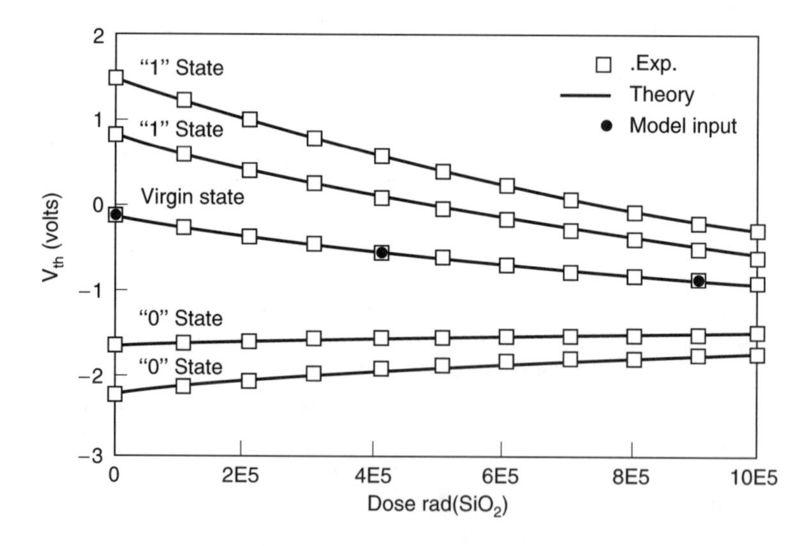

Figure 7.10 Agreement between modeled and experimental changes in threshold voltage for an SNOS memory device.

Irradiating a device at two radiation levels, such as 250 KRads and 1 MRad, allows the constants A and B to be obtained. At this point, Eq. (7.13) can be used to accurately model the time and radiation performance of a SNOS memory device. Unfortunately, the integral is not available in closed form, and a numerical integration must be performed. Again, excellent agreement between this model and the experimental results is obtained, as shown in Fig. 7.10 [7.15].

7.3.3 Response to Dose Rate

The SNOS memory device stores such a large amount of charge at the nitride–oxide interface that the ionization in most prompt pulses of interest cannot degrade the memory. However, the dose-rate upset threshold of a SNOS memory will be determined by the associated read and write circuitry.

When SNOS devices are properly designed into computer systems, they provide EEPROMs with excellent radiation hardness characteristics. The design must provide circumvention to protect the memory from invalid write or erase commands during a prompt radiation pulse, and software to identify and ignore misreads during a radiation pulse. These two requirements are often simultaneously met by the circumvention subsystem.

7.3.4 Response to Neutron Fluence

SNOS memories have a high degree of inherent hardness to ionizing radiation, and so their response to neutrons usually depends only on displacement damage effects. Since they are majority carrier devices, the only effect of lifetime degradation is a small increase in leakage currents. The displacement damage effects of concern are an increase in resistivity and a decrease in channel mobility. These effects were discussed in Section 7.1.2, and the data provided can be used to predict the degradation in transconductance and switching speed, as well as the increase in channel resistance. The failure threshold is usually between 10^{14} and $10^{15}\,n/cm^2$. The response of the transconductance, q_m, can be obtained as

$$g_m = g_{mo} \exp -(\phi/K) \approx g_{mo}(1 - K\phi) \tag{7.16}$$

where K is a function of the SNOS channel resistivity [7.1]. The increase in channel resistance is obtained in a similar fashion, as is the reduction in cutoff frequency, $f_c = g_m/2\pi C_g$,

$$f_c \approx f_{co}(1 - K'\phi) \tag{7.17}$$

where K' is the sum of K and the mobility degradation constant for the appropriate channel resistivity. Although the radiation effects can be approximately calculated from Eqs. (7.16) and (7.17), K and K' are often obtained as fitted constants from a simple radiation experiment [7.1].

7.3.5 Hardening of SNOS Memories

The hardening of SNOS memories has proceeded by optimizing the manufacturing process to improve the inherent hardness of the nitride–oxide gate dielectric. Since these devices are used in hardened processors and computers, a large effort has been required to harden the associated microcircuits to a level consistent with the SNOS devices. A circumvention procedure is usually required. A detector is used to sense the dose rate at a level substantially below that which would cause an upset. This activates circuits that clamp the memory to prevent miswrites or erasures and activates a software procedure that identifies and ignores any misreads that might have occurred during the pulse. It then restores normal function from a valid start point. This provides continuous operation with only a very short interrupt. For most tactical applications, circumvention is probably not necessary.

7.4. OTHER NONVOLATILE RAMs

EEPROM requirements in radiation environments have also been satisfied using magnetic memories and hardened RAMs with battery backup. Recently, ferroelectric memories have been proposed for hardened nonvolatile applications.

7.4.1 Magnetic RAMs

Magnetic memories include core memories, which have received extensive application, and plated wire memories which so far have received limited application due to their relatively high cost. Bubble memories have also been used in some applications. All of these magneto-inductive devices are inherently hard for the normally encountered range of radiation environments. However, all of these memories must also be protected from invalid write commands and cannot be read in the presence of dose-rate environments approaching the upset level of the associated circuitry. Another type of memory based on the magneto-resistive effect has recently been proposed and is under development.

7.4.2 Static RAMs with Battery Backup

Perhaps the most formidable competitors for radiation-hardened nonvolatile applications are the new state-of-the-art CMOS-SOS and CMOS-SOI RAMs. These devices are not truly nonvolatile but can be designed with battery backup to handle temporary power supply failures. They have the same density, speed, and power requirements as the associated ICs.

7.4.3 Ferroelectric RAMs

Recently, ferroelectric memories have appeared as candidates for radiation-hard nonvolatile memories. The basic cell promises to be inherently hard for environments of practical interest, but so far the problems of integrating their manufacture with the process for silicon ICs has not been completely solved. However, useful hardened devices are expected in the foreseeable future [7.16].

References

[7.1] G. C. Messenger, and M. S. Ash, *The Effects of Radiation on Electronic Systems*, Van Nostrand Reinhold, 1986.

[7.2] S. Glasstone, and P. Dolan, *The Effects of Nuclear Weapons*, U.S. Departments of Defense and Department of Energy, 1977.

[7.3] A. E. Profio, *Radiation Shielding and Dosimetry*, John Wiley, New York, 1979.

[7.4] N. Rudie, *Principles and Techniques of Radiation Hardening*, vol. 1, *Western Periodicals*, pp. 2–40, 1980.

[7.5] J. Adams, Jr., "The ionizing particle environment near earth," NRL Report presented at AIAA Aerospace Science Meeting—41–82–107, p. 5.

[7.6] Nuclear IEEE Standards, ANSI/IEEE Standard 382-1972, Part 3, Design Basis Accident Environment Simulation for BWR and PWR, vol. 1, October 1978.

[7.7] R. J. Chaffin, *Microwave Semiconductor Devices*, John Wiley, New York, pp. 120–122, 1973.

[7.8] T. P. Ma and P. V. Dressendorfor, *Ionizing Radiation Effects in MOS Devices & Circuits*, John Wiley, New York, 1989.

[7.9] James R. Murray, "Design considerations for a radiation hardened non-volatile memory," *IEEE Trans. Nucl. Sci.*, vol. NS-40, no. 6, pp. 1610–1618, 1993.

[7.10] Dick Wellekens et al., "Single poly cell as the best choice for radiation-hard floating gate EEPROM technology," *IEEE Trans. Nucl. Sci.*, vol. 40, no. 6, pp. 1619–1627, 1993.

[7.11] E. S. Snyder et al., "Radiation response of floating gate EEPROM memory cells," *IEEE Trans. Nuc. Sci.*, vol. NS-36, pp. 2131–2139, 1989.

[7.12] D. Sampson, "Time and total dose response of non-volatile UVPROMS," *IEEE Trans. Nuc. Sci.*, vol. NS-35, pp. 1542–1546, 1988.

[7.13] P. J. McWhorter et al., "Retention characteristics of SNOS nonvolatile devices in a radiation environment," *IEEE Trans. Nucl. Sci.*, vol. NS-34, pp. 1652–1657, 1987.

[7.14] P. J. McWhorter et. al., "Radiation response of SNOS nonvolatile transistors," *IEEE Trans. Nucl. Sci.*, vol. NS-33, pp. 1414–1419, 1986.

[7.15] Blanford et al., "Cosmic ray induced permanent damage in MNOS EAROM's," *IEEE Trans. Nucl. Sci.*, vol. NS-31, pp. 1568–1570, 1984.

[7.16] G. C. Messenger and F. Coppage, "Ferroelectric memories: a possible answer to the hardened nonvolatile memory question," *IEEE Trans. Nuc. Sci.*, vol. NS-35, pp. 1461–1466, 1988.

Chapter 8

David Sweetman

Procurement Considerations

8.0. INTRODUCTION

When procuring a microcircuit, many considerations are important. These may be divided into two categories: administrative and technical. Administrative issues, such as price, delivery, service, and product assurance, are not discussed here; instead, technical issues, such as performance, quality, and reliability, are addressed.

In order to ensure that the manufacturer properly executes technical considerations, the customer should verify the existence and adequacy of the manufacturer's product assurance system. Recognized industry standards for quality systems, such as ISO-9001 or EIA-599, can be used for evaluation. With the reduction of the DOD's microcircuit standardization program and the emphasis on purchasing commercial products, there is no longer an independent body to develop or assess microcircuit standards and performance. The user is therefore cautioned to carefully assess any given microcircuit's data sheet and actual performance for compliance with application requirements. There are neither formal industry standards nor documented best commercial practices that clearly define or assess device specifications or performance. Although ISO-9001 registration will assure the adequacy of a company's quality systems, there is no assurance of the adequacy of either the commercial data sheet or device performance to meet application requirements. "Best commercial practices" are perfectly adequate for many applications; however, application requirements, for example, temperature range, endurance, and data retention, for reprogrammable nonvolatile memories often necessitate that the user clearly specify and review device conformance.

The performance of a microcircuit is evaluated on the basis of parameters specified in a data sheet. Quality is a measure of conformance to specification, typically expressed in parts per million (PPM) nonconforming, reference EIA Standard 554. Quality levels, as maximum PPM lot acceptance guarantees, are usually provided in manufacturers' warranty policies. Reliability is an expectation of quality over time, typically expressed as a failure rate in %/1000 hours or failures in time (FITs), or as a mean time to failures (MTTF). Reliability expectations are usually provided in manufacturers' reliability reports for device or process families.

Reprogrammable nonvolatile semiconductor memories have performance values, data sheets, and quality levels similar to other microcircuits. Reliability expectations are complicated by considerations of endurance and data retention, which have failure mode characteristics that are not applicable to other microcircuits. Thus, reprogrammable nonvolatile memories have additional procurement considerations that affect manufacturing methods, data sheet specifications, quality levels, and reliability.

The following sections discuss the elements necessary for a complete data sheet, critical parameters for evaluating the quality of a device, and a reliability evaluation methodology.

8.1. COMMERCIAL SPECIFICATION PRACTICES

The industry uses three primary documents to specify microcircuits: data sheets, reliability documents, and warranty policies. The following sections describe the desired contents of each of these documents.

8.1.1 Data Sheets

Microcircuit electrical performance is specified by a data sheet. Data sheets may be very simple expressions of "nominal" performance levels or an extensive listing of minimum and maximum performance levels under all allowed conditions. The magnitude of testing by the manufacturer usually correlates with the stringency of the specification. The most thorough and detailed device specifications were the Standardized Microcircuit Drawings (SMDs) provided by the U.S. government. Currently, no industry body generates or validates device specifications, that is, data sheet format, specifications, or criteria.

The device data sheet contains a description of how the device performs its intended function, for example, how to write to and read from a memory; and a list of parameters with performance limits and conditions that define and specify how each function is implemented. Testing validates that the circuit performs as specified. Data sheet limits and values are based on considerations of device performance, system requirements, and test capability. The ability to test parameter performance is critical in understanding various data sheet limits and values.

Reprogrammable nonvolatile memories have more complicated features and functions than most memories, due to the complexities and limitations of the technologies used to erase and program the memory cell. The wide variety of technologies used to implement the various features and functions can be confusing or conflicting. Although some features, for example, pin-out, page-write, and Software Data Protection, have been standardized according to reference JEDEC Standard 21-C, there still may be significant differences in performance (e.g., endurance) between manufacturers for supposedly the same device type. In addition, major functional differences exist between devices depending on the intended application. For example, interfacing the memory to the system may be serial or parallel; the number of power supplies required may be one or more; the duration of time to erase and program may take nanoseconds to minutes; and the minimum erase/program element size may be bits, bytes, pages, sectors, or the entire chip. The user must carefully consider the performance of all features and functions when designing with reprogrammable nonvolatile memories.

Device performance will predictably change as a function of environmental parameters. For example, with MOS devices, read speed will be slower at high temperatures, and power supply current will be higher at low temperatures. Therefore, the environment is specified. Device testing by the manufacturer is performed at worst case conditions to assure the user of device conformance to the data sheet under all allowable conditions. The power supply level directly affects the performance of the device; a lower value is worst case for most parameters because of the poorer conductivity of the MOS transistor at lower voltages. This is most evident for read access times, which are generally slower at low V_{cc} and high temperature.

The temperature range is typically specified as commercial, 0 to 70°C; industrial, –40 to 85°C; or military, –55 to 125°C, reflecting the severity of the expected application. Commercial and industrial temperatures are ambient temperatures; therefore, the junction temperature of an operating device will be higher than the ambient, and the device must be tested at other than the specified limits. The test temperature can be calculated as a function of the specified temperature, the device power dissipation at the temperature limit, and the thermal resistance of the package. Military temperatures are case temperatures; therefore, the junction temperature will be the same as the case test temperature. For example, the test temperature, and the specified case temperature are the same, excluding guardbands. Ambient humidity is a test concern because, at low temperatures, frozen condensation can lock the equipment or cause leakage paths.

Devices are typically specified with $\pm 10\%$ power supplies; however, both extremes need not be tested for all parameters. A worst case power supply condition, based on characterization, can be stipulated for most parameters. Thus, the worst case is the only condition that would be tested in production.

Reprogrammable nonvolatile memory test parameters fall into three categories: AC or timing, DC or parametric, and functional. AC parameters are the maximum or minimum timing conditions needed for the device to function, for example, address setup times, data hold times, as well as the minimum/maximum responses from the device, such as read access times. DC parameters are the static levels on each pin in the various operational modes, including power supply current, input levels, and output leakage. Functional parameters include those timing and static conditions (i.e., AC and DC parameters) necessary for the device to function in a given mode. Specific parameter conditions are usually included in the data sheet and should reflect how the device is tested.

Conventions for measuring parameters should be clearly stated, because device performance can change significantly for apparently minor differences in measuring methods. The most critical reference points are for AC or timing measurements. Timing limits are minimum or maximum timing values for each parameter. Input requirements for the device are specified from the external system point of view, that is, what the system must provide to the device. Responses from the device are specified from the device point of view, that is, what the device provides to the system. Timing edge reference points are required to differentiate between the provided input pulse levels or output levels and their respective (as recognized by the device) valid trip points. Edge reference points must be specified relative to where the device recognizes a valid signal level. Actual test specification edge reference points generally indicate those points that the tester recognizes as the beginning of a transition for timing measurement, not where the device recognizes a valid signal. The difference provides a built-in guardband between test conditions and actual performance requirements. The levels specified for DC or static performance may not be the same levels specified for AC or timing performance.

Integral to the testing of an integrated circuit is the test philosophy utilized to determine the parameters that are tested and in what manner. Included in the philosophy are guardbanding methodologies, interface hardware design rules, test routine algorithms, and characterization requirements.

Guardbanding is the off-setting of a parameter, condition, or attribute acceptance level from the specified value. Guardbanding is done to account for variability in equipment and device performance or to make test programs more efficient and effective. Machine guardbands, implemented in the forcing, measured, or external conditions, are required to account for the accuracy and precision capabilities of testers, interface hardware, and handlers. Device guardbands are implemented where device performance greatly exceeds the parameter limit; thus, an early warning of a change in performance is available. Test program guardbands are implemented to speed up device testing, where worst case conditions can be applied based on predictable device behavior, for example, for pattern sensitivities.

Parameter conformance to specification can be measured in a variety of ways. In variate testing, which is generally used only for characterization, the actual value of the parameter is determined. For DC parameters, this is relatively simple because the measurement is effectively the output of a voltmeter or ammeter. For AC parameters, this can be very complex, depending on the timing signal measured, because a narrow strobe must be continuously repositioned until the desired transition is detected or the reference edge must be continuously repositioned until the desired output is obtained. Various addressing and data patterns can be utilized, which can significantly affect test time, for example, a galpat versus an address increment addressing pattern. Traditional RAM read/write patterns are not usually employed because most reprogrammable nonvolatile memories have much longer write times compared with their read times. Variate testing, whether on an automated tester or a bench setup, is used primarily to validate the design and performance models for initial device release or after design changes.

Attribute testing is the comparison of a measured parameter under given conditions to a specified limit. The tested parameter then either passes or fails—go/no-go. Some parameters are directly compared with the limit, whereas others must be "tested by inference" or "tested by the application of specified signals and conditions." Tested by inference is the validation of the performance of a parameter by the measurement of the correct performance of a correlated parameter or function. Tested by inference also applies when testing a worst case condition; therefore, all other conditions need not be tested. Tested by the application of specified signals and conditions is the applying of input parameters at their specified minimum or maximum and measuring the correct performance of a dependent parameter or function. Parameters that are outputs from the device are measured or compared with standards. Inputs to the device are tested by inference or application of specified signals and conditions.

Data sheets usually reference mechanical specifications for the packages containing the microcircuits. Most packages conform with JEDEC Publication 95. Otherwise, the manufacturer should have a similar type of specification that provides all dimensional and material requirements. Visual and mechanical quality characteristics are usually inspected for compliance to the applicable specification by the manufacturer before shipment of the devices. Dimensions, such as package thickness and lead spacing, may be critical for automatic insertion equipment operation. Composition, such as lead finish, may be critical for solderability. Explicit methodology and criteria for validating mechanical and visual performance are contained in JEDEC Standard 22 or MIL-STD-883 for plastic and hermetic devices, respectively.

Data sheets are typically divided into several sections. The following lists and describes appropriate sections for reprogrammable nonvolatile memories:

Product Description: A prose description of each device function and feature.

Package Pinouts: A drawing of each offered package, with pin assignments.

Block Diagram: A drawing of the overall logical architecture of the device, identifying external control signals and major logic blocks.

Pin Description: A tabular description of each pin assignment, including symbol, name, and function.

Operation Mode Tables: A tabular description of each functional mode of the device, including the status of each externally applied signal and internally generated signal.

Reliability Characteristics: A tabular listing of reliability parameters, including endurance, data retention, electrostatic discharge (ESD) immunity, and latchup that identifies the specification limit, units, and test method.

Absolute Maximum Stress Ratings: A list of those parameters and limits, which if exceeded, could cause permanent damage to the device.

Operating Range: A table of the temperature and voltage operating range limits.

DC Operating Characteristics: A table listing DC parameters, limits, units, and test conditions. Note: "Typicals" should not be listed in this table, but provided in the form of a plot of the parameter as a function of key external variables, that is, voltage, temperature.

AC Characteristics: A table listing AC parameters, limits, units, and test conditions. Often, separate tables will be provided for the read and write functions. Note: "Typicals" should not be listed in this table, but provided in the form of a plot of the parameter as a function of key external variables, that is, voltage, temperature.

AC Conditions of Test: A table and drawing defining the input and output conditions and loading during AC measurements.

Timing Diagrams: A graphical description of each function, showing the level and sequence of all signals.

Flowcharts: A graphical flowchart of the sequence of steps to perform each device function.

Device Ordering Information: A table listing the information necessary to generate the applicable part number for each device option, for example, package, speed, temperature range, and voltage.

8.1.2 Reliability Parameters

Reliability evaluation may be divided into two categories: mechanical and electrical. Reprogrammable nonvolatile semiconductor memories are assembled in packages using similar materials and processes as other microcircuits; thus, the mechanical reliability of the package is the same. Package stresses can be used to evaluate thermal and mechanical stress effects on the reprogrammable nonvolatile memory die. Of particular concern is data retention, which can be adversely affected by thermal or mechanical stresses induced during the package assembly or board assembly processes. Adherence to recommended soldering profiles is very important to avoid potential adverse effects on data retention. Mechanical reliability evaluations typically use JEDEC Standard 22 or MIL-STD-883 for test methods.

Electrical reliability for reprogrammable nonvolatile semiconductor memories is different from reliability for other microcircuits. Reprogrammable nonvolatile memory reliability is the summation of the factors of operating life (read), data retention, and endurance [see Eq. (8.1)]; see reference JEDEC Standards 29, 34, and 47. The specific failure modes and mechanisms are described in Chapter 6. The manufacturer's reliability assurance program for screening, qualification, and product monitoring must address the specific concerns for each technology, for example, floating gate, floating trap.

$$\text{Failure Rate (FR) device} = \text{FR read} + \text{FR endurance} + \text{FR data retention} \quad (8.1)$$

The read, endurance, and data retention failure rates are thermally accelerated; therefore, failure rates must be given stating the temperature, confidence interval, and apparent activation energy (Arrhenius acceleration) or alternative deceleration technique.

Endurance is the most important failure rate because the actual endurance is a direct function of the application, that is, the number of times the device is rewritten during system operation. In other words, the total system life can be compromised by the endurance capability of the reprogammable nonvolatile memory. Some device technologies have known endurance wearout mechanisms, which is not a factor in normal operation, but can affect system performance if the specified number of endurance cycles is greatly exceeded.

Endurance is defined as "The measure of the ability of a nonvolatile memory device to meet its data sheet specifications as a function of accumulated nonvolatile data changes," per IEEE Std 1005-1991, Definitions and Characterization of Floating Gate Semiconductor Arrays. The data sheet specifications include write functionality, data retention, and read access time. Typically, a nonvolatile data change is the completion of an erase/program cycle for each byte, that is, transferring charge to and from the storage node in the memory cell.

Endurance, in floating gate or floating trap MOS technologies, has two primary failure mechanisms: charge trapping or oxide damage. Three failure modes can result—data retention degradation, access time degradation, or loss of write functionality (including various "disturb" mechanisms). The charge is transferred to and from the storage node through an oxide, resulting in the failure mechanisms of oxide damage and charge trapping. These mechanisms are caused by the cumulative effects of passing a current through a nominal insulator and placing a high electric field across an oxide. Thicker oxides have a greater likelihood of measurable charge trapping. Thinner oxides require greater care in processing to reduce initial oxide defects, which cause yield loss. Endurance cycling over the lifetime of the system will cause random oxide damage and charge trapping at some constant low level. Design and processing by the manufacturer must be such as to minimize initial defects and reduce generated defects to the lowest possible level. Stressing and testing must be performed to separate devices with various levels of endurance performance.

When a high number of endurance cycles (before the onset of wearout) or a very low endurance cycle failure rate is desired, error correction is suitable for random failures. Bit or byte error correction methods may be used to improve the endurance of devices whose dominant failure mode is random damage to the oxides which insulate the storage node. Error correction is not practical for uniform charge degradation mechanisms, for example, charge-trapping induced failures. Note that low failure rates during the stipulated useful life region, suitable for most applications, are achieved by proper design, processing, and screening without the use of error correction.

Endurance follows the "bathtub" curve, with an infant mortality region governed by latent defects, a useful life region governed by the intrinsic integrity of the design and process (assuming infant mortality has been eliminated by screening), and a predictable wearout region governed by the cumulative effects of transferring charge through an oxide. The manufacturer eliminates infant mortality during screening and testing. The failure rate level in the useful life region is assessed by way of product monitors. The onset of wearout is determined by extended endurance cycling, including stressing devices past the initial failure. Endurance cycling, as a periodic qualification test, has historically been considered the preferred means of verifying capability because cycling can be performed in real time and is the actual operating mode of the device.

Data retention has infant mortality, which must be screened in the manufacturing flow. There is a useful life region that is governed by the intrinsic integrity of the design and process. Wearout does not occur (in the sense that permanent, nonreversible degradation is present) because the storage node may be refreshed. Intrinsic data retention, the time the storage node is capable of retaining charge independent of the application, may vary by device design and process technology, but for most technologies it is essentially very long compared with real-world operating conditions. The extrinsic data retention is a function of endurance—that is, data retention degradation introduced by erase/program cycling. Endurance failure rate expectations should contain the extrinsic data retention failure rate induced by endurance. The intrinsic data retention failure rate should be considered independent of endurance.

Test Method 1033 of MIL-STD-883 describes the procedures to be used when performing endurance cycling for screening or endurance performance verification. Various methods exist to eliminate infant mortality. These methods are a function of product design and the dominant failure mechanisms of the process. For example, some devices use endurance cycling, and others use a margin test to screen out infant mortality. Most device manufacturing flows contain an infant mortality data retention unbiased bake screen, for example, per JESD-22-A103 High Temperature Storage Life. Military requirements contain a periodic Quality Conformance Inspection (QCI) that must be performed on SMD or 883 compliant byte-alterable EEPROMs to verify operating life, data retention, and endurance. A similar requirement could be added for other reprogrammable nonvolatile devices in the applicable procurement specification or in the manufacturer's internal manufacturing documentation.

Random defects, which occur naturally in the wafer fabrication process, cause infant mortality endurance or data retention failures. In neither case is there an explicit relationship that correlates infant mortality with device performance in the useful life or wearout regions. To improve yields, manufacturers may include redundant memory in the device which is used to repair initial or infant mortality failures. For large and complex memory arrays, for example, RAMs, EPROMs, or EEPROMs, few device types are shipped that do not include some level of redundancy repair. Due to the localized nature of the random defects that cause initial or infant mortality failures, the reliability of repaired and nonrepaired devices is equivalent.

The endurance failure rate of the reprogrammable nonvolatile memories in a system increases in importance as a function of the number of times the system rewrites the memory during the system's life. System reliability is a function of the failure rate in the specified useful life region of the device, not of the time when the onset of wearout occurs. Given that the operating life failure rate of an MOS memory is on the order of 100 FITs (0.01%/1000 hours) or less, the endurance and intrinsic data retention failure rate contributions to the total failure rate should be an order of magnitude or more lower than 100 FITs, that is, 10 FITs or less [see Eq. (8.1)].

8.1.3 Warranty Policies

All microcircuit manufacturers provide warranty policies or terms and conditions of sale. These documents are typically divided into three categories: the warranty, the guaranty, and the applicable conditions.

The warranty typically states that any nonconforming device may be returned to the manufacturer for credit or replacement. Conformance is to the data sheet or other applicable procurement specification and is usually for a term of one year from date of shipment.

The guaranty is for lot acceptance and typically states that any lot that fails the lot acceptance sampling plan per the applicable specification may be returned to the manufacturer for credit or replacement, within one year from shipment.

The warranty policy defines those conditions under which devices may be returned, including administrative and technical requirements. Administrative requirements define the logistics and methodology of documenting and returning the affected devices, so that credit may be applied to the correct purchase order. Technical requirements include: defining the condition of returned devices (e.g., must be testable) and the amount of correlation needed to validate nonconformance.

Lot acceptance guarantees, specified per the applicable lot acceptance sampling plan, define guaranteed quality levels. Typically, the quality level applies to all data sheet electrical parameters and the applicable mechanical/visual parameters, but does not apply to reliability expectations. Data retention and endurance may be treated as quality parameters; thus, they often have lot acceptance guaranty. These parameters are conventionally considered reliability expectations; however, the number of years for data retention and the number of cycles for endurance are typically specified in the data sheet. The relatively quick and simple measurement of these parameters allows specific lot acceptance requirements to be stipulated.

Quality levels, measured in PPM nonconforming, are estimates of the average outgoing quality (AOQ) of the manufacturers' production line, after all screening, testing, and acceptance sampling have been performed. EIA Standard 554 defines how to assess AOQ in PPM for microcircuit manufacturing. Transformation of acceptable quality level (AQL) or lot tolerant percent defective (LTPD) sampling plans and lot guaranty levels to AOQ values is treated in standard texts on acceptance sampling.

The warranty policy should contain a definition and statement of guaranty for endurance and data retention. An example is as follows:

> Endurance is the measure of the ability of a reprogrammable nonvolatile memory device to meet its data sheet specifications as a function of accumulated erase/program cycles. An erase/program cycle is the act of changing data from original (e.g., erased) to opposite (e.g., programmed) back to original for all bits of the memory array.

The memory shall be capable of the specified number of erase/program cycles per specified memory element, e.g., byte, sector, page, independent of the erase or program method, e.g., byte, page, sector, chip.
Data retention is the measure of the integrity of the stored data as a function of time. Data retention time is the time from data storage to the time at which a repeatable data error is detected.
The device shall be capable of the specified number of years of data retention. This applies across the operating temperature range and after the specified minimum number of endurance cycles.
The memory has a lot acceptance guaranty of a 1% AOQL (LTPD 5/1) for the specified number of endurance cycles and data retention years, as verified by the specified test methodology (see section 8.4.2).

8.2. CRITICAL DEVICE PARAMETERS

Some device characteristics are more important for the operation of the user's system than others. The following sections describe those parameters that are deemed most critical and require careful evaluation.

8.2.1 Electrical

All electrical parameters are important for the correct functioning of a device in an application; however, a few parameters tend to be more visible because they are the ones that most often appear to fail.

Critical DC parameters are input/output leakage, for example, I_{il}, I_{ih}, and power supply currents, for example, I_{cc}, I_{pp}, I_{sb}. High input/output leakage levels, typically caused by electro-static discharge (ESD) or electrical over-stress (EOS) will cause nonfunctionality by address lines, control pins, or outputs being unable to go to correct levels. High power supply currents, either active or standby, typically caused by EOS, may overload supply lines and damage other components. Logic test methods such as I_{ddq} and built-in self test (BIST) are generally not applicable to large memory arrays because these test methods are not sensitive or complete enough to detect subtle memory cell defects or variations.

Critical AC parameters are access timing values (e.g., T_{aa}, T_{ce}), and input/output level conditions (e.g., V_{ih}/V_{il}, V_{oh}/V_{ol}). Access times are sensitive to data patterns, including checkerboard, diagonal, and address patterns, such as address complement and diagonal galpat. If the device is inadequately tested by the manufacturer (i.e., not using worst case data and address patterns), it may occasionally read incorrectly in the application. Input/output level test conditions differ widely from device to device and from manufacturer to manufacturer. Timing values are extremely sensitive to the applied input/output levels. Thus, devices with supposedly the same timing value may function differently in the application because of different levels used during manufacturers' testing.

All parameters should be controlled by the manufacturer's internal documentation for how they are tested. Some parameters, such as capacitance, are only tested initially and after a design change that affects the parameter. Others should indicate whether testing is by inference (e.g., $V_{ih\text{-}max}$), or by application of specified signals and conditions (e.g., setup/hold times). The address and data patterns used for verifying write and read functionality, as well as appropriate machine, test, or device guardbands, should be included in the manufacturer's internal documentation.

8.2.2 Mechanical/Visual

Critical mechanical parameters are the package dimensions of thickness and lead spacing, which can affect how devices interact with automatic insertion equipment or the dimensions of the application; and solderability, which affects the mechanical, thermal, and electrical connection of the device to the application.

Critical visual parameters are the marking of the device and the marking permanency. These clearly and permanently identify the device, as to part number, date code, orientation, and the like.

8.2.3 Reliability

Critical reliability parameters include endurance, data retention, and package integrity. Endurance is application dependent (i.e., how often the device is rewritten) and affects the overall system failure rate. Data retention is application sensitive (i.e., the intrinsic or extrinsic data retention failure rate of some devices or technologies may preclude some applications). Standard package integrity may be application sensitive (i.e., concerns with hermeticity, especially for glass-sealed packages, and concerns with cumulative exposure to temperature and humidity for plastic packages). Plastic package recommended soldering profiles must be carefully adhered to in order to prevent thermal or mechanical stress-induced data retention degradation.

Two other issues, though not exactly normal application reliability concerns, may be of concern when using nonvolatile memories: radiation tolerance and declassification ability. Radiation tolerance is a measure of how much radiation a device may receive and continue to function. In some cases, for similar technologies, radiation tolerance correlates with reliability performance but is not always a means for comparing different reprogrammable nonvolatile technologies for reliability. Declassification ability is a measure of the difficulty or possibility of recovering information that has been supposedly removed (i.e., erased) from the device.

8.3. MANUFACTURER'S SCREENING

The manufacturer of microcircuits must include, as part of the test and inspection flow, various screens to eliminate latent or generated defects that could cause the device to become nonconforming to specification. The following sections describe some of the more common screening practices for reprogrammable nonvolatile memories.

8.3.1 Electrical

The manufacturing of microcircuits consists of four major steps: fabrication, assembly, screening/test, and quality assurance. Fabrication consists of various physical, chemical, photolithography, and inspection operations to form die on the microcircuit wafer; assembly consists of placing microcircuit die in a package for connection to other elements; screening/test consists of applicable stressing of devices, then identification of the conformance level of each microcircuit (and removing nonconforming devices); and quality assurance of regular sampling to verify quality and reliability performance.

Normal microcircuit manufacturing practices include one or more 100% electrical tests; for example, each device is tested for DC, AC, and functional parameters, separating passing devices by performance level. For complex microcircuits such as reprogrammable nonvolatile memories, electrical testing before assembly (e.g., wafer sort) is done at room temperature, and electrical testing post assembly is performed at high or cold temperature or both. Military devices require testing at high, low, and room temperatures. Commercial devices usually have a single insertion at high temperature and are guardbanded for parameters that are adversely affected by lower temperatures. For example, MOS devices are worst case at high temperature for speed; however, I_{cc} is worst case at low temperature. Therefore, I_{cc} is guardbanded in the high-temperature test.

Complex devices are tested using automated testers (automated test equipment—ATE) and test handlers with suitable interface hardware. The ATE is controlled by software, called a test program, which contains the various algorithms for testing a device. These include the forced and measured values, guardbands, data patterns, and address patterns in a sequence sufficient to exercise all functions at applicable data sheet limits. In addition, manufacturer's test programs typically include special modes that allow operation of the device in a nondata sheet specified manner to improve test effectiveness or efficiency, for example, apply an accelerating stress for reduced test time. Handlers control the ambient temperature for test and segregate devices into various bins by test results.

Users should verify that the manufacturer has fully documented the test program, the interface hardware, the accuracy and precision of the test setup, and the operating procedures for performing tests.

8.3.2 Reliability

Although microcircuits are designed for reliability, variability in manufacturing processes could result in degradation of performance sooner than expected. This infant mortality is usually the result of random latent defects in manufacturing material or processes and may be detected by accelerated stresses.

Reprogrammable nonvolatile memories have two major reliability parameters, endurance and data retention, both of which require evaluation and possible additional screening to remove infant mortalities. Screening of other reliability parameters should be consistent with that of other microcircuits fabricated with similar processes and assembled in similar packages. Test methods should reference MIL-STD-883 for hermetic devices or JEDEC Standards for plastic devices.

The manufacturer's test flow should include screens for endurance and data retention. Typical endurance screening is the performance of some number of endurance (erase/program) cycles or other oxide stress procedure to accelerate latent defects in the charge transmission oxide. Endurance stressing also is used to accelerate defects that can cause various disturb failures—for example, read-disturb and program-disturb. The number and type of disturbs is a function of the process technology and circuit design. Endurance screening is usually followed by a data retention stress, such as a high temperature unbiased bake, for accelerating latent data retention failures. The manufacturer should be able to provide a methodology and data to support whatever endurance and data retention screens are used. Test Method 1033 of MIL-STD-883 provides a format for defining the requirements for an endurance and data retention screen.

8.4. QUALIFICATION TESTING

In order to transfer a microcircuit to production or ship to a customer, manufacturers perform some level of qualification activities. The following sections describe some of these basic qualification activities.

8.4.1 Characterization

Upon identifying a potential device for an application, the devices's adequacy for the application must be verified. Given that the application constraints are known, this usually consists of characterizing the electrical and reliability performance of the proposed device. Before embarking on the very expensive effort of device characterization, several other activities should be performed.

The manufacturer should be audited by the user or users' representative, for example, the National Supervising Inspectorate for the ISO-9000 series Quality Systems (NSI) for general capability to consistently manufacture a device that conforms to all applicable specifications. This includes system capability as well as specific technical abilities. The manufacturer may provide information to the user that details the manufacturing technology, device performance specifications, and reliability expectations.

Once the user is satisfied that the manufacturer's quality system and the device's performance comply with the application requirements, the user should obtain some devices for validation of expected results. Testing performed by the user should verify device performance to the data sheet or other applicable specification. This testing can be used as the basis for establishing a correlation between the manufacturer's inspection and the user's application. Critical electrical and mechanical/visual parameters may require extra attention to assure consistency in measurement and repeatability in results.

Reliability parameters should receive characterization, especially endurance and data retention. Erase/program cycling and data retention bake should continue until the onset of endurance wearout, that is, where the endurance failure rate increases with additional cycles. Verification of the intrinsic data retention failure rate should establish, using standard deceleration techniques, that the MTTF of the reprogrammable nonvolatile memory is greater than the application's required storage time. Other reliability parameters, such as life test and package stresses, may use data provided by the manufacturer.

Typical endurance and data retention characterization tests consist of choosing two samples. Subject the first sample to the number of endurance cycles at the temperature used in the application, followed by a data retention bake that correlates to the storage time required of the application. Subject the second sample to increasing numbers of erase/program cycles, interweaved periodically with short data retention bakes, until a statistically significant fraction, for example, 20%, of the devices have failed two or more bits. Analysis of these data in conjunction with the manufacturer's supplied data should validate the feasibility of the proposed device in the application.

8.4.2 Verification

After characterization verifies that the microcircuit is capable of meeting the application requirements, some ongoing testing may be required to assure that production deliveries continue to conform to specification. Commonly used methods include source inspection, incoming inspection, regular audits, and periodic monitors.

Source inspection requires the user or designate to witness critical manufacturing or quality assurance operations in order to verify that the shipped product conforms to the applicable specification. Incoming inspection accomplishes a similar purpose by having the user inspect production material upon receipt.

In regular audits, the user periodically assesses the manufacturer to ensure that manufacturing methods, systems, procedures, and specifications are adhered to in the ongoing production of the purchased microcircuits. Periodic monitoring subjects a sample to an incoming test or a characterization evaluation.

Incoming inspection is probably the least productive of the various verification methodologies. Regular audits, combined with periodic monitors, are likely the most effective, but require a skilled and trained audit and evaluation team for implementation. Source inspection is the simplest method for the user to directly verify product conformance to specification. The method that is most appropriate to a given application depends on the relationship of manufacturer and user, the manufacturing and quality system capabilities of both parties, and any other overriding considerations, for example, government requirements. In all cases, regular communication between manufacturer and user is vital to ensure ongoing performance improvement.

The user should monitor reliability parameters by periodically reviewing the manufacturer's reliability reports, using the following methodology:

An endurance test, reference Method 1033 of MIL-STD-883, shall be added before performing the steady state life test and extended data retention test. Cycling may be chip, sector, block, byte, or page on finished devices. The following conditions shall be met:

(1) All bytes shall be cycled for a minimum of the specified number of cycles at equipment room ambient.
(2) Perform parametric, functional, and timing tests at room (high) temperature, after cycling. Devices having bits not in the proper state after functional testing shall constitute a device failure. Separate the devices into two groups for extended data retention and steady state life test, then write correct data patterns.
(3) Perform extended data retention, consisting of high temperature unbiased storage for 1000 hours minimum at 150°C minimum. The storage time may be accelerated by using a higher temperature according to the Arrhenius relationship and an apparent activation energy of 0.6 eV. The maximum storage temperature in a nitrogen environment shall not exceed 175°C for hermetic or 160°C for plastic devices. All devices shall be written with a charge on all memory cells in each device, such that a loss of charge can be detected (e.g., worst case pattern).
(4) Read the data retention pattern and perform parametric, functional, and timing tests at room (high) temperature, after cycling and bake. Devices having bits not in the proper state after functional testing shall constitute a device failure.
(5) Perform steady state life, reference method 1005 condition D of MIL-STD-883, for 1000 hours at 125°C in a nitrogen environment. The steady state life time may be accelerated by using an Arrhenius relationship and an apparent activation energy of 0.4 eV. The maximum operating junction temperature shall not exceed 160°C. All devices shall be written with a checkerboard or equivalent topological alternating bit pattern.
(6) Read the steady state life pattern and perform parametric, functional, and timing tests at room (high) temperature, after cycling and steady state life. Devices having bits not in the proper state after functional testing shall constitute a device failure.
(7) The endurance, data retention, and steady state life tests shall individually pass a sample plan to an LTPD of 5/1 (sample size = 77, accept = 1), equivalent to an AOQL = 1%.

8.5. SUMMARY

Reprogrammable nonvolatile memories should be procured with the same care as other microcircuits. The additional application-dependent reliability parameters of endurance and data retention require careful consideration of device design, manufacturing methodology, reliability criteria, and documented performance. The features and functions of various technology and device alternatives require careful consideration before final system design approval.

The preceding sections have outlined major areas in which microcircuits, in general, and reprogrammable nonvolatile memories, in particular, require special procurement considerations.

References

[8.1] *Testing Semiconductor Memories, Theory & Practice*, A. J. van de Goor, Wiley, 1989.

[8.2] *Introduction to Component Testing*, Anthony K. Stevens, Addison-Wesley, 1986.

[8.3] *Automatic Testing & Evaluation of Digital Integrated Circuits*, James T. Healy, Reston Publishing, 1981.

[8.4] "Guardbanding VLSI EEPROM test programs," David Sweetman, IEEE VLSI Test Symposium, 1991.

[8.5] MIL-STD-883,Test Methods and Procedures for Microelectronics.

[8.6] EIA-625, Requirements for Handling Electro-Static Sensitive (ESDS) Devices.

[8.7] EIA-554, Assessment of Outgoing Nonconforming Levels in Part Per Million (PPM).

[8.8] EIA-557, Statistical Process Control Systems.

[8.9] EIA-599-1, Process Certification Standard for Semiconductor Device Assemblers.

[8.10] ANSI/ASQC M1, American National Standard for Calibration Systems.

[8.11] JESD-26,General Specification for Plastic Encapsulated Microcircuits for Use in Rugged Applications.

[8.12] JESD-29,Failure Mechanism Driven Reliability Monitoring of Silicon Devices.

[8.13] JESD-34,Failure Mechanism Driven Reliability Qualification of Silicon Devices.

[8.14] JESD-47,Stress Test Driven Qualification of Integrated Circuits.

[8.15] JEP-122, Failure Mechanisms and Model for Silicon Semiconductor Devices.

[8.16] EIA-599, National Electronic Process Certification Standard.

[8.17] ISO-9000, Quality Management and Quality Assurance Standards—Guidelines for Selection and Use.

[8.18] ISO-9001, Quality Systems—Model for Quality Assurance in Design/Development, Production, Installation, and Servicing.

[8.19] EIA-670, Quality System Assessment.
[8.20] EIA-672, Guideline for User Notification of Product/Process Changes by Semiconductor Suppliers.
[8.21] JESD-99, Glossary of Microelectronic Terms, Definitions, and Symbols.
[8.22] JESD-100, Terms, Definitions, and Letter Symbols for Microcomputers and Memory Integrated Circuits.
[8.23] IEEE Std-1005-1991, IEEE Standard Definitions and Characterization of Floating Gate Semiconductor Arrays.
[8.24] JESD-21-C, Configurations for Solid State Memories.
[8.25] JEP-95, Registered and Standard Outlines for Solid State and Related Products.

Note: A wide variety of MIL-STD-883 and JESD-22-XX Test Methods are applicable to the specification, qualification, and reliability monitoring of reprogrammable nonvolatile memories. Some of particular interest are:
MIL-STD-883, Test Method 1033, Endurance Life.
JESD-22-A103, High Temperature Storage Life.
JESD-22-A112, Moisture-Induced Stress Sensitivity for Plastic Surface Mount Devices.
JESD-22-A113, Preconditioning of Plastic Surface-Mount Devices Prior to Reliability Testing.

W. D. Brown

Bibliography

The following is a list of references by year composed of most of those listed at the end of each chapter and others that have been included because of either their direct or indirect relevance to the subject matter of the book.

1967

Budenstein, P. P., and P. J. Hayes, "Breakdown conduction in Al-SiO-Al capacitors," *J. Appl. Phys.*, vol. 38, p. 2837, 1967.

Chu, T. L., et al., "The preparation and C-V characteristics of Si-Si_3N_4 and Si-SiO_2-Si_3N_4 structures," *Sol. St. Electr.*, vol. 10, p. 897, 1967.

Hu, S. M., et al., "Evidence of hole trapping and injecting effects in the nitride films prepared by reactive sputtering," *Appl. Phys. Lett.*, vol. 10, p. 97, 1967.

Kahng, D., and S. M. Sze, "A floating gate and its application to memory devices," *Bell Syst. Tech. J.*, vol. 46, p. 1288, 1967.

Sze, S. M., "Current transport and maximum dielectric strength of silicon nitride films," *J. Appl. Phys.*, vol. 38, p. 2951, 1967.

Wegener, H. A. R., et al., "The variable threshold transistor, a new electrically alterable, non-destructive read-only storage device," *IEDM Tech. Dig.*, 1967.

1969

Dill, H. G., et al., "Anomalous behavior in stacked-gate MOS tetrodes," *IEEE ISSCC Tech. Dig.*, p. 44, 1969.

Frohman-Bentchkowsky, D., and M. Lenzlinger, "Charge transport and storage in metal–nitride–oxide–silicon (MNOS) structures," *J. Appl. Phys.*, vol. 40, no. 8, p. 3307, 1969.

Lenzlinger, M., and E. H. Snow, "Fowler-Nordheim tunneling in thermally grown SiO_2," *J. Appl. Phys.*, vol. 40, p. 278, 1969.

Ross, E. C., and J. T. Wallmark, "Theory of the switching behavior of MIS memory transistors," *RCA Rev.*, p. 366, June 1969.

Wallmark, J. T., and T. H. Scott, "Switching and storage characteristics of MIS memory transistors," *RCA Rev.*, vol. 30, p. 335, 1969.

1970

Agrawala, B. N., et al., "Dependence of electromigration induced failure time on length and width of aluminum thin-film conductors," *J. Appl. Phys.*, 1970.

Frohman-Bentchkowsky, D., "The metal–nitride–oxide–silicon (MNOS) transistor—characteristics and applications," Proc. IEEE, vol. 58, p. 1207, 1970.

Ross, E. C., A. M. Goodman, and M. T. Duffy, "Operational dependence of the direct-tunneling mode MNOS memory transistor on the SiO_2 layer thickness," *RCA Rev.*, p. 467, September 1970.

1971

Frohman-Bentchkowsky, D., "A fully decoded 2048 bit electrically programmable FAMOS read-only memory," *IEEE J. Sol. St. Cir.*, vol SC-6, p. 301, 1971.

Frohman-Bentchkowsky, D., "A fully decoded 2048-bit electrically programmable MOS-ROM," *IEEE ISSCC Tech. Dig.*, p. 80, 1971.

Frohman-Bentchkowsky, D., "Memory behaviour in a floating gate avalanche injection MOS (FAMOS) structure," *Appl. Phys. Lett.*, vol. 18, p. 332, 1971.

1972

Iizuka, H., et al., "Stacked gate avalanche injection type MOS (SAMOS) memory," Proc. 4th Conf. Sol. St. Dev., 1972.

Lundström, K. I., and C. M. Svensson, "Properties of MNOS structures," *IEEE Trans. Elect. Dev* ., vol. ED-19, no. 6, p. 826, 1972.

Tarui, Y., Y. Hayashi, and K. Nagai, "Electrically reprogrammable non-volatile semiconductor memory," *IEEE J. Sol. St. Cir.*, vol. SC-7, p. 369, 1972.

White, M. H., and J. R. Cricchi, "Characterization of thin-oxide MNOS memory transistors," *IEEE Trans. Elect. Dev.*, vol. ED-19, no. 12, p. 1280, 1972.

Woods, M. H., and J. W. Tuska, "Degradation of MNOS memory transistor characteristics and failure mechanism model," *Proc. IRPS*, p. 120, 1972.

1973

Bosselaar, C. A., "Charge injection into SiO_2 from reverse-biased junctions," *Sol. St. Electr.*, vol. 16, p. 648, 1973.

Chaffin, R. J., *Microwave Semiconductor Devices*, John Wiley, New York, p. 120, 1973.

Cricchi, J. R., F. C. Blaha, and M. D. Fitzpatrick, "The drain-source protected MNOS memory device and memory endurance," *IEDM Tech. Dig.*, p. 126, 1973.

DiStefano, T. H., "Dielectric breakdown induced by sodium in MOS structures," *J. Appl. Phys.*, vol. 44, p. 527, January 1973.

Iizuka, H., et al., "Stacked gate avalanche injection type MOS (SAMOS) memory," *J. Jap. Soc. Appl. Phys.*, vol. 42, p. 158, 1973.

Lundkvist, L., I. Lundstrom, and C. Svensson, "Discharge of MNOS structures," *Sol. St. Electr.*, vol. 16, p. 811, 1973.

Naber, C. T., and G. C. Lockwood, "Processing of non-volatile memories," *J. Electrochem. Soc.*, vol. 120, p. 401, 1973.

Sansbury, J. D., "MOS field threshold increase by phosphorus-implanted field," *IEEE Trans. Elect. Dev.*, vol. ED-20, no. 5, p. 473, 1973.

Simmons, J. G., et al., "Thermally stimulated currents in semiconductors and insulators having arbitrary trap distributions," *Phys. Rev.*, B7, p. 3714, 1973.

Yun, B. H., "Direct display of electron back tunneling in MNOS memory capacitors," *Appl. Phys. Lett.*, vol. 23, no. 3, p. 152, 1973.

1974

Chen, P. C. Y., "Interface instability of r.f. sputtered silicon nitride films on silicon," Thin Sol. Films, vol. 21, p. 245, 1974.

Chen, P. C. Y., et al., "Thin film memory transistors," *IEEE Trans. Elect. Dev.*, vol. ED-21, p. 740, 1974.

Cricchi, J. R., et al., Nonvolatile block-oriented RAM," *IEEE ISSCC Tech. Dig.*, p. 204, 1974.

Dennard, R. H., et al., "Design of ion-implanted MOSFET's with very small physical dimensions," *IEEE J. Sol. St. Cir.*, SC-9, p. 256, 1974.

DiStefano, T. H., and M. Shatzkes, "Impact ionization model for dielectric instability and breakdown," *Appl. Phys. Lett.*, vol. 25, p. 685, 1974.

Frohman-Bentchkowsky, D., "FAMOS – A new semiconductor charge storage device," *Sol. St. Electr.*, vol. 17, p. 517, 1974.

Grant, E. I., and R. S. Leavenworth, "Statistical Quality Control," Chapter 18, *Some Aspects of Life Testing and Reliability*, McGraw-Hill Book Co., New York, 4th ed., 1974.

Khang, D., et al., "Interfacial dopants for dual-dielectric charge-storage cells," *IEEE Trans. Elect. Dev.*, vol. ED-21, p. 740, 1974.

Koo, T. K., "The non-volatility characteristics of MNOS memory FET," IEEE NAECON, p. 37, 1974.

Ning, T. H., and H. N. Yu, "Optically-induced injection of hot electrons into SiO_2," *J. Appl. Phys.*, vol. 45, p. 5373, 1974.

Sato, S., and T. Yamaguchi, "Study of charge behaviour in metal–alumina–silicon dioxide–silicon (MAOS) field effect transistor," *Sol. St. Electr.*, vol. 17, p. 367, 1974.

Taylor, G. W., and J. G. Simmons, "Effects of bulk trapping on the memory characteristics of thick-oxide MNOS variable-threshold capacitors," *Sol. St. Electr.*, vol. 17, p. 1, 1974.

Yun, B. H., "Measurements of charge propagation in Si_3N_4 films," *Appl. Phys. Lett.*, vol. 25, no. 6, p. 340, 1974.

1975

Arnett, P. C., and B. H. Yun, "Silicon nitride trap properties as revealed by charge-centroid measurements on MNOS devices," *Appl. Phys. Lett.*, vol. 26, no. 3, p. 94, 1975.

DiMaria, D. J., and D. R. Kerr, "Interface effects and high conductivity in oxides grown from polycrystalline silicon," *Appl. Phys. Lett.*, vol. 27, p. 505, 1975.

Maes, H. E., and R. Van Overstraeten, "Simple technique for determination of the centroid of nitride charge in MNOS structures," *Appl. Phys. Lett.*, vol. 27, no. 5, p. 282, 1975.

Peck, D. S., "Practical applications of accelerated testing—introduction," Reliab. Phys., 13th Ann. Proc., p. 253, 1975.

Scoggan, G. A., et al., "Width dependence of electromigration life," *Proc. IRPS*, 1975.

Uchida, Y., et al., "A 1024-Bit MNOS RAM using avalanche-tunnel injection," *IEEE J. Sol. St. Cir.*, vol. SC–10, no. 5, p. 288, 1975.

Yun, B. H., "Electron and hole transport in CVD nitride films," *Appl. Phys. Lett.*, vol. 27, no. 4, p. 256, 1975.

1976

Barnes, J., J. Linden, and J. Edwards, "Operation and characterization of n-channel EPROM cell," *IEEE IEDM Tech. Dig.*, p. 173, 1976.

Brewer, J. E., et al., "Army/Navy MNOS BORAM," GOMAC, 1976.

Chang, J. J., "Effect of distributed charge in the nitride of an MNOS structure on the flat-band voltage," *Appl. Phys. Lett.*, vol. 29, no. 11, p. 742, 1976.
Dickson, J. F., "On-chip high-voltage generation in MNOS integrated circuits using an improved voltage multiplier technique," *IEEE J. Sol. St. Cir.*, vol. SC–11, p. 374, 1976.
Fagan, J. L., et al., "A high-density, read/write, non-volatile charge-addressed memory," *IEEE ISSCC Tech. Dig.*, p. 184, 1976.
Iizuka, H., et al., "Electrically alterable avalanche injection type MOS read-only memory with stacked gate structure," *IEEE Trans. Elect. Dev.*, vol. ED–23, p. 379, 1976.
Kamata, T., K. Tanabashi, and K. Kobayashi, "Substrate current due to impact ionization in MOSFET," *Jap. J. Appl. Phys.*, vol. 15, p. 1127, 1976.
Maes, H. E., and R. J. Van Overstraeten, "Memory loss in MNOS capacitors," *J. Appl. Phys.*, vol. 47, no. 2, p. 667, 1976.
Neugebauer, C. A., and J. F. Burgess, "Endurance and memory decay of MNOS devices," *J. Appl. Phys.*, vol. 47, no. 7, p. 3182, 1976.
Raider, S. I., et al., "Surface oxidation of silicon nitride films," *J. Electrochem. Soc.*, vol. 123, p. 560, 1976.
Saito, S., et al., "A 256 bit nonvolatile static random access memory with MNOS memory transistors," Proc. 7th Conf. Sol. St. Dev., *Suppl. Jap. J. Appl. Phys.*, vol. 15, p. 185, 1976.
Stein, H. J., "Hydrogen content and annealing of memory quality of silicon-oxynitride films," *J. Electr. Mat.*, vol. 5, p. 161, 1976.
Weinberg, Z. A., and A. Hartstein, "Photon-assisted tunneling from aluminum into silicon dioxide," *Sol. St. Comm.*, vol. 20, p. 179, 1976.
Williams, R. A., and D. K. Nichols, "Radiation-induced memory loss in thin-oxide MNOS devices," *IEEE Trans. Nucl. Sci.*, vol. NS–23, no. 6, p. 1554, 1976.

1977

Anderson, R. M., and D. R. Kerr, "Evidence for surface asperity mechanism of conductivity in oxide grown on polycrystalline silicon," *J. Appl. Phys.*, vol. 48, no. 11, p. 4834, 1977.
Chang, J. J., "Theory of MNOS memory transistor," *IEEE Trans. Elect. Dev.*, vol. ED–24, no. 5, p. 511, 1977.
Chen, P. C., "Threshold-alterable Si-gate MOS devices," *IEEE Trans. Elect. Dev.*, ED–24, no. 5, p. 584, 1977.
Cricchi, J. R., M. D. Fitzpatrick, F. C. Blaha, and B. T. Ahlport, "1 MRad hard MNOS structures," *IEEE Nucl. Sci.*, vol. NS–24, no. 6, p. 2185, 1977.
Glasstone, S., and P. Dolan, *The Effects of Nuclear Weapons*, U. S. Departments of Defense and Energy, 1977.
Guterman, D., et al., "An electrically alterable nonvolatile memory cell using a floating gate structure," *IEEE Trans. Elect. Dev.*, vol. ED–24, p. 806, 1977.

Hsia, Y., "MNOS LSI memory device data retention measurements and projections," *IEEE Trans. Elect. Dev.*, vol. ED–24, no. 5, p. 568, 1977.

Johannessen, J. S., et al., "Auger depth profiling of MNOS by ion sputtering," *IEEE Trans. Elect. Dev.*, vol. ED–24, no. 5, p. 547, 1977.

Lehovec, K., "Charge distribution in the nitride of MNOS memory devices," *J. Electr. Mats.*, vol. 6, no. 2, p. 77, 1977.

Mar, H. A., and J. G. Simmons, "A review of the techniques used to determine trap parameters in the MNOS structure," *IEEE Trans. Elect. Dev.*, vol. ED–24, no. 5, p. 540, 1977.

Nishi, Y., and H. Iizuka, "Invited: nonvolatile semiconductor memory," *Suppl. Jap. J. Appl. Phys.*, vol. 16, p. 191, 1977.

Rössler, B., and R. Müller, "Electrically erasable and reprogrammable read-only memory using the n-channel SIMOS one-transistor cell," *IEEE Trans. Elect. Dev.*, vol. ED–24, p. 806, 1977.

Solomon, P., "High-field electron trapping in SiO_2," *J. Appl. Phys.*, vol. 48, p. 3843, September 1977.

Stein, H. J., and H. A. R. Wegener, "Chemically bound hydrogen in CVD Si_3N_4: dependence on NH_3/SiH_4 ratio and on annealing," *J. Electrochem. Soc.*, vol. 124, p. 908, 1977.

Weinberg, Z. A., "Tunneling of electrons from Si into thermally grown silicon dioxide," *Sol. St. Elect.*, vol. 20, p. 11, 1977.

White, M. H., J. W. Dzimianski, and M. C. Peckerar, "Endurance of thin-oxide nonvolatile MNOS memory transistors," *IEEE Trans. Elect. Dev.*, vol. ED–24, no. 5, p. 577, 1977.

1978

Angle, R. L., and H. E. Talley, "Electrical and charge storage characteristics of the tantalum oxide–silicon dioxide device," *IEEE Trans. Elect. Dev.*, vol. ED–25, p. 1277, 1978.

Barnes, J. J., et al., "Operation and characterization of N-channel EPROM cells," *Sol. St. Electr.*, vol. 22, p. 521, 1978.

Endo, N., "Charge distributions in silicon nitride of MNOS devices," *Sol. St. Electr.*, vol. 21, p. 1153, 1978.

Gentil, P., and S. Chausse, "Measurement of the effect of write-erase cycling on noise in MNOS memory transistors," *IEEE Trans. Elect. Dev.*, vol. ED–25, no. 8, p. 1042, 1978.

Harari, E., "Dielectric breakdown in electrically stressed thin films of thermal SiO_2," *J. Appl. Phys.*, vol. 49, p. 2478, 1978.

Harari, E., et al., "A 256 bit non-volatile static RAM," *IEEE ISSCC Tech. Dig.*, p. 108, 1978.

Hezel, R., and E. W. Hearn, "Mechanical stress and electrical properties of MNOS devices as a function of the nitride deposition temperature," *J. Electrochem. Soc.*, vol. 125, no. 11, p. 1848, 1978.

Horne, M. A., and B. A. Brillhart, "A military grade 1024-bit nonvolatile semiconductor RAM," *IEEE Trans. Elect. Dev.*, vol. ED–25, no. 8, p. 1061, 1978.

Hsia, Y., "Cross gate MNOS memory device," *IEEE Trans. Elect. Dev.*, vol. ED–25, no. 8, p. 1071, 1978.

Hughes, R. C., "High field electronic properties of SiO_2," *Sol. St. Electr.*, vol. 21, p. 251, 1978.

Joh, D. Y., "Charge storage nonvolatile semiconductor device," Ph.D. diss., University of New Mexico, 1978.

Multani, J. S., B. B. Kosicki, and J. S. Sandhu, "In-process measurement of memory properties of MNOS devices," *IEEE Trans. Elect. Dev.*, vol. ED–25, no. 8, p. 1072, 1978.

Nuclear IEEE Standards, Volume 1, ANSI/IEEE Standard 382-1972, Part 3, Design Basis Accident Environment Simulation for BWR and PWR, October 1978.

Schauer, H., E. Arnold, and P. C. Murau, "Interface states and memory decay in MNOS capacitors," *IEEE Trans. Elect. Dev.*, vol. ED–25, no. 8, p. 1037, 1978.

Schols, G., et al., "High temperature hydrogen anneal of MNOS structures," *Revue de Phys. Appl.*, vol. 13, p. 825, 1978.

Schuermeyer, F. L., and C. R. Young, "Endurance studies on MNOS devices," *J. Appl. Phys.*, vol. 49, no. 8, p. 4556, 1978.

Topich, J. A., "Charge storage model for SNOS memory devices," Abs. 268, Proc. Electrochem. Soc., 1978.

Uchida, Y., et al., "1K-bit nonvolatile semiconductor read/write RAM," *IEEE Trans. Elect. Dev.*, vol. ED-25, no. 8, p. 1066, 1978.

Williams, R. A., and M. E. Beguwala, "The effect of electrical conduction of nitride on the discharge of MNOS memory transistors," *IEEE Trans. Elect. Dev.*, vol. ED–25, p. 1019, 1978.

1979

Anolick, E. S., and G. R. Nelson, "Low field time-dependent dielectric integrity," *Proc. IRPS*, p. 8, 1979.

Brown, W. D., "Effects of HCl annealing of memory oxides on MNOS capacitor memory window," *J. Electr. Mats.*, vol. 8, no. 2, p. 87, 1979.

Brown, W. D., "Metal–nitride–oxide–semiconductor transistor characterization using implanted MNOS capacitors," *Thin Sol. Films*, vol. 59, no. 1, p. 125, 1979.

Brown, W. D., "MNOS technology—will it survive?," *Sol. St. Technol.*, p. 77, July 1979.

Brown, W. D., "Retention and endurance characteristics of HCl-annealed and unannealed MNOS capacitors," *Sol. St. Electr.*, vol. 22, p. 373, 1979.

Cottrell, P. E., R. R. Troutman, and T. H. Ning, "Hot electron emission in n-channel IGFET's," *IEEE J. Sol. St. Cir.*, vol. SC–14, p. 442, 1979.

Crook, D. L., "Method of determining reliability screens for time-dependent dielectric breakdown," *Proc. IRPS*, p. 1, 1979.

Hampton, F. L., and J. R. Cricchi, "Space charge distribution limitations on scale down of MNOS memory devices," *IEEE IEDM Tech. Dig.*, p. 374, 1979.

Hampton, F. L., and J. R. Cricchi, "Steady-state electron and hole space charge distribution in LPCVD silicon nitride films," *Appl. Phys. Lett.*, vol. 35, no. 10, p. 802, 1979.

Hartstein, A., and Z. A. Weinberg, "Unified theory of internal photoemission and photon-assisted tunneling," *Phys. Rev. B*, vol. 20, no. 4, p. 1335, 1979.

Hezel, R., "High temperature annealing of MNOS devices and its effect on Si-nitride stress, interface charge density and memory properties," *J. Electr. Mat.*, vol. 8, no. 4, p. 459, 1979.

Hsia, Y., and K. L. Ngai, "MNOS traps and tailored trap distribution gate dielectric MNOS," Proc. 11th Conf. Sol. St. Dev., 1979.

Hu, C., "Lucky electron model of hot electron emission," *IEEE IEDM Tech. Dig.*, p. 22, 1979.

Ito, T., et al., "Low voltage alterable EAROM cells with nitride barrier avalanche injection MIS (NAMIS)," *IEEE Trans. Elect. Dev.*, vol. ED–26, p. 906, 1979.

Kirk, C. T., "Valence alternation pair model of charge storage in MNOS memory devices," *J. Appl. Phys.*, vol. 50, no. 6, p. 4190, 1979.

Klein, R., et al., "5-V-only, nonvolatile RAM owes it all to polysilicon," *Electronics*, p. 111, October 11, 1979.

Ngai, K. L., and C. T. White, "Reconstructing states at the Si-SiO_2 interface," *J. Vac. Sci. Technol.*, vol. 16, p. 1412, July 1979.

Peercy, P. S., et al., "Hydrogen concentration profiles and chemical bonding in silicon nitride," *J. Electr. Mat.*, vol. 8, p. 11, 1979.

Profio, A. E., *Radiation Shielding and Dosimetry*, John Wiley, New York, 1979.

Schauer, H., and E. Arnold, "Simple technique for charge centroid measurement in MNOS capacitors," *J. Appl. Phys.*, vol. 50, no. 11, p. 6956, 1979.

Schroder, D. K., and M. H. White, "Characterization of current transport in MNOS structures with complementary tunneling emitter bipolar transistors," *IEEE Trans. Elect. Dev.*, vol. ED–26, p. 899, 1979.

Suzuki, E., and Y. Hayashi, "Transport processes of electrons in MNOS structures," *J. Appl. Phys.*, vol. 50, p. 7001, 1979.

Suzuki, E., Y. Hayashi, and H. Yanai, "A model of degradation mechanisms in metal–nitride–oxide–semiconductor structures," *Appl. Phys. Lett.*, vol. 35, no. 10, p. 790, 1979.

Wang, S. T., "On the I-V characteristics of floating gate MOS transistors," *IEEE Trans. Elect. Dev.*, vol. ED–26, p. 1292, 1979.

Yatsuda, Y., et al., "N-channel Si-gate MNOS device for high speed EAROM," Proc. 10th Conf. Sol. St. Dev., p. 11, 1979.

1980

Aitken, J. M., et al., "Study of the physics of insulating films as related to the reliability of metal–oxide semiconductor devices," Interim Report No. RADC–Tr–79–280, January 1980.

Boccaletti, G., "Accelerated testing of non-volatile memories," 2nd Colloque International sur la Fiabilite et la Maintenabilite, Perros-Guirec, Tregastel, France, September 8–12, 1980, Publ. by CNET, Cent de Fiabilite, Lannion, France, p. 160.

Deal, B. E., "Standardized terminology for oxide charges associated with thermally oxidized silicon," *IEEE Trans. Elect. Dev.*, ED–27, p. 606, 1980.

DiMaria, D. J., R. Ghez, and D. W. Dong, "Charge trapping studies in SiO_2 using high current injection from Si-rich SiO_2 films," *J. Appl. Phys.*, vol. 51, p. 4830, 1980.

Gerber, B., and J. Fellrath, "Low voltage single supply CMOS electrically erasable read-only memory," *IEEE Trans. Elect. Dev.*, vol. ED–27, p. 1211, 1980.

Hagiwara, T., et al., "A 16kbit electrically erasable PROM using n-channel Si-gate MNOS technology," *IEEE J. Sol. St. Cir.*, vol. SC–15, p. 346, 1980.

Hsia, Y., and K. L. Ngai, "MNOS traps and tailored trap distribution gate dielectric MNOS," *Jap. J. Appl. Phys.*, vol. 19, Sup. 191, p. 245, 1980.

Huff, H. R., et al., "Experimental observations on conduction through polysilicon oxide," *J. Electrochem. Soc.*, vol. 127, no. 11, p. 2482, 1980.

Ito, T., et al., "Retardation of destructive breakdown of SiO_2 films annealed in ammonia gas," *J. Electrochem. Soc.*, vol. 127, no. 10, p. 2248, 1980.

Johnson, W., et al., "A 16Kb electrically erasable nonvolatile memory," *IEEE ISSCC Tech. Dig.*, p. 152, 1980.

Johnson, W. S., et al., "16-K EE-PROM relies on tunneling for byte-erasable program storage," *Electr.*, p. 113, February 1980.

Kapoor, V. J., and S. B. Bibyk, "Energy distribution of electron trapping defects in thick-oxide MNOS structures," *The Physics of MOS Insulators*, Eds. G. Lucovsky, S. Pantelides and G. Galeener, Pergamon, New York, p. 117, 1980.

Kupec, J., et al., "Triple level poly silicon E^2PROM with single transistor per bit," *IEEE IEDM Tech. Dig.*, p. 602, 1980.

Landers, G., "5-V only EEPROM mimics static RAM timing," *Electr.*, p. 127, June 30, 1980.

Li, S. P., J. Maserjian, and S. Prussin, "Model for MOS field-time-dependent breakdown," *NASA Tech. Briefs*, vol. 5, no. 2, p. 145, 1980.

Maes, H. E., and G. Heyns, "Influence of a high temperature hydrogen anneal on the memory characteristics of p-channel MNOS transistors," *J. Appl. Phys.*, vol. 51, p. 2706, 1980.

Peckerar, M. C., and N. Bluzer, "Hydrogen annealed nitride/oxide dielectric structures for radiation hardness," *IEEE Nucl. Sci.*, vol. NS–27, p. 1753, 1980.

Rosenberg, S., "E-PROM reliability: Part 2—tests and screens weed out failures, project rates of reliability," *Electr. Mag.*, p. 17, August 14, 1980.

Rosenberg, S., "Test and screens weed out failures, project rates of reliability," *Electr.*, p. 136, August 14, 1980.

Rudie, N., *Principles and Techniques of Radiation Hardening*, vol. 1, Western Periodicals, pp. 2–40, 1980.

Shiner, R. E., J. M. Caywood, and B. L. Euzent, "Data retention in EPROMs," Proc. IRPS, p. 238, 1980.

Wada, M., et al., "Limiting factors for programming EPROMs of reduced dimensions," *IEEE IEDM Tech. Dig.*, p. 38, 1980.

Wang, S. T., "Charge retention of floating gate transistors under applied bias conditions," *IEEE Trans. Elect. Dev.*, vol. ED–27, p. 297, 1980.

Yatsuda, Y., et al., "Effects of high temperature annealing on n-channel Si-gate MNOS devices," *Jap. J. Appl. Phys.*, vol. 19, S19–1, p. 219, 1980.

1981

Ajike, T., et al., "Temperature accelerated estimation of MNOS memory realiability," *Proc. IRPS*, p. 17, 1981.

Anolick, E. S., and L.-Y. Chen, "Application of step stress to time dependent breakdown," *Proc. IRPS*, p. 23, 1981.

Berman, A., "Time-zero dielectric reliability test by a ramp method," *Proc. IRPS*, p. 204, 1981.

Chaudhari, P. K., "Subthreshold degradation in MNOS technology," *J. Electrochem. Soc.*, vol. 128, no. 1, p. 170, 1981.

Drori, J., et al., "A single 5V supply non-volatile static RAM," *IEEE ISSCC Tech. Dig.*, p. 148, 1981.

Eitan, B., and D. Frohman-Bentchkowsky, "Hot electron injection into the oxide in n-channel MOS-devices," *IEEE Trans. Elect. Dev.*, vol. ED–28, p. 328, 1981.

Euzent, B., et al., "Reliability aspects of a floating gate EEPROM," *Proc. IRPS*, p. 11, 1981.

Hallberg, O., "NMOS voltage breakdown characteristics compared with accelerated life tests and field use data," *Proc. IRPS*, 1981.

Isagawa, M., H. Oniyama, and H. Azegami, "Memory retention life at various environmental and life tests," *Proc. IRPS*, p. 52, 1981.

Jacobs, E. P., and U. Schwabe, "N-channel Si-gate process for MNOS EEPROM transistors," *Sol. St. Electr.*, vol. 24, p. 517, 1981.

Jeng, C. S., et al., "High-field generation of electron traps and charge trapping in ultra-thin SiO_2," *IEEE IEDM Tech. Dig.*, p. 388, 1981.

Jones, R. V., and W. D. Brown, "Endurance and retention of MNOS devices over the temperature range from $-50°C$ to $+125°C$," *J. Electr. Mats.*, vol. 10, no. 6, p. 959, 1981.

Kapoor, V. J., R. S. Bailey, and S. R. Smith, "Impurities-related memory traps in silicon nitride thin films," *J. Vac. Sci. Technol.*, vol. 18, no. 2, p. 305, 1981.

Kapoor, V. J., and S. B. Bibyk, "Energy distribution of electron-trapping centers in low pressure chemically vapor-deposited Si_3N_4 films," *Thin Sol. Films*, vol. 78, p. 193, 1981.

Kapoor, V. J., and R. A. Turi, "Charge storage and distribution in the nitride layer of the metal–nitride–oxide–semiconductor structures," *J. Appl. Phys.*, vol. 52, no. 1, p. 311, 1981.

Ko, P., R. Müller, and C. Hu, "A unified model for hot electron currents in MOSFET's," *IEEE IEDM Tech. Dig.*, p. 600, 1981.

Liang, M. S., and C. Hu, "Electron trapping in very thin thermal silicon dioxides," *IEEE IEDM Tech. Dig.*, p. 396, 1981.

Maes, H. E., and S. Usmani, "Charge pumping measurements on stepped-gate MNOS memory transistors," *J. Appl. Phys.*, vol. 53, p. 7106, 1981.

Manning, R. W., and W. D. Brown, "Memory-window-size-temperature dependence of the metal–nitride–oxide–silicon (MNOS) structure," *Sol. St. Electr.*, vol. 24, no. 11, p. 1039, 1981.

Nishi, Y., and H. Iizuka, "Nonvolatile memories," *Appl. Sol. St. Sci.*, Suppl. 2, Pt. A, p. 121, 1981.

Nishi, Y., and H. Iizuka, *Silicon Integrated Circuits, Pt. A*, D. Kahng (Ed.), Academic Press, New York, p. 121, 1981.

Pryor, R. W., "A mechanism for endurance failure in metal–nitride–oxide–semiconductor device structures," *J. Appl. Phys.*, vol. 52, no. 5, p. 3702, 1981.

Rizzo, J., "2816 floating gate bring a revolution to non-volatile memory," WESCON 81 Conf. Record, 25/1/1-5, 1981.

Schatzkes, M., and M. Av-Ron, "Statistics of breakdown," *IBM J. Res. and Dev.*, vol. 25, no. 2–3, p. 167, 1981.

Schatzkes, M., and M. Av-Ron, "Statistics of defect related breakdown," *Proc. IRPS*, 1981.

Simko, R. T., "Substrate coupled floating gate memory cell," U. S. Patent 4,274,012, 1981.

Smith, K. C., "The prospects for multivalued logic: A technology and application view," *IEEE Trans. Computers*, vol. C-30, p. 619, 1981.

Suzuki, E., Y. Hayashi, and H. Yanai, "Degradation properties in metal–nitride–oxide–semiconductor structures," *J. Appl. Phys.*, vol. 52, no. 10, p. 6377, 1981.

Tanaka, S., and M. Ishikawa, "One-dimensional writing model of n-channel floating gate ionization-injection MOS (FIMOS)," *IEEE Trans. Elect. Dev.*, vol. ED–28, p. 1190, 1981.

Wiker, R. L., and R. Carter, "Accelerated testing of time related parameters in MNOS memories," *Proc. IRPS*, p. 111, 1981.

Williams, R. A., and M. M. Beguwala, "Reliability concerns for small geometry MOSFET's," *Sol. St. Technol.*, vol. 24, p. 65, 1981.

Yeargain, J., and K. Kuo, "A high density floating gate EEPROM cell," *IEEE IEDM Tech. Dig.*, p. 24, 1981.

1982

Anolick, E. S., and L.-Y. Chen, "Screening of time-dependent dielectric breakdown," *Proc. IRPS*, 1982.

Bailey, R. S., and V. J. Kapoor, "Variation in the stoichiometry of thin silicon nitride insulating films on silicon and its correlation with memory traps," *J. Vac. Sci. Technol.*, vol. 20, no. 3, p 484, 1982.

DiMaria, D., et al., "Electrically alterable read-only-memory using Si-rich SiO_2 injectors and a polycrystalline silicon storage layer," *J. Appl. Phys.*, vol. 52, p. 4825, 1982.

Donaldson, D. D., et. al., "SNOS 1k × 8 static nonvolatile RAM," *IEEE J. Sol. St. Cir.*, vol. SC–17, p. 847, 1982.

Edwards, D. G., "Testing for MOS IC failure modes," *IEEE Trans. Reliab.*, vol. R–31, no. 1, p. 9, 1982.

El-Dessouky, A., "Charge centroid in MIOS nonvolatile memory structures," *IEEE Trans. Elect. Dev.*, vol. ED–29, no. 5, p. 814, 1982.

Ellis, R. K., "Fowler–Nordheim emission from non-planar surfaces," *IEEE Elect. Dev. Lett.*, vol. EDL–3, p. 330, 1982.

Fowler, R. H., and L. Nordheim, "Electron emission in intense electric fields," Proc. Roy. Soc. London, Ser. A., vol. 119, p. 173, 1982.

Fujita, S., et al., "Variations of trap states and dangling bonds in CVD Si_3N_4 layer on Si substrate by NH_3/SiH_4 ratio," *J. Electr. Mats.*, vol. 11, no. 4, p. 795, 1982.

Gargini, P. A., et al., "Elimination of silicon electromigration in contacts by the use of an interposed barrier metal," *Proc. IRPS*, 1982.

Gee, L., et al., "An enhanced 16K EEPROM," *IEEE J. Sol. St. Cir.*, vol. SC–17, p. 828, 1982.

Gupta, A., et al., "A 5V-only 16K EEPROM utilizing oxynitride dielectrics and EPROM redundancy," *IEEE ISSCC Tech. Dig.*, p. 184, February 1982.

Heimann, P. A., S. P. Murarka, and T. T. Sheng, "Electrical conduction and breakdown in oxides of polycrystalline silicon and their correlation with interface texture," *J. Appl. Phys.*, vol. 53, p. 6240, 1982.

Hsia, Y., E. Mei, and K. L. Ngai, "MNOS retention-endurance characteristics enhancement using graded nitride dielectric," Proc. 14th Conf. Sol. St. Dev., 1982.

Hsia, Y., and K. L. Ngai, "Empirical study of the metal–nitride–oxide semiconductor device characteristics deduced from a microscopic model of memory traps," *Appl. Phys. Lett.*, vol. 41, no. 2, p. 159, 1982.

Jenq, C., et al., "Properties of thin oxynitride films used as floating gate tunneling dielectrics," *IEEE IEDM Tech. Dig.*, p. 309, 1982.

Kamiya, M., et al., "EPROM cell with high gate injection efficiency," *IEEE IEDM Tech. Dig.*, p. 741, 1982.

Knoll, M., D. Braunig, and W. R. Fahrner, "Comparative studies of tunnel injection and irradiation on metal oxide semiconductor structures," *J. Appl. Phys.*, vol. 53, p. 6946, 1982.

Maserjian, J., and N. Zamani, "Behavior of the Si/SiO_2 interface observed by Fowler–Nordheim tunneling," *J. Appl. Phys.*, p. 559, January 1982.
Maserjian, J., and N. Zamani, "Observation of positively charged state generation near the Si/SiO_2 interface during Fowler–Nordheim tunneling," *J. Vac. Sci. Tech.*, vol. 20, p. 743, 1982.
Matsukawa, N., et al., "Selective polysilicon oxidation technology for VLSI isolation," *IEEE Trans. Elect. Dev.*, vol. ED–29, p. 561, 1982.
Suciu, P. I., et al., "Cell model for EEPROM floating-gate memories," *IEEE IEDM Tech. Dig.*, p. 737, 1982.
Takeda, E., et al., "Submicrometer MOSFET structure for minimizing hot carrier generation," *IEEE Trans. Elect. Dev.*, vol. ED–29, p. 611, 1982.
Vaidya, S., "Electromigration induced leakage at shallow junction contacts metallized with aluminum/poly-silicon," *Proc. IRPS*, 1982.
Weinberg, Z. A., "On tunneling in metal–oxide–silicon structures," *J. Appl. Phys.*, vol. 53, p. 5052, 1982.
Yamamoto, H., H. Iwasawa, and A. Sasaki, "Discharging process by multiple tunnelings in thin-oxide MNOS structures," *IEEE Trans. Elect. Dev.*, vol. ED–29, no. 8, p. 1255, 1982.
Yaron, G. et al., "A 16K E^2PROM employing new array architecture and designed-in reliability features," *IEEE J. Sol. St. Cir.*, vol. SC–17, no. 5, p. 833, 1982.
Yatsuda, Y., et al., "An advanced MNOS memory device for highly integrated byte erasable 5V only EEPROMs," *IEEE IEDM Tech. Dig.*, p. 733, 1982.
Yatsuda, Y., et al., "Scaling down MNOS nonvolatile memory devices," *Jap. J. Appl. Phys.*, vol. 21, S21–1, p. 85, 1982.

1983

Argawal, A. K., et al., "On the transient and steady-state transport of electrons and holes in the MNOS and MONOS devices," *IEEE IEDM Tech. Dig.*, p. 400, 1983.
Arzigian, J. S., "Aluminum electromigration lifetime variations with linewidth: the effects of changing stress conditions," *Proc. IRPS*, 1983.
Bagula, M., and R. Wong, "A 5V self-adaptive microcomputer with 16Kb of E^2 program storage and security," *IEEE ISSCC Tech. Dig.*, p. 34, 1983.
Becker, N. J., et al., "A 5V-only 4K nonvolatile static RAM," *IEEE ISSCC Tech. Dig.*, p. 170, 1983.
Brown, D. K., and C. A. Barile, "Ramp breakdown study of double polysilicon RAMs as a function of fabrication parameters," *J. Electrochem. Soc.*, vol. 130, p. 1597, 1983.
Caywood, J. M., and B. L. Prickett, "Radiation-induced soft errors and floating gate memories," *Proc. IRPS*, p. 167, 1983.
Dham, V., et al., "A 5V-only E^2PROM using 1.5μm lithography," *IEEE ISSCC Tech. Dig.*, p. 166, 1983.

Dobbs, C. S., W. D. Brown, and J. R. Yeargan, "Charge loss in metal–nitride–oxide–semiconductor (MNOS) devices at high temperatures," *Sol. St. Electr.*, vol. 26, no. 5, p. 427, 1983.

Donaldson, D. D., E. H. Honnigford, and L. J. Toth, "+5V-only 32K EEPROM," *IEEE ISSCC Tech. Dig.*, p. 168, 1983.

Gale, R., "Hydrogen migration under avalanche injection of electrons in Si metal-oxide-semiconductor capacitors," *J. Appl. Phys.*, vol. 54, p. 6938, 1983.

Haken, et al., "An 18V double-level poly CMOS technology for nonvolatile memory and linear applications," IEEE Int. Sol. St. Cir. Conf., p. 90, 1983.

Hieda, K., et al., "Optimum design of dual control gate cell for high density EEPROM's," *IEEE IEDM Tech. Dig.*, p. 593, 1983.

Hillen, M. W., et al., "Charge build-up prior to breakdown in thin gate oxides," Insulat. Films on Semicond., Proc. Intl. Conf. INFOS 83, p. 274, 1983.

Hillen, M. W., et al., "Influence of charge build-up on breakdown and wear-out in thin SiO_2 layers on Si," Proc. 1st Intl. Conf. on Conduction and Breakdown in Sol. Dielect., p. 355, 1983.

Horiuchi, M., and H. Katto, "FCAT-II: A 50 NS/15-V alterable nonvolatile memory device. I.: experimental," *IEEE Trans. Elect. Dev.*, vol. ED–30, no. 10, p. 1369, 1983.

Hsia, Y., "Threshold referenced MNOS sense amplifier," U.S. Patent No. 4,376,987, March 15, 1983.

Hsia, Y., E. Mei, and K. L. Ngai, "MNOS retention-endurance characteristics enhancement using graded nitride dielectric," *Jap. J. Appl. Phys.*, vol. 22, Suppl. 22–1, p. 89, 1983.

Hu, C., "Hot electron effects in MOSFET's," *IEEE IEDM Tech. Dig.*, p. 176, 1983.

Jewell-Larsen, S., et al., "5-volt RAM-like triple polysilicon EEPROM," Conf. Proc. 2nd Ann. Phoenix Conf., p. 508, 1983.

Kapoor, V. J., R. S. Bailey, and H. J. Stein, "Hydrogen-related memory traps in thin silicon nitride films," *J. Vac. Sci. Technol.*, vol. A1, no. 2, p. 600, 1983.

Katz, L. E., *VLSI Technology*, S. M. Sze (Ed.), McGraw-Hill, 1983.

Knoll, M. G., T. A. Dellin, and R. V. Jones, "A radiation-hardened 16 kbit MNOS EAROM," *IEEE Nucl. Sci.*, vol. NS-30, p. 4224, 1983.

Lai, S. K., J. Lee, and V. K. Dham, "Electrical properties of nitrided-oxide systems for use in gate dielectrics and EEPROM," *IEEE IEDM Tech. Dig.*, p. 190, 1983.

Lancaster, A., et al., "A 5 V-only EEPROM with internal program/erase control," *IEEE ISSCC Tech. Dig.*, p. 164, 1983.

Lee, D. J., et al., "Control logic and cell design for a 4K NVRAM," *IEEE J. Sol. St. Cir.*, vol. SC–18, p. 525, 1983.

Lee, J., and V. K. Dham, "Design considerations for scaling FLOTOX EEPROM cell," *IEEE IEDM Tech. Dig.*, p. 589, 1983.

Liang, M.-S., et al., "Creation and termination of substrate deep depletion in thin oxide MOS capacitors by charge tunneling," *IEEE Elect. Dev. Lett.*, vol. EDL–4, p. 350, 1983.

Liang, M.-S., et al., "Hot carrier-induced degradation in thin gate oxide MOSFET's," *IEEE IEDM Tech. Dig.*, p. 186, 1983.

Lin, P. S. D., R. B. Marcus, and T. T. Sheng, "Leakage and breakdown in thin oxide capacitors—correlation with decorated stacking faults," *J. Electrochem. Soc.*, vol. 130, p. 1878, 1983.

Lucero, E. M., N. Challa, and J. Fields, "A 16 kbit smart 5V-only EEPROM with redundancy," *IEEE J. Sol. St. Cir.*, vol. SC–18, p. 539, 1983.

Maes, H. E., "Recent developments in non-volatile semiconductor memories," *Tech. Dig.*, ESSCIRC83, p. 1, 1983.

Maes, H. E., and G. Heyns, "Two-carrier conduction in amorphous chemically vapour deposited (CVD) silicon nitride layers," Proc. Int. Conf. Insul. Films on Semicond., North Holland Publ. Co., p. 215, 1983.

Maes, H. E., and J. Remmerie, "Effects of annealing in different ambients on the hydrogen and charge distributions in CVD silicon nitride films," Proc. Electrochem. Soc., vol. 83–8, Eds. V. J. Kapoor and H. Stein, p. 73, 1983.

Manzini, S., and A. Modelli, "Tunneling discharge of trapped holes in silicon dioxide," Insulating Films on Semicond., J. F. Verwey and D. R. Wolters (Ed.), Elsevier Science Publishers, p. 112, 1983.

Mielke, N., "New EPROM data-loss mechanisms," *Proc. IRPS*, p. 106, 1983.

Oto, D. H., et al., "High-voltage regulation and process considerations for high-density 5V-only EEPROM's," *IEEE J. Sol. St. Cir.*, vol. SC–18, p. 532, 1983.

Ricco, B., M. Ya. Azbel, and M. H. Brodsky, "Novel mechanism for tunneling and breakdown of thin SiO_2 films," *Phys. Rev. Lett.*, vol. 51, no. 19, p. 1795, 1983.

Schols, G., and H. E. Maes, "High temperature hydrogen anneal of MNOS structures," Proc. Electrochem. Soc., vol. 83–8, Eds. V. J. Kapoor and H. Stein, p. 94, 1983.

Shiner, R., N. Mielke, and R. Haq, "Characterization and screening of SiO_2 defects in EEPROM structures," *Proc. IRPS*, p. 248, 1983.

Sternhein, M., et al., "Properties of thermal oxides grown on phosphorus in-situ doped polysilicon," *J. Electrochem. Soc.*, vol. 130, p. 1735, 1983.

Suzuki, E., et al., "A low voltage alterable EEPROM with metal–oxide–nitride–oxide–semiconductor (MONOS) structure," *IEEE Trans. Elect. Dev.*, ED–30, p. 122, 1983.

Suzuki, E., et al., "Traps created at the interface between the nitride and the oxide on the nitride by thermal oxidation," *Appl. Phys. Lett.*, vol. 42, no. 7, p. 609, 1983.

Sze, S .M., ed., "VLSI Technology," Chapter 14, *Yield and Reliability*, by W. J. Bertram, McGraw-Hill, New York, 1983.

Takeda, E., and N. Suzuki, "An empirical model for device degradation due to hot-carrier injection," *IEEE Elect. Dev. Lett.*, vol. EDL–4, no. 4, April 1983.

Takeda, E., N. Suzuki, and T. Hagiwara, "Device performance degradation due to hot-carrier injection at energies below the Si-SiO_2 energy barrier," *IEEE IEDM Tech. Dig.*, p. 396, 1983.

Towner, J. M., "Aluminum electromigration under pulsed D.C. conditions," *Proc. IRPS*, 1983.

Vail, P., "Radiation hardened MNOS: a review," Proc. Electrochem. Soc., vol. 83–8, p. 207, 1983.

Van Buskirk, M., et al., "EPROMs graduate to 256-k density with scaled n-channel process," *Electronics*, February 24, 1983.

van der Schoot, J. J., and D. R. Wolters, "Current induced dielectric breakdown," Insulat. Films on Semicond., Proc. Intl. Conf. INFOS 83, p. 270, 1983.

Wolters, D. R., J. J. van der Schoot, and T. Poorter, "Damage caused by charge injection," Insulat. Films on Semicond., Proc. Intl. Conf. INFOS 83, p. 256, 1983.

Yamabe, K., K. Taniguchi, and Y. Matsushita, "Thickness dependence of dielectric breakdown failure of thermal SiO_2 films," *Proc. IRPS*, p. 184, 1983.

Yoshida, M., et al., "A 288K CMOS EPROM with redundancy," *IEEE J. Sol. St. Cir.*, vol. SC–18, p. 544, 1983.

1984

Alexander, K., J. Hicks, and T. Soukup, "Moisture resistive U.V. transmissive passivation for plastic encapsulated EPROM devices," *Proc. IRPS*, p. 218, 1984.

Baglee, D. A., "Characteristics and reliability of 100 Å oxides," *Proc. IRPS*, p. 152, 1984.

Bhattacharyya, A., "Modelling of write/erase and charge retention characteristics of floating gate EEPROM devices," *Sol. St. Electr.*, vol. 27, p. 899, 1984.

Bibyk, S. B., and V. J. Kapoor, "Trapping kinetics in high trap density silicon nitride insulators," *J. Appl. Phys.*, vol. 56, no. 4, p. 1070, 984.

Blanford, J. T., A. E. Waskiewicz, and J. C. Pickel, "Cosmic ray induced permanent damage in MNOS EAROMs," *IEEE Trans. Nucl. Sci.*, NS–31, p. 1568, 1984.

Bursky, D., "Nonvolatile memories—enroute to higher density, speed, and reliability—explore new processes, circuit refinements, and above all, CMOS," *Electron. Des. (USA)*, vol. 32, no. 17, p. 122, 1984.

Cuppens, R. et al., "An EEPROM for microprocessors and custom logic," IEEE *ISSCC Tech. Dig.*, p. 268, 1984.

De Keersmaecker, R. F., et al., "Breakdown and wearout of MOS gate oxides," Proc. 2nd Internat. Symp. VLSI Sci. and Tech.: Materials for High Speed/High Density Applications, p. 301, 1984.

Dettmer, R. "E^2PROMs: The quest for the nonvolatile RAM," *Electr. and Pow. (GB)*, vol. 30, no. 5, p. 359, 1984.

Domangue, E. R., R. Rivers, and C. Shepard, "Reliability predicting using large MOS capacitors," *Proc. IRPS*, p. 140, 1984.

Duthie, I., "ROMS to bubbles: the selection of nonvolatile memories," *Electron. and Pow. (GB)*, vol. 30, no. 11–12, p. 865, 1984.

Globig, J. E., "Designing with serial E^2PROMs," ELECTRO 84, Electron. Conventions, p. 708, 1984.

Groeseneken, G., et al., "A reliable approach to charge-pumping measurements in MOS transistors," *IEEE Trans. Elect. Dev.*, vol. ED–31, no. 1, p. 41, 1984.

Grossman, S. "EE-PROMs open new application areas to the design engineer," *Electr.*, vol. 57, no. 7, p. 129, 1984.

Hatfield, R. L., "Serial to bytewide NOVRAMs for every microprocessor application," MIDCON 84, Electron. Conventions, p. 468, 1984.

Hezel, R., K. Blumenstock, and R. Schorner, "Interface states and fixed charges in MNOS structures with APCVD and plasma silicon nitride," *J. Electrochem. Soc.*, vol. 131, no. 7, p. 1679, 1984.

Holland, S., et al., "On physical models for gate oxide breakdown," *IEEE Elect. Dev. Lett.*, vol. EDL–5, no. 6, p. 302, 1984.

Hsu, F.-C., and K.-Y. Chiu, "A comparative study of tunneling, substrate hot-electron, and channel hot-electron injection induced degradation in thin-gate MOSFET's," *IEEE IEDM Tech. Dig.*, p. 96, 1984.

Iyer, S. S., and C.-Y. Ting, "Electromigration study of the Al-Cu/Ti/Al-Cu system," *Proc. IRPS*, 1984.

Johnson, A., "Super recovery of total dose damage in MOS devices," *IEEE Trans. Nucl. Sci.*, vol. NS–31, p. 1568, 1984.

Jolly, R. D., H. R. Grinolds, and R. Groth, "A model for conduction in floating gate EEPROM's," *IEEE Trans. Elect. Dev.*, vol. ED–31, p. 767, 1984.

Kessler, R. A., and E. P. Endre, "Accelerated testing of nonvolatile memory retention," Electr. Reliabil. Conf., SAE Special Publications SP–573, p. 17, 1984.

Kuniyoshi, S., K. Itabashi, and K. Tanaka, "Charge analysis of MNOS memory by the thermally stimulated current," Proc. 17th Symp. Elect. Insul. Mats., p. 145, 1984.

Lai, S. K., et al., "Design of an E^2PROM memory cell less than 100 square microns using micron technology," *IEEE IEDM Tech. Dig.*, p. 468, 1984.

Liang, M.-S., et al., "MOSFET degradation due to stressing of thin oxide," *IEEE Trans. Elect. Dev.*, vol. ED–31, p. 1238, 1984.

Maes, H., and G. Groeseneken, "Conduction in thermal oxides grown on polysilicon and its influence on floating gate EEPROM degradation," *IEEE IEDM Tech. Dig.*, p. 476, 1984.

Masuoka, F., et al., "A new Flash EEPROM cell using triple polysilicon technology," *IEEE IEDM Tech. Dig.*, p. 464, 1984.

Matsukawa, N., S. Morita, and H. Nozawa, "High performance EEPROM using low barrier height tunnel oxide," Ext. Abstr. Int. Conf. Sol. St. Dev. and Mat., p. 261, 1984.

Mehrotra, S., et al., "A 64Kb CMOS EEPROM with on-chip EEC," *ESSCC Digest of Tech. Papers*, p. 142, 1984.

Mori, S., et al., "Polyoxide/nitride/oxide structure for highly reliable EPROM cells," *Tech. Dig. Symp. VLSI Tech.*, p. 38, 1984.

Moslehi, M. M., and K. C. Saraswat, "Studies of trapping and conduction in ultra-thin SiO_2 gate insulators," *IEEE IEDM Tech. Dig.*, p. 157, 1984.

Robertson, J., and M. J. Powell, "Gap states in silicon nitride," *Appl. Phys. Lett.*, vol. 44, no. 4, p. 415, 1984.

Shatzkes, M., M. Av-Ron, and K. V. Srikrishnan, "Determination of reliability from ramped voltage breakdown experiments; application to dual dielectric MIM capacitors," *Proc. IRPS*, p. 138, 1984.

Stewart, R., et al., "A shielded substrate injector MOS (SSIMOS) EEPROM cell," *IEEE IEDM Tech. Dig.*, p. 472, 1984.

Topich, J. A., "Long-term retention of SNOS nonvolatile memory devices," *IEEE Trans. Elect. Dev.*, vol. ED–31, no. 12, p. 1908, 1984.

Wegener, H. A. R., "Endurance model for textured poly floating gate memories," *IEEE IEDM Tech. Dig.*, p. 480, 1984.

Yoshikawa, K., et al., "Technology requirements for mega bit CMOS EPROMS," *IEEE IEDM Tech. Dig.*, p. 456, 1984.

1985

Agarwal, A. K., and M. W. White, "New results on electron injection, hole injection, and trapping in MNOS nonvolatile memory devices," *IEEE Trans. Elect. Dev.*, vol. ED–32, no. 5, p. 941, 1985.

Baglee, D. A., and M. C. Smayling, "The effects of write/erase cycling on data loss in EEPROMs," *IEEE IEDM Tech. Dig.*, p. 624, 1985.

Bill, C. S., et al., "A temperature and process tolerant 64K EEPROM," *IEEE J. Sol. St. Cir.*, vol. SC–20, no. 5, p. 979, 1985.

Brown, W. D., R. V. Jones, and R. D. Nasby, "The MONOS memory transistor: application in a radiation-hard nonvolatile RAM," *Sol. St. Electr.*, vol. 29, p. 877, 1985.

Chan, T. Y., P. K. Ko, and C. Hu, "Dependence of channel electric field on device scaling," *IEEE Elect. Dev. Lett.*, vol. EDL–6, p. 551, 1985.

Chen, I. C., S. E. Holland, and C. Hu, "Electrical breakdown in thin gate and tunneling oxides," *IEEE Trans. Elect. Dev.*, vol. ED–32, no. 2, p. 413, 1985.

Chen, I. C., S. Holland, and C. Hu, "A quantitative physical model for time-dependent breakdown in SiO_2," *Proc. IRPS*, p. 24, 1985.

Hu, Chenming, et al., "Hot electron induced MOSFET degradation—model, monitor and improvement," *IEEE Trans. Elect. Dev.*, vol. ED–32, no. 2, 1985.

Claassen, W. A. P., et al., "Influence of deposition temperature, gas pressure, gas phase composition and RF frequency on composition and mechanical stress of plasma silicon nitride layers," *J. Electrochem. Soc.*, vol. 132, no. 4, p. 894, 1985.

Cohen, Y. N., J. Shappir, and D. F. Bentchkowsky, "Dynamic model of trapping-detrapping in SiO_2," *J. Appl. Phys.*, vol. 58, p. 2252, 1985.

Cuppens, R., et al., "An EEPROM for microprocessors and custom logic," *IEEE J. Sol. St. Cir.*, vol. SC-20, no. 2, p. 603, 1985.

Duthie, I., "ROMs to bubbles: non-volatile memories, which-when-why?", *Microelectron. J. (GB)*, vol. 16, no. 3, p. 13, 1985.

Faraone, L., et al., "Characterization of thermally oxidized n^+ polycrystalline silicon," *IEEE Trans. Elec. Dev.*, vol. ED–32, p. 577, 1985.

Fujita, S., and A. Sasaki, "Dangling bonds in memory-quality silicon nitride films," *J. Electrochem. Soc.*, vol. 132, no. 2, p. 398, 1985.

Gerosa, G., et al., "A high performance CMOS technology for 256K/1Mb EPROMs," *IEEE IEDM Tech. Dig.*, p. 631, 1985.

Hata, T., et al., "Data retention test method of EPROMS," Proc. ISTFA 1985: Intl. Symp. for Testing and Failure Analysis, p. 104, 1985.

Heyns, G. L., and H. E. Maes, "New model for the discharge behavior of metal–nitride–oxide–silicon (MNOS) non-volatile memory devices," *Appl. Surf. Sci.*, vol. 30, no. 1–4, p. 153, 1985.

Hezel, R., and W. Bauch, "Charge storage properties of plasma-deposited silicon nitride films and the effect of interface states," *Electrochem. Soc. Ext. Abs.*, vol. 85–2, p. 305, 1985.

Hill, M. D., "55ns CMOS E^2PROMS that work in bipolar PROM sockets," Conf. Rec.: WESCON/85, pap. 13.2, 1985.

Hokari, Y., T. Baba, and N. Kawamura, "Reliability of 6–10 nm thermal SiO_2 films showing intrinsic dielectric integrity," *IEEE Trans. Elect. Dev.*, vol. ED–32, p. 2485, 1985.

Hoffmann, K. R., et al., "Hot-electron and hole emission effects in short n-channel MOSFET's," *IEEE Trans. Elect. Dev.*, vol. ED–32, p. 691, 1985.

Hong, C. C., and D. L. Crook, "Breakdown energy of metal (BEM)—A new technique for monitoring metallization," *Proc. IRPS*, 1985.

Horiguchi, S., et al., "Interface-trap generation modeling of Fowler–Nordheim tunnel injection into ultra-thin gate oxide," *J. Appl. Phys.*, vol. 58, p. 387, 1985.

Hosoi, T., M. Akizawa, and S. Matsumoto, "The effect of Fowler–Nordheim tunneling current on thin SiO_2 metal–oxide–semiconductor capacitors," *J. Appl. Phys.*, vol. 57, no. 6, p. 2072, 1985.

Hsu, F.-C., and K.-Y. Chiu, "Hot electron substrate current generation during switching transient," *IEEE Trans. Elect. Dev.*, vol. ED–32, p. 375, 1985.

Hu, C., "Thin oxide reliability," *IEEE IEDM Tech. Dig.*, p. 368, 1985.

Hu, C., et al., "Hot-electron-induced MOSFET degradation—model, monitor and improvement," *IEEE Trans. Elect. Dev.*, vol. ED–32, p. 375, 1985.

Jaffe, J. "Reliability of textured poly floating gate E^2PROMS," Midcon/85 Conf. Rec., vol. 15, no. 2, p. 1, 1985.

Krause, H., "Trap induction and breakdown mechanism in SiO_2 films," Phys. Status Solidi a (Germany), vol. 89, no. 1, p. 353, 1985.

Kuniyoshi, S., and K. Tanaka, "Charge analysis of MNOS memory devices by thermally stimulated current," Proc. 18th Symp. on Elect. Insul. Mats., p. 45, 1985.

Libsch, F. R., A. Roy, and M. H. White, "Amphoteric trap modeling of multidielectric scaled SONOS nonvolatile memory structures," *Appl. Sur. Sci.*, vol. 30, no. 1–4, p. 160, 1985.

Manzini, S., and F. Volonte, "Charge transport and trapping in silicon nitride-silicon dioxide dielectric double layers," *J. Appl. Phys.*, vol. 58, no. 11, p. 4300, 1985.

Masuoka, F., et al., "A 256K Flash EEPROM using triple polysilicon technology," *IEEE ISSCC Tech. Dig.*, p. 168, 1985.

McPherson, J. W., and D. A. Baglee, "Acceleration factors for thin gate oxide stressing," Proc. IRPS, p. 1, 1985.

Mizutani, Y., and K. Makita, "A new EPROM cell with a side-wall floating gate for high density and high performance device," *IEEE IEDM Tech. Dig.*, p. 63, 1985.

Mori, S., et al., "Reliable CVD inter-poly dielectrics for advanced E&EEPROM," 1985 Symp. VLSI Tech., *Tech. Dig.*, p. 16, 1985.

Muhkerjee, S., " A single transistor EEPROM cell and its implementation in a 512K CMOS EEPROM," *IEEE IEDM Tech. Dig.*, p. 616, 1985.

Nagai, K., T. Sekigawa, and Y. Hayashi, "Capacitance-voltage characteristics of semiconductor–insulator–semiconductor (SIS) structure," *Sol. St. Electr.*, vol. 28, no. 8, p. 789, 1985.

Nissan-Cohen, Y., J. Shappir, and D. Frohman-Bentchkowsky, "High-field and current-induced positive charge in thermal SiO_2 layers," *J. Appl. Phys.*, vol. 57, p. 2830, 1985.

Palhak, S., et al., "A 25-ns 16K CMOS PROM using a four-transistor cell and differential design techniques," *IEEE J. Sol. St. Cir.*, vol. SC–20, no. 5, p. 964, 1985.

Pryor, R. W., "Isotope effects in MNOS transistors," *IEEE Elect. Dev. Lett.*, vol. EDL–6, no. 1, p. 31, 1985.

Pye, K., "Ideal non-volatile memory," Conf. Rec.: WESCON/85, pap. 13.5, 1985.

Root, B. J., and T. Turner, "Wafer level electromigration tests for production monitoring," *Proc. IRPS*, 1985.

Saleh, N., A. El-Hennawy, and S. El-Hennawy, "Simulation of a non-avalanche injection based CMOS EEPROM memory cell compatible with scaling down trends," Conf. Proc.: IEEE ELECTRONICOM '85, p. 590, 1985.

Schafft, H. A., and T. C. Grant, "Electromigration and the current density dependence," *Proc. IRPS*, 1985.

Suciu, P. I., et al., "A 64K EEPROM with extended temperature range and page mode operation," *IEEE ISSCC Tech. Dig.*, p. 170, 1985.

Tesch, R., et al., "A 35-ns 64K EEPROM," *IEEE J. Sol. St. Cir.*, vol. SC–20, no. 5, p. 971, 1985.

Tzou, J. J., et al., "Hot-electron-induced MOSFET degradation at low temperatures," *IEEE Elect. Dev. Lett.*, EDL–6, no. 9, 1985.

Verwey, J. F., and E. J. Korma, "Oxide layers in EEPROM devices," Phys. of Semicond. Dev., Proc. 3rd Intl. Workshop, p. 38, 1985.

Wolters, D. R., and J. J. van der Schoot, "Dielectric breakdown in MOS devices. I. Defect-related and intrinsic breakdown," *Philips J. Res.*, vol. 40, no. 3, p. 115, 1985.

Wolters, D. R., and J. J. van der Schoot, "Dielectric breakdown in MOS devices. II. Conditions for the intrinsic breakdown," *Philips J. Res.*, vol. 40, no. 3, p. 137, 1985.

Wolters, D. R., and J. J. van der Schoot, "Dielectric breakdown in MOS devices. III. The damage leading to breakdown," *Philips J. Res.*, vol. 40, no. 3, p. 164, 1985.

Yamabe, K., and K. Taniguchi, "Time-dependent dielectric breakdown of thin thermally grown SiO_2 films," *IEEE Trans. Elect. Dev.*, vol. ED–32, p. 423, 1985.

Yatsuda, Y., et al., "Hi-MNOS II technology for a 64-kbit byte-erasable 5V-only EEPROM," *IEEE Trans. Elect. Dev.*, vol. ED–32, no. 2, p. 224, 1985.

Yatsuda, Y., et al., "Hi-MNOS II technology for a 64-Kbit byte-erasable 5V-only EEPROM," *IEEE J. Sol. St. Cir.*, vol. SC–20, no. 1, p. 144, 1985.

1986

Asai, S., "Semiconductor memory trends," *Proc. IEEE*, vol. 74, no. 12, p. 1623, 1986.

Bisschop, J., et al., "A model for the electrical conduction in polysilicon oxide," *IEEE Trans. Elect. Dev.*, vol. ED–33, p. 1809, 1986.

Chen, I. C., S. Holland, and C. Hu, "Oxide breakdown dependence on thickness and hole current-enhanced reliability of ultra thin oxides," *IEEE IEDM Tech. Dig.*, p. 660, 1986.

Chen, C.-F., and C.-Y. Wu, "A characterization model for constant current stressed voltage-time characteristics of thin thermal oxides grown on silicon substrates," *J. Appl. Phys.*, vol. 60, no. 11, p. 3926, 1986.

Chen, C. F., and C. Y. Wu, "A characterization model for ramp-voltage-stressed I-V characteristics of thin thermal oxide grown on silicon substrate," *Sol. St. Electr.*, vol. 29, p. 1059, October 1986.

Cole, B. C., "Exploding role of nonvolatile memory," *Electronics*, vol. 59, no. 29, p. 47, 1986.

De Almeida, A. A. M., "Effects of processing variations and endurance stress on the MNOS nonvolatile memory device," Ph.D. diss., University of Florida, 1986.

De Almeida, A. M., and S. S. Li, "Effects of varying the processing parameters on the interface-state density and retention characteristics of an MNOS capacitor," *Sol. St. Electr.*, vol. 29, no. 6, p. 619, 1986.

Demidova, G. N., and N. I. Gavrilin, "Spectrum of defects responsible for breakdown of poly–SiN–SiO_2–Si and Al–SiO_2–Si structures," Trans. in Sov. Microelectron. (USA), vol. 15, no. 6, p. 272, 1986.

Dixon, S., and P. Shah, "Stress-and-test procedure weeds out weak EPROMs," *Electron. Prod.*, vol. 28, no. 20, p. 47, 1986.

Esquivel, J., et al., "High density contactless self aligned EPROM cell array technology," *IEEE IEDM Tech. Dig.*, p. 592, 1986.

Faraone, L., "Endurance of 9.3 nm EEPROM tunnel oxide," Insulat. Films on Semicond., Proc. Intl. Conf. INFOS 85, p. 151, 1986.

Faraone, L., "Thermal SiO_2 Films on n^+ polycrystalline silicon: electrical conduction and breakdown," *IEEE Trans. Elect. Dev.*, vol. ED–33, p. 1785, 1986.

Fong, Y., et al., "Dynamic stressing on thin oxide," *IEEE IEDM Tech. Dig.*, p. 664, 1986.

Gadiyak, G. V., M. S. Obrecht, and S. P. Sinitsa, "Numerical simulation of bipolar injection and recombination in MNOS structures," *Intl. J. for Comput. and Math. in Elect. and Electr. Engr.*, vol. 5, no. 4, p. 227, 1986.

Gadiyak, G. V., and I. V. Travkov, "Injection and charge transfer in MNOS structures," Optoelectron. Instrum. Data Process., no. 5, p. 93, 1986.

Ghidini, G., and A. Modelli, "Processing dependence and wear-out characterization in thin SiO_2 films," Insulat. Films on Semicond., Proc. Intl. Conf. INFOS 85, p. 141, 1986.

Goetting, E., "EEPROM-based ASIC propels programmable logic to new levels of complexity," *Electronic Design*, vol. 34, no. 10, p. 201, 1986.

Goldberger, A., "Overview: electrically erasable memory technology," Electro/86 and Mini/Micro Northeast Conf. Rec., p. 12, 1986.

Groeseneken, G., "Programming behaviour and degradation phenomena in electrically erasable programmable floating gate memory devices," Ph.D. thesis, K. U. Leuven, 1986.

Groeseneken, G., and H. E. Maes, "A quantitative model for the conduction in oxides thermally grown from polycrystalline silicon," *IEEE Trans. Elect. Dev.*, vol. ED–33, p. 1028, 1986.

Guterman, D. C., et al., "New ultra-high density textured poly-Si floating gate EEPROM cell," *IEEE IEDM Tech. Dig.*, p. 826, 1986.

Guterman, D. C., "Nonvolatile electrically alterable memory," U.S. Patent 4,599,706, 1986.

Heremans, P., H. E. Maes, and N. Saks, "Evaluation of hot carrier degradation of n-channel MOSFET's with the charge pumping technique," *IEEE Elect. Dev. Lett.*, vol. EDL–7, p. 428, 1986.

Kolodny, A., et al., "Analysis and modeling of floating gate EEPROM cells," *IEEE Trans. Elect. Dev.*, vol. ED–33, p. 835, 1986.

Kusaka, T., Y. Ohji, and K. Mukai, "Breakdown characteristics of ultra thin silicon oxide," Ext. Abs. 18th Intl. Conf. Sol. St. Dev. and Mats., p. 463, 1986.

Lai, S. K., and V. Dham, "Comparison and trends in today's dominant E^2PROM technologies," *IEEE IEDM Tech. Dig.*, p. 580, 1986.

Lee, S.-S., et al., "Three transistor cell for high speed CMOS EPROM technology," *IEEE IEDM Tech. Dig.*, p. 588, 1986.

Lee, Y.-H., "Dual-carrier charge transport and damage formation in LPCVD nitride for nonvolatile memory devices," Ph.D. diss., Ohio State University, 1986.

Liang, M.-S., et al., "Degradation of very thin gate oxide MOS devices under dynamic high field/current stress," *IEEE IEDM Tech. Digest*, p. 394, 1986.

Libsch, F. R., A. Roy, and M. H. White, "A computer simulation program for erase/write characterization of ultra-thin nitride scaled SONOS/MONOS memory transistors," Ext. Abstr., Electrochem. Soc. Fall Mtg., vol. 86–2, 1986.

Madan, P., and G. Landers, "EEPROMs provide more memory solutions than you may think," Electro/86 and Mini/Micro Northeast Conf. Rec., p. 12, 1986.

Marin, K., "Four approaches to nonvolatile memory open to designers," Computer Design, vol. 25, no. 17, p. 30, 1986.

Matsubara, K., and T. Arai, "EEPROM on-chip single-chip microcomputer," *Hitachi Review*, vol. 35, no. 5, p. 237, 1986.

McGowan, J. E., "A field programmable logic array using EEPROM technology," M.S. thesis, Santa Clara University, 1986.

McWhorter, P. J., et al., "Radiation response of SNOS nonvolatile transistors," *IEEE Trans. Nucl. Sci.*, NS–33, p. 1414, 1986.

Messenger, G. C., and M. S. Ash, *The Effects of Radiation on Electronic Systems*, Van Nostrand Reinhold, 1986.

Miyamoto, J. I., et al., "An experimental 5V-only 256-kbit CMOS EEPROM with a high-performance single-polysilicon cell," *IEEE J. Sol. St. Cir.*, vol. SC–21, p. 852, 1986.

Mori, S., et al., "Reliability aspects of 100 Å inter-poly dielectrics for high density VLSIs," 1986 Symp. VLSI Tech., *Tech. Dig.*, 1986.

Nozaki, S., and R. V. Giridhar, "Study of carrier trapping in stacked dielectrics," *IEEE Elect. Dev. Lett.*, vol. EDL–7, no. 8, p. 486, 1986.

Nozawa, H., N. Matsukawa, and S. Morita, "An EEPROM cell using a low barrier height tunnel oxide," *IEEE Trans. Elect. Dev.*, vol. ED–33, no. 2, p. 275, 1986.

Parkerson, J. P., "High-endurance testing of nonvolatile-random-access-memory," M.S. thesis, University of Arkansas, 1986.

Righter, B., and D. Sur, "EEPROM technology revitalises PLAs," *Electron. Prod. Des.*, vol. 7, no. 2, p. 41, 1986.

Saks, N. S., et al., "Observation of hot-hole injection in nMOS transistors using a modified floating-gate technique," *IEEE Trans. Elect. Dev.*, vol. ED–33, p. 1529, 1986.

Schneider, B., C. Hansen, and G. Jorgensen, "E^2PROMS: a comparative evaluation into the performance of E^2PROMS under various extreme environmental and operating test conditions," Elektronikcentralen Rep ECR, p. 46, 1986.

Sekiya, K., et al., "Trench self-aligned EPROM technology," *Tech. Dig.*: Symp. on VLSI Tech., p. 87, 1986.

Shukuri, S., et al., "Novel EPROM device using focused boron ion beam implantation," Ext. Abs.: 1986 Intl. Conf. on Sol. St. Dev. and Mats., p. 327, 1986.

Slocombe, D., "Programming schemes change applications role of EPROMS," *EDN*, vol. 31, no. 22, p. 191, 1986.

Stewart, R., et al., "High density EPROM cell and array," *Tech. Dig*: 1986 Symp. on VLSI Tech., p. 87, 1986.

Suzuki, E., and Y. Hayashi, "On oxide–nitride interface traps by thermal oxidation of thin nitride in metal–oxide–nitride–oxide–semiconductor memory structures," *IEEE Trans. Elect. Dev.*, vol. ED–33, no. 2, p. 214, 1986.

Swartz, G. A., "Gate oxide integrity of MOS/SOS devices," *IEEE Trans. Elect. Dev.*, vol. ED–33, p. 119, 1986.

Verwey, J. F., and D. R. Wolters, *Insulating Films on Semiconductors*, p. 125, 1986.

Venkatesh, B., et al., "MOS 1MB EPROM," *IEEE ISSCC Tech. Dig.*, p. 40, 1986.

Vollebregt, F., J. F. Verwey, and D. Wolters, "The switching degradation in floating gate transistors," Insulat. Films on Semicond., Proc. Intl. Conf. INFOS 85, p. 137, 1986.

Wolters, D. R., and J. J. van der Schoot, "Breakdown by charge injection," Insulat. Films on Semicond., Proc. Intl. Conf. INFOS 85, p. 145, 1986.

Wu, A., et al., "A novel high-speed, 5-V programming EPROM structure with source-side injection," *IEEE IEDM Tech. Dig.*, p. 584, 1986.

Yau, L. D., "Determination of the Fowler–Nordheim tunneling barrier from nitride to oxide in oxide:nitride dual dielectric," *IEEE Elect. Dev. Lett.*, vol. EDL–7, no. 6, p. 365, 1986.

Yoshikawa, K., S. Mori, and N. Arai, "EPROM cell structure for EPLDs compatible with single poly gate process," Ext. Abs.: 1986 Intl. Conf. on Sol. St. Dev. and Mats., p. 323, 1986.

1987

Ali, S. B., et al., "50-ns 256K CMOS split-gate EPROM," *IEEE J. Sol. St. Cir.*, vol. SC–23, no. 1, p. 79, 1987.

Ali, S., et al., "50 ns 256K CMOS split gate EPROM," *Symp. VLSI Tech. Tech. Dig.*, p. 53, 1987.

Atsumi, S., et al., "A 120ns 4Mb CMOS EPROM," *IEEE ISSCC Tech. Dig.*, p. 74, 1987.

Baglee, D. A., et al., "Effects of processing on EEPROM reliability," *Proc. IRPS*, p. 93, 1987.

Bleiker, C., and H. Melchior, "Four-state EEPROM using floating-gate memory cells," *IEEE J. Sol. St. Cir.*, vol. SC–22, no. 3, p. 460, 1987.

Calzi, P., and J. Devin, "Anatomy of a 256-KBit-EPROM," *Elektronik*, vol. 36, no. 7, p. 82, 1987.

Cham, K. M., et al., "Self-limiting behavior of hot carrier degradation and its implication on the viability of lifetime extraction by accelerated stress," *Proc. IRPS*, 1987.

Chan, T. Y., K. K. Young, and C. Hu, "True single-transistor oxide–nitride–oxide EEPROM device," *IEEE Elect. Dev. Lett.*, vol. ED–8, no. 3, p. 93, 1987.

Chang, C., and J. Lien, "Corner-field induced drain leakage in thin oxide MOSFETs," *IEEE IEDM Tech. Dig.*, p. 714, 1987.

Chao, C. C., and M. H. White, "Characterization of charge injection and trapping in scaled SONOS/MONOS memory devices," *Sol. St. Electr.*, vol. 30, no. 3, p. 307, 1987.

Chen, I. C., J. Lee, and C. Hu, "Accelerated testing of silicon dioxide wearout," 1987 Symposium on VLSI Tech., Dig. Tech. Paps., p. 23, 1987.

Chen, C.-F., et al., "The dielectric reliability of intrinsic thin SiO_2 films thermally grown on a heavily doped Si substrate-characterization and modeling," *IEEE Trans. Elect. Dev.*, vol. ED–34, no. 7, p. 1540, 1987.

Chen, J., et al., "Subbreakdown drain leakage current in MOSFET," *IEEE Elect. Dev. Lett.*, vol. EDL–8, p. 515, 1987.

Chingchi, Y., et al., "Structure and frequency dependence of hot-carrier-induced degradation in CMOS VLSI," Proc. IRPS, 1987.

Chu, S. S.-D., and A. J. Steckl, "Effect of trench-gate-oxide structure on EPROM device operation," *IEEE Elect. Dev. Lett.*, vol. EDL–9, no. 6, p. 284, 1987.

Cioaca, D., et al., "A million-cycle CMOS 256K EEPROM," *IEEE J. Sol. St. Cir.*, vol. SC–22, no. 5, p. 684, 1987.

Coffman, T., et al., "A 1M CMOS EPROM with a 13.5μ^2 cell," *IEEE ISSCC Tech. Dig.*, p. 72, 1987.

Cole, B. C., "Catalyst's EEPROM needs a miserly 3 volts," *Electr.*, vol. 60, no. 14, p. 67, 1987.

Cole, B. C., "Changing face of nonvolatile memories," *Electr.*, vol. 60, no. 14, p. 61, 1987.

Cole, B. C., "Fast control stores can now use EEPROMs," *Electr.*, vol. 60, no. 9, p. 66, 1987.

Cole, B. C., "Waferscale's 256-K EPROM runs superfast," *Electr.*, vol. 60, no. 14, p. 65, 1987.

Cordan, E., and M. Eby, "EEPROM supercell for custom standard cell ICs," Future Transportation Technol. Conf. and Expos., pap. 871544, 1987.

De Almeida, A. M., and S. S. Li, "Retention-temperature and endurance characteristics of the MNOS capacitor with processing variations as parameter," *Sol. St. Electr.*, vol. 30, no. 9, p. 889, 1987.

Dellin, T. A., and P. J. McWhorter, "Scaling of MONOS nonvolatile memory transistors," Proc. Electrochem. Soc., vol. 87-10, Eds. V. J. Kapoor and K. T. Hankins, p. 3, 1987.

Elliott, N., "Road to high performance EEPROMS," *New Electronics*, vol. 20, no. 8, p. 46, 1987.

Esquivel, A., et al., "A 8.6μ^2 cell technology for a 35.5mm^2 Megabit EPROM," *IEEE IEDM Tech. Dig.*, p. 859, 1987.

Esquivel, A., et al., "Novel trench-isolated buried N^{++} FAMOS transistor suitable for high-density EPROMs," *IEEE Elect. Dev. Lett.*, vol. EDL–8, no. 4, p. 146, 1987.

Fong, Y., et al., "Oxides grown on textured single-crystal silicon for low programming voltage non-volatile memory applications," *IEEE IEDM Tech. Dig.*, p. 389, 1987.

Gadiyak, G. V., et al., "Frenkel-limited monopolar conductivity of MNOS structures," *Soviet Microelectr.*, vol. 16, no. 1, p. 27, 1987.

Gill, M., et al., "Contactless one-transistor cell for VLSI Flash EEPROM," *IEEE Trans. Elect. Dev.*, vol. ED–34, no. 11, p. 2372, 1987.

Gritsenko, V. A., et al., "Nonstationary transport of electrons and holes in the depolarized mode of MNOS devices: an experiment and numerical modeling," *Soviet Microelectr.*, vol. 16, no. 1, p. 19, 1987.

Gurtov, V. A., A. I. Nazarov, and V. Y. Uritskiy, "Effect of an external voltage on the charged state of MNOS memory elements during irradiation," *Sov. J. Commin. Technol. Electron.*, vol. 32, no. 1, p. 188, 1987.

Haddad, S., and M.-S. Liang, "The nature of charge trapping responsible for thin-oxide breakdown under a dynamic field stress," *IEEE Elect. Dev. Lett.*, vol. EDL–8, no. 11, p. 524, 1987.

Haifley, T., "Endurance of EEPROMs with on-chip error correction," *IEEE Trans. on Reliability*, vol. R-36, no. 2, p. 222, 1987.

Hansen, C., and G. Jorgensen, "Comparative evaluation into the performance of 128K, 256K, and 512K NMOS and CMOS UV-PROMs, including reliability and erasability," Elektronikcentralen ECR, p. 209, 1987.

Heyns, G., and H. E. Maes, "A new model for the discharge behavior of MNOS non-volatile memory devices," *Appl. Surf. Sci.*, vol. 30, p. 153, 1987.

Kendall, T., "Keeping data safe with nonvolatile memory," Chilton's Instrumentation and Control Systems, vol. 60, no. 9, p. 89, 1987.

Kume, H., et al., "A flash-erase EEPROM cell with an asymmetric source and drain structure," *IEEE IEDM Tech. Dig.*, p. 560, 1987.

Kusaka, T., Y. Ohji, and K. Mukai, "Time-dependent dielectric breakdown of ultra-thin silicon oxide," *IEEE Elect. Dev. Lett.*, vol. EDL–8, no. 2, p. 61, 1987.

Lassig, S., and M.-S. Liang, "Time-dependent degradation of thin gate oxide under post-oxidation high-temperature anneal," *IEEE Elect. Dev. Letts.*, vol. EDL–8, no. 4, p. 160, 1987.

Lee, J., et al., "Statistical modeling of silicon dioxide reliability," *Proc. IRPS*, p. 131, 1988.

Lee, S.-S., et al., "High speed EPROM process technology development," *Sol. St. Technol.*, vol. 30, no. 10, p. 149, 1987.

Libsch, F. R., A. Roy, and M. H. White, "Amphoteric trap modelling of multidielectric scaled SONOS nonvolatile memory structure," INFOS 87 Conf., Applied Solid-State Science Series of Elsevier Science, Publ. B. V., Eindhoven, The Netherlands, October 1987.

Libsch, F. R., A. Roy, and M. H. White, "True 5-V EEPROM cell for high-density NVSM," *IEEE Trans. Elect. Dev.*, vol. ED–34, no. 11, p. 2371, 1987.

Lim, H.-K., et al., "64K CMOS EEPROM with page mode," Proc. IEEE Region 10 Conference, Comput. and Comm. Tech. Toward 2000, p. 1067, 1987.

Maes, H. E., and E. Vandekerckhove, "Non-volatile memory characteristics of polysilicon oxynitride oxide silicon devices," Proc. Electrochem. Soc., vol. 87–10, Eds. V. J. Kapoor and K. T. Hankins, p. 28, 1987.

Maes, H. E., J. Witters, and G. Groeseneken, "Trends in non-volatile memory devices and technologies," Proc. 17th European Sol. St. Dev. Res. Conf., p. 743, 1987.

Maslovskii, V. M., "MNOS degradation mechanism," *Soviet Microelect.*, vol. 16, no. 4, p. 202, 1987.

Masuoka, F., et al., "A 256-kbit Flash E^2PROM using triple-polysilicon technology," *IEEE J. Sol. St. Cir.*, vol. SC–22, no. 4, p. 548, 1987.

Masuoka, F., et al., "New ultra high density EPROM and Flash EEPROM with NAND cell structure cell," *IEEE IEDM Tech. Dig.*, p. 552, 1987.

McWhorter, P. J., et al., "Retention characteristics of SNOS nonvolatile devices in a radiation environment," *IEEE Trans. Nucl. Sci.*, NS–34, p. 1652, 1987.

Mielke, N., A. Fazio, and H.-C. Liou, "Reliability comparison of flotox and textured-polysilicon E^2PROMs," *Proc. IRPS*, p. 85, 1987.

Mitchell, A. T., C. Huffman, and A. L. Esquivel, "New self-aligned planar array cell for ultra high density EPROMs," *IEEE IEDM Tech. Dig.*, p. 548, 1987.

Mizutani, Y., and K. Makita, "Characteristics of a new EPROM cell structure with a sidewall floating gate," *IEEE Trans. Elect. Dev.*, vol. ED–34, no. 6, p. 1297, 1987.

Mori, S., et al., "Novel process and device technologies for submicron 4MB CMOS EPROMs," *IEEE IEDM Tech. Dig.*, p. 556, 1987.

Nguyen, T. N., P. Olivo, and B. Ricco, "A new failure mode of very thin (< 50 Å) thermal SiO_2 films," *Proc. IRPS*, p. 66, 1987.

Ohtsuka, N., et al., "A 4-Mbit CMOS EPROM," *IEEE J. Sol. St. Cir.*, vol. SC–22, no. 5, p. 669–675, 1987.

Prall, K., W. Kinney, and J. Macro, "Characterization and suppression of drain coupling in submicrometer EPROM cells," *IEEE Trans. Elect. Dev.*, vol. ED–34, no. 12, p. 2463, 1987.

Remmerie, J., et al., "Two carrier transport in MNOS devices," Proc. Electrochem. Soc., vol. 87–10, Eds. V. J. Kapoor and K. T. Hankins, p. 93, 1987.

Roy, A., F. R. Libsch, and M. H. White, "Investigations on ultrathin silicon nitride and silicon dioxide films in nonvolatile semiconductor memory transistors," Proc. Electrochem. Soc., vol. 87–10, Eds. V. J. Kapoor and K. T. Hankins, p. 38, 1987.

Samachisa, G., et al., "A 128K flash EEPROM using double polysilicon technology," *IEEE ISSCC Tech. Dig.*, p. 76, 1987.

Samachisa, G., et al., "A 128K flash EEPROM using double polysilicon technology," *IEEE J. Sol. St. Cir.*, vol. SC–22, no. 5, p. 676, 1987.

Sato, K., et al., "Carrier conduction characteristics of Si gate MNOS memory device," *Electron. Commun. Jpn.*, Part 2, vol. 70, no. 10, p. 1, 1987.

Shukiri, S., et al., "A novel EPROM device fabricated using focused boron ion beam implantation," *IEEE Trans. Elect. Dev.*, vol. ED–34, p. 1264, 1987.

Suga, H., and K. Murai, "Effect of bulk defects in silicon on SiO_2 film breakdown," Emerging Semiconductor Tech., ASTM Special Tech. Pub. 960, 4th Intl. Symp. Semiconductor Processing, p. 336, 1987.

Sweetman, D., and T. Haifley, "Reliability enhancements—million cycle EEPROMS," Proc. Reliab. and Maintain. Symp., p. 399, 1987.

Terada, Y., et al., "New architecture for the NVRAM—an EEPROM backed-up dynamic RAM," *IEEE J. Sol. St. Cir.*, vol. SC–23, no. 1, p. 79, 1987.

Terada, Y., et al., "New architecture of NVRAM," Symp. VLSI Tech. *Tech. Dig.*, p. 22, 1987.

Tsai, H.-H., et al., "The effects of thermal nitridation conditions on the reliability of thin nitrided oxide films," *IEEE Elect. Dev. Lett.*, vol. EDL–8, p. 143, 1987.

Venkatachalam, D., "A 5V EEPROM using modified CMOS technology," M.S. thesis, University of Missouri at Rolla, 1987.

Witters, J. S., G. Groeseneken, and H. E. Maes, "Programming mode dependent degradation of tunnel oxide floating gate devices," *IEEE IEDM Tech. Dig.*, p. 544, 1987.

Wright, M., "High-speed EPROMs," *EDN*, vol. 32, no. 19, p. 132, 1987.

Wu, C. Y., and C. F. Chen, "Transport properties of thermal oxide films grown on polycrystalline silicon—modeling and experiments," *IEEE Trans. Elect. Dev.*, vol. ED–34, p. 1590, 1987.

Yoshida, M., et al., "80 ns address-data multiplex 1MB CMOS EPROM," *Proc. IEEE ISSCC*, p. 70, 1987.

1988

Ali, S., et al., "A 50ns 256K CMOS split-gate EPROM," *IEEE J. Sol. St. Cir.*, vol. SC–23, no. 1, p. 86, 1988.

Aminzadeh, M., S. Nozaki, and R. V. Giridhar, "Conduction and charge trapping in polysilicon–silicon nitride–oxide–silicon structures under positive gate bias," *IEEE Trans. Elect. Dev.*, vol. ED–35, no. 4, p. 459, 1988.

Arima, H., et al., "A novel process technology and cell structure for megabit EEPROM," *IEEE IEDM Tech. Dig.*, p. 420, 1988.

Baker, A., et al., "An in-system reprogrammable 32K × 8 CMOS Flash memory," *IEEE J. Sol. St. Cir.*, vol. SC–23, no. 5, p. 1157, 1988.

Ballay, N., and B. Baylac, "CAD MOSFET model for EPROM cells," *J. de Phys.*, suppl. C4 to no. 9, vol. 49, p. 681, 1988.

Bez, R., et al., "SPICE model for transient analysis of EEPROM cells," Proc. ESSDERC88, p. 677, 1988.

Brown, M. D., J. A. Small, and J. A. Vincent, "EBS-1—An EPROM-based sequencer ASIC," Proc. Custom Integ. Cir. Conf., p. 15.6/1–4, 1988.

Cacharelis, P. J., et al., "Modular 1 μm CMOS single polysilicon EPROM PLD technology," *IEEE IEDM Tech. Dig.*, p. 60, 1988.

Canepa, G., et al., "90 ns 4 Mb CMOS EPROM," *IEEE ISSCC Tech. Dig.*, p. 120, 1988.

Carney, B., et al., "Configurable EEPROMs for ASICs," Proc. Custom Integ. Cir. Conf., p. 4.2/1–4, 1988.

Chang, C., et al., "Drain-avalanche and hole trapping-induced gate leakage in thin-oxide MOS devices," *IEEE Elect. Dev. Letts.*, vol. EDL–9, no. 11, p. 588, 1988.

Chang, K. Y., et al., "Advanced high voltage CMOS process for custom logic circuits with embedded EEPROM," Proc. IEEE Custom Integ. Cir. Conf., p. 25.5/1–5, 1988.

Chen, I.-C., et al., "The effect of channel hot-carrier stressing on gate-oxide integrity on MOSFET's," *IEEE Trans. Elect. Dev.*, vol. 35, no. 12, p. 2253, 1988.

Chu, S., and A. Steckl, "The effect of trench-gate-oxide structure on EPROM device operation," *IEEE Elect. Dev. Lett.*, vol. EDL–9, no. 6, p. 284, 1988.

Do, J. Y., et al., "256K CMOS EEPROM with enhanced reliability and testability," 1988 Symp. VLSI Circuits Dig. Tech. Paps., p. 83, 1988.

Dua, S., "Flash memories: the logical choice," Wescon/88 Conf. Rec., p. 10.0/1–4, 1988.

Esquivel, A. L., et al., "Novel, shallow-trench-isolated, planar, N^+ SAG FAMOS transistor for high-density nonvolatile memories," *IEEE Trans. Elect. Dev.*, vol. ED–35, no. 12, p. 2437, 1988.

"ETOX Flash memory: the cost effective and reliable firmware management solution," Intel Corp., Order No. 296294-001, March 1988.

Evans, J., and R. Womack, "An experimental 512-bit nonvolatile memory with ferroelectric storage cell," *IEEE J. Sol. St. Cir.*, SC–23, no. 5, p. 1171, 1988.

Fukuda, M., et al., "55 ns 64 multiplied by 16 b CMOS EPROM," *IEEE ISSCC Tech. Dig.*, p. 122, 1988.

Gallant, J., "Conventional EEPROM and Flash EEPROM offer a spectrum of bit densities," EDN, vol. 33, no. 22, p. 89, 1988.

Gastaldi, R., et al., "1-Mbit CMOS EPROM with enhanced verification," *IEEE J. Sol. St. Cir.*, vol. SC–23, no. 5, p. 1150, 1988.

Gill, M., et al., "5-volt contactless array 256Kbit Flash EEPROM technology," *IEEE IEDM Tech. Dig.*, p. 428, 1988.

Heremans, P., et al., "Consistent model for the hot-carrier degradation in n-channel and p-channel MOSFET's," *IEEE Trans. Elect. Dev.*, vol. ED–35, p. 2194, 1988.

Hori, T., H. Iwasaki, and K. Tsuji, "Charge-trapping properties of ultrathin nitrided oxides prepared by rapid thermal annealing," *IEEE Trans. Elect. Dev.*, vol. ED–35, no. 7, p. 904, 1988.

Horiguchi, M., et al., "An experimental large-capacity semiconductor file memory using 16-level/cell storage," *IEEE J. Sol. St. Cir.*, vol. SC–23, no. 1, p. 27, 1988.

Hu, V. W., "EEPROM as analog storage device for neural networks," M. S. Research Report, University of California at Berkeley, 1988.

Hu, V., A. Kramer, and P. K. Ko, "EEPROMS as analog storage devices for neural nets," *Neural Networks*, vol. 1, no. 1, p. 385, SUPPL 1988.

"IEEE standard definitions and characterization of metal nitride oxide semiconductor arrays," ANSI/IEEE Std 641-1987, New York, 1988.

Isles, J., "Flash EEPROMs bridge the gap," *New Electronics*, vol. 21, no. 7, p. 42, 1988.

Johnson, M., "Flash EEPROM: a cost competitive user friendly non-volatile memory," Wescon/88 Conference Record, p. 10.4/1–7, 1988.

Kaga, T., and T. Hagiwara, "Short- and long-term reliability of nitrided oxide MISFETs," *IEEE Trans. Elect. Dev.*, vol. ED–35, no. 7, p. 929, 1988.

Karoly, C., "Development of microprocessor systems with RAM, EPROM emulator," Meres es Automatika, vol. 36, no. 11, p. 321, 1988.

Kazerounian, R., et al., "A 5 volt high density poly-poly erase Flash EPROM cell," *IEEE Tech. Dig. IEDM*, p. 436, 1988.

Khaliq, M. A., et al., "Memory characteristics of MNOS capacitors fabricated with PECVD silicon nitride," *Sol. St. Electr.*, vol. 31, no. 8, p. 1229, 1988.

Krick, D. T., P. M. Lenahan, and J. Kanicki, "Nature of the dominant deep trap in amorphous silicon nitride," *Phys. Rev. B*, vol. 38, 1988.

Kynett, V. N., et al., "An in-system reprogrammable 32K × 8 CMOS flash memory," *IEEE J. Sol. St. Cir.*, vol. SC–23, p. 1157, 1988.

Kynett, V. N., et al., "In-system reprogrammable 256K CMOS Flash memory," *IEEE ISSCC Tech. Dig.*, p. 132, 1988.

Lee, J., et al., "Statistical modeling of silicon dioxide reliability," *Proc. IRPS*, p. 131, 1988.

Lee, J. C., I.-C. Chen, and C. Hu, "Modeling and characterization of gate oxide reliability," *IEEE Trans. Elect. Dev.*, vol. 35, no. 12, p. 2268, 1988.

Maekawa, S., "EPROM keeps pace with need for more powerful microprocessors," *J. of Electr. Eng.*, vol. 25, no. 256, p. 47, April 1988.

Maruyama, T., et al., "Wide operating voltage range and low power consumption EPROM structure for consumer oriented ASIC applications," Proc. of the Custom Integ. Cir. Conf., p. 4.1/1–4, 1988.

Matsuoka, H., Y. Igura, and E. Takeda, "Device degradation due to band-to-band tunnelling," Proc. Intl. Conf. Sol. St. Dev. and Mats., Tokyo, p. 589, 1988.

Messenger, G. C., and F. Coppage, "Ferroelectric memories; a possible answer to the hardened nonvolatile memory question," *IEEE Trans. Nucl. Sci.*, vol. NS–35, p. 1466, 1988.

Miki, H., et al., "Electron and hole traps in SiO_2 films thermally grown on Si substrates in ultra-dry oxygen," *IEEE Trans. Elect. Dev.*, vol. 35, no. 12, p. 2245, 1988.

Minami, S., et al., "Improvement of written-state retentivity by scaling down MNOS memory devices," *Jap. J. of Appl. Phys.*, vol. 27, p. L2168, 1988.

Momodomi, M., et al., "New device technologies for 5V-only 4Mb EEPROM with NAND structure cell," *IEEE IEDM Tech. Dig.*, p. 412, 1988.

Nakayama, T., et al., "5V only 1 Tr. 256K EEPROM with page mode erase," *Symp. VLSI Circuits Tech. Dig.*, p. 81, 1988.

Naruke, K., S. Taguchi, and M. Wada, "Stress induced leakage current limiting to scale down EEPROM tunnel oxide thickness," *IEEE IEDM Tech. Dig.*, p. 424, 1988.

Nguyen, B. Y., et al., "Influences of processing chemistry of silicon nitride films on the charge trapping behavior of oxide/CVD–nitride/oxide capacitors," *J. Electrochem. Soc.*, vol. 135, no. 3, p. 776, 1988.

Norris, R., "FLASH technology: bridging the gap between EPROMS and EEPROMS," Midcon/88 Conf. Rec., p. 102, 1988.

Norris, R., "Flash technology: bridging the gap between EPROMS and EEPROMS," Wescon/88 Conference Record, p. 10.1/1–4, 1988.

Novosel, D., et al., "1 Mb CMOS EPROM with enhanced verification," *IEEE ISSCC Tech. Dig.*, p. 124, 1988.

Olivo, P., T. N. Nguyen, and B. Ricco, "High-field-induced degradation in ultra-thin SiO_2 films," *IEEE Trans. Elect. Dev.*, vol. 35, no. 12, p. 2259, 1988.

Ong, T.-C., et al., "Recovery of threshold voltage after hot-carrier stressing," *IEEE Trans. Elect. Dev.*, vol. ED–35, no. 7, p. 978, 1988.

Richmond, G., et al., "Advanced standard cell library provides high performance mixed digital, analog, and EEPROM single chip solutions," Proc. of the Custom Int. Cir. Conf., p. 24.3/1–6, 1988.

Sampson, D., "Time and total dose response of non-volatile UVPROMS," *IEEE Trans. Nucl. Sci.*, NS–35, p. 1542, 1988.

Shirota, R., et al., "New NAND cell for ultra high density 5v-only EEPROM," *Tech. Dig.*; Symp. VLSI Tech., p. 33, 1988.

Stoll, J. P., "Comparison of nonvolatile memory technologies for spacecraft applications," IEEE 1988 Aerospace Applications Conf. Dig., p. 10/1–14, 1988.

Tam, S., et al., "A high density CMOS 1-T electrically erasable non-volatile (flash) memory technology," Symp. on VLSI Tech., Dig. Tech. Paper, p. 31, 1988.

Terada, Y., et al., "A new architecture for the NVRAM—An EEPROM backed-up dynamic RAM," *IEEE J. Sol. St. Cir.*, vol. SC–23, no. 1, p. 86, 1988.

Ting, T.-K. J., et al., "A 50-ns CMOS 256K EEPROM," *IEEE J. Sol. St. Cir.*, vol. SC–23, no. 5, p. 1164, 1988.

Vancu, R., L. Chen, and G. Smarandoiu, "30 ns Fault tolerant 16k CMOS EEPROM," *IEEE ISSCC Tech. Dig.*, p. 128, 1988.

Verma, G., and N. Mielke, "Reliability performance of ETOX based Flash memories," *Proc. IRPS*, p. 167, 1988.

Yang, W., R. Jayaraman, and C. G. Sodini, "Optimization of low-pressure nitridation/reoxidation of SiO_2 for scaled MOS devices," *IEEE Trans. Elect. Dev.*, vol. ED–35, no. 7, p. 935, 1988.

Yasmauchi, Y., K. Tanaka, and K. Sakiyama, "Novel NVRAM cell technology for high density applications," *IEEE IEDM Tech. Dig.*, p. 416, 1988.

Yoshikawa, K., et al., "An asymmetrical lightly-doped source (ALDS) cell for virtual ground high density EPROMs," *IEEE IEDM Tech. Dig.*, p. 432, 1988.

Young, K. K., C. Hu, and W. G. Oldham, "Charge transport and trapping characteristics in thin nitride–oxide stacked films," *IEEE Elect. Dev. Lett.*, vol. EDL–9, no. 11, p. 616, 1988.

Zales, S., "Flash memories change system design fundamentals," Wescon/88 Conference Record, p. 10.2/1–6, 1988.

Zetterberg, C. T., "Design and programming considerations for flash memory devices," Wescon/88 Conference Record, p. 10.3/1–4, 1988.

1989

Aganin, A. P., V. M. Maslovskii, and A. P. Nagin, "Determination of the parameters of trapping centers in silicon nitride in a MNOS structure with electron injection from a field-emission electrode," *Soviet Microelect.*, vol. 17, no. 4, p. 200, 1989.

Ali, S., et al., "New staggered virtual ground array architecture implemented in a 4 Mb CMOS EPROM," Proc. Symp. VLSI Cir., p. 35, 1989.

Anderson, J., et al., "A 90-ns one-million erase/program cycle 1 Mbit Flash memory," *IEEE J. Sol. St. Cir.*, vol. SC–24, no. 5, p. 1259, 1989.

Anon, "EEPROM as a substitute for bubble memory," *Elektron (Johannesburg)*, vol. 6, no. 6, p. 21, February 1989.

Arreola, J. I., et al., "New charge gain failure mechanism of high performance EPROMs," 10th IEEE NVSM Workshop, Ext. Abs., p. 79, 1989.

D'Arrigo, S., et al., "5V-only 256 Kbit CMOS Flash EEPROM," *IEEE ISSCC Tech. Dig.*, p. 132, 1989.

Ashmore, B., et al., "20 ns 1 Mb CMOS burst mode EPROM," *IEEE ISSCC Tech. Dig.*, vol. 32, p. 40, 1989.

Bauch, W., K. Jaeger, and R. Hezel, "Effect of Cs contamination on the interface state density of MNOS capacitors," *Appl. Surf. Sci.*, vol. 39, no. 1–4, p. 356, 1989.

Bellezza, O., D. Laurenzi, and M. Melanotte, "A new self-aligned field oxide cell for multimegabit EPROMs," *IEEE IEDM Tech. Dig.*, p. 579, 1989.

Bergemont, A., et al., "A high performance CMOS process for submicron 16 Mb EPROM," *IEEE IEDM Tech. Dig.*, p. 591, 1989.

Berkman, J., G. Biran, and G. Miller, "A 2 volt secure serial EEPROM with a bipolar reference," 10th IEEE NVSM Workshop, Ext. Abs., p. 4, 1989.

Boyko, K. C., and D. L. Gerlach, "Time dependent dielectric breakdown of 210Å oxides," *Proc. IRPS*, p. 1, 1989.

Bez, R., D. Cantarelli, and P. Cappelletti, "Experimental transient analysis of tunnel current in EEPROM cells," 10th IEEE NVSM Workshop, Ext. Abs., p. 48, 1989.

Card, H. C., and W. R. Moore, "EEPROM synapses exhibiting pseudo-hebbian plasticity," *Electr. Letts.*, vol. 25, no. 12, p. 805, 1989.

Carley, R. L., "Trimming analog circuits using floating-gate analog MOS memory," *IEEE ISSCC Tech. Dig.*, vol. 32, p. 202, 1989.

Cernea, R. A., et al., "A 1Mb Flash EEPROM," *IEEE ISSCC Tech. Dig.*, p. 138, 1989.

Chang, C. C., et al., "A 20ns low power 128K CMOS EPROM," 10th IEEE NVSM Workshop, Ext. Abs., p. 61, 1989.

Chen, I. C., C. Kaya, and J. Paterson, "Band-to-band tunneling induced substrate hot-electron (BBISHE) injection: a new programming mechanism for nonvolatile memory devices," *IEEE IEDM Tech. Dig.*, p. 263, 1989.

Chen, Y.-Z., and T.-W. Tang, "Computer simulation of hot-carrier effects in asymmetric LDD and LDS MOSFET devices," *IEEE Trans. Elect. Dev.*, vol. ED–36, no. 11, pt. 1, p. 2492, 1989.

Ciampolini, P., et al., "Realistic device simulation in three dimensions (EPROM cell)," *IEEE IEDM Tech. Dig.*, p. 131, 1989.

Ciampolini, P., et al., "Three-dimensional simulation of a floating-gate EPROM cell," 3rd Ann. European Computer Conf., p. 5/55–57, 1989.

Dimmler, K., et al., "50ns 256K SNOS EEPROM," 10th IEEE NVSM Workshop, Ext. Abs., p. 3, 1989.

Do, J. Y., et al., "256K CMOS EEPROM with enhanced reliability and testability," 1988 Symp. VLSI Circuits Tech. Dig., p. 83, 1989.

Domashevskaya, E. P., et al., "Effect of composition and exposure to external factors on the electronic structure of amorphous silicon nitride in memory devices," *Microelectr. J.*, vol. 20, no. 6, p. 11, 1989.

Endoh, T., et al., "New design technology for EEPROM memory cells with 10 million write/erase cycling endurance," *IEEE IEDM Tech. Dig.*, p. 599, 1989.

Frake, S., et al., "9 ns, low standby power CMOS PLD with a single-poly EPROM cell," *IEEE ISSCC Tech. Dig.*, vol. 32, p. 230, 1989.

French, M. L., "Memory window studies of nonvolatile silicon–oxide–nitride–oxide–silicon (SONOS) memory devices," M.S. thesis, Lehigh University, 1987.

Furuyama, T., et al., "An experimental 2-bit/cell storage DRAM for macrocell or memory-on-logic applications," *IEEE J. Solid-State Circuits*, vol. SC–24, no. 2, p. 388, 1989.

Gaeta, I. S., and K. J. Wu, "Improved EPROM moisture performance using spin-on-glass (SOG) for passivation planarization," 27th Ann. Proc. Reliab. Phys., p. 122, 1989.

Gill, M., et al., "5-volt only Flash EEPROM technology for high density memory and system IC applications," Proc. IEEE 1989 Custom Integ. Cir. Conf., p. 18.4/1–4, 1989.

Gill, M., et al., "Reliability of ACEE-based 256Kb 5V-only Flash EEPROM," 10th IEEE NVSM Workshop, Ext. Abs., p. 83, 1989.

Gosain, D. P., et al., 'Nonvolatile memory based on reversible phase transition phenomena in telluride glasses," *Japan. J. Appl. Phys.*, Part 1: Regular Papers and Short Notes, vol. 28, no. 6, p. 1013, 1989.

Gowni, S. P., et al., "12 ns, CMOS programmable logic device for combinatorial applications," Proc. of the Custom Integ. Cir. Conf., p. 5.5/1–4, 1989.

Haddad, S., et al., "Degradation due to hole trapping in Flash memory cells," *IEEE Elect. Dev. Letts.*, vol. EDL–10, no. 3, p. 117, 1989.

Hart, M. J., P. J. Cacharelis, and M. H. Manley, "Mechanism for program disturb in EPROMs," 10th IEEE NVSM Workshop, Ext. Abs., p. 75, 1989.

Hisamune, Y. S., et al., "A 3.6 μm^2 memory cell structure for 16 Mb EPROMs," *IEEE IEDM Tech. Dig.*, p. 583, 1989.

Heremans, P., et al., "Analysis of the charge pumping technique and its application for the evaluation of MOSFET degradation," *IEEE Trans. Elect. Dev.*, vol. ED–36, no. 7, July 1989.

Hoe, D. H. K., et al., "Cell and circuit design for single-poly EPROM," *IEEE J. Sol. St. Cir.*, vol. SC–24, no. 4, p. 1153, 1989.

Hofmann, F., and W. Krautschneider, "MOSFET interface state densities of different technologies," Proc. IEEE Int. Conf. on Microelect. Test Struct., vol. 2, p. 109, 1989.

Hoff, D., et al., "23 ns 256 K EPROM with double-layer metal and address transition detection," *IEEE J. Sol. St. Cir.*, vol. SC–24, no. 5, p. 1250, 1989.

Igura, Y., H. Matsuoka, and E. Takeda, "New device degradation due to "cold" carriers created by band-to-band tunneling," *IEEE Elect. Dev. Letts.*, vol. EDL–10, no. 5, p. 227, 1989.

Imamiya, K., et al., "68ns 4 Mbit CMOS EPROM with high noise immunity design," Proc. Symp. VLSI Circuits, p. 37, 1989.

Intel Reliability Report, RR-60, "ETOX™ Flash memory reliability data summary," Order No. 293002-006, October 1989.

Ishihara, H., K. Tanaka, and K. Sakiyama, "A single poly silicon flip flop EEPROM for ASIC application," 10th IEEE NVSM Workshop, Ext. Abs., p. 40, 1989.

Ishihara, H., K. Tanaka, and K. Sakiyama, "Single-polysilicon flip-flop EEPROM for ASIC application," Proc. 2nd Ann. IEEE ASIC Sem. and Exh., p. 3.3/1–4, 1989.

Itoh, Y., et al., "An experimental 4-Mbit CMOS EEPROM with a NAND structured cell," *IEEE J. Sol. St. Cir.*, vol. SC–24, no. 5, p. 1238, 1989.

Kamigaki, Y., et al., "Highly reliable 256-Kb EEPROM," 10th IEEE NVSM Workshop, Ext. Abs., p. 1, 1989.

Kamigaki, Y., et al., "New scaling guidelines for MNOS memory devices," 10th IEEE NVSM Workshop, Ext. Abs., p. 42, 1989.

Kamigaki, Y., et al., "Yield and reliability of MNOS EEPROM products," *IEEE J. Sol. St. Cir.*, vol. SC–24, no. 6, p. 1714, 1989.

Kobayashi, K., et al., "120-ns 128K × 8-bit/64K × 16-bit CMOS EEPROM," *IEEE J. Sol. St. Cir.*, vol. SC–24, no. 5, p. 1238, 1989.

Kobayashi, K., et al., "Self-timed dynamic sensing scheme for 5V only multi-Mb Flash E^2PROMs," Symp. VLSI Cir., p. 39, 1989.

Kobayashi, K., and H. Arima, "120ns 1M EEPROM," Mitsubishi Electric. Advance., vol. 49, p. 18, 1989.

Kramer, A., et al., "EEPROM device as a reconfigurable analog element for neural networks," *IEEE IEDM Tech. Dig.*, p. 259, 1989.

Kynett, V., et al., "A 90ns 100K erase/program cycle megabit flash memory," *IEEE ISSCC Tech. Dig.*, p. 140, 1989.

Kynett, V., et al., "A 90-ns one-million erase/program cycle 1-Mbit flash memory," *IEEE J. Sol. St. Cir.*, vol. SC–24, no. 10, p. 1259, 1989.

Lambertson, R. T., and D. C. Goodman, "A comprehensive graphical method for evaluating textured polysilicon tunnel voltages and EEPROM cell operation," 10th IEEE NVSM Workshop, Ext. Abs., p. 14, 1989.

Lanzoni, M., P. Olivo, and B. Ricco, "Testing technique to characterize E^2PROM's aging and endurance," 20th Intl. Test Conf., p. 391, 1989.

Libsch, F. R., "Physics, technology, and electrical aspects of scaled MONOS/SONOS devices for low voltage nonvolatile semiconductor memories," Ph.D. diss., Lehigh University, 1989.

Libsch, F., A. Roy, and M. H. White, "Scaling reliability requirements in SONOS/MONOS EEPROM devices," 10th IEEE NVSM Workshop, Ext. Abs., p. 45, 1989.

Liew, B. K., et al., "Electromigration interconnect lifetime under AC and pulsed DC stress," *Proc. IRPS*, 1989.

Linton, T. D., Jr., and P. A. Blakey, "Fast, general three-dimensional device simulator and its application in a submicron EPROM design study," *IEEE Trans. Computer-Aided Design of Integ. Cir. and Sys.*, vol. 8, no. 5, p. 508, 1989.

Liou, T.-I., et al., "Single-poly CMOS process merging analog capacitors, bipolar and EPROM devices," 9th Symp. on VLSI Tech., Dig. of Tech. Pap., p. 37, 1989.

Ma, T. P., and P. V. Dressendorfor, *Ionizing Radiation Effects in MOS Devices & Circuits*, John Wiley, New York, 1989.

Maes, H. E., et al., "Trends in semiconductor memories," *Microelectron. J.*, vol. 20, p. 9, 1989.

Maiz, J. A., "Characterization of electromigration under bidirectional (BC) and pulsed unidirectional (PDC) currents," *Proc. IRPS*, 1989.

McWhorter, P. J., S. L. Miller, and T. A. Dellin, "Simple model for predicting the data retention characteristics of SNOS devices in a varying thermal scenario," 10th IEEE NVSM Workshop, Ext. Abs., p. 26, 1989.

Miller, W. M., and S. L. Miller, "Quantitative determination of SNOS reliability," 10th IEEE NVSM Workshop, Ext. Abs., p. 88, 1989.

Momodomi, M., et al., "High density NAND EEPROM with block-page programming for microcomputer applications," Proc. IEEE Custom Integ. Cir. Conf., p. 10.1/1–4, 1989.

Momodomi, M., et al., "Experimental 4-Mbit CMOS EEPROM with a NAND-structured cell," *IEEE J. Sol. St. Cir.*, vol. SC–24, no. 10, p. 1238, 1989.

Mori, S., et al., "Scaled EPROM cell technology in 0.6 μm regime," 10th IEEE NVSM Workshop, Ext. Abs., p. 32, 1989.

Nakayama, T., et al., "A 5V-only one-transistor 256K EEPROM with page-mode erase," *IEEE J. Sol. St. Cir.*, vol. SC–24, no. 4, p. 911, 1989.

Naruke, K., et al., "A new Flash-erase EEPROM cell with a sidewall select-gate on its source side," *IEEE IEDM Tech. Dig.*, p. 603, 1989.

Nolan, J., and P. Boydston, "Application for a family of EEPROM logic and memory cells for analog and digital standard cell ASICs," 10th IEEE NVSM Workshop, Ext. Abs., p. 51, 1989.

Nordlund, B., "High performance intelligent EPROM programmer," Wescon/89; Conf. Rec., p. 496, 1989.

Norris, R., "FLASH technology: bridging the gap between EPROMS and EEPROMS," Southcon/89; Conf. Rec., p. 135, 1989.

Ohtsuka, N., et al., "Fast programming in snap-back region (snap-back programming for sub-μm EPROM cells," 10th IEEE NVSM Workshop, Ext. Abs., p. 20, 1989.

Ong, T.-C., P. K. Tong, and C. Hu, "EEPROM as an analog memory drive," *IEEE Trans. Elect. Dev.*, vol. ED–36, no. 9, p. 1840, 1989.

Owen, W. H., and W. E. Tchon, "E^2PROM product issues and technology trends," 3rd Ann. European Computer Conf. p. 1/17–19, 1989.

Park, C. S., and Y. C. Lin, "High voltage switches for CMOS sub-micron EPROM," Conf. Rec.: 23rd Asilomar Conf. on Sigs., Syst. and Comput., vol. 2, p. 970, 1989.

Pashley, R. D., and S. K. Lai, "Flash memories: the best of two worlds," *IEEE Spectrum*, vol. 26, no. 12, p. 30, 1989.

Paterson, J., "Adding analog EPROM and EEPROM modules to CMOS logic technology: how modular?", *IEEE IEDM Tech. Dig.*, p. 413, 1989.

Rauh, G., "Flash EEPROMs can take the terror out of surface mounting," *Electronic Products*, vol. 32, no. 3, p. 30, 1989.

Roy, A., "Characterization and modeling of charge trapping and retention in novel multi-dielectric nonvolatile semiconductor memory devices," Ph.D. diss., Lehigh University, 1989.

Roy, A., F. R. Libsch, and M. H. White, "Electron tunneling from polysilicon asperities into polyoxides," *Sol. St. Elect.*, vol. 32, no. 8, p. 655, 1989.

Saeki, Y., et al., "Low-power consumption and low-voltage operation PLA (L**2-PLA) using 1.2 μm double poly-silicon CMOS E^2PROM technology," Midwest Symp. on Cir. and Sys., p. 1061, 1989.

Sage, J. P., R. S. Withers, and K. E. Thompson, "MNOS/CCD circuits for neural network implementations," Proc. IEEE Intl. Symp. on Cir. and Sys., vol. 2, p. 1207, 1989.

Sato, K., et al., "Tunnel oxide breakdown characteristics of floating-gate-type EEPROM," Electr. and Comm. in Japan, Part II: Electronics, vol. 72, no. 10, p. 1, 1989.

Schultz, K. J., et al., "Microprogrammable processor using single poly EPROM," *Integration, the VLSI J.*, vol. 8, no. 2, p. 189, 1989.

Shams, Q. A., and W. D. Brown, "Effects of nitrogen and argon as carrier gases and annealing ambients on the physical properties of PECVD silicon nitride," *Microelectr. J.*, vol. 28, no. 6, p. 49, 1989.

Shibata, Y., et al., "Influence of Pt ions in SiO_2 films on nonvolatile memory effects in Pt-diffused MOS devices," Electr. and Comm. in Jap., Part 2: Electronics, vol. 72, no. 1, p. 106, 1989.

Sharma, U., and M. H. White, "Ionization radiation induced degradation of MOSFET channel frequency response," *IEEE Trans. Elect. Dev.*, vol. ED–36, p. 1359, 1989.

Shimabukuro, R. L., P. A. Shoemaker, and M. E. Stewart, "Circuitry for artificial neural networks with non-volatile analog memories," Proc. IEEE Intl. Symp. on Cir. and Sys., vol. 2, p. 1217, 1989.

Shmidt, T. V., V. A. Gurtov, and V. A. Laleko, "Time characteristics of breakdown in films of silicon dioxide and silicon nitride," *Soviet Microelectr.*, vol. 17, no. 3, p. 139, 1989.

Snell, A. J., et al., "Application of vanadium-doped SiO_2 to EEPROM devices," *Appl. Surf. Sci.*, vol. 39, no. 1–4, p. 368, 1989.

Snyder, E. S., et al., "Radiation response of floating gate EEPROM memory cells," *IEEE Trans. Nucl. Sci.*, NS–36, p. 2131, 1989.

Suzuki, E., et al., "Hole and electron current transport in metal–oxide–nitride–oxide–silicon memory structures," *IEEE Trans. Elect. Dev.*, vol. ED–36, no. 6, p. 1145, 1989.

Terada, Y., et al., "120ns 128k $\times$ 8b/64k $\times$ 16b CMOS EEPROM's," *IEEE ISSCC Tech. Dig.*, p. 136, 1989.

Terada, Y., et al., "120-ns 128K multiplied by 8-bit/64K multiplied by 16-bit CMOS EEPROMs," *IEEE J. Sol. St. Cir.*, vol. SC–24, no. 5, p. 1244, 1989.

Tsao, R. R., and S. H. Chiao, "Flash E^2PROM role in NVM applications and technology," Conf. Rec.: 23rd Asilomar Conf. on Sigs., Syst. and Comput., vol. 2, p. 965, 1989.

Turner, J., "A family of high performance CMOS programmable logic devices using a single polysilicon electrically erasable floating gate cell," 10th IEEE NVSM Workshop, Ext. Abs., p. 53, 1989.

VHSIC/VHSIC-Liek reliability prediction modeling, RADC-TR-89-177, Final Tech. Rpt., October 1989.

Vollebregt, F., et al., "A new E(E)PROM technology with a $TiSi_2$ control gate," *IEEE IEDM Tech. Dig.*, p. 607, 1989.

Wada, Y., et al., "1.7 volts operating CMOS 64 kbit E^2PROM," Symp. VLSI Cir., p. 41, 1989.

Weiner, A., et al., "A 64K non-volatile SRAM," 10th IEEE NVSM Workshop, Ext. Abs., p. 59, 1989.

White, A., and B. J. Frost, "Channelled gate arrays," *Electr. & Wireless World*, vol. 96, no. 1639, p. 499, 1989.

Witters, J., "Characteristics and reliability of thin oxide floating gate memory transistors and their supporting programming circuits," Ph.D. thesis, K. U. Leuven, 1989.

Witters, J. S., G. Groeseneken, and H. E. Maes, "Degradation of tunnel-oxide floating-gate EEPROM devices and the correlation with high field-current-induced degradation of thin gate oxides," *IEEE Trans. Elect. Dev.*, vol. ED–36, no. 9, p. 1663, 1989.

Witters, J. S., and H. E. Maes, "Analysis and modeling of on-chip high-voltage generator circuits for use in EEPROM circuits," *IEEE J. Sol. St. Cir.*, vol. SC–24, no. 5, p. 1372, 1989.

Wright, P. J., et al., "Hot electron immunity of SiO_2 dielectrics with fluorine incorporation," *IEEE Elect. Dev. Lett.*, vol. EDL–10, p. 347, 1989.

Wrobel, T. F., "Radiation characterization of a 28C256 EEPROM," *IEEE Trans. Nucl. Sci.*, vol. NS–36, p. 2247, 1989.

Wu, A. T., et al., "Gate bias polarity dependence of charge trapping and time-dependent dielectric breakdown in nitrided and reoxidized nitrided oxides," *IEEE Elect. Dev. Lett.*, vol. EDL–10, no. 10, p. 443, 1989.

Yamauchi, Y., K. Tanaka, and K. Sakiyama, "A new cell technology for implementation in high-density NVRAMs," 10th IEEE NVSM Workshop, Ext. Abs., p. 37, 1989.

Yamauchi, Y., et al., "A versatile stacked storage capacitor on FLOTOX cell for megabit NVRAM applications," *IEEE IEDM Tech. Dig.*, p. 595, 1989.

Yoshikawa, K., et al., "0.6 μm EPROM cell design based on a new scaling scenario," *IEEE IEDM Tech. Dig.*, p. 587, 1989.

Zang, D. Y., and V. M. Ristic, "All-optical erasable programmable read-only memory (EPROM) based on $LiNbO_3$ channel waveguides," *IEEE Photonics Tech. Letts.*, vol. 1, no. 10, p. 323, 1989.

1990

Ajika, N., et al., "A 5 volt only 16M bit flash EEPROM cell with a simple stacked gate structure," *IEEE IEDM Tech. Dig.*, p. 115, 1990.

Ajika, N., et al., "A high density high performance cell for 4 M bit full feature EEPROM," Ext. Abs.; 22nd Intl. Conf. on Sol. St. Dev. and Mats., p. 397, 1990.

Ajika, N., et al., "Optimization of nitridation and re-oxidation conditions for an EEPROM tunneling dielectric," Ext. Abs.; 22nd Intl. Conf. on Sol. St. Dev. and Mat., p. 171, 1990.

Amin, A. A. M., "A novel flash erase EEPROM memory cell with asperities aided erase," ESSDERC 90; 20th European Sol. St. Dev. Res. Conf., Eds. W. Eccleston, P. J. Bristol Rosser, UK: Adam Hilger, p. 177, 1990.

Aritome, S., et al., "Extended data retention characteristics after more than 10^4 write and erase cycles in EEPROMs," *Proc. IRPS*, p. 259, 1990.

Aritome, S., et al., "A reliable bi-polarity write/erase technology in flash EEPROMs," *IEEE IEDM Tech. Dig.*, p. 111, 1990.

Atsumi, S., et al., "A 16 ns 1 Mb CMOS EPROM," *IEEE ISSCC Tech. Dig.*, p. 58, 1990.

Baglee, D. A., L. Nannemann, and C. Huang, "Building reliability into EPROMs," *Proc. IRPS*, p. 12, 1990.

Bez, R., D. Cantarelli, and P. Cappelletti, "Experimental transient analysis of the tunnel current in EEPROM cells," *IEEE Trans. Elect. Dev.*, vol. ED–37, no. 4, p. 1081, 1990.

Bez, R., et al., "A novel method for the experimental determination of the coupling ratios in submicron EPROM and flash EEPROM cells," *IEEE IEDM Tech. Dig.*, p. 99, 1990.

Bez, R., et al., "Series resistance effects on EPROM programming," ESSDERC 90; 20th European Sol. St. Dev. Res. Conf., Eds. W. Eccleston, P. J. Bristol Rosser, UK: Adam Hilger, p. 165, 1990.

Cacharelis, P. J., et al., "A fully modular 1 μm CMOS technology incorporating EEPROM, EPROM and interpoly capacitors," ESSDERC 90; 20th European Sol. St. Dev. Res. Conf., Eds. W. Eccleston, P. J. Bristol Rosser, UK: Adam Hilger, p. 547, 1990.

Cassi, D., and B. Ricco, "An analytical model of the energy distribution of hot electrons," *IEEE Trans. Elect. Dev.*, vol. ED–37, p. 1514, 1990.

Crisenza, G., et al., "Charge loss in EPROM due to ion generation and transport in interlevel dielectric," *IEEE IEDM Tech. Dig.*, p. 107, 1990.

Crook, D. L., "Evolution of VLSI reliability engineering," *Proc. IRPS*, p. 2, 1990.

Crowell, C. R., C.-C. Shih, and V. C. Tyree, "Simulation and testing of temperature distribution and resistance versus power for SWEAT and related joule-heated metal-on-insulator structures," *Proc. IRPS*, 1990.

Dejenfelt, A. T., "Analytical model for the internal electric field in submicrometer MOSFET's," *IEEE Trans. Elect. Dev.*, vol. ED–37, no. 5, p. 1352, 1990.

Doyle, B., et al., "Interface state creation and charge trapping in the medium-to-high gate voltage range ($Vd/2 < Vg < Vd$) during hot-carrier stressing of n-MOS transistors," *IEEE Trans. Elect. Dev.*, vol. ED–37, no. 3, p. 744, 1990.

Dutoit, M., et al., "Thin nitrided SiO_2 films for EEPROMs," Ext. Abs.; 22nd Intl. Conf. on Sol. St. Dev. and Mat., p. 175, 1990.

Endoh, T., et al., "An accurate model of subbreakdown due to band-to-band tunnelling and some applications," *IEEE Trans. Elect. Dev.*, vol. ED–37, no. 1, p. 290, 1990.

Fiegna, C., et al., "Simulation of EPROM writing," ESSDERC 90; 20th European Sol. St. Dev. Res. Conf., Eds. W. Eccleston, P. J. Bristol Rosser, UK: Adam Hilger, p. 527, 1990.

Fishbein, B. J., and D. B. Jackson, "Performance degradation of n-channel MOS transistors during DC and pulsed Fowler–Nordheim stress," *Proc. IRPS*, 1990.

Fong, Y., A. T.-T. Wu, and C. Hu, "Oxides grown on textured single-crystal silicon—dependence on process and application in EEPROMs," *IEEE Trans. Elect. Dev.*, vol. ED–37, no. 3, p. 583, 1990.

Gigon, F., "Modeling and simulation of the 16 megabit EPROM cell for write/read operation with a compact SPICE model," *IEEE IEDM Tech. Dig.*, p. 205, 1990.

Gill, M., et al., "A novel sublithographic tunnel diode-based 5-volt Flash memory," *IEEE IEDM Tech. Dig.*, p. 119, 1990.

Gupta, A., et al., "A user configurable gate array using CMOS-EPROM technology," Proc. IEEE 1990 Custom Integ. Cirs. Conf., p. 31.7/1–4, 1990.

Haddad, S., et al., "An investigation of erase-mode dependent hole trapping in flash EEPROM memory cell," *IEEE Elect. Dev. Lett.*, vol. EDL–11, no. 11, p. 514, 1990.

Haifley, T., and D. Sowards, "The endurance of EEPROMs/utilizing fault tolerant memory cells," *Proc. IRPS*, p. 378, 1990.

Halg, B., "On a micro-electro-mechanical nonvolatile memory cell," *IEEE Trans. Elect. Dev.*, vol. ED–37, no. 10, p. 2230, 1990.

Higuchi, M., et al., "An 85 ns 16 Mb CMOS EPROM with alterable word organization," *IEEE ISSCC Tech. Dig.*, p. 56, 1990.

Huber, L., D. Borojevic, and N. Burany, "Digital implementation of the space vector modulator for forced commutated cycloconverters," 4th Intl. Conf. on Power Elect. and Variable-Speed Drives, IEE Conference Publication no. 324, p. 63, 1990.

Ishijima, T., et al., "A deep-submicron isolation technology with T-shaped oxide (TSO) structure," *IEDM Tech. Dig.*, p. 257, 1990.

Ikeda, N., et al., "Single chip microcontroller with internal EPROMs," National Tech. Rpt. (Matsushita Electric Industry Company), vol. 36, no. 3, p. 295, 1990.

Imamiya, K., et al., "A 68-ns 4 Mbit CMOS EPROM with high-noise-immunity design," *IEEE J. Sol. St. Cir.*, vol. SC–25, no. 1, p. 72, 1990.

Iwata, Y., et al., "High-density NAND EEPROM with block-page programming for microcomputer applications," *IEEE J. Sol. St. Cir.*, vol. SC–25, no. 2, p. 417, 1990.

Joshi, A. B., et al., "Polarity dependence of charge to breakdown and interface state generation of oxynitride gate dielectrics prepared by rapid thermal processing," *Elect. Letts.*, vol. 26, no. 21, p. 1741, 1990.

Kamigaki, Y., et al., "A highly reliable 256-Kb MNOS EEPROM: memory cell design technology," Electr. and Comm. in Jap. Part 2 (Electronics), vol. 73, no. 9, p. 79, 1990.

Kang, J.-K., "Design of a CMOS temperature sensor with EEPROM error correction scheme," M.S. thesis, New Jersey Institute of Technology, 1990.

Kapoor, V.J., R. S. Bailey, and R. A. Turi, "Chemical composition, charge trapping, and memory properties of oxynitride films for MNOS devices," *J. Electrochem. Soc.*, vol. 137, no. 11, p. 3589, 1990.

Keeney, S., et al., "Complete transient simulation of flash EEPROM devices," *IEEE IEDM Tech. Dig.*, p. 201, 1990.

Kirisawa, R., et al., "A NAND structured cell with a new programming technology for highly reliable 5 V-only flash EEPROM," 1990 Symp. on VLSI Tech. Tech. Dig., p. 129, 1990.

Kobayashi, K., et al., "A high-speed parallel sensing architecture for multi-megabit Flash E^2PROM's," *IEEE J. Sol. St. Cir.*, vol. SC–25, no. 1, p. 79, 1990.

Kuriyama, M., et al., "A 16-ns 1-Mb CMOS EPROM," *IEEE J. Sol. St. Cir.*, vol. SC–25, no. 5, p. 1141, 1990.

Lahti, W., and D. McCarron, "Store data in a flash (flash-memory ICs)," *BYTE*, vol. 15, no. 12, p. 311, 1990.

Lanzoni, M., et al., "Testing of E^2PROM aging and endurance: a case study," European Trans. on Telecommunications and Related Technologies, vol. 2, no. 1, p. 201, 1990.

Leemann, R., and A. Birolini, "New experimental results in the qualification tests of large EPROMs," Annales des Telecommunications, vol. 45, no. 11–12, p. 591, 1990.

Libsch, F. R., A. Roy, and M. H. White, "Charge transport and storage of low programming voltage SONOS/MONOS memory devices," *Sol. St. Electr.*, vol. 33, no. 1, p. 105, 1990.

Lo, G. Q., et al., "Charge trapping properties in thin oxynitride gate dielectrics prepared by rapid thermal processing," *Appl. Phys. Lett.*, vol. 56, no. 5, p. 979, 1990.

Maes, H. E., G. Groeseneken, and J. Witters, "New developments in nonvolatile semiconductor memory technologies and devices," Proc. of the ESA Electr. Comp. Conf., p. 31, 1990.

Manos, P., and C. Hart, "A self-aligned EPROM structure with superior data retention," *IEEE Elect. Dev. Lett.*, vol. EDL–11, no. 7, p. 309, 1990.

Mantey, J. T., "Degradation of thin silicon dioxide films and EEPROM cells," Ph.D. thesis, Ecole Polytechnique Federale de Lausanne, 1990.

Matsukawa, N., K. Masuda, and J. Miyamoto, "A bipolar-EPROM (Bi-EPROM) structure for 3.3 V operation and high speed application," *IEEE IEDM Tech. Dig.*, p. 313, 1990.

Matsukawa, N., et al., "Process technologies for a 16 ns high speed 1 Mb CMOS EPROM," 1990 Symp. VLSI Tech. Tech. Dig., p. 127, 1990.

Mengucci, P., et al., "3D statistics from TEM observations of TPFG EEPROM memory cells," *Microcopy, Microanalysis, Microstructures*, vol. 1, no. 3, p. 215, 1990.

Miller, T., S. Illyes, and D. A. Baglee, "Charge loss associated with program disturb stresses in EPROMs," Proc. IRPS, p. 154, 1990.

Minami, S., and Y. Kamigaki, "New scaling guidelines for MNOS nonvolatile memory devices," Proc. Sol. St. Dev. and Mats., Japan, p. 9, 1990.

Miyawaki, Y., et al., "A new erasing and row decoding scheme for low supply voltage operation 16 Mb/64 Mb Flash EEPROMS," Tech. Dig. of the Sym. VLSI Tech., p. 85, 1990.

Mori, S., et al., "Reliability study of thin inter-poly dielectrics for non-volatile memory application," Proc. IRPS, p. 132, 1990.

Nakai, H., et al., "A 36 ns 1 Mbit CMOS EPROM with new data sensing technique," Symp. VLSI Cirs. Tech. Dig., p. 95, 1990.

Nakamura, Y., et al., "A 55 ns 4 Mb EPROM with 1-second programming time," *IEEE ISSCC Tech. Dig.*, p. 62, 1990.

Nozaki, T., et al., "A 1 Mbit EEPROM with MONOS memory cell for semiconductor disk application," Symp. VLSI Cirs. Tech. Dig., p. 101, 1990.

Offenberg, M., et al., "Role of surface passivation in the integrated processing of MOS structures," Symp. VLSI Tech. Dig. Tech. Papers, p. 117, 1990.

Ohshima, Y., et al., "Process and device technologies for 16 Mbit EPROMs with large-tilt-angle implanted p-pocket cell," *IEEE IEDM Tech. Dig.*, p. 95, 1990.

Oto, D., R. Habitzreiter, and K. Venkateswaran, "A high speed EEPROM configurable digital to analog convertor array," Proc. IEEE 1990 Custom Integ. Cir. Conf., p. 6.6/1–5, 1990.

Palm, R., "X88C64: improving microcontroller performance with a new E^2PROM," WESCON/90 Conf. Rec., p. 479, 1990.

Pan, C.-S., et al., "A scaling methodology for oxide–nitride–oxide interpoly dielectric for EPROM applications," *IEEE Trans. Elect. Dev.*, vol. ED–37, no. 6, pt. 1, p. 1439, 1990.

Polman, H. L. A., and J. J. Mul, "A new method for the fast determination of the data retention lifetime of UV EPROMs," *Annales des Telecommunications*, vol. 45, no. 11–12, p. 599, 1990.

Polman, H. L. A., and J. J. Mul, "A new method for the fast determination of the data retention life time of UV EPROMs," Proc. 7th Intl. Conf. on Reliability and Maintainability, p. 463, 1990.

Rakkhit, R., et al., "Drain-avalanche induced hole injection and generation of interface traps in thin oxide MOS devices," *Proc. IRPS*, p. 150, 1990.

Regtien, P. P. L., and P. J. Trimp, "Dynamic calibration of sensors using EEPROMs," Sensors and Actuators A (Physical), vol. A22, no. 1–3, p. 615, 1990.

Riemenschneider, B., et al., "A process technology for a 5-volt only 4 Mb flash EEPROM with an 8.6 μm^2 cell," 1990 Symp. VLSI Tech. Tech. Dig., p. 125, 1990.

Robinson, K., "Trends in flash memory system design," WESCON/90 Conf. Rec., p. 468, 1990.

Roome, C., "Implementing high performance EEPROM designs," WESCON/90 Conf. Rec., vol. 34, p. 484, 1990.

Santo, B., and K. T. Chen, "Technology '90: Solid State," IEEE Spectrum, vol. 27, no. 1, p. 41, 1990.

Seki, K., et al., "An 80-ns 1-Mb flash memory with on-chip erase/erase-verify controller," *IEEE J. Sol. St. Cir.*, vol. SC–25, no. 5, p. 1147, 1990.

Seki, K., et al., "An 80-ns 1-Mb flash memory with on-chip erase/erase-verify controller," *IEEE ISSCC Tech. Dig.*, p. 60, 1990.

Shah, P., "Flash memories help system design of the 1990s," WESCON/90 Conf. Rec., p. 473, 1990.

Shah, P., and G. Armstrong, "Programmable memory trends in the automotive industry," Vehicle Electronics in the 90's: Proc. of the Intl. Congress on Transportation Electronics, p. 143, 1990.

Shams, Q. A., and W. D. Brown, "Physical and electrical properties of memory quality PECVD silicon oxynitride," *J. Electrochem. Soc.*, vol. 137, no. 4, p. 1244, 1990.

Shirota, R., et al., "A 2.3 μm^2 memory cell structure for 16 Mb NAND EEPROMs," *IEEE IEDM Tech. Dig.*, p. 103, 1990.

Shohji, K., et al., "A novel automatic erase technique using an internal voltage generator for 1 Mbit flash EEPROM," Symp. VLSI Cirs. Tech. Dig., p. 99, 1990.

Smith, D., et al., "A 3.6 ns 1 Kb ECL I/O BiCMOS UV EPROM," 1990 IEEE Intl. Symp. on Cirs. and Sys., vol. 3, p. 1987, 1990.

Smith, D. L., et al., "Reduction of charge injection into PECVD SiN_xH_y by control of deposition chemistry," *J. Electr. Mats.*, vol. 19, no. 1, p. 19, 1990.

Stiegler, H., et al., "A 4 Mb 5 V-only flash EEPROM with sector erase," Symp. VLSI Cirs. Tech. Dig., p. 103, 1990.

Taffe, N. P., "Eliminate shadow RAM with fast EPROMs," Conf. Rec.: WESCON '90, vol. 34, p. 488, 1990.

Tanaka, T., et al., "A 4-Mbit NAND-EEPROM with tight programmed V_t distribution," Symp. VLSI Cirs. Tech. Dig., p. 105, 1990.

Terada, Y., et al., "High speed page mode sensing scheme for EPROMs and flash EEPROMs using divided bit line architecture," Symp. VLSI Cirs. Tech. Dig., p. 97, 1990.

Urai, T., et al., "Simulation of EPROM programming characteristics," *Electr. Letts.*, vol. 26, no. 11, p. 716, 1990.

Vancu, R., et al., "A 35ns 256K CMOS EEPROM with error correcting circuitry," *IEEE ISSCC Tech. Dig.*, p. 64, 1990.

Vorontsov, V. V., V. M. Efimov, and S. P. Sinitsa, "Hot electron capture in Si_3N_4," Soviet Microelect. (English translation of Mikroelektronika), vol. 18, no. 3, p. 127, 1990.

Weber, S., "Look out EPROM, here comes Flash," *Electronics*, vol. 63, no. 11, 1990.

Weinberg, Z. A., et al., "Ultrathin oxide–nitride–oxide films," *Appl. Phys. Lett.*, vol. 57, no. 12, p. 1248, 1990.

Woo, B. J., et al., "A novel memory cell using flash array contactless EPROM (FACE) technology," *IEEE IEDM Tech. Dig.*, p. 91, 1990.

Wu, K., et al., "A model for EPROM intrinsic charge loss through oxide–nitride–oxide (ONO) interpoly dielectric," *Proc. IRPS*, p. 145, 1990.

Yamauchi, Y., et al., "A 4 M bit NVRAM technology using a novel stacked capacitor on selectively self-aligned FLOTOX cell structure," *IEEE IEDM Tech. Dig.*, p. 931, 1990.

Yoneda, K., et al., "Reliability degradation mechanism of the ultra thin tunneling oxide by the post annealing," Dig. Symp. VLSI Tech., p. 121, 1990.

Yoshikawa, K., S. Mori, and N. Arai, "EPROM cell structure for EPLD's compatible with single poly-Si gate process," *IEEE Trans. Elect. Dev.*, vol. ED–37, no. 3, pt. 1, p. 675, 1990.

Yoshikawa, K., et al., "A flash EEPROM cell scaling including tunnel oxide limitations," ESSDERC 90; 20th European Sol. St. Dev. Res. Conf., Eds. W. Eccleston, P. J. W. Bristol Rosser, UK: Adam Hilger, p. 169, 1990.

Yoshikawa, K., M. Sato, and Y. Ohshima, "Reliable profiled lightly doped drain (PLD) cell for high-density submicrometer EPROMs and Flash EEPROMs," *IEEE Trans. Elect. Dev.*, vol. ED–37, no. 4, p. 999, 1990.

Yoshikawa, K., et al., "Asymmetrical lightly doped source cell for virtual ground high-density EPROMs," *IEEE Trans. Elect. Dev.*, vol. ED–37, no. 4, p. 1046, 1990.

Yoshikawa, K., et al., "Lucky-hole injection induced by band-to-band tunneling leakage in stacked gate transistors," *IEEE IEDM Tech. Dig.*, p. 577, 1990.

Zales, S., and D. Elbert, "Intel flash EPROM for in-system reprogrammable non-volatile storage," *Microprocessors and Microsystems*, vol. 14, no. 8, p. 543, 1990.

1991

Ajika, N., et al., "A novel cell structure for 4 MBit full feature EEPROM and beyond," *IEDM Tech. Dig.*, p. 295, 1991.

Amin, A. A. M., "A novel flash erase EEPROM memory cell with reversed poly roles," Proc. 6th Mediterranean Electrotechnical Conf., vol. 1, p. 311, 1991.

Amin, A., and M. Aref, "An intelligent EPROM silicon compiler," Proc. 91 IEEE Pacific Rim Conf. Commun. Comput. Signal Process., p. 757, 1991.

Arima, H., et al., "A high density performance cell for 4M bit full feature electrically erasable/programmable read-only memory," *Jap. J. Appl. Phys.* Part 2 (Letters), vol. 30, no. 3A, p. L334, 1991.

Arima, H., et al., "Optimization of nitridation and reoxidation conditions for an EEPROM tunneling dielectric," *Jap. J. Appl. Phys.*, Part 2 (Letters), vol. 30, no. 3A, p. L398, 1991.

Asquith, J. E., "A two-dimensional numerical model of an EEPROM transistor for VLSI," M.S. thesis, University of Alabama at Huntsville, 1991.

Barker, S. A., "Effects of carbon on charge loss in EPROM structures," *Proc. IRPS*, p. 171, 1991.

Bergonzoni, C., E. Camerlenghi, and P. Caprara, "Device simulations for EPROM cells scaling down," *Microelect. Eng.*, vol. 15, no. 1–4, p. 625, 1991.

Cappelletti, P., et al., "Accelerated current test for fast tunnel oxide evaluation (of EPROMs)," Proc. 1991 Intl. Conf. on Microelect. Test Structures, p. 81, 1991.

Chang, K.-M., S. Cheng, and C. Kuo, "Modular flash EEPROM technology for 0.8 μm high speed logic circuits," Proc. IEEE 1991 Custom Integ. Cirs. Conf., p. 18, 1991.

Ciampolini, P., A. Pierantoni, and G. Baccarani, "Efficient 3-D simulation of complex structures," IEEE Trans. Computer-Aided Des. of Int. Cir. and Sys., vol. 10, no. 9, p. 1141, 1991.

Crowell, C. R., C.-C. Shih, and V. C. Tyree, "SWEAT structure design and test procedure criteria based upon TEARS characterization and spacial distribution in iterated structures," *Proc. IRPS*, 1991.

Cummings, M., "Alias EPROM?," *Electronics World + Wireless World*, vol. 97, no. 1661, p. 254, 1991.

Dumin, D. J., N. B. Heilemann, and N. Husain, "Test structures to investigate thin insulator dielectric wearout and breakdown," Proc. 1991 Intl. Conf. on Microelect. Test Structures, p. 61, 1991.

Eitan, B., R. Kazerounian, and A. Bergemont, "Alternate metal virtual ground (AMG)—a new scaling concept for very high-density EPROMs," *IEEE Elect. Dev. Lett.*, vol. EDL–12, no. 8, p. 450, 1991.

Euzent, B. L., T. J. Maloney, and J. C. Donner, "Reducing field failure rate with improved EOS/ESD design EPROM," II Electr. Overstress/Electrostatic Discharge Symp. Proc. 1991, p. 59, 1991.

Fiegna, C., et al., "Simple and efficient modeling of EPROM writing," *IEEE Trans. Elect. Dev.*, vol. ED–38, no. 3, p. 603, 1991.

Fukumoto, K., et al., "A 256 K-bit non-volatile PSRAM with page recall and chip store," 91 Symp. VLSI Tech. Dig., p. 91, 1991.

Fukuda, H., et al., "Novel N_2O-oxynitridation technology for forming highly reliable EEPROM tunnel oxide films," *IEEE Elect. Dev. Lett.*, vol. EDL–12, no. 11, p. 587, 1991.

Gigon, F., "Compact spice model of EEPROM memory cell for writing/erasing/read operation," *Microelect. Eng.*, vol. 15, no. 1–4, p. 629, 1991.

Greenwood, C. J., "Construction and characterization of an ASIC EEPROM," Tenth Microelect. Conf. Proc., no. 91, pt. 5, p. 58, 1991.

Harari, E., "Flash EEPROM memory system having multi-level storage cells," U.S. Patent #5,043,940, 1991.

Hart, M. J., et al., "A back-biased 0.65 μm L_{effn} CMOS EEPROM technology for next-generation sub 7 ns programmable logic devices," *Microelect. Eng.*, vol. 15, no. 1–4, p. 613, 1991.

Hayashikoshi, M., et al., "A dual-mode sensing scheme of capacitor coupled EEPROM cell for super high endurance," 1991 Symp. on VLSI Cirs.; Dig. of Tech. Paps., p. 89, 1991.

Hemink, G. J., et al., "Modeling of VIPMOS hot electron gate currents," *Microelect. Eng.*, vol. 15, no. 1–4, p. 65, 1991.

Higuchi, M., et al., "An 85ns 16 Mb CMOS EPROM," *IEICE Trans.*, vol. E74, no. 4, p. 896, 1991.

Huang, J., and T. Liu, "Study on interface trap distribution of MNOS structures using TSC spectrum," *Chinese J. Semicond.*, vol. 12, no. 12, p. 728, 1991.

Jex, J. "Flash memory BIOS for PC and notebook computers," IEEE Pacific Rim Conf. on Communications, Computers and Signal Processing, vol. 2, p. 692, May 1991.

Jin-Yeong Kang, "Fabrication and operational stability of inverted floating gate E^2PROM (electrically erasable programmable read only memory)," *Japan. J. Appl. Phys.*, Part 1 (Regular Papers & Short Notes), vol. 30, no. 4, p. 627, 1991.

Joyner, A. V., "Design of a compilable ASIC EEPROM," Microelect. Conf. 1991: Enabling Technology, no. 91, pt. 5, p. 35, 1991.

Kammerer, W., et al., "A new virtual ground array architecture for very high speed, high density EPROMs," 1991 Symp. on VLSI Cirs.; Dig. of Tech. Paps., p. 83, 1991.

Kim, T., and K. Ohnishi, "Properties of SiN films deposited by photo CVD and memory characteristics of MNOS structure," Electr. & Comm. in Jap., Part II: Electr., vol. 74, no. 9, p. 63, 1991.

Kitazawa, S., and T. Harada, "Low detecting bias and its influence on nonvolatile memory data access," IEICE Trans., vol. E74, no. 4, p. 885, 1991.

Kramer, A., et al., "Compact EEPROM-base weight functions," Adv. in Neural Info. Proc. Systems 3, San Mateo, Calif., Morgan Kaufmann, p. 1001, 1991.

Kodama, N., et al., "A 5V only 16Mbit Flash EEPROM cell using highly reliable write/erase technologies," 91 Symp. VLSI Tech. Dig., p. 75, 1991.

Kodama, N., et al., "A symmetrical side wall (SSW)-DSA Cell for a 64Mbit Flash memory," *IEEE IEDM Tech. Dig.*, p. 303, 1991.

Kumagai, J., S. Sawada, and K. Toita, "Novel measurement technique for trapped charge centroid in gate insulator," Proc. of the 1991 Intl. Conf. on Microelect. Test Struct., p. 87, 1991.

Kume, H., et al., "A 3.42 μm^2 flash memory cell technology conformable to a sector erase," 1991 Symp. on VLSI Tech.; Dig. of Tech. Paps., p. 77, 1991.

Kuo, C., et al., "512 Kb flash EEPROM for a 32 bit microcontroller," 1991 Symp. on VLSI Cirs.; Dig. of Tech. Paps., p. 87, 1991.

Lai, S. K., "Oxide and interface issues in nonvolatile memory," Santa Clara Valley Section of the IEEE Elect. Dev. Soc. Symp.: Adv. in Semicond. Tech., p. 1, 1991.

Lai, S., et al., "Highly reliable E^2PROM cell fabricated with ETOX flash process," 1991 Symp. on VLSI Tech.; Dig. of Tech. Paps., p. 59, 1991.

Lalvani, D. D., "1 megabit burst mode EPROM provides unlimited 15 ns sequential access capability," Electro Intl.; Conf. Rec., p. 152, 1991.

Lambertson, R., et al., "A high-density dual polysilicon 5-volt-only EEPROM cell," *IEEE IEDM Tech. Dig.*, p. 299, 1991.

Lanzoni, M., et al., "Evaluation of E2PROM data retention by field acceleration," Qual. and Reliab. Eng. Intl., vol. 7, no. 4, p. 293, 1991.

Leblebici, Y., et al., "Numerical simulation of hot-carrier induced damages in VLSI circuits," Proc. IEEE 1991 Custom Integ. Cir. Conf., p. 29, 1991.

Lee, P., "Designing serial EEPROMs into microcontroller systems," Electro Intl.; Conf. Rec., p. 32, 1991.

Leibson, S. H., "Nonvolatile, in-circuit-reprogrammable memories," *EDN*, vol. 36, no. 1, p. 89, 1991.

Letourneau, P., et al., "Effect of nitrogen profile on electrical characteristics of ultrathin SiO_2 films nitrided by RTP," *Microelect. Eng.*, vol. 15, no. 1–4, p. 483, 1991.

Liu, Z. H., P. T. Lai, and Y. C. Cheng, "Characterization of charge trapping and high-field endurance for 15-nm thermally nitrided oxides," *IEEE Trans. Elect. Dev.*, vol. ED–38, no. 2, p. 344, 1991.

Lo, G. Q., et al., "Improved performance and reliability of MOSFETs with ultrathin gate oxides prepared by conventional furnace oxidation of Si in pure N_2O ambient," Symp. VLSI Dig. Tech. Papers, p. 43, 1991.

Maes, H. E., Chapter VI, *LPCVD Oxynitrides*, Ed. F. Habraken, 1991.

Mannem, R. R., "A BiCMOS EEPROM design using pipeline architecture," M.S. thesis, Arizona State University, 1991.

Martin, F., and X. Aymerich, "Transient analysis of charge transport in the nitride of MNOS devices under Fowler–Nordheim injection conditions," *Microelect. J.*, vol. 22, no. 7–8, p. 5, 1991.

Martin, J., "Aerospace applications of ferroelectric memory technology," IEEE Proc. National Aerospace and Electronics Conf., vol. 1, p. 282, 1991.

Masuhara, T., et al., "VLSI memories: present status and future prospects," *IEICE Trans.*, vol. E74, no. 1, p. 130, 1991.

Masuoka, F., R. Shirota, and K. Sakui, "Review and prospects of non-volatile semiconductor memories," *IEICE Trans.*, vol. E74, no. 4, p. 868, 1991.

McConnell, M., et al., "An experimental 4 Mb flash EEPROM with sector erase," *IEEE J. Sol. St. Cir.*, vol. SC–26, no. 4, p. 484, 1991.

Melanotte, M., R. Bez, and G. Crisenza, "Non-volatile memories-status and emerging trends," *Microelect. Eng.*, vol. 15, no. 1–4, p. 603, 1991.

Minami, S., and Y. Kamigaki, "New scaling guidelines for MNOS nonvolatile memory devices," *IEEE Trans. Elect. Dev.*, vol. ED–38, no. 11, p. 2519, 1991.

Minami, S., and Y. Kamigaki, "Tunnel oxide thickness optimization for high-performance MNOS nonvolatile memory devices," *IEICE Trans.*, vol. E74, no. 4, p. 875, 1991.

Miyashita, M., et al., "Dependence of thin oxide films quality on surface microroughness," Symp. VLSI Tech. Dig. Tech. Papers, p. 45, 1991.

Miyawaki, Y., et al., "A new erasing and row decoding scheme for low supply voltage operation 16 Mb/64 Mb flash EEPROMs," 1991 Symp. on VLSI Cirs.; Dig. of Tech. Paps., p. 85, 1991.

Momodomi, M., et al., "A 4 Mb NAND EEPROM with tight programmed Vt distribution," *IEEE J. Sol. St. Cir.*, vol. SC–26, no. 4, p. 492, 1991.

Mori, S., et al., "ONO inter-poly dielectric scaling for nonvolatile memory applications," *IEEE Trans. Elect. Dev.*, vol. ED–38, no. 2, p. 386, 1991.

Mori, S., et al., "Polyoxide thinning limitation and superior ONO interpoly dielectric for nonvolatile memory devices," *IEEE Trans. Elect. Dev.*, vol. ED–38, no. 2, p. 270, 1991.

Mori, S., et al., "Threshold voltage instability and charge retention in nonvolatile memory cell with nitride/oxide double-layered inter-poly dielectric," *Proc. IRPS*, p. 175, 1991.

Murray, K. G. D., "New developments in EEPROM technology for ASICs," *J. Semicustom ICs*, vol. 9, no. 1, p. 3, 1991.

Nachtwei, G., et al., "Density of states of the two-dimensional electron system at Si-MOS and Si-MNOS devices in the quantum well regime," *Surf. Sci.*, vol. 250, no. 1–3, p. 243, 1991.

Nakayama, T., et al., "A 60-ns 16-Mb flash EEPROM with program and erase sequence controller," *IEEE J. Sol. St. Cir.*, vol. SC–26, no. 11, p. 1600, 1991.

Nishida, T., and S. E. Thompson, "Oxide field and temperature dependent gate oxide degradation by substrate hot electron injection," *Proc. IRPS*, p. 310, 1991.

Nozaki, T., et al., "A 1 Mb EEPROM with MONOS memory cell for semiconductor disk application," *IEEE J. Sol. St. Cir.*, vol. SC–26, no. 4, p. 497, 1991.

Nughin, A., A. Multsev, and A. Milosheysky, "N-channel 256Kb and 1Mb EEPROMs," IEEE Intl. Sol. St. Conf., p. 228, 1991.

Olivo, P., et al., "Charge trapping and retention in ultra-thin oxide–nitride–oxide structures," *Sol. St. Electr.*, vol. 34, no. 6, p. 609, 1991.

Ozaki, T., et al., "Analysis of band-to-band tunnelling leakage current in trench-capacitor DRAM cells," *IEEE Elect. Dev. Letts.*, vol. EDL–12, no. 3, p. 95, 1991.

Pan, C.-S., K. Wu, and G. Sery, "Physical origin of long-term charge loss in floating-gate EPROM with an interpoly oxide–nitride–oxide stacked dielectric," *IEEE Elect. Dev. Lett.*, vol. ED–12, no. 2, p. 51, 1991.

Pan, C.-S., et al., "High-temperature charge loss mechanism in a floating-gate EPROM with an oxide–nitride–oxide (ONO) interpoly stacked dielectric," *IEEE Elect. Dev. Lett.*, vol. EDL–12, no. 9, p. 506, 1991.

Papadas, C., et al., "Programming window degradation in FLOTOX EEPROM cells," *Microelect. Eng.*, vol. 15, no. 1–4, p. 621, 1991.

Peters, B., "Non-volatile memory technologies hold promise for today and tomorrow," *Defense Electronics*, vol. 23, no. 10, p. 79, 1991.

Raikerus, P. A., and V. A. Gurtov, "Depolarization of MNOS structures in the regime of dispersion transport," *Sol. St. Electr.*, vol. 34, no. 1, p. 63, 1991.

Ramaswami, R., and H. C. Lin, "Simulation of time-dependent tunneling characteristics of MOS structures," *Sol. St. Electr.*, vol. 34, no. 3, p. 291, 1991.

Renan, T., "Electrical characterization of polysilicon surface roughness in double polysilicon EPROMS," Proc. 9th Bienn. Univ. Gov. Ind. Microelectron. Symp., p. 79, 1991.

Rofan, R., and C. Hu, "Stress-induced oxide leakage," *IEEE Elect. Dev. Letts.*, vol. EDL–12, p. 632, 1991.

Rosenbaum, E., and C. Hu, "High-frequency time-dependent breakdown of SiO_2," *IEEE Elect. Dev. Letts.*, vol. EDL–12, p. 267, 1991.

Rosenbaum, E., Z. Liu, and C. Hu, "The effect of oxide stress waveform on MOSFET performance," *IEEE IEDM Tech. Dig.*, p. 719, 1991.

Roy, A., and M. H. White, "Determination of the trapped charge distribution in scaled silicon nitride MNOS nonvolatile memory devices by tunneling spectroscopy," *Sol. St. Electr.*, vol. 34, no. 10, p. 1083, 1991.

Shiba, K., et al., "A composite process for EPROM and EEPROM and its reliability," Electr. and Comm. in Jap., Part 2 (Electronics), vol. 74, no. 8, p. 68, 1991.

Simard, R., et al., "An EPROM-based PWM modulator for a three-phase soft commutated inverter," IEEE Trans. Industrial Electronics, vol. IE–38, no. 1, p. 79, 1991.

Sune, J., et al., "Transient simulation of the erase cycle of floating gate EEPROM's," *IEEE IEDM Tech. Dig.*, p. 905, 1991.

Thomsen, A., and M. A. Brooke, "A floating-gate MOSFET with tunneling injector fabricated using a standard double-polysilicon CMOS process," *IEEE Elect. Dev. Letts.*, vol. EDL–12, no. 3, p. 111, 1991.

Ting, W., J. H. Ahn, and D. L. Kwong, "Ultrathin stacked Si_3N_4/SiO_2 gate dielectrics prepared by rapid thermal processing," *Electr. Letts.*, vol. 27, no. 12, p. 1046, 1991.

Tjulkin, V., and V. A. Miloshevsky, "A 4kb nMOS static NVRAM with extended 16 kb nonvolatile memory," *IEEE ISSCC Tech. Dig.*, p. 226, 1991.

Tomita, N., et al., "A 60-ns 16-Mb CMOS EPROM with voltage stress relaxation technique," *IEEE J. Sol. St. Cir.*, vol. SC–26, no. 11, p. 1593, 1991.

Turkman, R., "Electrical characterization of polysilicon surface roughness in double polysilicon EPROMS," Proc. of the 9th Biennial Univ./Gov./Ind. Microelect. Symp., p. 79, 1991.

Uraoka, Y., et al., "Evaluation of gate oxide reliability using luminescence method," Proc. 1991 Intl. Conf. on Microelect. Test Structures, p. 69, 1991.

Ushiyama, M., et al., "Two dimensionally inhomogeneous structure at gate electrode/gate insulator interface causing Fowler–Nordheim current deviation in nonvolatile memory," *Proc. IRPS*, p. 331, 1991.

Van Houdt, J., et al., "The high injection MOS cell: a novel 5 V-only flash EEPROM concept with a 1 μsec programming time," *Microelect. Eng.*, vol. 15, no. 1–4, p. 617, 1991.

Varhue, W. J., et al., "Electron trapping and chemical composition in radio frequency glow discharge a-SiN:H," *J. Vac. Sci. Technol.*, vol. A9, no. 6, p. 3076, 1991.

Wijburg, R. C., et al., "VIPMOS—a novel buried injector structure for EPROM applications," *IEEE Trans. Elect. Dev.*, vol. ED–38, no. 1, p. 111, 1991.

Wolf, K. A., "Flash memory technology to design where no nonvolatile memory has gone before," Electro Intl.; Conf. Rec., p. 223, 1991.

Woo, B. J., T. C. Ong, and S. Lai, "A poly-buffered FACE technology for high density flash memories," 1991 Symp. on VLSI Tech.; *Dig. of Tech. Paps.*, p. 73, 1991.

Yamada, S., K. Naruke, and M. Wada, "Modified impact-ionization recombination model under dynamic stress of thin oxide (of EEPROM)," Ext. Abs. 1991 Intl. Conf. on Sol. St. Dev. and Mats., p. 32, 1991.

Yamada, S., et al., "A self-convergence erasing scheme for a simple stacked gate Flash EEPROM," *IEEE IEDM Tech. Dig.*, p. 307, 1991.

Yamauchi, Y., et al., "A 5V-only virtual ground Flash cell with an auxillary gate for high density and high speed applications," *IEEE IEDM Tech. Dig.*, p. 319, 1991.

Yeh, B., "Side wall contact in a nonvolatile electrically alterable memory cell," U.S. Patent 5,023,694, 1991.

Ynegawa, K., et al., "Low power one chip microcomputer with 64Kbit EEPROM for smart card," *Shapu Giho/Sharp Tech. J.*, no. 51, p. 67, 1991.

Yoshikawa, K., et al., "Flash EEPROM cell scaling based on tunnel oxide thinning limitations," 1991 Symp. on VLSI Tech.; Dig. of Tech. Paps., p. 79, 1991.

1992

Akulov, A. P., V. V. Aphanasjev, and V. A. Gurtov, "Determination of the radiation-induced charge centroid in a MIS dielectric," *Sol. St. Elect.*, vol. 35, no. 9, p. 1353, 1992.

Amin, A. A. M., "Design, selection and implementation of flash erase EEPROM memory cells," IEE Proceedings G (Circuits, Devices and Systems), vol. 139, no. 3, p. 370, June 1992.

Amin, A. M. A., "A novel flash erase EEPROM memory cell with reversed poly roles," 6th Mediterranean Electrotechnical Conf., p. 311, 1992.

Baker, M., et al., "Non-volatile memory for fast, reliable file systems," Intl. Conf. on Architect. Support for Prog. Lang. and Oper. Sys.—ASPLOS, vol. 27, no. 9, p. 10, 1992.

Barlow, K. J., and B. S. Bold, "Optimised drain structures for improved hot carrier reliability in submicron NMOS transistors," IEE Colloquium on 'Sub-Micron VLSI Reliability' (Digest No. 002), p. 2, 1992.

Chang, M.-B., U. Sharma, and S. K. Cheng, "Improved model for the erase operation of a FLOTOX EEPROM cell," *Sol. St. Electr.*, vol. 35, no. 10, p. 1513, 1992.

Chang, M.-B., et al., "New scalable floating-gate EEPROM cell," *Sol. St. Electr.*, vol. 35, no. 10, p. 1521, 1992.

Concannon, A., et al., "Analysis of transient behaviour of floating gate EEPROMs," *Microelectr. Eng.*, vol. 19, no. 1–4, p. 773, 1992.

Crisenza, G., et al., "Floating gate memories reliability," Quality and Reliab. Eng. Intl., vol. 8, no. 3, p. 177, 1992.

Deleonibus, S., "GIGABIT scalable SILO field isolation using rapid thermal nitridation (RTN) of silicon," *Microelectr. Eng.*, vol. 19, no. 1–4, pp. 75, 1992.

El-Hennawy, A., "Design and simulation of a high reliability non-volatile CMOS EEPROM memory cell compatible with scaling-down trends," *Int. J. Elect.*, vol. 72, no. 1, p. 73, 1992.

Endoh, T., et al., "New write/erase operation technology for Flash EEPROM cells to improve the read disturb characteristics," *IEEE IEDM Tech. Dig.*, p. 603, 1992.

Fukuda, H., T. Arakawa, and S. Ohno, "Thin-gate SiO_2 films formed by in-situ multiple rapid thermal processing," *IEEE Trans. Elect. Dev.*, vol. ED–39, no. 1, p. 127, 1992.

Fukuda, H., M. Yasuda, and T. Iwabuchi, "Novel single-step oxynitridation (SS-RTON) technology for forming highly reliabile EEPROM tunnel oxide films," Ext. Abs.; 1992 Intl. Conf. on Sol. St. Dev. and Mats., p. 425, 1992.

Fukuda, H., et al., "Heavy oxynitridation technology for forming highly reliable Flash-type EEPROM tunnel oxide films," *Electr. Letts.*, vol. 28, no. 19, p. 1781, 1992.

Fukuda, H., et al., "High performance scaled Flash-type EEPROMs with heavily oxynitrided tunnel oxide films," *IEEE IEDM Tech. Dig.*, p. 465, 1992.

Gosch, J., "IC merges 32-kbyte flash EPROM with 16-bit micro," *Electronic Design*, vol. 40, no. 10, p. 99, May 14, 1992.

Hayashikoshi, M., et al., "A dual-mode sensing scheme of capacitor-coupled EEPROM cell," *IEEE J. Sol. St. Cir.*, vol. SC–27, no. 4, p. 569, 1992.

Herdt, C. E., and C. A. Paz de Araujo, "Analysis, measurement, and simulation of dynamic write inhibit in an nvSRAM cell," *IEEE Trans. Elect. Dev.*, vol. ED–39, no. 5, p. 1191, 1992.

Hori, T., et al., "A MOSFET with Si-implanted gate-SiO_2 insulator for nonvolatile memory applications," *IEEE IEDM Tech. Dig.*, p. 17.7.1–4, 1992.

Houdt, J. V., G. Groeseneken, and H. E. Maes, "Analytical model for the optimization of high injection MOS Flash E^2PROM devices," *Microelectr. Eng.*, vol. 19, no. 1–4, p. 257, 1992.

Hsu, C. C.-H., et al., "High speed, low power p-channel Flash EEPROM using silicon rich oxide as tunneling dielectric," Ext. Abs.; 1992 Intl. Conf. on Sol. St. Dev. and Mats., p. 140, 1992.

Hu, Y., and M. H. White, "A new buried-channel EEPROM device," *IEEE Trans. Elect. Dev.*, vol. ED–39, no. 11, p. 2670, 1992.

Inoue, H., and S. Ito, "High-speed 16-bit single-chip microcomputer with EPROM," *Mitsubishi Electric Advance*, vol. 59, p. 12, 1992.

Jacunski, M., et al., "Radiation hardened nonvolatile memory for space and strategic applications," GOMAC, p. 449, 1992.

Jinbo, T., et al., "A 5V-only 16 Mb Flash memory with sector-erase mode," *IEEE ISSCC Tech. Dig.*, p. 154, 1992.

Kapoor, V. J., D. Xu, and R. S. Bailey, "The combined effect of hydrogen and oxygen impurities in the silicon nitride film of MNOS devices," *J. Electrochem. Soc.*, vol. 139, no. 3, p. 915, 1992.

Kaya, C., et al., "Buried source-side injection (BSSI) for Flash EPROM programming," *IEEE Elect. Dev. Letts.*, vol. EDL–13, no. 9, p. 465, 1992.

Keeney, S., et al., "Complete transient simulation of Flash EEPROM devices," *IEEE Trans. Elect. Dev.*, vol. ED–39, no. 12, p. 2750, 1992.

Keeney, S., et al., "Simulation of enhanced injection split gate Flash EEPROM device programming," *Microelectr. Eng.*, vol. 18, no. 3, p. 253, 1992.

Keeney, S., et al., "Simulation of EPROM device programming using the hydrodynamic model," *Microelectr. Eng.*, vol. 19, no. 1–4, p. 261, 1992.

Koyama, S., "A novel cell structure for giga-bit EPROMs and Flash memories using polysilicon thin film transistors," Tech. Dig. of the Sym. VLSI Tech., p. 44, 1992.

Kuki, S. "Speed of Flash memories sparks maker interest," JEE, p. 40, November 1992.

Kume, H., et al., "A 1.28 μm^2 contactless memory cell technology for a 3V-only 64 Mbit EEPROM," *IEEE IEDM Tech. Dig.*, p. 991, 1992.

Kuo, C., et al., "A 512-kb flash EEPROM embedded in a 32-b microcontroller," *IEEE J. Sol. St. Cir.*, vol. SC–27, no. 4, p. 574, 1992.

Lai, S. K., "Oxide/silicon interface effects in E^2PROMs and ETOX™ Flash," Proc. SRC Topical Research Conference on Floating Gate Non-Volatile Memory Research, October 1992.

Lin, J. K., et al., "Simple model for the switching behavior of a SONOS EEPROM device," Ext. Abs.; 1992 Intl. Conf. on Sol. St. Dev. and Mats., p. 143, 1992.

Linton, T. D., Jr., P. A. Blakey, and D. P. Neikirk, "The impact of three-dimensional effects on EEPROM cell performance," *IEEE Trans. Elect. Dev.*, vol. ED–39, no. 4, p. 843, April 1992.

Maloberti, F., et al., "CMOS data output buffer for integrated memories," *Elect. Letts.*, vol. 28, no. 21, p. 1946, 1992.

Martin, F., and X. Aymerich, "Characterization of the spatial distribution of traps in Si_3N_4 by field-assisted discharge of metal–nitride–oxide–semiconductor devices," Thin Solid Films, vol. 221, no. 1–2, p. 147, 1992.

Masuoka, F., "Technology trends of Flash EEPROMs," VLSI Symp. Dig. Tech. Papers, p. 6, 1992.

Miyamoto, J., et al., "Multi-step stress test for yield improvement of 16 Mbit EPROMs with redundancy scheme," Dig. of Pap.—Intl. Test Conf., p. 540, 1992.

Miyawaki, Y., et al., "A new erasing and row decoding scheme for low supply voltage operation 16-Mb/64-Mb flash memories," *IEEE J. Sol. St. Cir.*, vol. SC-27, no. 4, p. 583, April 1992.

Moazzami, R., and C. Hu, "Stress-induced current in thin silicon dioxide films," *IEEE IEDM Tech. Dig.*, p. 139, 1992.

Mori, S., et al., "Bottom-oxide scaling for thin nitride/oxide interpoly dielectric in stacked-gate nonvolatile memory cells," *IEEE Trans. Elect. Dev.*, vol. EDL–39, no. 2, p. 283, 1992.

Ohmi, T., et al., "Wafer quality specification for future sub-half micron ULSI devices," Symp. VLSI. Tech. Dig. Tech. Papers, p. 24, 1992.

Ohnishi, K., and K. Miura, "Electron injection characteristics of MNOS nonvolatile semiconductor memory structure using the reverse breakdown of the PN junction," Electr. & Comm. in Jap., Part II: Electronics, vol. 75, no. 4, p. 52, 1992.

Ong, T. C., et al., "Instability of erase threshold voltage in ETOX Flash memory array," Proc. of SRC Topical Res. Conf. on Floating Gate Non-Volatile Memory Research, October 1992.

Onoda, H., et al., "A novel cell structure suitable for a 3V operation, sector erase Flash memory," *IEEE IEDM Tech. Dig.*, p. 599, 1992.

Oyama, K., et al., "A novel erasing technology for 3.3V Flash memory with 64Mb capacity and beyond," *IEEE IEDM Tech. Dig.*, p. 607, 1992.

Papadas, C., et al., "Influence of tunnel oxide thickness variation on the programmed window of FLOTOX EEPROM cells," *Sol. St. Electr.*, vol. 35, no. 8, p. 1195, 1992.

Papadas, C., et al., "Model for programming window degradation in FLOTOX EEPROM cells," *IEEE Elect. Dev. Lett.*, vol. EDL–13, no. 2, p. 89, 1992.

Patrikar, R. M., and R. Lal, "New method for assessing dielectric integrity of MOS oxides," Microelect. and Reliabil., vol. 32, no. 7, p. 961, 1992.

Paulsen, R. E., et al., "Observation of near-interface oxide traps with the charge-pumping technique," *IEEE Trans. Elect. Dev.*, vol. ED–39, no. 12, p. 627, 1992.

Rollins, J. G., V. Axelrad, and S. J. Motzny, "Carrier temperature dependent gate current modeling for EEPROM simulation," *Microelectr. Eng.*, vol. 19, no. 1–4, p. 265, 1992.

Roy, A., et al., "Substrate injection induced program disturb—a new reliability consideration for Flash-EPROM arrays," *Proc. IRPS*, p. 68, 1992.

San, T. K., and T.-P. Ma, "Determination of trapped oxide charge in Flash EPROM's and MOSFET's with thin oxides," *IEEE Elect. Dev. Letts.*, vol. EDL–13, no. 8, p. 439, 1992.

San, K. T., et al., "A new technique for determining the capacitive coupling coefficients in flash EPROMs," *IEEE Elect. Dev. Letts.*, vol. EDL–13, no. 6, p. 328, 1992.

Sato, K., et al., "Selective erasable and programmable NAND-type MNOS memory," Elect. & Comm. in Jap., Part II: Electronics, vol. 75, no. 10, p. 93, 1992.

Sethi, R. B., et al., "Electron barrier height change and its influence on EEPROM cells," *IEEE Elect. Dev. Lett.*, vol. EDL–13, no. 5, p. 244, 1992.

Shiba, K., and K. Kubota, "Downscaling of floating-gate EEPROM modules for ASIC applications," Elect. & Comm. in Jap. Part II: Electronics, vol. 75, no. 12, p. 67, 1992.

Sin, C.-K., et al., "EEPROM as an analog storage device, with particular applications in neutral networks," *IEEE Trans. Elect. Dev.*, vol. ED–39, no. 6, p. 1410, 1992.

Singh, R., "Growth of thin thermal silicon dioxide films with low defect density," *Microelectr. J.*, vol. 23, no. 4, p. 273, July 1992.

Takata, K., T. Kure, and T. Okawa, "Observation of deep contact holes and conductive components underlying insulator in a memory cell by tunneling acoustic microscopy," *Appl. Phys. Lett.*, vol. 60, no. 4, p. 575, 1992.

Tang, Y., et al., "Differentiating impacts of hole trapping vs. interface states on TDDB reduction in MOS transistors," Proc. SRC Top. Res. Conf. on Floating Gate Non-Volatile Memory Research, October 1992.

Umezawa, A., et al., "A 5-V-only operation 0.6-μm Flash EEPROM with row decoder scheme in triple-well structure," *IEEE J. Sol. St. Cir.*, vol. SC–27, no. 11, p. 1540, 1992.

Van Houdt, J., et al., "Analysis of the enhanced hot-electron injection in split-gate transistors useful for EEPROM applications," *IEEE Trans. Elect. Dev.*, vol. ED–39, no. 5, p. 1150, 1992.

Watts, T., "Technology steps to build a 4 Mbit EPROM," *Electronic Product Design*, vol. 13, no. 1, p. 38, 1992.

Wei, S. J., and H. C. Lin, "Multivalued SRAM cell using resonant tunnelling diodes," *IEEE J. Sol. St. Cir.*, vol. SC–27, no. 2, p. 212, 1992.

Wolstenholme, G. R., and A. Bergemont, "Novel isolation scheme for implementation in very high density AMG EPROM and FLASH EEPROM arrays," *Microelectr. Eng.*, vol. 19, no. 1–4, p. 253, 1992.

Wu, C.-Y., C.-F. Chen, and Hsin-Chu, "Physical model for characterizing and simulating a FLOTOX EEPROM device," *Sol. St. Electr.*, vol. 35, no. 5, p. 705, 1992.

Wu, C.-Y., et al., "Channel hot-carrier programming in EEPROM devices," Conf. on Sol. St. Dev. and Mats., p. 173, 1992.

Yamauchi, Y., et al., "A versatile stacked storage capacitor on FLOTOX cell for megabit NVRAM's," *IEEE Trans. Elect. Dev.*, vol. ED–39, no. 12, p. 2791, 1992.

Yoon, S., "Reliability of thin gate insulators for submicron MOSFET and EEPROM devices," Ph.D. diss., Lehigh University, 1992.

Yoshida, M., "Flash memories provide alternative method to standard data storage," *J. Electr., Eng.*, vol. 29, no. 308, p. 34, 1992.

Yoshikawa, K., et al., "Comparison of current Flash EEPROM erasing methods: stability and how to control," *IEEE IEDM Tech. Dig.*, p. 595, 1992.

Zhang, K., E. Zheng, and C. Peng, "An EPROM-based heart-rate meter with wide range and multifunctions," *J. Biomedical Eng.*, vol. 14, no. 2, p. 159, March 1992.

1993

Amin, A. A. M., "Design and analysis of a high-speed sense amplifier for single-transistor nonvolatile memory," IEE Proc., Part G, Circuits, Devices and Systems, vol. 140, no. 2, p. 117, 1993.

Amin, A. A. M., "Speed optimized array architecture for Flash EEPROMs", IEE Proc., Part G: Circuits, Devices and Systems, vol. 140, no. 3, p. 177, 1993.

Aritome, S., et al., "Reliability issues of Flash memory cells," *Proc. IEEE*, vol. 81, no. 5, p. 776, 1993.

Banerjee, A. K., et al., "An automated SONOS NVSM dynamic characterization system," Proc. 5th Nonvolatile Memory Tech. Rev. Conf., Linthicum Heights, Md., p. 78, 1993.

Bellafiore, N., F. Pio, and C. Riva, "Thin oxide nitridation in N_2O by RTP for non-volatile memories," *Microelectr. J.*, vol. 24, no. 4, p. 453, 1993.

Bergemont, A., "Status, trends, comparison, and evolution of FLASH technology for memory card applications," IC Card Conf., 9 pages, 1993.

Bergemont, A., et al., "NOR virtual ground (NVG)—a new scaling concept for very high density Flash EEPROM and its implementation in a 0.5μm process," *IEEE IEDM Tech. Dig.*, p. 15, 1993.

Bipert, B., and L. Hebert, "Flash memory goes mainstream," IEEE Spectrum, p. 48, October 1993.

Brand, A., et al., "Novel read disturb failure mechanism induced by FLASH cycling," *Proc. IRPS*, p. 127, 1993.

Castro, H. A., S. M. Tam, and M. A. Holler, "Implementation and performance of an analog nonvolatile neural network," Anal. Integ. Cir. and Sig. Process., vol. 4, no. 2, p. 97, 1993.

Chen, D., et al., "Modeling of the charge balance condition on floating gates and simulation of EEPROM's," IEEE Trans. on Computer-Aided Design of Integ. Cir. and Sys., vol. 12, no. 10, p. 1499, 1993.

Chi Chang, "Flash memory reliability," a tutorial at IRPS 1993.

Concannon, A., et al., "Two-dimensional numerical analysis of floating-gate EEPROM devices," *IEEE Trans. Elect. Dev.*, vol. ED–40, no. 7, p. 1258, 1993.

Dipert, B., and M. Levy, *Designing With Flash Memory*, Annabooks, 1993.

Dong, J., and Y. Peng, "Method of storing C-language programs in an erasable programmable read-only memory (EPROM)," Xibei Gongye Daxue Xuebao/ J. of Northwestern Polytech. Univ., vol. 11, no. 4, p. 431, 1993.

Dori, L., et al., "Optimized silicon-rich oxide (SRO) deposition process for 5V-only Flash EEPROM applications," *IEEE Elect. Dev. Letts.*, vol. EDL–14, no. 6, p. 283, 1993.

Dumin, D., and J. R. Mattux, "Correlation of stress-induced leakage current in thin oxides with trap generation inside the oxides," *IEEE Trans. Elect. Dev.*, vol. ED–40, no. 5, p. 986, 1993.

Dutoit, M., et al., "Optimization of thin Si oxynitride films produced by rapid thermal processing for applications in EEPROMs," *J. Electrochem. Soc.*, vol. 140, no. 2, p. 549, 1993.

Emel'yanov, A. M., "Charge phenomenon in MNOS dielectric films under avalanche charge carrier injection from silicon," *Mikroelektronika*, no. 2, p. 31, 1993.

Fazan, P. C., and V. K. Mathews, "A highly manufacturable trench isolation process for deep submicron DRAMs," *IEEE IEDM Tech. Dig.*, p. 57, 1993.

Fernholz, G., and H. T. Benz, "Dielectric properties of polyoxides for EEPROM," *Microelectr. J.*, vol. 24, no. 4, p. 435, 1993.

French, M., H. Sathianathan, and M. H. White, "A SONOS nonvolatile memory cell for semiconductor disk application," Proc. 5th Nonvolatile Memory Tech. Rev. Conf., Linthicum Heights, Md., p. 38, 1993.

Fukuda, H., M. Yasuda, and T. Iwabuchi, "Novel single-step rapid thermal oxynitridation technology for forming highly reliable electrically erasable programmable read-only memory tunnel oxide films," *Japan. J. Appl. Phys.*, Pt. 1: vol. 32, no. 1B, p. 447, 1993.

Fukuda, H., et al., "Determination of trapped oxide charge in flash-type EEPROMs with heavily oxynitrided tunnel oxide films," *Elect. Letts.*, vol. 29, no. 11, p. 947, 1993.

Gigon, F., et al., "Reliability simulations of the endurance performance of FLOTOX EEPROM cells using SPICE," Qual. and Reliab. Eng. Intl., vol. 9, no. 4, p. 347, 1993.

Griswold, M. D., et al., "Characterization of PECVD process-induced degradation of EEPROM reliability," *Proc. SPIE*, p. 34, 1993.

Gurtov, V. A., et al., "Radiation-induced charge in MNOS structures with nitrified silicon oxide," *Mikroelektronika*, no. 2, p. 43, 1993.

Hayashi, T., et al., "High performance scaled Flash-type EEPROMs fabricated by in-situ multiple rapid thermal processing," *Electr. Letts.*, vol. 29, no. 25, p. 2178, 1993.

Herdt, C. E., "Nonvolatile SRAM—the next generation," Proc. 5th Nonvolatile Memory Tech. Rev. Conf., Linthicum Heights, Md., p. 28, 1993.

Hisamune, Y. S., et al., "A high capacitive coupling ratio (HiCR) cell for 3V only 64 Mbit and future flash memories," *IEEE IEDM Tech. Dig.*, p. 19, 1993.

Holmes, A. J., et al., "Use of a Si:H memory devices for non-volatile weight storage in artificial neural networks," *J. of Non-Crystal. Sol.*, vol. 164–66, pt. 2, p. 817, 1993.

Hu, Y., and M. H. White, "Charge retention in scaled SONOS nonvolatile semiconductor memory devices—modeling and characterization," *Sol. St. Elect.*, vol. 36, no. 10, p. 1401, 1993.

Jardine, A. P., and J. S. Madsen, "Fabrication of "smart" ferroelastic-ferroelectric heterostructures," Proc. Smart Structures and Materials 1993: Smart Materials, p. 384, 1993.

Kim, Y.-S., et al., "Low-defect-density and high-reliability FETMOS EEPROM's fabricated using furnace N_2O oxynitridation," *IEEE Elect. Dev. Letts.*, vol. EDL–14, no. 7, p. 342, 1993.

Kong, S. O., and C. Y. Kwok, "Study of trapezoidal programming waveform for the FLOTOX EEPROM," *Sol. St. Electr.*, vol. 36, no. 8, p. 1093, 1993.

Koyama, S., and T. Jinbo, "Novel cell structure for EPROMs and Flash memories using polysilicon TFT," NEC Res. & Develop., vol. 34, no. 1, p. 70, 1993.

Lai, Stefan, "Flash memory reliability," a tutorial at IRPS 1993.

Lanzoni, M., et al., "Advanced electrical-level modeling of EEPROM cells," *IEEE Trans. Elect. Dev.*, vol. ED–40, no. 5, p. 951, 1993.

Lin, J.-K., et al., "Charge loss due to AC program disturbance stresses in EPROMs," *Jap. J. of Appl. Phys.*, Pt. 1, vol. 32, no. 9A, p. 3748, 1993.

Lin, J.-K., et al., "Transient and steady state carrier transport under high field stresses in SONOS EEPROM device," *Jap. J. of Appl. Phys.*, pt. 1, vol. 32, no. 6A, p. 2748, 1993.

Liong, L. C., and P. Liu, "A theoretical model for the current-voltage characteristics of a floating-gate EEPROM cell," *IEEE Trans. Elect. Dev.*, vol. ED–40, no. 1, p. 146, 1993.

Martin, F., and X. Aymerich, "Interface and bulk traps in oxide-nitride stacked films," Mat. Res. Soc. Symp. Proc., vol. 284, p. 141, 1993.

Matsumoto, O., et al., "1.5V high speed read operation and low power consumption circuit technology for EPROM and Flash-EEPROM," Proc. of the Custom Integ. Cir. Conf., p. 25.4.1–25.4.4, 1993.

Minami, S.-I., and Y. Kamigaki, "Novel MONOS nonvolatile memory device ensuring 10-year data retention after 10^7 erase/write cycles," *IEEE Trans. Elect. Dev.*, vol. ED–40, no. 11, p. 2011, 1993.

Minpour, M., et al., "Reliability issues of a crack-resistant passivation layer process for sub-micron non-volatile memory technology," Mats. Res. Soc. Symp. Proc., vol. 309, p. 41, 1993.

Moison, B., et al., "New method for the extraction of the coupling ratios in FLOTOX EEPROM cells," *IEEE Trans. Elect. Dev.*, vol. ED–40, no. 10, p. 1870, 1993.

Murray, J. R., "Design considerations for a radiation hardened nonvolatile memory," *IEEE Trans. Nucl. Sci.*, vol. NS–40, no. 6, pt. 1, p. 1610, 1993.

Noguchi, K., and H. Arai, "Block erasable high-speed 4Mb Flash memory," *Mitsubishi Electric Advance*, vol. 63, p. 22, 1993.

Ohi, M., et al., "An asymmetrical offset source/drain structure for virtual ground array Flash memory with DINOR operation," *IEEE IEDM Tech. Dig.*, p. 57, 1993.

Ohsaki, K., N. Asamoto, and S. Takagaki, "Planar type EEPROM cell structure by standard CMOS process for integration with gate array, standard cell, microprocessor and for neural chips," Proc. of the Custom Integ. Cir. Conf., p. 23.6.1–23.6.4, 1993.

Papadas, C., et al., "Analytical modelling of the programmed window in FLOTOX EEPROM cells," *Electr. Letts.*, vol. 29, no. 1, p. 122, 1993.

Papadas, C., et al., "Impact of reactive ion etching using O_2 plus CHF_3 plasma on the endurance performance of FLOTOX EEPROM cells," *IEEE Trans. Elect. Dev.*, vol. ED–40, no. 8, p. 1549, 1993.

Papadas, C., et al., "Influence of rapid thermal nitridation of process in N_2O ambient on the endurance performance of FLOTOX EEPROM cells," *Electr. Letts.*, vol. 29, no. 2, p. 242, 1993.

Papadas, C., et al., "New method for the experimental determination of the control gate and drain coupling ratios in FLOTOX EEPROM cells," Proc. of the IEEE Intl. Conf. on Microelectr. Test Struct., p. 293, 1993.

Papadas, C., et al., "Numerical transient simulation of the programming window degradation in FLOTOX EEPROM cells," *Sol. St. Electr.*, vol. 36, no. 9, p. 1303, 1993.

Papadas, C., et al., "On the endurance performance of FLOTOX EEPROM cells with WSi_2 overcoated floating gate electrode," *Microelectr. J.*, vol. 24, no. 4, p. 395, 1993.

Papadas, C., et al., "Reliability issues of silicon-dioxide structures—application to FLOTOX EEPROM cells," *Microelectr. and Reliab.*, vol. 33, no. 11–12, p. 1867, 1993.

Patel, R., et al., "10NS, 4000 gate, 160 pin CMOS EPLD developed on a 0.8 μm process," Proc. of the Custom Integ. Cir. Conf., p. 7.6.1–7.6.5, 1993.

Pein, H. B., and J. D. Plummer, "Performance of the 3-D sidewall Flash EEPROM cell," *IEEE IEDM Tech. Dig.*, p. 11, 1993.

Pein, H., and J. D. Plummer, "3-D sidewall Flash EPROM cell and memory array," *IEEE Elect. Dev. Letts.*, vol. EDL–14, no. 8, p. 415, 1993.

Sathianathan, H., "A single-supply, low-voltage, programmable SONOS memory array for high density EEPROM and semiconductor disk applications," M.S. thesis, Lehigh University, 1993.

Sharma, U., et al., "A novel technology for megabit density, low power, high speed NVRAMs," VLSI Symp., Japan, p. 53, 1993.

Sethi, R. B., et al., "Robust LPCVD nitride integrated process for high density nonvolatile EPROM memories," Rapid Thermal and Integrated Processing II; MRS Symp. Proc., vol. 303, p. 383, 1993.

Smayling, M., et al., "Modular merged technology process including submicron CMOS logic, nonvolatile memories, linear functions, and power components," Proc. of the Custom Integ. Cir. Conf. p. 24.5.1–24.5.4, 1993.

Staudinger, T., "Flash EPROM controller SAB 88C166 sets new standards for real-time applications," Siemens Components, vol. 28, no. 4, p. 15, 1993.

Takahashi, Y., and K. Ohnishi, "Estimation of insulation layer conductance in MNOS structure," *IEEE Trans. Elect. Dev.*, vol. ED–40, no. 11, p. 2006, 1993.

Tarui, Y., "Flash memory features simple structure, superior integration," *J. Electr. Eng.*, vol. 30, no. 321, p. 84, 1993.

Umezawa, A., et al., "New self-data refresh scheme for a sector erasable 16-Mb Flash EEPROM," 1993 Symp. on VLSI Cir., Dig. Tech. Pap., p. 99, 1993.

Van Houdt, J., et al., "A 5V/3.3V—compatible Flash E^2PROM cell with a 400 ns/70μs programming time for embedded memory applications," Proc. 5th Biennial Nonvolatile Memory Technology Review, Linthicum Heights, Md., p. 54, June 1993.

Van Houdt, J., et al., "HIMOS-a high efficiency Flash EEPROM cell for embedded memory applications," *IEEE Trans. Elect. Dev.*, vol. ED–40, p. 2255, 1993.

van Steenwijk, G., K. Hoen, and H. Wallinga, "Nonvolatile analog programmable voltage source using the VIPMOS EEPROM structure," *IEEE J. of Sol. St. Cir.*, vol. 28, no. 7, p. 784, 1993.

Vasilenko, V. S., and Ya. O. Roizin, "Novel techniques of charge centroid evaluation in dielectric layers," *J. Electrostatics*, vol. 30, p. 355, 1993.

Wellekens, D., et al., "On the total dose radiation hardness of floating gate EEPROM cells," Proc. NVMTR, IEEE Cat. #93TH0547-0, p. 54, 1993.

Wellekens, D., et al., "Single poly floating gate cell as the best choice for radiation-hard floating gate EEPROM technology," *IEEE Trans. Nucl. Sci.*, vol. NS–40, no. 6, pt. 1, p. 1619, 1993.

Wiker, R. L., "High density memory technology in advanced signal, data control processors, and other memory systems," Proc. 5th Nonvolatile Memory Tech. Rev. Conf., Linthicum Heights, Md., p. 38, 1993.

White, W., and P. Holden, "Marginal programming triggers failures in EPROM memories," *Evaluation Engineering*, vol. 32, no. 8, p. 29, 1993.

Wu, S. L., et al., "Tunnel oxide prepared by thermal oxidation of thin polysilicon film on silicon," *IEEE Elect. Dev. Lett.*, vol. EDL–14, p. 379, 1993.

Yabe, T., "High-speed circuit techniques for 1 to 5 V operating memories," *IEICE Trans. on Electr.*, vol. E76–C, no. 5, p. 708, 1993.

Yamada, S., et al., "Degradation mechanism of flash EEPROM progamming after programming cycles," *IEEE IEDM Tech. Dig.*, p. 23, 1993.

Yamaguchi, Y., et al., "ONO interpoly dielectric scaling limit for non-volatile memory devices," Symp. VLSI Tech. Dig. Tech. Papers, p. 85, 1993.

Yuan, X.-J., J. S. Marsland, and W. Eccleston, "Substrate hot electron injection modelling based on lucky drift theory," *Microelectr. Eng.*, vol. 22, no. 1–4, p. 261, 1993.

Zolotov, M. V., V. A. Gurtov, and I. N. Surikov, "Influence of the dose rate and temperature of irradiation upon radiation effects in MNOS structures," *Mikroelektronika*, no. 4, p. 82, 1993.

1994

Abdul-Kader, R., and H. A. Ahmed, "Charge trapping in MIOS memory structures under positive and negative biasing," *Intl. J. of Electr.*, vol. 76, no. 2, p. 249, 1994.

Aoki, M., et al., "Triple density DRAM cell with silicon selective growth channel and NAND structure," *IEEE IEDM Tech. Dig.*, p. 631, 1994.

Aritome, S., et al., "A 0.67 μm^2 self-aligned shallow trench isolation cell, [SA-STI cell] for 3V-only 256 Mbit NAND EPROMs," *IEEE IEDM Tech. Dig.*, p. 61, 1994.

Aritome, S., et al., "Advanced NAND-structure cell technology for reliable 3.3 V 64 Mb electrically erasable and programmable read only memories (EEPROMs)," *Jap. J. Appl. Phys.*, pt. 1: Regular Papers and Short Notes and Review Papers, vol. 33, no. 1B, p. 524, 1994.

Aritome, S., et al., "Data retention characteristics of Flash memory cells after write and erase cycling," *IEICE Trans. on Electr.*, vol. E77–C, no. 8, p. 1287, 1994.

Atsumi, S., et al., "A 16-Mb Flash EEPROM with a new self data refresh scheme for a sector erase operation," *IEEE J. Sol. St. Cir.*, vol. SC–29, no. 4, p. 454, 1994.

Baker, A., et al., "A 3.3 V 16Mb Flash memory with advanced write automation," *IEEE ISSCC Tech. Dig.*, p. 146, 1994.

Bhattacharyya, A., "Effects of the variation in the dose of the injector implant on the endurance characteristics of floating gate electrically erasable programmable read only memory devices," *Jap. J. Appl. Phys.*, pt. 1: Regular Papers and Short Notes and Review Papers, vol. 33, no. 4A, p. 1793, 1994.

Botros, N. M., and M. Abdul-Aziz, "Hardware implementation of an artificial neural network using field programmable gate arrays (FPGAs)," *IEEE Trans. Indust. Elect.*, vol. 41, no. 6, p. 665, 1994.

Bursky, D., "Flash memory interfaces simplify systems," *EDN*, p. 168, November 7, 1994.

Cacharelis, P., et al., "A 0.8 μm CMOS, double polysilicon EEPROM technology module optimized for minimum wafer cost," Proc. 24th European Solid State Device Research Conference, p. 195, 1994.

Cao, M., et al., "A simple EEPROM cell using twin polysilicon thin film transistors," *IEEE Elect. Dev. Letts.*, vol. EDL–15, no. 8, p. 304, 1994.

Chern, H. N., et al., "Improvement of polysilicon oxide characteristics by fluorine incorporation," *IEEE Elect. Dev. Lett.*, vol. EDL–15, p. 181, 1994.

Chester, A. J., et al., "Experimental investigation of EEPROM reliability issues using the progressional offset technique," IEEE Intl. Conf. Microelect. Test Structures, p. 135, 1994.

Chioffi, E., "High-speed, low-switching noise CMOS memory data output buffer," *IEEE J. Sol. St. Cir.*, vol. SC–29, no. 11, p. 1359, 1994.

Chiou, Y. L., E. Sheybani, J. P. Gambino, and P. J. Tsang, "Extraction of Fowler–Nordheim tunneling parameters from I-V characteristics of thin gate oxides,"

Silicon Nitride and Silicon Oxide Thin Insulating Films, Eds. V. J. Kapoor and W. D. Brown, Electrochemical Society, Pennington, N.J., p. 612, 1994.

Choi, W. L., and D. M. Kim, "A new technique for measuring coupling coefficients and 3-D capacitance characterization of floating gate devices," *IEEE Trans. Elect. Dev.*, vol. ED–41, no. 12, p. 2337, 1994.

Ciao, M., et al., "A simple EEPROM cell using twin polysilicon thin film transistors," *IEEE Elect. Dev. Lett.*, vol. EDL–15, no. 8 p. 304, 1994.

Concannon, A., et al., "Applications of a novel hot carrier injection model in flash EEPROM design," Proc. 24th European Solid State Device Research Conference, p. 503, 1994.

Concannon, A., et al., "Model for hot-electron and hot hole injection in flash EEPROM programming," *Microelect. J.*, vol. 25, no. 7, p. 469, 1994.

Depas, M., et al., "Ultra thin gate oxide yield and reliability," Symp. VLSI, Tech. Dig. Tech. Papers, p. 23, 1994.

Devore, J., et al., "Monolithic power IC uses nonvolatile memory to increase system flexibility," *Power Conversion and Intelligent Motion*, vol. 20, no. 5, 6 p., 1994.

Dunn, C., et al., "Flash EPROM disturb mechanisms," Ann. Proc.—Reliab. Phys. (Symp.), p. 299, 1994.

Fang Hao, et al., "Plasma-induced in-line charging and damage in non-volatile memory devices," *IEEE IEDM Tech. Dig.*, p. 467, 1994.

Fonash, S. J., "A survey of damage effects in plasma etching," *Solid State Tech.*, p. 99, July 1994.

Franhani, M. M., et al., "Conventional contact interconnect technology as an alternative to contact plug (W) technology for 0.85 μm CMOS EPROM IC devices," *IEEE Trans. Semicond. Manufac.*, vol. 7, no. 1, p. 79, 1994.

French, M., et al., "Design and scaling of a SONOS multidielectric device for nonvolatile memory applications," IEEE Trans. on CPMT, pt. A, vol. 17, no. 3, p. 390, 1994.

Fujii, M., and D. W. Fine, "Flash memories illustrate progress in technology uses," *J. Electr. Eng.*, vol. 31, no. 326, p. 98, 1994.

Gardner, M., J. Fulford, M. Bhat, and D. L. Kwong, "RPT N_2O oxides for nonvolatile memory applications," *Silicon Nitride and Silicon Oxide Thin Insulating Films*, Eds. V. J. Kapoor and W. D. Brown, The Electrochemical Society, Pennington, N.J., p. 357, 1994.

Gardner, M., et al., "Hydrogen denudation for enhanced thin oxide quality, device performance, and potential epitaxial elimination," VLSI Tech. Dig. Papers, p. 111, 1994.

Gow, R., "Thin SiO_2 films nitrided by rapid thermal processing in NH_3 or N_2O for applications in EEPROMs," *Microelect. J.*, vol. 25, no. 7, p. 539, 1994.

Hermann, M. R., et al., "Long term charge loss in EPROMs with ONO interpoly dielectric," Ann. Proc.—Reliab. Phys. (Symp.), p. 368, 1994.

Hoffman, R. D., "Speciality memories gain currency," *EBN*, p. 25, April 18, 1994.

Hoffstetter, D. M., and M. H. Manley, "Systematic test methodology for identifying defect-related failure mechanisms in an EEPROM technology," IEEE Intl. Conf. Microelect. Test Structures, p. 114, 1994.

Hori, T., "A 0.1 μm CMOS technology with Tilt-Implanted Punchthrough Stopper (TIPS)," *IEEE IEDM Tech. Dig.*, p. 75, 1994.

Hu, G. J., et al., "Advanced CMOS EPROM technology for high/speed/high density programmable logic devices and memory applications," Proc. Cust. Cir. Conf., p. 488, 1994.

Kapoor, V. J., et al., "MNOS memory technology with oxynitride thin films," IEEE Trans. CPMT, Pt. A, vol. 17, no. 3, p. 367, 1994.

Kato, M., et al., "A 0.4-μm^2self-aligned contactless memory cell technology suitable for 256-Mbit Flash memories," *IEEE IEDM Tech. Dig.* p. 921, 1994.

Kianian, S., et al., "A novel 3 volts-only, small sector erase, high density Flash E^2PROM," Tech. Dig. Symp. VLSI Tech., p. 71, 1994.

Kobayashi, S., et al., "Memory array architecture and decoding scheme for 3V-only sector erasable DINOR Flash memory," *IEEE J. Sol. St. Cir.*, vol. SC-29, no. 4, p. 461, 1994.

Kuo, D., et al., "TEFET—A high density, low erase voltage, trench Flash EEPROM," Tech. Dig. Symp. VLSI Tech., p. 51, 1994.

Kwok, C. Y., et al., "Effects of controlled texturization of the crystalline silicon surface on the SiO_2/Si effective barrier height," *IEEE Elect. Dev. Lett.*, vol. EDL–15, p. 513, 1994.

Lanzoni, M., et al., "A novel approach to controlled programming of tunnel based floating gate MOSFETs," *IEEE J. Solid State Cir.*, vol. SC–29, p. 147, 1994.

Lee, D. J., et al., "An 18 Mb serial Flash EEPROM for solid-state disk applications," Tech. Dig. Symp. VLSI Cir., p. 59, 1994.

Lin, J.-K., et al., "New polysilicon–oxide–nitride–oxide–silicon electrically-erasable programmable read-only memory device approach for eliminating off-cell leakage current," *Jap. J. Appl. Phys.*, Pt. 1: Regular Papers and Short Notes and Review Papers, vol. 33, no. 5A, p. 2513, 1994.

Link, V. W., "Ferroelectric memories for identification systems, Part I: system overview and applications," Frequenz, vol. 48, no. 1–2, p. 37, 1994.

Ma, Y., et al., "A novel high density contactless Flash memory array using split-gate source-side-injection cell for 5V-only applications," Tech. Dig. Symp. VLSI Tech., p. 49, 1994.

McKearney, B., "UV EPROM is dead. Long live the mask ROM?", *New Electronics*, vol. 27, no 4, p. 32, April 1994.

Mercer, D. A., "A 14-b 2.5 MSPS pipelined ADC with on chip EPROM," Proc. Bipolar/BiCMOS Circuits and Technology Meeting, p. 15, 1994.

Mihara, M., et al., "Row-redundancy scheme for high-density Flash memory," *IEEE ISSCC Tech. Dig.*, p. 150, 1994.

Minami, S.-I., et al., "3 volt 1 Mbit full-featured EEPROM using a highly-reliable MONOS device technology," *IEICE Trans. on Electr.*, vol. E77–C, no. 8, p. 1260, 1994.

Miyawaki, Y., et al., "An over-erasure detection technique for tightening Vth distribution for low voltage operation NOR type Flash memory," Tech. Dig. Symp. VLSI Cir., p. 63, 1994.

Mockridge, D., "EPROM emulation," Dr. Dobb's J. of Software Tools for Professional Programmer, vol. 19, no. 10, p. 56, 1994.

Mori, S., et al., "High speed sub-half micron Flash memory technology with simple stacked gate structure cell," Tech. Dig. Symp. VLSI Tech., p. 53, 1994.
Nakanishi, T., et al., "Improvement in MOS reliability by oxidation in ozone," Symp. VLSI Tech. Dig. Tech. Papers, p. 45, 1994.
Nakao, Y., et al., "Study of Pb-based ferroelectric thin films prepared by sol-gel method for memory application," *Jap. J. Appl. Phys.*, pt. 1: Regular Papers and Short Notes and Review Papers, vol. 33, no. 9B, p. 5265, 1994.
Oda, H., et al., "New buried channel Flash memory cell with symmetrical source/drain structure for 64-Mbit or beyond," Tech. Dig. Symp. VLSI Tech., p. 69, 1994.
Ohmi, K., et al., "Hydrogen radical balanced steam oxidation for growing ultra-thin high reliability gate oxide films," Symp. VLSI Tech. Dig. Tech. Papers, p. 109, 1994.
Ohsaki, K., N. Asamoto, and S. Takagaki, "A singly poly EEPROM cell structure for use in standard CMOS processes," *IEEE J. Solid State Cir.*, vol. SC–29, p. 311, 1994.
Ohsaki, K., et al., "SIPPOS (single poly pure CMOS) EEPROM embedded FPGA by news ring interconnection and highway path," Proc. Cust. Integ. Cir. Conf., p. 189, 1994.
Ohzone, T., and T. Hori, "MOSFET with Si-implanted gate-SiO_2 structure for analog-storage EEPROM applications," *Solid-State Electr.*, vol. 37, no. 10, p. 1771, 1994.
Ohzone, T., and T. Hori, "Test structure of a MOSFET with Si-implanted gate-SiO_2 for EEPROM applications," IEEE Intl. Conf. Microelect. Test Structures, p. 135, 1994.
Okada, Y., et al., "Gate oxynitride grown in nitric oxide," Symp. VLSI Tech. Dig. Tech. Papers, p. 105, 1994.
Onishi, S., "A half-micron ferroelectric memory cell technology with stacked capacitor structure," *IEEE IEDM Tech. Dig.* p. 843, 1994.
Peng, J., et al., "Flash EPROM endurance simulation using physics-based models," *IEEE IEDM Tech. Dig.*, p. 295, 1994.
Peng, J. Z., et al., "Accurate simulation of EPROM hot-carrier induced degradation using physics based interface and oxide charge generation models," Ann. Proc.—Reliab. Phys. (Symp.), p. 154, 1994.
Peng, J. Z., et al., "Integrated efficient method for deep submicron EPROM/flash device simulation using energy transport model," *IEICE Trans. on Electr.*, vol. E77–C, no. 2, p. 166, 1994.
Raymond, D. W., et al., "Nonvolatile programmable devices and in-circuit test," Proc. IEEE Intl. Test Conf., p. 817, 1994.
Rinaldi, D., et al., "Electron tunnelling from rough surfaces: an application to TPFG EEPROM cells," Semicond. Sci. and Tech., vol. 9, no. 7, p. 1414, 1994.
Sakui, K., and F. Masouka, "Sub-half micron Flash memory technologies," *IEICE Trans. Electron*, vol. E-77, no. 8, p. 1251, 1994.
Samukawa, S., and D. Terada, "Pulse time modulated ECR plasma etching for highly selective, highly anisotropic and less-charging poly-Si gate patterning," VLSI Tech. Dig. Tech. Papers, p. 27, 1994.

Sato, S., et al., "An ultra-thin fully depleted floating gate technology for 64 Mb Flash and beyond," Tech. Dig. Symp. VLSI Tech., p. 65, 1994.

Satoh, J., et al., "Optical beam induced current technique as a failure analysis tool of EPROMs," *IEICE Trans. Electr.*, vol. E77–C, no. 4, p. 574, 1994.

Sexton, F. W., et al., "SEU and SEL response of the Westinghouse 64K E^2 PROM, Analog Devices AD7876 12-bit ADC, and the Intel 82527 serial communications controller," Workshop Record—IEEE Radiation Effects Data Workshop, p. 55, 1994.

Shacham, E., et al., "Novel test structures for process development and monitoring of stack etches for high density FLASH and EPROM memories," IEEE Intl. Conf. Microelect. Test Structures, p. 90, 1994.

Strass, H., "Data systems for PCMCIA flash memories," *Elektronik*, vol. 43, no. 22, p. 146, 1994.

Su, H. P., et al., "Superthin O/N/O stacked dielectrics formed by oxidizing thin nitrides in low pressure oxygen for high-density memory devices," *IEEE Elect. Dev. Lett.*, vol. EDL–15, no. 11, p. 440, 1994.

Takeshima, T., et al., "A 3.3V single-power-supply 64Mb Flash memory with dynamic bit-line latch (DBL) programming scheme," *IEEE ISSCC Tech. Dig.*, p. 148, 1994.

Tan, C., M. Xu, and Y. Wang, "Application of the difference subthreshold swing analysis to study generation interface trap in MOS structure due to Fowler–Nordheim aging," *IEEE Elect. Dev. Letts.*, vol. EDL–15, no. 7, p. 257, 1994.

Tanaka, T., et al., "High speed programming and program-verify methods suitable for low voltage Flash memories," Tech. Dig. of the Symp. VLSI Cir., p. 61, 1994.

Tanzawa, T., et al., "A quick boosting charge pump circuit for high density and low voltage Flash memories," Tech. Dig. Symp. VLSI Cir., p. 65, 1994.

Thomsen, A., and M. A. Brooke, "Low control voltage programming of floating gate MOSFETs and applications," IEEE Trans. Cir. and Sys., I: Fundamental Theory and Applications, vol. 41, no. 6, p. 443, 1994.

Van Houdt, J., et al., "5V-compatible Flash EEPROM cell with microsecond programming time for embedded memory applications," IEEE Trans. on CPMT, pt. A, vol. 17, no. 3, p. 380, 1994.

Verkasalo, R., "Effect of test method on the failure dose of SEEQ 28C256 EEPROM," *IEEE Trans. Nucl. Sci.*, vol. 41, no. 6, pt. 1, p. 2600, 1994.

Watanabe, H., et al., "Scaling of tunnel oxide thickness for Flash EEPROMS realizing stress-induced leakage current reduction," Tech. Dig. Symp. VLSI Tech., p. 47, 1994.

Westmont, C., and B. Wall, "Reprogrammability eases PCMCIA designs," *Electronic Design*, p. 110, April 18, 1994.

Xie, W., et al., "A vertically integrated bipolar storage cell in 6H silicon carbide for nonvolatile memory applications," *IEEE Elect. Dev. Letts.*, vol. EDL–15, no. 6, p. 212, 1994.

Yano, K., et al., "Room-temperature single-electron memory," *IEEE Trans. Elect. Dev.*, vol. ED–41, no. 9, p. 1628, 1994.

Yeh, B., and C. Jeng, "Super tech drives 3.3V superflash," *Australian Electronics Engineering*, vol. 27, no. 6, p. 32, 1994.

Yoon, G. W., et al., "Formation of high quality storage capacitor dielectrics by in-situ rapid thermal reoxidation of Si_3N_4 films in N_2O ambient," *IEEE Elect. Dev. Lett.*, vol. 15, no. 8, p. 266, 1994.

Yu, T.-K., et al., "AN EEPROM model for low power circuit design and simulation," *IEEE IEDM Tech. Dig.*, p. 157, 1994.

Zaleski, A., et al., "Design and performance of a new Flash EEPROM on SOI(SIMOX) substrates," IEEE Intl. SOI Conf. Proc., p. 13, 1994.

1995

Adan, A. O., et al., "A scaled 0.6 μm high speed PLD technology using single-poly EEPROM's," Proc. IEEE Custom Integr. Cir. Conf., p. 55, 1995.

Barsan, R., et al., "Advanced salicided EECMOS Technology for complex PLDS," Proc. 14th NVSM Workshop, Monterey, Calif., 1995.

Baumann, A., "Testing EEPROM cell endurance versus write and erase programming cycles during IC fabrication," *TI Tech. Journal*, vol. 12, no. 5, p. 88, September/October 1995.

Bergemont, A., M. Chi, and H. Haggag, "Process technology and device built-in reliability in NOR virtual ground (NVG^{TM}) Flash memories," Proc. 14th NVSM Workshop, Monterey, Calif., 1995.

Bhanumurthy, G., et al., "A practical EPROM-based extended memory code conversion system," *Microelectronics Journal*, vol. 26, no. 6, p. 621, 1995.

Bhat, M., et al., "Electrical properties and reliability of MOSFETs with rapid thermal NO-nitrided SiO_2 gate dielectrics," *IEEE Trans. Elect. Dev.*, vol. ED–42, p. 907, 1995.

Bottini, R., et al., "Passivation scheme impact on retention reliability of nonvolatile memory cells," IEEE Intl. Integ. Reliab. Workshop Final Report, p. 18, 1995.

Bude, J. D., et al., "EEPROM/Flash sub 3.0 V drain-source bias hot carrier writing," *IEEE IEDM Tech. Dig.*, p. 989, 1995.

Burns, S. G., et al., "Design and fabrication of alpha-Si:H-based EEPROM cells," Proc. Second Symp. Thin Film Transistor Technologies, p. 370, 1995.

Calligaro, C., et al., "Dichotomic current-mode serial sensing methodology for multistorage non-volatile memories," Proc. 38th Midwest Symp. Cir. and Sys., vol. 1, p. 302, 1995.

Calligaro, C., et al., "A new serial sensing approach for multistorage non-volatile memories," IEEE Intl. Workshop on Memory Technol., Design and Testing, p. 21, 1995.

Caprara, P., et al., "A 70ns CMOS double metal 16Mbit EPROM with hierarchical word line decoder," Proc. 14th NVSM Workshop, Monterey, Calif., 1995.

Cernea, R., et al., "A 34 Mb 3.3 V serial Flash EEPROM for solid-state disk applications," *IEEE ISSCC Dig. Tech.* Pap., p. 126, 1995.

Chang, K. T., et al., "Implementation of split gate Flash EEPROM with improved programming characteristics," Proc. 14th NVSM Workshop, Monterey, Calif., 1995.

Chang, Y., et al., "Related reliability issues of band-to-band tunneling induced hot hole injection in Flash EEPROMs," Proc. 14th NVSM Workshop, Monterey, Calif., 1995.

Chen, C., and T. P. Ma, "A new source-side erase algorithm to reduce disturb problem in Flash EPROM," International Symposium on VLSI Technology, Systems, and Applications, Proc. Tech. Papers, p. 321, 1995.

Chen, J. C., et al., "Short channel enhanced degradation during discharge of Flash EEPROM memory cell," *IEEE IEDM Tech. Dig.*, p. 331, 1995.

Chi, M.-H. and A. Bergemont, "Multi-level Flash/EPROM memories: new self-convergent programming methods for low-voltage applications," *IEEE IEDM Tech. Dig.*, p. 271, 1995.

Chi, M.-H., and A. Bergemont, "Programming and erase with floating-body for high density low voltage Flash EEPROM fabricated on SOI wafers," 1995 IEEE Intl. SOI Conf. Proc., p. 129, 1995.

Chi, M., A. Bergemont, and M. Liang, "Steady-state analysis and comparison of erase V_T tightening methods for high density Flash EEPROM," Proc. 14th NVSM Workshop, Monterey, Calif., 1995.

Chi, M.-H., H. Haggag, and A. Bergemont, "A new erase VT distribution model for reliability design in high density Flash EEPROM," Intl. Symp. VLSI Technol., Sys., and Appl. Proc. Tech. Pap., p. 326, 1995.

Concannon, A., et al., "The numerical simulation of hot carrier processes in Flash EEPROM programming," Proc. 14th NVSM Workshop, Monterey, Calif., 1995.

DiMaria, D. J., "Hole trapping, substrate currents, and breakdown in thin silicon dioxide films," *IEEE Elect. Dev. Letts.*, vol. EDL–16, no. 5, p. 184, 1995.

Dorval, D., et al., "Improved lifetime determination of deep submicron n-channel MOSFETs using charge pumping technique and drain current degradation modeling," *Proc. IRPS*, p. 51, 1995.

Dumin, D. J., et al., "Characterizing wearout, breakdown, and trap generation in thin silicon oxide," *J. Vac. Sci. Technol. B*, vol. 13, no. 4, p. 178, 1995.

Dumin, D. J., et al., "High field related thin oxide wearout and breakdown," *IEEE Trans. Elect. Dev.*, vol. ED–42, no. 4, p. 760, 1995.

Duncan, M., and P. Pansana, "Optimising the performance of advanced non-volatile memories using differentiated cell source and drain implants," *Proc. SPIE*, p. 143, 1995.

Fukumoto, T., et al., "2 V 120 nsec 8/16-bit microcontroller with embedded flash EEPROM," Proc. IEEE Custom Integrated Circuits Conference, p. 155, 1995.

Fratin, L., et al., "Degradation mechanism of Flash cell induced by parasitic drain stress conditions," Proc. 14th NVSM Workshop, Monterey, Calif., 1995.

Ghezzi, P., A. Kramer, and P. Zabberoni, "Novel Flash cell for neural network application," Proc. 14th NVSM Workshop, Monterey, Calif., 1995.

Ghidini, G., et al., "Sidewall sealing effects on Flash EEPROM," Proc. 14th NVSM Workshop, Monterey, Calif., 1995.

Gotou, H., "New operation mode for stacked-gate Flash memory cell," *IEEE Elect. Dev. Letts.*, vol. EDL–16, no. 3, p. 121, 1995.

Habas, P., "Modeling study of the experimental techniques for the characterization of MOSFET hot-carrier aging," Microelectron. Reliab., vol. 35, no. 3, p. 481, 1995.

Han, Jaechun, et al., "Programming characteristics of single-poly EEPROM," *J. Korean Instit. of Telematics and Electronics*, vol. 33A, no. 2, p. 131, 1995.

Hazama, H., et al., "A new testing methodology for Flash EEPROM devices," Proc. 14th NVSM Workshop, Monterey, Calif., 1995.

Henry, P., "Flash memories: Evaluating density vs. architecture," *Electronic Products*, p. 27, August 1995.

Herdt, C., et al., "A 256K nonvolatile static RAM," Proc. 14th NVSM Workshop, Monterey, Calif., 1995.

Herth, R., "A super-flash EEPROM," *Elektronik Industrie*, vol. 26, no. 3, p. 40, 1995.

Himeno, T., et al., "A new technique for measuring threshold voltage distribution in Flash EEPROM devices," Proc. Intl. Conf. on Microelectronic Test Structures, p. 283, 1995.

Himeno, T., et al., "A novel threshold voltage distribution measuring technique for Flash EEPROM devices," *IEICE Trans. Electr.*, vol. E79–C, no. 2, p. 145, 1995.

Hsu, J., and S. Shumway, "Comparison of over erase susceptibility and cycling reliability between channel erase and bitline erase in Flash EEPROM," Proc. 14th NVSM Workshop, Monterey, Calif., 1995.

Hsu, J.-T. and S. Shumway, "Erasure enhancement technique in Flash EEPROM by pulsed gate-drain erasure (PGDE)," *Electr. Lett.*, vol. 31, no. 14, p. 1195, 1995.

Hsu, J., and S. Shumway, "Novel erasing enhancement technique in Flash EEPROM by pulsed gate-drain erase (PGDE)," Proc. 14th NVSM Workshop, Monterey, Calif., 1995.

Hu, C.-Y., et al., "Analysis of substrate-bias-enhanced hot electron injection for self-convergence of over-erased Flash cells," Proc. 14th NVSM Workshop, Monterey, Calif., 1995.

Huang, C., and T. Wang, "Transient simulation of EPROM writing characteristics with a novel hot electron injection model," Personal Engineering and Instrumentation News, vol. 12, no. 2, p. 59, February 1995.

Huang, C., and T. Wang, "Transient simulation of EPROM writing characteristics, with a novel hot electron injection model," *Sol. St. Electr.*, vol. 38, no. 2, p. 461, 1995.

Huang, C., et al., "Characterization and simulation of hot carrier effect on erasing gate current in Flash EEPROM's," Proc. IRPS, p. 61, 1995.

Huh, Y., et al., "Hot-carrier-induced circuit degradation in actual DRAM," *Proc. IRPS*, p. 72, 1995.

Hwang, H., et al., "Impact of velocity saturation region on MOSFET's hot carrier reliability at elevated temperatures," *Proc. IRPS*, p. 48, 1995.

Imamiya, K., et al., "A 35 ns-cycle-time 3.3V-only 32 Mb NAND Flash EEPROM," *IEEE ISSCC Tech. Dig.*, p. 130, 1995.

Iwata, Y., et al., "A 35 ns cycle time 3.3 V only 32 Mb NAND Flash EEPROM," *IEEE J Sol. St. Circ.*, vol. SC–30, no. 11, p. 1157, 1995.

Iwata, Y., et al., "The internal voltage system of the 32 Mbit NAND Flash EEPROM for low-voltage power supply," 14th IEEE NVSM Workshop, paper #4-2, 1995.

Kamran, S., "Flash memory," *Advanced Packaging*, p. 18, November/December 1995.

Kanamori, K., Y. Hisamune, and T. Okazawa, "A new source-side injection EPROM (SIEPROM) cell for 3V only Flash memory," Proc. 14th NVSM Workshop, Monterey, Calif., 1995.

Kaya, C., et al., "Low-level gate current injections in Flash memories initiated by minority carrier collection action of floating terminals," *IEEE Trans. Elect. Dev.*, vol. ED–42, no. 12, p. 2131, 1995.

Kelsey, J., "Intel Flash devices offer flexible mapping, noncontiguous memory spaces," Personal Engineering and Instrumentation News, vol. 12, no. 2, p. 59, 1995.

Kim, D. M., et al., "Characterization of split-gate Flash memory devices: reliability, gate-disturbance and capacitive coupling coefficents," IEEE Region 10 Intl. Conf. Microelectronics and VLSI, Asia-Pacific Microelectronics 2000, Proc., p. 460, 1995.

Kjelso, M., S. Jones, and H. G. Baker, "Memory management in Flash-memory disks with data compression," Proc. Memory Management International Workshop, p. 399, 1995.

Klemm, T., and M. Hamma, "Electronically trimmed resistance for proximity switches using EEPROM-stored settings," *Elektronik*, vol. 44, no. 17, p. 90, 1995.

Kong, S. O., C. Y. Kwok, and S. P. Wong, "An EPROM cell with a magnesium electronic injector," IEEE TENCON. IEEE Region 10 International Conference on Microelectronics and VLSI, Asia-Pacific Microelectronics 2000 Proceedings, p. 456, 1995.

Kosaka, H., et al., "Excellent weight-updating-linearity EEPROM synapse memory cell for self-learning neuron-MOS neural networks," *IEEE Trans. Elect. Dev.*, vol. ED–42, no. 1, p. 135, 1995.

Kotov, A. E., "Reliability aspects of thin dielectric films used in NVM ICs," Materials Reliability in Microelectronics V, MRS Symp. Proc., vol. 391, p. 27, 1995.

Lanzoni, M., and B. Ricco, "Experimental characterization of circuits for controlled programming of floating gate MOSFETs," *IEEE J. Solid State Cir.*, vol. SC–30, p. 706, 1995.

Leduc, H., "16 Megabit EPROM of compact chip layout," *Elektronik Industrie*, vol. 26, no. 9, p. 44, 1995.

Lee, C. H., et al., "Simulation of a long term memory device with a full bandstructure Monte Carlo approach," *IEEE Elect. Dev. Lett.*, vol. EDL–16, no. 8, p. 360, 1995.

Licciardello, M., "Super FLASH mixed memory technology," *Electronic Product Design*, vol. 17, no. 6, p. 21, 1995.

Ma., T. P., and C. Chen, "Combined effect of bitline and wordline disturb in Flash EPROM," Proc. 14th NVSM Workshop, Monterey, Calif., 1995.

Madson, G., et al., "Building reliability into an EPROM cell using in-line WLR monitors," IEEE Intl. Integr. Reliab. Workshop, p. 40, 1995.

Matsumoto, O., et al., "1.5 V high speed read operation and low power consumption circuit technology for EPROM and flash-EEPROM," Proc. IEEE Custom Integ. Cir. Conf., p. 25.4.1, 1995.

Mausi, S., K. Sawada, and Y. Sugawara, "A charge pump circuit for EEPROMs with a power supply voltage below 2V," Proc. 14th NVSM Workshop, Monterey, Calif., 1995.

Mercer, D. A., "A 14-b 2.5 MSPS pipelined ADC with on-chip EPROM," *IEEE J. Sol. St. Circ.*, vol. SC–31, no. 1, p. 70, 1995.

Mondon, F., et al., "Electron trapping/detrapping and current leakage in Flash E^2 cells using nitride-oxide or oxide–nitride–oxide interpoly dielectric," Proc. 14th NVSM Workshop, Monterey, Calif., 1995.

Muramatsu, S., et al., "The analysis of over erase mechanism for Flash memory cell—the effect of substrate and floating gate quality," Proc. 14th NVSM Workshop, Monterey, Calif., 1995.

Myrvaagnes, R., "Flash memory: single-transistor cells promise cheaper storage for battery systems," *Electronic Products*, p. 18, December 1995.

Niijima, H., "Design of a solid-state file using Flash EEPROM," IBM J. Res. Develop., vol. 39, no. 5, p. 531, 1995.

Nishimoto, T., et al., "High density 'AND' cell technology for 32Mbit Flash memory," Proc. 14th NVSM Workshop, Monterey, Calif., 1995.

Nozoe, A., et al., "A 3.3 V high density AND Flash memory with 1 ms/512B erase and program time," *IEEE ISSCC Tech. Dig.*, p. 124, 1995.

Olivo, M., L. Bedarida, and G. Fusillo, "A 1 Mbit embedded Flash memory for a high performance 16 bits microcontroller family," Proc. 14th NVSM Workshop, Monterey, Calif., 1995.

Ookawa, T., et al., "A 32 bit 4Mb boot-block Flash EEPROM," Proc. 14th NVSM Workshop, Monterey, Calif., 1995.

Papadas, C., et al., "Modeling of the intrinsic retention characteristics of FLOTOX EEPROM cells under elevated temperature conditions," *IEEE Trans. Elect. Dev.*, vol. ED–42, no. 4, p. 678, 1995.

Park, J.-H., J.-W. Park, and T.-S. Jung, "A simple and efficient model of NAND-Flash EEPROM cell for accurate programming characteristics simulation," Proc. 14th NVSM Workshop, Monterey, Calif., 1995.

Park, K.-H., "Effects of source bias on the programming characteristics of submicron EPROMs/Flash EEPROMs," Proc. 14th NVSM Workshop, Monterey, Calif., 1995.

Pein, H., and J. D. Plummer, "Performance of the 3-D PENCIL Flash EPROM cell and memory array," *IEEE Trans. Elect. Dev.*, vol. ED–42, no. 11, p. 1982, 1995.

Ravazzi, L., et al., "The Flash E^2PROM cell with boron p-pocket architecture: advantages and limitations," Proc. 14th NVSM Workshop, Monterey, Calif., 1995.

San, K. T., et al., "Effects of erase source bias on Flash EPROM device reliability," *IEEE Trans. Elect. Dev.*, vol. ED–42, no. 1, p. 150, 1995.

Shen, S.-J., C.-S. Wei, and C.-H. Hsu, "A new observation on initial charge loss in EPROM," Intl. Symp. VLSI Tech., Systems, and Applications, Proc. Tech. Papers, p. 317, 1995.

Shimizu, K., T. Endoh, and H. Iizuka, "Mechanism of ac-stress-induced leakage current in EEPROM tunnel oxides," *Proc. IRPS*, p. 56, 1995.

Shirota, R., et al., "A new programming method for multi-level NAND Flash memories," Proc. 14th NVSM Workshop, Monterey, Calif., 1995.

Shone, F., et al., "Characterization and improvements of Flash EEPROM endurance performance by integrating optimized process modules in advanced flash technology," Intl. Symp. VLSI Tech., Sys., and Appl. Proc. Tech. Pap. p. 314, 1995.

Shum, D., et al., "Effective in-process plasma damage screening of tunnel oxide for nonvolatile memory applications," Proc. 14th NVSM Workshop, Monterey, Calif., 1995.

Shum, D. P., "A highly robust process integration with scaled ONO interpoly dielectrics for embedded nonvolatile memory applications," *Trans. Elect. Dev.*, vol. ED–42, no. 7, p. 1376, 1995.

Su, H. P., et al., "Novel tunneling dielectric prepared by oxidation of ultrathin rugged polysilicon for 5-V-only nonvolatile memories," *IEEE Elect. Dev. Letts.*, vol. EDL–16, no. 6, p. 250, 1995.

Suh, K.-D., et al., "A 3.3 V 32 Mb NAND Flash memory with incremental step pulse programming scheme," *IEEE ISSCC Tech. Dig.*, p. 128, 1995.

Sun, W., E. Rosenbaum, and S.-M. Kang, "Fast timing simulation for submicron hot-carrier degradation," *Proc. IRPS*, p. 65, 1995.

Sweetman, D., "Floating gate reprogrammable non-volatile EEPROM technology," *Electr. Engr.*, vol. 67, no. 819, p. 99, 1995.

Sweha, S., et al., "A smart auto-configuring selectable-voltage 16 Mb Flash memory," Proc. 14th NVSM Workshop, Monterey, Calif., 1995.

Takebuchi, M., et al., "A novel high-density EEPROM cell using a polysilicon-gate hole (POLE) structure suitable for low-power applications," *Jap. J. Appl. Phys.*, pt. 1: Regular Papers & Short Notes, vol. 35, no. 2B, p. 797, 1995.

Turner, J., et al., "A fault tolerant in-system programmable 560 macrocell EEPROM based PLD," Proc. 14th NVSM Workshop, Monterey, Calif., 1995.

Van Houdt, J., et al., "A 25ns/byte-programmable SSI Flash array with a new low-voltage erase scheme for embedded memory applications," Proc. 14th NVSM Workshop, Monterey, Calif., 1995.

Van Houdt, J. F., et al., "An analytical model for the optimization of source-side injection Flash EEPROM device," *IEEE Trans. Elect. Dev.*, vol. ED–42, no. 7, p. 1314, 1995.

Van Houdt, J. F., et al., "Investigation of the soft-write mechanism in source-side injection Flash EEPROM device," *IEEE Elect. Dev. Letts.*, vol. EDL–16, no. 5, p. 181, 1995.
Wang, C.-K., et al., "Investigation of a high quality and ultraviolet-light transparent plasma-enhanced chemical vapor deposition silicon nitride film for non-volatile memory application," *Jap. J. Appl. Phys.*, pt. 1: Regular Papers & Short Notes, vol. 34, no. 9A, p. 4736, 1995.
Wellekens, D., et al., "Write/erase degradation in source side injection Flash EEPROM's: characterization techniques and wearout mechanisms," *IEEE Trans. Elect. Dev.*, vol. ED–42, no. 11, p. 1992, 1995.
Wellekens, D., et al., "Write/erase degradation and disturb effects in source-side injection flash EEPROM devices," Qual. Reliab. Eng. Int. (UK), Quality and Reliability Engineering International, vol. 11, no. 4, p. 239, 1995.
Wen, K.-S., and C.-Y. Wu, "A simple and accurate simulation technique for Flash EEPROM writing and its reliability issue," *Sol. St. Electr.*, vol. 38, no. 7, p. 1373, 1995.
Wett, T., and S. Levy, "Flash-the memory technology of the future that's here today," Proc. IEEE National Aerospace and Electronics Conference, p. 359, 1995.
Wilkie, D., and M. Hensen, "A study of EEPROM endurance correlation with wafer level reliability data," IEEE Intl. Integr. Reliab. Workshop Final Report, p. 55, 1995.
Wolf, S., *Silicon Processing for the VLSI era—Submicron MOSFET*, vol. 3, chap. 7, p. 495, Lattice Press, 1995.
Yang, C. K., C. L. Lee, and T. F. Lei, "Enhanced H_2-plasma effects on polysilicon thin-film transistors with thin ONO gate-dielectrics," *IEEE Elect. Dev. Letts.*, vol. EDL–16, no. 6, p. 228, 1995.
Yeoh, T.-S. and S.-J. Hu, "Wafer level sort programming-impact on EPROM memory retention," IEEE Trans. on Semiconductor Manufacturing, vol. 9, no. 2, p. 278, 1995.
Yoshida, T., K. Sato, and T. Ono, "3.3 V single power supply 16 Mbit flash EEPROM," Oki Technical Review, vol. 62, no. 155, p. 13, 1995.

1996

Butler, J. F., et al., "Improved EEPROM Tunnel- and Gate-Oxide Quality by Integration of a Low-Temperature Pre-Tunnel-Oxide RCA SC-1 Clean," IEEE Trans. Semicond. Manufact., vol. 9, no. 3, p. 471, 1996.
de Cari, G., "New flash memory devices from Intel," Elettronica Oggi, p. 93, 1996.
Degraeve, R., et al., "On the field dependence of intrinsic and extrinsic time-dependent dielectric breakdown," *Proc. IRPS*, p. 44, 1996.
Hanafi, H. I., S. Tiwari, and I. Khan, "Fast and Long Retention-Time Nano-Crystal Memory," *IEEE Trans. Elect. Dev.*, vol. 43, no. 9, pp. 1553–1558, 1996.

Hemink, G. J., et al., "Trapped hole enhanced stress-induced leakage currents in NAND EEPROM tunnel oxides," *Proc. IRPS*, p. 117, 1996.

Kubota, T., K. Ando, and S. Muramatsu, "The effect of the floating gate/tunnel SiO_2 interface on Flash memory data retention reliability," *Proc. IRPS*, p. 12, 1996.

Lin, C.-J., et al., "Enhanced Tunneling Characteristics of PECVD Silicon–Rich–Oxide (SRO) for the Application in Low Voltage Flash EEPROM," *IEEE Trans. Elect. Dev.*, vol. 43, no. 11, p. 2021, 1996.

Maiti, B., et al., "Highly reliable furnace-grown N_2O tunnel oxide for microcontroller with embedded Flash EEPROM," *Proc. IRPS*, p. 55, 1996.

Martin, A., et al., "A new oxide degradation mechanism for stresses in the Fowler–Nordheim tunneling regime," *Proc. IRPS*, p. 67, 1996.

Miwa, T., et al., "A 1 Mb 5-transistor/bit non-volatile CAM based on flash-memory technologies," IEEE ISSCC Dig. Tech. Pap., p. 414, 1996.

Mori, S., et al., "Thickness scaling limitation factors of ONO interpoly dielectric for nonvolatile memory devices," *IEEE Trans. Elect. Dev.*, vol. 43, no. 1, p. 47, 1996.

Pagey, M., et al., "Unified model for n-channel hot-carrier degradation under different degradation mechanisms," *Proc. IRPS*, p. 289, 1996.

Runnion, E. F., et al., "Limitations on oxide thicknesses in Flash EEPROM applications," *Proc. IRPS*, p. 93, 1996.

Schlund, B., et al., "A new physics-based model for time-dependent-dielectric-breakdown," *Proc. IRPS*, p. 84, 1996.

Teramoto, A., et al., "Excess current induced by hot-hole injection and F–N stress in thin SiO_2 films," *Proc. IRPS*, p. 113, 1996.

Van Tran, H., et al., "A 2.5 V 256-level non-volatile analog storage device using EEPROM technology," IEEE ISSCC Dig. Tech. Pap., p. 512, 1996.

Yamada, S., et al., "Non-uniform current flow through thin oxide after Fowler–Nordheim current stress," *Proc. IRPS*, p. 108, 1996.

Yamada, S., et al., "A Self-Convergence Erase for NOR Flash EEPROM Using Avalanche Hot Carrier Injection," *IEEE Trans. Elect. Dev.*, vol. 43, no. 11, p. 1937, 1996.

Yeoh, T.-S., and S.-J. Hu, "Wafer level sort programming—impact on EPROM memory retention," IEEE Trans. Semicond. Manufact., vol. 9, no. 2, p. 278, 1996.

Young, N. D., et al., "The Fabrication and Characterization of EEPROM Arrays on Glass Using a Low-Temperature Poly-Si TFT Process," *IEEE Trans. Elect. Dev.*, vol. 43, no. 11, p. 1930, 1996.

1997

Aritome, S., et al., "A Side-Wall Transfer-Transistor Cell (SWATT Cell) for Highly Reliable Multi-Level NAND EEPROM's," *IEEE Trans. Elect. Dev.*, vol. 44, no. 1, p. 145, 1997.

Cappelletti, P., et al., "CAST: An Electrical Stress Test to Monitor Single Bit Failures in Flash-EEPROM Structures," *Microelectron. Reliab.*, vol. 37, no. 3, p. 473, 1997.

Farahani, M. M., J. F. Butler, and S. Garg, Limitation of the TiN/Ti Layer Formed by the Rapid Thermal Heat Treatment of Pure Ti Films in an NH_3 Ambient in Fabrication of Submicrometer CMOS Flash EPROM ICs," *IEEE Trans. Semicond. Manufacturing*, vol. 10, no. 1, p. 147, 1997.

Wilkie, D., et al., "A Design of Experiment Analysis of Serial EEPROM Endurance," *Microelectron. Reliab.*, vol. 37, no. 3, p. 487, 1997.

Index

D

E

F

G

H

I

S

T

X

Y

Editors' Biographies

William D. Brown is currently a university professor and head of the Department of Electrical Engineering at the University of Arkansas. As a Member of the Technical Staff at Sandia Laboratories in Albuquerque, New Mexico, from 1969–1977, he initiated Sandia's research and development effort on metal-nitride–oxide-silicon (MNOS) device technology. After joining the faculty at the University of Arkansas in 1977, his nonvolatile semiconductor memory research concentrated on the synthesis and characterization of plasma-enhanced chemical vapor deposited (PECVD) silicon nitride for application in MNOS and SNOS memory devices. He has also served on IEEE NVM standards committees and is an organizer and participant in the IEEE International Nonvolatile Memory Technology Conference and its precursor conferences.

Joe E. Brewer is currently a Senior Advisory Engineer at the Northrop Grumman Corporation Electronic Sensors and Systems Division located near Baltimore, MD. Throughout his 36-year engineering career he has been engaged in the development of state-of-the-art microelectronic technology. Mr. Brewer has been both a developer and user of NVSM devices. He has had extensive experiences with MNOS block-oriented devices and storage systems, as well as a variety of SONOS devices. He has served on IEEE NVM standards committees and is an organizer and participant in the IEEE International Nonvolatile Memory Technology Conference and its precursor conferences.